英汉·汉英
自动控制与仪表词汇

ENGLISH–CHINESE / CHINESE–ENGLISH
DICTIONARY OF
AUTOMATIC CONTROL AND INSTRUMENT

刘振全　陈　浩　主编

化学工业出版社
·北京·

本书收集自动控制类、仪器仪表类基本词汇、常用词汇及最新专业词汇，同时包括电工电子类、机械类、计算机类部分常用词汇，共收词 50000 余条。分为"英汉部分"和"汉英部分"，以满足英译中和中译英的不同需求。

本书重点是两种文字的对应释义，不解决定义、概念和应用方面的问题，这样能在相同的篇幅中包含更多的词汇，以满足读者专业文献翻译、阅读、写作等需要。

本书可供从事自动控制、仪器仪表相关研究、开发、教学、培训的技术人员参考。

图书在版编目（CIP）数据

英汉·汉英自动控制与仪表词汇/刘振全，陈浩主编. —北京：化学工业出版社，2012.6
ISBN 978-7-122-13870-5

Ⅰ. 英… Ⅱ. ①刘…②陈… Ⅲ. ①自动控制-词汇-英、汉 ②自动化仪表-词汇-英、汉 Ⅳ. ①TP273-61 ②TH82-61

中国版本图书馆 CIP 数据核字（2012）第 057459 号

责任编辑：宋　辉		文字编辑：丁建华	
责任校对：蒋　宇		装帧设计：关　飞	

出版发行　化学工业出版社（北京市东城区青年湖南街 13 号　邮政编码 100011）
印　　刷　北京永鑫印刷有限责任公司
装　　订　三河市万龙印装有限公司
850mm×1168mm　1/32　印张 21¾　字数 1135 千字
2012 年 8 月北京第 1 版第 1 次印刷

购书咨询：010-64518888（传真：010-64519686）　　售后服务：010-64518899
网　　址：http://www.cip.com.cn
凡购买本书，如有缺损质量问题，本社销售中心负责调换。

定　价：88.00 元　　　　　　　　　　　　　　　　　版权所有　违者必究

前 言

随着科学技术的发展，自动控制与仪器仪表方面的研究、开发、应用日新月异，自动化与仪器仪表等方面的技术越来越受到人们的重视，与之相关的新的专业名词和缩略语大量涌现，令人应接不暇。国内外技术的交流和科研、开发、应用等工作，都要求专业技术人员接触大量的英文专业文献。广大学生和工程技术人员由于技术交流、课题研究、产品研发、国外产品应用、专业学习、课程设计、论文撰写、文献互译等多方面的需要，经常遇到专业词汇的英汉互译，许多专业词汇在一般英汉词典上难以查到，或者释义过多需要读者进一步分析，很不方便。为方便专业读者，尤其为方便科研及工程技术人员、广大师生在阅读、翻译和学习有关书刊、文献资料时查阅，特编写本书。

本书收集自动控制类、仪器仪表类基本词汇、常用词汇及最新专业词汇，同时包括电工电子类、机械类、计算机类部分常用词汇，共收词50000余条。分为"英汉部分"和"汉英部分"。本书重点放在两种文字的对应释义，不解决定义、概念和应用方面的问题，这样能节省篇幅，以满足读者专业文献翻译、阅读、写作等需要。希望能够为从事相关研究、开发、教学、培训的技术人员提供一本有用的工具书。

本书具有以下特点：

（1）收词量丰富，多而不陈，以专业基本词、常用词、新词为主，对于过时的、不常用的词汇尽量不收录，涉及的复合词汇收词专而精；

（2）言简意赅，一目了然，不做过多的解释，以保证每页涵盖更多的词汇；

（3）对于缩略词在缩略和全拼两处均给出解释，以便读者

查阅;

(4) 收词具有时代性、科学性、实用性,反映自动控制与仪表领域的最新进展,确保新颖性和相对稳定性,力求从实用的角度,采用较小篇幅,给读者以更多的词汇和信息量;

(5) 以查阅频率高的词汇为主,收录的基本词条是名词词条,兼收个别其他词条。

本书由刘振全、陈浩主编,杨世凤、刘振德、王汉芝副主编,参与本书编写和校对工作的还有:刘静、孙海霞、刘伟、薛薇、陈晓艳、刘东伟、贾红艳、张葆璐、白瑞祥、黄华芳、彭一准、张晓平、贺庆、戴凤智等。王德进教授审阅了全稿并提出了建议和修改意见,在此表示感谢。

限于编者水平,书中难免存在不妥和疏漏之处,欢迎读者不吝指正。

编者

凡 例

(1) 本书收集自动控制类、仪器仪表类基本词汇、常用词汇及最新专业词汇，同时包括电工电子类、机械类、计算机类部分常用词汇。分英汉和汉英两部分。

(2) 英汉部分按照英文字母排序；汉英部分按照汉语拼音顺序排序。

(3) 英汉部分排列顺序不分字母大写和小写，缩略语按照缩写以后的字母顺序排列。

(4) 汉英部分的同音字，按照四声（阴平 -、阳平 ˊ、上声 ˇ、去声 ˋ）顺序排列。

(5) 汉英部分，以非汉字开头的词条集中在最后的"其他"部分，按照阿拉伯数字、罗马数字、英文字母和希腊字母的顺序排列。

(6) 复合词或者固定搭配的词组，如 multi-channel 和 multichannel 等，均可见于英文文献，故不作统一。

目 录

凡例
英汉部分 ……………………………………………… 1
汉英部分 ……………………………………………… 361

英汉部分

A

A/D and D/A conversion 模/数与数/模转换
A/D conversion 模拟/数字转换,模/数转换
A/D converter 模拟/数字转换器,模/数转换器
A/D encoder 模拟/数字译码器,模/数译码器
AA (absolute apparatus) 绝对仪器
AA (active addressing) 主动寻址
AA (artificial antenna) 仿真天线
AAS (atomic absorption spectrophotometry) 原子吸收分光光度法
AAS (atomic absorption spectroscopy) 原子吸收光谱学
AASP (advanced automated sample processor) 半自动样品制备系统
AB (audio bandwidth) 音频带宽
Abe metro-scope 阿贝测长仪
aberration 色差
abnormal condition 异常情况,异常条件,异常状态
abnormal condition handler 异常条件处理程序
abnormal glow 不规则辉光放电,反常辉光放电
abnormal handling 异常处理
abnormal operating condition 异常运转工况,不正常运行状态,异常运行状态
abnormal release condition 异常释放条件
abnormal termination 异常终止,异常结束
abort routine 异常终止程序
abort 终止,停止
abrasion 磨损,磨耗,擦伤
abrasion hardness 磨损硬度,耐磨硬度,研磨硬度
abrasive 研磨料,粗糙的,有研磨作用的,砂轮
abrasive cut-off machine 磨切机,砂轮切断机
abrasive disk 砂轮,磨盘,研磨盘
abrasive grain 磨粒,抛光粉
abrasive grinding machine 磨机,火石机
abrasive particle 磨粒
abrasive tool 研磨工具,磨具,砂磨工具
abrasive tooth wear 齿面研磨磨损
abscissa 横线,横坐标
abscissa axis 横坐标,横轴,横坐标轴
absence of offset 零静差
absence-of-ground search selector 无接地搜索选择器,未接地寻线器
absent 不在的,缺少的
absolute 绝对的,完全的
absolute address 绝对地址,绝对位址
absolute alarm 绝对值报警
absolute apparatus (AA) 绝对仪器
absolute cell address 绝对单元地址,绝对单元位址
absolute derivative algorithm 导数绝对值程序
absolute deviation 绝对偏差,实际弹着偏差
absolute deviation integral 绝对偏差积分
absolute dimension 绝对尺寸
absolute dimension measurement 全尺寸测量
absolute dimensional factor 绝对尺寸系数
absolute electrometer 绝对静电计
absolute error 绝对误差
absolute error criterion 绝对误差判据
absolute frequency meter 绝对频率计
absolute gravimeter 绝对重力仪
absolute humidity 绝对湿度
absolute measurement 绝对测量

3

absolute motion　绝对运动
absolute permeability　绝对磁导率，绝对渗透性
absolute pressure　绝对压强
absolute pressure indicator　绝对压力表
absolute pressure sensor　绝对压力传感器
absolute pressure transducer　绝对压力传感器
absolute pressure transmitter　绝对压力变送器
absolute resolution　绝对分辨率
absolute salinity　绝对盐度
absolute scale　绝对温标，开氏温标
absolute size　绝对尺寸
absolute stability　绝对稳定性
absolute stability of a linear control system　线性控制系统的绝对稳定性
absolute static pressure of the fluid　流体绝对静压
absolute tangential velocity　绝对切向速度
absolute temperature　绝对温度
absolute temperature scale　绝对温标，开氏温标
absolute track address　绝对磁道地址
absolute value　绝对值
absolute velocity　绝对速度
absolute zero　绝对零点，绝对零度
absorber　吸光器，吸收体，减震器
absorber circuit　吸收电路
absorptance　吸收比，吸收率
absorptiometer　吸收比色计，吸收计
absorption　吸收，吸收过程，吸收作用
absorption band　吸收带
absorption cell　吸收池
absorption coefficient　吸收系数
absorption correction　吸收修正
absorption edge　吸收边沿，吸收限
absorption factor　吸收系数，吸收率
absorption flaw detector　吸收式探伤仪
absorption hygrometer　吸收湿度表
absorption infrared spectrometer　吸收式红外光谱仪
absorption of vibration　震动阻尼，减震
absorption probe analyzer　吸收式探头分析仪
absorption spectrum　吸收光谱，吸收频谱
absorption type monitor　吸收式监测器
absorption X-ray spectrometry　吸收X射线度谱术，吸收X射线光谱法
absorptivity　吸收率，吸收能力，吸收性
absorptivity of an absorbing　吸引材料的吸收率
abstract　摘要，抽象
abstract system　抽象系统
abundance sensitivity　丰度灵敏度
abutment　邻接，桥台，拱座，接合点，接合器
abutting end　邻接端，对接端
AC and DC network analyzer　交直流网络分析仪
AC circuit　交流电路
AC circuit constant voltage regulator　交流电路稳压器
AC commutator machine　交流整流子电机，交流换向器电机
AC conductivity　交流电导率
AC converter machine　交流变换器
AC differential transformer displacement transducer　交流差动变压器式位移传感器
AC digital voltmeter　交流数字电压表，交流数字伏特计
AC flip-flop　交流触发器
AC galvanometer　交流检流计
AC machine　交流电机
AC motor　交流电动机
AC power controller　交流调功电路，交流功率控制器，交流电源控制器
AC power electronic switch　交流电力电子开关
AC regulated power source　交流稳压电源
AC regulated power supply　交流稳压电源
AC relay　交流继电器
AC tachometer generator　交流测速发电机
AC to digital converter　交流-数字转换器

AC voltage controller 交流调压电路
AC voltage stabilizer 交流稳压器
AC voltage-stabilized source 交流稳压电源
AC/DC differential voltmeter 交直流差动电压表，交直流差动式伏特计
ACB (auxiliary controller bus) 辅助控制器总线
accelerated graphics port (AGP) 图形加速接口
accelerated test 加速试验
accelerating chamber 加速室，加速箱，加速腔，加速真空箱
accelerating key 加速键
accelerating tube 加速管
acceleration 加速，加速度
acceleration analysis 加速度分析
acceleration diagram 加速度曲线
acceleration error coefficient 加速度误差系数
acceleration of gravity 重力加速度
acceleration pedal 加速器踏板
acceleration response 加速器响应
acceleration sensitive element 加速器敏感元件
acceleration sensor 加速度传感器
acceleration simulator 加速度仿真器
acceleration transducer 加速度传感器
acceleration voltage 加速电压
accelerator 油门，加速器，催化剂
accelerometer 加速计，加速度仪，加速表
acceptable quality level 可接受的质量标准
acceptance criteria 接受准则，验收准则
acceptance inspection 验收检验
acceptance of the mass filter 滤质器的接收容限
acceptance testing 验收测试
accepted goods 良品
accepted parts 良品
acceptor 接收器，接受者
acceptor circuit 谐振电路，接收电路
access 进入，通路
access authority 存取权限

access code 访问码，存取码
access control 存取控制
access control field 存取控制字段
access control register 存取控制寄存器
access coupler 通路耦合器
access door 检修门，通道门
access environment 访问环境
access hole 余隙孔
access lane 进出路径
access level 存取级别
access point 入口处，出入通道处
access ramp 入口坡道，斜通道
access road 通路，通道
access shaft 竖井通道
access spiral loop 螺旋式回旋通道
access time 存取时间
accessibility 易接近性，可及性
accessories of testing machine 试验机附件
accessory 附件，配件
accessory hardware 附属硬件
accessory instrument 附属仪表，辅助仪表
accessory of limited interchangeability 有限互换附件
accessory slot 扩展槽
accessory time 辅助时间
accident 事故，意外
accidental collapse 意外坍塌
accidental error 随机误差，偶然误差
accidental loading 偶然荷载，偶然负荷
accommodate 装设，容纳
accretion 增加物，冲积层，炉瘤
accumulate 累加
accumulated error 累积误差
accumulated time difference 累积时差
accumulation of electric energy 电能储存
accumulative raingauge 累积雨量器
accumulator 储压器，蓄电池，累加器
accumulator cell 蓄电池
accuracy 精确度，准确性
accuracy class 精确度等级
accuracy limit 准确度极限
accuracy of measurement 测量精确度

5

accuracy of the wavelength 波长精确度
accuracy rating 精度标称值
accurate 精确
accurate die casting 精密压铸
accurate measuring instrument 精度测量仪表
acetate 醋酸盐
acetate wire 醋酸绝缘线
acetone 丙酮
acetylene 乙炔,电石气
acetylene cylinder 乙炔汽缸(气缸)
acetylene pressure gauge 乙炔压力表
acetylene regulator 乙炔减压器
AC-frequency converter 交流变频器
acid 酸
acid converter 酸性转炉
acid lining cupola 酸性熔铁炉
acid open-hearth furnace 酸性平炉
acid plant 酸洗设备,酸洗机
acid proof cast iron 耐酸铸铁
acid pump 酸液泵
acid solvent 酸性溶剂
acid tank 酸液缸
acid-base indicator 酸碱指示剂
acid-base titrations 酸碱滴定法
ACIDP (advanced control interface point data) 先进的控制接口数据点
acid-resistant water purifier 耐酸滤水器
acknowledge 确认
acknowledge interrupt 确认中断,中断应答
acknowledge operation 确认操作
acme thread form 梯形螺纹
acorn nuts 盖形螺母
acoustic amplifier 声放大器,声波放大器
acoustic amplitude logger 声波幅度记录器
acoustic analyzer 声学分析仪
acoustic and light signals 声光信号
acoustic beacon 水声信标
acoustic couplant 声耦合剂
acoustic coupler 声耦合器
acoustic current meter 声学海流计
acoustic Doppler system 声学多普勒定位系统

acoustic element 声学元件
acoustic emission 声发射
acoustic emission amplitude 声发射振幅
acoustic emission analysis system 声发射分析系统
acoustic emission detection system 声发射检测系统
acoustic emission detector 声发射检测仪
acoustic emission energy 声发射能量
acoustic emission event 声发射事件
acoustic emission preamplifier 声发射前置放大器
acoustic emission pulser 声发射脉冲发生器
acoustic emission rate 声发射率
acoustic emission signal processor 声发射信号处理器
acoustic emission source 声发射源
acoustic emission source location and analysis system 声发射源定位及分析系统
acoustic emission source location system 声发射源定位系统
acoustic emission spectrum 声发射光谱
acoustic emission technique 声发射技术
acoustic emission transducer 声发射换能器
acoustic fatigue 声疲劳,噪声疲劳
acoustic frequency generator 声频发生器
acoustic impedance 声阻抗
acoustic interferometer 声干涉仪
acoustic line 声传输线
acoustic logging instrument 声波测井仪
acoustic malfunction 声失效
acoustic matching layer 声匹配层
acoustic oscillograph 声波示波器
acoustic radiometer 声辐射计
acoustic ratio 声学比,声强比,声波比
acoustic releaser 声释放器
acoustic resistance 声阻

acoustic scanner 声扫描器
acoustic screen 隔声屏
acoustic thermometer 声学温度计，声波温度表
acoustic tide gauge 回声验潮仪
acoustic transducer 声换能器
acoustic transponder 声应答器
acoustical hologram 声全息图
acoustical holography 声全息术
acoustical holography by electron-beam scanning 电子束扫描声全息
acoustical holography by laser scanning 激光束扫描声全息
acoustical holography by mechanical scanning 机械扫查声全息
acoustical imaging by Bragg diffraction 布拉格衍射声成像
acoustical impedance method 声阻法
acoustical interferometer 声干涉仪
acoustical lens 声透镜
acoustical light modulator 声光调制器
acoustical meter 比声计
acoustically transparent pressure vessel 透声压力容器
ACP（advanced control program） 高级控制程序
ACPR（adjacent-channel power ratio） 邻频功率比
acquisition time 收集时间
acrylic 丙烯酸的，亚克力
acrylic paint 水性漆，丙烯酸涂料
acrylic resin 丙烯酸树脂，亚克力树脂
acrylic sheet 丙烯胶片，丙烯酸片
actinometer 化学光度计，光线强度计，曝光计，露光计
action 动作
activation 激活，活化作用
activator 活化剂
active 活的，有效的
active addressing（AA） 主动寻址
active attitude stabilization 主动姿态稳定
active block 有源组件
active circuit elements 有源电路元件
active compensation 有效补偿
active component 活性组分，有效元件，有功分量
active control 有效控制
active corrosion 活性腐蚀
active device 有源器件
active earth pressure 主动土压力
active electric circuit 有源电路
active element 活性元素，有源元件
active energy meter 有功电度表，瓦时计
active filter 有源滤波器
active four-terminal network 有源四端网络
active gauge length 有效基长
active gauge width 有效基宽
active in respect to 相对呈阳性
active linear two-port network 有源线性二端网络
active load 有功负载
active metal indicated electrode 活性金属指示电极
active narrow band suspension 有效窄带悬挂
active noise control 有效噪声控制
active oxidation 活性氧化
active plate 活动板
active power 有功功率
active remote sensing 主动遥感
active transducer 有源传感器
activity coefficient 活度系数
activity selection 作用选择
actual higher measuring range value 实际（测量范围）上限值
actual line of action 实际啮合线
actual lower measuring range value 实际（测量范围）下限值
actual material calibration 实物校准
actual time of observation 实际观测时间
actual transformation ratio of current transformer 电流互感器的变压系数
actual value 实际值
actual voltage ratio 实际电压比
actuating cable 控制电缆
actuating element 执行机构
actuating error signal 动作偏差信号
actuating motor 伺服电机，制动

马达
actuating signal　驱动信号
actuator　执行机构，激励器，促动器
actuator bellows　执行机构波纹管
actuator load　执行机构负载
actuator power unit　执行机构动力部件
actuator sensor interface（ASI）　执行器传感器接口
actuator shaft　执行机构输出轴
actuator spring　执行机构弹簧
actuator stem　执行机构输出杆
actuator stem force　执行机构刚度
actuator travel characteristic　执行机构行程特性
acylation　酰基化，酰化作用
adaline　线性适应元
adamantane　金刚烷
adapt　改装
adapt meter　匹配测量仪
adaptation　适应，适配
adaptation layer　适应层
adapter　适配器，衔接口，转接器
adapter check　适配器检验
adapter control block　适配器控制块
adaptive　自适应的
adaptive algorithm　自适应算法
adaptive array　自适应矩阵
adaptive control　自适应控制
adaptive control system　自适应控制系统
adaptive controller　自适应控制器
adaptive correlation　自相关
adaptive digital filter　自适应数字滤波器
adaptive equalization　自适应均衡，配接等化，均一化调整
adaptive equalizer　自适应补偿器
adaptive filter　自适应滤波器
adaptive fuzzy control　自适应模糊控制
adaptive fuzzy system　自适应模糊系统
adaptive generic model control（AGMC）　自适应一般模型控制
adaptive law　自适应规则
adaptive prediction　自适应预报
adaptive system　自适应系统
adaptive telemeter system　自适应遥测系统
adaptive tuning　自适应整定
adaptor　适配器，适配接头，转接器，接合器
ADC（analog to digital converter）　模拟-数字转换器
add lubricating oil　加润滑油
add operation　加（法）运算，加（法）操作
addendum　齿顶高
addendum circle　齿顶圆
addendum surface　上齿面
adder　加法器
adder-subtracter　加减器
addition　添加，增加物
addition method　叠加法
addition system　附加系统
additional building works　增补建筑工程
additional correction　补充修正
additional horizontal force　额外横向力
additional LCN module　附加局部控制网络模件
additional network　附加网络
additional plan　增补计划
additional terminals for pulse width output　脉冲宽度输出附件端子
additional vent　加设通风口
additive　添加剂，添加物
additive operation　加法运算
additive white Gaussian noise（AWGN）　附加的白高斯噪声
additivity of mass spectra　质谱的可叠加性
address buffer　地址缓冲器
address bus　地址总线
address code　地址代码
address comparator　地址比较器
address decoder　地址解码器
address map configuration　地址变换组态
address read wire　地址读出线
address register　地址寄存器
address space　地址空间
address wire　地址线
address　地址

addressable fire monitor system　可寻址火灾监视系统
addressable location addressing　可定位寻址
addressable portable test instrument　可寻址的便携测试仪表
add-subtract time　加减时间
adhesive　黏结剂，胶黏剂
adhesive coated dielectric film　涂胶黏剂绝缘薄膜
adhesive coated foil　涂胶铜箔
adhesive face　胶黏剂面
adhesive force　黏附力
adhesive wear　黏着磨损
adhesive-coated catalyzed laminate　涂胶催化层压板
adhesive-coated uncatalyzed laminate　涂胶无催层压板
adiabatic scanning calorimeter　绝热扫描热量计
ADIMM（advanced dual in-line memory modules）　高级双重内嵌式内存模块
adit　入口，通路，坑道口
adjacency　接近，相邻
adjacent channel　相邻信道，相邻通道
adjacent level　相邻水平
adjacent-channel interference　相邻信道干扰
adjacent-channel power ratio（ACPR）　邻频功率比
adjoint operator　伴随算子
adjoint variable　伴随变量
adjust buffer total ion strength　总离子强度调节缓冲剂
adjust　调整，校正
adjustability　可调性
adjustability coefficients　可调系数
adjustable　可调整的
adjustable capacitor　可变电容器
adjustable cistern barometer　动槽气压表，调槽式气压表
adjustable condenser　可变电容器，可调电容器
adjustable gain　可调增益
adjustable higher measuring range limit　可调测量范围限
adjustable impulse counter　可调脉冲计数器
adjustable length gauge　可调式长度量规
adjustable lower measuring range limit　可调测量范围下限
adjustable pliers　可调手钳
adjustable relative humidity range　相对湿度可调范围
adjustable resistance　可变电阻
adjustable resistor　可变电阻器
adjustable spanner　活动扳手
adjustable speed motors　调速电动机
adjustable starting rheostat　可调启动变阻器
adjustable temperature range　温度可调范围
adjustable wrench　活扳手
adjustable-speed motor　变速电动机
adjusted retention time　调整保留时间，调整保持时间
adjusted retention volume　调整保留体积
adjuster　调整机构，调节器
adjusting　调整，调整的，校准
adjusting pin　校正针
adjusting wedge　可调型楔块
adjusting screw　调整螺钉
adjustment　调校，调整，调节
adjustment bellows　调节波纹管
adjustment device　调节器装置
administration unit　管理单元
admissible error　容许误差
admittance　导纳
admittance measuring instrument　导纳测量仪
admixture　掺合剂，外加剂
adsorbent　吸附剂
adsorption　吸附
adsorption chromatography　吸附色谱法
advance directional sign　前置指路标志，方向预告标志
advance earthworks　前期土方工程
advance warning sign　前置警告标志
advance works　前期工程

advanced 高级的，先进的
advanced automated sample processor (AASP) 半自动样品制备系统
advanced control 先进控制，先行控制
advanced control function 先行控制功能，超前控制功能
advanced control interface data point (ACIDP) 先行控制接口数据点，先进的控制接口数据点
advanced control point ID 先行控制点标志
advanced control program (ACP) 高级控制程序
advanced control system 先行控制系统
advanced control technique 先行控制技术
advanced diskette operating system 先进磁盘操作系统
advanced dual in-line memory modules (ADIMM) 高级双重内嵌式内存模块
advanced mobile phone system (AMPS) 高级移动电话系统
advanced multifunction controller (AMC) 高级多功能控制器
advanced operation 先进操作
advanced process control (APC) 高级过程控制
advanced process manager (APM) 高级过程管理综合控制器，高级过程管理站
advancing 超前，超前的
AE pulse analyzer 声发射脉冲分析仪
AED (atomic emission detection) 原子发射检测
aeration 曝气，通风，充气
aeration tank 曝气池
aerator 通风装置
aerial 天线，天线的，架空的
aerial camera 航空照相机
aerial change-over switch 天线转换开关
aerial coil 天线线圈
aerial feeder 天线馈线
aerial line 架空线
aerial remote sensing 航空遥感

aerial surveying camera 航摄仪
aerial view 鸟瞰图，航测图
aerial wire 天线
aerodynamic balance 空气动力天平
aerodynamic noise 气体动力噪声
aerofoil 机翼，翼型
aerograph 航空气象记录仪
aerogravity survey 航空重力测量
aerometeorograph 高空气象计
aerometer 量气计，气体比重计
aeronautical and astronautical science and technology 航空宇航科学与技术
aerosol 悬浮微粒，气溶胶，喷雾
aerospace 航空航天空间
aerospace computer control 航空计算机控制
aerospace control 航空控制
aerospace engineering 航空工程
aerospace propulsion theory and engineering 航空宇航推进理论与工程
aerospace trajectory 航空轨迹
AES (atomic emission spectrometry) 原子发射光谱术
AES (Auger electron spectroscopy) 俄歇电子光谱学
AF measuring instrument 声频测量仪表
AFC amplifier 自动频率控制放大器
AFD (alkali flame detection) 碱火焰检测
affinity chromatography 亲和色谱法
AFID (alkali flame ionization detector) 碱焰离子化检测器
A-frame A形骨架，金字塔形建筑物
AFS (atomic fluorescence spectroscopy) 原子荧光光谱学
after service 售后服务
aftercooler 二次冷却器
afterfilter 二次过滤器，补充过滤器
AGC (automatic gain control) 自动增益控制
age hardening 时效硬化
ageing 老化处理
aggregate 骨料，集料，碎石
aggregate area 总面积
aggregate superficial area 表面总面积

aggregation matrix 集结矩阵
agile 敏捷，灵巧
agile control 灵巧控制
agile manufacturing 敏捷制造
aging of column 柱老化
aging property tester 老化性能测定仪
agitator 搅拌器，搅动机
AGMC（adaptive generic model control） 自适应一般模型控制
AGP（accelerated graphics port） 图形加速接口
agricultural analyzer 农用分析仪
agricultural biological environmental and energy engineering 农业生物环境与能源工程
agricultural electrification and automation 农业电气化与自动化
agricultural engineering 农业工程
agricultural mechanization engineering 农业机械化工程
agricultural water-soil engineering 农业水土工程
agriculture 农业
AHP（analytic hierarchy process） 层次分析法
AI（analog input） 模拟输入，模拟量输入
AI（artificial intelligence） 人工智能
aided design 辅助设计
air 空气，大气
air activity monitor 空气活度监测器
air bleeding 放气
air blower 鼓风机
air brake 气压制动器
air break switch 空气断路器
air capacitor 空气电容器
air circuit 空气回路
air circuit breaker 空气断路器，空气自动断路器
air cleaner 空气过滤器
air compressor 空气压缩机，空压机，压缩机
air compressor governor 空气压缩机调压器
air condition area 空调区
air conditioned location 空调场所
air conditioner 空调设备

air conditioning 空气调节
air consumption 耗气量
air cooled chiller 风冷式冷冻机
air cooler 空气冷却器
air cooling system 空气冷却系统，风冷系统
air core 空心，空气磁芯［心］
air coupling valve 空气联结阀
air curtain fan 风帘风扇
air cushion plate 气垫板
air cylinder 气缸，气筒
air damper 风闸，气流调节器，空气阻尼器
air damping 空气阻尼
air distribution system 配气系统
air distributor 空气分配器
air dryer 空气干燥器
air duct 通风管道，气槽
air electric connecting plug 气电插座
air entrained cement 加气水泥，伴水泥
air entraining agent 加气剂
air exhaust 排气口
air exhaust fan 排气扇，抽风机
air filter 空气滤器，风隔，隔尘网
air filter chamber 空气过滤室
air flowmeter 风量计，空气流量计
air gap 气隙
air grill 空气格栅
air gun 空气枪
air hardening 气体硬化
air header 集气管
air humidity indicator 空气湿度指示器
air inlet 进风口，进气孔
air inlet louver 进气百叶窗
air inlet port 进气口
air intake 进风口，进气孔，入气口
air intake duct 进气槽
air intake filter 进气过滤器
air isolating cock 空气隔断旋塞
air line breathing apparatus 气喉型呼吸器具
air line strainer 进气管道隔滤器
air lock 气锁
air micrometer 气体测微计
air operated electrical switch 气动电

开关
air outlet 回风口，出风口
air outlet grille 空气出口栅格
air particle 空气粒子
air passage 风道
air patenting 空气韧化
air permeability 空气磁导率，空气渗透性
air permeability test 透气性试验
air pipe 气管
air piston 空气活塞
air pollution 空气污染
air pressure 气压，风压
air pressure balance 空气压力天平
air pressure gauge 气压表
air pressure regulator-filter 空气过滤调压器
air pressure switch 气动开关，气压开关
air pressure test 空气压力试验
air pressurization system 空气加压系统
air receiver 空气储存器
air reservoir 储气缸
air restrictor 空气节流器
air safety valve 空气安全阀
air set 空气中凝固，常温自硬
air set mold 常温自硬铸模
air shuttle valve 空气换向阀
air sleeve 风向袋
air sprayer 喷涂器，空气喷雾器
air spring 空气弹簧，气垫
air strainer 空气隔滤器，空气滤网
air supply outlet 供气出口
air supply valve 供气阀
air switch 空气开关
air tank 空气箱
air temperature 气温
air terminal 航空集散站
air tight instrument 气密式仪表
air to close 气关
air to open 气开
air traffic control 空中交通控制
air valve 进出气阀，放气阀
air valves 空气阀门
air vent 排气道，通气孔
air vent cock 通风管旋塞

air vent valve 通气阀
air ventilator 空气通风器，通风孔，通风管
air wire 天线架空线
airborne flux-gate magnetometer 航空磁通门磁力仪
airborne gamma radiometer 航空伽玛辐射仪
airborne gamma spectrometer 航空伽玛能谱仪
air-borne gravimeter 航空重力仪
airborne infrared spectroradiometer 机载红外光谱辐射计
airborne optical pumping magnetometer 航空光泵磁力仪
airborne proton magnetometer 航空质子磁力仪
air-control 气控
air-core coil 空心线圈
aircraft 航空飞行器
aircraft control 飞行器控制
aircraft operation 飞行器操纵
air-cushion eject-rod 气垫顶杆，安全气囊顶杆
air-deployable buoy 空投式极地浮标
air-drop automatic station 空投自动气象站
airfoil flow meter 翼式流量计
air-gap field 气隙磁场
air-gap flux 气隙磁通
air-gap flux distribution 气隙磁通分布
air-gap line 气隙磁化线
airgun controller 喷雾控制器
air-lock device 锁气装置
airmeter 气流表
air-operated damper 气动风闸
airshaft 通风竖井
airtight cover 气密盖
air-tight instrument 气密式仪器仪表
airy dry varnish 风干清漆
AL (application layer) 应用层
alarm 警报，警报器
alarm annunciator display 报警灯显示画面
alarm bell 警钟
alarm buzzer 警报器

alarm console 报警控制台
alarm cutout 报警切除
alarm device 报警信号装置
alarm enunciator 报警信号器
alarm high 高限报警
alarm indicator 报警指示器
alarm lamp 报警信号灯
alarm logger package 报警记录程序包
alarm low 低限报警
alarm priority 报警优先级
alarm priority change 改变报警优先级
alarm process 报警处理
alarm relay 报警信号继电器
alarm signal 报警信号
alarm status 报警状态
alarm summery panel 报警汇总画面
alarm switch 报警信号开关
alarm system 报警系统
alarm type 报警类型
alarm unit 报警组件，报警单元
alarming 信号报警
alcohol thermometer 酒精温度表
aldehyde 醛，乙醛
alga 水藻
algebraic 代数的
algebraic approach 代数方法
algebraic Riccati equation 代数黎卡提方程
algebraic selection 代数选择
algebraic systems theory 代数系统理论
algorithm 算法
algorithmic 算法的
algorithmic language 算法语言
alias 替换名，别名，化名
alias point 不同位号的点
alidade 照准仪
align 匹配，定位，对准，调整
aligner 调整器，汽车的转向轮安装角测定仪，前轴定位器
alignment 对齐，对准，找正，调整
alignment capacitor 微调电容器
alignment instrument 准线仪
alignment oscillator 校准振荡器
alignment wire 准线
ALIVH (any layer inner via hole) 任意层内部导通孔

alkali 碱，碱性的
alkali flame detection (AFD) 碱火焰检测
alkali flame ionization 碱火焰离子化
alkali flame ionization detector (AFID) 碱焰离子化检测器
alkalidipping 脱脂
alkaline aggregate reaction 碱性集料反应
alkaline error 碱误差
alkaline pump 碱液泵
alkaline tank 碱液缸
alkalinity 碱度，碱性
alkalinity of seawater 海水碱度
alkalinization 碱化
alkali-silica reaction 碱硅反应
alkaloid 生物碱
alkylation 烷基化，烃化
all band 全波段
all core molding 集合式铸模
all digital analyzer 全数字分析器
all drilled hole 全部钻孔
all pass 全通
all pass element 全通元件
all pass filter 全通滤波器
all round die holder 通用模座
all sidedly 全面地
all wave 全波的
allen wrench 通用扳手
all-metal prevailing torque type nuts 金属预置扭矩式螺帽
all-mine pig iron 不用冶炼的生铁
allocate 分配
allocated logical storage 已分配的逻辑存储器
allocated physical storage 已分配的物理存储器
allocation 定位
allocation problem 配置问题，分配问题
allowable amount of unbalance 许用不平衡量
allowable current 容许电流
allowable error 容许误差
allowable load 容许负载
allowable load impedance 允许的负载阻抗

allowable pressure angle　许用压力角
allowable pressure differential　允许压差
allowable stress　允许应力
allowable unbalance　许用不平衡量
allowable value　容许值
alloy　合金
alloy cast iron　合金铸铁
alloy iron　铁合金
alloy steel bar　合金钢筋条
alloy steel nuts　合金钢螺帽
alloy structural steel wire　合金结构钢丝
alloy tool steel　合金工具钢
allround die holder　通用模型
all-sky camera　全天空照相机
all-wave band　全波段
all-wave filter　全波滤波器
all-weather wind vane and anemometer　全天候风向风速计
alpha　字母
alphabetic　字母的
alphabetic character　字母字符
alphabetic character set　字母字符集
alphabetic code　字母代码
alphabetic shift　字母换挡（键）
alphabetic string　字符串
alphameric field　字母数字字段
alphameric keyboard　字母数字键盘
alphanumeric　字母数字的
alpha-numeric　字数组
alphanumeric keyboard　字母数字键盘
alteration　更改，改建，改动
alternate blade cutter　双面刀盘
alternate pulse generator　交替脉冲发生器
alternating current　交流电
alternating current bridge　交流电桥
alternating current circuit　交流电路
alternating current generator　交流发电机
alternating current mains　交流电力网
alternating current motor　交流电动机
alternating current resistance　交流电阻
alternating current transmission　交流输电
alternating load　交变负荷
alternating voltage　交流电压
alternating magnetic field　交变磁场
alternating-current commutator motor　交流换向器电动机
alternating-current relay　交流继电器
alternative design　替代设计
alternative route　替代路线
altimeter　高度计
altitude angle　高度角
altitude indicator　高度指示器
altitude meter　测高仪
ALU（arithmetic logic unit）　算术逻辑运算单元
aluminium（Al）　铝
aluminium alloy　铝合金
aluminium alloy wire　铝合金线
aluminium clad wire　包铝钢丝（双金属线）
aluminium tape　铝带
aluminium wire　铝线
aluminium-plastic composite panel　铝塑复合板
aluminum continuous melting and holding furnaces　连续溶解保温炉
aluminum nuts　铝螺帽
alumite wire　防蚀铝线
AM（amplitude modulation）　幅度调制，调幅
AM/FM function generation　调幅/调频函数发生器
amalgamation method　汞齐化法
amber flashing light　黄色闪灯
ambient humidity range　环境湿度范围
ambient pressure　环境压力，周围压力
ambient pressure error　环境压力误差
ambient temperature　环境温度，周围温度
ambient temperature range　环境温度范围
ambient vibration　环境振动
ambient　周围的，环境的
ambient noise　环境噪声
ambiguity error　模糊误差

AMC (advanced multifunction controller) 高级多功能控制器
amendment 修订
American National Standards Institute (ANSI) 美国国家标准学会
American Public Transit Association 美国大众运输协会
American Standard Code for Information Interchange (ASCII) 美国信息交换标准码
americium (Am) 镅
ammeter 电流表,安培计
amorphous polymer 非晶态聚合物
amortisseur 阻尼器,阻尼绕组
amount of precipitation 降水量,降雨量
amount of unbalance 不平衡量
amount of unbalance indicator 不平衡量指示器
ampere 安培(电流单位)
ampere turn 安(培)匝(数)
Ampere's circuital law 安培环路定律
Ampere's law 安培定律
Ampere's right-handed screw rule 安培右手螺旋定则
ampere-turns 安(培)匝(数)
amperometric titration 电流滴定法
amphoteric solvent 两性溶剂
amplidyne 电机放大机,交磁放大机
amplification 放大,增强,增益
amplification control 增益控制
amplifier 放大器,扩音器
amplifier inverter 倒相放大器
amplifier stage 放大级
amplifier system 放大器系统
amplifier transformer 放大器变压器
amplify 放大,增高
amplifying element 放大环节
amplitude 幅度,幅值,振幅,最大值
amplitude control 幅度控制
amplitude detector module 振幅检测组件
amplitude discriminator 检波器,振幅鉴别器
amplitude distortion 幅值失真,畸变
amplitude error 振幅误差

amplitude filter 振幅滤波器
amplitude limiter 限幅器
amplitude locus 幅值轨迹
amplitude lopper 限幅器
amplitude modulation (AM) 幅度调制,调幅,幅值调制
amplitude modulation detector 调幅检波器
amplitude modulator 调幅器
amplitude of first harmonic 基波振幅
amplitude of vibration 振幅
amplitude quantization error 幅值量化误差
amplitude ratio 幅值比
amplitude response 幅值响应
amplitude sampler 振幅取样器
amplitude shift keyed infra-red (ASKIR) 长波形可移动输入红外线
amplitude shift keying (ASK) 幅移键控
amplitude stabilizer 稳幅器
amplitude-frequency characteristic 幅频特性
amplitude-modulated 调幅的
amplitude-phase error 幅相误差
AMPS (advanced mobile phone system) 高级移动电话系统
AMU (atomic mass unit) 原子质量单位
anacamptometer 反射计
analog channel 模拟通道
analog computer 模拟计算机
analog computer control 模拟计算机控制
analog computing element 模拟计算元件
analog control 模拟控制
analog controller 模拟量控制器,模拟量调节器
analog curve plotter 模拟曲线描绘器
analog data 模拟数据
analog deep-level seismograph 模拟深层地震仪
analog dial 模拟刻度盘
analog digital adapter 模拟数字适配器

analog digital converter 模拟数字转换器
analog divider 模拟除法器
analog faceplate block 模拟面板功能块
analog input (AI) 模拟输入，模拟量输入
analog input data point 模拟量输入数据点
analog input/output 模拟输入/输出
analog integrator 模拟积分器
analog magnetic tape record type strong-motion instrument 模拟磁带记录强震仪
analog model 模拟模型
analog multiplexed 模拟多路转换器
analog multiplier 模拟乘法器，模拟多路开关
analog operation blocks 模拟操作功能块
analog output (AO) 模拟输出，模拟量输出
analog output data point 模拟输出数据点
analog point 模拟点
analog regulator 模拟调节仪表
analog seismograph tape recorder 模拟磁带地震记录仪
analog signal 模拟信号
analog simulation 模拟仿真
analog superconduction magnetometer 模拟式超导磁力仪
analog switch 模拟开关
analog system 模拟系统
analog telemetering system 模拟遥测系统
analog to digital transient recorder 模拟-数字瞬态记录仪
analog transducer 模拟传感器
analog unit (AU) 模拟单元
analog 模拟
analog-digital conversion 模-数转换
analog-mode device 类模器
analog-to-digital conversion accuracy 模-数转换精确度
analog-to-digital conversion rate 模-数转换速度

analog-to-digital converter (ADC) 模拟-数字转换器
analogue comparing calculating hardware 模拟比较计算仪表
analogue comparing control hardware 模拟比较控制仪表
analogue computer 模拟计算器，模拟计算机
analogue computing unit 模拟计算单元
analogue date 模拟数据
analogue indication 模拟示值，模拟读数
analogue measuring instrument 模拟测量仪表
analogue measuring instrument with digital presentation 数字显示模拟测量仪表
analogue measuring instrument with semi-digital presentation 半数字显示模拟测量仪表
analogue read-out 模拟示值，模拟读数
analogue representation of a physical quantity 物理量的模拟表示
analogue signal 模拟信号
analogue-digital converter (ADC) 模拟-数字转换器
analogue-to-digital conversion 模-数转换
analogue-to-pulse converter 模拟-脉冲变换器
analogy 模拟，类比
analyse 分析
analyse mechanics 分析力学
analyser 分析仪
analysis 分析，分解
analysis of mechanism 机构分析
analysis of simulation experiment 仿真实验分析
analysis of variance 方差分析
analytic approximation 解析近似
analytic hierarchy process (AHP) 层次分析法
analytical balance 分析天平
analytical chemistry 分析化学
analytical design 解析设计

analytical electron microscope 分析型电子显微镜
analytical gap 分析间隙
analytical instrument 分析仪器
analytical line 分析线
analytical method 分析法
analytical plotter 解析测图仪
analytical predictor (AP) 分析预估器
analytical quality control (AQC) 分析质量管理
analyzer 分析仪,解析器
analyzer for clinic medicine concentration 临床药物浓度仪
analyzer tube 分析管
anchor 锚
anchor bearing 锚承,锚座
anchor bolt 锚栓,地脚螺栓,锚定螺栓
anchor nuts 壁虎螺帽
anchor pin 锚销
anchor plate 锚定板
anchorage 锚碇,碇泊区,抛锚区,锚固
anchorage length 锚固长度
anchoring spur 盘址
anchoring strength 锚固强度
anchoring wire 锚索
ancillary facilities 附属设施
AND 与
AND circuit 与门,与电路
AND element 与元件
AND gate 与门
AND NOT gate 与非门
AND operation 与操作运算
AND OR NOT gate 与或非门
anechoic chamber 消声室,无回音室
anechoic tank 消声水池
anemograph 风速计
anemometer 风速表
anemometer tower 测风塔
aneroid barograph 膜盒气压表
aneroid barometer 空盒气压表,空盒气压计
aneroidograph 空盒气压计
anesthetic equipment 麻醉机
angle 角度
angle beam technique 斜角法

angle beam testing 角钢梁测试
angle cutter 角铣刀,角铁切割机
angle ejector rod 斜顶杆
angle external threaded block valve 角形外螺纹切断阀
angle form 角型
angle iron 角钢,角铁
angle iron bracket 角铁支架,角铁托架
angle of advance 超前角,导前角
angle of attach 冲角
angle of contact 接触角
angle of departure 出射角,分离角
angle of field of view 视场角
angle of incidence 入射角
angle of inclination 倾斜角,倾角
angle of refraction 折射角
angle of spread 指向角,半扩散角
angle of view of telescope 望远镜视场角
angle of X-ray projection X射线辐射圆锥角
angle pin 斜导边,斜针,角度针
angle probe 斜探头
angle resolved electron spectroscopy (ARES) 角分辨电子谱法
angle sensor 角度传感器
angle stop valves 角形断流阀
angle strain 角应变
angle throttle valves 角式节流阀
angle type globe valves 角式截止阀
angle valve 角阀,角形阀
angle welding 角焊
angle-attack sensor 迎角传感器
angle-attack transducer 迎角传感器
angular 角度
angular acceleration 角加速度
angular acceleration sensor 角加速度传感器
angular acceleration transducer 角加速度传感器
angular backlash 角侧隙
angular bevel gears 斜交锥齿轮
angular contact bearing 角接触轴承
angular contact radial bearing 角接触向心轴承
angular contact thrust bearing 角接触

推力轴承
angular deviation 角偏差,角偏移
angular displacement 角位移
angular displacement grating 角位移光栅
angular displacement transducer 角位移传感器
angular encoder 角编码器
angular frequency 角频率
angular momentum 角动量,动量矩
angular pin 角销,倾斜销
angular pitch 角节距
angular position 角位置
angular sensitivity 角灵敏度
angular testing machine 可调角度试验机
angular velocity 角速度
angular velocity ratio 角速比
angular velocity sensor 角速度传感器
angular velocity transducer 角速度传感器
anhydride 酸酐,脱水物
anion 阴离子
ANN(artificial neural network) 人工神经网络
anneal 退火
annealed aluminum wire 退火铝线,软铝线
annealed copper foil 退火铜箔
annealed copper wire 软铜线
annealing 退火
announciator 报警器
annual load curve 年负荷曲线
annual load factor 年负荷系数
annual load variation 年负荷变化
annular chamber 环室
annular coil clearance 环形线圈间隙
annular pad 环形盘
annular ring 孔环
annular space 环形间隙
annular spring 环形弹簧
annunciation lamp 警示灯
annunciator 信号器,电铃指示装置
annunciator message 报警器信息
anode 阳极,正极
anode effect 阳极效应
anode voltage 阳极电压

anodize 阳极电镀
anodizing 阳极氧化处理
ANSI (American National Standards Institute) 美国国家标准学会
answerback 回答,应答信号
answering 应答
antenna 天线
antenna down-lead 天线引下线
antenna matching 天线匹配
antenna pointing control 天线指向控制
antenna system 天线系统
anti reset windup 防复位终止
anti skid device 防滑装置
anti-burglary glazing 防盗玻璃窗
anti-cavitation valve 防空化阀
anticlockwise 逆时针方向的
anti-contamination device 防污染装置
anti-corrosion paint 防蚀漆
anti-coupling bi-frequency induced polarization instrument 抗耦双频激电仪
antifriction cast iron 抗磨铸铁
anti-friction quality 减摩性
anti-interference 抗干扰
anti-interference equipment 抗干扰设备
anti-interference filter 抗干扰滤波器
antijamming capability 抗干扰能力
anti-lift roller 防升滚轮
antilock braking system 防锁相制动系统
anti-magnetized varistor 消磁电压敏电阻器
antimatter 反物质
antimony (Sb) 锑
anti-oxidizing paint 抗氧化漆
antiparticle 反粒子
antiproton 反质子
antiresonance 反共振,并联谐振,电流谐振
antiresonance frequency 反共振频率,并联谐振频率,电流谐振频率
anti-rust paint 防锈漆
antiskid 防滑的,防滑装置
antiskid control 防滑控制

anti-skid dressing 防滑钢砂
anti-skid material 防滑物料
anti-slip 防滑动，防空转，防打滑
anti-spin regulation 防自旋调节，防自转调节
anti-static chain 抗静电链
anti-static tyre 抗静电轮胎
anti-stripping agent 防剥剂
anti-syphonage pipe 反虹吸管
anti-tip roller 防倾侧滚轮
anti-vibration mounting 防震
anti-vibration pad 防震垫
anti-wheelspin control 防打滑控制
anti-windup 防饱和，抗饱和，防积分饱和，防止结束
anvil 铁砧
any layer inner via hole (ALIVH) 任意层内部导通孔
any layer inner via hole multilayer printed board 层间全内导通多层印制板
AO (analog output) 模拟输出
AOCS (attitude and orbit control system) 姿态轨道控制系统
AOQ (average output quality) 平均出厂品质
AP (analytical predictor) 分析预估器
APC (advanced process control) 高级过程控制
aperiodic damping 非周期阻尼，过阻尼
aperiodic decomposition 非周期分解
aperiodic speed fluctuation 非周期性速度波动
aperiodic vibration 非周期振动
aperture 隙缝，孔口，光圈
aperture of pressure difference 压差光圈
aperture photographic method 针孔摄影法
aperture time 空隙时间
API (atmospheric pressure ionization) 常压电离，大气压电离，大气压离子化
APM (advanced process manager) 高级过程管理综合控制器，高级过程

管理站
apparatus 仪器，设备，器材
apparatus dew point temperature 设备露点温度
apparent 表观的，视在的
apparent impedance 视在阻抗
apparent pH 表观 pH 值
apparent power 表观功率，视在功率
apparent power consumption 视在功率消耗
apparent temperature 表观温度
appearance potential spectrometer (APS) 出现电热谱法，出现电热谱仪
appearance potential 出现电位
appliance 器具，器械，装置
applicable medium 适用介质
applicable temperature 适用温度
application 应用
application builder and executive 应用编码程序和执行
application control language 应用控制语言
application drawing 操作图
application factor 工况
application layer (AL) 应用层
application layer protocol specification 应用层协议规范
application layer service definition 应用室服务定义
application module 应用组件，应用模块
application module redundancy test 应用模块冗余测试
application module status display 应用模块状态显示画面
application software 应用软件
application processor 应用处理器
application program 应用程序，请求应用
application-specific integrated circuit (ASIC) 专用集成电路
applied biomechanics 应用生物力学
applied chemistry 应用化学
applied covering 外加覆盖物
applied load 外加负荷
applied neural control 应用神经控制

applied voltage　外加电压
apposable　并列的
approach　接近，靠近，方法
approach action　啮入
approach channel　进港航道，引渠
approach distance　接近距离
approach lock　接近锁定
approach ramp　引道坡
approach road　引道，进路
approach speed　进场速度
approach viaduct　高架引道
approval　批准
approval examine and verify　审核
approved by　核准
approved material　经批准的物料
approved plan　经批准的图则，经批准的计划；经审定的图纸
approximate absolute temperature scale　近似绝对温标
approximate analysis　近似分析
approximate equivalent circuit　近似等效电路
approximate reasoning　近似推理
approximate value　近似值
appurtenance　附属物
apron　跳板，护板，停机坪
aprotic solvent　无质子溶剂
APS (appearance potential spectrometer)　出现电热谱法，出现电热谱仪
AQC (analytical quality control)　分析质量管理
aqua fortis　硝酸
aqueous ultrasonic cleaning systems　大型超声波清洗机
aqueous vapour　水汽
arbitrary waveform generator　任意波形发生器
arbor　柄轴，心轴，刀轴
arbor assembly　心轴组件
arbor distance　心轴距
arc contact　电弧触点
arc discharge　电弧放电
arc of approach　啮入弧
arc of recess　啮出弧，渐远弧
arc resistance　电弧电阻，弧阻
arc suppressing resistor　消弧电压敏电阻器
arc welding　电弧焊接
arc　电弧，弧光
arch　拱，弓形，拱门
architectural history and theory　建筑历史与理论
archive replay module　档案重访模块
archive time period　归档时间周期
arctan　反正切
arctic buoy　极地浮标
area alarm summary display　区域报警总貌画面
area change　区域改变
area data base　区域数据库
area display　分区显示
area graphic display　分区流程图显示
area location　区域定位
area name　区域名
area name configuration　区域名称组态
area of cross section of the main air flow　主送风方向横截面积
area status display　分区状态画面
area traffic control system　区域交通控制系统
area trend display　分区趋势画面
area　面积，区域
ARES (angle resolved electron spectroscopy)　角分辨电子谱法
argon (Ar)　氩
argon arc welding　氩气焊，氩弧焊接
argon ionization detector　氩电离检测器
argon welding　氩焊
argon-ion gun　氩离子枪
arithmetic　算术，运算
arithmetic algorithm　算术，算法
arithmetic logic unit (ALU)　算术逻辑运算单元
arithmetic mean　算术平均值
arithmetic weighted mean　算术加权平均值
arm error　不等臂误差
ARMA model　自回归移动平均模型
ARMA parameter estimation　自回归移动平均参数估计
armament launch theory and technology　兵器发射理论与技术

armament science and technology 兵器科学与技术
armature 电枢，衔铁线圈
armature circuit 电枢电路
armature coil 电枢线圈
armature core 电枢铁芯
armature current 电枢电流
armature field 电枢磁场
armature leakage inductance 电枢漏磁电感
armature leakage reactance 电枢漏磁电抗
armature ohmic loss 电枢铜损
armature reaction 电枢反应
armature resistance 电枢电阻
armature stroke 衔铁行程
armature winding 电枢绕组
armored thermocouple 铠装热电偶
armoured cable wire 铠装电缆钢丝
armouring wire 铠装（电缆用）钢丝
aromatic polyamide paper 聚芳酰胺纤维纸
arrangement 排列，布置
arrangement on the back of instrument panel-board 仪表盘盘后布置图
array 数组，排列，阵列
array configuration 阵排列
array filter 组合滤波器
array point 数组点
array processor 数组处理机，阵列处理机，排列信息处理装置
arrester 避雷器
arrester brake 制动器
arrester varistor 防雷用电压敏电阻器
arsenic (As) 砷
arsenic iron 砷化铁
articulated robot 关节型机器人
artificial 人工的，仿真的
artificial antenna (AA) 仿真天线
artificial atmospheric phenomena simulator 人工气候室
artificial defect 人工缺陷
artificial environment 人工环境
artificial intelligence (AI) 人工智能
artificial intelligent work station 人工智能工作站
artificial lighting 人工照明
artificial load 仿真负载
artificial neural network (ANN) 人工神经网络
artillery, automatic gun and ammunition engineering 火炮、自动武器与弹药工程
as this article is fragile be sure to put enough padding 此物易碎，请务必装入足够的填充物
asbestos abatement works 石棉拆除工程
asbestos cement 石棉水泥
asbestos covered wire 石棉包线，石棉绝缘线
asbestos gasket 石棉垫片
asbestos-cement corrugated sheet and ridge tile 石棉水泥波纹瓦及其脊瓦
asbestos-cement pipe for well casing 石棉水泥井管
asbestos-cement pressure pipe for gas transmission 石棉水泥输煤气管
as-built drawing 竣工图纸
ascending development 上行展开法
ASCII (American Standard Code for Information Interchange) 美国信息交换标准码
ash pit 排渣槽，灰坑
ASI (actuator sensor interface) 执行器传感器接口
ASIC (application-specific integrated circuit) 专用集成电路
ASK (amplitude shift keying) 幅移键控
ASKIR (amplitude shift keyed infrared) 长波形可移动输入红外线
asphalt 沥青
asphalt distributor 沥青喷洒机
asphalt paver 沥青摊铺机
asphalt roofing 沥青屋面
asphaltic coating 沥青涂层
asphaltic concrete 沥青混凝土
aspirating valves 吸气阀，抽气阀
aspirator 吸气器
assemblage 装配
assemble line 装配线，生产线
assembler 汇编程序，汇编机，装

配工
assembler language 汇编语言
assembly 组合,组装,汇编
assembly condition 装配条件
assembly drawing 装配图
assembly language 汇编语言
assembly line 组装线
assembly mark 铸造合模记号
assembly parts 部件
assembly robot 装配机器人
assessment 评估
assignment problem 配置问题,分配问题
associate display 关联画面
associate equipment 关联设备
associate storage 内容定址存储器
associated works 相关工程,相关设施
associative memory model 联想记忆模型
associatron 联想机
assume 假定
Assur group 阿苏尔杆组,杆组
A-stage resin 甲阶树脂
astatic action 无定位作用
astatic control 无静差控制
asymmetric 不对称的,不平衡的
asymmetrical 非对称的
asymmetrical deformation vibration 不对称变形振动
asymmetrical polyphase current 不对称多相电流
asymmetrical stretching vibration 不对称伸缩振动
asymmetry potential 不对称电位
asymptote 渐近线
asymptote centroid 渐近中心
asymptotic 渐近的
asymptotic analysis 渐近分析
asymptotic approximation 渐近近似,渐近逼近
asymptotic property 渐近性
asymptotic stability 渐近稳定性
asynchronous 异步的
asynchronous circuit 异步电路
asynchronous communication 异步通信
asynchronous communication interface adapter 异步通信接口适配器

asynchronous computer 异步计算机
asynchronous data transfer 异步数据传递
asynchronous flow 异步数据流
asynchronous frequency changer 异步变频器
asynchronous input 异步输入
asynchronous machine 异步电机
asynchronous modulation 异步调制
asynchronous motor 异步电动机
asynchronous multiplexed 异步多路转换器
asynchronous operation 异步运行
asynchronous receive 异步接收
asynchronous send 异步发送
asynchronous sequential 异步序列
asynchronous sequential logic 异步序列逻辑
asynchronous starting 异步启动
asynchronous transfer mode (ATM) 异步传输模式
asynchronous transmission 异步传输
Atbas metal 镍铬钢
ATC (automatic train control) 列车自动控制装置
ATIM (automatic ticket issuing machine) 自动售票机
ATM (asynchronous transfer mode) 异步传输模式
atmidometer 蒸发仪,蒸发表
atmometer 汽化表
atmospheric distillation 常压蒸馏
atmospheric electricity 大气电
atmospheric opacity 大气不透明度
atmospheric pressure 气压,常压,大气压力
atmospheric pressure altimeter 气压高度计
atmospheric pressure ionization (API) 大气压电离,大气压离子化
atmospheric temperature 大气温度,常温
atmospheric-pressure ionization (API) 常压电离
atmospherics 天电,大气干扰,天电干扰
atom 原子

atom bomb 原子弹
atom force microscope 原子力显微镜
atomic absorption spectrometry 原子吸收光谱，原子吸收光谱法
atomic absorption spectrophotometry (AAS) 原子吸收分光光度法
atomic absorption spectroscopy (AAS) 原子吸收光谱学
atomic beam magnetic resonance apparatus 原子束磁共振仪
atomic boiler 原子锅炉
atomic emission detection (AED) 原子发射检测
atomic emission spectrometry (AES) 原子发射光谱术
atomic emission spectroscopy 原子发射光谱学
atomic emission spectrophotometry 原子发射分光光度法
atomic fluorescence spectrometry 原子荧光光谱法
atomic fluorescence spectrophotometer 原子荧光光度计
atomic fluorescence spectrophotometry 原子荧光分光光度法
atomic fluorescence spectroscopy (AFS) 原子荧光光谱学
atomic fluorometry 原子荧光测定术
atomic mass 原子质量
atomic mass unit (AMU) 原子质量单位
atomic number 原子序数
atomic number correction 原子序数修正
atomic power 原子能
atomic resonance magnetometer 原子共振磁强计
atomic spectroscopy 原子光谱法
atomic spectrum 原子光谱
atomic weight 原子量
atomic-absorption spectrophotometer 原子吸收分光光度计
atomization 原子化
atomization air fan 雾化空气风扇
atomizer 喷雾器，雾化器
atomizing 雾化
attachment 附件，附属物

attack resisting glazing 防盗玻璃窗
attained pose drift 实际位姿漂移
attemperator 温度调节器，保温器
attention 注意
attenuation 衰减
attenuation coefficient 衰减系数
attenuation correction 衰减校正
attenuation length 衰减长度
attenuation observation 衰减观察器
attenuation ratio 衰减比
attenuator 衰减器
attitude 姿态
attitude acquisition 姿态捕获
attitude algorithm 姿态算法
attitude and orbit control system (AOCS) 姿态轨道控制系统
attitude angular velocity 姿态角速度
attitude control 姿态控制
attitude disturbance 姿态扰动
attitude maneuver 姿态机动
attitude sensor 姿态传感器
attitude transducer 姿态传感器
attraction 收紧，引力
attractor 吸引子，牵引机
attribute 属性
attribute byte 属性字节
attribute label 属性标号
AU (analog unit) 模拟单元
audible cab indicator 驾驶室音响指示器
audible frequency 音频，声频
audible signal 音响信号
audio amplifier 音频放大器
audio band amplifier 声频带通放大器
audio bandwidth (AB) 音频带宽
audio equipment 音频设备
audio frequency track circuit 音频轨道电路
audio monitor 监听器
audio noise meter 噪声计
audio signal 音频信号
audioformer 声频变压器
audio-frequency 音频，声频
audio-frequency amplifier 音频放大器
audio-frequency generator 音频发生

器,音频振荡器
audio-frequency reception 音频接收
audio-frequency spectrograph 声频频谱仪
audio-frequency transformer 音频变压器
audiphone 助听器
Auger electron image 俄歇电子像
Auger electron spectrometer 俄歇电子能谱仪
Auger electron spectroscopy (AES) 俄歇电子光谱学
augment ability 可扩充性
augmented system 增广系统
aural detector 音频检波器
aurora 极光
austempering 等温淬火,沃斯回火法,奥氏体回火法
austenite 奥氏体
austenitic cast iron 奥氏体铸铁
austenitic steel 奥氏体钢
authentication 证实,鉴定
authoritarian checking 授权检查
authoritarian code 授权代码
authoritarian message 授权信息
authority 权限,特许
auto 自动
auto correlation 自相关
auto correlation function 自相关函数
auto manual station algorithm 自动手动控制站算法
auto mode loop status switching key 自动方式回路状态转换键
auto selector 自动选择器
auto selector system 自动选择系统
auto station 自动站
auto synchronous motor 自动同步电动机
auto test equipment 自动测试设备
auto tuner 自动整定,调谐
auto reclose 自动重合闸
auto-bias 自给偏压
auto-car racing 汽车竞赛
autoclaved aerated concrete blocks 蒸压加气混凝土砌块
autoclaved aerated concrete slabs 蒸压加气混凝土板

autoclaved lime-sand brick 蒸压灰砂砖
autoclaves sterilizers 高压灭菌器,高压灭菌锅
autocollimator 自动准直机
auto-compensation logging instrument 电子自动测井仪
auto-compound current transformer 自耦式混合绕组电流互感器
autoconduction 自感
autocontrol 自动控制
autoconverter 自耦变压器
autocorrection 自动校正
autocorrelation function 自相关函数
auto-excitation 自激振荡,自激励
autoformer 自耦变压器
autographic apparatus 自动绘图仪
automata 自动装置,机器人
automata controller 自动控制器
automata frequency control 自动频域控制
automata gain control 自动增益控制
automata operation 自动操作
automata process control 自动过程控制
automata recognition 自动辨识
automata regulator 自动调节器
automata restart 自动重复启动
automata sequence control 自动顺序控制
automata testing 自动测试
automata theory 自动机理论
automate 使自动
automated guided vehicle 自动导向机车
automated inspection 自动化检验
automatic 自动化的
automatic adaptive equalizer 自适应均衡器
automatic air circuit breaker 自动空气断路器
automatic analyzer for microbes 微生物自动分析系统
automatic and hand operated change-over switch 自动手动转换开关
automatic assembly system 自动化装配系统

automatic balancing circuit 自动平衡电路
automatic balancing machine 自动平衡机
automatic bias 自动偏压
automatic blood cell analyzer 全自动血细胞分析仪
automatic break 自动切断
automatic channel switch 自动频道开关
automatic check 自动校验
automatic circuit breaker 自动断路器
automatic coding 自动编码
automatic control 自动控制
automatic control source of vacuum 真空自动控制电源
automatic control switch 自动控制开关
automatic control system 自动控制系统
automatic data processing 自动数据处理
automatic diagnostic program 自动诊断程序
automatic engineering 自动化工程
automatic exposure device 自动曝光装置
automatic fare collection system 自动收费系统
automatic feeder for brine 盐水溶液自动补给器
automatic focus and stigmator 自动调焦和消像散
automatic following 自动跟踪
automatic gain control (AGC) 自动增益控制
automatic gate 自动闸门
automatic generation control 自动发电控制
automatic ignition device 自动点火装置
automatic interlocking device 自动联锁装置
automatic lathe 自动车床
automatic level controller 自动电平控制器
automatic manual station 自动-手动操作器
automatic measurement 自动测量

automatic message recording 自动抄表
automatic mode 自动模式
automatic modulation control 自动调节控制
automatic null-balancing potentiometer 自动零平衡电位计
automatic operation 自动操作
automatic packer 自动包装机
automatic phase shifter 自动移相器
automatic phase synchronization 自动相位同步
automatic power control 自动功率控制
automatic pressure reducing valve 自动减压阀
automatic program interrupt 自动程序中断
automatic programming 自动程序设计
automatic radio wind wane and anemometer 无线电自动风向风速仪
automatic reclosing 自动重合闸
automatic regulation 自动调节
automatic regulator 自动调节器
automatic release 自动脱扣
automatic reset relay 自动复位继电器
automatic scanning 自动扫查
automatic screwdriver 电动螺丝刀，电动螺钉旋具，电动起子
automatic send-receive 自动收发机
automatic spark ignition device 自动火花点火装置
automatic sprinkle system 自动洒水系统
automatic start 自动启动
automatic starter 自动启动器
automatic station 无人值守电站，自动操作站
automatic switch 自动开关
automatic synchronized system 自同步系统
automatic system 自动化系统
automatic teleswitch 自动遥控开关
automatic temperature recorder 温度自动记录器
automatic testing machine 自动试验机
automatic ticket issuing machine

(ATIM) 自动售票机
automatic titrator 自动滴定仪
automatic tracking 自动跟踪，自动跟踪仪
automatic train control (ATC) 列车自动控制装置
automatic train protection speed limit 列车自动保护速限
automatic vertical index 竖直度盘指标补偿器
automatic voltage regulator (AVR) 自动调压器，自动电压调整器
automatic weather station 自动气象站
automatic with bias station 带偏置自动操作器
automatic 自动
automation 自动化
automation and drives 自动化与驱动
automation bootstrap loader 自动引导装入程序
automation equipment 自动化设备
automation output station 自动输出操作站
automation system interaction 自动化交互系统
automation system interface 自动化系统接口
automaton 自动机
automobile 汽车
automobile industry 汽车制造工业
automorphism 自同构
automotive 汽车的，自动的
automotive body 汽车车身
automotive chassis 汽车底盘
automotive control 自动控制
automotive electrical equipment 汽车电气设备
automotive emission 自动发射
automotive engine 汽车发动机
autonomous 自治的，自主的
autonomous control 自治控制，自主控制
autonomous mobile robot 自主移动机器人
autonomous system 自治系统
autonomous vehicle 自主机车

autopneumatic circuit breaker 自动气动断路器
auto-polarization compensator 自动极化补偿器
autoprotolysis constant 质子自递常数
autoprotolysis reaction 质子自递反应
autoradiography 放射自显影法
autoregressive model 自回归模型
autoregressive moving-average 自回归移动平均
autosyn 自整角机，自动同步机
autotransductor 自耦磁放大器
auto-transformer 自耦变压器
auxiliary attachment 辅件
auxiliary calculation blocks 辅助计算功能块
auxiliary console 辅助控制台
auxiliary contact 辅助触点，联锁触点
auxiliary control panel 辅助操纵板，辅助控制盘
auxiliary controller bus (ACB) 辅助控制器总线
auxiliary device 辅助装置
auxiliary equation 辅助方程
auxiliary equipment 辅助设备
auxiliary feedwater pump 辅助给水泵
auxiliary feedwater tank 辅助给水箱
auxiliary function 辅助功能
auxiliary gas 辅助气体
auxiliary loop 辅助回路
auxiliary motor 辅助电动机
auxiliary output signal 辅助输出信号
auxiliary polynomial 辅助多项式
auxiliary power supply 辅助电源，自备电源
auxiliary servomotor 辅助伺服电动机
auxiliary storage 辅助存储器
auxiliary switch 辅助开关
auxiliary system 辅助系统
auxiliary terminal 辅助端
auxiliary type gravimeter 助动型重力仪
auxiliary unit 辅助单元
auxiliary valves 副阀，辅助阀
auxiliary water pump 辅助水泵

auxiliary winding　辅助绕组
auxiliary window　辅助窗口
auxochrome　助色团（用以使染料固着在织物上）
availability　可用性
available material　可用材料
available time　可用时间，有效时间
avalanche　雪崩
avalanche diode　雪崩二极管
avalanche transistor　雪崩三极管
avalanche-type photodiode　雪崩型光电二极管
average absolute value　平均绝对值
average availability　平均可用度
average compressive stress　平均压应力
average current　平均电流
average cutter diameter　平均刀尖直径
average error　平均误差
average field intensity　平均场强
average output quality (AOQ)　平均出厂品质
average output quality level　平均出厂品质水平
average power　平均功率
average sound level　平均声级
average strength　平均强度
average stress　平均应力
average value　平均值
average value detector　平均值检波器
average velocity　平均速度
average voltage　平均电压
average wind speed　平均风速
averaging control　平均值控制，均化控制
AVOmeter　安伏欧计，万用电表
AVR (automatic voltage regulator)　自动调压器，自动电压调整器
AWGN (additive white gaussian noise)　附加的白高斯噪声
axes　轴线
axial clearance　轴向间隙
axial contact bearing　轴向接触轴承
axial current flow method　轴向通电法
axial direction　轴向
axial displacement　轴向位移
axial factor　轴向
axial fan　轴流式风扇，轴流式通风机
axial force　轴向力
axial internal clearance　轴向游隙
axial load　轴向载荷
axial load factor　轴向载荷系数
axial locating surface　轴向定位面
axial pitch　轴向齿距
axial plane　轴向平面
axial pump　轴流泵
axial rake angle　轴向倾角
axial sensitivity　轴向灵敏度
axial stress　轴向应力
axial thrust　轴向推力
axial thrust load　轴向分力
axial tooth profile　轴向齿廓
axial vibration　轴向振动
axial-flow pump　轴流泵
axiom　原理，定理
axis　轴，轴线
axis guide　轴套
axis of abscissas　横轴
axis of ordinates　纵轴
axis of reference　参考轴，基准轴
axis of rotation　摆轴，旋转轴
axis of strain gauge　应变计［片］轴线
axle　车轴
axle bearing　轴承
axle testing machine　传动桥试验机

B

back angle 背锥角
back angle distance 背角距
back born wire 反生线
back cone 背锥
back cone distance 背锥距
back cone element 背锥母线
back connected switch 背面接线开关
back contact 后触点,背触点
back feed 反馈,回授
back flushing 反冲,反吹
back ionospheric sounder 反向电离层探测器
back layering 回流层现象
back panel 背板,后面板
back propagation (BP) 反向传播
back shaft 支撑轴
back pressure 背压
back up 支持,备用
back wash 反冲洗
back-coupled generator 反馈振荡器
backfill 回填
backfill material 回填物料
back-fire 逆火
back-forward counter 双向计数器
background 后台,背景,本底
background color 背景颜色,底色
background current 基流
background job 后台作业
background mass spectrum 本底质谱
background noise 本底噪声,背景噪声
background plate 背景板
background processing 后台处理
background program 后台程序
background register 辅助寄存器,后备寄存器
backing 垫料,衬垫
backing coil 补偿绕组
backing plate 垫板,背板
backing sand 背砂
bcklash 间隙,回差,间断,回跳
backlash characteristics 间隙特性

backlash tolerance 侧隙公差
backlash variation 侧隙变量
backlash variation tolerance 侧隙变量公差
backplane 总线板,背板
back-pressure valve 反压阀
backpropagation algorithm 反向传播算法
back-resistance 反向电阻计
backscattered electron image 背散射电子像
backspace 空格,回退
backspace character 退格符
back-to-back 背靠背的,紧接的
back-to-back arrangement 背对背安装
backtracking 反向跟踪
backup 备用,备份,拷贝,后备
backup cascade 后备串级
backup controller 备用控制器
backup copy 副本
backup diskette 备份软盘
backup file 备份文件
backup panel for console 控制台用备用仪表板
backup power source 后备电源
backup system 备用系统
backward channel 反向信道
backward coupler 反向耦合器
backward difference 反向差分
backward diode 反向二极管
backward kinematics 反向运动学
backward wave tube 反波管
backward-travelling waves 向后传播的波
backward-wave parametric amplifier 回波参数放大器
bad data 不良数据
bad data identification 不良数据辨识
bad value 坏值
baffle 隔板,挡板
baffle plate 挡块,遮挡板
baffle wall 隔板,遮挡墙

baffle-plate converter 挡板变换器
bag tie wire 捆扎用丝（锁口丝）
bag filter 除尘布袋
bainite 贝氏体
Baker clamping circuit 贝克钳位电路
balance 平衡，天平，余额，平衡称
balance condition 平衡状态
balance converter 平衡变换器
balance gas 平衡气
balance indicator 平衡指示器
balance level 衡准仪，水准仪
balance of machinery 机械平衡
balance of mechanism 机构平衡
balance of rotor 转子平衡
balance of rotors 回转体平衡
balance of shaking force 惯性力平衡
balance of voltage 电压平衡
balance out 抵消
balance output 对称输出
balance quality of rotor 转子平衡精度
balance state 平衡状态
balance tank 调节池，均衡槽
balance valves 平衡阀
balance weight 平衡锤，平衡块
balanced 平衡的，均衡的
balanced amplifier 平衡放大器，对称放大器
balanced bridge 平衡电桥
balanced bridge sampler 平衡桥式取样器
balanced demodulator 平衡解调器
balanced detector 平衡检波器
balanced level-type bell gauge 杠杆平衡式浮钟压力计
balanced load 平衡载重，平衡负载，对称负载
balanced modulator 平衡调制器，对称调制器
balanced noise limiter 平衡式限噪器
balanced relay 差动继电器
balanced supply 平衡电源
balanced three-phase system 对称三相系统
balanced three-wire system 对称三线制
balanced-pressure rotameter 压力补偿式转子流量计

balancer 平衡器
balances 天平
balance-type differential pressure recorder 平衡式差压记录仪
balancing 平衡
balancing equipment 平衡设备
balancing machine 平衡机
balancing machine sensitivity 平衡机灵敏度
balancing mass 平衡质量
balancing motor 平衡电动机
balancing pipeline 平衡水管
balancing quality 平衡品质
balancing relay 平衡继电器
balancing speed 平衡转速
balata spring 橡胶弹簧
ball 球，球芯
ball bearing 滚珠轴承，球轴承
ball screw 滚珠丝杠
ball screw assembly 滚珠丝杠装配
ball seat 密封圈
ball slider 球塞滑块
ball valve 球阀
ball valve for air supply 气源球阀
ball-and-socket joint 球窝接头
ballast 镇流器
ballast impedance 镇流阻抗
ballast resistance 镇流电阻
ball-bearing wire 滚珠轴承钢丝，滚珠用钢丝
ballistic galvanometer 冲击电流计
band 波段，频带，带，区域
band analyzer 带宽分析仪
band brake 带式制动器
band broadening 谱带扩展
band clamp 带夹
band elimination 带阻滤波器
band heater 环带状的电热器
band meter 波长计
band pass amplifier 带通放大器
band pass filter 带通滤波器
band rejection filter 带阻滤波器
band selector 波段选择器，波段开关
band switch 波段转换开关
band width 带宽，通带宽度，频带宽度
band width allocation 带宽分配

band width of video amplifier 视频放大器频宽
banded structure 条纹状组织
bandwidth coaxial probe 带宽同轴探头
bandwidth delayed sweep oscilloscope 带宽延迟扫描示波器
bandwidth electrical pulse 带宽电脉冲
bandwidth measurement 带宽测量
bandwidth meter 带宽测量仪
bandwidth minimization problem 带宽极小化问题
bandwidth voice network 声宽网络
bang-bang control 开关控制
bar 条,棒
bar generator 条状信号发生器
bar graph 棒图
bar graph indicator 棒图指示器
bar magnet 条形磁铁,磁棒
bar primary bushing type current transformer 棒形电流互感器
barbell 杠铃
barcode 条形码,条码技术
barcode scanner 条码扫描器
bare board 裸板,空板,裸印制板
bare copper wire 裸铜线
bare electrode 光焊条,裸露电极
bare metal 金属裸露
bare thermocouple temperature transducer 裸露式热电偶温度传感器
bare wire 光焊丝,裸线
barium 钡
barograph 气压计
barometer 气压表,气压计
barometer relay 气压继电器
barometric correction 气压修正
barostat 恒压器
barothermograph 气压温度计
barrel 滚筒,管筒,芯管
barrel distortion 桶形畸变,负畸变
barrel plating 滚镀
barrel tumbling 滚筒打光
barreling 滚光加工
barrette 方形桩
barretter 镇流电阻器
barricade 路障,障碍物
barrier 栏栅,护栏,障碍物,屏障

barrier block 路障
barrier gate 路闸
barrier layer 阻挡层
barrier plate 阻挡板
BAS (base application software) 基本应用软件
base application software (BAS) 基本应用软件
base circle 基圆
base cone 基圆锥
base coordinate system 基座坐标系
base cylinder 基圆柱
base diameter 基圆直径
base electrode 基极
base frame 基架
base insulator 基架绝缘器
base line 基线
base material 基材
base peak 基峰
base pitch 基圆齿距,基节
base plate 底板,垫板
base quantity 基本量
base radius 基圆半径
base register 基址寄存器
base sealing 底部密封胶
base signal generator 基极信号发生器
base slab 平底板
base spiral angle 基圆螺旋角
base support 底座支架
base unit 基础单元
base unit of measurement 基本测量单位
baseband coaxial cable 基带同轴电缆
baseline 基线,底线,基准线
baseline drift 基线漂移
baseline noise 基线噪声
baseline programme 基线计划
baseline 基准线
basement 地库,地窖,地下室
basic 基本的,主要的
basic arithmetic module 基本算法模块
basic circuit 基本电路,原理电路
basic controller 基本控制器
basic dynamic axial load rating 轴向基本额定动载荷
basic input subsystem 基本输入子系统

basic iron 碱性铁
basic NMR frequency 基本核磁共振频率
basic operation station 基本操作站
basic output subsystem 基本输出子系统
basic physics 基本物理量测定
basic processing cycle 基本处理周期
basic processing unit 基本处理单元
basic rack 基准齿条
basic solvent 碱性溶剂
basic standard 基础标准
basic static axial load rating 轴向基本额定静载荷
basic static radial load rating 径向基本额定静载荷
basic terminal 基本型端子
basis material 基体材料
batch 批次，批处理
batch control 批量控制，间歇控制
batch control station 批量控制站
batch controller 批量控制器，批量调节器
batch data acquisition unit 批量数据采集器
batch data set unit terminal 批量数据设定单元型端子
batch history data point 批量历史数据点
batch history prototype data point 批量历史模型数据点
batch inlet 分批进样
batch mode 间歇模式
batch process 间歇过程
batch process control 间歇过程控制
batch processing 批处理，成批处理
batch processing simulation 批处理仿真
batch reactor 间歇反应器
batch set 批量设定
batch set station 批量设定器
batch set unit 批量设定单元，批量设定器
batch status indicator 批量状态指示器
batch trend 批量趋势
batching plant 混凝土混合机，配料厂

batter pile 斜桩
battery 蓄电池
battery acid level 电池酸位
battery backup unit (BBU) 电池备用单元，电池后备单元
battery cell volt 蓄电池电压
battery charger 电池充电器
battery electric locomotive 电力机车
battery electrolyte 电池电解液
battery voltage meter 电瓶电压表
battery operated receiver 电磁供电接收机
battery/electric locomotive 电力电瓶双电源机车
baud 波特
baud rate 波特率
baud rate generator 波特率产生器
Bayes classifier 贝叶斯分类器
bayonet gauge 插入式测量仪表
BBU (battery backup unit) 电池备用单元，电池后备单元
BCD (binary coded decimal) 二-十进制编码
be put in storage 入库
beacon 警示灯，闪光指示灯
bead 焊珠
bead chain 滚子链条
bead glazing 压条装配玻璃法
beading stress 弯曲应力
beaker 烧杯
beam 横梁，声速
beam deflector 电子束偏转器
beam finder 电子束探测器
beam path distance 声程
beam ratio 声束比
beam-deflection ultrasonic flowmeter 声速偏转式超声流量计
beam-loading thermobalance 水平式热天平
bearing 支座，轴承
bearing alignment 方位对准
bearing alloy 轴承合金
bearing axis 轴承中心线
bearing block 轴承座
bearing bore diameter 轴承内径
bearing bush 轴瓦、轴承衬
bearing capacity 承载力，承载能力

bearing detector 轴承侦测器
bearing fittings 轴承配件
bearing force 承重能力，承载能力
bearing height 轴承高度
bearing life 轴承寿命
bearing outside diameter 轴承外径
bearing pad 支承垫片，承重垫片
bearing pile 支承桩
bearing pin 支承栓钉
bearing plate 支承垫板
bearing preload 轴承预负荷
bearing processing equipment 轴承加工机
bearing ring 轴承套圈
bearing stress 支承应力
bearing surface 支承面
bearing width 轴承宽度
beat 拍，拍
beat frequency 差频，拍频
beat frequency oscillator 拍频振荡器，差频振荡器
beat indicator 差频指示器
beat method 差拍法
beat method of measurement 差拍测量法
beat receiver 外差式收音机
beatening 敲打成形
bechtop 超净工作台
Beckman differential thermometer 贝克曼温度计
bed 机座
bed type milling machines 床身式铣床
bedding 底层，层理
bedrock 基层岩
Beer law 比尔定律
beginner all-purpose symbolic instruction code 初学者通用符号指令码
beginning of contact 起始啮合点
behavior 属性
behaviour 性能，状况
behavioural science 行为科学
Belisha beacon 人行道指示灯
bell button 电铃按钮
bell manometer 钟罩压力计
bell metal 钟铜
bell prover 钟罩校准器
bell 铃

belleville spring 碟形弹簧
belling 压凸，压凸加工
bellow pot 气囊，气囊筒
bellow type 波纹管式
bellows 膜盒，波纹管
bellows absolute-pressure gauge 膜盒式绝对压力表
bellows differential flowmeter 膜盒式差式流量计
bellows pressure gauge 波纹管压力表
bellows seal bonnet 波纹管密封型上盖
bellows valve 波纹管阀
bellows-type orifice meter 膜盒式孔板流量计
belt conveyor 带式输送机
belt drive 皮带传动
belt guard 皮带护罩
belt pulley 带轮
belt sander 带式打磨机
belt tension 皮带拉力
belt weigher 皮带秤
belt 带，皮带
bench hook 垫板，台垫
bench mark 水准点
bench type hand winding machine 台式手绕线机
benchmark 基准
benchmark example 基准例子，标准例子
bend 弯曲
bending 弯曲度
bending block 折刀
bending fatigue 弯曲疲劳
bending force 弯曲力
bending machine 弯曲机，折床
bending machines 弯曲机
bending moment 弯矩，弯曲力矩
bending strength 弯曲强度
bending stress 弯曲应力
bending vibration 弯曲振动
Benkelman beam test 贝克曼梁试验
bent stem earth thermometer 曲管地温表
bentonite 膨润土，膨土岩
benzalkonium bromide 苯扎溴铵
benzene 苯

benzoic acid and sodium benzoate 苯甲酸及其钠盐
BER（bit error rate） 误码率
berkelium（Bk） 锫
beryllium（Be） 铍
Bessel function 贝塞尔函数
Besson nephoscope 贝森测云器
betatron 电子回旋加速器，电子感应加速器
bevatron 质子加速器
bevel 斜角，斜面
bevel gear 锥齿轮，伞形齿轮
bevel gears 锥齿轮，圆锥齿轮机构
bezel 斜视规
bezel panel 面板
B-H curve 磁化曲线，B-H 曲线
B-H loop 磁滞回线
bias 偏差，偏压，偏流，偏置
bias control potentiometer 偏压调整电位计
bias error 偏置误差
bias in 内对角接触
bias out 外对角接触
bias resistance 偏压电阻，偏流电阻
bias voltage 偏置电压
biased circuit 偏压电路，偏流电路
biased relay 极化继电器，带制动的继电器
biased-rectifier amplifier 偏压整流放大器
BIBO（bounded input and bounded output） 有界输入有界输出
bidirectional pulse train 双向脉冲列
bi-directional triode thyristor 双向晶闸管
bi-directional vane 双向风向标，风信标
bifilar 双线的，双股的
bifilar coil 双线无感线圈
bifilar winding 双线无感绕组，双线无感绕法
bifurcate 分叉，分支，分路
big size nuts 大尺寸螺帽
bilateral antenna 双向天线
bilateral circuit 双向电路
bilateral current stabilizer 双向稳流器

bilateral diode 双向二极管
bilateral servomechanism 双向伺服机构
bilateral switch 双向开关
bilinear 双线性
bilinear control 双线性控制
bilinear system 双线性系统
bilinear transformation 双线性变换
bill of material 物料清单
billet 坯料
bills receivable 应收票据
bimetal 双金属
bimetal relay 双金属继电器
bimetallic 双金属的
bimetallic element 双金属元件
bimetallic instrument 双金属式仪表
bimetallic rotor 双金属转子
bimetallic temperature transducer 双金属温度传感器
bimetallic thermometer 双金属温度计
bimetallic wire 双金属线
bimotored 双马达的
binary 二进制的，二元的，二态的
binary arithmetic 二进制运算
binary array 双元阵
binary bit 二进位
binary card 二进制卡片，二进制号码
binary coded decimal 二进制编码的十进制
binary coded decimal（BCD） 二-十进制编码
binary control 二进制控制
binary digit 二进制数
binary digital 二进制数字
binary elastic scattering event 双弹性散射过程
binary elastic scattering peak 双弹性散射峰
binary element 二进制元件
binary image 双映像
binary logic system 双逻辑系统
binary phase shift keying（BPSK） 二相位键控
binary search tree 双搜索树
binary signal 二进制信号
binary storage element 二值存储元件

33

binary system 二进制系统
binder 黏剂，黏合料，黏结剂
binder plate 压板
binding 连接，联结
binding energy 结合能
bioautography 生物自显影法
biochemical analysis 生物技术分析
biochemical analyzer 生化分析仪
biochemical engineering 生物化工
biochemical quantity sensor 生化量传感器
biochemical quantity transducer 生化量传感器
biocybernetics 生物控制论
biohazard safety equipment 生物安全柜
biological feedback system 生物反馈系统
biological quantity sensor 生物量传感器
biological quantity transducer 生物量传感器
biomedical 生物医学
biomedical analyzer 生物医学分析仪
biomedical control 生物医学控制
biomedical engineering 生物医学工程
biomedical system 生物医学系统
biomolecular interaction analysis system 生物分子间相互作用分析系统
bionics 仿生学
bio-reactor 生物反应器
bio-safety cabinet 生物柜
biosensor 生物传感器，生物感应器
biotechnology 生物工艺学
bio-technology related instruments 生物技术关连仪器
biphase 双相的
biphenyl 联苯，联二苯
bipolar junction transistor (BJT) 双极结型晶体管
bipolar motor 双极电动机
bipolar transistor 双极型晶体管
birefringent polarizer 双折射偏光器
bisector 平分线
bismaleimide-triazine resin 双马来酰亚胺三嗪树脂
bismuth (Bi) 铋
bismuth mold 铋铸模

bispectrum 双频谱
bispectrum estimation 双频谱估计
bistability 双稳性
bistability device 双稳定装置
bistable circuit 双稳态电路
bistable element 双稳态元件
bistable logic multivibrator 双稳态逻辑多谐振荡器
bistable multivibrator 双稳态多谐振荡器
bistable trigger element 双稳态触发元件
bit 比特，位
bit error rate (BER) 误码率
bit parallel 位并行
bit per inch 位/英寸
bit per second (bps) 位/秒，每秒传送位数
bit serial 位串行
bit vector 位向量
bite 咬入
bit-serial highway 位串行信息公路
bitumen 沥青
bitumen coating 沥青外搪层，沥青外衬
bitumen lining 沥青内搪层，沥青衬里
bituminous concrete 沥青混凝土
bituminous waterproof membrane 沥青防水膜
bivalent 二价的
bivane 双向风向标，双风信标
BJT (bipolar junction transistor) 双极结型晶体管
black 黑色
black box 未知框，黑匣，黑箱
black box modeling 黑箱模型化
black box testing approach 黑箱测试法
black light crack detector 黑光探伤器
black light filter 透过紫外线的滤光片，黑光滤光片
black sheet iron 黑钢皮
black smoke 黑烟
black wire 黑色导线
blackbody 黑体
blackbody chamber 黑体腔

blackbody furnace 黑体炉
black-heart malleable iron 黑心可锻铸铁
blacking hole 涂料孔（铸疵）
blacking scab 涂料疤
blade 剪刀，叶片，刀形开关
blade angle 刀齿齿廓角
blade edge radius 刀尖圆角半径
blade letter 刀尖凸角代号
blade life 刀尖寿命
blade point width 刀顶宽
blade saw 锯片
blade wheel type flowmeter 叶轮式流量计
blades 刀片
bland position 毛坯位置
blank 空白，间隔
blank character 间隔符
blank flange 盲板法兰，盲板凸缘，管口盖板
blank form 空白格式
blank holder 防皱压板
blank semi-graphic instrument panelboard 空白半模拟仪表盘
blank through dies 漏件式落料模
blanking die 冲裁模，下料模，切口冲模
blanking plate 封板
blast furnace 鼓风炉
blast pressure gauge 风压表
blast wave 冲击波
blastfurnace 高炉，鼓风炉
blast-furnace cast iron 高炉生铁
blast-furnace slag cement 炉渣水泥
blasting 爆石，爆破
blazed grating 闪耀光栅
bleed nipple 放气嘴，减压嘴
bleed screw 放气螺钉，减压螺钉
bleeder 分压器，分泄电阻，附加电阻
bleeder circuit 分压电路，泄放电路
blend 混合
blend ration control 混合比例控制
blended cement 混合水泥
blender 试料混合器
blending 混合
blending control 混合控制
blending controller 混合调节器

blending PI controller 混合PI调节器
blending system 混合系统，调和系统
blind hole 盲孔
blind power 无功功率，电抗功率
blind search 盲目搜索
blind set plug 设定盲插头
blind via 盲孔
blinding 碎石料
blinding plate 盲板
blinker lamp 闪光信号灯
blinker light 闪烁灯光
blinker signal equipment 闪光信号装置
blister 气泡，水泡
blister steel 浸碳钢
block 块体，字块，字组，均温块
block check 块检验
block check character 块检验字符
block diagonalization 块对角化
block diagram 方块图，方框图
block gauge 块规
block length 字块长度
block push-button 联锁按钮
block relay 联锁继电器
block transfer 块传递
block word data 字数据块
blocker 雏形锻模
blocking 闭塞，粗胚锻件
blocking condenser 隔直流电容器，耦合电容器
blocking hammer 落锤
blocking layer 阻挡层，闭锁层
blocking tube oscillator 电子管间歇振荡器
blood calcium ion transducer 血钙传感器
blood carbon dioxide transducer 血液二氧化碳传感器
blood chlorine ion transducer 血氯传感器
blood electrolyte transducer 血液电解质传感器
blood flow transducer 血流传感器
blood gas transducer 血气传感器
blood oxygen transducer 血氧传感器
blood pH transducer 血液pH传感器
blood potassium ion transducer 血钾

传感器
blood sodium ion transducer　血钠传感器
blood sugar analyzer　血糖分析仪
blood transfusion set　输血器
blood-gas analyzer　血气分析仪
blood-group immune transducer　免疫血型传感器
blood-pressure transducer　血压传感器
blood-volume transducer　血容量传感器
bloomery iron　熟铁块
blow down　吹倒
blow down valve　排水阀
blow hole　破孔，铸件气孔
blow moulding　吹膜
blow　吹
blower　吹风机，鼓风机
blue shift　蓝移
blue shortness　青熟脆性
blue smoke　蓝烟
blue　蓝色
bluff body　阻流体，非流线形体，不良流线体
blushing　雾浊
board　板
board drop hammer　板落锤
board tester　电路板测试仪
bobbin　绕线管
Bode diagram　伯德图
body　本体，机壳
body guidance mechanism　刚体导引机构
body temperature transducer　体温传感器
body wetting before glazing　补水
body wrinkle　侧壁皱纹
bogie　转向架
bogie washer　转向架清洗设备
boiler room　锅炉房
boiler　锅炉
boiler water gauge　锅炉水位表
boiling point　沸点
bold line　粗线
bollard　护柱，系船柱
bollard plinth　护柱柱基，护柱基座
bolometer　辐射热计，热辐射仪
bolster　垫板，承枕，横撑，上下模板

bolster bogie　有承梁转向架
bolsterless bogie　无承梁转向架
bolt wire　螺栓钢条
bolt　螺栓，拧螺丝
bolts small hexagon head with fit neck　小六角头导颈螺栓
bolts small hexagon head with fit neck and hole through the shank　小六角头螺杆带孔导颈螺栓
bolts small hexagon head with hole through the shank　小六角头螺杆带孔螺栓
bolts small hexagon head with holes through the head　小六角头头部带孔螺栓
Boltzman machine　玻耳兹曼机
bomb head tray　弹头托盘
bombardment　轰击
bond　耦合，结合，连接器
bond coat　黏合层
bond strength　黏合强度
bond stress　黏合应力
bond test　黏结力试验
bond tester　接头电阻测试仪
bonded phase chromatography　化学键合相色谱法
bonded strain gauge　粘贴式应变计
bonded wire　黏合漆包线
bonderized steel sheet　邦酸防蚀钢板
bonderizing　磷酸盐皮膜处理，磷化处理
bonding agent　结合剂
bonding layer　黏结层
bonding phase column　键合相柱
bonding sheet　黏结片
bonding strength　黏合强度
bonding wire　接合线，焊线
bonnet　阀盖
Boolean algebra　布尔代数
Boolean operation　布尔运算
Boolean alarm　布尔报警
Boolean function　布尔函数
Boolean logic　布尔逻辑
boom　吊杆
boomerang grab　自返式取样器，回飞棒
boomerang gravity corer　自返式深海

取样管，重力式取样管
boost 升高，加强，加速，增压，提高
boost chopper 升压斩波电路
boost converter 升压变换器
boost transformer 增压变压器
boost pump 升压泵
booster 增强器
booster amplifier 升压放大器
booster diode 升压二极管
booster pump 增压泵
booster pumping station 增压抽水站
booster relay 升压继电器
booster transformer 升压变压器，增压变压器
booster water pump 增压水泵，增压抽水机
bootload 引导装入
bootstrap circuit 自举电路，启动电路
borated water storage tank 含硼水贮存箱
borazon 氮化硼立方晶
bore 孔，腔，镗，钻孔，内孔，钻削，镗削
bore check 精密小孔测定器
bored pile 螺旋钻孔桩
bored tunnel 钻挖的隧道
borehole acoustic television logger 超声电视测井仪
borehole compensated sonic logger 补偿声波测井仪
borehole gravimeter 井中重力仪
borehole log 钻孔记录，测井曲线
borehole thermometer 井温仪
boric acid 硼酸
boring 镗，镗孔，钻探，冲孔
boring heads 镗孔头
boring machine 镗床，镗孔机，钻探机
boring machines 镗床
borneol 冰片
boron 硼
bottom block 下垫脚
bottom board 浇注底板
bottom clearance 顶隙
bottom echo 底面反射波
bottom flange 下盖
bottom heave 底部隆起

bottom land 齿槽底面
bottom layer 底层
bottom plate 下托板，底板，下固定板
bottom pour mold 底浇铸模
bottom pouring 浇注
bottom slide press 下传动式压力机
bottom surface 底面
bottom 底部
bottoming 浓度饱和，到达底部
bottom-up development 自下而上开发
boulder 巨砾
bound 结合
bound charge 束缚电荷
bound data point 连接数据点
bound electron 束缚电子
boundary 分界线，界线
boundary condition 边界条件
boundary detection 边界搜索
boundary dimension 外形尺寸
boundary element method 边值原理方法
boundary integral formulation 边值积分公式
boundary layer 边界层
boundary lubrication 界面润滑
boundary surface 分界面
boundary value analysis 边界值分析
boundary-layer type mass flowmeter 边界层流型质量流量计
boundary-value problem 边值问题，极限值问题
bounded 有界的
bounded disturbance 有界扰动
bounded input and bounded output (BIBO) 有界输入有界输出
bounded noise 有界噪声
bounding method 约束方法
Bourdon pressure sensor 弹簧管压力检测元件
Bourdon tube 弹簧管，波登管
Bourdon tube pressure gauge 弹簧管压力表
Bourdon-tube manometer 弹簧管压力表
Bourdon-tube manometer with electric contacts 电接点弹簧压力表
box 盒，箱

box annealing 箱型退火
box bridge 盒式电桥，电阻箱电桥
box carburizing 封箱渗碳
box control state 箱控制状态
box girder 箱形大梁
box iron 槽钢 槽块
box spanner 管钳子
box status display 箱状态显示画面
BP (back-propagation) 反向传播
B-power 乙电源，阳极电源
bps (bit per second) 位/秒，每秒传送位数
BPSK (binary phase shift keying) 二相位键控
brace 撑杆，支撑
bracing 支撑
bracing structure 支撑结构
bracing wire 拉索，拉线，拉铁丝
bracket 支架，托架，括号
Bragg's equation 布拉格方程
braided cable 编织电缆，屏蔽电缆
braided wire 编（织）线
brain model 脑模型
brake 制动器，闸，刹车
brake horse power 制动马力
brake lining 制动器摩擦衬片
brake magnet 制动电磁铁
brake motor 制动电动机
brake pedal 刹车踏脚板
brake power 制动功率
brake switch 制动开关
brake system 制动系统
brake test 制动器试验
brake tester 制动系统测试器
brake valves 制动阀，闸阀
brake protecting relay 制动保护继电器
braking 制动
braking distance 制动距离，刹车距离
braking element 制动元件
braking magnet 制动电磁铁
braking relay 制动继电器
braking time 制动时间
branch 分支，支路，转移
branch admittance matrix 支路导纳矩阵
branch and link 转移和连接
branch box 分线盒

branch cable 支线电缆
branch circuit 分支电路
branch current 支路电流
branch current method 支路电流法
branch impedance matrix 支路阻抗矩阵
branch line 支线
branch pipe 支管
branch pipework 支管
branch switch 分路开关
branching 分支，支路
branching cable 分支电缆
brand 品牌
brass 黄铜
brass gate valve 黄铜闸门阀
brass nuts 铜螺帽
brass wire brush 铜丝刷
brass-plated steel wire 镀黄铜钢丝
Braun-tube 布劳恩管，阴极射线管
brazing wire 铜焊线
breadth-first search 宽度优先搜索，横向搜索
break frequency 拐点频率
break pressure tank 减压配水缸箱，水压调节池
break 断开，断路
break contact 常闭触点，动断触点
breakage 破裂
break-and-make switch 断续开关
breakaway force 起步阻力
break-away panel 可断拼板
breakaway point 分离点
breakdown diode 击穿二极管
breakdown maintenance 故障维修
breakdown slip 停转转差率，极限转差率
breakdown switch 故障开关
breakdown torque 极限转矩
breakdown voltage 击穿电压，破坏电压
breakdown voltage rating 绝缘强度
breakdown voltage testing 绝缘强度测试
breaker 轧碎机，碎石机，隔断器，闸，断路器
breaker switch 断路开关
breaker coil 跳闸线路

breaking 断开，切断
breaking length 断裂长度
breaking of contact 触点断开
breaking point control 断点控制
breaking step 失步
breaking strength 抗断强度
breaking-down test 击穿试验，破坏试验
break-off core 缩颈砂芯
breakpoint 断点，折点
breakpoint function 中断点功能
breakpoint switch 停止点开关，折点开关
breakthrough 击穿，突破
breather 换气，通气孔，呼吸器
breather valve 通气阀
breathing 排气
breathing apparatus 呼吸器具
breeder reactor 增殖反应堆
B-register 变址寄存器
bremsstrahlung 轫致辐射
brick 砖
brick molding 砌箱造模法
bridge 桥，电桥，桥接线，桥型网络，桥接器
bridge abutment 桥台
bridge and tunnel engineering 桥梁与隧道工程
bridge arm 电桥臂
bridge arm ratio 电桥比
bridge balance 电桥平衡
bridge calibration 电桥校准
bridge circuit 桥接电路
bridge connection 桥接，桥式接线
bridge crane 桥式吊机
bridge cranes 桥式起重机，桥式吊车
bridge deck 桥面板，桥板，桥面
bridge for measuring temperature 测温电桥
bridge girder 桥大梁
bridge megger 桥式高阻表，绝缘测量器
bridge method 电桥法
bridge pier 桥墩
bridge rectifier 桥式整流器
bridge resistance 桥路电阻，电桥电阻
bridge reversible chopper 桥式可逆斩波电路
bridge-contact 桥式断点
bridged contact pattern 桥型接触斑点
bridged impedance 桥接阻抗
bridge's balance range 电桥平衡范围
bridge-type frequency meter 电桥式频率计
bridgeworks 桥梁工程
bridle joint 啮接
bridle wire 绝缘跨接线，跳线
bright annealed wire 光亮退火钢丝
bright electroplating 辉面电镀
bright field electron image 明场电子像
bright heat treatment 光辉热处理
bright wire 光亮钢丝光面线
brightness 亮度
brightness meter 亮度计
brightness regulator 亮度调节器
brilliance 亮度，辉度，光泽
Brinell hardness 布氏硬度
Brinell hardness number 布氏硬度值
Brinell hardness penetrator 布氏硬度压头
Brinell hardness test 布氏硬度试验
Brinell hardness tester 布氏硬度计
bring into step 整步，使同步
bring to rest 停车，停止运行
brittleness material 脆性材料
broach 钻头，凿子
broaching 拉孔，拉刀切削
broaching machine 拉床，剥孔机，铰孔机
broad band 宽频带，宽波段
broad band electro-optic modulator 宽带光电调制器
broad band millivoltmeter 宽带毫伏计
broad band spectrum 宽波段
broadband LAN 定带局域网，宽带局域网
broad-band operation 宽带运行，宽波段运行
broad-band random vibration 宽带随机振动
broadcast 广播
broadcast band 广播波段

39

broadcast interference 广播干扰
broadcast network 广播网
broadening correction 加宽校正
broadening correction factor 加宽校正因子
brominated epoxy resin 溴化环氧树脂
bromine (Br) 溴
bromine method 溴量法
bromophenol blue 溴酚蓝
bronze 青铜
brown 棕色
Brownian motion 布朗运动
brush 电刷,刷子
brush contact 电刷触点
brush motor 整流子电动机,换向器电动机
brush spark 电刷火花
brushless 无触点运动
bubble 水准泡,膜泡,气泡
bubble accumulator 气泡贮存器
bubble chamber 气泡室
bubble-tube 吹气管
bubble-tube pressure sensing device 吹气式压力计
Buchholz relay 气体继电器,瓦斯继电器
buck chopper 降压斩波电路
buck converter 降压变换器
bucket 斗,吊斗,桶,存储桶
bucket conveyor 斗式输送机
bucket thermometer 表层温度表
bucking coil 补偿绕组
buckle 使弯曲
buckling 压曲,压弯,纵弯曲
buckling load 压曲临界荷载
buffer 备件,备品,缓冲,缓冲器,减震器,阻尼器
buffer amplifier 缓冲放大器
buffer area 缓冲地区
buffer circuit 缓冲电路,阻尼电路
buffer solution 缓冲溶液
buffer stage 缓冲级
buffer storage 缓冲存储器
buffing 抛光
buffing wheel 抛光轮,砂轮
bug 故障,程序错误
build 建立

build up 增加,组合,提升
builder 生成
builder's lift 施工用升降机
building automation and control net 建筑物自动化和控制网络
building management system 智能建筑管理系统
building site 建筑工地
building technology science 建筑技术科学
build-up multilayer printed board 积层多层印制板
build-up printed board 积层印制板
build-up time 建立时间
built-in check 自动校验,内部校验
built-in galvanometer 内装式检流计
built-in storage 内存储器
bulb 电灯泡
bulging 撑压加工
bulk cargo 散装货
bulk storage memory 大容量存储器
bulk supply substation 主变电站
bulk type semiconductor strain gauge 体型半导体应变计
bulk zinc oxide varistor 体型氧化锌电压敏电阻器
bulldozer 推土机,铲泥车
bullet wire 中碳钢丝
bump 连续冲击,碰,撞击
bump test 连续冲击试验,颠簸试验
bump testing machine 连续冲击台,碰撞试验台
bumper 缓冲器,防撞器,防撞杠
bumper block 缓冲块
bumpless start 无扰动启动
buna-n rubber 丁钠橡胶
bunch 群,串
bunch wire 绞合线,多绞线
bundle of electrons 电子束
Bunsen burner 本生灯
buoy 浮标
buoy array 浮标阵
buoy float 浮标体
buoy motion package 浮标运动监测
buoy station 浮标站
buoyancy correction 浮力修正
buoyancy displacer level transmitter

浮力沉筒式液位变送器
buoyancy level measuring device 浮力液位测量仪表
buoyancy type level measuring transmitter 浮力式液位变送器
buoyant force 浮力
Burdon pressure sensor 波登管压力传感器
burette 滴定管
burglar alarm 防窃报警器，防盗自动警铃
burglar alarm system 防盗警钟系统
buried 埋藏的
buried cable 地下电缆
buried concrete 埋入地下的混凝土
buried resistance board 埋电阻板
buried via hole 埋孔
buried wire 埋地电线
burner 燃烧器，炉头
burner control system 燃烧器控制系统
burner management system 燃烧器管理系统
burning method 燃烧法
burnishing 抛光
burnishing die 挤光模
burnt iron 过烧钢
burr 毛边
burring 冲缘加工，毛口磨光
burst 爆炸，爆发
burst pressure 破裂压力
bursting 爆裂
bus 母线，总线
bus converter 总线转换器
bus line 总线
bus master 总线主设备
bus mother board 总线母板
bus multiprocess 总线多处理过程
bus network 总线网
bus repeater 总线转发器
bus slave 总线从设备，总线受控
bus topology 总线拓扑
bus type current transformer 母线式电流互感器
busbar 母线，汇流排

bus-bar 汇流条
bus-coupler 母线耦合器
bushing 衬套，护线帽，套筒
bushing block 衬套
bushing type current transformer 套管式电流互感器
busy 忙，占线
busy state 忙碌状态
busy-back signal 占线信号，忙音信号
butane 丁烷
butt fusion welding 对头熔接
butt joint 对接合，对接榫
butt mitred joint 对接角榫
butt welding 对焊，对接焊，对接焊接
butterfly cock 蝶形旋阀
butterfly gate 蝶形闸门，蝶阀
butterfly tuner 蝶式调谐器
butterfly valve 蝶阀
butterfly valves with gear actuator 蜗轮传动蝶阀
Butterworth filter 巴特沃斯滤波器，蝶值滤波器，最大平坦响应滤波器
button 按键，按钮
button configuration 键组态
button die 镶入式圆形凹模
button switch 按钮开关
buttress 支墩
buttress thread form 锯齿形螺纹
butt welding 对焊
buttwelding valves 对焊连接阀
buzzer 蜂音器，蜂鸣器
buzzle 蜂鸣器
bypass 旁路
by-pass 旁路，支路
bypass heat treatment 旁路热处理
by-pass injector 旁通进样器
by-pass joint 旁路接头
bypass manifold 旁路集管
bypass type flowmeter 旁通管式流量计
bypass valve 旁通阀
by-pass valve 旁通阀
Byram anemometer 拜拉姆风速表
byte serial 字节串行
byte 字节（八位）

C

cab 小室，驾驶室
cab composite material 复合材料
cab heater 驾驶室加温器
cabinet 橱柜，机箱、柜
cabinet grounding 机柜接地
cable 电缆，多芯导线
cable armor 电缆铠装
cable box 分线盒
cable bracket 电缆托架
cable break 电缆断线事故
cable bridge 电缆桥架
cable channel 电缆沟，电缆槽
cable column 卷缆柱
cable conduit 电缆管
cable coupler 电缆耦合器
cable cranes 缆索起重机
cable draw pit 电缆沙井，铺缆井
cable duct 电缆管道，电缆汇线槽
cable fault 电缆损伤，电缆故障
cable fault locator 电缆故障定位器
cable for communication 通信电缆
cable gland 电缆密封套
cable joint 电缆接头
cable laying wagon 电缆敷设车
cable lead 电缆引线
cable loss 电缆损失
cable making tools 造线机
cable noise 电缆噪声
cable paper 电缆纸
cable route 电缆路线
cable routing 电缆路由选择
cable sheath 电缆包皮层
cable supported viaduct 悬索高架桥
cable suspension wire 电缆悬挂线
cable terminal 电缆接头
cable trench 电缆槽
cable trough 电缆坑
cable trunk 电缆干线，电缆管道
cable tunnel 电缆隧道
cable type current transformer 电缆式电流互感器
cable wire 钢丝绳

cable-joint box 电缆接头盒
cable-tension transducer 电缆张力传感器
cableway system 架空电缆系统
cabling requirements 电缆安装要求
cache memory 高速缓冲存储器
CACSD (computer-aided control system design) 计算机辅助控制系统设计
CAD (computer aided design) 计算机辅助设计
CADFC (computer-aided design of feedback controller) 计算机辅助反馈控制器设计
cadmium 镉
cadmium cell 镉电池
cadmium copper wire 镉铜合金线，强抗张力导线
cadmium photocell 镉光电池
cadmium-nickel accumulator 镉镍蓄电池
CAE (computer aided engineering) 计算机辅助工程
CAE (computer-aided engineering) 计算机辅助工程
caesium (Cs) 铯
cage 套筒，潜水罐笼
cage guiding 套筒导向
cage rotor 笼形转子
cage valve 套管阀
cage-type float-operated gauge 箱式浮子液位计
cake adhesive retention meter 泥饼黏滞性测定仪
calcium (Ca) 钙
calcium silicate insulation 硅酸钙绝热制品
calculate 计算
calculated bending moment 计算弯矩
calculated maximum flow coefficient 最大计算流量
calculated normal flow coefficient 正常计算流量系数

calculated results data point（CRDP） 计算结果数据点
calculated value 计算值
calculation 计算
calculation block 计算功能块
calculation function 计算功能
calculation sheet 计算书
calculation unit 算术运算器
calculation value 计算值
calculator 计算器
calculator algorithm 计算器算法
calculus 计算，演算，微积分学
calendaring molding 压延成形
caliber gauge 测径规，外径卡规
caliber 口径
calibrate 校正，调整，校准，定标
calibrated attenuator 校正衰减器
calibrated signal generator 校准信号发生器
calibrating devices for instruments 仪表检验设备
calibrating gas 校正气
calibrating period 校准周期
calibrating voltage 校准电压
calibration battery 校准用电池
calibration block 标准试块
calibration characteristics 校准特性，分度特性
calibration coefficient of wave height 波高校正系数
calibration component 校准组分
calibration curve 校准曲线，校正曲线
calibration cycle 校准周期，校验周期，校准循环
calibration equation 校准公式，分度公式
calibration equipment of reversing thermometers 颠倒温度表检定设备
calibration gas mixture 校准混合气
calibration hierarchy 校准层次
calibration instrument 校准用测量仪表
calibration point 校准点，分度点
calibration quantity 校准量
calibration record 校准记录
calibration regulator 校准用调节器
calibration report 校验报告

calibration rotor 标定转子
calibration solution 校准液
calibration table 校准工作台
calibration traceability 跟踪校准
calibration 校准，标定
calibrator 口径测量器，校验器
calibrator above ice-point 零上检定器
calibrator below ice-point 零下检定器
calibrator for ice-point 零点检定器
calibrator for oxygen pressure gauge 氧气压力表校验仪
calibrator for pressure gauges 标准压力表
calibrator for vacuum gauges 真空表校验表
californium（Cf） 锎
caliper gauge 卡规
caliper measure 测径
caliper profiler 纸张厚度计
calipers 卡规
calking tool 密缝錾
call error 调用错误
calling 呼叫
calling device 呼叫设备
calling relay 呼叫继电器
calling signal 呼叫信号
calliper gauge 孔径规
calomel electrode 甘汞电极
calomel half-cell 甘汞半电池
calorie（cal） 卡（热量单位）
calorific capacity 热容（量）
calorific value 热值
calorifier 加热器
calorimeter 热量计
calorimeter instrument 热量计式测试仪表
cam 凸轮
cam bezel ring 卡口式盖环
cam block 滑块
cam die bending 凸轮弯曲加工
cam mechanism 凸轮机构
cam operated switch 凸轮开关
cam profile 实际廓线，凸轮廓线
cam set controller 凸轮设定控制器
cam with oscillating follower 摆动从动件凸轮机构
CAM（computer aided manufacturing）

计算机辅助制造
CAMAC (computer automated measurement and control) 计算机自动测量和控制
camber 拱度,电弧弯曲
camera 照相机,摄影机,暗箱
camera length 相机长度
camera signal 影像信号
Campbell bridge 坎贝尔电桥
Campbell-Stokes sunshine recorder 聚集日照计,坎贝尔-斯托克斯日照计
cancel character 作废字符
cancel 取消、省略
cancellation 抵消,相约
candela (cd) 烛光(发光强度单位)
canonical state variable 规范化状态变量
cantilever 悬臂,悬臂梁
cantilever beam 悬臂梁
cantilever bridge 悬臂桥
cantilever crane 悬臂吊机
cantilever footing 悬臂基脚
cantilever foundation 悬臂地基
cantilever structure 悬臂结构
cantilever support 悬臂支架
cap 帽,盖,管帽
cap nut 盖螺母,盖形螺母
capacitance balance 电容平衡
capacitance bridge 电容电桥
capacitance coupling 电容耦合
capacitance diaphragm manometer 电容式膜片压力计
capacitance effect 电容效应
capacitance hygrometer 电容湿度计
capacitance measurement instrument 电容测量仪表
capacitance meter 电容计
capacitance micrometer 电容式测微计
capacitance pressure transducer 电容式压力传感器
capacitance relay 电容继电器
capacitance type pressure sensor 电容式压力传感器
capacitance 容量,电容
capacitance-resistance coupling 阻容耦合
capacitive 电容的,电容性的
capacitive bolometer 电容式辐射热测定器
capacitive compensation 电容补偿
capacitive coupling 电容耦合
capacitive current 电容电流
capacitive displacement transducer 电容式位移传感器
capacitive divider 电容分压器
capacitive feedback 电容反馈
capacitive load 电容负载
capacitive reactance 容抗
capacitive susceptance 容纳
capacitively loaded 电容负载
capacitor 电容器
capacitor activated transducer 电容触发传感器
capacitor filter 电容滤波器
capacitor microphone 电容话筒,微音器
capacitor motor 电容启动电动机
capacitor start 电容启动
capacitor storage 电容器存储器
capacity 容量
capacity control valve 容量控制阀
capacity correction 容量修正
capacity factor 容量因子
capacity ground 接地电容
capadyne 电致伸缩继电器
capillary column 毛细管柱
capillary column gas chromatography 毛细管柱气相色谱法
capillary electrochromatography 毛细管电色谱法
capillary electrophoresis 毛细管电泳,毛细管电泳法
capillary gas chromatograph 毛细管气相色谱仪
capillary gas chromatography 毛细管气相色谱法
capillary gel electrophoresis 毛细管凝胶电泳
capillary ion analysis 毛细管离子分析
capillary isoelectric focusing 毛细管等电聚焦
capillary phenomenon 毛细现象
capillary supercritical-fluid chromatog-

raphy 毛细管超临界流体色谱法
capillary viscometer 毛细管黏度计
capillary zone electrophoresis（CZE）毛细管区带电泳法
capital works 基本建设工程
capping ends 顶盖末端
capsule 膜盒
capsule aneroid 膜盒气压计
capsule platinum resistance thermometer 套管式铂电阻温度计
capsule pressure gauge 膜盒压力表
capsule type manometer 膜盒式压力计
capsule-type micromanometer 膜盒式微压计
captance 容抗
captive chains calibration 链码校准
capture 俘获
car wheel lathe 车轮车床
carat balance 克拉天平
carbide 碳化合金，碳化物
carbon（C） 碳
carbon and hydrogen analysis meter 碳氢元素分析仪
carbon brush 碳刷
carbon copy 打字副本
carbon disulfide 二硫化碳
carbon humidity-dependent resistor 碳湿敏电阻器
carbon monoxide 一氧化碳
carbon pole 碳素电极
carbon resistor 碳电阻器
carbon ring 炭环，炭精环
carbon steel 碳钢
carbon strip 炭条
carbon structural wire 碳结构钢丝
carbon tool steel 碳素工具钢
carbonation 碳化
carbonation depth 碳化深度
carbonation process 碳化过程
carbon-filament lamp 碳丝灯泡
carbonization 碳化
carbonyl iron 羰基铁
carborundum 碳化硅，金刚砂
carburettor 化油器，汽化器
carburization 渗碳
carburized case depth 渗碳层深度
carburizing 渗碳

card cage instrumentation 插件箱式仪器
card connector 印版插座，卡件插座
card installation procedure 卡件安装程序
card punch 卡片穿孔机
card punch unit 卡片穿孔装置
card read unit 卡片阅读装置
card reader 卡片阅读机
card reproducer 卡片复制机
card slot 卡槽
card test extender 卡片测试延伸板，功能板测试延伸器
card test module 功能板测试转接器
card type indicator 图表式指示器
cardan shaft 万向轴
car-door electric contact 车厢门触点
career card 履历卡
cargo collection 揽货
cargo handling area 货物装卸区
Carlson type strain gauge 卡尔逊应变计
carrier 载波，载流子
carrier and sideband 载波和边带
carrier band LAN 载波带局域网络
carrier detector 载波检波器
carrier equipment 载波设备
carrier foil 载体箔
carrier frequency 载波频率
carrier frequency amplifier 载波频率放大器
carrier gas 载气
carrier generator 载波发生器
carrier level 载波电平
carrier magnetic amplifier 载波磁放大器
carrier ring 承载圈，垫圈
carrier sense 载波侦听
carrier sense multiple access detect 线路监听多次存取检测
carrier sense multiple collision detect 线路监听多次碰撞检测
carrier sync 载波同步
carrier wave 载波
carrier wire 载波电缆，载波线
carrier-current communication 载波通信

carry　进位
carry storage　进位存储
carry storage register　进位寄存器
carrying capacity　运载量，载重量，承载能力
car-switch control　开关控制，转接控制
Cartesian　笛卡儿
Cartesian coordinate manipulator　直角坐标操作器
Cartesian coordinates　笛卡儿坐标
Cartesian manipulator　笛卡儿算子
Cartesian robot　直角坐标型机器人
cartography and geographic information engineering　地图制图学与地理信息工程
carton box　纸箱
carton　纸箱，纸板箱
cartridge　子弹，弹药筒，盒式磁带机
cartridge disk　盒式磁盘
cartridge disk drive　盒式磁盘机
cartridge heater　发热管
cartridge operated tool　弹药推动的工具
cartridge paper　图画纸
cartridge tape　盒式磁带
cartridge type respirator　滤罐型呼吸器，筒型防毒面具
carts and laboratory table attachments　实验台用附属器具
cascade　串级，串联，级，级联
cascade amplifier　串级放大器，级联放大器
cascade amplitude limiter　串级限幅器
cascade compensation　级联补偿
cascade compensation network　串联校正网络
cascade connection　级联，串联
cascade control　串级控制，串级调节
cascade control module　串级控制组件框
cascade control system　串级控制系统
cascade controller　串级调节器
cascade converter　级联变换器
cascade electrooptic modulator　级联光电调制器
cascade exciter　串级激励，级联激励
cascade recording controller　串级记录

cascade set　串级设定
cascade setpoint　串级设定点
cascade speed control　串级调速
cascade system　串级系统
cascade voltage transformer　级联式电压互感器，感应式电压互感器
case　箱
case crushing　齿面塌陷
case-based design　基于实例设计
casein　酪蛋白，干酪素
cash counting and bagging equipment　钱币计数及袋装设备
casing　外壳，套管，箱
casserole　勺皿
cassette　盒式磁带，卡式磁带，暗盒
cast aluminium　铸铝
cast blade　铸造叶片
cast copper　铸铜
cast gray pig iron　铸灰口铁
cast iron　铸铁，生铁
cast iron conductor　铸铁导管
cast iron pipe　铸铁管（生铁管）
cast steel　铸钢，生钢
cast-aluminum rotor　铸铝转子
cast-in anchorage　浇注锚固
casting　铸造，铸件
casting aluminium brass　铸铝黄铜
casting aluminium bronze　铸铝青铜
casting basin　预制件工场
casting flange　铸造凸缘
casting ladle　浇注包
casting on flat　水平铸造
cast-in-place　现场浇铸
casualty　人身事故、伤亡、故障
CAT (computer-aided test)　计算机辅助测试
catadioptric telescope　折反射望远镜
catalog　目录，编目录
cataloged data set　编目数据集
cataloging system　编目系统
catalogue　目录，为……编目录
catalysis　催化作用
catalysis element　催化元件
catalyst　催化剂
catalytic action　催化作用
catalytic analyzer　催化分析器

catalytic chromatography 催化色谱法
catalytic gas transducer 催化式气体传感器
catalytic reactor 催化反应器
catalyzed board coated catalyzed laminate 催化板材
catastrophe theory 突变论，失败理论
catch 捕获，挡片，制止器，门扣
catch fan 扇形防护网架
catch fence 拦截围墙
catch platform 坠台
catching diode 钳位二极管
catchment area 下游区，排水区
catchpit 排水井，集水坑，聚泥坑
catchwater channel 集水槽
categorical data 分类数据
catenary wire 吊索
cathode 阴极，负极
cathode coupling 阴极耦合
cathode follower 阴极耦合器，阴极跟踪器
cathode follower amplifier 阴极输出放大器
cathode lead wire 阴极引线
cathode of electron gun 电子枪阴极
cathode ray null indicator 阴极射线指零仪
cathode ray tube (CRT) 阴极射线管
cathode ray tube display 阴极射线管显示器
cathode-ray oscillograph 阴极射线示波器
cathodic protection 阴极保护
cation 阳离子
catwalk 跳板，轻便梯，轻便栈桥
caulk 堵缝
caulking compound 填隙料
caulking material 填隙料
caulking metal 填隙合金
cause analysis 原因分析
cause description 原因说明
caustic potash 苛性钾
caustic soda 苛性钠
caution sign 警告标志
caution 注意
cavern 洞穴
cavitation 气穴现象，空穴作用

cavitation corrosion 气蚀，空化腐蚀
cavitation noise 空穴噪声
cavity 中空部分，穴，型腔，母模，模穴
cavity insert 上内模
cavity retainer plate 模穴托板
cavity-resonator frequency meter 谐振腔频率计
cavity-stabilized oscillator 空腔稳频振荡器
CCCS (current-controlled current source) 电流控制电流源
CCD (charge coupled device) 电荷耦合元件
CCD camera 电荷耦合摄像机
CCD (computer controlled display) 计算机控制显示
CCITT (Consultative Committee International Telegraph and Telephone) 国际电报电话咨询委员会
CCVS (current-controlled voltage source) 电流控制电压源
CD Rom 光盘只读存储器
CDMA (code-division multiple access) 码分多址
CDPD (cellular digital packet data) 蜂窝数字包数据，蜂窝状数字式分组数据
ceiling 天花板，上限
ceiling slab 天花板
ceiling suspension hook 天花吊钩
ceiling switch 拉线开关
ceilometer 云高计，云幂仪
cell 电池，传感器，气泡
cell constant 电池常数
cell potential sensor 细胞电位传感器
cell potential transducer 细胞电位传感器
cell scalar analyzer 细胞计数分析仪
cell vital analyzer 细胞生死判别系统
cellular 细胞的，单元的
cellular array processor 单元阵列处理机
cellular automation 单元自动化
cellular digital packet data (CDPD) 蜂窝数字包数据，蜂窝状数字式分组数据

47

cellular logic 单元逻辑
cellular neural network 单元神经网络
cellular office 分格式办公室
cellulose acetate 醋酸纤维素
cellulose acetate butyrate 醋酸丁酸纤维素
Celsius 摄氏，摄氏温度计
Celsius temperature 摄氏温度
Celsius temperature scale 摄氏温标
cement 水泥
cement content 水泥含量
cement mortar 水泥砂浆
cement plaster 水泥灰泥
cement sand mix 水泥砂浆
cementation 渗碳
cementite 渗碳体，碳化铁，碳化三铁体
cementitious content 水泥质成分
center 中心
center buckle 表面中部波皱
center distance 中心距
center distance change 中心距变动
center of mass 质心
center of pressure 压力中心
center pin 中心轴，中心销
center-gated mold 中心浇口式模具
centering 定中心
centering potentiometer 中心调整电位器
centi 厘，百分之一
centimeter-wave calibrating instrument 厘米波校准仪
central computer 中央计算机
central conductor method 中心导体法, 电流贯通法
central divider 中央分隔栏
central dividing strip 中央分隔带
central gear 中心轮
central line 中线
central median 中央分隔带
central power-driven machine 中央动力机械
central principal inertia axis 中心主惯性轴
central processing unit (CPU) 中央处理单元, 中央处理机, 中央处理装置
central processor 中央处理器
central reserve 中央预留带
central span 中跨距
central control room 中控室
central processing unit 中央处理器
centrality 集中性
centralization 集中化
centralize 集中
centralized control 集中控制
centralized data processing 集中数据处理
centralized intelligence 集中智能
centralized management system 集中管理系统
centralized network 集中式网络
centralized operation 集中操作
centralized process control computer 集中型过程控制计算机
centralized traffic control code wire 调度集中电码线
centre line 中心线
centre of mass 质量中心
centreless 无中心的
centrifugal balancing machine 离心力式平衡机
centrifugal clutch 离心式离合器
centrifugal development 离心展开
centrifugal filter 离心过滤器
centrifugal force 离心力，向心力
centrifugal freeze dryers 离心型冻结干燥器，离心型冻结干燥机器
centrifugal governor 离心调速器
centrifugal integrator 离心式积分器
centrifugal load 离心荷载
centrifugal pump 离心泵
centrifugal pumps for refinery chemical and petrochemical processes 炼厂、化工及石油化工流程用离心泵
centrifugal seal 离心密封
centrifugal starting switch 离心式启动开关
centrifugal stress 离心应力
centrifugal switch 离心开关
centrifugal tachometer 离心式转速表
centrifugal 离心的
centrifugal fan 离心风机
centrifuge 离心机

centripetal development 向心展开
centripetal force 向心力
ceramic filter 陶瓷滤波器
ceramic metal 陶瓷金属
ceramic microphone 陶瓷传声器
ceramic substrate printed board 陶瓷印制板
ceramic tile 瓷砖
ceramic type vibration displacement meter 陶瓷式振动位移计
ceramics 陶瓷
ceramics base copper-clad laminates 陶瓷基覆铜箔板
ceraunograph 雷电计
cereals, oils and vegetable protein engineering 粮食、油脂及植物蛋白工程
cerebellar model articulation computer (CMAC) 小脑模型连接计算机
cerium (Ce) 铈
cerium sulphate method 硫酸铈法
cermet potentiometer 金属陶瓷电位计
certainty 可靠性
certificate 证明书
certificate of conformity 合格证书
certificate of control 控制证书
certificate of registration 注册证明书,登记证明书
certification 核证,认证
certification specifications 认可规范
certification system 认证体系
certified standard material 有证标准物质
cesium (Cs) 铯
cesspool 污水池
CG (computer graphics) 计算机图形学
chain 链,链式,链条
chain block 起重机
chain code 链式码
chain dotted line 点划线
chain drive 链传动
chain gearing 星形轮
chain making tools 造链机
chain printer 链式打印机
chain reaction 链式反应
chain wheel 链轮,滑轮

chainage 链测长度
chained aggregation 链式集结
chained list 链式表
chamber 小室,间隔
chamfer 斜面,凹槽,去角,挖槽,斜切
chamfering 倒角,去角斜切
chamfering machine 倒角机
chamfering tool 去角刀具
chamotte sand 烧磨砂
chance failure 偶发故障
change detect 变化检查器
change gear 变速齿轮
change of temperature test 温度变化试验
change storage option 改变存储选择
change wheel 变齿轮,变速[变向]齿轮
change 改变
change of direction 方向变化
change of phase 相位变化
change of polarity 极性变化
change-over contact 转换触点
change-over switch 转换开关,双向开关
change-over valve 转换阀
change-pole motor 变极电动机
changer 变换器,换能器
change-tune switch 波段调整开关
channel 信道,通道,频道,槽,沟渠,线槽,渠道
channel adapter 通道适配器
channel address word 通道地址字
channel bandwidth 通道带宽
channel bases 沟渠基底
channel bus controller 通道总线控制器
channel cover 槽盖,管箱盖
channel demodulator 通道解调器
channel iron 槽钢,铁
channel multiplexer 通道多路转换器
channel range 通道范围
channel selector 频道选择器,频道转换开关
channel shifter 信道移频器
channel span 通道量程
channel transistor 沟道晶体管

channeling	沟道作用
channelizing line	导行线
chaos	混沌
chaos theory	混沌理论
chaotic behaviour	混沌特性
chaotic control	混沌控制
chaotic system	混沌系统
chaotic time series	混沌时间序列
character boundary	字符边界
character code	字符码
character die	字模
character printer	字符打印机
character recognition	字符识别
character set	字符集，字符组
character	字符，符号，信息，特性，性质
character-at-time printer	串行打印机
characteristic	特性，特性的，特性曲线，特性曲线的
characteristic admittance	特性导纳
characteristic curve	特性曲线
characteristic curve tracer	特性曲线图示仪
characteristic equation	特性方程式
characteristic frequency	特征频率
characteristic impedance	特性阻抗
characteristic locus	特征轨迹
characteristic polynomial	特征多项式
characteristic root	特征根
characteristic strength	特征强度
characteristic time	特征时间
characteristic vector	特征矢量
characteristic X-ray	特征 X 射线
characteristics	特性
characterizer	特征化运算器
characters	字符串
charactron	显像管
charge	电荷，充电
charge amplifier	电荷放大器
charge balance	电荷平衡
charge balance equation	电荷平衡式
charge conservation	电荷守恒
charge coupled device (CCD)	电荷耦合摄像机
charge density	电荷密度
charge neutralization	电荷中和
charge sensitivity	电荷灵敏度
charge transfer	电荷转移
charge up	充电
charge indicator	验电器、带电指示器
charge-coupled device	电荷耦合器件
charged	带电荷的，已充电的
charged body	带电体
charger	充电器
charger switch	充电开关
charging current	充电电流
charging equipment	充电设备
charging hopper	加料漏斗
charging period	充电时间
charging voltage	充电电压
charging wiring diagram	充电接线图
Charpy impact test	沙尔皮冲击试验，摆锤式冲击试验
chart	记录纸
chart datum	海图基准面
chart driving mechanism	传纸机构，记录纸驱动机构
chart lines	记录纸分度线
chart recorder	图表记录器
chart scale length	记录纸标度尺长度
chart speed	记录纸速度
chasis	车身底盘
chassis control	起落架控制
chassis	基座，底座，底盘
chassis earth	机壳接地
chattering	颤动，振动
check	检验，校验，查核，检查
check block	挡块
check for bad	坏值检查器
check gauge	校对规
check mechanism	制动
check meter	校验仪
check plate	垫板，挡板
check point	查核点
check pointing	检查指针，校验点
check rail	护轮轨
check register	校验寄存器
check screw	止动螺钉
check value	校验检查值
check valve	止回阀
checker	检验器
checkered iron	网纹钢
checkpoint	监测点，检查点，检验点
checksum	校验和，检查和

chelate compound 螯合物
chemical action 化学作用
chemical analysis 化学分析
chemical cell 化学电池
chemical dosing 化学剂量
chemical double layer 化学双电层
chemical engineering 化学工程
chemical engineering and technology 化学工程与技术
chemical grout 化学灌浆
chemical industry 化学工业
chemical ionization 化学电离，化学离子化
chemical ionization source 化学离子源
chemical microsensor 化学微型传感器
chemical plating 化学电镀
chemical process equipment 化工过程机械
chemical processing engineering of forest products 林产化学加工工程
chemical property 化学特性
chemical propulsion 化学推进
chemical reactor 化学反应器
chemical refuse 化学垃圾
chemical sensor 化学传感器
chemical shift 化学位移
chemical technology 化学工艺
chemical test 化学测试
chemical vapor deposition 化学蒸镀
chemical variables control 化工变量控制
chemically bonded phase packing 化学键合相填充剂
chemicals 化学品
chemiluminescence apparatus 化学发光仪
chemiluminescence detection 化学发光检测
chemometrics 化学计量学
chequered plate 网纹板
chill mold 冷硬用铸模
chill plate 冷却板
chilled air fan 冷风风扇
chilled cast iron 冷硬铸铁
chilled water pump 冷冻水泵
chiller 冷冻机
chiller plant 致冷设备，制冷设备

chimney coping 烟囱盖顶
chimney effect 烟囱效应
chimney 烟沟，烟囱，烟道
china 陶瓷，瓷器
China Power 中国电力
chip 芯片，晶片
chip conveyor 排屑输送机
chip enable 片选，组件选通
chip microprocessor 单片微处理器，单片微处理机
chip on board 载芯片板
chip testing 芯片测试
chipping 修整表面缺陷
chippings 碎屑，破片
chiral chromatography 手性色谱法
chiral derivatization reagent 手性衍生试剂
chiral separation chromatography 手性分离色谱
chiral stationary phase 手性固定相，手性固定化
chiralion pair complex 手性离子对络合剂
chirping laser 调频激光器
chisel 錾，凿，扁錾
chloride 氯化物
chloride content 氯化物含量
chloride diffusion 氯化物扩散
chloride extraction 除氯
chloride ion 氯离子
chloride ion content 氯离子含量
chlorinated polyethylene plastic sheets for waterproofing 氯化聚乙烯防水卷材
chlorinated water 加有氯气的水
chlorinator 加氯器
chlorine (Cl) 氯
chloroform 氯仿
CHN Analysis 环境成分分析仪
choke 抑止，抗流圈，扼流圈
choke circuit 抗流电路
choke filter 扼流滤波器
choke transformer 扼流变压器
choke turn 扼流圈
choke winding 扼流线圈
chokon 高频隔直流电容器
chopper 斩波器，断路器

chopper amplifier 斩波放大器，截断放大器
chopper circuit 斩波电路
chopper voltage stabilizer 斩波稳压器
chopping 截断
chopping astable multivibrator 断续非稳态多谐振荡器
chopping signal 斩波信号，断路信号
chordal addendum 弦齿高
chordal thickness 弦齿厚
chromate 铬酸处理，铬酸盐
chromatogram 色谱图，色层分离谱
chromatographic analysis 色谱分析
chromatographic column 色谱柱
chromatographic optimization function 色谱优化函数
chromatographic peak 色谱峰
chromatographic response function 色谱响应函数
chromatography 色谱法（层析法）
chromatron 彩色电视显像管，色标管
chrome bronze 铬铜
chrome wire 铬线
chromel alloy 镍铬合金
chromel-alumel thermocouple 镍铬-镍铝热电偶
chromel-constantan thermocouple 镍铬-康铜热电偶
chromel-copel extension wire 镍铬-考铜补偿导线
chromel-nisiloy thermocouple 镍铬-镍硅热电偶
chrome-nickel wire 铬镍线
chromic iron 铬铁矿
chrominance demodulator 彩色信号解调器
chrominance modulator 彩色信号调制器
chrominance signal carrier 彩色信号载波
chromium (Cr) 铬
chromium irons 铬铁合金
chromium stainless steel 铬不锈钢
chromium-copper wire 铬铜线，铬铜合金线
chromium-molybdenum steel 铬钼钢

chromium-molybdenum-vanadium steel 铬钼钒钢
chromophore 生色团
chromoscope 彩色显像管
chronometer 精密计时表
chronoscope 计时器，极微时间测定器
chuck 夹具，卡盘
chute 溜槽，滑道，槽管
CIMS (computer integrated manufacture system) 计算机集成制造系统
cinder pig iron 夹渣生铁
CIPS (computer integrated process system) 计算机集成过程系统
circuit 电路，线路
circuit alarmer 线路故障报警器
circuit analysis 电路分析
circuit analyzer and tester 电路分析测试仪
circuit brake 电路制动器
circuit branch 支路
circuit break 断路
circuit breaker 断路器，保护断路器，断路保险掣，电路断路器
circuit breaker cabinet 断路器柜
circuit calculation 电路计算
circuit card 电路插卡，电路插件板
circuit components 电路元件
circuit design 电路设计
circuit diagram 电路图，线路图
circuit efficiency 电路效率
circuit element 电路元件
circuit equation 电路方程式
circuit impedance 电路阻抗
circuit loss 电路损耗
circuit model 电路模型
circuit parameters 电路参数
circuit performance 电路性能
circuit schematic diagram 电路原理图解
circuit simulation 电路仿真，电路模拟
circuit switched network 电路开关网络
circuit theory 电路理论，电路原理
circuit transformation 电路变换
circuit with distribution parameters

分布参数电路
circuitous 旁路，绕行的
circuitry 电路，电路系统
circuits and systems 电路与系统
circular buffering 循环缓冲
circular current 环流
circular development 环形展开
circular face-mill 圆盘端面铣刀
circular file 循环文件
circular footing 圆基脚
circular frequency 角频率，角速度
circular function 循环功能
circular gear 圆形齿轮
circular measuring dial 圆形测量刻度盘
circular peripheral-mill 圆盘铣刀
circular pitch 齿距，周节
circular prestressed concrete pole 环形预应力混凝土电杆
circular reinforced concrete pole 环形钢筋混凝土电杆
circular thickness 弧齿厚，圆弧齿厚
circulating current 环流，平衡电流
circulating power load 循环功率流
circulating register 循环寄存器
circulating water pump 循环水泵
circulating 循环
circulation mode 循环模式
circulator 循环器，回转器
civil engineering 土木工程
civil speed limit 轨道速限
civil works 土木工程，土建工程，建筑工程
CL block 控制语言程序块
CL (control language) 控制语言
clad sheet 被覆板
clad weld 被覆熔接
cladding 覆层，骨架外墙，覆盖层
claim 索赔，声称，专利申请，权利要求，索偿
clamp 夹具，钳，夹钳，夹紧，固定
clamp pulse generator 钳位脉冲发生器
clamp valves 对夹式阀门
clamper 钳位器
clamping 夹紧
clamping block 锁定块

clamping circuit 钳位电路
clamping diode 钳位二极管
clamping plates for fixing steel wire ropes 钢丝绳用压板
clamping systems 夹具系统
clamp-off 铸件凹痕
clamp-on alternating-current milliammeter 钳位交流毫安表
clamp-on flowmeter 钳形流量计
claplock cable clamp 拍扣式电缆线夹
class 类，等级，程度
class A amplification 甲类放大
class A amplifier 甲类放大器
class A insulation A级绝缘
class A modulation 甲类调制
class A operation 甲类工作状态
class B amplification 乙类放大
class B amplifier 乙类放大器
class B modulation 乙类调制
class C circuit 丙类电路
class of accuracy 准确度等级
class of insulation 绝缘等级
classical control theory 经典控制理论
classical information pattern 经典信息模式
classical Venturi tube 经典（古典）文丘里管
classification 分类，整理
classification for fire retardancy 防火性能分级
classification of process variable 过程变量分类
classifier 分类器
clay 黏土
clay field pipe 瓦管
clean bomb 干净核弹，低污染核弹
clean ovens 洁净恒温器
clean 清洁的，纯净的
clean supply 无干扰供电
cleaning eye 清理孔，通管孔
cleaning of casting 铸件清理
cleaning rod 清理棒
cleanness 清扫
cleanse 净化，洗净，消毒
cleansing 洁净
clear 清除
clear display 清晰显示

clear effective length	净有效长度	closed chain mechanism	闭链机构
clear height	净高	closed circuit	闭路，通路
clear opening	净开口	closed circuit television	闭路电视
clear space	净空间	closed core	闭合［口］铁芯，闭式铁芯
clear span	净跨距，净孔		
clear width	内径，净宽度	closed core transformer	闭口铁芯变压器
clearance	顶隙，间隙，径向间隙		
clearance angle	后角，留隙角	closed end	闭端
clearance gauge	间隙规	closed kinematic chain	闭式运动链
clearance goods	清仓品，结关货物	closed loop	闭环
clearance space	间隙空间	closed loop control	闭环调节，闭环控制
clearing of fault	故障清除		
clearness	清晰度	closed loop control system	闭环控制系统
cleat	夹具，楔子，栓		
clevis	U形夹，连接叉	closed loop controller	闭环调节器，闭环控制器
climb form technique	提升模板技术		
clinch nuts	齿花螺帽	closed loop gain	闭环增益
clinical control system	临床控制系统	closed loop identification	闭环辨识
clinometer	测斜仪	closed loop phase angle	闭环相角
clip	小夹	closed loop pole	闭环极点
clipping	削波，限幅	closed loop stabilization	闭环稳定
clipping circuit	削波电路，限幅电路，限幅器	closed loop system	闭环系统
		closed loop transfer function	闭环传递函数
clip-type probe	线夹型探头		
clock	时钟	closed loop zero	闭环零点
clock cycle	脉冲周期	closed magnetic circuit	闭合磁路
clock exerciser	时钟练习程序	closed position	停止位置，闭合位置
clock frequency	时钟脉冲频率	closed queuing network	闭合排队网络
clock pulse	时钟脉冲		
clock switch	时钟开关	closed system	封闭系统，闭合式系统
clock synchronization	时钟同步	closed type track circuit	闭路式轨道电路
clock system repeater	时钟系统重复器		
		closed-die forging	合模锻造
clock-controlled governor	时钟控制调节器	closed-loop design principle	闭路设计原理
clocking	计时	closed-loop frequency response	闭环频率响应
clockwise	顺时针，右旋		
clogged	障碍，塞满，粘注	closed-loop transfer function	闭环传递函数
close coupling	紧密耦合		
close fitting cover	紧合封盖	closed-loop	闭环
close fittings	紧合配件	close-up	接通，闭合
close type socket joint	封闭式插接	closing contact	闭合触点
close	关闭	closing magnet	合闸磁铁
close-boarded platform	密合封板平台	closing of contact	触点闭合
closed	闭合的，接通的	closing relay	合闸继电器
closed antenna	闭路天线	closure	封闭
closed area	禁区	closure member	截流件

cloud amount 云量
cloud base 云底
cloud chamber 云室，云雾室
cloud detection radar 测云雷达
cloud direction 云向
cloud height indicator 云高指示器
cloud height meter 云高仪
cloud of electrons 电子云
cloud searchlight 探照灯
cloud speed 云速
cloud top 云顶
cloud-base recorder 云底记录仪
cloud-drop sampler 云滴取样器
cloudiness radiometer 云辐射仪
cloverleaf buoy 三叶浮标
cloverleaf interchange 四叶式交汇处，蝶式交汇处
cluster analysis 聚类分析
cluster automatic switch 装置式自动开关
clutch 离合器
clutch boss 离合器轮壳
clutch brake 离合器制动器
clutch lining 离合器覆盖
CM（computing module） 计算模件，计算模块
CMAC（cerebellar model articulation computer） 小脑模型连接计算机
CMC（critical micelle concentration） 临界胶束浓度
CMRR（common mode rejection ratio） 共模抑制比
CNC（computerized numerical control） 电脑数值控制，计算机数控，数控
CNC 5-coordinate milling-boring machine 五坐标数控铣镗床
CNC abrasive belt surface-grinding machine 数控砂带平面磨床
CNC abrasive line cut-off machine 数控砂线切割机床
CNC abrasive wire sawing machine 数控砂线切割机床
CNC angular approach cylindrical grinding machine 数控端面外圆磨床
CNC automatic multi-slide lathe 数控自动多轨车床

CNC automatic riveting machine 数控自动铆接机
CNC automatic turning machine 数控自动车床
CNC bar loading magazine 数控杆料进料匣
CNC bed-type horizontal milling machine 数控工作台不升降卧式铣床
CNC bed-type milling machine with travelling column 数控立柱移动工作台不升降铣床
CNC bed-type vertical milling machine 数控工作台不升降立式铣床
CNC bending presses 电脑数控压弯机
CNC boring machines 电脑数控镗床，电脑数控钻孔机
CNC camshaft milling machine 数控凸轮轴铣床
CNC chucking lathe 数控卡盘车床
CNC controller 数控控制器
CNC coordinate measuring machine 数控坐标测量机
CNC copy milling machine 数控仿形铣床
CNC deep hole boring machine 数控高效深孔镗床
CNC deep-hole drilling and boring machine 数控深孔钻镗床
CNC double column jig boring machine 数控双柱坐标镗床
CNC double spindle lathe 双轴数控车床
CNC double-column vertical turning & boring mill 数控双柱立式车床
CNC EDM wire-cutting machines 电脑数控电火花线切削机
CNC electric discharge machines 电脑数控电火花机
CNC electrical discharge machine 数控电火花加工机床
CNC electrode grinding machine 数控电极磨床
CNC electrolytic tool and cutter grinder 数控电解工具磨床
CNC engraving machines 电脑数控雕刻机

CNC external cylindrical angular plunge-cut grinder 数控斜角全面进磨式外圆磨床
CNC floor-type milling and boring machine 落地式数控铣床
CNC formed cylindrical grinder 数控成形外圆磨床
CNC front loaded machine 数控前载机车床
CNC gear hard finishing machine 数控齿轮精加工机床
CNC grinder for external thread 数控外螺纹磨床
CNC grinder for internal thread 数控内螺纹磨床
CNC grinding machine 数控磨床
CNC heavy turning machine 数控重型车床
CNC heavy-duty horizontal lathe 数控重型卧式车床
CNC high speed double-wheel angular approach cylindrical grinding machine 数控高速双砂轮端面外圆磨床
CNC high speed turret milling machine 数控高速转塔铣床
CNC horizontal broaching machine 数控卧式拉床
CNC horizontal milling and boring machine 数控卧式铣镗床
CNC horizontal surface grinding machine with rectangular table 数控卧式矩（形）台平面磨床
CNC hydraulic press brake 数控液压折弯机
CNC internal grinding machine for bearing ring 数控轴承套圈内圆磨床
CNC internal thread grinding machine 数控内圆端面磨床
CNC jig grinding machine 数控坐标磨床
CNC knee type horizontal milling machine 数控卧式升降台铣床
CNC knee type vertical milling machine 数控立式升降台铣床
CNC laser processing machine 数控激光加工机
CNC lathe 计算机数控机床，数控车床，数控车床加工
CNC machine tool fittings 电脑数控机床配件
CNC milling cutter 数控铣刀
CNC milling machines 电脑数控铣床
CNC movable single column vertical turning and boring mill 数控单柱移动式立式车床
CNC oblique cutting cylindrical special grinding machine 数控斜切外圆专用磨床
CNC planer horizontal milling and boring machine 数控刨台卧式铣镗床
CNC planer type milling machine 数控龙门铣床
CNC precision guideway grinder 数控精密导轨磨床
CNC race grinding machine for ball bearing inner ring 数控轴承内圈沟磨床
CNC race grinding machine for ball bearing outer ring 数控轴承外圈沟磨床
CNC ram-type milling machine 数控滑枕式铣床
CNC roll grinder 数控轧辊磨床
CNC shearing machines 电脑数控剪切机
CNC single column jig boring machine 数控单柱坐标镗床
CNC single column vertical turning and boring mill 数控单柱立式车床
CNC slant-bed turning machine 数控斜式车床
CNC sliding headstock auto-lathe 数控纵切车床
CNC special purpose milling machine 数控专用铣床
CNC tap grinding machine 数控丝锥磨床
CNC three dimensional measuring machine 数控三维测量机
CNC tool profile grinding machine 数控工具曲线磨床
CNC tube bending machine 数控弯

管机
CNC turning machine 数控车床
CNC turning tool 数控车刀
CNC turret punch press 数控冲模回转头压力机
CNC turret punching press 数控冲模回转头压力机
CNC turret vertical miller 转塔式数控立式铣床
CNC twin-spindle chucker-type turning machine 数控双轴卡盘式车床
CNC universal milling machine 数控万能铣床
CNC universal tool milling machine 数控万能工具铣床
CNC vertical external broaching machine 数控立式外拉床
CNC vertical turning machine 数控立式车床
CNC water cutting machine 数控水力切割机
CNC wire-cut electric discharge machine 数控电火花线切割机床
CNC wire-cut machine 数控线切割机床
CNC wire-cutting machines 电脑数控线切削机
coal-tar epoxy 环氧煤焦油
coarse aggregate 粗集料，粗骨料
coarse indented cut plating iron 粗牙槽刨刀
coarse pitch 大节距，大螺距
coarse screening 粗筛
coarse thread 粗牙螺纹
coarse vacuum 粗真空，前级真空
coarse wire 粗拔钢丝
coarse 粗的，不精确的
coarse-fine 粗糙-精细
coarse-fine control 粗-精控制，粗糙-精细控制
coarse-fine control system 粗糙-精细控制系统
coarse-fine relay 粗糙-精细继电器
coarse-fine switch 粗糙-精细开关
coarsening 结晶粒粗大化
coast side 不工作齿侧
coastal zone color scanner (CZCS) 海岸带水色扫描仪
coating 覆盖层，保护层，涂层，涂布被覆
coating material 涂层材料
coaxial 共轴的，同轴的
coaxial bolometer 同轴测辐射热计
coaxial cable 同轴电缆
coaxial line transformer 同轴线变换器
coaxial matched taper 同轴匹配渐变器
coaxial relay 同轴继电器
coaxial tuner 同轴调谐器
coaxiality 同轴度
coaxial-to-waveguide transformer 同轴-波导变换器
cobalt (Co) 钴
cobweb model 蛛网模型
co-channel rejection ratio 同频抑制比
cock 二通，旋塞，旋
code input 代码输入
code of practice 工作守则，操作守则
code reader 读码器
code sequence generator 码序发生器
code wire 隔离导线，标准线号导线
code 码，编码，代码，符号，标记
code conversion 码变换
code converter 代码转换器，译码器
code output 代码输出
code translator 译码器
coded address 编码地址
coded circle 编码度盘
coded modulation 编码调制
coded-decimal notation 十进制编码
code-division multiple access (CDMA) 码分多址
coder 编码器
code-transparent data communication 代码透明的数据通信
coding scheme 编码方案
coefficient 系数
coefficient matrix 系数矩阵
coefficient of amplification 放大系数
coefficient of attenuation 衰减系数
coefficient of chromatic aberration 色差系数
coefficient of correction 修正系数

coefficient of coupling　耦合系数
coefficient of expansion　膨胀系数
coefficient of friction　摩擦系数
coefficient of linear expansion　线性膨胀系数
coefficient of mutual inductance　互感系数
coefficient of radial distortion　径向畸变系数
coefficient of rotational distortion　旋转畸变系数
coefficient of self-inductance　自感系数
coefficient of speed fluctuation　速度不均匀系数，机械运转不均匀系数
coefficient of spherical aberration　球差系数
coefficient of stability　稳定系数
coefficient of stabilization　稳定系数
coefficient of temperature　温度系数
coefficient of travel speed variation　行程速度变化系数
coefficient of variation　变异系数
coefficient of velocity fluctuation　运转不均匀系数
coercive force　矫顽力
coercivity meter　矫顽力计
cofactor　辅因子，余子式
cofferdam　围堰坝
cognitive　认知
cognitive science　认知科学
cognitive system　认知系统
cognitron　认知机
Cohen and Coon controller setting　科恩-库恩控制器整定
cohere　相关，相干
coherent　相关的，相干的，耦合的
coherent detector　相干检测
coherent of measurement　测量单位
coherent of unit　一贯单位
coherent system　单调关联系统，相干系统
coherent system of measurement　测量单位制
coherent system of unit　一贯单位制
cohesive force　黏合力，凝聚力
coil　线圈，卷，簧圈，盘管

coil aerial　环形电线
coil arrangement　线圈布置
coil car　带卷升降运输机
coil cradle　卷材进料
coil galvanometer　线圈式振动子
coil insulator tester　线圈绝缘测试器
coil method　线圈法
coil reel stand　钢材卷料架
coil spring　螺旋弹簧，卷簧，盘簧
coil stock　卷料
coil winding　线圈绕组
coincide in phase with　与……同相
coincidence　一致，符合
coincidence discrimination　符合鉴别
coincident points　重合点
coining　压印加工
coke pig iron　焦炭生铁
cold　冷，冷启动
cold blast pig iron　冷风生铁
cold chamber die casting　冷式压铸
cold forging　冷锻，锤锻，自由锻
cold hobbing　冷挤压制模
cold junction　冷端
cold junction reference　冷端参考点
cold machining　冷加工
cold punched nut　冷冲螺母
cold reduced steel wire　冷轧钢丝
cold resistance　冷态电阻
cold runner　冷流道
cold shortness　低温脆性
cold slag　冷料渣
cold slug　注塑冷料
cold solvent welding　冷冻溶剂焊接
cold storage　冷藏库
cold test　常温试验，不通电试验
cold treatment　冷处理，冰冷处理
cold work die steel　冷锻模用钢
cold-cathode source　冷阴极离子源
cold-heading wire　冷镦钢丝
cold-rolled steel wire　冷轧钢丝
collapse　塌陷，坍塌
collapsible cantilever platform　可折叠悬臂平台
collar　护圈，束套，套环
collateral　并联的，平行的
collateral contact　并联触点
collect　收集

collective 集中的
collective control 集中控制
collector 集电器，集电极
collector ring 集电环
collector slit 接收器狭缝
collet 套爪，筒夹
collimation axis 视轴
collimation line 视准线
collision 冲突，碰撞
collision ionization 碰撞电离
collisional activation 碰撞激活，活化作用
collisional activation mass spectrometer 碰撞激活质谱计
color agent 显色剂
color comparator 比色器
color display 彩色显示器
color hard copy unit 彩色硬拷贝器
color ink-jet printer 彩色喷墨式打印机
color masterbatch 色母料
color matching 调色
color mottle 色斑
color video copier 彩色视频拷贝机，彩色视频复印机
colorant 着色剂
colorimeter 比色计，色度计
colour 颜色，色彩，彩色
colour balance 彩色平衡
colour bar 色带
colour change interval 变色范围
colour code 色码
colour coder 彩色编码器
colour contrast 彩色对比度
colour demodulator 彩色解调器
colour filter 颜色滤光片
colour kinescope 彩色显像管
colour light signal 颜色灯号
colour meter 水色计
colour pencil 颜色笔
colour pick-up tube 彩色摄像管
colour picture tube 彩色显像管
colour television 彩色电视
colour television receiver 彩色电视机
coloured cement 颜色水泥
coloured mark for gas cylinders 气瓶颜色标记

coloured noise 有色噪声
column 塔，列，圆柱
column bleeding 柱流失
column cap 柱帽
column capacity 柱效能
column chromatography 柱色谱法
column footing 柱基脚
column frame 柱架
column head 柱头
column life 柱寿命
column liquid-solid extraction 液固柱萃取
column packing 柱填充剂
column switching 柱切换
column-parity field 列奇偶校验字段
comb filter 多通带滤波器，梳齿型滤波器
comb plate 梳板
combination 组合，合成作用
combination decision 组合决策
combination digital logger 数字式综合记录仪
combination electrode 复合电极
combination in parallel 并联式组合
combination in series 串联式组合
combination logging instrument 组合测井仪
combination pH electrode 复合pH电极
combination valves 组合阀
combination water meter 复式水表
combinational 组合，复合
combinational circuit 组合电路，复合电路
combinational logic element 组合逻辑元件
combinational mathematic 组合数学
combinational network 组合网络，复合网络
combinational switching 组合开关
combinations 合谱；制品
combinatorial explosion 组合爆炸
combinatory logic 组合逻辑
combined pressure and vacuum gauge 压力真空表
combined column 复合柱
combined dead load 组合恒载

combined effect 混合效应
combined footing 联合基脚
combined lighting 综合照明
combined load 合并载重
combined load testing machine 综合载荷试验机
combined mechanism 组合机构
combined preload 综合预负荷
combined pressure and vacuum gauge 压力真空表
combined stress 复合应力
combined test 综合试验
combined test cabinet 综合试验箱
combined type digital voltmeter 复合式数字电压表
combined voltage current transformer 电压、电流互感器
combustibility 燃烧性,可燃性
combustible 可燃的,易燃的,可燃物
combustible component 可燃成分
combustible gas 可燃气
combustible goods 可燃物品
combustible material 可燃烧物料
combustion 燃烧
combustion chamber 燃烧室
combustion property tester 燃烧性能测定仪
coming into step 进入同步
command accepted 命令接受
command and control system 指令和控制系统
command code 指令码,操作码
command control 指令控制
command data dictionary 命令字典
command disagree 命令不一致,命令不符
command message 命令报文,命令消息,指令讯息
command monitoring decoder 指令监视译码器
command operation 命令操作
command output 命令输出
command variable 指令信号
command 命令,指挥,指令
commencement of operation 开始操作
comment 注释,注解

comment column 注解列
comment line 注释行,注解行
commercial hardware and software 商用硬件与软件
commissioning 投运,启用,试车,试运转
commissioning test 运行试验
commitment 承担
commix 混合,混合物
common 公共的,共有的
common apex of cone 锥顶
common area 公共区
common block 公共块
common business-oriented language 面向商业的通用语言
common card file assembly 通用卡存放器组件
common control signals 公共控制信号
common equipment 常用设备
common magnet galvanometer 共磁式振动子
common memory test program 公共存储器测试程序
common mode interference 共模干扰
common mode rejection 共模抑制
common mode rejection ratio (CMRR) 共模抑制比
common mode signal 共模信号
common mode voltage 共模电压
common normal line 公法线
common operation and monitoring function 通用操作和监视功能
common software 通用软件
common use 共同使用
common waste pipe 共用废水管
common-base 共基极
common-collector 共集电极
common-mode noise 普通模式噪声
common-strength steel wire 普通强度钢丝
communication and information systems 通信与信息系统
communication and transportation engineering 交通运输工程
communication antenna television 公用天线电视

communication buffer 通信缓冲器
communication card 通信卡
communication channel 信道
communication control application program 通信控制应用程序
communication control protocol 通信控制协议
communication control unit 通信控制器，通信控制单元
communication environment 通信环境
communication function 通信功能
communication gateway unit 通信门单元
communication interface adapter 通信接口适配器
communication laser 通信激光器
communication line 通信线路
communication line interface 通信线接口
communication mode control 通信方式控制
communication module 通信模块
communication multiplexer 通信多路转接器
communication net 通信网
communication network 通信网络
communication processor (CP) 通信处理机，通信处理器
communication protocol 通信协议
communication satellite 通信卫星
communication subnet 通信子网
communication system 通信系统，通讯系统
communication zone 通信区域
communication 通信，通讯
communications terminal 通信终端
commutate 换向，整流
commutating coil 换向线圈
commutating field 换向磁场
commutating pole 换向极
commutating winding 换向绕组
commutation 换向，整流，换流
commutation condition 换向状况
commutation error 换码误差，转码误差
commutation loss 换向损耗
commutative law 交换律，互易律

commutator 换向器，整流子，整流器
commutator frequency changer 换向器变频机，整流子频率变换器
commutator induction motor 换向器感应电动机
commutator motor 换向器电动机
commutator-brush combination 换向器-电刷总线
commutatorless machine 无换向器电机
compact 小型的
compact disk read-only memory 只读光碟
compact FCS 小型现场控制站
compact sleek design 紧凑、精巧设计
compact spectra 压缩频谱，压缩光谱
compacted concrete 压实混凝土
compacting molding 粉末压出成形
compaction 压实，夯实
compaction test 压实测试
compact-stranded wire 压紧多股绞合线
companion matrix 伴随矩阵
company standard 企业标准
comparative 比较的
comparative read-out 比较读出
comparator 比较器，比测仪
comparator coil 比较线圈
compare 比较，对照
compare and swap 比较和交换
compare file difference 比较文件
compare logic 比较逻辑
comparing element 比较元件
comparison bridge 比较电桥
comparison calibration 比较法校准
comparison method 比较法
comparison method of calibrating thermocouple 热电偶比较检定法
comparison method of calibration 比较法标定
comparison method of measurement 比较测量法
comparison standard 比较标准器
comparison value 比较值
compartment 分隔室，隔室

compartment wall 分隔墙
compass 罗盘，两角规
compass theodolite 罗盘经纬仪
compatibility 兼容性
compatible 兼容的，相容的
compatible hardware 兼容硬件
compatible software 兼容软件
compatible time-sharing system 兼容分时系统
compensate 补偿
compensated induction motor 有补偿的感应电动机
compensated micromanometer 补偿微压计
compensated scale barometer 定槽水银气压表
compensated temperatures 补偿温度
compensating 补偿，补偿的
compensating bellows 补偿波纹管
compensating element 补偿元件，补偿环节
compensating error of automatic vertical index 竖直度盘指标补偿误差
compensating extension lead 补偿型延长导线
compensating feedback 补偿反馈，反馈补偿
compensating feedforward 补偿前馈，前馈补偿
compensating gauge 补偿计，补偿片
compensating network 补偿网络
compensating resistance 补偿电阻
compensating valve 补偿阀，补气阀
compensating winding 补偿绕组
compensation 补偿，校正，补偿金
compensation coil 补偿线圈
compensation density logger 补偿密度测井仪
compensation equipment 补偿装置
compensation method 补偿法
compensation output action 补偿输出信号作用
compensation output signal 补偿输出信号
compensation set 补偿设定
compensation theorem 补偿定理
compensation type airborne electromagnetic instrument 补偿式航电仪
compensation-type gas flowmeter 补偿式气体流量计
compensator 补偿器
compensator level 自动安平水准仪
compensator starter 自耦变压器启动器
compensatory leads 补偿导线
competing ions 竞争离子
compile 编译
compile and link a program 编译和连接一个程序
compile and link in background 在后台编译和连接
compiler 编译程序，自动编码器，程序编制器
compiler diagnostics 编译程序的诊断程序
compiler generator 编译程序的生成程序
compiler optimization 编译器优化
complement 补码，补充设备
complementarity problem 互补问题
complementary 互补的，补充的，余的
complementary code 补码
complementary crown gears 互补冠状齿轮
complementary feedback 辅助反馈
complementary formulation 附加公式，辅助公式
complementary function 余函数
complementary gas 附加气
complementary measurement 互补测量
complementary metal oxide semiconductor 互补金属氧化物半导体
complementary symmetry circuit 互补对称电路
complementary transistor logic circuit 互补晶体管逻辑电路
complementary unijunction transistor 互补单结晶体管
complete 完全的
complete controllability 完全可控性，完全能控性
complete decoupling 完全解耦
complete failure 完全失效，完

故障
complete function test　全面功能试验
complete fusion　完全熔接, 助熔剂
complete observability　完全可观测性, 完全能观测性
complete overhaul　全面大修
complete response　全响应
complete solution　全解, 全面的方案
completing cycle　全工序循环
complex　复变量, 复数, 复数的, 复杂的
complex admittance　复数导纳
complex conjugate　复共轭
complex conjugate root　共轭复根
complex form　复数形式
complex impedance　复数阻抗
complex mechanism　复杂机构
complex number　复数
complex perturbation　复杂扰动
complex plane　复平面
complex power　复数功率, 复功率
complex radiation　复电阻率仪
complex s plane　复s平面
complex system　复杂系统
complex variable　复变量
complex z plane　复z平面
complexation chromatography　络合色谱法
compleximetry　配位滴定法, 络合滴定
compliance　顺应, 柔度
component　部件, 零件, 组件, 元件
component balance　组分平衡
component density　元件密度
component electromagnetic flowmeter　二组分电磁流量计
component hole　元件孔
component of a symmetrical system　对称系统的分量
component positioning　元件安置
component proton magnetometer　分量质子磁力仪
component side　元件面
component superconducting magnetometer　超导分量磁力仪
component under test　被测元件
composite　复合的, 组成的
composite action　复合作用
composite block　复合块
composite building　综合用途建筑物
composite cable　复合电缆
composite control　组合控制
composite control systems　复合控制系统
composite dies　复合模具
composite error　合成误差
composite laminate　复合层压板
composite metallic material　复合金属材料
composite pile　混合桩
composite point　复合点
composite sandwich construction　复合夹层结构
composite signal　复合信号, 混合信号
composite steel plate　复合钢板
composite sync generator　复合同步信号发生器
composite synchronizing signal　复合同步信号
composite test　组合试验
composite tooth form　组合齿形
composite wire　双金属丝
composite-temperature-compensation strain gauge　组合温度补偿应变计
composition　成分
composition amplifier　组合式放大器
composition control　组成控制, 成分控制
composition deviation transmitter　成分偏差变送器
composition sensor　成分传感器
compound　混合料
compound action　组合作用
compound combining　复合式组合
compound control　复合控制
compound control system　复合控制系统
compound controller　复合控制器
compound die　复合模
compound dynamo　复励发电机
compound excitation　复励
compound flat belt　复合平带
compound flowmeter　复式流量计
compound gear train　复合轮系, 混合轮系

compound generator 复励发电机
compound hinge 复合铰链
compound molding 复合成形
compound motor 复励电动机
compound oxide series gas sensor 复合氧化物系气敏元件
compound relay 复合继电器
compound screw mechanism 复式螺旋机构
compound semiconductor 复合半导体,合成半导体
compound system 复合系统
compound twisted wire 合成双绞线
compound winding 复励绕组
compounded 复励
compounding 复励,复合
compounding feedback 复式反馈
compounding feedforward 复式前馈
compound-wound 复励式,复绕式
compound-wound current transformer 混合绕组电流互感器
comprehensive development area 综合发展区
comprehensive transport interchange facilities 综合交通交汇设施
comprehensive transport study 整体运输研究
compressed air 压缩空气
compressed air tunnelling method 压缩空气开挖隧道法
compressed gas 压缩气体
compressibility factor 压缩因子
compressing 压缩
compressing tool 压挤工具
compression 压缩
compression bending 压弯曲加工
compression force 压缩力
compression joint 承压接缝
compression load 压缩荷载
compression molding 压缩成形
compression plate 压板
compression strength 抗压强度
compression test 抗压测试
compression testing machine 压力试验机
compressive failure 压缩塌毁,压缩毁坏

compressive strength 抗压强度,压缩强度
compressive stress 压应力,抗压应力
compressor 压缩机
compressors classification 压缩机分类
computable general equilibrium model 可计算一般均衡模型
computational expression 计算表达式
computational method 计算方法
compute 计算
computed torque control 计算力矩控制
computer 计算机
computer aided analysis 计算机辅助分析
computer aided analysis for circuit 计算机辅助电路分析
computer aided circuit design 计算机辅助电路设计
computer aided control engineering 计算机辅助控制工程
computer aided debugging 计算机辅助故障诊断
computer aided design (CAD) 计算机辅助设计
computer aided design of control system 控制系统计算机辅助设计
computer aided diagnosis 计算机辅助诊断
computer aided drawing 计算机辅助制图
computer aided engineering (CAE) 计算机辅助工程
computer aided instruction 计算机辅助教学,计算机辅助指令
computer aided manufacturing (CAM) 计算机辅助制造
computer aided planning 计算机辅助规划
computer aided production planning 计算机辅助生产计划
computer aided quality control 计算机辅助质量控制
computer aided quality management 计算机辅助质量管理
computer aided researching and develo-

ping 计算机辅助研究开发
computer aided testing 计算机辅助测试
computer and computing hardware 计算机和计算硬件
computer application 计算机应用
computer applied technology 计算机应用技术
computer architecture 计算机结构
computer assisted management 计算机辅助管理
computer auto backup 计算机自动备用
computer auto-manual station 计算机自动/手动操作器
computer automated measurement and control (CAMAC) 计算机自动测量和控制
computer automatic-manual station 计算机自动-手动操作器
computer code 计算机编码
computer communication network 计算机通信网络
computer control 计算机控制
computer control system 计算机控制系统
computer controlled display (CCD) 计算机控制显示
computer controlled system 计算机控制系统
computer experiment 计算机试验，计算机实验
computer gateway 计算机接口
computer graphics (CG) 计算机图形学
computer hardware 计算机硬件
computer integrated manufacturing system (CIMS) 计算机集成制造系统
computer integrated process system (CIPS) 计算机集成过程系统
computer interface 计算机接口
computer interface station 计算机接口操作器
computer interface unit 计算机接口单元
computer management instruction 计算机管理说明
computer network 计算机网络
computer program 计算机程序
computer programming 计算机编程，计算机程序设计
computer recreation 计算机创造
computer science and technology 计算机科学与技术
computer set station 计算机设定操作器
computer simulation 计算机仿真
computer software 计算机软件
computer software and theory 计算机软件与理论
computer subroutine 计算机子程序
computer system 计算机系统
computer systems organization 计算机系统结构
computer tomography 计算机断层摄影，计算机X射线断层术
computer vision 计算机视觉
computer-aided circuit analysis 计算机辅助电路分析
computer-aided control system design (CACSD) 计算机辅助控制系统设计
computer-aided design of feedback controller (CADFC) 计算机辅助反馈控制器设计
computer-aided engineering (CAE) 计算机辅助工程
computer-aided software engineering 计算机辅助软件工程
computer-aided system design 计算机辅助系统设计
computer-aided test (CAT) 计算机辅助测试
computer-aided testing 计算机辅助测试
computer-aided work 计算机辅助工作
computer-integrated 计算机集成
computer-integrated manufacturing 计算机集成制造
computerized automatic concrete cube crushing machine 电脑化混凝土立方块压力试验机
computerized numerical control (CNC) 电脑数值控制，计算机数控，数控

computer-reset-hold switch 计算-复位-保持三用开关
computing amplifier 运算放大器
computing element 计算元件，计算单元
computing instrument 计算器，计算仪表
computing linkage 计算连接
computing module (CM) 计算模件，计算模块
computing module status display 计算模块状态显示画面
computing relay 计算继电器
computing station 计算操作器
computing system 计算系统
computing unit 计算装置
concatenated motor 串级电动机，级联电动机
concave 凹形
concave cutter 凹面铣刀，凹锣刀片
concave grating 凹面光栅
concave side 凹面
concavity 凹面，凹度
concealed installation 隐藏式安装，暗装
concealed piping 隐蔽喉管，隐藏喉管，暗管敷设
concentrated load 集中载重，集中负荷
concentrating column 浓缩柱
concentrating flux plate 浓缩导磁板
concentrating flux sleeve 浓缩导磁套
concentrating thin layer plate 浓缩薄层板
concentration 浓度
concentration cell 浓差电池
concentration of tracer 示踪物浓度
concentration sensitive detector 浓度敏感型检测器
concentrator 集中器，集线器，集中者
concentric cable 同心电缆，同轴电缆
concentric line 同轴电缆，同心线，公共线
concentric orifice plate 同心孔板，同心锐孔隔板
concentric orifice plate flowmeter 同心孔板流量计

concentric winding 同心绕组
concentricity 同心度
concentricity tester 同心度检查仪
concept design 方案设计，概念设计
conceptual development model 初样，初步开发样机
conceptual representation 概念表达式，概念说明，概念模型
conclusion 结论
concrete 混凝土，三合土
concrete admixture 混凝土外加剂，混凝土掺和料
concrete and reinforced concrete drainage and sewer pipes 混凝土和钢筋混凝土排水管
concrete block seawall 混凝土海堤
concrete buffer 混凝土缓冲
concrete core 混凝土芯，土芯
concrete cover 混凝土保护层
concrete cube 混凝土立方块
concrete durability 混凝土耐用程度
concrete foundation 混凝土基础
concrete grade 混凝土标号，混凝土等级
concrete lining 混凝土搪层，混凝土衬砌
concrete mix 混凝土混合物，混凝土拌和料
concrete mixer 混凝土混合机，混凝土搅拌机
concrete mixing plant 混凝土拌和厂
concrete pile 混凝土桩
concrete pipe 混凝管，水泥管
concrete pour works 混凝土浇灌工程
concrete re-alkalization 混凝土再碱性化
concrete sample 混凝土样本
concrete slab 混凝土板
concrete sleeper 混凝土轨枕
concrete spalling 混凝土剥落
concrete strength 混凝土强度
concrete stress 混凝土应力
concrete structure 混凝土结构，具体结构，混凝土建造物
concrete technology 混凝土科技，混凝土工艺，混凝土工艺学
concrete test 混凝土测试

concrete vibrator 混凝土振动器，混凝土路面振动整实器
concrete wall 混凝土墙，堡垒墙
concurred design 并行设计
concurrency 并行，并行并发性，同时发生
concurrency control 并行控制
concurrent 并行的
concurrent architecture 并行结构
concurrent engineering 同步工程，并行工程
concurrent engineering on network 网上协作工程
concurrent operation 同时操作
concurrent program 并行程序
concurrent search 并行搜索
concurrent system 并行系统
condensable gas 可凝性气体
condensance 容抗，电容器的电抗，电容阻抗
condensate pot 冷凝容器
condensate 冷凝，使凝结
condensation 冷凝，凝结，压缩
condenser 冷凝器，电容器，聚光器
condenser bank 电容器组
condenser coupling 电容器耦合
condenser excitation 电容器励磁
condenser lens 聚光透镜，聚束透镜，聚光镜，聚焦透镜
condenser microphone 电容传声器
condenser motor 电容电动机
condition code 条件码，特征码，状态码
condition number 条件数，性态数
condition of self-locking 自锁条件
condition signal 条件信号
condition 条件，状况，环境
conditional 条件的
conditional behavior 条件属性
conditional instability 条件不稳定性
conditional potential 条件电位
conditional probability 条件概率
conditional stability 条件稳定
conditional stability of a linear control system 线性控制系统的条件稳定性
conditional stability of a linear system 线性系统的条件稳定性

conditional stable system 条件稳定系统
conditioning 调节
conditioning amplifier 调节放大器
conduct 传导
conductance 电导，电导量，导率
conductance relay 导纳继电器
conductimetry 电导分析法
conducting electrical instrument 传导类电法仪器
conducting medium 导电介质
conducting wire 导线，传导线
conduction 导电，传导
conduction angle 导通角，传导角
conduction current 传导电流
conduction holes 导电空穴
conduction of heat 热传导
conductive 导电的，传导的
conductive coupling 电导耦合
conductive foil 导电箔
conductive gas sensor 电导式气体传感器
conductive gas transducer 电导式气体传感器
conductive level sensor 电导式物位传感器
conductive level transducer 电导式物位传感器
conductive part 导电部分
conductive pattern 导电图形
conductivity 传导率，传导性，导电性，电导率
conductivity cell 传导单元
conductivity detection 电导检测
conductivity instrument 电导率仪器
conductivity meter 导电计，电导仪，导热计
conductivity recorder 电导率记录仪
conductivity sensor 电导率传感器
conductivity transmitter 电导率变送器
conductometer 电导计，热导计
conductometric analysis 电导分析法
conductometric analyzer 电导式分析器
conductometric titration 电导滴定，电导滴定法
conductometry 电导分析法，电导测

定法
conductor 导体，导线
conductor cross-section 导线截面
conductor element 导体元件
conductor layer 导体层，导线层
conductor resistance 导线电阻
conductor side 导线面
conductor spacing 导线距离
conductor trace line 导体痕迹线
conductor width 导线宽度
conduit 管道，导管
conduit entry 导管引入
cone 锥形筒
cone angle 圆锥角
cone distance 锥距
cone-plate viscometer 锥-板黏度计
confidence 置信度
confidence interval 置信区间，可靠区间
confidence level 置信级，可信度，信赖水准，置信水平
confidence limits 置信区间，信赖界限，可信限
configurable 可组态的，可配置的，结构的
configurable control function 可组态的控制功能
configuration 配置，组态，构造，结构，形态，配置
configuration control 组态控制，组合控制
configuration identification table 组态标识表
configuration list 组态表
configuration management 组态管理，组合管理
configuration of engineering functions 工程功能的组态
configuration of FCS control functions FCS 控制功能的组态
configuration of hardware 硬件的配置
configuration of operation and monitoring functions 操作和监视功能的组态
configuration space 组合空间
configuration space enable 可分配

空间
configuration specification 组态说明书
configuration stability 结构稳定性
confined lysimeter 封闭式渗漏测定计
confined space 密闭空间，有限空间
confining stress 局限应力，侧限应力，围压应力
confirmatory test 证实试验
conflicting routes 路径冲突
conformal mapping 保角（保形）映射，共形映象
conformal mapping technique 保形映射技术，保形变换技术
conformal surfaces 共型表面
conformal transformation 保角变换，保角映射，保形变换，共形变换
conformance test 一致性测试，符合性测试
conformity 一致性
conformity certification 合格认证
conformity error 一致性误差
congealer 冷却器，冷冻器
conic 双曲线，二次曲线
conical bearing 圆锥轴承，锥形轴承
conical entrance orifice plate 圆锥入口孔板，锥形入口孔板
conjugate 共轭的
conjugate cam 共轭凸轮
conjugate complex number 共轭复数
conjugate curves 共轭曲线
conjugate gradient method 共轭梯度法
conjugate match 共轭匹配
conjugate point 共轭点
conjugate profile 共轭齿廓
conjugate roots 共轭根
conjugate tooth 共轭齿轮
conjugate yoke radial cam 等径凸轮
conjunction 连接
connect to earth 接地
connect to neutral 接零，接中线
connect up 接上，连接
connect 连接，接通
connected in series 串联的
connected parallel 并行连接

connected parallel computer 并行连接计算机
connecting 连接
connecting format 连接形式
connecting in parallel 并联
connecting in series 串联
connecting link 连接杆,可拆链环,连接环节,连接链节,连接片
connecting pipeline 连接管线
connecting rod 连杆
connecting wire error 连接线误差
connection 连接,接线,接线图,引线,通信线,接驳,接驳处
connection box 连接匣,接线板,连接器
connection point 接点
connection screw 连杆调节螺钉
connection test 检查接线
connection to subsystem 连接到子系统
connectionism 联结主义,连接机制
connections of circuits 电路的连接
connections of polyphase circuits 多相电路的连接
connective 接通的,连接的
connective instability 不稳定连接
connective stability 稳定连接
connectivity 连接性,连通性,联络性
connector 接插件,接头,插头,接驳器,连接器,接线盒
connector plug 连接器插头,插头钳
connector socket 插座
conoid helical-coil compression spring 圆锥螺旋扭转弹簧
consecutive test result 连续测试结果
consequent 后续的
conservation 清洁,守恒
conservation of energy 能量守恒,节约能源
conservation of matter 物质守恒
conservation of mechanical energy 机械能守恒
conservation principle 收敛原理
conservative system 保守系
conservatory 温室
consistency 相容性,一致性
consistency check 一致性检验,连续检查
consistency transmitter 浓度变送器,稠度变送器
console 操纵台,控制台
console display 操纵台显示器,控制台显示器
console message processor 操纵台信号处理机,控制台信号处理机
console monitor 操纵台监视器,控制台监视器
console printer 控制台打印机
console scope 控制台显示器
console status display 操纵台状态显示画面
console subsystem 操纵台子系统,操作站子系统
console switch 操纵台开关,操作盘开关
console with wardrobe 直形橱式操纵台
consolidate 加固
consolidation 沉积,固结
consolidation coefficient 固结系数
conspicuous place 显眼地方
conspicuous position 显眼位置
const 常数,不变的
constac 自动稳压器
constant 常数,常量,不变的,恒定的
constant acceleration and deceleration motion 等加等减速运动规律
constant bandwidth filter 恒定带宽滤波器
constant current adjustable signal source 恒流可调信号源
constant current power supply 恒流电源
constant error 恒定误差
constant frequency oscillator 固定频率振荡器
constant frequency power supply 稳频电源
constant head flowmeter 恒定压头流量计
constant load 固定负载
constant low temperature facilities 低温室

constant of a measuring instrument 测量仪表的常数
constant of inertia 惯性常数
constant pressure line 等压线
constant state 恒定状态,恒定常态
constant temperature and humidity chambers 恒温恒湿器
constant temperature bath 恒温池
constant temperature circulator 恒温循环泵
constant temperature devices 恒温装置
constant temperature drying ovens 恒温干燥箱
constant temperature incubators 恒温培养器
constant temperature water baths 恒温水槽
constant velocity motion 等速运动
constant velocity testing machine 常速试验机
constant voltage 恒定电压,定压
constant voltage constant frequency (CVCF) 恒压恒频
constant voltage diode 稳压二极管
constant voltage power supply 恒压电源
constant voltage source 恒压源
constantan 康铜
constant-breadth cam 等宽凸轮
constant-current characteristic 恒流特性
constant-current power supply 稳流电源
constant-current regulation 恒流调节
constant-current source 恒流源
constant-current stabilizer 稳流器,恒流稳定器
constant-current transformer 恒流变压器
constant-frequency oscillator 恒频振荡器
constant-level head tank 恒液位压头容器
constant-rate injection method 恒定速率注入法
constant-temperature method of reference junction 参比端恒温法

constituent 成分
constitution of mechanism 机构组成
constrain 强制
constrained 约束的
constrained current operation 强制励磁
constrained optimization 约束优化
constrained parameter 约束参数
constrained pole 约束极点
constraining force 约束力
constraint 约束,强制
constraint condition 约束条件
constraint satisfaction 约束补偿
constraint satisfaction problem 约束补偿问题
constraints on control 约束控制
constriction resistance 集中电阻,接触电阻
construction 结构,构成,组成,建造,建筑结构,构造
construction joint 施工接缝,施工缝
construction site vehicle 建筑工地车辆
construction winch 建筑卷扬机
construction works 建筑工程,建造工程,施工工程
constructional hardware 结构硬件
consultation 咨询,磋商,会诊,讨论会
Consultative Committee International Telegraph and Telephone (CCITT) 国际电报电话咨询委员会
consumer 用户
consumer electronics 消费性电子产品,家用电子产品
consumer package 销售包装
consumption 消耗
consumption function 消费函数
contact 触点,接触,接触触点,接触联系
contact anemometer 接触式风速表,电接风速计
contact controlled electric actuator 有触点电动执行机构
contact current 触点电流
contact drop 接触点电压降
contact electromotive force 接触电

动势
contact fatigue 接触疲劳，接触疲惫
contact input 触点输入，接点输入
contact mercury barometer 电接点水银气压表
contact output 触点输出，接点输出
contact pads 接触垫
contact pattern 接触图形，（齿轮的）接触斑点
contact points 联络点，接触点，啮合点
contact process 接触过程，接触法
contact ratio 重合度，接触比，啮合系数
contact resistance 接触电阻
contact seal 接触式密封
contact sense module 接触感测模块
contact stress 接触应力，接面应力
contact switch 接触开关
contact terminal 触头
contact voltage 接触电压
contact to earth 接地，触地，碰地
contactless 无触点的，不接触的，遥控的
contactless pickup 无触点传感器
contactless switch 无触点开关
contactor 开关，电流接触器
contactor control 接触器控制
contactor control systems 继电控制系统
container 贮存器，货柜，集装箱，容器
container cranes 集装箱起重机
container function 集装箱功能
containment 包含，牵制，容量，密闭度
contaminate 污染
contamination 沾染，污染
contamination monitoring system 污染监测系统
content uniformity 含量均匀度
context state record 上下文状态记录
context-free grammar 上下文无关文法
contingency plan 应急计划，意外事件计划
continued fraction 连分数，连分式

continued fraction expansion 连续展开，连续空间
continuity 连续性
continuity test 通路测试，连续性试验
continuous 连续的，不断的
continuous action 连续作用
continuous barrier 连续护栏
continuous beam 连续梁
continuous control 连续控制
continuous control block 连续控制功能块
continuous control function 连续控制功能
continuous control system 连续控制系统
continuous controller 连续作用控制器，连续作用调节器
continuous current 恒向电流，直流，连续流
continuous current generator 直流发电机
continuous cycling method 连续循环方法
continuous data point 连续数据点
continuous development 连续展开
continuous discrete event hybrid system simulation 连续离散事件混合系统仿真
continuous duty 连续运行，连续负荷，连续使用，连续工作
continuous footing 连续基脚，条型基础
continuous index 连续分度
continuous load 持续负载，连续荷载
continuous operation 连续操作，连续运算
continuous oscillation 等幅振荡
continuous overload limit 持续过范围限
continuous overrange limit 持续过范围限
continuous path control 连续轨道控制
continuous phase modulation 连续相位调制
continuous process 连续过程

71

continuous signal 连续信号
continuous speech recognition 连续语音识别
continuous stirred tank reactor (CSTR) 连续带搅拌反应器，连续搅拌釜［槽］式反应器
continuous system 连续系统，连续式，连续生产系统
continuous time filter 连续时间滤波器
continuous time signal 连续时间信号
continuous time system 连续时间系统
continuous to discrete-time conversion 连续到离散时间变换
continuous transformation 连续变换
continuous variable 连续变化，连续变量
contour 轮廓
contour line 等高线，等值线，轮廓线
contour map 等高线图，等值线图
contouring machine 仿形机床，轮廓锯床，靠模机床
contraction coefficient 收缩系数
contraction crack 缩裂，收缩裂缝
contraction joint 收缩接缝，收缩节理
contra-flow 回流，对流
contrast 对比
contrast law 对比定律
contrast sensibility 对比灵敏度
contravention 违反
control 控制，操纵，调节
control apparatus 控制电器
control accuracy 控制精度，调节精度
control action 控制作用，控制动作，调节作用
control algorithm 控制算法
control appliance 控制设备，控制装置
control application 控制应用
control applications of computer 计算机控制应用
control board 控制板，仪器板，控制台，控制屏
control cabinet 控制柜，操纵室

control cable 操纵索，控制电缆
control center 调度中心，控制中心
control characteristic 控制特性，调节特性
control characteristic of a control system 控制系统的控制特性
control circuit 控制电路，控制线路
control circuit closed-loop 闭环控制电路
control combination 控制组合
control command word 控制命令字
control computation 控制运算
control computer 控制用计算机，电脑控制
control configuration 控制组态，控制结构
control console 操纵台，控制台
control console body 控制箱体
control contact 控制触点
control contactor 控制接触点
control counter 控制计数器
control current 控制电流
control desk 控制台，操纵台
control device 控制装置，控制器
control drawing 控制图
control drawing definition 控制图定义
control drawing example 控制图举例
control drawing window 控制图窗口
control electrode 控制电极
control engineering 控制工程
control equation 控制方程
control equipment 控制设备，控制仪器
control error 控制差错，调节误差
control floating action 浮点控制作用
control function 控制作用，控制函数
control gear 自动调整仪，控制机构，操纵装置
control group display block 控制组显示块
control group panel 分组画面
control group panel definition 分组画面定义
control handle 控制旋钮，控制手柄，操纵手柄

control hardware 控制硬件
control hierarchy 控制等级，控制层次
control input 控制输入
control input source 控制输入源
control instant 瞬态控制
control instrument 控制仪表
control interaction factor 控制相关因数
control key 控制键
control kiosk 控制站
control language (CL) 控制语言
control language for application module 应用模块使用的控制语言
control language for continuous data points 用于连续控制点的控制语言
control language for process manager 过程管理机的控制语言
control language for the multifunction controller 用于多功能控制器的控制语言
control law 控制率，控制规则
control loop 控制回路，调节回路，操纵系统
control loop guideline 控制回路导向图
control loop interaction 控制回路交互作用
control loop troubleshooting 控制回路故障排除
control mark 控制标志
control message display 控制信号显示
control mode 控制方式，调节方式
control model 控制方式
control module redundancy 控制模件冗余
control moment gyro 控制力矩陀螺
control monitor 控制监视器
control motor 控制电动机
control motor actuator 电机执行机构
control nonlinearity 非线性控制
control object 控制目标
control of phase-sequence 相序控制
control of total plant 整体工厂控制
control oriented model 原模型控制

control output 控制输出
control output destination 控制输出指定站
control output module 控制输出组件，调节输出组件
control panel 控制盘，控制台，控制面板，控制屏
control parameter 控制参数
control point 控制点
control point deletion 删除控制点
control precision 控制精度
control pressure of hydraulic choke 液动节流控制压力
control processor 控制处理机
control range 控制范围，调节范围
control relay 控制继电器，监测继电器
control requirement 控制要求
control room 控制室，仪表室，配电室
control room area 控制室区
control room design 控制室设计
control room environment 控制室环境
control room management system 控制室管理系统
control scheme 控制方案，控制元件，控制环节
control science and engineering 控制科学与工程
control signal 控制信号，操纵信号
control station 调度站，控制站，控制点，控制台，操作器
control status display function 控制状态显示功能
control strategy 控制策略
control structure 控制结构，内部控制结构
control switch 控制开关，主令开关
control synchro 控制式自整角机
control system 控制系统，调节系统
control system analysis 控制系统分析
control system design 控制系统设计
control system instrumentation 控制系统仪表化
control system synthesis 控制系统综合

control technology 控制技术
control theory and control engineering 控制理论与控制工程
control time horizon 控制时间范围，控制时间层
control transformer 控制变压器
control unit 控制器，控制部件
control unit definition 控制单元定义
control unit start 控制单元的启动
control unit status display panel 控制单元状态显示画面
control unit stop 控制单元的停止
control value 控制值
control valve 控制阀，调节阀
control valve actuator 阀控传动机构
control variable 控制变量，决策变量
control voltage 控制电压
control winding 控制绕组，控制线圈
control wire 操纵索，操纵线，控制线，操纵钢绳
control with fixed set point 定值调节，定值控制
controllability 可控性
controllability index 可控指数
controllable canonical form 可控规范型
controlled area 控制区，监管区，放射性区域，信号灯控制区
controlled atmosphere 受控大气，人造大气
controlled avalanche rectifier 可控雪崩整流器
controlled condition 控制条件
controlled system 控制系统，被控系统
controlled variable 被控量，控制量，被控变量
controlled wellhead number 控制井口数量
controller 控制器，调节器
controller design 控制器设计
controller gain 控制器增益
controller modulator 调制控制器
controller output 控制器输出
controller saturation 控制器饱和
controller setting 控制器整定
controller subsystem 控制器子系统
controller vehicle 车厢控制器
controlling 控制
controlling device 控制装置，操纵装置，调节装置
controlling element 控制因子，控制环节，控制元件
controlling instrument 控制仪表
controlling machine 控制机械
controlling system 施控系统
controlling unit 调节单元
controlling value 控制值
convection 对流
conventional control 常规控制
conventional mechanism 常用机构
conventional operation 常规操作
conventional true value 约定真值
convergence 收敛
convergence analysis 收敛分析
convergence factor 收敛因子
convergence of numerical method 收敛数值方法
convergence proof 收敛证明
convergent 收敛的，非周期衰减的
convergent control 衰减控制
convergent series 收敛级数，收敛数列
conversational monitoring system 会话监督系统
conversion 转换，变换，换算
conversion equipment 转换设备，转换器
conversion factor 换算因子，转换因子，变换因数
conversion iron 炼钢生铁
conversion of signal 信号变换
convert 转换，使转变
converter 整流器，转化器，变流器
convertor 转炉，变流器，转化器
convex 凸的，凸面体
convex cutter 凸形铣刀，凸锣刀片
convex optimization 凸规划，凸面优化
convex programming 凸规划，凸编程
convex projection 凸计划
convex side 凸面
convey 输送
conveyer 输送机，运输器
conveyer belt 输送带

conveying chains　输送链
conveyor　传送带，输送机，输送器
convolution integral　卷积积分，褶合积分
convolution model　卷积模型
convolution spectrometry　褶合光谱法
convolution transform (CT)　褶合变换，对合变换
cooktop　炉灶面
cool　冷的
coolant　冷却液，冷却剂
coolant temperature gauge　冷却水温度表
cooler　冷却器
cooling　冷却
cooling agent　冷却剂
cooling and dehumidifying coil　冷却除湿盘管
cooling coil　冷却盘管，冷却旋管
cooling fan　冷却风扇，冷却风机
cooling fan blade　散热扇片
cooling fin　散热片，防裂片
cooling fluid　冷却液，冷却流体
cooling liquid circulators　冷却液体循环器
cooling pipe　冷却管
cooling pond　凉水池
cooling rib　冷却肋片
cooling spiral　螺旋冷却栓，冷却螺管
cooling system　冷却系统
cooling tank　冷却罐
cooling tower　冷却塔，冷却水塔
cooling trap　寒流捕获器
cooling water circulators　冷却水循环器，冷却水外部循环器
cooling water pump　冷却水泵
co-operation　协作，协同
co-operative control　协调控制
cooperative game　合作对策
co-ordinate　坐标系
co-ordinate time　等同时间
co-ordinate transformation　等同变换
coordinated control system　协调控制系统
co-ordination　调整，协调，同等
coordination strategy　协调策略

co-ordinator　协调器，统筹人
coordinator　协调器
coping　盖顶
copolymer　共聚合体，共聚物
copper (Cu)　铜
copper binding wire　铜包线
copper contact　铜触点
copper core　铜芯线，铜芯电力电缆
copper damper　铜阻尼器
copper electrode　铜电极
copper foil　铜箔
copper glazing　铜条嵌镶玻璃
copper loss　铜损
copper loss of rotor　转子铜损
copper sheet　铜片，铜皮，薄铜板
copper tube　铜管
copper weld steel wire　铜焊钢线，铜焊钢丝
copper-clad aluminium conductor　包铜铝导体
copper-clad covered steel wire　包铜钢线
copper-clad laminate　覆铜箔层压板
copper-clad surface　铜箔面
coprocessor　协处理器，协调处理器
coprocessor daughter board　协处理器子板
copy　复制
copy a file　复制一个文件
copy grinding machine　仿形磨床
copy lathe　仿形车床
copy milling machine　仿形铣床
copyright　版权
cord bracket　夹线板
cord switch　拉线开关
core area　铁芯面积
core data bank　土芯样本资料库
core diameter　内径，焊条直径
core drilling　钻取土芯，取芯钻探
core iron　芯铁
core loss　铁损，磁芯损耗
core material　型芯材料，裂变物质，内层芯版，磁芯材料
core memory driver　磁芯存储器驱动器
core module test system　磁芯模件测试系统

core pin 塑孔栓，中心销，心形销，后模镶针
core sample 矿样，岩心样本，土芯样本
core template 砂芯模板
core testing 岩心试验，岩心测试，中心抽样检验
core transistor logic 磁芯晶体管逻辑
core vent 砂芯排气孔
core volume 体积
core wire 心线，焊条芯
core 核心，要点，磁芯
coring 钻取样本，钻探抽样
corner 角位
corner effect 锐角效应，角隅作用
corner frequency 转折频率，转角频率
corner gate 压边浇口，角形控制极
corner shear drop 直角压陷
corona 电晕
corona discharge 电晕放电
corona effect 电晕效应
corona loss 电晕损失
corporate strategy 自治策略，公司策略
correct 校准，修正，正确的，改正
corrected result 修正结果
correcting 校正，校正的
correcting condition 校正条件
correcting element 调节机构，校正元件，执行元件
correcting feedback 反馈校正
correcting feedforward 前馈校正
correcting plane 平衡平面，校正平面
correcting range 校正范围，操纵范围
correcting unit 校正单元
correcting variable 校正变量
correction 修正值，修正，改正
correction coefficient 校正系数
correction curve 修正曲线，补偿曲线
correction curve of a measuring instrument 测量仪表的修正值曲线
correction data 校正数据
correction factor 校正因子，修正因数

correction rate 校正率
correction time 校正时间，补偿时间
correction value 校正值
corrective action 校正动作
corrective action report 校正动作报告
corrective maintenance 故障检修，维修保养，出错维修，安全改进维护
correlation absorption band 相关吸收带
correlation coefficient 相关系数，关联系数
correlation interferometer 相关干涉仪
correlation length measuring instrument 相关式测长仪
correlative method input system 相关对比法感应脉冲瞬变系统
correlator 相关器，相关因子
correspondence theorem 相似定理
corridor 通路，走廊
corrosion 腐蚀，侵蚀，点状腐蚀
corrosion failure 腐蚀破裂
corrosion inhibiting grease 抗蚀油脂
corrosion preventative 防腐蚀
corrosion resistance 耐蚀性，耐腐蚀性
corrosion testing machine 腐蚀试验机
corrosion-proof instrument 防腐式仪器仪表
corrosion-resistant material 防蚀物料
corrosive air 腐蚀性空气
corrosive atmospheres test 腐蚀性大气试验
corrosive fume 腐蚀性烟气
corrosive liquid 腐蚀液
corrosive salt 腐蚀性盐
corrosive substance 腐蚀性物质
corrosive wear 腐蚀性磨损
corrugated iron 陨铁，波状钢
corrugated roof glazing 瓦楞玻璃屋面
corrugated tool 阶梯刨刀
corrugation grinding machine 波纹研磨机
corrugation pad 波形垫
cosine (cos) 余弦
cosine acceleration motion 余弦加速

度运动
cosine collector 余弦收集器
cosine law 余弦定理
cosine radiator 余弦辐射体
cosine transform 余弦变换
cosmetic defect 外观不良
cosmetic inspect 外观检查
cosmic rays 宇宙线，宇宙射线
cost matrix 费用矩阵
cost of electric energy 电能成本
cost of labor 人工费
cost performance 性能价格比，成本效益
cost of fuel 燃料费用
costate variable 共态变量
cotter 开口销
cotton covered enamel wire 纱包漆包线
cotton-covered wire 纱包线
Coulomb 库仑，库（电量电位）
Coulomb damping 库仑阻尼，干摩擦阻尼
Coulomb friction 库仑摩擦
Coulomb interaction 库仑作用
Coulomb's law 库仑定律
coulometer 库仑表，库仑计，电量计
coulometric analysis 电量分析，电量分析法
coulometric analyzer 电量分析器
coulometric titration 电量滴定，库仑滴定法
coulometry 电量分析
count circuit 计数电路
count impulse 计数脉冲
count of acoustic emission event 声发射事件计数
countdown frequency divider 脉冲分频器
counter 计数器，计算器，相反的，反作用
counter electromotive force 反电动势
counter electromotive force relay 反电动势继电器
counter extraction 反萃取
counter lock 止口镶嵌方式，平衡锁口

counter punch 反凸模
counter register 计数寄存器
counter switch 计数开关
counter type superconduction magnetometer 计数式超导磁力仪
counter voltage 反电压
counterbalance valve 反平衡阀，背压阀，配衡阀
counterbore 镗（沉头）孔，埋头孔，扩孔
counter-clockwise 逆时针方向的
counterclockwise 逆时针方向的
countermeasures against noise 抗噪声对策
countermeasures against static electricity 抗静电措施
countersink 埋头孔，暗钉眼，钻（沉头）孔
countersunk socket set screw 埋头螺丝
counterweight 配重，平衡重，平衡锤
counting anemometer 计数风速表
counting rate 计数率
counting rate meter 计数率计
counting 计数器
counting-shallow-layer seismograph 计数型浅层地震仪
couplant 耦合剂，耦合介质
couple 力偶
couple unbalance 力偶不平衡
coupled 耦合的，连接的
coupled chamber method reciprocity calibration 耦合腔法互易校准
coupled circuit 耦合电路
coupled device 耦合装置
coupled mode analysis 耦合模型分析
coupled mode theory 耦合模型理论
coupled modes 耦合方式
coupled oscillatory circuit 耦合振荡电路
coupled resonator 耦合谐振器
coupled system 耦合系统
coupled vibration 耦合振动
coupler 耦合器，耦合腔，联结器，联轴器
coupler-curve 连杆曲线
coupling 管箍，接头，联结，耦合，

匹配，配合，联轴器
coupling capacitor 耦合电容器
coupling capacity 耦合电容
coupling coefficient 耦合系数
coupling element 耦合元件
coupling factor 耦合系数
coupling filter 耦合滤波器
coupling flange 联结翼板
coupling function 耦合函数
coupling impedance 耦合阻抗
coupling inductance 耦合电感
coupling joint 联轴器连接，活节联轴器
coupling loss 耦合损失
coupling model 耦合模型
coupling of orbit and attitude 轨道和姿态耦合
coupling reactance 耦合电抗
coupling transformer 耦合变压器
covalent bond 共价键
covariance 协方差
covariance matrix 协方差矩阵
cover 盖板，外盖，遮盖，封盖，保护层
cover center section 盖板中心部分
cover layer 覆盖层
cover plate 电机活动盖板，荧光屏前面防护玻璃
cover tile 盖瓦
covered channel 有盖排水槽，暗槽
covered wire 绝缘电线，绝缘线被覆线，包线
cowcatcher 排障器
CP (communication processor) 通信处理机，通信处理器
CPU (central processing unit) 中央处理单元，中央处理机，中央处理器
crack 裂缝，裂痕
crack propagation strain gauge 裂纹扩展应变计
cracking pattern 裂纹图形，裂缝分布图
cradle 摇台，送料架
cramp iron 铁搭，钢筋，两爪钉，扒钉
crane 吊机，起重机

crane controller 起重机控制器
crane motor 起重电动机
crane trolley wire 起重机接触导线
crane weigher 吊车秤
crank 曲柄
crank balancing machine 曲轴平衡机
crank pin bearing 曲柄销轴承
crank plate 曲板
crank shaft 曲轴，曲柄轴，曲柄转轴
crankless 无曲柄式
crank-rocker mechanism 曲柄摇杆机构
crank-slider mechanism 曲柄滑块机构
crash 粉碎，破碎
crash barrier 防撞栏，护栏
crate 机箱
crate address 机箱地址
crate controller 机箱控制器
crator 焊疤
CRC (cyclic redundancy check) 循环冗余校验
CRDP (calculated results data point) 计算结果数据点
creat command file 建立命令文件
creat directory 建立目录
creation design 创新设计
creep 蠕变，潜变，恢复，徐变
creep rupture strength 蠕变断裂强度
creep rupture strength testing machine 持久强度试验机
creep strength 潜变强度，抗蠕变强度
creep testing machine 蠕变试验机
creeping discharge 蠕缓放电，沿面放电
creeping motion 蠕动
creeping speed 蠕变速度，最低航速，爬坡速率
cresol resin 甲酚树脂
crest factor 波峰因素，峰值系数，振幅因数
crest meter 峰值计
crest voltage 巅值电压，峰值电压
crest voltmeter 峰值伏特计
crimping tools 卷边工具，机械压线钳
criterion 判据，准则
criterion for stability 稳定度判据

criterion function 标准函数
critesister 热敏电阻
critical 临界的，极限的
critical area 临界区域，关键部位
critical clearing time 极限切除时间
critical condition 临界条件，临界状态
critical current density 临界电流密度
critical damped 临界衰减
critical damping 临界衰减，临界阻尼
critical defect 临界缺陷
critical element 关键构件，临界构件
critical evaporator 临界点干燥器
critical flow 临界流
critical flow measurement 临界流量测量
critical flow nozzle 临界流喷嘴
critical load 临界荷载
critical mass 临界物质
critical micelle concentration（CMC）临界胶束浓度
critical path analysis 关键路线分析，临界通道分析
critical point 临界点
critical pressure differential 临界压差
critical pressure differential ratio 临界压差比
critical pressure ratio 临界压力比
critical process controller 关键过程控制器
critical resistance 临界电阻
critical speed 临界速度，临界转速
critical stability 临界稳定性
critical stable state 临界稳定状态
critical state model 临界状态模式
critical stimulus 临界激励
critical temperature coefficient thermistor 临界温度系数热敏电阻器
critical temperature thermistor 临界温度热敏电阻器
critical viscous damping 临界黏性阻尼
critical voltage 临界电压
cross 交叉，跨越，十字接头
cross assembling 交叉汇编，交叉编译
cross compiling 交叉编译
cross crank 横向曲轴
cross grain 横纹
cross head 十字头，丁字头

cross joint 十字接头，四通
cross over filter 交叠滤波器
cross point 交叉点
cross product 叉积，矢量积
cross road 十字路口，交路口
cross section 横截面
cross sectional area 横截面积，横切面积，横切面面积
cross tie 枕木，横木，横向堆放层
cross wire 十字交叉线，十字丝
cross-beam 横梁
cross-belt drive 交叉带传动
cross-correlation flowmeter 互相关流量计，交叉相关流量计
cross-correlation function 互相关函数，交叉相关函数
cross-country vehicle 越野汽车
crossed helical gears 交错轴斜齿轮
crossfall 横斜度，横向坡，横向坡度
crosshead 十字结联轴节
crossing point 交错点
crosslinking column 交联柱
crossover 跨交，跨接
crossover frequency 交越频率，交叉频率，分频频率，跨越频率，穿越频率
crossover of electron gun 电子枪交叉点
crossover of load characteristic 负载特性交叠
cross-phase modulation 交叉相位解调
cross-section ionization detector 离子截面积检测器
crosstalk 串扰，串音，串话，交扰，交调失真
crosstalk interference 交叉干扰，串话干扰，串音干扰
crowbar 铁撬棍，铁笔
crowd loading 群众荷载，集束荷载
crown 齿冠，路拱，拱顶
crown circle 锥齿轮冠圆，外端齿顶圆
crown gear 冠状齿轮，差动器侧面伞齿轮
crowned teeth 鼓形齿
CRT（cathode ray tube）阴极射线管
CRT display device CRT显示器
CRT display terminal CRT显示终端

CRT terminal CRT 终端
crucible 坩埚
crude petroleum 原油
cruise control 恒速操纵器，巡航控制
crushed stone 碎石
crusher 破碎机，碎石机
crushing strength 压碎强度，抗碎强度，破碎强度
cryogenic current standard 低温电流标准器
cryogenic device 低温器件
cryogenic temperature 冷冻温度
cryostat 低温恒温器
cryptometer 遮盖力计，遮盖力测定仪
crystal 振子，晶片，晶体
crystal amplifier 晶体放大器
crystal cell 晶体光电池
crystal clock 晶体钟
crystal converter 晶体变换器
crystal cut 晶体切片
crystal diode 晶体二极管
crystal frequency 晶体频率
crystal grating 晶体光栅
crystal lattice 晶格
crystal microphone 晶体话筒，晶体传声器
crystal mixer 晶体混频器
crystal oscillator 晶体振荡器
crystal photoelement 晶体光电元件
crystal pressure transducer 晶体压力传感器
crystal rectifier 晶体整流器
crystal speaker 晶体扬声器
crystal violet 结晶紫
crystalline 晶体的
crystalline electrode 晶体电极
C-scope C 型显示
CSTR（continuous stirred tank reactor） 连续带搅拌反应器，连续搅釜（槽）式反应器
CT（convolution transform） 褶合变换，对合变换
CTN carton 卡通箱
cubic metre 立方米（体积单位）
cull 残料废品
cultural aspects of automation 自动控制文化概念

culvert 涵洞，阴沟，电缆管道
cumulative average unit 区间平均运算器
cumulative compound excitation 积复励
cumulative compound generator 积复励发电机
cumulative compound motor 积复励电动机
cumulative compound winding 积复励绕组
cumulative error 累计误差
cup anemometer 转杯风速表
cup flow test 杯模式流动度试验
cupping testing machine 杯突试验机
cuprum alloy 铜合金
curdle 凝固
Curie 居里
Curie temperature 居里温度，居里点
curing 养护
curing agent 固化剂
curing compound 养护剂
curing condition 养护环境
curing time 硬化时间，固化时间
curium（Cm） 锔
curl bending 卷边弯曲加工
curling 卷曲加工
current 电流，当前的，现行的，水流，当前，气流
current amplifier 电流放大器
current balance 电流平衡
current bias source 偏流源
current bottle 漂流瓶
current by phase 每相电流
current carrier 载流子
current carrying capacity 载流容量（等于最大允许电流），电流容许量
current center 电流中心
current collecting device 集电设备
current collecting methods 集电方式
current comparators 电流比较器
current constant 电流常数
current consumption 电流消耗
current converter 换流器
current coupling 电流耦合
current damper 电流阻尼器
current decay 电流衰减，电流损耗

current demand　电流需求量
current density　电流密度
current direction　海流方向，电流方向
current distribution　电流分布
current divider　分流器
current error　电流误差
current feedback　电流反馈
current file　当前文件，现行文件
current gain　电流增益，增益
current in middle wire　中线电流
current indicator　电流指示器
current intensity　电流强度
current limiter　限流器
current limiting reactor　限流电抗器
current limiting resistor　限流电阻，限流电阻器
current limiting threshold　限流阈（值）
current limiting transistor　限流晶体管
current line　当前行，流线
current loading　电流负载
current loss　电流损失
current loss regulator　电流损耗调节器
current matching transformer　电流匹配互感器
current meter　电流表，电流计，流速计
current modulation　电流调制
current multiplication type transistor　电流倍增型晶体管
current of commutation　换向电流
current output　电流输出
current output module　电流输出模块
current overload　电流过载
current path　电流路径
current phase　电流相位
current pole　测流标杆
current protection　电流保护装置
current range　电流范围
current rating　额定电流
current ratio　电流比
current reducing resistor　减流电阻器
current regulator　电流调节器，稳流器
current relay　电流继电器
current resonance　电流谐振，并联谐振
current reversible chopper　电流可逆斩波电路
current saturation　电流饱和
current sensitivity　电流灵敏度
current setting　电流整定值
current source　电流源
current source type inverter　电流型逆变电路
current stability　电流稳定度
current stabilizer　电流稳定器，镇流器，稳流器
current suppressing resistor　抑流电阻器
current transformer　电流互感器，电流转换，电流转变
current triangle　电流三角形
current value buffer　当前值缓冲区
current velocity　流速
current voltage characteristic　电流电压特性，伏安特性
current-carrying capacity　载流量
current-carrying conductor　载流导体
current-controlled current source（CCCS）　电流控制电流源
current-controlled voltage source（CCVS）　电流控制电压源
currentless　无电流的
current-limit relay　限流继电器
cursor　光标，游标
cursor control key　光标控制键
cursor key　光标键
cursor position　光标位置
curtain wall　幕墙，护墙
curtain wall supports　幕墙承托物
curvature　曲率
curvature of field　像场弯曲，场曲
curve　曲线，图标
curve generator　曲线发生器
curve matching　曲线拼接
curve of magnetization　磁化曲线
curved snips　弯铁剪
curved-shoe follower　曲面从动件
curvilinear　曲线的
curvilinear motion　曲线运动
curvilinear ordinates recording instrument　曲线坐标记录仪
cushion　缓冲，软垫
cushion pin　缓冲销
cushioning material　垫承物料

cushioning spring	缓冲弹簧
custom data point	用户数据点
custom data segment	用户数据段
custom graphic display	用户图形显示
cut	切割
cut and fill design	挖填设计
cut and paste	剪切和粘贴
cut into operation	接入继电器
cut off	断开，截止
cut off key	断开电键
cut set	割集
cut set matrix	割集矩阵
cut slope	削土斜坡，切削斜坡
cut to size panel	剪切板
cut-in	接入，接通
cut-off frequency	截止频率
cut-off of supply	停止供电
cut-off rate	截止速率
cut-off state	截止状态
cut-off voltage	截止电压
cut-out	断开，切断，切口，开孔，断电器，熔断器
cut-out switch	切断开关
cut-over	切换，转换
cutter	刀具，刀盘，切割器，切削工具
cutter axial	刀盘的轴向位置
cutter axial plane	刀盘轴向平面
cutter axis	刀盘轴线
cutter diameter	刀盘直径
cutter edge radius	刀刃圆角半径
cutter head	刀盘体
cutter number	刀号
cutter point diameter	刀尖直径
cutter point radius	刀尖半径
cutter point width	刀顶距
cutter spindle	刀盘主轴
cutter spindle rotation angle	刀盘主轴转角
cutters	刀具
cutting	切削，切削加工
cutting depth	切削深度，切削厚度，掘槽深度，切削度，开挖深度
cutting die	冲裁模
cutting distance	切齿安装距
cutting edge clearance	刃口余隙角
cutting mills	粉碎器
cutting opening	切孔
cutting out	切断，断开
cutting-off machines	切断机
CVCF (constant voltage constant frequency)	恒压恒频
cybernetic system	控制论系统
cybernetics	控制论
cycle	循环，周期，周波
cycle convertor	周波变流器
cycle count	循环计数
cycle counter	周波表，循环计数器
cycle length	循环时间，周期
cycle of magnetization	磁化循环
cycle of motion	运动周期
cycle per second	赫兹
cycle time	循环时间，周期
cycle track	单车径
cyclic	循环的，周期的
cyclic damp heat test	循环湿热试验
cyclic load	循环负荷
cyclic loading	循环荷载，周期性负荷，周期荷载
cyclic memory	循环存储器
cyclic redundancy check (CRC)	循环冗余校验
cyclic remote control	循环遥控
cyclic strain	循环应变
cyclic stress	循环应力
cyclic voltammetry	循环伏安法
cycling life	循环寿命，周期数
cycling redundancy check	循环冗余校验
cycling solenoid valve	周期电磁阀
cyclodextrin	环糊精
cyclodextrin chromatography	环糊精色谱法
cyclodextrins electrokinetic chromatography	环糊精动电色谱
cycloidal	摆线的
cycloidal gear	摆线齿轮
cycloidal mass spectrometer	摆线质谱计
cycloidal motion	摆线运动规律
cycloidal profile	摆线轮廓
cycloidal tooth profile	摆线齿形
cycloidal-pin wheel	摆线针轮
cyclotron	回旋加速器

cylinder 汽缸［气缸］，圆筒，圆柱状物，柱面
cylinder address 磁道柱面地址
cylinder crushing strength 圆柱体抗压强度
cylinder efficiency 汽缸［气缸］效率
cylinder square 圆筒直尺
cylindrical cam 圆柱凸轮
cylindrical coordinate manipulator 圆柱坐标操作器
cylindrical gear 圆柱齿轮
cylindrical grinding machine 外圆磨床
cylindrical lathe cutting 外圆车削
cylindrical robot 圆柱坐标型机器人
cylindrical roller 圆柱滚子，（输送机）圆筒形滚柱
cylindrical roller bearing 圆柱滚子轴承
cylindrical throat Venturi nozzle 圆筒形喉部文丘里喷嘴
cylindrical worm 圆柱蜗杆，柱形蜗杆
cylindroid helical-coil compression spring 圆柱螺旋压缩弹簧
cylindroid helical-coil extension spring 圆柱螺旋拉伸弹簧
cylindroid helical-coil torsion spring 圆柱螺旋扭转弹簧
cymometer 频率表，频率计
CZCS (coastal zone color scanner) 海岸带水色扫描仪
CZE (capillary zone electrophoresis) 毛细管区域电泳

D

D controller 微分调节器,微分控制器
D/A converter 数/模转换器
D/A (digital-to-analog) 数/模转换
DAC (distance amplitude compensation) 距离振幅补偿
DAD (photodiode array detector) 光电二极管检测器,光二极管阵列检测器
dado 墙裙,护壁
daemon declaration file 说明文件
daemon declaration program 说明程序
Dahlin's algorithm 达林算法
Dahlin's controller 达林控制器
daily load 日负荷
daily load curve 日负荷曲线
daily report 日报
daisy chain 菊花链
daisy chain bus 菊花链总线
Daly detector 戴利检测器
damage 损坏,破坏,损毁,损害
damp 阻尼,衰减
damp proofing 防潮
damped 阻尼的,衰减的
damped alternating current 衰减交流
damped coefficient 阻尼系数
damped frequency 阻尼频率
damped natural frequency 阻尼固有频率
damped oscillation 阻尼振荡,衰减振荡
damped oscillator 阻尼振荡器
damped wave 阻尼波,衰减波
damped winding 阻尼绕组
damper 气闸,气流调节器,减震器,防火闸,阻尼器
damping 阻尼
damping action 阻尼作用
damping characteristic 阻尼特性
damping coefficient 阻尼系数,衰减系数
damping coil 阻尼线圈
damping constant 阻尼常数
damping curve 阻尼曲线
damping factor 阻尼因数,阻尼因子
damping oscillation 阻尼振荡
damping ratio 阻尼比
damping time 阻尼时间
damping torque 阻尼力矩
damping torque coefficient 阻尼力矩系数
damping winding 阻尼绕组
damping wire 减振拉筋,阻尼拉筋
dampness 湿度,潮湿
damp-proof 防潮的
damp-proof insulation 防潮绝缘
danger light 危险信号灯
danger signal 危险信号,危险标志
danger 危险,危险物
danger zone 危险区
dangerous articles package 危险品包装
dangerous building 危险建筑物
dangerous goods 危险品
dangerous goods store 危险品仓库
dangerous 危险的
dark field electron image 暗场电子像
dark noise 暗噪声
dashpot 缓冲器,减震器,阻尼器
data access 数据存取
data acquisition 数据采集
data acquisition algorithm 数据采集算法
data acquisition equipment 数据采集设备
data acquisition station 数据采集站
data acquisition system 数据采集与处理系统
data adapter unit 数据适配器
data alarm 数字警报器
data and time setting panel 数据和时间设定画面
data base (DB) 数据库
data base management system 数据库

管理系统
data binding functions 数据捆绑功能
data buoy system 数据浮标系统
data bus 数据总线
data channel 数据通道
data circuit 数据电路,数据线路
data circuit-terminating equipment (DCE) 数据电路终端设备
data communication 数据通信
data communication system 数据传输系统
data compression 数据压缩
data compression algorithm 数据压缩算法
data concentration 数据集中
data concentrator 数据汇集器,数据集中分配器
data connection 数据连接
data definition table 数据定义表
data display module 数据显示模块
data distributor 数据分配器
data driven 数据驱动
data encoding system 数据译码系统
data encryption 数据加密
data entity builder (DEB) 数据实体编制程序,数据实体编码程序
data entry panel 数据输入面板
data fitting 数据拟合
data flow 数据流
data flow analysis 数据流分析
data flow diagram (DFD) 数据流图,数据流程图
data generator 数据发生器
data handling 数据处理
data handling system 数据处理系统
data highway (DHW) 数据公路,数据总线,高速数据通道
data highway interface 数据总线接口
data highway port (DHP) 高速数据通道子系统接口,数据总线子系统接口
data hold 数据保持
data integrity 数据完整性
data link 数据链路
data link command indicator 数据传输指令指示器
data link layer (DLL) 数据链路层

data link protocol specification 数据链路协议规范
data link service definition 数据链路服务定义
data logger 巡检测,巡回检测,数据记录,数据记录装置
data logging 数据记录,数据资料记录
data model 数据模型,数据模式
data network 数据网络
data owner 数据所有者
data point (DP) 数据点
data point mode 数据点方式
data point mode attribute 数据点方式属性
data pointer 数据指示器
data pool 数据库,数据源
data potentiometer 数据输出电位计
data preprocessing 数据预处理,资料预处理
data privacy 数据保密
data processing 数据处理
data processing system 数据处理器,数据处理机
data processor 数据处理器
data processor subsystem 数据处理机子系统
data reconciliation 数据协调
data recorder 数据记录
data reduction 数据简化,数据减缩,数据整理,资料精编
data reduction system 数据简化系统
data replication 数据复制,数据拷贝
data sampling switch 数据采样开关
data segment description 数据段说明
data set 数传机,数据传输转换器,数据集
data set switch 数据设定开关
data set unit 数据设定器
data set unit with input indicator 带输入指示的常数设定器
data signalling rate 数据传信率,数据信号传输率
data sink 数据接收器
data source 数据源,数据发送器
data station 数据站

data storage 数据储存
data stream 数据流
data structure 数据结构
data symbol 数据符
data symbol transmission 数据转移
data terminal equipment (DTE) 数据终端设备
data transfer 数据传递
data transfer rate 数据传送率,数据传输速率
data transmission 数据传输,数据输送,数据传送
data transmission interface 数据传输接口
data type 数据类型
data upgrading 数据更新,数据升级
data word 数据字
data word echo 数据字回波,数据字回信
data 数据
database 资料库
database structure 数据库结构
database system 数据库系统
data-in 输入数据
data-out 输出数据
data-taking equipment 数据读出设备
datum error 基准误差
datum mark 基准记号
datum tooth 基准齿
daylight savings time 夏时制
DB (data base) 数据库
dB-loss 分贝衰减
DC (direct current) 直流,直流电
DC milliammeter 直流毫安表
DC reflecting galvanometer 直流复射式检流计
DC standard resistance 直流标准电阻
DC ammeter 直流电流表
DC bridge for measuring high resistance 直流高阻电桥
DC bridge for measuring resistance 测量电阻用的直流电桥
DC chopping 直流斩波
DC chopping circuit 直流斩波电路
DC circuit 直流电路
DC comparator potentiometer 直流比较仪式电位差计

DC comparator type bridge 直流比较仪式电桥
DC component 直流分量
DC compound generator 直流复励电动机
DC control circuit 直流控制回路
DC control system 直流控制系统
DC current transformer 直流电流互感器
DC distributing equipment 直流配电装置
DC drive 直流传动,直流拖动
DC electrical source 直流电源
DC exciting-winding 直流励磁绕组
DC generator 直流发电机
DC generator-motor set drive 直流发电机-电动机组传动
DC high voltage transmission 直流高压输电
DC integrating meter 直流积量表
DC machine 直流电机
DC motor 直流电动机
DC potentiometer 直流电位差计
DC power voltage ripple 直流电源电压纹波
DC resistance box 直流电阻箱
DC resistor volt ratio box 直流电阻分压箱
DC series motor 直流串励电动机
DC signal 直流信号
DC source 直流电源
DC standard voltage generator 直流标准电压发生器
DC supply 直流供电
DC voltage calibrator 直流电压校准器
DC welding generator 直流电焊发电机
DC-AC converter 直流-交流变换器
DC-AC-DC converter 直交直电路
DC-DC converter 直流-直流变换器
DCE (data circuit-terminating equipment) 数据电路终端设备
DCI (desorption chemical ionization) 解吸化学电离
DCS (distributed control system) 集散控制系统,分布式控制系统
DDC (direct digital control) 直接数字控制

DDS（drug delivery system） 释药系统，药物传递系统
deactivate 使无效
deactivation 减活化作用，钝化作用
dead band 死区，非灵敏区，无控制作用区
dead band error 死区误差
dead end 尽头处
dead file 停用文件
dead halt 完全停机
dead layer 死层
dead load 静荷载，恒荷载，恒载量
dead lock 死锁
dead point 死点
dead section 盲段，空段（备用段）
dead time 停滞时间，窝工时间，时滞，纯滞后
dead time compensation unit 滞后时间补偿器
dead time unit 滞后时间器
dead time-delay compensation 纯滞后补偿
dead volume 死体积，静容量
dead weight tester 活塞式压力计，活塞压力器，净重测试仪
dead wire 不载电导线
dead zone 盲区，死区
dead zone error 死区误差
deadbeat control 无差控制
deadbeat control algorithm 非振荡控制算法
deadline 限期
dead-stop titration 永停滴定法
dead-time compensation 纯滞后补偿
DEB（data entity builder） 数据实体编制程序，数据实体编码程序
debris 瓦砾，碎屑，泥石
debug 排除故障，调试
debugger 调试器，调试程序
debugging 调试
debugging aid 调用辅助程序
debugging package 调试程序包
debugging routine 调试程序
debugging utility 调试实用程序
deburr 去毛刺
deburring punch 压毛边冲子
decad type resistance box 十进位电阻箱
decade resistance box 十进电阻箱
decarbonizing 脱碳
decarburization 脱碳，脱碳处理
decarburizing 脱碳退火
decatron 十进制计数管
decay 衰减
decay ratio 衰减比
deceleration lane 减速车道
deceleration valves 减速阀门，减速阀
deceleration 减速，减速度
decentrality 分散性
decentralization 分散化
decentralized 分散
decentralized automation 分布式自动化系统
decentralized control 分散控制
decentralized control system 分散控制系统
decentralized model 分散模型
decentralized robust control 分散鲁棒控制
decentralized stochastic control 分散随机控制
decentralized system 分散系统
deci 十分之一
decibel（dB） 分贝（声音强度单位）
decimal 十进制的
decimal number 十进制数
decipher 译码
deciphering machine 译码器
decision 决策，判决
decision analysis 决策分析
decision circuit 决策电路
decision feedback 决策反馈
decision feedback equalization 等效决策
decision fusion 决策合成
decision items 决议事项
decision making 决策
decision model 决策模型
decision program 决策程序
decision space 决策空间
decision support system 决策支持系统
decision table 决策表，判定表
decision theory 决策理论
decision tree 决策树
deck 桥面，层面，露天平台

declaration 说明，声明
declarator 说明符
declination 倾斜
decode 译码，译码
decoder 译码器，解码器
decoding relay 译码继电器
decommissioning 解除运作，停止运作，关闭
decomposable searching 可分解搜索
decomposable searching problem 可分解搜索问题
decomposition 分解，变质
decomposition method 分解方法
decomposition theorem 分解定理，分解理论
decomposition voltage 分解电压，分解电势
decomposition-aggregation approach 分解集结法
decompression chamber 减压室，降压室
deconvolution 反褶积，去卷积
decorations 装饰，装修
decorative finish 饰面，装饰性修整
decorative lighting 装饰照明
decorrelation 去相关，解相关
decouple 去耦，解耦，分离
decoupled subsystem 去耦合子系统
decoupling 去耦的，解耦的
decoupling circuit 去耦电路
decoupling control 解耦控制
decoupling filter 去耦滤波器
decoupling parameter 解耦参数
decoupling problem 解耦问题
decoupling zero 零解耦
decrease 减少，降低
decrement key 减少键
decrement ratio 减幅比
decrepitation 老化
dedendum 齿根高
dedendum angle 齿根角
dedendum circle 齿根圆
dedendum surface 下齿面
dedicated 专用的
dedicated cabinet 专用箱
dedicated channel 专用通道
dedicated instrument 专用仪表

dedicated line 专用线
dedicated storage 专用存储区
dedication 专用
deductive-inductive hybrid modeling method 演绎与归纳混合建模法
deemphasis 去加重，减加重
deemphasis circuit 去加重电路
deep bore well pump 深钻井泵
deep compaction 深层压实
deep foundation 深层地基，深基础
deep groove 深槽
deep groove ball bearing 深槽滚珠轴承
deep sea instrument capsule 深海仪器舱
deep trench excavation 深沟挖掘
deep vibration compaction 深层振荡式压实
deep well water pump 深井水泵
deep 深度，深的，深
deep-slot squirrel-cage induction motor 深槽式笼形感应电动机
de-excitation 去励，灭励，反励
deface 污损
default 缺省，缺席，过失，默认值，缺省值
default area 默认区域
default option 缺省选择
default value 补缺值，缺省值
defect 缺陷，缺点，故障，毛病，欠妥之处
defective 欠妥，错误的，有故障的，有缺陷的
defective product 不良品
defective product box 不良品箱
defective product label 不良标签
defective to staking 铆合不良
defectoscope 探伤仪
deficient manufacturing procedure 制程不良
deficient purchase 来料不良
defining fixed point 定义固定点
definite 确定的，规定的
definite corrective action 固定校正作用
definite integral 定积分
definite sequence 定顺序
definite time relay 定时继电器

definition 定义,清晰度
definitional domain 定义域
deflagration 爆燃过程
deflecting 偏转,偏移
deflecting angle 偏转角
deflecting torque 偏转力矩
deflecting voltage 偏转电压
deflection 偏向,偏差,挠曲
deflection coefficient 挠度系数,偏转系数
deflection method 偏位法
deflection period 摆动周期
deflection test 挠曲试验
deflection testing machine 挠曲试验机
deformation 变形
deformation characteristic 变形特征
defrost 融雪,解冻
degassing 脱气,除气,放气
degate 打浇口
degausser 去磁器,消磁器
degaussing 退磁,去磁
degeneration 退化,衰减,负反馈
degeneration feedback amplifier 负反馈放大器
degenerative feedback 负反馈
degenerative feedback amplifier 负反馈放大器
degradation 降级,退化,退降
degrease 脱脂
degrease solvent 去脂溶剂
degree 程度,度,度数,级
degree Celsius 摄氏度(温度单位)
degree Fahrenheit 华氏度(温度单位)
degree of accuracy 准确度
degree of attenuation 衰减程度
degree of compensation 补偿度
degree of cross linking 交联度
degree of freedom 自由度
degree of freedom mobility 自由度
degree of modulation 调制程度,调制深度
degree of reliability 可靠度
degree of sensitivity 灵敏度
degree of stability 稳定度
DEH (digital electro-hydraulic) control system 数字电液控制系统

dehumidification 除湿
dehumidifier 除湿器,抽湿机
dehydrating cartridge 干燥剂筒
dehydration 脱水
dehydrator 干燥机,脱水器
deion 消电离
delamination 层状剥落,分层,分叶
delay 时滞,纯时滞,延迟,阻延,延时
delay action 延时动作
delay amplifier 延时放大器
delay analysis 时滞分析
delay automatic gain control 延迟自动增益控制
delay cable 延时电缆
delay circuit 时延电路,延迟电路
delay compensation 时滞补偿
delay demodulation 时滞解调
delay distance 滞后距离
delay element 时滞环节,时滞元件
delay estimation 时滞估计
delay flip-flop 延迟触发器
delay line 延迟线
delay spread 时滞延伸
delay spread modulation 延时调制
delay switch 延时开关
delay time 延迟时间,滞后时间
delay-action 延时动作
delay-action circuit-breaker 延时断路器,延时开关
delayed echo 迟到反射波
delayed telemetry 延时遥测
delayed transformation 延迟变换
delayer 延迟器,延时器
delete 删除
delete a file 删除一个文件
delete command file 删除命令文件
delete function 删除功能
deleterious substance 有害杂质
delimiter bit 定界符
delimiter byte 定界字符
delivery system 交付系统,发射系统
delta connection 三角形接法,△接法
delta function 脉冲函数,δ函数
delta-T 时间差,时差
delta-T timing unit 时差计时单元

deluge valve 集水阀，雨淋阀，涌流阀
demagnetization 去磁，消磁，退磁
demagnetize 退磁
demagnetizer 去磁器
demagnetizing coil 消磁线圈
demand 请求
demand and supply 需求
demand archive 需要归档
demodulation 检波，反调制，解调制
demodulator 解调器
demolition 清拆，拆卸
demolition works 拆卸工程
demonstrate 证明
deniermeter 纤度计
denominator 分母
densely-installed I/O module 密集安装式输入/输出模件
densitometer 密度计，黑度计，光密度计
density 密度
density and water content 密度和含水率
density correction 密度修正
density measurement 密度测量
density meter 密度计
density of heat flow 热流密度
density sensor 密度传感器
density transducer 密度传感器
dents 压痕
department director 部长
departure 偏离
dependability 可靠性
dependent source 受控源
dependent variable 因变量，他变数，应变数
dephosphorized pig iron 脱磷生铁
deploy 配置
deposit 淤积物，沉积物，押金，定金，保证金
depot 车厂，仓库，车站
depth bellows 深波纹管
depth controller 深度控制器
depth gauge 测深规，深度规，深度计
depth of cut 切齿深度，切削深度
depth of field 视野深度
depth of focus 焦深，景深

depth of hardening 硬化深层，硬化深度，淬硬深度，淬火深度
depth of penetration 穿透深度
depth of snow 雪深
derailer 脱轨器
derivation 求导
derivation gain 微分增益
derivation time 微分时间
derivation tree 导出树
derivative 衍生物，导数，微分
derivative absorption spectrum 导数吸收光谱
derivative action 微分作用
derivative action coefficient 微分作用系数
derivative action gain 微分增益，微分作用增益
derivative action time 微分时间，微分作用时间
derivative action time constant 微分作用时间常数
derivative control 微分调节，微分控制
derivative control action 微分调节作用，微分控制作用
derivative controller 微分调节器，微分控制器
derivative differential thermal analysis 导数差示热分析
derivative differential thermal curve 导数差示热曲线
derivative dilatometry 导数膨胀法
derivative element 微分环节，微分元件
derivative feedback 微分反馈
derivative thermogravimetric curve 导数热重曲线
derivative thermogravimetry 导数热重法
derivative time constant 微分时间常数
derivative unit 微分器，微分单元
derived quantity 导出量
derived unit of measurement 测量的导出单位
derrick crane 桅杆起重机，吊臂起重机
derust 除锈

derusting machine 除锈机
descale 清除氧化皮
descending development 下行展开，下行法展开
describing function 描述函数
descriminator 鉴别器，鉴频器
description 品名，说明，说明，描述
descriptive geometry 画法几何学
descriptor 描述符，描述词，说明词
descriptor system 描述系统
desensitization 降低灵敏度
design 设计，计划，计算
design assumption 设计假定
design automation 设计自动化
design calculation 设计计算资料
design code 设计准则，设计规范
design condition 设计状态
design constant 设计常数
design constraints 设计约束，设计制约，设计限制，设计约束条件
design data sheet 设计数据表
design database 设计数据库
design distance 设计距离
design earth pressure 设计土压力
design flow 设计流量
design for environment 绿色设计
design for product's life cycle 面向产品生命周期设计
design life 设计使用年限
design load 设计载重，设计荷载
design methodology 设计方法学
design modification 设计变化
design of feedback control 反馈控制设计
design of simulation experiment 仿真实验设计
design origin 设计原点
design pressure 设计压力
design quality assurance 设计品质保证
design rule checking 设计规则检查
design rules for cranes 起重机设计规范
design speed 设计速度
design stress 设计应力
design system 设计系统

design variable 设计变量
designator 指示符，命令符
designed immersion depth 设计浸入深度
design-for-manufacturability 可制造性设计
desilting sand pit 隔沙池，沉沙池
desilting sand trap 隔沙池，沉沙池
desired value 预期值，期望值，希望值，目标值
desk grinding wheel machine 台式砂轮机
desk top 便携电源适配器，台式计算机
desktop equipment 台式设备
desktop management interface (DMI) 桌面管理接口
desk-top type recorder 台式记录器
desorption chemical ionization (DCI) 解吸化学电离
despinner 消旋体
destination 目的站
destructive pitting 破坏性点蚀
destructive test 破坏性测试
destructive wear 破坏性磨损
desynchronizing 失步
detail display 细目显示，详细显示画面
detail file 细目文件
detail key 细目显示调出键
detailed description 详细描述
detailed drawing 详图，细节图，明细图
detailed plan 详图
details drawing 零件详图，详细图
detect 探测，检波，整流，发现，检定
detect amplifier 检波放大器
detectability 可检测性，能检测性，检测能力
detecting device 检出器，敏感元件
detecting element 检测元件，探测元件
detecting instrument 检出器，检测元件，检波器
detection 检测，探测
detection algorithm 检测算法，探测算法

detection system 检测系统，探测系统
detection technology and automatic equipment 检测技术与自动化装置
detector 探测器，检波器，检出器，检测器，鉴别器
detector diode 检波二极管
detector performance 检测器性能，探测器性能
detector saturation 探测器饱和，检测器饱和
detectors 多探测器，多检测器
detergent 清洁剂
deterioration 退化，恶化
determinate error 可定误差
determination of mass per unit area 单位面积质量的测定
determination of stability 稳定性的确定
determination of steady-state thermal resistance 稳态热阻的测定
deterministic 确定的
deterministic automaton 确定性自动机
deterministic behaviour 固有特性
deterministic system 确定性系统
detrimental resistance 有害阻力
detuner 解调器
deuterium 氘
deuteron 氘核
developed setting 试切调整
developed winding diagram 绕组展开图
developer 开发者，显影剂
developing solvent 展开剂，显影溶剂
developing tank 展开槽，显影罐
development 展开
development chamber 展开室
deviate 使偏离，脱离
deviation 偏差，误差，背离
deviation alarm 偏差报警，偏差警器
deviation alarm sensor 偏差报警检测器
deviation check 偏差检查
deviation from linearity 偏离线性度
deviation high alarm trip point 偏差高限报警点

deviation limit 偏差限值
deviation low alarm trip point 偏差低限报警点
deviation variable 偏差变量
device 设备，装置，器件，仪器
device commutation 器件换流
device control point 设备控制点
device degradation 设计算法
device driver 设备驱动程序，设备驱动器
device simulation 装置仿真
device simulator 装置仿真器
dew cell 湿敏元件
dew point 露点
dew point hygrometer 露点湿度计，露点湿度表
dew point sensor 露点传感器
dew point temperature 露点温度
dew point transducer 露点传感器
dewatering 降低地下水位，脱水作用
dew-point 露点
dew-point hygrograph 露点计
dew-point meter 露点仪
DFD (data flow diagram) 数据流图，数据流程图
DHP (data highway port) 高速数据通道子系统接口，数据总线子系统接口
DHW (data highway) 高速数据通道，数据总线
DI (digital input) 数字输入
diagnosis 诊断
diagnostic 诊断的，特征的
diagnostic error handler 诊断错误的处理程序
diagnostic function 诊断功能
diagnostic function test 诊断功能测试程序
diagnostic inference 诊断推理
diagnostic model 诊断模型
diagnostic program 诊断程序，自诊断程序
diagnostic routine 诊断程序
diagnostic test 诊断测试
diagnostor 诊断程序
diagonal beam 斜束
diagonal bridge 对角电桥

diagonal dominance 对角优势
diagonal line 对角线
diagonal voltage 对角电压
diagonally dominant matrix 对角主导矩阵
diagram 图表，简图，图解
diagram method 图解法
diagrammatic plan 图解平面，示意平面
dial 转盘，表盘上的指针，标度盘
dial bore gauge 内径千分尺
dial exchange 自动交换区
dial feed 刻度盘进给装置
dial flowmeter 指针式流量计
dial indicator 刻度盘式指示表，千分表
dial instrument 表盘式指示仪表
dial terminal 拨号终端
dial up 拨号，标度
diamagnetic 抗磁的，反磁的
diamagnetic effect 抗磁效应，反磁效应
diamagnetic material 抗磁材料，反磁材料
diamagnetism 抗磁性，反磁性
diameter 直径
diameter ratio 直径比，内外径比，直径螺距比
diameter series 直径系列
diametral 直径的
diametral pitch 径节
diametral quotient 蜗杆直径系数，直径系数
diamond 钻石，金刚石，菱形，方块牌
diamond array 菱形阵
diamond cutters 金刚石切割器，钻石刀具
diamond pad 菱形盘，金刚石镶嵌瓣
diaphragm 膜片，隔膜，保护膜
diaphragm actuator 薄膜执行机构，膜盒式制动器
diaphragm and diaphragm capsule 膜片和膜盒
diaphragm capsule 膜盒
diaphragm control valve 气动薄膜调节阀，气动薄膜控制阀

diaphragm gas meter 膜盒式煤气表
diaphragm gate 隔膜浇口，盘形浇口
diaphragm gauge 膜片式压力表
diaphragm plate 横隔板
diaphragm pressure gauge 膜片压力表
diaphragm pressure span 膜片压力量程
diaphragm seal 密封膜片
diaphragm sealed manometer 隔膜压力表
diaphragm strain gauge 膜片式应变计
diaphragm type pressure sensor 膜片式压力传感器
diaphragm valve 隔膜阀
diaphragm wall 膜壁，隔墙
diaphragm-box element 膜盒元件
diaphragm-seal pressure gauge 隔膜压力表
diastereoisomer 非对映异构体
diazotization reaction 重氮化反应
diazotization titration 重氮化滴定法
DICEPS programming language DICEPS编程序语言
dicing saws 切割机，切割锯，钻石轮划片机，晶圆切割机
dicyandiamide 双氰胺
dicyclopentadienyl iron 二茂铁
die 模具
die approach 模口角度，模头料道，拉拔模入口
die assembly 合模
die bed 型底
die block 模块，滑块
die body 铸模座
die button 冲模母模，模具叶状模槽
die casting dies 压铸冲模
die casting machines 压铸机
die change 换模
die clamper 夹模器
die cushion 模具缓冲垫，模具缓冲器，模垫（压机）
die engineering 模具工程
die fastener 模具固定用零件
die height 闭模高度，冲压闭合高度
die holder 模型夹头，下模座，凹模固定板，板牙绞手
die insert 模衬，压模嵌入件

die life 模具寿命
die lifter 举模器,起模装置
die lip 模唇
die locker 锁模器
die opening 母模逃孔,模具孔,模距
die pad 下垫板,冲模垫,芯片安装面积
die plate 模板,拉模板
die repair 模修
die set 模组,成套冲模,置模器
die spotting machine 合模机
die structure dwg 模具结构图
die worker 模工
dielectric 电介质,绝缘材料,介质,绝缘的
dielectric amplifier 介质放大器
dielectric amplitude induction logging instrument 幅度介电感应测井仪
dielectric conductance 介质电导
dielectric constant 介质常数
dielectric loss 介质损耗
dielectric loss angle 介质损耗角
dielectric phase induction logger 相位介电感应测井仪
dielectric resistance 绝缘电阻
dielectric sheets 介电原片
dielectric strength 绝缘强度
dielectric wire 电介质线
diesel engine 柴油引擎,柴油机
diesel fuel 柴油
diesel hydraulic locomotive 液压传动柴油机车
diesel locomotive 柴油机车
diesel oil 柴油
diesel pile hammer 筒式柴油打桩锤
diesel generator 柴油发电机
diesel-electric locomotive 柴油电力机车头
diesel-hydraulic locomotive 柴油液动机车头
difference 差,微分,分歧,差距
difference absorption spectrum 差式吸引光谱
difference amplifier 差分放大器
difference analysis 差分分析
difference equation 差分方程
difference equation model 差分方程模型
difference galvanometer 差动检流计
difference input 差分输入
difference quantity 差异量
differential 差动的,微分的,微分,差速器,差别的
differential action 差动作用
differential amplifier 差分放大器,减法电路放大器,推挽式放大器
differential amplifier multiplier 差分放大乘法器
differential analyzer 微分分析器
differential chromatography 差示色谱法
differential circuit 差动电路
differential coil 差动线圈
differential comparator 差动比较器,差分比较器
differential compound 差复励
differential compound motor 差复励电动机
differential connection 差动接法
differential detection 差分检测
differential detector 微分型检测器
differential dilatometry 差示热膨胀法
differential dynamical system 微分动力学系统
differential equation 微分方程式
differential equations 微分方程
differential error of the slope 斜率的微分误差
differential excitation 差励
differential field rotor 差动磁场转子
differential gain 差分增益,微分增益
differential galvanometer 差动检流计
differential game 差额博弈,微分对策
differential gap 差隙,不可调间隙,切换差
differential gap control 差隙控制
differential gap controller 差隙控制器,两位式调节器
differential gap pneumatic controller 间隙差式气动控制器
differential gear 差速齿轮
differential gear train 差动轮系

differential geometric method 差分几何方法
differential geometry 微分几何
differential input 差动输入，差分输入
differential input resistance 差动输入电阻
differential Manchester encoding 差分曼彻斯特编码
differential measurement 微差测量
differential measuring instrument 差动测量仪表，差动式测量仪器
differential method of calibrating thermocouple 热电偶微差检定法
differential method of measurement 微差测量法
differential mode interference 串模干扰
differential movement 差异移动
differential pair 微分对
differential piston 差动活塞
differential preamplifier 差动前置放大器
differential pressure 差压
differential pressure device 差压装置
differential pressure devices 差压装置
differential pressure flow sensor 差压流量传感器
differential pressure flow transducer 差压流量传感器
differential pressure flowmeter 差压流量计
differential pressure gauge 差压压力表
differential pressure gauge with double bellows 双波纹管差压计
differential pressure level meter 差压液位计
differential pressure level sensor 差压（式）物位传感器
differential pressure level transducer 差压（式）物位传感器
differential pressure ratio 差压比
differential pressure sensor 差压传感器
differential pressure transducer 差压传感器
differential pressure transmitter 压力变送器，压力传送器
differential pressure type flowmeter 差压式流量计
differential read-out 差动读出
differential refraction detector 示差折光检测器
differential refractive index detector 示差折光检测器
differential regulator 差动调节器
differential relay 差动继电器
differential scanning calorimeter 示差扫描量热仪
differential scanning calorimetry (DSC) 差示扫描量热法
differential screw mechanism 微动螺旋机构，差动螺旋机构
differential thermal analysis (DTA) 差示热分析
differential thermal analysis curve 示差热曲线，差热分析曲线，差热曲线
differential thermal analysis meter 差示热分析仪
differential thermal analyzer 差热分析仪
differential thermal curve 差示热曲线，DTA 曲线
differential thermocouple 差分热电偶，差示热电偶
differential thermocouple voltmeter 差动热电偶式电压表
differential thermometric titration 差示温度滴定法
differential transducer 差动传感器
differential transformer 差动式换能器，差接变压器，差动变压器
differential transformer displacement transducer 差动变压器式位移传感器
differential transformer pressure transducer 差动变压器式压力传感器
differential-excited generator 差励发电机
differentiating action 差动作用
differentiating effect 区分效应，辨别效应

differentiating element 差动元件
differentiating solvent 辨别溶剂，区分溶剂
differentiation 微分
differentiation element 微分环节
differentiator 微分器，微分电路
diffraction grating 衍射光栅
diffraction lens 衍射透镜
diffraction resolution 衍射分辨力
diffuse field 扩散声场
diffused junction transistor 扩散面结型晶体管
diffused silicon semiconductor force meter 扩散硅式测力计
diffused silicon semiconductor force transducer 扩散硅式力传感器
diffuse-field response of microphone 传声器扩散场响应
diffuse-field sensitivity of microphone 传声器扩散场灵敏度
diffuser 散光罩，扩散器
diffusion 扩散，漫射
diffusion annealing 扩散退火
diffusion coefficient 扩散系数，扩散率
diffusion current 扩散电流
diffusion pump 扩散泵
digit 数字
digital 数字，数字的，数字量
digital ammeter 数字电流表
digital analogy 数模
digital burette 数字式滴定管
digital circuit 数字电路
digital communication 数字通信
digital compensation 数字补偿
digital composite data point 数字混合数据点
digital composite point 数字复合点
digital computer 数字计算机
digital computer application 数字计算机应用
digital computer compensator 数字补偿器，数字计算机校正网络
digital computer control 数字计算机控制
digital control 数字控制
digital control algorithm 数字控制算法
digital control system 数字控制系统
digital controller 数字控制器
digital conversion technique 数字变换技术，数字转换技术
digital converter 数字转换器
digital counter 数字计数器
digital data 数字数据
digital data acquisition system 数字数据采集系统
digital data processing system 数字数据处理系统
digital deep-level seismograph 数字深层地震仪
digital differential analyser 数字微分分析器，数字差分分析器
digital differential circuit 数字微分电路
digital differential integrator 数字式微分积分器
digital displacement measuring instrument 数字式位移测量仪
digital displacement transducer 数字式位移传感器
digital display instrument 数字式显示仪表
digital electric actuator 数字式电动执行机构
digital electro-hydraulic (DEH) control system 数字电液控制系统
digital equipment corporation 数字设备公司
digital feedback control loop 数字反馈控制回路
digital feedback system 数字反馈系统
digital film room 数字胶片室
digital filter 数字滤波，数字滤波器
digital filter processor 数字滤波处理器
digital filter structure 数字滤波器结构
digital fluxmeter 数字磁通表
digital frequency meter 数字频率表
digital gaussmeter 数字式高斯计
digital I/O interface 数字 I/O 接口
digital I/O module 数字量输入/输出模件

digital image 数字图像
digital indication 数字示值，数字读数
digital information processing system 数字信息处理系统
digital input (DI) 数字输入
digital input data point 数字输入数据点
digital input-sequence of event 事件序列开关量输入
digital integrated circuit tester 数字集成电路测试仪
digital integrating circuit 数字积分电路
digital integrating fluxmeter 数字积分式磁通表
digital integrator 数字积分器
digital logging instrument 数字测井仪
digital magnetic tape record type strong-motion instrument 数字磁带记录强震仪
digital measuring instrument 数字测量仪表，数字式测量仪器仪表
digital micrometer 数位式测微计
digital mobile radio 数字车用无线电
digital multimeter 数字万用表，数字复用表
digital network architecture 数字网络体系，数字网络结构
digital ohmmeter 数字电阻表
digital output (DO) 数字输出
digital output data point 数字输出数据点
digital pattern 数字模式
digital phase meter 数字相位表
digital photometer 数字式光度计
digital position transmitter 数字式位置发送器
digital positioner 数字式定位器
digital power driver 数字功率表
digital pressure gauge 数字压力表
digital pulse duration modulation 数字脉宽调制
digital radio 数字式无线电
digital readout 数字读出
digital read-out 数字示值，数字读数
digital recording instrument 数字记录仪表

digital representation of a physical quantity 物理量的数字表示
digital seismic recording system 数字地震仪
digital sensor 数字传感器
digital servomechanism 数字伺服机构
digital signal 数字信号
digital signal analyzer 数字信号分析仪
digital signal processing 数字信号处理
digital signal processor 数字信号处理器
digital simulation 数字仿真
digital simulation computer 数字仿真计算机
digital simulator 数字仿真器
digital sine wave generator 正弦波数字信号发生器
digital strain indicator 数字应变仪
digital system 数字系统
digital telemetering system 数字遥测系统
digital tension sensor 数字应变传感器
digital thermal infrared scanner 数字式热红外扫描仪
digital thermometer 数字温度计
digital transducer 数字传感器
digital valve 数字阀门
digital vibration meter 数字震动表
digital viscometer 数字黏度计
digital voltmeter 数字电压表
digital input 数字输入
digital output 数字输出
digital-analog conversion 数字-模拟转换
digital-analog converter 数字-模拟转换器
digital-analog decoder 数字-模拟译码器
digital-analog hybrid computer 数字模拟混合计算机
digital-analog simulator 数字模拟仿真器
digitalization error 数字化误差
digital-to-analog (D/A) 数/模转换
digitiser 数字转换器
digitization 数字化
digitization error 数字化误差

digitizer　数字转换器
digitizing　数字化
dilapidation　破旧，崩塌
dilatometry　膨胀法
diluent gas　稀释气
dilute　冲淡，稀释
dilution factor　稀释因数
dilution methods　稀释法
dilution rate　稀释速率，稀释比例
dilution ratio　稀释比
dim transformer　光暗变压器
dimension　尺度，尺寸，量纲
dimension sensor　尺度传感器
dimension series　尺寸系列
dimension survey　尺寸测量
dimension transducer　尺度传感器
dimensional　量纲的，尺寸的，因次的
dimensional measuring instrument　长度测量工具
dimensional system　量纲系统
dimensional tolerance　尺寸公差
dimensional transfer function　量纲转换函数
dimensioned location　用尺寸标明的位置
dimensioned plan　度量图
dimensionless quantity　无量纲量
dimensions　外形尺寸
dimmer　车辆调光器，光暗控制器，衰减器
dimorphism　双晶现象
Dines anemometer　丹斯风速计
diode　二极管
diode amplifier　二极管放大器
diode capacity meter　二极管电容器
diode clamping　二极管钳位
diode clipper　二极管削波仪
diode limiter　二极管限幅器
diode-transistor-logic　二极管晶体管逻辑电路
diopter　屈光度
dip　倾角
dip gauge　垂度规，垂度计
dip logger　地层倾角测井仪
dip mold　下压模
dip tube　液面探测管

dipole　偶极子，偶极天线
direct　直接，控制，引导
direct acting actuator　正作用执行机构
direct acting actuator controller　正作用调节器，正作用调节控制器
direct acting actuator measuring instrument　直接作用测量仪表
direct acting instrument　直接作用仪表
direct acting recording instrument　直流作用记录仪
direct action　正作用
direct action solenoid valve　直动式电磁阀
direct actuator　正作用执行机构
direct axis　直轴
direct axis transient time constant　直轴瞬变时间常数
direct brake　直接制动器
direct brake valve　直接制动阀
direct component　直流部分
direct connection　直接连接
direct control　直接控制
direct control layer　直接控制层
direct coordination　直接协调
direct coupled　直接耦合的，直接连接的
direct coupled amplifier　直接耦合放大器
direct current (DC)　直流，直流电
direct current circuit　直流电路
direct current control　直接电流控制
direct current feed control panel　直流馈电屏
direct current plasma emission spectrometer　直流等离子体发射光谱仪
direct current power supply　直流电源
direct digital control (DDC)　直接数字控制
direct digital control station　直接数字控制站
direct dynamic problem　直接动力学问题
direct Fourier reconstruction　直接傅

里叶重构
direct frequency modulation 直接频率调制
direct gate 直接浇口
direct heated type thermistor 直热式热敏电阻器
direct indication 直接示值
direct injection 直接喷射，直接注入，直接射出法
direct injection burner 直接注入燃烧器
direct injector 直流进样器
direct kinematic problem 直接动力学问题
direct kinematics 正向运动学
direct memory access 直接存储器（内存）访问，直接存储器（内存）存取
direct method of measurement 直接测量法
direct mounting gauge 直接安装压力表
direct overwrite 直接重写
direct potentiometry 直接电位测定法
direct probe inlet 直接探头进样
direct purging 直接驱气
direct reading audio-frequency meter 直读式声频计
direct reading ceramic condenser tester 直读式陶瓷电容器测试仪
direct reading current meter 直读式测流计
direct reading instrument 直读式仪器
direct record strong-motion instrument 直接记录式强震仪
direct resistance heating 直接电阻加热
direct seawater cooling system 海水直接冷却系统
direct synthesis method 直接综合法
direct-comparison method of measurement 直接比较测量法
direct-current 直流
direct-current contactless potentiometer 直流无触点电位计
direct-current main 直流电源
direct-current moving coil meter 直流动圈式仪表
direct-current transmission 直流输电，直流传输
direct-drive robot 直接驱动机器人
directed graph 直接图形
direct-fired vaporizer 明火直热式汽化器
direct-imaging mass analyser 直接成像质量分析仪
direction arrow plate 方向指示板
direction finding 探向
direction focusing 方向聚焦
direction indicator 中心指示器
direction mark meter 方位标仪
direction of feed 进给方向
direction of main movement 主运动方向
direction of resultant movement of cutting 合成切削运动方向
direction of vector 矢量方向
directional coil 定向线圈
directional frequency response of microphone 传声器指向性频率响应
directional interchange 定向道路交汇处
directional pattern of microphone 传声器指向性图案
directional radiation 定向发射
directional reception 定向接收
directivity 指向性
directivity index of microphone 传声器指向性指数
directly controlled system 直接被控系统，直接被控制系统
directly controlled variable 直接被控变量
directly-coupled transistor circuit 直接耦合晶体管电路
direct-on starting 直接启动
direct-operated regulator 直接作用式调节，自力式调节
director 指挥站
directory 目录
directory file 目录文件
direct-reading instrument 直读式仪表
direct-recording 直接记录式仪表
disable 禁止

disaggregation 解裂
disappearing-filament optical pyrometer 隐丝式光学高温计
disassembly 拆散，拆卸
disaster 事故，故障
disaster shutdown 事故停机
disc 板，瓣，磁盘
disc brake 碟式制动器，圆盘制动器，碟式（编辑）
disc drive 磁盘驱动器
disc friction clutch 圆盘摩擦离合器
disc plug 盘形芯
disc recorder 圆盘（形）记录仪
disc type positive displacement flowmeter 圆盘式容积流量计
discharge 排水，卸载，流量，放电，放电的
discharge capacity 放电容量
discharge characteristic 放电特性曲线
discharge circuit 放电电路
discharge coefficient 流量系数，放电系数
discharge condition 放电状态
discharge energy 放电能量
discharge lamp 放电灯
discharge process 放电过程
discharge valve 卸料阀，排出阀
disc-mill cuter 盘铣刀
discoloration 变色
discoloring agent 脱色剂
discolouration 变色，褪色
disconnect 截断，断路，断开分离
disconnect amplifier 隔离放大器
disconnect switch 隔离开关
disconnecting trap 隔气弯管，存水隔气弯管
disconnection 断开，开路，截断
disconnector 隔离器，隔离，断路器，切断开关，隔离开关
discontinue 中止
discontinuity 不连续性
discontinuous 离散的，不连续的
discontinuous action 非连续作用，断续作用
discontinuous control 非连续控制，不连续控制
discontinuous control system 不连续控制系统
discontinuous digital dynamic control 离散数字动态控制
discontinuous Fourier transformer 离散傅里叶变换
discontinuous measurement 离散测量
discontinuous process 非连续过程
discontinuous simultaneous techniques 间歇联用技术，不连续同时串用技术
discontinuous system 离散系统
discontinuous time 离散时间
discontinuous time detection 离散时间检测
discontinuous time system 离散时间系统
discontinuous-event dynamic system 离散事件动态系统
discontinuous-event system 离散事件系统
discoordination 失协调
discount 折扣
discrete 离散，不连续，离散的，不连续的
discrete control system 离散控制系统
discrete cosine transformer 离散余弦变换
discrete event dynamic system 离散事件动态系统
discrete integrator 离散积分仪
discrete signal 离散信号
discrete state 离散状态
discrete system 离散系统
discrete system model 离散系统模型
discrete system simulation 离散系统仿真
discrete system simulation language 离散系统仿真语言
discrete type 离散类型
discrete wire block 分离接线块
discrete wiring board 散线印制板
discrete input 离散量输入
discrete output 离散量输出
discrete-time model 离散时间模型
discrete-time response 离散时间响应
discrete-time signal 离散时间信号

discrete-time system 离散时间系统
discretization 离散化
discretization of partial differential equation 偏微分方程离散化
discriminant analysis 识别分析，求解分析
discriminant function 判别函数
discriminate 鉴别，辨别
discrimination 辨识，区分，鉴别力
discrimination threshold 鉴别阈，鉴别灵敏度
discriminator 鉴别器，抑制器
discus 铁饼
dish angle 凹角
disinfectant tank 消毒箱
disintegration 衰变，蜕变
disk 磁盘，磁碟
disk cam 盘形凸轮
disk capacity 磁盘容量
disk drive 磁盘驱动器
disk file index 磁盘文件索引
disk memory 磁盘存储器
disk pack 磁盘组
diskette 磁盘，磁碟，软盘，软磁盘
disk-like rotor 盘形转子
dismantle the die 拆模
dismantling 拆卸，拆除
dispersant 分散剂，化油剂
disperse reinforcement 分散性强化复合材料
dispersion 弥散
dispersion dose 散射剂量
dispersion gate 扩散闸，与非门
dispersion meter 色散计
dispersive crystal 分光晶体
dispersive infra-red gas analyzer 色散红外线气体分析器
dispersive power 色散本领
displace 位移，移动
displaced phase 位移相
displacement 移离原位，位移，替换
displacement cascade 位移串级
displacement current 位移电流
displacement diagram 位移曲线
displacement error 位移误差
displacement factor 位移因数
displacement flowmeter of duplex rotor pattern 双转子式容积式流量计
displacement flowmeter of semi-rotary pattern 半旋转式容积式流量计
displacement flux 位移通量
displacement law 位移定律
displacement measuring instrument 位移测量仪表
displacement pickup 位移传感器
displacement sensor 位移传感器
displacement transducer 位移传感器
displacement vibration amplitude transducer 位移振幅传感器
display 显示，显示屏，显示画面，显示器，指示器
display attribute 显示属性
display board 显示板
display console 显示控制台
display data base 显示数据库
display default directory 显示缺省目录
display device 显示器，显示设备
display element 显示元件
display generator 显示生成程序
display instrument 显示仪表
display lamp 指示灯
display log 显示记录
display lower 显示器下盖
display request 显示画面请求
display schematic 显示画面
display standard 标准显示画面
display stem 显示器支撑杆
display tube 显示管
display unit 显示单元
displaying unit 显示单元
disposable infusion set 一次性使用输液器
disposable sterile injector 一次性无菌注射针
disposable toolholder bits 舍弃式刀头
disposable venous infusion needle 一次性静脉输液针
disposal 处置
disposed goods 处理品
disposed products 处理品
disposition 排列，配置
dispute 纠纷，争议
disruptive 击穿，破坏

disruptive voltage 击穿电压
dissipation 分配，分发
dissipation constant 耗散常数，电离常数
dissipation power 耗散功率
dissipative structure 耗散结构
dissociation 离解
dissolution 分解，溶解
dissolved oxygen analyzer 溶解氧分析器
dissolved oxygen analyzer for seawater 海水溶解氧测定仪
dissolved oxygen of seawater 海水中的溶解氧
dissymmetry 不对称，不平衡
distance amplitude compensation (DAC) 距离振幅补偿
distance amplitude curve 距离振幅曲线
distance between cable 电缆之间的距离
distance between conductors 线间距离
distance constant 距离常数
distance gauge 测距仪，测距规
distance marker 距离刻度
distance meter 测距仪
distance of centre of gravity 重心距
distance transformation 距离变换
distance velocity lag 距离速度滞后，距离速度延迟
distance 距离，间隔
distant control 遥控
distant radio communication 远距离无线电通信
distant regular 遥控调节器
distant-action instrument 遥感仪表
distillation 蒸馏
distillation column 精馏柱
distillation control 精馏控制
distilled water 蒸馏水
distort 失真，畸变
distorted alternating current 非正弦交流电，失真交变电流
distorted peak 畸峰
distorted wave 失真波，畸变波
distortion 失真，畸变，扭曲变形

distortion factor 失真系数，畸变系数
distortion factor meter 失真度测量仪
distortion pad 畸变衰减器
distortion power 畸变功率
distortion tester 失真测试仪
distortionless 无失真的，无畸变的
distribute 分布，分配
distributed 分布，分散，分布式的，分散型
distributed amplifier 分布式放大器
distributed artificial intelligence 分布式人工智能
distributed capacitance 分布电容
distributed circuit 分布参数电路
distributed communication architecture 分布式通信体系，分散型通信体系
distributed computer computer control system 分散型计算机控制系统，分布式计算机控制系统
distributed computer-control SDAS 分布式计算机控制数字地震仪
distributed control 分散控制，分布控制
distributed control system (DCS) 集散控制系统，分布式控制系统
distributed database 分布式数据库
distributed detection 分散式检测
distributed feedback 分散式反馈
distributed inductance 分布电感
distributed intelligence 分布式智能
distributed load 分布负载，分布载重，分布荷载
distributed model 分布式模型
distributed network 分布式网络，分布式通信网络，分散型通信网络
distributed non-linear element 分散式非线性环节，分散式非线性元件
distributed parameter 分布参数
distributed parameter control system 分布参数控制系统
distributed parameter system 分布参数系统
distributed processing unit 分布式处理单元
distributed simulation 分散式仿真
distributed telemetry SDAS 分布式遥

测型数字地震仪
distributer 配电器
distributing board 配电盘
distribution 分布，分配，分散
distribution automation 分布自动化
distribution board 配电屏，配电板，配电盘
distribution box 接线盒，配电箱
distribution coefficient 分布系数
distribution control 分散控制
distribution equipment 配电装置
distribution feeder 配电线路，配电馈线
distribution line 配电线
distribution network 分散网络，配电网
distribution of electric charges 电荷分布
distribution panel 配电屏，配电板，配电盘
distribution ratio 分配比
distribution readout system 分散读出系统
distribution system 分散系统，配电系统
distribution temperature 分布温度
distribution transformer 配电变压器
distribution winding 分布绕组
distribution wire 配电线路
distributor 分配器，配电器
district heating 集中供热，区域供暖
disturbance 干扰，骚扰，扰动
disturbance compensation 扰动补偿
disturbance current 干扰电流
disturbance effect 干扰效应
disturbance localization 本地扰动
disturbance parameter 扰动参数
disturbance rejection 抗干扰，扰动
disturbance signal 扰动信号
disturbance variable 扰动量，扰动变量
ditch 明沟
ditching machine 挖沟机
dither 颤振
dithering 颤动
diurnal variation 日变化，日变程，日际变化，昼夜变化
diverging three way valve 三通分流阀
diversion kerb 导流
diversion of water pipe 迁移水管
diversion of water main 迁移水管
diversity 多样性
diverter 分流器，避雷针，换向器，转向器
diverter switch 转换开关
divided circuit 分流电路
divider 分配器，除法器
dividing circuit 除法电路
diving bell 潜水钟
divisibility 可分性
division 分，除法，部门
division indicator 分度指示器
DLL (data link layer) 数据链路层
DLL (down line loading) 离线装载数据，向下加载
DMC (dynamic matrix control) 动态矩阵控制
DMI (desktop management interface) 桌面管理接口
DNA sequencers DNA 测序仪
DO (digital output) 数字输出
dock 船坞，码头
document 文件，文本，文献，资料
document folder 文件夹
document sorter 文件分类机
document structure 文件结构，文献结构
documentary evidence 文件证据
documentation 文件，文本
dog chuck 爪牙夹头，爪形夹盘
dog iron 两爪铁扣，铁钩
dog spike 钩头道钉
domain 磁畴，领域，区域
domain analysis 区域分析
domain knowledge 领域知识
domain wall 畴壁
dome 圆顶，凸圆
domestic package 内销包装
domestic sub-contracting system 内部自行分包制
dominant 主要的，显著的
dominant frequency 优势频率，主频率

dominant point　要点
dominant pole　主导极点
dominant poles　主导极点
dominant route　主根
dominant time constant　主时间常数
donor　施主
donor impurity　施主杂质
door contact　门联锁触点
door frame　门框
door interlock　门联锁装置
door sill　门槛
door switch　门开关
dope additive　掺杂剂
Doppler current meter　多普勒海流计
Doppler effect　多普勒效应
Doppler flowmeter　多普勒流量计
Doppler frequency tracker　多普勒频率跟踪仪
Doppler radar　多普勒雷达
Doppler sonar　多普勒声呐
Doppler sonar navigator　多普勒声呐导航器
dose　剂量
dose rate　剂量率
dose rate meter　剂量率计
dosemeter　放射量测定器
dose-response model　剂量反应模型
dosing pump　剂量泵
dot　点
dot matrix　点阵
dot matrix printer　点阵打印机,点阵式打印机
dot printer　点阵印刷机,点阵打印机
dot-bar generator　点-条信号发生器
dotted line　虚线
dotted line recorder　断续线记录仪
dotting strip chart recorder　打点式长图记录仪
dotting time　打点时间
double acting positioner　双作用定位器
double amperometric titration　双电流滴定法(双安培滴定法)
double beam spectrum radiator　双光束光谱辐射计
double bounce technique　二次反射法

double bridge　双臂电桥,凯尔文电桥
double cage motor　双笼式电动机
double collectors　双接收器
double commutator motor　双换向器式电动机
double crank mechanism　双曲柄机构
double crank press　双曲柄轴冲床,双曲轴压力机
double crystal probe　双振子探头
double current generator　交直流发电机
double disc　双闸板
double end converter　双端电路
double focusing　双聚焦
double focusing analyzer　双聚焦分析器
double focusing at all masses　全质量双聚焦
double glazing glass　双层玻璃
double housing planer　龙门刨床
double index　双分度
double indicator titration　双指示剂滴定法
double inlet system　双进样系统
double insulation　双重绝缘
double layer　双层的
double liquid balance manometer　双液柱平衡压力计
double opening exhaust valves　双口排气阀
double pan balance　双盘天平
double pole double throw　双刀双掷
double probe technique　双探头法,热电偶双极检定法
double range voltmeter　双量程电压表
double refined iron　二重精炼铁
double rocker mechanism　双摇杆机构
double roll　双向滚动
double roller chain coupling　双滚子链联轴器
double sided board　双面电路板
double squirrel cage motor　双笼式电动机
double stack mold　双层模具
double star connection　双星形接法,双星形连接

double track 双轨
double treated foil 双面处理铜箔
double tube mercury manometer 双管水银压力表
double tube thermometer 套管温度表
double universal joint 双万向联轴节
double vibration amplitude 双振幅
double 两倍的,双重的
double-base diode 双基型二极管
double-beam mass spectrometer 双束质谱计
double-beam torsion balance 双杆扭秤
double-bellows differential element 双波纹管差压元件
double-column transformer 双绕组变压器
double-cone viscometer 双锥黏度计
double-contact wires 双接触导线
double-cotton-covered wire 双纱包[铜]线
double-crystal diffractometer 双晶体衍射计
double-delta connection 双三角接法,双三角接线
double-density format 双倍密度格式
double-direction thrust bearing 双向推力轴承
double-dry calorimeter 双干式热量计
double-ended grinder 双头砂轮机
double-ended spanner 双头扳子
double-exponential filter 双指数滤波器
double-fed 双馈
double-fed repulsion motor 双馈推斥电动机
double-focus X-ray tube 双焦点X射线管
double-focusing mass spectroscope 双聚焦质谱仪器
double-glazing 双层玻璃
double-image tacheometer 双像速测仪
double-length 双倍字长
double-magnification imaging 双放大倍率成像法
double-pass internal reflection element 双通内反射元件
double-path ratio thermometer 双道比色温度计
double-polarity method for calibrating thermocouple 热电偶双极标定法
double-pole double-throw switch 双刀双掷
double-ported globe valve 双座阀
double-row bearing 双列轴承
double-sided copper-clad laminate 双面覆铜箔层压板
double-sided printed board 双面印制板
double-silk covered wire 双丝包线
double-slider mechanism 双滑块机构
doublet 双峰
doublet impulse 双向脉冲
dovetail joint 鸠尾榫,鱼尾榫,燕尾榫
dovetail saw 燕尾锯
dowel 暗钉,销钉,定位销,暗销
dowel bar 传力杆,暗销杆
dowel hole 导套孔
dowel joint 榫钉接缝,暗钉接合
dowel pin 销子,管钉,合模销,固定销,定位销
down line loading (DLL) 离线装载数据,向下加载
down pipe 水落管
down roll 向下滚动
down stream 下游
down time 停机时间,故障时间,停工时间
down 向下的,向下
downhole instrument 下井仪器
download 下载
downstream valve 下游阀
downtime 故障停机时间
downward 向下
downward force 下向力
downward radiation 向下辐射
downward terrestrial radiation 大气向下辐射
downward total radiation 向下全辐射
dozzle 辅助浇口
DP (data point) 数据点
D-port D型端口
drafting machine 绘图仪
drag coefficient 阻力系数,曳引系数

drag cup motor 托杯形电机
drag wire 阻力张线
drain 疏水，排放，漏电，排水渠
drain cock 放水旋塞，排气阀，排液旋塞
drain current 漏电流
drain hole 排水孔
drain holes 排泄孔
drain pipe 排泄管，疏水管
drain plug 排水塞，排放丝堵
drain tube 排泄管，疏水管
drain valve 限流阀
drain wire 加蔽线，排扰线
drainage blanket 疏水层，排水层，排水砂层
drainage connection system 排水接驳系统
drainage inlet 排水入口
drainage layer 排水层
drainage material 排水物料
drainage measure 排水措施
drainage pump 排水泵
drainage system 渠道系统，排水系统
drainage valves 排水阀
drainage works 渠务工程，排水工程
DRAM (dynamic random access memory) 动态随机存储器
draught drying cabinet 电热鼓风干燥箱
draw bar 起模杆，拉杆
draw bead 张力调整杆，拉深压边筋，拉道
draw box 拉线箱
draw hole 拉孔
draw out 拉拔，锻造拔长
draw pit 拉线井
draw plate 起模板
draw spike 起模长针
drawbridge 吊桥，开闭的吊桥
drawing 绘图，制图，画图；图则，图样；拉，抽，牵引
drawing board 绘图板，画板
drawing force 拉力
drawing format 绘图格式
drawing frame 图框
drawing instruments 绘图仪器
drawing machines 拔丝机，拉伸机

drawing materials 绘图材料
drawing paper 绘图纸
drawing pencil 绘图铅笔
drawing technique 绘图技巧
drawing-down 锤尖法，锻延
drawing-in wire 电缆牵引线
drawn wire 拉制钢丝
draw-off tap 放水龙头
dredge 挖泥机，挖泥船，拖曳式采样器
dredging works 挖沙工程，挖泥工程
drencher 喷淋器
drencher curtain 水帘
drencher head 喷淋头
drencher system 水帘系统，大型灭火洒水器
dresser 砂轮整修机
drier 干燥器
drift 漂移，偏移
drift bottle 漂流瓶
drift card 漂流卡片
drift carrier 漂移载流子
drift corrected amplifier 漂移校正放大器
drift current 漂移电流
drift mobility 漂移迁移率
drift rate 漂移率
drift velocity 漂移速度
drifting buoy 漂流浮标
drill 钻，钻机，钻孔，钻头，钻床
drill bit 钻头
drill drawing 钻孔图
drill stand 钻台
drilling 钻孔，钻取，钻探
drilling machine 钻床，钻孔机
drilling machines 钻床
drilling machines bench 钻床工作台
drilling machines high-speed 高速钻床
drilling machines multi-spindle 多轴钻床
drilling machines radial 摇臂钻床
drilling machines vertical 立式钻床
drills 钻头
drive 驱动，拖动，强迫
drive amplifier 驱动放大器
drive axle 传动轴

drive bearing 传动轴承
drive belt 传动皮带
drive chain 传动链
drive file 驱动文件
drive gear 传动齿轮
drive magnet 驱动磁铁
drive motor 传动电机,传动马达,驱动电动机
drive pin 传动销
drive pinion 传动小齿轮
drive point 驱动点,策动点
drive screw 传动螺杆
drive side 工作齿侧,主动侧
drive wire 激励线圈
driven gear 从动轮
driven link follower 从动件
driven pulley 从动带轮
driven system 传动系统
driver 驱动器
driver behavior 驱动器特性
driver model 驱动器模式
driveway 车道
driving 驱动
driving cab 驾驶室
driving chart paper 走纸偏差
driving chart swaying 走纸偏差
driving circuit 驱动电路
driving console 驾驶控制台
driving device 传动装置
driving force 驱动力
driving gear 主动齿轮
driving gearbox 驱动齿轮箱
driving link 原动件,主动件
driving moment 驱动力矩
driving motor 驱动电动机,主动电动机
driving pulley 主动带轮
driving torque 驱动力矩,驱动转矩
driving voltage 驱动电压
driving winch 驱动绞车
droop characteristics 下降特性
droop rate 下降率
drooping characteristic 下降特性
drop 站
drop forging 锤锻法,冲锻法,模锻法
drop point 落点

drop size meter 滴谱仪
drop test 跌落试验
drop wire 引入线,接户线
droplet 滴,微滴
drop-out of step 失步
dropping mercury electrode 滴汞电极
dropsonde 下投式探空仪
drosometer 露量表,露量计
drug delivery system (DDS) 释药系统,药物传递系统
drum armature 鼓形电枢
drum brake 鼓式制动器
drum controller 鼓形控制器
drum memory 磁鼓存储器
drum recorder 鼓形记录仪
drum water meter 转桶式水表
dry 干,干燥
dry air 干空气
dry battery 干电池
dry bulb 干球(温度计)
dry cell 干电池
dry density 干密度
dry friction 干摩擦
dry gas meter 干式气体表
dry glazing 干法施釉
dry glazing dispenser 干法施釉器
dry heat test 干热试验
dry meter 干燥计
dry powder type fire extinguisher 干粉型灭火器
dry-bulb thermometer 干球温度计
dryer 烘干机,干燥机,干燥剂
drying oven 干燥箱
drying shrinkage 干水收缩,干收缩,干燥失重
drying sterilizers 干热灭菌器
DSC (differential scanning calorimetry) 差示扫描量热法
DTA (differential thermal analysis) 差示热分析
DTA range DTA 范围,差热分析范围
DTE (data terminal equipment) 数据终端设备
dual absolute alarm card 双通道绝对值报警卡
dual carriageway 双程分隔车道

dual composition control 双组分控制
dual computer system 双并列计算机系统，复式计算机系统
dual edge trimming iron 双片边刨刨刀
dual indicator 双针指示器
dual meter 两用表
dual modulation telemetering system 双重调制遥测系统
dual principle 对偶原理
dual purpose voltage transformer 双重用途电压互感器
dual sealer card 双通道标度变换卡
dual signal selector 双信号选择器
dual spin stabilization 双自旋稳定
dual temperature thermostat 双温恒温器
dual triode 双三极管
dual-beam synchroscope 双线同步检定器
dual-channel high-sensitive voltmeter 双通道高灵敏度电压表
dual-computer system 双计算机系统
dual-feed 双端馈电，双路馈电
dual-flame ionization detector 双火焰离子化检测器
duality 双重性，对偶性
dual-mode control 双重模式控制
dual-mode control system 双重方式调节系统，双重方式控制系统
duct 风道，管道
ductile cast iron 球墨铸铁
ductile iron 球墨铸铁
ductile iron pipe 球墨铸铁管
ductility 延度
dug iron 熟铁，锻铁
dummy antenna 仿真天线
dummy test 模拟测试
dummy thermocouple 补偿热电偶
dummying 预锻
dump 转储，转储数据
dump valve 倾泻阀，安全阀，应急排放阀，事故排放阀
dumper 倾泥车，泥头车，倾卸车
dumping area 卸泥区
duoplasmatron ion source 双等离子体离子源

duplex 双重双面法
duplex air gauge 双计气压表
duplex cable 双芯电缆，双股电缆
duplex control 双重控制
duplex deviation alarm card 双偏差报警卡
duplex difference alarm card 双差值报警卡
duplex gauge 双工压力表，双针压力表
duplex helical 双重螺旋法
duplex pressure gauge 双工压力表，双针压力表
duplex rotor type flowmeter 双转子式流量计
duplex spread blade 双重双面刀齿
duplex transmission 双工传输
duplex wire 双芯导线
duplexed computer system 双计算机系统
duplicated cavity plate 复板模
duplicating milling machines 仿形铣床
durability 耐用度，耐久性
durability factor 耐久性系数
durable material 耐久物料
duration 持续时间
duration of cycle 周期时间
duration of load 负荷保持时间
dust 灰尘，尘埃，尘土
dust analyzer 尘量分析仪
dust counter 计尘器
dust sampling 粉末取样，粉尘取样
dust screen 隔尘网
dust seal 防尘封
dust catcher 除尘器，吸尘器
dustproof instrument 防尘式仪器仪表
dustproof packaging 防尘包装
dust-proof solenoid valve 防尘型电磁阀
duty 责任，负载，功率，工作状态
duty cycle 工作周期，占空度，忙闲度
duty factor 占空因数，线圈间隙因数，工作因数，工作系数
duty factor control system 工作比控制系统

duty ratio 负载比
dwell 停歇，保压
dwell time 停留时间
dye penetrant 染料渗透剂
dye penetrant inspection 染料渗透检查
dye penetrant test 染料渗透试验
dye-penetrant testing method 着色渗透探伤法
dynamic 动态，动态的，动力学，动力学的
dynamic accumulation error 动态累计误差
dynamic analysis design 动态分析设计
dynamic analysis of machinery 机械动力分析
dynamic balance 动平衡
dynamic balancing machine 动态平衡机
dynamic behavior of various system 各种系统动态特性
dynamic behaviour 动力学特性
dynamic bias control 动态偏置控制
dynamic brake 动力制动器
dynamic braking 能耗制动
dynamic calibrator 动态校准器
dynamic capacity 动态电容
dynamic channel assignment 动态信道分配
dynamic characteristic 动态特性
dynamic characteristic calibrater 动态特性校准仪
dynamic characteristics 动态特性
dynamic coil 动圈
dynamic compensation 动态补偿
dynamic compensator 动态补偿器
dynamic curve 动态曲线
dynamic decoupling 动态解耦
dynamic degradation 动态降解，动态衰退
dynamic design of machinery 机械动力设计
dynamic deviation 动态偏差
dynamic display image 动态显示图像
dynamic effect 动力效应
dynamic element 动态元件

dynamic energy 动能
dynamic equations 动态方程
dynamic equivalent axial load 轴向当量动载荷
dynamic equivalent radial load 径向当量动载荷
dynamic error 动态误差
dynamic error coefficient 动态误差系数
dynamic exactness 动态吻合性
dynamic flex board 动态挠性板
dynamic gain 动态增益
dynamic gauging 动态容积测量法
dynamic inductance 动态电感
dynamic input-output model 动态输入-输出模型
dynamic load 动负荷，动载荷
dynamic lubrication 动力润滑
dynamic magnetization curve 动态磁化曲线
dynamic mass spectrometer instruments 动态质谱仪器
dynamic matrix 动态矩阵
dynamic matrix control（DMC） 动态矩阵控制
dynamic measurement 动态测量
dynamic memory 动态存储器
dynamic memory allocation 动态存储器分配
dynamic microphone 电动传声器
dynamic model 动态模型，动态模式
dynamic modelling 动态模型
dynamic noise suppresser 动态噪声抑制器
dynamic optimization model 动态优化模型
dynamic output feedback 动态输出反馈
dynamic parameter 动态参数
dynamic performance analysis 动态性能分析
dynamic pressure 动压
dynamic pressure of fluid element 流体单元动压
dynamic pressure sensor 动态压力传感器

dynamic pressure transducer 动态压力传感器
dynamic programming 动态规划
dynamic property 动态性质,动态性能
dynamic RAM 动态随机存储器
dynamic random access memory (DRAM) 动态随机存储器
dynamic range 动态范围
dynamic range of microphone 传声器动态范围
dynamic reaction 动态反应
dynamic resistance 动态电阻
dynamic resistance strain gauge 动电阻应变仪
dynamic resolution 动态分辨力
dynamic response 动态响应
dynamic sensitivity 动态灵敏度
dynamic SIMS 动态二次离子质谱法
dynamic specification 动态指标
dynamic stability 动态稳定,动态稳定度
dynamic standard strain device 动态标准应变
dynamic state 动态
dynamic stiffness 动刚度
dynamic stiffness of the moving element suspension 运动部件悬挂动刚度
dynamic stiffness ratio 动刚度比
dynamic storage allocation 动态存储分配
dynamic strain 动应变
dynamic strain indicator 动态应变仪
dynamic system 动态系统
dynamic test 动力测试,动态测试,动态试验
dynamic thermomechanical analysis 动态热机械分析
dynamic thermomechanical analysis apparatus 动态热机械分析仪
dynamic two-plane balancing machine 动态双面平衡机
dynamic unbalance 动态不平衡
dynamic vane bias 风向标的动力偏幅
dynamic viscosity 动力黏度
dynamic water tank 动水槽
dynamic weighing method 动态称重法
dynamic windows set 动态窗口设定
dynamically equivalent model 等效动力学模型
dynamics 动力学
dynamics of feedback loop 反馈回路动态
dynamics of machinery 机械动力学
dynamics of measuring instrument 测量仪表动态
dynamic-state operation 动态运行
dynamite 黄炸药,甘油炸药
dynamo 发电机,直流发电机
dynamo magneto 永磁发电机
dynamometer 测功器,功率表,测力计,功率计
dynamometric system 测力系统
dynamotor 电动发电机,发电机
dynatron effect 负阻效应
dysprosium (Dy) 镝

E

EAF (effective attenuation factor) 有效衰减因数
ear muffler 护耳罩
ear plug 耳塞
early failure 早期故障,早期失效,过早损坏
early return bar 复位键,提前回杆
early warning system 预警系统,预先警报系统
earphone 耳机
earth 地,大地,接地,泥土,地线
earth bus 接地母线
earth capacity 对地电容
earth copper strap 接地铜织带
earth current 地电流,泄地电流
earth electrode 接地体,接地电极
earth leakage circuit breaker (ELCB) 通地泄漏电路断器,漏电保护断路器,接地漏电断路器
earth leakage detector 接地漏电检示器
earth loop impedance test 接地阻抗测试
earth moving machinery 土方机械
earth plate 接地板,接地导板
earth potential 地电位
earth pressure 土压力,地压
earth resistance 接地电阻,大地电阻
earth resistance meter 接地电阻表
earth resource technology satellite (ERTS) 地球资源技术卫星
earth rod 接地棒
earth tag 接地片
earth tape 接地带
earth wire 地线,接地线
earth connector 接地线
earth fault 接地故障
earth lead 接地线
earthed circuit 接地电路
earthed conductor 接地导线
earthed input 接地输入
earthed neutral conductor 接地中线
earthed output 接地输出
earthed point 接地点
earthed system 接地系统
earthed voltage transformer 接地型电压互感器
earthenware pipe 陶管
earth-fault protection 接地保护装置
earthing 接地
earthing autotransformer 接地自耦变压器
earthing bus 接地母线
earthing cable 接地线,接地电缆
earthing jumper 接地跨线
earthing of casing 外壳接地
earthing switch 接地转换开关
earthing system 接地系统
earthing terminal 接地终端
earthing twisted pair 接地绞线
earthometer 接地测量仪,兆欧计,兆欧表,高阻表
earthquake loading 地震荷载
earth-resistance meter 接地电阻表
earth-retaining structure 挡土构筑物,护土结构
earthworks 土方工程
EAS (equivalent air speed) 指示空速
ease of ignition 易起燃性
easily damaged parts 易损件
easy setting dual jumper (ESDJ) 简化双重跳线法
eaves gutter 檐沟
EB (exception build) 例外制作
EB (expansion bus) 扩展总线
EBT (electronic bathythermograph) 电子深度温度计
eccentric 偏心,偏心盘
eccentric angle 偏心角
eccentric load 偏心载荷
eccentric mass 偏心质量
eccentric orifice plate 偏心孔板
eccentric shaft 偏心轴
eccentricity 偏心,扰度,偏心度,

111

偏心距
eccentricity of rotor 转子偏心距
eccentricity ratio 偏心率
ECD (electron capture detector) 电子捕获检测器
ECG (electrocardiograph) 心电图（仪）
echelon grating 阶梯光栅
echo 反射波，回波，反应
echo height 反射波高度
echo sounder 回声测深仪
echo sounding receiver 回声测探接收器
EC-link EC链
ECN (engineering change notice) 工程变更通知
ecology 生态学
economic control 经济控制
economic control theory 经济控制理论
economic cybernetics 经济控制论
economic data 经济数据
economic design 经济性设计
economic model 经济模型
economic system model 经济系统模型
economics in process control 过程控制经济
economizer 节热器，省煤器
ECS (expert control system) 专家控制系统
ECU (electromagnetic control unit) 电磁控制单元
EDA (electric design automation) 电子设计自动化
eddy 涡流，旋涡
eddy current 涡流
eddy current analysis 涡流分析
eddy current inspection instrument 涡流探伤仪
eddy current loss 涡流损耗
eddy current problem 涡流问题
eddy current technique 涡流技术
eddy current testing method 涡流探伤法
eddy current thickness meter 电涡流厚度计
eddy diffusion 涡流扩散

eddy velocity 涡动速度
eddy-current 涡流
eddy-current flowmeter 涡流流量计
eddy-current revolution counter 涡流式转速表
eddy-current thickness meter 电涡流厚度计
eddy-current type transducer 涡流式传感器
edge 切边碎片
edge connector 边缘连接器
edge crack 裂边，（平板玻璃）板边裂纹
edge effect 边缘效应
edge file 刃用锉刀
edge finder 巡边器，边缘查找器
edge iron 角钢，角铁
edge joint 端接接头，边缘连接
edge radius 刀尖圆角半径
edge triggered flip-flop 边沿触发的触发器
edge-board contact 板边插头
edging 边缘处理
edit 编辑
edit a file 编辑一个文件
edit command file 编辑命令文件
edit table of contents 目录编辑表
editor 编辑程序，编辑器
EDM (electric discharge machine) 放电机
EDTA (ethylene diamine tetraacetic acid) 乙二胺四乙酸
EED (electron-diffraction method) 电子衍射法
EEG (electroencephalogram) 脑电图
EELS (electron energy loss spectroscopy) 电子能量损失谱学
effect 效应，作用，影响，效果
effect device power 特效装置电源
effect devices 特效装置
effect transistor structure 特效晶体管结构
effective 有效，有效的，有作用的
effective aperture 有效孔径
effective area 有效面积
effective attenuation factor (EAF) 有效衰减系数

effective bandwidth 有效带宽
effective bearing spacing 轴承有效间距
effective channel length 有效波道长度
effective circle force 有效圆周力
effective current 有效电流
effective cut-off wavelength 有效截止波长
effective data transfer rate 有效数据传送率
effective deadtime 有效纯滞后，等效纯滞后
effective diaphragm area 膜片有效面积
effective emissivity 有效发射率
effective enhancement 效率提高
effective excitation force 有效激振力
effective face width 有效宽度（有效齿宽）
effective impedance 有效阻抗
effective inductance 有效电感
effective magnetic field 有效磁场
effective mass 有效质量，等效质量
effective mass of the moving element 运动部件有效质量
effective path length 有效光程长度
effective radiation exitance 有效辐（射）出（射）度
effective range 有效范围，有效量限，有效测量范围
effective reactance 有效电抗
effective resistance 工作阻力，有效电阻
effective resistance moment 工作阻力矩
effective sound pressure 有效声压
effective tension 有效拉力
effective time constant 有效时间常数
effective values 有效值
effectiveness 有效性
effectiveness theory 效益理论
effects of saturation 饱和效应
efficiency 效率，功能率，能力
efficiency curve 效率曲线
efficiency diode 高效二极管，阻尼二极管
efficient 有效的，因子
efficient algorithm 有效算法

efficient engineering 有效工程
efficient evaluation 有效评估
effluent 污水
effluent outfall 污水出口管，污水排放管
efflux viscometer 流出式黏度计
EGA (evolved gas analysis) 逸出气分析
EGD (evolved gas detection) 逸出气检测
EID (electron impact desorption) 电子轰击解吸
EID (electron induced desorption) 电子诱导解吸
eigen frequency 特征频率
eigen structure assignment 本征结构测定
eigenfunction 本征函数，特征函数
eigenmode analysis 本征模式分析
eigenvalue 特征值
eigenvalue assignment 特征值赋值
eigenvalue lower bound 特征值低限
eigenvalue problem 特征值问题
eigenvalue replacement 特征值替换
eigenvector 特征向量
EIP (event initiated processing) 事件触发处理
EISA (enhanced industry standard architecture) 增强形工业标准架构
eject pin 顶出针
ejected 喷射，驱逐，被放出的
ejection pad 顶出衬垫
ejection 弹出，排出，喷出，喷射
ejection pad 顶出衬垫
ejector 喷射器，脱模器
ejector guide pin 顶出导销
ejector leader busher 顶出导销衬套
ejector pad 顶出垫
ejector pin 顶出针，顶出销
ejector plate 顶出板
ejector pump 喷射泵
ejector rod 顶出杆
ejector sleeve 顶出衬套
ejector valve 顶出阀
Ekman current meter 埃克曼流速仪
elapsed time controller 消逝时间调节器
elastic after-effect 弹性后效

elastic analysis 弹性分析
elastic coefficient 弹性系数
elastic coupling 弹性联轴器
elastic deformation 弹性变形
elastic force 弹力
elastic foundation 弹性地基，弹性基础
elastic limit 弹性极限
elastic modulus 弹性系数，弹性模数
elastic pressure gauge 弹性元件压力计
elastic scatter 弹性散射
elastic support 弹性支座
elastic system 弹性系统
elasticity 弹性
elasticity of demand 需求弹性
elasticity sliding motion 弹性滑动
elastomer 弹性体
elbow 弯通，弯管接头，弯头，肘形弯管
ELCB (earth leakage circuit breaker) 通地泄漏电路断电器，漏电保护断路器，接地漏电断路器
electret microphone 驻极体传声器
electric 电的，电气的，电力的，导电的
electric accumulator 蓄电池，累加器，积聚者
electric actuated stop valves 电动截止阀
electric actuated wedge gate valves 电动楔式闸阀
electric actuator 电动执行机构，电动执行器
electric angle 电角度
electric apparatus 电气设备
electric appliance 电器
electric automatization 电气自动化
electric axis 电轴
electric bell 电铃
electric blower 电吹风
electric braking 电气制动
electric cable 电缆
electric car 电车
electric charge 电荷，电费
electric charge time constant 充电时间常数
electric circuit 电路

electric communication 电信
electric condenser 电容器
electric conductivity 电导率
electric conductivity meter 电导仪
electric contact liquid-in-grass thermometer 电接点玻璃温度计
electric contact set 电接点
electric control 电动控制，电气控制
electric current 电流
electric current meter 电流表
electric current sensor 电流传感器
electric current transducer 电流传感器
electric degree 电角度
electric design automation (EDA) 电子设计自动化
electric differential pressure transmitter 电动差压传感器
electric discharge 放电
electric discharge machine (EDM) 放电机
electric discharge machining 放电加工
electric discharge time constant 放电时间常数
electric displacement 电位移
electric domain 电畴
electric double disk parallel gate valves 电动平行式双闸板闸阀
electric doublet 电偶极子
electric drill 电钻
electric drive 电气传动，电力驱动
electric drive control 电力拖动控制
electric drive control gear 电动传动控制设备
electric drying 电气干燥
electric drying oven force-air circulation 电热干燥器强制空气循环
electric drying oven with forced convection 电热鼓风干燥箱
electric dynamometer 电功率表
electric elevator 电梯，电升降机
electric energy 电能
electric excavator 电动挖掘机
electric field 电场
electric field controller 电场控制仪
electric field intensity 电场强度
electric field meter 电场强度计

electric field sensor 电场传感器
electric field strength sensor 电场强度传感器
electric field strength transducer 电场强度传感器
electric filter 电滤波器
electric flux 电通量
electric flux density 电通密度,电位移
electric furnace 电炉
electric generator 发电机
electric heat tracing 电伴随加热
electric heater 电热器,电暖气
electric hoist 电动起重机,电葫芦,电动卷扬机
electric hydraulic converter 电-液转换器
electric inertia 电气惯性
electric installation 电气装置,电气设备
electric instrument 电测仪表,电表
electric iron 电熨斗
electric lamp 电灯
electric lift 电梯,电升降机
electric lighting 电气照明
electric line of force 电力线
electric locomotive 电气机车
electric logger 电测井仪器
electric machine 电机,电动机械
electric machines and electric apparatus 电机与电器
electric material 电工材料
electric measurement technique of strain gauge 应变计电测技术
electric meter 电度表
electric motor 电动机
electric motor drive 电动机拖动
electric network 电力网
electric operating station 电动操作台
electric overhead line 架空电缆
electric pig iron 电炼生铁
electric pneumatic converter 电-气转换器
electric potential 电位
electric potential difference 电位差
electric power 电功率
electric power distribution 电力分配,配电
electric power economy 电力经济
electric power generation 发电
electric power industry 电力工业
electric power pool 电力网系统,联合供电网
electric power supply 电力供应,供电
electric power system 电力系统,供电系统
electric power tools 电动刀具
electric power transmission 电力输送,输电
electric pulse 电脉冲
electric pump 电动泵,电动抽水机
electric quantity sensor 电量传感器
electric quantity transducer 电量传感器
electric railway 电气铁道
electric rate 电费率
electric refrigerator 电冰箱
electric regulator 电气调节器,电动调节器
electric relay 继电器
electric resistance 电阻
electric resonance 谐振
electric screw driver 电动起子,电动螺丝刀,电动螺丝起子,电动螺钉旋具
electric shaft 电气联动
electric signalling 电气信号
electric soldering 电焊
electric soldering iron 电烙铁
electric spark machining 电火花加工,放电加工
electric steam boiler 电热式蒸汽锅炉
electric substation 变电站
electric test 电气试验
electric thermometer 电测温度计
electric throttle control 电力扼流控制
electric train 电气列车
electric tramway 有轨电车
electric travelling crane 移动式起重机,行车
electric vehicle 电力机车
electric voltage 电压

electric welding set 电焊机
electric wire 电线
electric failure 触电
electrical 电的，电动的，电气的
electrical activity 电活动
electrical aluminium wire 导电铝线，铝导线
electrical angle 电角
electrical apparatus 电气设备
electrical appliance 电器，电气器具，电气设备，电气用品
electrical behaviour 电力动态
electrical breakdown 停电，绝缘破坏，电击穿
electrical capacitance level measuring device 电容物位测量仪表
electrical center 电气中心
electrical characteristic 电力特性
electrical conductance level measuring device 电导物位测量仪表
electrical conduction 电导，电子传导
electrical conductivity 电导率
electrical conductivity detector 电导检测器，导电性检测器
electrical conductivity test 电导率测试
electrical connection 电气连接
electrical contact 电接触，电插头
electrical control valve 电动调节阀
electrical cut-off switch 电力断路开关，电力断路接触器
electrical device 电气设备
electrical discharge wire-cutting 电火花线切割加工
electrical efficiency 电效率
electrical energy 电能，电力
electrical engineering 电机工程，电气工程
electrical engineering handbook 电工手册
electrical equipment 电气设备
electrical feedback 电反馈
electrical hygrometer 电气湿度计，电测湿度表
electrical impedance 电阻抗
electrical infrastructure 电气基础设施

electrical installation 电气安装技术
electrical interlock 电气连锁
electrical isolation 电气隔离
electrical load 电力负载
electrical machine 电机，电动机械
electrical measurement method 电测法
electrical measurement method of optical pyrometer 光学高温计电测法
electrical measuring instrument 电工测量仪器仪表，电法勘探仪器
electrical network 电力网络
electrical panel 配电盘，配电板
electrical plug 电插头
electrical power consumption 耗电量
electrical power system 电力系统
electrical products 电器产品
electrical property 电力性质
electrical property tester 电性能测定仪
electrical pulse 电脉冲
electrical quadripole 四端网络
electrical resonance frequency of the moving element 运动部件电谐振频率
electrical shock 电击
electrical signal 电信号
electrical socket outlet 电源插座
electrical sparkle 电火花
electrical stimulation 电刺激，电激发
electrical supplier 供电商
electrical technician 电力工程技术员
electrical thermometer 电测温度表
electrical wind vane and anemometer 电传风向风速仪
electrical wiring 电气布线
electrical works 电气工程
electrical zero 电零位，电零点
electrical zero adjuster 电零位调节器，电零点调整器
electrical service 供电设施
electrically conductive path 导电通路
electrically heated drying cabinet 电热干燥箱
electrically-erasable ROM 电子可擦只读存储器

electrical-mail 电子邮件
electrical-music industry 电子音乐工业
electric-charge density 电荷密度
electric-hydraulic control 电-液控制
electrician 电工,电气技术员,电业工匠
electricity 电力,电流
electricity interruption 电力干扰,电力中断,电力故障
electricity meter box 电表箱
electricity transmission 输电
electric-pneumatic transducer 电-气转换器
electric-resistance wire strain gauge 电阻丝应变仪
electrification 电气化,带电
electrified railroad 电气化铁路
electro magnetic interference (EMI) 电磁干扰
electro microscopy 电子显微镜
electroacoustic transducer 电声换能器
electroacoustical reciprocity theorem 电声互易定理
electro-arc contact machining 接触放电加工
electrocardiograph (ECG) 心电图
electro-cardiography sensor 心电图传感器
electro-cardiography transducer 心电图传感器
electrochemical analysis 电化学分析
electrochemical analyzer 电化学式分析器
electrochemical corrosion 电化学腐蚀
electrochemical detection 电化学检测,电化学检测器
electrochemical detector 电化学检测器
electro-chemical machining 电化学加工
electrochemical sensor 电化学式传感器
electrochemical transducer 电化学式传感器
electroconductive paste printed board 导电胶印制板
electro-copper glazing 铜条嵌镶玻璃
electrocution 触电,电毙

electrode 电极,电焊条
electrode polarity 焊条极性
electrode potential 电极电位,金属电极电位
electrode signal 电极信号
electrode type salinometer 电极式盐度计
electrode wire 焊条钢丝
electrode with a mobile carrier 流动载体电极
electrodeless-discharge lamp 无极放电灯
electrodeposited copper foil 电解铜箔
electrodialysis method for desalination 电渗析淡化法
electrodynamic 电动力的,电动力学的
electrodynamic bridge 电动式电桥
electrodynamic force 电动力
electrodynamic instrument 电动系仪表,电动力式仪表
electrodynamic meter 电动系电度表
electrodynamic vibrator 电动振动器
electroencephalogram (EEG) 脑电图
electroence-phalographic sensor 脑电图传感器
electroence-phalographic transducer 脑电图传感器
electroformed mold 电铸成形模
electro-fusion coupler 电熔套管
electro-fusion welding 电熔焊
electro-galvanized steel wire 电镀锌钢丝
electro-gravimetric analysis 电重量分析(法)
electrogravimetry 电重量分析法
electro-hydraulic actuator 电动-液压执行机构
electrohydraulic control 电液控制
electrohydraulic servo valve 电液伺服阀
electro-hydraulic servocontrolled fatigue testing machine 电液伺服疲劳试验机
electro-hydraulic system 电-液系统
electrokinetic injection 电动进样
electrolysis 电解

electrolysis humidity sensor 电解式湿度传感器
electrolysis humidity transducer 电解式湿度传感器
electrolyte 电解质，电解液
electrolytic 电解的
electrolytic action 电解作用
electrolytic analysis method 电解法
electrolytic analyzer 电解质分析仪
electrolytic capacitor 电解电容器
electrolytic cell 电解池，电解电池
electrolytic condenser 电解电容器
electrolytic grinding 电解研磨
electrolytic hardening 电解淬火
electrolytic hygrometer 电解湿度计
electrolytic iron 电解铁
electromagnet 电磁铁
electromagnet damping galvanometer 电磁阻尼振动子
electromagnet fluid damping galvanometer 电磁液体阻尼振动子
electromagnetic 电磁的
electromagnetic application 电磁应用
electromagnetic brake 电磁闸，电磁制动器
electromagnetic braking 电磁制动
electromagnetic clutch 电磁离合器
electromagnetic control unit (ECU) 电磁控制单元
electromagnetic core 电磁铁芯
electromagnetic counter 电磁计数器
electromagnetic crane 电磁铁起重机
electromagnetic damper 电磁阻尼器
electromagnetic deflector alignment system 电磁偏转对中系统
electromagnetic densitometer 电磁密度计
electromagnetic device 电磁装置
electromagnetic distance meter 电磁波测距仪
electromagnetic element 电磁元件
electromagnetic energy 电磁能
electromagnetic field 电磁场
electromagnetic field and microwave technology 电磁场与微波技术
electromagnetic field problem 电磁场问题
electromagnetic flow sensor 电磁流量传感器
electromagnetic flow transducer 电磁流量传感器
electromagnetic flowmeter 电磁流量计
electromagnetic force 电磁力
electromagnetic gun 电磁枪
electromagnetic induction 电磁感应
electromagnetic inertia 电磁惯性
electromagnetic instrument 电磁式仪表
electromagnetic interference 电磁干扰，电磁感应法仪器
electromagnetic lens 电磁透镜
electromagnetic mail 电磁信件
electromagnetic method instrument 电磁法仪器
electromagnetic methods 电磁法
electromagnetic mode 电磁模式，电磁模型
electromagnetic pulse 电磁脉冲
electromagnetic pump 电磁泵
electromagnetic radiation 电磁辐射
electromagnetic scattering 电磁散射
electromagnetic scattering problem 电磁散射问题
electromagnetic screen 电磁屏蔽
electromagnetic sensor 电磁式传感器
electromagnetic signal 电磁信号
electromagnetic spectrum 电磁波谱
electromagnetic switch 电磁开关
electromagnetic system 电磁系统
electromagnetic torque 电磁转矩
electromagnetic transducer 电磁式传感器
electromagnetic transient 电磁瞬态过程
electromagnetic transmission 电磁发射
electromagnetic type relay 电磁继电器
electromagnetic unit 电磁单元
electromagnetic vibrator 电磁振动器
electromagnetic wave 电磁波
electromagnetic wave propagation logging instrument 电磁波传播测井仪
electromechanical 电动机械的，机电的
electromechanical liquid level indicator

机电式液位指示器
electromechanical regulator 机电调节器
electrometer 静电表,静电计
electromobile 电动汽车
electromotive force (EMF) 电动势,电池电动势
electromotor 电动机
electromyographic sensor 肌电图传感器
electromyographic transducer 肌电图传感器
electromyography (EMG) 肌电图
electron 电子
electron accelerator 电子加速器
electron avalanche 电子雪崩
electron beam 电子束
electron beam exposure apparatus 电子束曝光机
electron beam processing machine 电子束加工机
electron capture detection 电子捕获检测
electron capture detector (ECD) 电子捕获检测器
electron channelling pattern 电子通道花样
electron cloud 电子云
electron concentration 电子浓度,电子密度
electron device 电子装置
electron diffraction image 电子衍射像
electron diffractometer 电子衍射谱仪
electron donating group 供电子取代基
electron emission 电子发射
electron energy loss spectroscopy (EELS) 电子能量损失谱学
electron gun 电子枪
electron gun alignment adjustment 电子枪对中调节
electron hole 电子空穴
electron image intensifier 电子像增强器
electron impact 电子碰撞
electron impact desorption (EID) 电子轰击解吸

electron impact ion source 电子轰击离子源
electron impact ionization 电子轰击离子化
electron induced desorption (EID) 电子诱导解吸
electron lens 电子透镜
electron microprobe 电子微探针
electron microscope 电子显微镜
electron mobility detector 电子迁移率检测器
electron operation desk of EPMA 电子探针的电子操纵台
electron optical system of EPMA 电子探针的电子光学系统
electron optics 电子光学
electron paramagnetic resonance (EPR) 电子顺磁共振
electron paramagnetic resonance spectrometer 电子顺磁共振波谱仪
electron probe 电子探针
electron probe micro-analysis (EPMA) 电子探针微分析
electron probe X-ray microanalyzer 电子探针X射线微分析仪
electron rays 电子射线,电子束,电磁波
electron relay 电子继电器
electron spectrometer 电子能谱仪
electron theory 电子理论
electron type rock ore densimeter 电子式岩矿密度仪
electron volt 电子伏特
electron wave length 电子波长
electron-beam atomizer 电子束原子化器
electron-diffraction method (EED) 电子衍射法
electron-energy analyzer 电子能量分析器
electron-gun 电子枪
electron-hole pairs 电子—空穴对
electronic 电子的,电子学的
electronic AC voltage stabilizer 电子交流稳压器
electronic amplifier 电子管式放大器
electronic analog and simulation equip-

ment 电子模拟仿真设备
electronic analog multiplier 电子模拟乘法器
electronic analogue-to-digital converter 电子模/数转［变］换器
electronic analyzer 电子分析仪
electronic automatic balancer 电子自动平衡仪
electronic automatic compensator 电子自动补偿仪
electronic balance 电子天平
electronic batching scale 电子配料秤
electronic bathythermograph (EBT) 电子深温计
electronic belt conveyor scale 电子皮带秤
electronic control 电子控制
electronic controller 电子调节器，电动调节器
electronic controlling element for current 电流的电子主控元件
electronic counter 电子计数器
electronic counting scale 电子计数秤
electronic data processing centre 电子数据处理中心
electronic device 电子器件
electronic distance-meter theodolite 电子测距光学经纬仪
electronic fluxmeter 电子磁通表
electronic galvanometer table model 台式电子检流计
electronic hanging scale 电子吊秤
electronic hoist scale 电子吊秤
electronic hopper scale 电子料斗秤
electronic indicator 电子指示仪，电动指示仪
electronic instrument 电子仪表
electronic integrating fluxmeter 电子积分式磁通表
electronic level 电子水准仪
electronic magnetism inspect 高磁测试
electronic measuring instrument 电子测量仪表，电子测量仪器仪表
electronic null-balance recorder 电子零位平衡记录仪
electronic plane table equipment 电子平板仪
electronic platform scale 电子平台秤
electronic railway scale 电子轨道衡
electronic regulator 电子调节器
electronic relay 电子继电器
electronic sampling switch 电子采样开关
electronic self-balance instrument 电子自动平衡仪表
electronic shell 电子层
electronic switch 电子开关
electronic tacheometer 电子速测仪
electronic temperature contact controller 接触式电子调温计
electronic testing machine 电子式试验机
electronic theodolite 电子经纬仪
electronic tilt sensor 电子测倾器
electronic transmitter 电子变送器，电动变送器
electronic trunk scale 电子汽车秤
electronic tube 电子管
electronic valve 电子阀
electronic weigher 电子秤
electronically-controlled 电子控制的
electronically-controlled transmission 电子控制发射，电子控制传送
electronics 电子学的
electronics module (EM) 电子模块
electronics science and technology 电子科学与技术
electron-tube photodetector 电子管光电探测器
electron-withdrawing group 吸电子取代基
electro-oculogram (EOG) 眼电图
electro-optical distance meter 光电测距仪
electroosmosis 电渗
electrophoresis 电泳，电泳法
electrophoresis meter 电泳仪
electrophoresis system 电泳系统
electroplating 电镀
electro-pneumatic brake 电动气动制动器
electro-pneumatic breaker 电动气动断路器

electropneumatic positioner 电动气动定位器，电气定位器
electro-pneumatic positioner 电动气动定位器，电气定位器
electro-pneumatic valve 电动气动阀门
electropolarized relay 极化继电器
electropolisher 电解抛光机
electropositive 正电性的，阳电性的
electroretinographic sensor 视网膜电图传感器
electroretinographic transducer 视网膜电图传感器
electroscope 验电器
electrosensitive printer 电灼式印刷机
electro-servo control 电气随动控制，电气伺服控制
electrospray 电喷雾
electrostatic 静电，静电的
electrostatic actuator 静电激发器
electrostatic analyzer 静电分析器
electrostatic charge 静电荷
electrostatic coupling 静电耦合
electrostatic deflection 静电偏转
electrostatic discharge 防静电腕带
electro-static discharge (ESD) 静电排放
electrostatic display recorder 静电显示记录仪
electrostatic electron microscope 静电电子显微镜
electrostatic emanometer 静电计式射气仪
electrostatic field interference 静电场，静电场干扰
electrostatic filter 静电过滤器
electrostatic flux 电位移通量
electrostatic force 静电力
electrostatic induction 静电感应
electrostatic instrument 静电系仪表，静电式仪表
electrostatic lens 静电透镜
electrostatic octupole lens 静电八极透镜
electrostatic printer 静电印刷机
electrostatic quadrupole lens 静电四极透镜
electrostatic screen 静电屏蔽

electrostatic wattmeter 静电功率表
electrostriction 电致伸缩
electrotechnical measurement 电工测量
electrotechnics 电工技术，电工学
electrotechnology 电工技术，电工学
electrothermal 电热的
electro-thermal damper 电热防火闸
electrothermic instrument 电热式仪表
electro-volumetric analysis 电容量分析（法）
element 元素，元件，单元，母线，要素，零件
element analysis 元素分析，组成分析
elemental error 单元误差
elevated approach road 高架引道
elevated structure 高架结构
elevated trackway 高架轨道
elevated vehicular link 架空行车路
elevated-zero range 零点提升范围
elevation 正面图，立视图，标高，高程，负迁移，仰角
elevation of zero point of barometer 气压表零点高度
elevator 电梯，升降机
ELISA (enzyme-linked immunosorbent assay) 酶联免疫吸附测定
elliptical polarization instrument 椭圆极化仪
elliptical vibration 椭圆振动
elongated glass bulb 移液管
elongation 伸长率
elongation rate 延伸率
eluate 流出液，洗出液
elution 洗脱（淋洗）
elution chromatography 洗脱色谱法
eluvial 渗蚀层
EM (electronics module) 电子模块
e-mail 电子邮件
emanation survey 射气测量
emanation thermal analysis 放射热分析，放射性热分析
emanation thermal analysis apparatus 放射热分析仪
emanometer 测氡仪，射气仪
embedded core 加装砂芯
embedded lump 镶头

embedded metal 内置金属
embedded strain gauge 埋入式应变计
embedded system 嵌入系统
embedded thermometer 嵌入式温度计
emboss 装饰,浮雕(图案);压花
embossing 浮花压制加工,压花,压花加工
embossing iron 压花铁
emerg 紧急情况
emergency 紧急事件,紧急情况
emergency access 紧急通道
emergency alarm 紧急报警
emergency brake 紧急制动器
emergency button 事故按钮,备用按钮
emergency cable 事故电缆,备用电缆
emergency channel 紧急频道
emergency condition handler 紧急停车条件处理程序
emergency control centre 紧急事故控制中心
emergency cut-off valves 紧急切断阀
emergency exercise 应急演习
emergency exit 紧急出口
emergency exit door 紧急出口门
emergency generator 紧急发电机,后备发电机,应急发电机
emergency hand-drive 事故手动装置
emergency hand-winding equipment 紧急人手绞动器
emergency light 事故照明器
emergency lighting 紧急照明设备
emergency manual 紧急事故手册
emergency plan 应急计划
emergency power supply 事故备用电源
emergency preparedness 应急准备
emergency push button 紧急按钮
emergency push-button switch 事故备用按钮
emergency shutdown 紧急停车
emergency standby 应急待命
emergency stop 异常停止
emergency stop button 紧急止动按钮
emergency stop protection 紧急停止保护装置
emergency stop switch 应急停机开关,紧急停机开关
emergency supply source 应急电源,备用电源
emergency switch 应急开关
emergency switching-off 事故切断
emergency trip system 紧急跳闸系统
emergency valve 安全阀
emergency-off 应急断电
emery 金刚砂
emery cloth 砂布,金刚砂布
EMF (electromotive force) 电动势,电池电动势
EMF to current transmitter 电动势-电流变送器
EMF to pneumatic transmitter 电动势-气动变送器
EMG (electromyography) 肌电图
EMI (electro magnetic interference) 电磁干扰
emission 发射
emission electron microscope 发射电子显微镜
emission spectrum 发射光谱
emission X-ray spectrometry 发射X射线谱法
emissivity 发射率
emittance of the earth's surface 地表面辐射
emitter 发射管,放射器,发射极,发射体
emitter current 发射极电流
emitter follower 射极输出器,发射极输出(放大)器
emitter junction 发射结
emitter resistance 射极电阻
emitter voltage 射极电压
empirical distribution 经验分布
empirical evidence 经验数据
empirical model 经验模型
empty 排空
emulsifier 乳化剂,黏合剂
emulsion 乳剂,乳化液
emulsion paint 乳胶漆
enable 使能够,允许
enamel covered wire 漆包线
enamel insulated wire 漆包绝缘线

enamel paint 磁漆，瓷漆
enamel silk-covered wire 丝包漆包线
enamelled fire clay 釉瓷耐火黏土
enamelled iron 搪瓷铁
enamelled wire 漆包线
enantiomer 对映（导构）体
encapsulation 浇封
encapsulation molding 低压封装成型，注入成形
encircling coil 环形线圈
enclose 包围
enclosed motor 封闭式电动机
enclosed relay 封闭式继电器
enclosed switch 封闭式开关
enclosed-scale liquid-in-glass thermometer 内标式玻璃温度计
enclosed-scale thermometer 内标式温度计
enclosure 密封，外壳，包围，围封部分，附件
encode 编码
encoder 编码器
encoding 编码
end absorption 末端吸收
end bearing 端支承
end byte 结束字节
end cap 后盖
end capping 封尾，封顶，遮盖
end connection 连接端
end connector 终端接头
end fittings 管端配件
end grain 端面晶粒
end missing 断经
end movement 轴向移动
end of contact 终止啮合点
end of file (EOF) 文件结束
end of file indicator 文件结束指示器
end off block 字块结束，信息组结束
end point control 终点控制
end ring 端环
end span 端跨
end user 最终用户
end view 端视图，侧视图
end 末端，终结
end cover 端盖
end-effector 末端执行器

endless grinding belt 循环式研磨带
endless wire 无端铜网
endogenous variable 内生变量
endothermic peak 吸热峰
endothermic reaction 吸热反应
end-points 端点
endurance life 耐久寿命
endurance limit 耐久极限
endurance test 耐久性试验
endurance training 耐力训练
energization 通电
energized part 带电部分
energizing frequency 激励频率
energizing voltage 激励电压
energy 能，能源，能量
energy balance 能量平衡
energy conservation 能量守恒，节能
energy control 能量控制，发动机控制
energy converter 电能转换器，换能器
energy crisis 能源危机
energy dependence 能量相关
energy disperse spectroscopy 能谱仪
energy dispersion 能量色散
energy dissipation 能量损耗
energy distribution 能量分布
energy dynamometer 发动机功率计
energy efficiency 发动机效率，能源效率
energy equivalent 能当量
energy expenditure 能源消耗
energy filter 能量过滤器
energy level 能级
energy level diagram 能级图
energy loss 能量损耗，能量损失
energy loss of electron spectrometer 电子能量损失谱仪
energy loss spectrometer 能量损失谱仪
energy management 机车管理
energy management system 能源管理系统
energy meter 电度表，电能表
energy modelling 发动机模型
energy processor module 能量处理组件

energy spectra　能谱
energy spread　能量分散
energy storage　能量储存
energy system　发动机系统
energy transformation　能量转换
energy weighted acquisition　能量加权搜索
engaged switch　接通开关
engagement　啮合
engaging-in　啮入
engaging-out　啮出
engine　引擎，发动机
engine driven　机动
engine driven pump　机动抽水机，机动泵
engine exhaust system　引擎排气系统
engine oil　机油
engine torque　发动机扭矩
engineer　工程技术人员，工程师
engineer keyboard　工程师键盘
engineer personality　工程师属性
engineering change notice (ECN)　工程变更通知
engineering change order　技术更改指令，工程变更次序
engineering custom engineering proposal　用户技术建议
engineering cybernetics　工程控制论
engineering description　技术说明书
engineering design automation　工程设计自动化
engineering drawing　工程图
engineering functions　工程功能
engineering mechanics　工程力学
engineering model　初样
engineering personality　工程师属性，工程师特性
engineering plastics　工程塑胶
engineering project difficulty　工程瓶颈
engineering prototype　正样，工程样机
engineering simulator　工程仿真器
engineering standardization　工程标准
engineering statistics　工程统计
engineering system simulation　工程系统仿真

engineering thermophysics　工程热物理
engineering unit (EU)　工程单位
engineering working station　工程师工作站
engineerings　工程技术站
engineer's operating station　工程师操作站
engineer's transit　工程经纬仪
Engler viscosity　恩氏黏度
engraving machine　雕刻机
enhanced industry standard architecture (EISA)　增强型工业标准架构
enhanced operator station　增强型操作站
enhanced operator station Ⅲ (EOS Ⅲ)　Ⅲ型高性能操作站
enhancement　提高，增强
enhancement effect　增强效应
enriched uranium　浓缩铀
enrichment　浓缩
enterprise　企业
enterprise expansion projects　企业扩建计划
enterprise integration　企业集成
enterprise management　企业管理
enterprise modelling　企业模型
enterprise plan　企划
enterprise resource planning (ERP)　企业资源规划
enterprise technology solutions　企业技术解决
enthalpy relaxation　焓衰减
entire depth　整个深度
entity　实体
entrance　入口，入口处
entropy　熵
entry　输入，入口，记载事项
enumeration　列举
envelop curve　包络线
envelope　包络，封皮，包络线
enveloped thermistor　密封型热敏电阻器
enveloping　包络
environment architectures　环境建筑学
environmental　环境
environmental area　环境区域

environmental condition 环境条件
environmental engineering 环境工程,环境工程学,环境保护工程
environmental error 环境误差
environmental factor 环境因素
environmental gas analyzer 环境气体分析仪
environmental impact assessment 环境影响评估
environmental influence 环境条件影响,环境影响
environmental location 环境区域,环境场所
environmental monitor station 环境监测站
environmental noise 环境噪声
environmental parameter 环境参数
environmental pollution 环境污染
environmental regulation 环境调节
environmental science 环境科学
environmental science and engineering 环境科学与工程
environmental specification 环境规范
environmental stability 环境稳定
environmental stress cracking test 环境应力龟裂试验
environmental test 环境试验
enzyme electrodes 酶电极
enzyme immunoassay 酶免疫测定
enzyme substrate electrode 酶敏电极
enzyme-linked immunosorbent assay (ELISA) 酶联免疫吸附测定
enzyme-multiplied immunoassay technique 酶多联免疫测定技术
EOF (end of file) 文件结束
EOG (electro-oculogram) 眼电图
EOS Ⅲ (enhanced operator station Ⅲ) Ⅲ型高性能操作站
ephedrine 麻黄碱
epicyclic gear train 周转轮系
EPMA (electron probe micro-analysis) 电子探针微分析
epoch angle 初相角
epoxide cellulose paper copper-clad laminates 环氧纸质覆铜箔板
epoxide cellulose paper core glass cloth surfaces copper-clad laminates 环氧玻璃布纸复合覆铜箔板
epoxide synthetic fiber fabric copper-clad laminates 环氧合成纤维布覆铜箔板
epoxide woven glass fabric copper-clad laminates 环氧玻璃布基覆铜箔板
epoxy adhesive 环氧树脂黏合剂
epoxy glass substrate 环氧玻璃基板
epoxy glue 环氧树脂胶
epoxy novolac 环氧酚醛
epoxy paint 环氧漆
epoxy resin 环氧树脂
epoxy resin coat 环氧树脂搪层
epoxy resin grout 环氧树脂浆
epoxy value 环氧值
epoxy-coated reinforcement 环氧树脂密封钢筋
EPR (electron paramagnetic resonance) 电子顺磁共振
EPROM 电可编程只读存储器
equal addendum teeth 等齿顶高齿
equal and opposite in direction 大小相等且方向相反
equal percent 等百分比
equal precision measurement 等精密度测量
equal-height sleeves 等高套筒
equalization 平衡,相等,均等,调平均,均衡
equalizer 均衡器
equalizing orifice 平衡孔
equally divided scale 等分刻度
equally loaded 均衡负载的
equal-percentage valve 等百分比阀
equal-phase 同相的
equation 方程式
equation of network 网络方程
equation of state of gas 气体状态方程
equations set 方程组
equilibrium 平衡,均衡,稳定,力平衡
equilibrium growth 均衡增长
equilibrium point 平衡点
equilibrium state 平衡状态
equipment 设备,设施,装备,器具,工具

equipotential 等电位
equipotential line 等电位线
equipotential point 等电位点
equipotential zone 等电位区域
equivalence partitioning 等价类划分
equivalent 等效量，等效于
equivalent AC resistance 等效交流电阻
equivalent air speed (EAS) 指示空速
equivalent air volume 等效空气容积
equivalent capacity 等效电容
equivalent circuit 等效电路
equivalent coefficient of friction 当量摩擦系数
equivalent conductance 当量电导
equivalent force 等效力
equivalent gain 等效增益
equivalent gear ratio 当量传动比
equivalent generator 等效发电机
equivalent inductance 等效电感
equivalent input impedance 等效输入阻抗
equivalent link 等效构件
equivalent load 当量载荷，等效载重
equivalent mass 等效质量
equivalent mechanism 替代机构
equivalent moment of force 等效力矩
equivalent moment of inertia 等效转动惯量
equivalent number of teeth 当量齿数
equivalent pitch radius 当量节圆半径
equivalent potential screen 等电位屏蔽
equivalent rack 当量齿条
equivalent resistance 等效电阻
equivalent sine wave 等效正弦波
equivalent spur gear 当量直齿轮
equivalent spur gear of the bevel gear 锥齿轮的当量直齿轮
equivalent spur gear of the helical gear 斜齿轮的当量直齿轮
equivalent T circuit 等效T形电路
equivalent teeth number 当量齿数
equivalent uniform roughness 等效均匀粗糙度
equivalent value wave impedance 等值波阻抗

erasable programmable read-only memory 可擦可编程的只读存储器
erase panel 消隐画面
erbium (Er) 铒
erecting telescope 正像望远镜
erection 架设，安装，建立，竖设
ergonomics 人机工程学，工效学
erosion control 侵蚀防治
erosion protection works 防止侵蚀工程
ERP (enterprise resource planning) 企业资源规划
erratic motion of the moving element 运动部件的漂移运动
error 错误，误差
error analysis 误差分析
error bandwidth 误差带宽
error checking and correcting 错误检查和校正
error coefficient 误差系数
error compensation 误差校正，误差补偿
error control 误差控制
error correction 误差校正
error criteria 误差判据，误差准则
error curve of a measuring instrument 测量仪表的误差曲线
error detection 误差检测
error detection and correction 错误检测及校正
error detector 误差检测器
error estimation 误差估计
error expressed as a percentage of the fiducial value 引用误差，百分数引用误差
error in observation 观测误差
error of measurement 测量误差
error of observation 观察误差
error probability 误差率，误码率
error rate 出错率
error rate performance 差错率指标
error signal 偏差信号，误差信号
error transfer function 误差传递函数
error-correcting code 误差校正码
error-correction parsing 纠错剖析
error-detecting code 误差检测码
error-free 无误差，正常的
error-sensing element 误差敏感元件，

误差传感器
ERTS (earth resource technology satellite) 地球资源技术卫星
Esaki diode 隧道二极管，江崎二极管
escalator 电动扶梯，行人电梯，行人自动梯，自动梯
ESCD (extended system configuration data) 可扩展系统配置数据
ESD (electro-static discharge) 静电排放
ESDJ (easy setting dual jumper) 简化双重跳线法
establish 建立实体
estimate 估计量
estimation 估计
estimation algorithms 估计算法
estimation parameter 估计参数
estimation theory 估计理论
estimator 估计器
etalon 标准器
etamsylate 酚磺乙胺
etching 表面蚀刻，浸蚀
etching machines 蚀刻机
Ethernet 以太网
Ethernet interface 以太网接口
ethyl cellulose 乙基纤维素
ethylene diamine tetraacetic acid (EDTA) 乙二胺四乙酸
EU (engineering unit) 工程单位
Euclidean distance 欧几里得距离
europium (Eu) 铕
evacuation 抽空，疏散
evaluation 评价，求值，评估
evaluation and decision 评价与决策
evaluation standard 评定标准
evaluation technique 评价技术
evaporating dish 蒸发皿
evaporative cooling system 蒸发冷却系统
evaporative light scattering detector 蒸发光散射检测器
evaporativity 蒸发度
evaporator 蒸发器
evaporator coil 蒸发器旋管，蒸发器蛇形管
evaporators 浓缩器
evaporimeter 蒸发仪，蒸发计

evaporograph 蒸发成像仪
evapotranspirometer 蒸散表
even cooling 均匀冷却
even electron 含偶数个电子
evenly 均匀地
event 事件
event chain 事件链
event deletion 删除事件
event history 历史事件
event initiated point 事件初始点
event initiated processing (EIP) 事件触发处理
event logger package 事件记录程序包
event marker 标记
event pulse 事件脉冲
event recorder 事故记录仪
event option 事件选项
evolutionary system 进化系统
evolved gas analysis (EGA) 逸出气分析
evolved gas analysis apparatus 逸出气分析仪
evolved gas detection (EGD) 逸出气检测
evolved gas detection apparatus 逸出气检测器
examine 检查，查看，检验
excavation permit 挖掘许可证
excavation works 挖掘工程
excavator 挖土机
exceed 超过
exceeded variation 回差大
excellent computer 增强型计算机
excellent field control station 增强型现场控制站
excellent field gateway 增强型现场接口
excellent field monitoring station 增强型现场监视站
excellent gateway unit 增强型接口单元
excellent operator console 增强型操作台
excellent operator station 增强型操作站
exception build (EB) 例外制作
excess current 过电流
excess energy meter 超量电度表

excess flow device 溢流装置
excess flow protection device 溢流保护装置
excess metal 多余金属
excess voltage 过电压
excess voltage protection 过电压保护
excess 超过，过度
excessive defects 过多的缺陷
excessive gap 间隙过大
excessive temperature sensor 过热感应器
exchange data with NT applications 和 NT 应用软件交换数据
exchange capacity 交换容量
exchanging data with Windows programs 和 Windows 程序交换数据
excitation 激励，励磁，激发，刺激
excitation control 激励控制
excitation force 激振力
excitation index 激发指数
excitation spectrum 激发光谱
excitation system 励磁系统
excitation voltage 励磁电压
excitation winding 激励绕组，励磁绕组
excite 励磁，激发
excited by 励磁
exciter 励磁机，激振器
exciter circuit 励磁电路
exciting circuit 励磁电路
exciting coil 励磁线圈
exciting current 激磁电流，励磁电流
exciting voltage 励磁电压
exciting winding 励磁绕组
excitor 励磁器
exclusion limit 排除极限
execution 执行
execution 进行
execution time 执行时间
execution unit 执行单元
executive component 执行元件
executive link 执行构件
executive routine 执行程序
exhaust 废气，排气口
exhaust baffle 排气隔板
exhaust blower 排气机，排气风箱

exhaust duct 排气管道
exhaust elbow 排气肘管，排气弯管
exhaust emission control system 排气排放物控制系统
exhaust fan 排气扇，抽气扇
exhaust fume collecting hood 集烟罩
exhaust gas 废气
exhaust gas recirculation 废气循环，排气循环
exhaust manifold 排废管汇，排气集管
exhaust muffler 排气消声器，排气管减声器
exhaust pipe 排气喉，排气管
exhaust pollution of motor vehicle 汽车排气污染
exhaust purification 排气净化
exhaust silencer 排气消音器，排气管减声器
exhaust smoke 排气烟度
exhaust stack 排气管
exhaust system 排气系统
exhaust valve 排气阀
exhauster 排气器，抽风机
exit 出口，太平门
exit route 出口路线
exit switch 救生掣，出口开关
exogenous variable 外生变量
exothermic peak 放热峰
exothermic reaction 放热反应
exothermic welding 放热式焊接
expanded graphite 膨胀石墨
expanded memory 扩展存储器
expanded perlite insulation 膨胀珍珠岩绝热制品
expanded scale instrument 扩展标度尺仪表
expanded scale 扩展标度
expander 扩展器，膨胀器
expander die 扩径模
expanding agent 膨胀剂
expanding arbor 胀缩心轴，扩管器，胀杆
expansibility factor 可膨胀性
expansion 膨胀
expansion bolt 伸缩栓，伸缩螺栓，膨胀螺栓
expansion bus (EB) 扩展总线

expansion coefficient 膨胀系数
expansion dwg 展开图
expansion factor 溶胀因子,膨胀系数,膨胀率
expansion joint 伸缩接口,伸缩接缝,伸缩接头
expected characteristics 希望特性
expected long term stability 预期的长期稳定性
expendable bathythermograph 投弃式探海温度测量器
expenditure pattern 开支模式
experimental determination of time constant 时间常数实验测定
experimental intensity of scattered ion 散射离子的实验强度
experimental modeling 实验模型
experimental standard deviation 实验标准偏差
experimental standard deviation of the mean 平均值的实验标准偏差
experimental temperature scale 经验温标
experimental testing 实验测试
experimental variance 实验方差
expert adaptive controller tuning 专家自适应控制器整定
expert control system (ECS) 专家控制系统
expert system 专家系统
explanatory drawing 说明图则/绘图
explanatory guide 阐释指引
explanatory plan 说明图则/绘图
exploded drawing 爆炸图,零件分散图
exploding wire 爆丝
exploration 勘探
exploratory works 勘探工程
explosion 爆炸,爆裂
explosion proof 防爆型的
explosion proof electric machine 防爆型电机
explosion proof freezer and refrigerator 防爆冷藏柜
explosion proof instrument 防爆型仪表
explosion proof motor 防爆型电动机
explosion relief measure 防爆泄压设施
explosion-proof area 防爆区域
explosion-proof electric actuator 防爆型电动执行机构
explosion-proof instrument 防爆型仪器仪表
explosion-proof socket 防爆插座
explosion-proof solenoid valve 防爆型电磁阀
explosive 炸药,爆炸品
explosive limit of flammable gas 可燃气体的爆炸限
explosive sound source 爆炸声源
explosive store 炸药贮藏室
exponential filter 指数滤波器
exponential function 指数函数
exponential lag 指数滞后
export 出口
export package 出口包装
exposed face 外露面
exposed junction 外露端
exposed junction type sheathed thermocouple 露端型铠装热电偶
exposed metal 金属裸露
exposure 暴露,曝光,外露,曝光量
exposure chart 曝光曲线图
exposure time 曝光时间
expression 表达式
extend 延伸
extended controller 扩展控制器
extended industry standard architecture 扩充的工业标准结构
extended Kalman filter 扩展卡尔曼滤波器
extended network 扩展网络
extended rating current 扩展的额定电流
extended rating type current transformer 扩展的额定型电流互感器
extended system configuration data (ESCD) 可扩展系统配置数据
extendible barrier 可延伸护栏
extending the V net 扩展V网
extension 延展部分,扩建部分,扩建物,伸长
extension dwg 展开图

extension jib 伸臂
extension lead 延长导线
extension lead method 延长导线法
extension of time 延期
extension of time limit 延展时限
extension sleeve 伸缩套筒
extension socket 拖苏板，拖板
extension wing 延伸翼
extension wire 补偿导线
extensometer 引伸计，延伸仪
exterior package 外包装
external 外部的，外来的，表面的
external armature circuit 电枢外电路
external bracing 外部支撑
external characteristic 外特性
external circuit 外电路
external conversion 外转换
external crack 外表裂缝
external critical damping resistance 外部临界电阻
external critical resistance 外临界电阻
external diameter 外直径
external distributor road 外干路
external disturbance 外扰
external feedback 外反馈
external force 外力
external gear 外齿轮
external grinding 外圆磨削
external indicator 外指示剂
external interface 外部界面
external joint 外部接缝
external loads 工作载荷
external lock signal 外锁信号
external mode switching option state 外部方式键选择，外部方式键允许状态
external protection 外保护层
external reference sample 外参比试样
external road 外围道路
external shell 外壳
external standard method 外标法
external standardization 外标准化
external store 外存储器
external strengthening works 外部巩固工程
external structural bracing frame 外部支撑结构架
external summing 外部加法
external surface 外层，外表面
external wall 外墙
external works 外部工程
external-convection column sensitive element 柱状外对流敏感元件
external-convection ring sensitive element 环状外对流敏感元件
external-scale liquid-in-glass thermometer 外标式玻璃温度计
external-scale thermometer 外标式温度计
extinction coil 消弧线圈
extinction of arc 灭弧
extinguisher 灭火器
extra high voltage 超高压
extra-column effect 柱外效应
extracorporeal shock wave lithotrite 体外冲击碎石机
extract 挖取
extract fan 抽气扇
extraction lens 引出透镜
extractor 提取器，拔出器
extra-high voltage 超高压
extrapolated onset 外推起始点，外延始点
extreme position 极限位置
extreme pressure lubricant 耐特压润滑剂
extreme temperature 极端温度
extreme value 极限值
extruded bead sealing 压出粒涂层法
extrusion 挤压，挤制加工
extrusion die 挤出模
extrusion molding 挤出成形
eye bolt 环首螺栓
eye nuts 环首螺帽
eye screen 护眼屏
eye shield 护眼罩
eyelet wire 带环线
eyeshade 遮光眼罩

F

FA-(factory automation) 工厂自动化
FAB-(fast atom bombardment) 快原子轰击,快速原子轰击
fabrication 制造,建造,装配
fabrication drawing 制造图纸,制作图
fabrication process 制造工艺
face angle 顶锥角,面锥角
face angle distance 顶锥角距
face apex 顶锥顶
face apex beyond crossing point 顶锥顶至相错点距离
face cone 顶锥
face cone element 齿顶圆锥母线,面锥母线
face contact ratio 齿长重合度,轴向重合度,面接触率
face shield 面罩,护面罩
face to face dimension 端面距尺寸
face width 表面宽度,齿面宽
faceplate block function 面板功能块功能
faceplate blocks 面板功能块
face-to-face arrangement 面对面安装
facilities 设施
facilities planning 设施规划
facing decorations 面层装饰
facsimile 传真,复制
facsimile seismograph 传真式地震仪
facsimile telegraph 传真电报
fact base 事实
factor 因数
factor meter 功率因数表
factor of stress concentration 应力集中
factored moment 计算力矩
factorization 因式分解
factorization method 因式分解方法
factory automation (FA) 工厂自动化
factory communication 工厂通信
factory diagnosis and improvement method 工厂诊断与改善方法
factory illumination 工厂照明
factory information protocol 工厂信息协议
factory 工厂,制造厂
fading 退色
Fahrenheit 华氏度
Fahrenheit temperature scale 华氏温标
fail 故障
fail safe 故障保护,失效安全
fail to start shutdowns 起动失败停机
fail tree analysis 失效树分析
fail-safe system 故障保险系统
failure 故障,停机,失败,中断,失效
failure cable 流线型拖缆
failure detection 故障检测
failure diagnosis 失效诊断,故障诊断
failure isolation 故障隔离
failure mechanism 失效机理
failure mode 失效率
failure model effectiveness analysis (FMEA) 失效模式分析
failure monitor 故障检测器
failure recognition 失效识别
failure trouble 故障
failure valve position 断源位置,阀断源位置
fall of potential 电压降
fall out of step 失步
fall time 下降时间
faller wire 下垂线
falling ball impact test 落球冲击试验
falling coaxial cylinder viscometer 同轴圆筒下落黏度计
falling into step 进入同步
falling sphere viscometer 落球黏度计
false ceiling 假天花板
false strain 虚假应变
false 假的,错误的
falsework 脚手架,临时支架
family mold 多腔铸型,集成塑模
family of curves 曲线族
fan blade 扇叶

fan ducting　风扇通风槽
fan gate　扇形浇口
fan heater　风扇式空气加热器，暖风机
fan motor　风扇电动机
fan room　通风机房
fan　风扇，风机
fanlight　楣窗，扇形窗
fanning mill anemometer　叶轮式风速表
fantail die　扇尾形模具
far field　远场
far infra-red radiant element　远红外辐射元件
far infrared radiation　远红外辐射
farad　法［拉］（电容单位）
Faraday cage　法拉第笼，静电屏蔽
Faraday effect　法拉第效应
Faraday electromagnetic induction law　法拉第电磁感应定律
Faraday's law　法拉第定律
far-infrared interferometer　远红外干涉仪
far-infrared spectrophotometer　远红外分光光度计
fashioned iron　型钢
fast　快的，迅速的
fast acting fuse　快速熔断器
fast atom bombardment (FAB)　快原子轰击，快速原子轰击
fast Fourier transforms　快速傅里叶变换
fast Kalman algorithms　快速卡尔曼算法
fast lane　快行车线，快行车道
fast mode　快变模态
fast parallel algorithms　快速并行算法
fast processor　快速处理器
fast recovery diode　快恢复二极管
fast recovery epitaxial diodes　快恢复外延二极管
fast relay　快速继电器
fast switching thyristor　快速晶闸管
fast timing method　快速定时法
fasten　锁紧（螺丝）
fastener　扣件，锁定器，紧固件，系固件，扣件

faster-than-real-time simulation　超实时仿真
fast-setting epoxy adhesive　快凝固环氧胶黏剂
fatigue　疲劳
fatigue breakage　疲劳破裂
fatigue characteristic　疲劳特性
fatigue failure　疲劳破裂，疲劳失效
fatigue life　疲劳寿命
fatigue limit　疲劳极限
fatigue strain gauge　疲劳应变计
fatigue strength　疲劳强度
fatigue testing machine　疲劳试验机
fatique test　疲劳测试，疲劳试验
fatty acid　脂肪酸
fault　故障，缺陷，过失，失灵
fault clearance　故障清除
fault clearing　排除故障
fault clearing time　故障切除时间
fault containment　故障抑制
fault detection　故障检测，探伤
fault detection and isolation (FDI)　故障诊断与分离
fault diagnosis　故障诊断
fault distribution　故障分布
fault identification　故障识别
fault impedance　故障阻抗
fault indication　故障指示灯
fault isolation　故障隔离
fault location　故障定位
fault log　故障记录
fault point　故障点
fault section　故障区间
fault tolerance　耐故障性，容错
fault tolerant　容错
fault tolerant design　容错设计
fault tolerant software　容错软件
fault tolerant system　容错系统
fault tree analysis　故障树分析
faultless　没有缺陷，完美的
faulty switching　开关误操作
faulty operation　误操作
FB (field bus)　现场总线
FBC (frame buffer cache)　帧缓冲缓存器
FBF (fuzzy basis functions)　模糊基函

数
FCS (field bus control system) 现场总线控制系统
FCS status display window FCS 状态显示窗口
FD (field desorption) 场解吸法
FDD (floppy disk drive) 软磁盘机,软盘驱动器
FDI (fault detection and isolation) 故障诊断与分离
feasibility 可行性
feasibility study 可行性研究
feasible coordination 可行协调
feasible region 可行域
feather key 滑键,导向键
feather length 毛圈长
feature change 特性变更
feature detection 特征检测
feature die 公母模
feature extraction 特征抽取
features 特点
fed-batch operation 补料间歇操作
fed-batch processes optimization 补料过程优化
feed 供料,馈,供给
feed cam 进给凸轮
feed control instrument 反馈控制仪表
feed control instrument display block 反馈控制仪表显示块
feed gears 进给齿轮
feed length 送料长度
feed level 送料高度
feed motion 进给运动,合闸动作
feed movement 进给运动
feed pipe 供料管
feed stock 原料
feed system 供电系统
feed through capacitor 旁路电容
feed water check valve 给水止回阀
feed water pump 给水泵
feed water treatment reservoir 给水处理池
feedback 反馈
feedback amplifier 反馈放大器
feedback bellows 反馈波纹管
feedback capacity 反馈能力
feedback channel 反馈通道
feedback circuit 反馈电路
feedback coil 反馈线圈
feedback combining 反馈式组合
feedback compensation 反馈补偿
feedback component 反馈元件
feedback control 反馈控制,反馈调节
feedback control systems 反馈控制系统
feedback controller 反馈控制器,反馈调节器
feedback controller design 反馈控制器设计
feedback element 反馈元件
feedback factor 反馈系数
feedback gain 反馈增益
feedback laser 反馈激光器
feedback linearization 反馈线性化
feedback loop 反馈回路
feedback method 反馈控制方法
feedback modifier 反馈校正器
feedback moment 反馈力矩
feedback path 反馈通道,反馈回路
feedback signal 反馈信号
feedback stabilization 反馈稳定
feedback system 反馈系统
feedback variable 反馈变量,反馈量
feedback voltage 反馈电压
feedback winding 反馈绕组
feeder 送料机
feeder automation 馈线自动化
feeder cable 馈电电缆,馈电线
feeder line 馈线
feeder messenger wire 馈电吊线,馈电悬缆线
feeder panel 馈电盘
feedforward 前馈
feedforward compensation 前馈补偿
feedforward control 前馈控制,前馈调节
feedforward controller design 前馈控制器设计,前馈调节器设计
feedforward network 前馈网络,前向网络
feedforward path 前馈通路
feedforward signal 前馈信号
feedforward-feedback control (FFC)

前馈-反馈控制
feeding 供电，馈电
feeding is not in place 送料不到位
feeler arm 探针臂
feeler gauge 测隙规，厚薄规
felt ring 毡圈
felt ring seal 毡圈密封
felt strip 毡条
FEM (field emission microscope) 场发射显微镜
female die 母模（凹模）
female plug 插座
fencing 围网，铁网，围栏
fender pile 防撞桩，护桩，护舷桩
fender system 防撞系统，保护装置
FEP (fluorinated ethylene propylene) 氟化乙丙烯
FEP (fluoro-ethylene polymer) 氟化乙烯离聚物
fermentation 发酵
fermentation engineering 发酵工程
fermentation processes 发酵过程
fermenter 发酵罐
fermium (Fm) 镄
Ferraris instrument 费拉里感应测试仪器
Ferraris motor 两相感应电动机，费拉里电动机
ferrite 铁素体，铁氧体，纯铁体，肥粒铁
ferrite core 铁氧体磁芯
ferrite-core memory 铁氧体磁芯存储器
ferritic 铁素体的
ferrofluid seal 铁磁流体密封
ferromagnetic 铁磁性的，铁磁的
ferromagnetic material 铁磁物质
ferromagnetism 铁磁性
ferro-resonance 铁磁谐振
ferrous metallurgy 钢铁冶金
ferrule 套圈
ferrule-type 卡套式
fertile element 可转换元素
FES (functional electrical stimulation) 功能电刺激
FET gas sensor 场效应（管）气体传感器

FET gas transducer 场效应（管）气体传感器
FET (field effect transistor) 场效应管，场效应晶体管，场效应管斩波器
FF (foundation fieldbus) 基金会现场总线，FF总线
FF protocol intelligent temperature transmitter FF协议智能温度变送器
FFC (feedforward-feedback control) 前馈-反馈控制
FFC (flexible flat cable) 挠性扁平电缆
FFL (free format log) 自由格式报表
FI (field ionization) 场致电离
fiber communication 光纤通信
fiber optic 光导纤维，光纤
fiber optic clock receiver 光纤时钟接收器
fiber optic clock transmitter 光纤时钟传送器
fiber optic image cable 光纤图像电缆
fiber reinforcement 纤维强化热固性，纤维强化复合材料
fiber-glass braided wire 玻璃丝编织线
fibre amplifier 光纤放大器
fibre conduction velocity 光纤通导速度
fibre connector 光纤连接器
fibre coupler 光纤耦合器
fibre glass 玻璃纤维
fibre interferometer 光纤干涉仪
fibre network 光纤网络
fibre optics 纤维光学
fibre preamplifier 光纤前置放大器
fibre-insulated wire 纤维绝缘线
fibrous iron 纤维断口铁
fibrous plaster 纤维灰泥
FID (flame ionization detector) 火焰离子化检测器
fidelity 保真度
fiducial error 引用误差，基准误差
fiducial value 引用值，基值
field 字段，现场

field mounting 现场安装
field balancing 现场平衡
field balancing equipment 现场平衡设备
field bus (FB) 现场总线
field bus control system (FCS) 现场总线控制系统
field bus module 现场总线模块
field cable area 现场电缆区域
field coil 励磁线圈
field control station 现场控制站
field control system 现场控制系统
field control unit 现场控制单元
field controlled thyristor 现场控制的晶闸管
field controller 磁场控制器
field current 励磁电流
field data 现场数据
field desorption (FD) 场解吸法
field discharge 励磁放电
field effect 场效应,电场效应
field effect transistor (FET) 场效应管,场效应晶体管,场效应管斩波器
field emission 场致发射,自动发射,静电发射,电场发射
field emission electron image 场发射电子像
field emission gun 场发射电子枪
field emission microscope (FEM) 场发射显微镜
field instrument 携带式仪表,野外作业仪器
field intensity meter 场强计
field investigation 实地勘测
field ion emission microscope 场离子发射显微镜,场致发射显微镜
field ionization (FI) 场致电离
field of view 视野
field reliability test 现场可靠性试验
field rheostat 磁场变阻器,励磁变阻器
field signal 现场信号
field stop 视场光阑,场阑
field strength meter 场强计
field surveying 实地测量
field sweeping 场扫描

field termination assemblies 现场端子组件
field test 现场试验
field testing 实地测试,工地测试
field tuning 现场整定
field winding 磁场绕组,励磁绕组
field wire 绕组,被覆线
field-failure protection 磁场失效保护装置
field-frequency lock 场频锁
figured iron 型铁,型钢
filament 单丝
filament image 灯丝像
filar suspended galvanometer 悬丝式检流计
file 文件,文卷,文件箱,锉刀
file builder 文件编制程序,文件编码程序
file compare 比较同步
file dust 锉屑
file folder 资料夹
file maintenance 文卷维护
filing 锉刀修润,锉削加工;文件归档
filings 铁屑,锉屑
fill 填土,填料,填塞,装填
fill factor 占空因数,填充因数
fill in the form 填表式
fill material 填土材料
fill slope 填土斜坡
fill valve 灌注阀
filled system thermometer 压力式温度计
filled thermal system 充灌式感温系统
filler 垫板,填料,充填剂
filler rod 焊条
fillet 镶,嵌边,齿根圆角
fillet curve 齿根过渡曲线
fillet radius 圆角半径,齿根圆角半径
fillet weld 角焊接
fillet welding 角焊
filling core 埋入砂芯
filling in 填砂
film adhesive 胶膜,胶纸,黏结膜
film blowing 薄膜吹制法

film gate 薄膜浇口
film recording thermograph 照相温度(表)
film strength 液膜强度
film type bolometer 薄膜式测辐射热计
film varistor 膜式压敏电阻器
film viewer 底片观察用光源
filter 滤光计,过滤器,滤波器,滤光板,滤线板,过滤层,滤管,滤器,隔网,滤网
filter amplifier 滤波放大器
filter bank 系列滤波器
filter capacitor 滤波电容,滤波电容器
filter cell 滤波器元件
filter characteristic 滤波器特性
filter circuit 滤波器电路
filter design 滤波器设计
filter device 滤波装置
filter element 过滤器滤芯
filter stability 滤波器稳定性
filter stop band 滤波器阻带,滤光器不透明带
filter transmission band 滤波器通带,滤光器透射带
filter differential pressure 滤网压差
filtered electron image 过滤电子像
filtered electron lens 过滤电子透镜
filtering 滤波
filtering problem 滤波问题
filtering technique 滤波技术
filtering theory 滤波理论
final 最后的,确定的
final amplifier 末级放大器
final audit 期末审计
final contact 终止啮合点
final control element 最终控制元件
final controlling element 终端控制元件,执行器
final inspection 终检
final mold design 正式模图设计
final percent output sent to control element 最后的百分比输出送到调节机构
final quality control (FQC) 终点品质管制人员
final state 终态

final temperature 终了温度,终止温度
final value 终值
final value theorem 终值原理
financial system 金融系统
fine aggregate 细集料,幼骨料,细骨料
fine blanking 精密冲裁,精密下料加工
fine blanking press 精密下料冲床
fine clay 幼黏土,细黏土
fine dust 细粉尘
fine screening 精筛
fine steel casting iron 优质碳素钢铸件
fine threads 细牙螺纹
fine turning resistance box 微调电阻箱
Fineman nephoscope 法因曼测云器
finger guard 指形护罩
finger pin 指形销
finish 饰面,终饰,抛光,修整
finish blanking 光制下料加工,刃口冲裁模
finish machining 精加工
finish paint 面漆
finished products 成品
finisher 精切机床,修整机,整面机
finishing 精整加工
finishing allowance 加工余量
finishing coat 面漆,最后一道涂工
finishing machines 整面机,研磨机,精整加工线,精加工机床
finishing slag 炼后熔渣
finishing works 终饰工程
finite 有限的
finite analysis 有限元分析
finite arc segment 有限线段
finite automata 有限自动化
finite automaton 有限自动机
finite computation 有限元计算
finite difference 有限差分
finite element 有限元,有限元件
finite field 有限场,有限域
finite field simulation 有限元场仿真,有限场模拟

finite impulse response (FIR) 有限脉冲响应
finite state machine 有限状态机
finite step response (FSR) 有限阶跃响应
FIR (finite impulse response) 有限脉冲响应
fire alarm system 火灾报警系统
fire barriers 防火间隔
fire behaviour 着火性能，火势
fire console 消防控制台
fire detection 火灾探测，火警探测，火警侦察器
fire escape 走火通道
fire extinguisher 灭火筒，灭火器
fire fighting engine 消防车
fire fighting equipment 消防设备
fire fighting system 消防系统
fire hydrant 消防街井，消防龙头
fire integrity 整体着火性
fire main 消防水管
fire proof concrete 耐火混凝土
fire proof machine 防爆式电机
fire protection equipment 防火设备
fire public device 消防设施
fire pump 消防泵，消防水泵
fire pump room 消防泵房
fire resistance 耐火性
fire resistant 耐高温的
fire resistant material 耐火材料
fire resisting construction 耐火结构
fire resisting shield 防火障板，防火挡板
fire roller shutter 消防卷闸
fire safety gas valve 消防气阀
fire services equipment 消防设备
fire services inspection 消防设备检查
fire services installation 消防装置
fire stability 对火稳定性
fire station 消防局，消防站，消防队
fire wall 防火墙
fire wire 火线，带电线
fire 燃烧，火焰
fire-extinguisher 灭火器
fireman's emergency switch 消防员紧急开关

fireman's lift 消防员升降机
fireproof 防火的，阻燃的
fire-resistant wire 耐火导线
fire-retarding glazing 耐火玻璃
firmware 固件，固体
first aid box 急救箱
first aider 急救人员
first article assurance 首件确认
first article inspection 新品首件检查
first harmonic 基波，一次谐波
first order lag unit 一阶滞后器
first order lead unit 一阶微分器
first order predicate logic 一阶谓词逻辑
first order spectrum 一级光谱
first principles model 第一机理模型
first stage annealing 第一段退火
first-order circuit 一阶电路
first-order hold element 一阶保持元件
first-order lag 一阶滞后
first-order lag system 一阶滞后系统
first-order lead 一阶超前
first-order plus time delay 一阶加纯滞后
first-order system 一阶系统
fish eye 鱼眼
fish finder 鱼探仪
fishing wire 牵引线
fishtail 鱼尾形
fishtail die 鱼尾形模具
fission 裂变
fission products 裂变产物
fissionable material 裂变物质
fissure 裂纹
fit together 组装在一起
fit tolerance 配合公差
fitting 管件
fitting model parameter 模型参数拟合
fitting out works 装修工程
fittings 设备
five-component borehole magnetometer 井中五分量磁测井仪
five-hole pneumatic socket 五孔气插座
fix 固定
fixed bearing 固定支座
fixed bias 固定偏压

fixed bolster plate　　固定侧模板
fixed capacitor　　固定电容器
fixed capacity　　固定电容
fixed coil　　固定线圈
fixed command control　　固定命令控制
fixed contact　　固定触点
fixed core　　定铁芯
fixed error　　固定误差
fixed format log　　固定格式报表
fixed frequency　　固定频率
fixed instrument　　固定式仪表
fixed level　　固定水平
fixed link　　固定构件
fixed measuring instrument　　固定式测量仪表
fixed point　　固定点
fixed points method of calibration　　定点法标定
fixed program computer　　固定程序计算机
fixed pump installation　　固定水泵装置
fixed resistance　　固定电阻
fixed resistance input type volt ratio box　　定阻输入式分压箱
fixed resistance output type volt ratio box　　定阻输出式分压箱
fixed restrictor　　恒节流孔
fixed sequence manipulator　　固定顺序机械手
fixed series capacitor compensation　　固定串联电容补偿
fixed set point control　　定值控制，定值控制系统
fixed set-point control system　　恒值控制系统
fixed setting　　固定安装法
fixed size pulse　　脉冲发生器
fixed storage　　固定存储器
fixed time lag　　固定时间间隔
fixed voltage　　固定电压
fixed Winchester disk driver　　固定的温彻斯特磁盘驱动器
fixed　　固定的，固定，确定，保护屏
fixed-based natural frequency　　固定基础固有频率
fixed-coil indicator　　定圈式指示器

fixing bolt　　固定螺栓
fixture　　夹具
fixtures　　固定，固定附物，固定装置
flag　　标记
flag data point　　特征数据点，标记数据点
flag point　　旗标点
flag variable　　标志变量
flagman　　旗号员，信号旗手
flame cutting　　火焰切割
flame detector　　火焰探测器
flame emission spectrometry　　火焰发射光谱法
flame failure alarm　　火焰警报器
flame failure protection　　熄火保险
flame hardening　　火焰硬化
flame ionization detection　　火焰离子化检测
flame ionization detector (FID)　　火焰离子化检测器
flame monitor　　火焰监测器
flame photometric detection　　火焰光度检测
flame photometric detector (FPD)　　火焰光度检测器
flame proof enclosure　　隔爆外壳
flame propagation　　火焰传播
flame temperature detector　　火焰温度检测器
flame thermionic detection　　火焰热离子检测
flame treatment　　火焰处理
flame-proof electrical fittings　　防火电气配件
flameproof enclosure　　隔爆外壳，防火外壳
flammable liquid　　易燃液体
flammable products　　易燃产品
flammable vapour　　易燃蒸气
flange　　法兰，凸缘
flange ball valves　　法兰球阀
flange bush　　凸缘套筒，法兰套筒
flange coupling　　凸缘喉套，凸缘联轴器，法兰式联轴节
flange gate valves　　法兰闸阀
flange globe valves　　法兰截止阀
flange joint　　凸缘接头，凸缘接驳位，

法兰接头
flange nylon insert lock nuts 凸缘尼龙盖帽
flange pressure tappings 法兰取压口
flange wrinkle 凸缘起皱
flanged diaphragm sealed manometer 法兰膜片隔离式压力表
flanged ends 法兰连接端
flanged pipe 凸缘管
flangeless ends 无法兰连接端
flangeless valve 无法兰阀
flanging 凸缘加工
flank 侧面,侧翼,下齿面
flanking 下齿面加工
flapper 挡板
flapper valve 片状阀,铰链阀
flash back arrester 回火制止器
flash butt weld 闪光焊
flash butt welding machine 闪光对接焊机,闪光电焊机
flash light 闪光灯
flash mold 溢流式模具,溢料式模具
flash plate 闪熔镀层,锚链垫板
flash 闪光,闪烁,闪蒸
flash lamp 闪光灯
flasher relay 闪弧继电器,闪光继电器,闪灯继电器
flashing 闪,防水盖片
flashing annunciator 闪光信号报警器
flashing beacon 闪动标灯
flashing lantern 闪动灯具
flashing light 闪灯,闪光,闪光信号
flashover 闪络,飞弧,跳火,抢火
flask 烧瓶,上箱
flask molding 有箱造模法
flat belt 平带
flat belt driving 平带传动
flat bus bar 扁母线,汇流条
flat cable 扁平电缆,带状电缆,排线
flat file 扁锉
flat gate valves 平板闸阀
flat hoop iron 平箍钢
flat leaf spring 板簧

flat pair 平面副
flat steel 扁钢
flat throttling valve 平面节流阀
flat wire 扁钢丝
flat-face follower 平底从动件
flattened wire 压扁丝
flaw 刮伤
flaw echo 缺陷回波,缺陷的回波信号,缺陷回声
flaw resolution 缺陷分辨力
flaw sensitivity 缺陷灵敏度
flex-hone 软磨石
flexible 柔性
flexible arm 柔性臂
flexible automation 柔性自动化
flexible cable 软电缆
flexible carriageway 柔性行车道
flexible conduit 软管
flexible copper-clad dielectric film 挠性覆铜箔绝缘薄膜
flexible cord 软电线,软线
flexible coupling 弹性联轴器
flexible disc 弹性闸板
flexible disk 软磁盘
flexible disk driver 软盘驱动器
flexible double-sided printed board 挠性双面印制板
flexible flat cable (FFC) 挠性扁平电缆
flexible frame 弹性构架,柔性构架
flexible gear 柔轮
flexible hose 软管
flexible impulse 柔性冲击
flexible joint 弹性接缝,柔性接缝
flexible manufacturing system (FMS) 柔性制造系统
flexible multilayer printed board 挠性多层印制板
flexible pavement 柔性路面
flexible pipe 软管
flexible printed board 挠性印制板
flexible printed circuit (FPC) 挠性印制电路
flexible printed wiring 挠性印制线路
flexible rigidity 弯曲刚也
flexible rotor 挠性转子,柔性转子
flexible shaft 软轴
flexible shaft wire 软轴用钢丝

flexible single-sided printed board 挠性单面印制板
flexible stranded wire 软性绞合线
flexible waveguide 可弯曲波导管，柔性波导管
flexible wire 软线，花线
flexural critical speed 挠曲临界转速
flexural principl mode 挠曲主振型
flexural properties 抗弯性能
flexural strength 抗弯强度，抗挠强度
flexural stress 弯曲应力，挠曲应力
flexure 簧片
flicker 闪烁，闪光，脉动器
flicker relay 闪光继电器
flight control 飞行控制
flight vehicle design 飞行器设计
flint glazing machine 磨光机
flint gun 火石枪，火石点火器
flip and flop generator 双稳态触发器
flip-chip 后滚翻
flip-flop 触发器，双稳态触发器，复振器，正反器
flip-flop circuit 触发电路，双稳态多谐振荡电路
float 浮标，浮子
float adjusting valve 浮球调节阀
float and cable level measuring device 浮标和缆索式液位测量装置，浮标和缆索式物位测量仪表
float barograph 浮子气压计
float flowmeter 浮子流量计，盘塞式流量计
float level measuring device 浮子液位测量，浮标式液位测量仪表
float level regulator 浮子型液位调节
float operated valve 浮球阀，浮动阀
float switch 浮控开关，浮子开关
float tide gauge 浮子式验潮仪
float-actuated recording liquid-level instrument 浮球式液位记录仪
float-area-type flowmeter 浮子面积式流量计
float-charge 浮充电
floating 浮点，浮动
floating accelerometer 重力式测波仪
floating action 无定位作用

floating ball 浮球
floating control 无差调节，浮点控制
floating control action 无差调节，浮点控制
floating controller 无定位控制器
floating cranes 浮式起重机
floating disc 浮环，浮动盘
floating input 浮空输入，浮置输入
floating output 浮空输出，浮置输出
floating platen 活动模板
floating point 浮点
floating point computation 浮点计算
floating punch 浮动冲头
floating speed 无定位速度
float-type differential pressure recorder 浮子式差压记录仪
flood level 溢流水位
flooding 水浸
floodlight 泛光灯
floodlight mast 泛光灯柱
floor iron 底铁
floor load 楼面荷载
floor model 落地式
floor structure 基底结构
floppy 磁碟片
floppy disk 软盘
floppy disk drive（FDD） 软磁盘机，软盘驱动器
flow board 流水板
flow characteristic 流量特性
flow characteristic curve 流量特性曲线
flow characteristic liner 流量特性直线
flow chart 流程图，信号流图，流程图表，流程表单
flow coefficient 流量系数
flow compensation algorithm 流量补偿算法
flow conditioner 流动调整器
flow control 流量控制
flow control device 流量控制
flow control valve 流量控制阀，流量调节阀
flow corrector 流量修正器，流量控制器
flow cytometer 流式细胞仪

flow diagram 流程图，信号流图
flow elbow 流量弯管
flow heterogeneity 多相流
flow indicator controller 流量指示控制器
flow injection analysis 流动注射分析
flow instrument 流量计
flow integrator 流量积算仪，流量积分仪
flow mark 流痕
flow measurement 流量测量
flow measurement calibration 流量测量校准
flow measuring device 流量测量仪器
flow meter 流量计，流速计
flow monitoring 流量监测
flow nozzle 流量喷嘴
flow profile 流动剖面
flow rate 流量，瞬时流量
flow rate indicator 流量指示仪
flow rate of a fluid through a cross section of a conduit 流经管道横截面的流体流量
flow rate of mobile phase 流动相流速
flow ratio 流量比
flow sensor 流量传感器
flow sheet 流程图
flow signal 流量信号
flow stabilizer 流量稳定器
flow straightener 流动整直器
flow switch 流量开关，气流换向器
flow to close 流关
flow to open 流开
flow transducer 流量传感器
flow transmitter 流量变送器
flow 流量，流动
flowchart 工艺流程图
flowmeter 流量计
flowmeter primary device 流量计一次仪表
flowmeter secondary device 流量计二次仪表
flow-rate 流量
flow-rate range 流量范围
flow-up control 随动控制
fluctuating circulating stress 脉动循环应力

fluctuating load 变动负载，脉动载荷
fluctuating power 波动功率
fluctuation 波动，不稳定，涨落
fluctuation of load 负载波动
fluctuation of speed 转速波动
flue 废气道，烟道，通气道
flue gas desulphurization 废气脱硫
flue pipe 烟道排气管，烟筒
flue gas 烟气
fluid 流体，流动的，液体
fluid clutch 液压离合器，液压驱动泵，液压联轴节
fluid damping galvanometer 液体阻尼振动子
fluid drive mechanism 液压传动机构
fluid dynamics 流体动力学
fluid friction 流体摩擦
fluid level controller 液位控制器
fluid level measuring instruments 液位测量仪表
fluid machinery and engineering 流体机械及工程
fluid mechanics 流体力学
fluidic flowmeter 射流流量计
fluorescence 荧光，荧光修正
fluorescence detection 荧光检测
fluorescence detector 荧光检测器
fluorescence effect 荧光效应
fluorescence efficiency 荧光效率
fluorescence life time 荧光寿命
fluorescence polarization immunoassay 荧光偏振免疫测定法
fluorescence quantum yield 荧光量子产率
fluorescence quenching method 荧光熄灭法
fluorescence spectrum 荧光光谱
fluorescence thin layer plate 荧光薄层板
fluorescent finish paint 荧光面漆
fluorescent image 荧光像
fluorescent lamp tube 荧光灯管，光管
fluorescent light tube 荧光灯管，光管
fluorescent magnetic particle inspection

machine 荧光磁粉探伤机
fluorescent penetrant testing method 荧光渗透探伤法
fluorescent screen 荧光屏
fluorinated ethylene propylene (FEP) 氟化乙丙烯
fluorinated ethylene-propylene copolymer film 聚全氟乙烯丙烯薄膜
fluorine (F) 氟
fluorine plastic 氟塑料
fluorine rubber 氟橡胶
fluoro-ethylene polymer (FEP) 氟化乙烯离聚物
fluorometer 荧光计
fluorometry 荧光分析法
fluorous rubber 氟橡胶
flurescent magnetic particle 荧光磁粉
flush 冲洗
flush conductor 齐平导线
flush mounted gauge 嵌装表
flush mounted pressure gauge 嵌装压力表
flush printed board 齐平印制板
flush switch 嵌入开关
flush water booster pump 冲水增压泵
flush water pump 冲水泵,冲洗水泵
flushing pipe 冲厕喉管
flushing water 冲洗水,冲厕水,冲厕用水
flush-type instrument 嵌入式仪表
fluted bar iron 凹面方钢
flutter 颤振
flux 流量,焊剂,磁通,通量,磁力线
flux constant 磁通常数
flux density 通量密度
flux distribution 磁通分布
flux guide 磁通量控制器
flux linkage 磁链,磁通链
flux meter 磁通表
flux of radiation 辐射通量
flux space vector 通量空间矢量
fluxgate compass 磁通门罗盘
fluxgate magnetometer 磁通门磁力仪,磁通门磁强计
fluxmeter 磁通表,磁通计
fluxmeter calibrator 磁通表校验仪

fluxoid 全磁通
flyback converter 反激电流
flyback transformer 反馈变压器
flywheel 飞轮
flywheel brake 飞轮制动器
FM (frequency modulation) 调频,频率调制
FM exciter 调频激励机
FM receiver 调频接收机
FM wave 调频波
FM-AM multiplier 调频-调幅倍增器
FMEA (failure model effectiveness analysis) 失效模式分析
FMS (flexible manufacturing system) 柔性制造系统
FNN (fuzzy neural network) 模糊神经网络
foam concrete 泡沫混凝土
foaming agent 发泡剂
focal distance 焦距
focal plane 焦平面
focus 焦点
focus size 焦点尺寸
focusing 聚焦
focusing regulator 聚焦调整器
focusing type probe 聚焦探头
focus-to-film distance 焦距
fog-gauge 雾量器
foil profile 箔剖面轮廓
foil removal surface 去铜箔面
foil strain gauge 箔式应变计
fold of packaging belt 打包带褶皱
folded block 折弯块
folding 折叠加工,折边弯曲加工
folding bar 折铁器
folding chart 折叠式记录纸
folding paper 折叠式记录纸
folding rule 折尺
follow 跟踪,追随
follow-control 随动调节,随动控制
follower 跟随器,输出器,随动机构
follower dwell 从动件停歇
follower motion 从动件运动规律
follow-up 随动,跟踪
follow-up control 随动调节,随动控制
follow-up control system 跟随控制系统
follow-up device 随动器,跟踪器

follow-up pointer 从动针
food analysis instruments 食品分析仪器
food analyzer 食品分析仪
food processing 食品工业
food science 食品科学
food science and engineering 食品科学与工程
foot press 脚踏冲床
foot pump 脚踏泵
foot rest 踏脚板,脚踏
foot switch 踏脚
footing 基脚,底脚,墙基
forbid 禁止
force 力,力度,强制
force balance 力平衡
force balance proportional controller 力平衡式比例调节器
force control 力控制
force convection 强制对流
force measuring instrument 力测量仪表
force oscillation 强迫振荡
force polygon 力多边形
force sensor 力传感器
force standard machine 力标准机
force transducer 力传感器
force-balance acceleration transducer 力平衡式加速度传感器
force-balance accelerometer 力平衡式加速度计
force-balance principle 力平衡原理
force-balance transmitter 力平衡变送器
force-bar 主杠杆
force-closed cam mechanism 力封闭型凸轮机构
forced commutation 强迫换流
forced convection constant temperature drying ovens 送风定温干燥器
forced convection constant temperature ovens 送风定温恒温器
forced convection constant temperature ovens with air curtain 空气幕送风定温恒温器
forced draft 压力气流
forced draught fan 压风机,压力抽风机,鼓风式风扇,送风机
forced oscillation 强制振荡
forced vibration 强迫振动,受迫振动

force-drive cam mechanism 力封闭型凸轮机构
forcing function 强迫函数,强制函数
forecast 预测报告
forefinger 食指
foreground 前台
foreground processing 前台处理
foreground program 前台程序
foreign exchange 外汇
foreign matter 异物,杂质
forestry engineering 林业工程
forge 锻造
forging aluminium 锻铝
forging die steel 锻造模用钢
forging dies 锻模
forging machine 锻造机
forging roll 轧锻,辊锻机
forklift 叉车
fork-lift truck 叉式起重车,铲车
form block 折刀
form cutting 仿形法
form factor 齿形,波形系数,波形因数
form grinding machine 成形磨床
form of structure 结构形式
form tool 成形刀
form 形式,形状,形成,构成
formal 正式的,合法的
formal insulated wire 聚乙烯绝缘线
formal language 正规语言
formal language theory 形式语言理论
formal method 正规方法
formal neuron 形式神经元
formal specification 正式说明规格
formal verification 合法证明
formaldehyde 甲醛
formale copper wire 聚乙烯铜线
format 形式,格式,格式化
formation 平整,开拓
formation vibration 变形振动
former 齿廓样板,靠模
forming 成型,成形加工,在磨具内挤压成型
forming die 成型模
formwork 模板,模架
Forster bridge 福斯特电桥
Fortin barometer 福丁气压表
forward 向前,向前的,正向的

forward channel 正向信道，正向通道，单向通道
forward construction 正向设置
forward control 前向控制
forward controlling element 正向主控元件
forward converter 正向变流器
forward current 正向电流
forward difference 前向差分
forward element 正向环节，正向元件
forward kinematics 正向运动学
forward path 前向通路，正向通路
forward reasoning 正向推理
forward resistance 正向电阻
forward signal 前向信号
forward transfer function 正向传递函数
forward voltage 正向电压
forward-travelling waves 向前传播的波
foucus 聚焦
foucusing 聚焦
foucusing regulator 聚焦调整器
foul air 浊气
foul water 污水，脏水
foul water pump 污水泵
found 铸造
foundation 基础，地基，底座，根据
foundation fieldbus (FF) 基金会现场总线，FF总线
foundation grouting technique 基础灌浆技术
foundation plan 基础图则
foundry iron 铸造生铁
foundry equipment 铸造设备
four core cable 四芯电缆
four terminal network 四端口网络
four-bar linkage 四连杆机构
four-core wire 四芯线
fourdrinier wire 长网线
Fourier 傅里叶
Fourier analysis 傅里叶分析
Fourier analyzer 傅里叶分析仪，频谱仪
Fourier coefficients 傅里叶系数
Fourier expansion 傅里叶展开式
Fourier optics 傅里叶镜片

Fourier transform 傅里叶变换
Fourier transform infrared spectrometry 傅里叶变换红外光谱法
Fourier transform near infrared spectrometer 傅里叶变换近红外分光仪
Fourier transform spectrometer 傅里叶变换光谱仪
Fourier's series 傅里叶级数
four-layer 四层
four-terminal device 四端器件
four-terminal standard resistor 四端（钮）标准电阻器
fourth-generation language 第四代语言
four-wheel 四轮
four-wheel drive 四轮驱动
four-wheel steering 四轮操纵
FPC (flexible printed circuit) 挠性印制电路
FPD (flame photometric detector) 火焰光度检测器
FQC (final quality control) 终点品质管制人员
fraction 分数，百分率，部分
fraction collector 馏分收集器，部分收集器
fractional harmonic 部分谐波
fractional-horsepower motor 分马力电动机
fractionation 分馏
fracture 裂面，断裂
fracture toughness 断裂韧性
fragile 易碎
fragment ion 碎片离子
fragmentation 碎裂过程
fragmentation pattern 碎裂图形
fraising 绞孔
frame 帧，框架，构架，支架，床身机架
frame buffer 帧缓冲器
frame buffer cache (FBC) 帧缓冲缓存器
frame construction 框架结构
frame fixed link 机架
frame stiffness 框架劲度
frame synchronization 帧同步
frame type instrument panel 框架式仪表盘

frame-check sequence 机架检查程序
framed joint 构架接合,构架榫
framework 框架,构架
francium (Fr) 钫
free 游离
free charge 自由电荷
free cutting steel wire 易切削钢丝
free electron 自由电子
free field 自由声场
free field correction curves 自由场修正曲线
free field reciprocity calibration 自由声场互易校准
free float type steam trap 浮球式疏水阀
free flow operating speed 无阻行车速度
free flow through route 无阻直通路线
free format log (FFL) 自由格式报表
free fuse breaker 无熔丝断路器
free hand cutting 徒手切割
free induction decay signal 自由感应衰减信号
free iron 游离铁
free motion 自由运动
free oscillating period 自由振动周期
free oscillation 自由振荡
free signal flow between control drawing 控制图之间的自由信号流
free standing 独立式
free swing of pendulum 摆锤空击
free vehicle respirometer 活动式海底生物呼吸测量器
free vibration 自由振动
freedom from bias of measurement 无偏置测量
free-field frequency response of microphone 传声器自由声场频率响应
free-field sensitivity of microphone 传声器自由声场灵敏度
freeweight 负重
freeze dryers 冻结干燥器
freeze drying equipment 冻干机
freezing heat 凝固热
freezing point 凝固点
freight handling 货物处理
freight marshalling facilities 货运调车设施
freight yard 货运场
french chalk 滑石粉
freon 氟利昂,氟氯烷
frequency 频率
frequency analysis 频谱分析,频率分析
frequency analyzer 频率分析仪
frequency band 频带
frequency calibration 频率校准
frequency change-over switch 频率转换开关
frequency changer 变频器
frequency channel 频道
frequency characteristic 频率特性
frequency compensator 频率补偿器
frequency component 频率成分
frequency control 频率控制
frequency control of motor speed 变频调速
frequency conversion 频率变换
frequency converters 变频器
frequency convertor 变频机
frequency correction 频率校正
frequency demultiplier 分频器
frequency difference 频差
frequency discrimination 鉴频
frequency dispersion 频率偏移,频率漂移
frequency distortion 频率失真
frequency distrbution 频率分布
frequency divider 分频器,分频仪
frequency division 分频
frequency division multiple access 频分多址
frequency division multiplexing 频分多路传输
frequency domain 频域
frequency domain analysis 频域分析
frequency domain method 频域法
frequency domain model reduction method 频域模型降阶法
frequency doubler 倍频器
frequency estimation 频率估计
frequency index 频率指数
frequency locus 频率轨迹
frequency measurement 频率测量
frequency measurement by comparison

with time scale 时标比较法测频
frequency measurement by digital meter 用数字频率计测频
frequency measurement by Lissajou's figure 用利萨如图形测频
frequency measurement by stroboscope 闪光测频
frequency meter 测频计，频率计，频率表
frequency modulation (FM) 调频，频率调制
frequency modulation inspect 调频测试
frequency modulator 调频器
frequency of the natural hydraulic mode 液压固有频率
frequency of vibration 振动频率
frequency output 频率输出
frequency range 频率范围
frequency regulation 频率调整
frequency response 频率响应
frequency response characteristic 频率响应特性
frequency response curve 频率响应曲线
frequency response locus 频率响应轨迹图
frequency response method 频率响应法
frequency response of microphone 传声器频率响应
frequency response range 频率响应范围
frequency response tracer 频率响应显示仪
frequency selection 频率选择
frequency shift keying (FSK) 频移键控
frequency shift magnetometer 频移磁强计
frequency signal analysis 频率信号分析
frequency sounding instrument 频率测深仪
frequency spectra induced polarization instrument 频谱激电仪
frequency spectrograph 频谱仪
frequency spectrum 频谱
frequency stability 频率稳定性，频率稳定度

frequency stabilization 频域稳定，稳频，频率稳定
frequency stabilizer 稳频器
frequency standard 频率标准
frequency sweeping 频率扫描
frequency tracking 频率跟踪
frequency transducer with integrator 频率转换积算器，频率转换积分器
frequency transformation 频率变换
frequency transformer 变频器
frequency-dependent 频率相关
frequency-dependent characteristic 频率相关特性
frequency-domain analysis 频域分析
frequency-domain design 频域设计
frequency-modulation broad-casting 调频广播
frequency-phase characteristic 相频特性
frequency-response method 频率响应方法
frequency-temperature coefficient 频率-温度系数
fresh air supply fan 鲜风供应风扇
fresh concrete 新浇混凝土
fresh iron 初熔铁
fresh water 淡水，食水
fresh water cooling tower 淡水冷却水塔
fresh water main 食水管
fresh water pipe 食水管
fresh water pump 食水抽水机，淡水泵，清水泵
fresh water pumping station 食水抽水站
Fresnel diffraction string 菲涅耳衍射条纹
friction 摩擦，摩阻，摩擦力
friction angle 摩擦角
friction brake 摩擦刹车，摩擦制动器
friction circle 摩擦圆
friction coefficient 摩擦系数
friction course 防滑面层
friction disc 摩擦圆碟
friction error 摩擦误差
friction force 摩擦力
friction glazing 摩擦轧光
friction load 摩擦负荷
friction moment 摩擦力矩

friction pile 摩擦桩
friction velocity 摩擦速度
frictional error 摩擦误差
frictional force 摩擦力
frictional resistance 摩擦抗力,摩擦阻力
from help menu of builder 来自系统生成菜单的帮助
from windows start menu 来自窗口启动菜单
front angle 前角
front cone 前锥
front crown 前锥齿冠
front crown to crossing point 前锥齿冠至交错点
front end 前端,高频端,调谐器
front end processor 前端处理机
front plate 前板
front side bus (FSB) 前置总线,即外部总线
front view 正面图
frontal chromatography 迎头色谱法,前端分析
frost point hygrometer 霜点湿度计(表)
FSB (front side bus) 前置总线,即外部总线
FSK (frequency shift keying) 频移键控
FSP (full-screen processing) 全屏幕处理
FSR (finite step response) 有限阶跃响应
FSSS (furnace safety supervision system) 锅炉炉膛安全监控系统,炉膛监控系统
FT-IR 傅里叶变换红外光谱学
fuel 燃油,燃料
fuel battery 燃料电池
fuel cell 燃料电池
fuel consumption 耗油量,燃料消耗量
fuel control 燃料控制
fuel control injection 燃料喷射控制
fuel dispensing pump 燃油分配泵
fuel emergency cut-off lever 燃油紧急截断杆
fuel filter 燃油过滤器,滤油器
fuel filter element 滤油器滤芯
fuel gauge 燃油表

fuel injection 燃油喷射
fuel injection pump 燃油喷射泵
fuel injector 燃油喷射器
fuel intake device 燃油吸入装置
fuel level 油位
fuel level control switch 燃油液位控制开关
fuel lift pump 升油泵
fuel line 燃油管道
fuel meter 油表
fuel pipe 燃油管道
fuel pump 燃油泵
fuel safety system 燃料安全系统
fuel supply system 燃油供应系统
fuel supply tap 燃油供应
fuel tank 油箱,油缸
fuel transfer control system 燃油运送控制系统
fuel transfer pump 燃油输送泵
fuel valve 燃油阀
fuel-air ratio control 燃-空比控制
fulcrum 支点,支轴
fulcrum pin 支轴销
full 充满,完全
full annealing 完全退火
full automatic 全自动的
full balance of shaking force 惯性力完全平衡
full bridge converter 全桥电路
full bridge measurement 全桥测量
full bridge rectifier 全桥整流电路
full capacity trim 全容量内件
full directional interchange 全面定向道路交汇处
full graphic display 全图形显示器
full load 满载
full load test 最高载重测试,满载测试
full order observer 全阶观测器
full penetration 完全焊透
full point form 全点
full scale 满量程
full scale flow rate 满标度流量
full sewage treatment plant 全套程序污水处理厂
full span 全跨度
full wave 全波
full wave analysis 全波分析

full wave discontinuity 不连续波
full wave rectifier 全波整流器
full welding 完全焊透
full speed 额定频率
full-depth teeth 全齿高齿
full-featured batch control package 全特色批量控制软件包
full-load current 满载电流
full-load test 满载试验
full-load torque 满载转矩
full-screen editing 全屏幕编辑
full-screen mode 全屏模式
full-screen processing (FSP) 全屏幕处理
full-subtracter 全减器
full-wave logger 声波全波列测井仪
full-wave rectification 全波整流
fully automatic operation 全自动操作
fully enclosed motor 全封闭式电动机
fully insulated current transformer 全绝缘电流互感器
fully rough turbulent flow 充分混杂湍流（紊流）
fully 充分的，完全的
fully-galvanized wire 全镀锌钢丝
fume 烟气，烟雾，烟，冒烟
fume hoods 通风柜
function 函数，功能
function access level 操作级别
function analyses design 功能分析设计
function analysis 功能分析
function approximation 功能近似，函数近似
function block 功能块
function block detail definition 功能块细目定义
function card 功能卡，功能件
function class module 功能级模块
function fitter 折线函数发生器
function generation 函数生成
function generator 函数发生器
function key 功能键
function key definition 功能件定义
function module 功能模块
function relation 函数关系
function switch 函数开关
function type optic-fibre temperature transducer 功能型光纤温度传感器

functional 功能的，函数的
functional adjustment 功能整定
functional block 功能块，功能框，功能器件
functional chain 功能链
functional decomposition 功能分解
functional electrical stimulation (FES) 功能电刺激
functional element 执行元件
functional insulation 功能绝缘
functional lift 复合抬举机
functional similarity 功能相似
functional simulation 功能仿真
functional specification 功能规格
fundamental 基本的，固有的
fundamental constant 固有常数
fundamental current 基波电流
fundamental factor 基波因数
fundamental frequency 基本频率
fundamental harmonic 基波
fundamental matrix 固有矩阵
fundamental mechanism 基础机构
fundamental method of measurement 基础测量法
fundamental natural mode of vibration 基本固有振型
fundamental period 基本周期
fundamental process 基本过程
fundamental relation 固有关系
fundamental wave 基波
funnel-shaped mud viscometer 漏斗式泥浆黏度计
furnace 熔炉
furnace bottom ash 炉底灰
furnace for reproduction of fixed points 定点炉
furnace for verification use 检定炉
furnace safety supervision system (FSSS) 锅炉炉膛安全监控系统，炉膛监控系统
fuse 熔断器，保险丝，熔线，熔丝，导火线
fuse link 熔断片，熔线
fuse machine 热熔机
fuse together 熔合
fuse wire 熔丝，保险丝
fuse holder 保险盒

fused silica open tubular column 熔融石英开管柱
fuselage truss wire 机身构架拉线
fuse-switch 熔丝开关，熔线开关
fusible cutout 熔断器
fusible link 保险连杆
fusible plug for gas cylinders 气瓶用易熔合金塞
fusing 熔断，烧熔
fusing current 熔断电流
fusing time 熔断时间
fusion 熔化，聚变
fusion melting 熔解
fusion welding 熔焊，熔焊接
fuzzification 模糊化
fuzziness 模糊
fuzzy 模糊的
fuzzy adaptive filter 模糊自适应滤波
fuzzy basis functions (FBF) 模糊基函数
fuzzy control 模糊控制
fuzzy controller 模糊控制器
fuzzy data 模糊数据
fuzzy decision 模糊决策
fuzzy evaluation 模糊评价
fuzzy expert system 模糊专家系统
fuzzy game 模糊对策
fuzzy hybrid system 模糊混杂系统
fuzzy identifier 模糊辨识器
fuzzy inference 模糊推理
fuzzy information 模糊信息
fuzzy input 模糊输入
fuzzy logic 模糊逻辑
fuzzy logic controller 模糊逻辑控制器
fuzzy logic system 模糊逻辑系统
fuzzy model 模糊模型
fuzzy modelling 模糊模型化
fuzzy neural network (FNN) 模糊神经网络
fuzzy output 模糊输出
fuzzy reasoning 模糊推理
fuzzy reasoning matrix 模糊推理矩阵
fuzzy relation 模糊关系
fuzzy rule 模糊规则
fuzzy rule base 模糊规则库
fuzzy sensor 模糊传感器
fuzzy set 模糊集
fuzzy subset 模糊子集
fuzzy supervision 模糊监控
fuzzy system 模糊系统
fuzzy theory and application 模糊理论与应用
fuzzy-neural network 模糊神经网络
fuzzy-set theory 模糊集理论

G

GA (genetic algorithm) 遗传算法
gadolinium (Gd) 钆
gage glass 液位玻璃管
gain 增益,放大
gain characteristic 增益特性
gain constant 放大系数
gain control 增益控制
gain controller 增益控制器
gain crossover frequency 增益交越频率
gain cut-off frequency 截止频率增益
gain dynamics 增益动力学
gain enhancement method 增益增强方法
gain margin 增益裕度,增益裕量,增益范围
gain modulation 幅值调制
gain saturation 增益饱和
gain scheduling 增益规划
gain setting divider 增益调节分压器
gain suppression 增益抑制
gain time control 增益时间控制
gain-crossover frequency 增益交越频率
gain-switching amplifier 增益转换放大器
gallium (Ga) 镓
galvanic 直流电的
galvanic cell 原电池
galvanic corrosion 电化腐蚀,电蚀
galvanic couple 电偶
galvanization 电镀,镀锌
galvanize 电镀,镀锌
galvanized 镀锌的
galvanized iron 锌铁,镀锌铁,白铁
galvanized iron pipe 镀锌铁管
galvanized iron tube 镀锌铁管
galvanized iron wire 镀锌铁丝
galvanized metal 镀锌金属
galvanized mild steel plate 镀锌软钢板
galvanized sheet iron 镀锌铁片,镀锌铁皮
galvanized steel 镀锌钢
galvanized steel sheet 镀锌钢板
galvanized stranded wire 镀锌钢绞线
galvanizing 镀锌
galvanomagnetism 电磁,电磁学
galvanometer 检流计,电流计
galvanometer record type strong-motion instrument 电流计记录式强震仪
galvanometer with optical point 光点式检流计
galvano-voltmeter 伏安表
game theory 对策论,博弈论,博弈树
game tree 对策树
gamma directioned radiometer γ定向辐射仪
gamma processing image γ处理像
gamma radiometer in borehole 井中γ辐射仪
gamma ray level measuring device γ射线物位测量仪表,伽马射线液位测量
gamma ray logger γ测井仪
gamma ray spectrometer γ能谱仪
gamma rays γ射线
gamma sampling radiometer of banded screen 带屏γ取样辐射仪
gamma scintillator radiometer γ闪烁辐射仪
gamma spectrometer in borehole 井下γ能谱仪
gamma-ray detection apparatus γ射线探伤机
gamma-rays γ射线
gamma-spectrometry γ光谱测量
gang control 同轴控制,联动控制
gang dies 复合模
gang switch 同轴开关,联动开关
gangway 通道
gantry 起重龙门,地下支架,门架
gantry crane 龙门起重机

Gantt chart 甘特图，线条图
gap 气隙，间隙，缝隙
gap action controller 间隙调节器，间隙控制器
gap element 间隙元件
gap gauge 间隙规
gap measurement 间隙测量
gap reluctance 气隙磁阻
gap shear 凹口剪床
gap transient torque 间隙动态力矩
GAP (generalized analytical predictor) 广义分析预估器
garage 车房，车库
garbage 无用数据
garbage bag 垃圾袋
garnett wire 锯齿钢丝，钢刺条
gas 气体，天然气
gas analysis 气体分析
gas analyzer 气体分析器
gas appliance 煤气用具
gas bearing 气体轴承
gas cell alarm 充气光电管报警器
gas chromatograph 气相色谱仪
gas chromatograph-mass spectrometer (GC-MS) 气相色谱-质谱联用仪
gas chromatography 气相色谱法
gas chromatorgaphy-mass 气相色谱-质谱法
gas consumption 煤气耗用量
gas cushion 气垫
gas cylinder regulator 气瓶减压器
gas densitometer 气体密度计
gas density gauge 气体密度计
gas detector relay 气体检测继电器
gas diode 充气二极管
gas discharge 气体放电
gas filled thermal system 充气热系统
gas flow computer 气体流量计算器
gas generator 气体发生器
gas geyser 煤气热水炉
gas insulated 气体隔离
gas insulated substations 气体分配站
gas insulated switchgear 气体绝缘设备，空气开关
gas laser 气体激光器
gas main 煤气总管
gas main laying works 铺设煤气总管工程
gas mark 焦痕
gas meter 煤气表，气体表
gas pipe 煤气喉管
gas plasma asher 气体等离子灰化机
gas plasma dry cleaner 气体等离子清洗机
gas plasma etcher 气体等离子蚀刻机
gas port 气口
gas proportional detector 气体正比检测器
gas pump thermocouple 抽气式热电偶
gas relay 瓦斯继电器，气体继电器
gas replacement vacuum furnaces 真空气体置换炉
gas riser 煤气竖管，煤气立管
gas sensing electrode 气敏电极
gas sensor 气敏元件，气体传感器
gas shield 气体遮蔽
gas thermometer 气体温度计
gas transducer 气体传感器
gas turbines 燃气轮机，燃气涡轮
gas vent 气孔
gas welding 气焊
gas-discharge source 气体放电源
gas-enclosed pressure gauge 气密式压力计
gaseous cyaniding 气体氧化法
gaseous form 气态
gas-filled bellows 充气膜盒
gas-filled thermometer 充气温度计
gasification 气化
gasket 垫圈，密封垫圈，软垫，衬垫，胶边，垫片
gasket joint 接口垫片
gasket seal 垫片密封
gas-liquid chromatography 气液色谱法
gas-liquid separator 气液分离器
gasoline gauge 汽油表
gas-sensitive element 气敏元件
gas-solid chromatography 气固色谱法
gas-tight 气密，不透气
gastrointestinal inner pressure sensor 胃肠内压传感器
gastrointestinal inner pressure transducer 胃肠内压传感器
gate 闸，闸门，浇口

gate array 门阵列，门阵
gate circuit 门电路
gate location 进入位
gate size 水口大小
gate turn-off thyristor (GTO) 可关断晶闸管
gate type 水口形式
gate valve 闸用阀，闸式阀
gate voltage 栅电压，触发电压
gate-type flowmeter 闸门式流量计
gateway 网关
gateway processor 接口处理器
gauge 计量规，仪表，表，计，尺，仪
gauge block 量块
gauge board 样板，模板，规准尺，仪表板
gauge circuit 应变计［片］电路
gauge connector 压力表接头
gauge factor 应变计灵敏系数，应变系数，量规因数
gauge length 标距长度
gauge multiport valve 多路阀
gauge plate 量规定位板
gauge pointer 仪表指针
gauge pressure 表压
gauge pressure sensor 表压传感器
gauge pressure transducer 表压传感器
gauge resistance 应变计（片）电阻
gauge valves 仪表阀
Gauss 高斯（磁通密度单位）
Gauss optics 高斯光学
Gauss theorem 高斯定理
Gaussian 高斯的
Gaussian distribution 高斯分布
Gaussian function 高斯函数
Gaussian integration method 高斯求积法
Gaussian noise 高斯噪声
Gaussian process 高斯过程
Gauss-Markov source 高斯-马尔科夫源
Gauss-meter 高斯计
gauze 金属丝网
gauze strainer 网状滤器
gauze wire 细目丝网
GC-MS (gas chromatograph-mass spectrometer) 气相色谱-质谱联用仪

GDF (global description file) 全局说明文件
gear 齿轮，齿轮机构
gear axial displacement 齿轮轴向位移
gear axial plane 齿轮轴向平面
gear axis 齿轮轴线
gear box 变速箱，齿轮箱，波箱
gear center 齿轮中心
gear combination 齿轮组合
gear cone 大轮锥距
gear coupling 齿轮联轴器
gear cutting machines 齿轮切削机，齿轮加工机床
gear machining 齿轮加工
gear manufacturing summary 齿轮加工调整卡
gear marking compound 检查齿轮啮合涂色剂
gear measuring wire 齿轮测量线
gear member 大轮
gear planer 成型刨齿机
gear ratio 齿数比
gear reducer 减速器，齿轮减速箱，齿轮减速装置
gear rougher 齿轮粗切机床
gear shaper 插齿机，刨齿机
gear shaping 插齿，滚齿，刨齿
gear tipping 齿轮倾斜
gear tooth vernier gauge 齿厚游标卡尺
gear train 轮系，齿轮传动链
gear train with fixed axes 定轴齿轮系，普通齿轮系
gear wheel 齿轮，大齿轮
gear pump 齿轮泵
gear shift housing 变速箱
gearbox 变速箱，齿轮箱
gearbox casing 变速箱体
geared index 齿轮系分度
gears 齿轮，齿轮组
Geiger type vibrograph 盖格尔式测振仪
GEK (geomagnetic electrokinetograph) 地磁动电计
gel 凝胶体
gel chromatography 凝胶色谱法
gel filtration chromatography (GFC) 凝胶过滤色谱法

gel permeation chromatograph 凝胶渗透色谱仪
gel permeation chromatography (GPC) 凝胶渗透色谱法
gene pattern analyzer 基因检查仪器
general 通用的
general and fundamental mechanics 一般力学与力学基础
general arrangement 总体布置，总体设计
general bilinear transformation 通用广义双线性变换
general constraint 公共约束
general description 一般描述
general equilibrium theory 一般均衡理论
general failure error 一般故障错误
general identification test 一般鉴别试验
general input 一般输入
general layout plan 总平面图
general linearization algorithm 通用线性化算法
general machining centers 通用加工中心
general nonperiodic wave 通用非周期波
general output 一般输出
general purpose computer 通用计算机
general purpose computer interface 通用计算机接口
general purpose diode 普通二极管
general purpose inputs 普通操作输入
general purpose interface bus 通用接口总线
general purpose strain gauge 常温应变计
general purpose thimbles for use with steel wire ropes 钢丝绳用普通套环
general requirements for test methods of aerated concrete 加气混凝土性能试验方法总则
general simulator 通用仿真器
general specification 一般规格
general stability criterion 广义稳定判据
general system theory 一般系统理论
general tag 一般工位号
general technical requirements 通用技术条件
general technical requirements for radial gates 弧形闸门通用技术条件
general control panel 总控制屏
generalization 通用化
generalized 一般化，推广，普及
generalized analytical predictor (GAP) 广义分析预估器
generalized connection network 广义连接网络
generalized coordinate 广义坐标
generalized error coefficients 广义误差系数
generalized kinematic chain 一般化运动链
generalized least squares estimation 广义最小二乘估计
generalized linear system 广义线性系统
generalized modeling 广义建模
generalized predictive control (GPC) 广义预测控制
generalized predictive controller 广义预测控制器
generalized quantizer 通用数字转换器，通用数字量化器
generalized state space 广义状态空间
general-purpose gauge 一般压力表
general-purpose pressure gauge 一般压力表
generate 发电，发生
generated energy 发电量
generated gear 展成法齿轮
generated Lyapunov function 广义李雅普诺夫函数
generating 发电，发生，展成法
generating cam 展成凸轮
generating cutting 范成法
generating equipment 发电设备
generating gear 展成齿轮
generating line 发生线
generating line of involute 渐开线发生线

generating plane 发生面
generating plant 发电厂
generating pressure angle 产形轮压力角
generating train 展成传动键
generation 产生，形成，展成
generation function 生成函数
generation lifetime 形成生命时间
generation mechanism 产生机制
generation method 范成法
generator 发电机，振荡器，发生器
generator capacity 发电机容量
generator circuit 发电机电路
generator control panel 发电机控制面板
generator excitation 发电机励磁
generator exciter 发电机励磁机
generator exciting winding 发电机励磁绕组
generator field control 发电机磁场控制
generator group 发电机组
generator lead wire 发电机导线
generator operation 发电机运行
generator protection 发电机保护
generator terminal 发电机线接头
generator tripping 切机
generator voltage 发电机电压
generic cabling 综合布线
generic model control (GMC) 一般模型控制
genetic algorithm (GA) 遗传算法
geneva index 星形轮分度，槽轮分度
geneva mechanism 槽轮机构
geneva numerate 槽数
geneva wheel 槽轮，马尔特机构间歇传动轮
genset running hour meter 机组运行小时表
geodesy and survey engineering 大地测量学与测量工程
geodetic instrument 大地测量仪器
geodetic survey 大地测量，测地学测量
geologic compass 地质罗盘仪
geological condition 地质状况
geological engineering 地质工程
geological resources and geological engineering 地质资源与地质工程
geological stereometer 地质立体量测仪
geology 地质
geomagnetic 地磁的
geomagnetic electrokinetograph (GEK) 地磁动电计
geomagnetic field 地磁场
geomagnetic torque 地磁力矩
geomagnetism 地磁
geometric 几何学的
geometric aberration 几何像差
geometric approach 几何学方法
geometric centre of the dial 度盘几何中心
geometric code 几何编码
geometric distribution 几何分布
geometric property 几何特性
geometric similarity 几何相似
geometrical 几何的
geometrical position 几何位置
geometrical theory 几何理论
geometry 几何学
geostationary meteorological satellite 地球同步气象卫星
geotechnical assessment 岩土评估
geotechnical data 岩土数据
geotechnical engineering 岩土工程
geotechnical record 岩土纪录
geotechnical survey 土力测量
germanium (Ge) 锗
germanium diode 锗二极管
germanium thermometer 锗温度计
germanium triode 锗三极管
Gershum tube 格森管
GFC (gel filtration chromatography) 凝胶过滤色谱法
ghost peak 假峰
giant transistor (GTR) 电力晶体管
giga (G) 千兆（十亿）
gimbal 万向接头
girder 大梁
girder iron 梁钢，梁铁
given 给定的
glare 眩光
glass beading 玻璃珠
glass bulb 玻璃泡

glass circle 玻璃电极
glass fabric 玻璃布
glass fiber 玻璃纤维
glass fiber reinforced plastics 玻璃纤维增强塑料
glass fiber reinforced polyester corrugated panels 玻璃纤维增强聚酯波纹板
glass fibre insulator 玻璃纤维绝缘器
glass filter 滤光镜，滤光片
glass funnel 玻璃漏斗
glass hydrometer 玻璃管式比重计
glass mats 玻璃纤维垫
glass panel 玻璃嵌板
glass piston pressure gauge 玻璃活塞式压力表
glass reinforced plastic pipe 玻璃强化胶管，玻璃钢管
glass rod 玻棒
glass type rotameter 玻璃转子流量计
glass-paper 砂纸
glass-tube level gauge 玻璃管液面计
glass-tube rotameter 玻璃管式转子流量计
glassware 玻璃器皿
glazed 镶有玻璃
glazed brick 釉面砖
glazed earthenware 釉面陶土
glazed porcelain 玻璃瓷
glazed tile 玻璃瓦，玻璃砖，釉面砖
glazed ware 釉面物料
glazing 光滑，玻璃装配，上釉，釉化，装配玻璃
glazing agent 上光剂
glazing bar 玻璃格条
glazing brad 玻璃钉
glazing by dipping 汤浸釉
glazing by immersion 汤蘸釉
glazing by rinsing 汤釉
glazing by splashing 泼釉
glazing by sufflation 吹釉
glazing calender 擦光机
glazing clip 玻璃卡子
glazing color 釉色
glazing felt 上毛毯
glazing industry 玻璃窗装配行业
glazing kiln 上釉窑

glazing machine 上光机，研光机，抛光机
glazing mill 电子管密封玻璃管制造机
glazing paint 发光漆，上光涂料
glazing pottery 上釉陶器
glazing rebate 玻璃槽口
glazing spray gun 喷釉枪
glazing sprig 玻璃钉
glazing tape 玻璃密封条
glazing wheel 研磨轮
glazing with putty 油灰镶玻璃法
glazing without frame 无窗框安装法
glazy pig iron 高硅生铁
global 全局的，总的，整体的
global analysis 整体分析
global asymptotic stability 全局渐进稳定性
global data base 全球数据库
global description file (GDF) 全局说明文件
global optimization 全局优化，整体优化
global optimum 全局最优
global overview 全球概览
global position system (GPS) 全球定位系统
global positioning system 全局定位系统
global radiation 总辐射
global stability 全局稳定
global system for mobile communications 全球移动通信系统
globe valve 球形，球阀
globe valve with sealing bushing 卡套式球阀
globular cementite 球状碳化铁
gloss 光泽
gloves 手套
glow discharge 辉光放电
glow-tube 辉光放电管
glucose enzyme sensor 酶（式）葡萄糖传感器
glucose enzyme transducer 酶（式）葡萄糖传感器
glucuronides 葡糖苷酸
glutamate microbial transducer 微生物谷氨酸传感器

glutamic acid microbial sensor 微生物谷氨酸传感器
glycerine 甘油
GMC（generic model control） 一般模型控制
goal coordination method 目标协调法
goal 目的，目标
goggle 护目镜
gold（Au） 金
gold bonding wire 金连接线，金键合线
goldclad wire 镀金导线
gold-plated kovar wire 镀金科伐线
good parts 良品
good product 良品
good stability 稳定性好
goods lift 载货升降机
gooseneck 鹅颈管
gouge 沟槽，凿槽，半圆凿
governor 调节器，调压器，调速器
governor motor 调速器电动机
governor switch 调速器开关
GPC（gel permeation chromatography） 凝胶渗透色谱法
GPC（generalized predictive control） 广义预测控制
GPS（global position system） 全球定位系统
grab iron 铁撬棍
grace period 宽限期
grade 级，等，级别，等级
grade Ⅰ standard dynamometer 一等标准测力计
grade Ⅱ standard load calibrating machine 二等标准测力机
grade Ⅲ standard dynamometer 三等标准测力计
grade of protection 防护等级
grade strength 等级强度
gradient 斜度，坡度，倾斜度，梯度
gradient development 梯度展开
gradient elution 梯度洗脱，等度洗脱
gradient flux-gate magnetometer 磁通门磁力梯度仪
gradient magnetometer 梯度磁强计
gradient method 梯度法，最速下降法
gradient superconducting magnetometer 超导磁力梯度仪
gradient thin layer plate 梯度薄层板
gradients 梯度，斜率
gradiometer 梯度计
graduated circle 分度圈
graduated cylinder 量筒
graduated dial 标度盘，刻度盘
graduated flask 量筒量杯
graduated pipettes 刻度吸管
graduated range 刻度范围
graduated release valve 缓解阀，阶段放气阀
graduation 定标，分度，刻度，分度号
graduation of scale 刻度
grain 细粒，磨粒
grain size 结晶粒度
gram atom 克原子
grammatical inference 文法推断
granular material 颗粒材
granularity 间隔尺寸
granule 细粒，颗粒料
graph 图，图表
graph search 图搜索
graph theory 图论
graphic 图解，图解的
graphic calculation 图解计算法
graphic definition 流程图定义
graphic display 图形显示
graphic display file 图形显示文件
graphic display panel 图形显示板
graphic library 图形库
graphic method 图解法
graphic object 流程图对象
graphic panel 图示板，全模拟盘，流程图画面，图形板
graphic presentation technique 图形表达技巧
graphic printer 图解打印
graphic search 图搜索
graphic window 流程图窗口
graphical communication 图像传意
graphical dispaly 图形显示
graphical kernel system 图形核心系统
graphical symbols 图形符号
graphite 石墨
graphite contraction allowance 电极

缩小余量
graphite grease 石墨润滑脂
graphite holder 电极夹座
graphite machine 石墨加工机
graphite mould 石墨模子
graphitic pig iron 灰口铁
graphs 图形，图像，图示
grating 光栅
grating displacement transducer 光栅式位移传感器
grating monochromator 光栅单色仪
grating spectrograph 光栅摄谱仪
gravimeter 重力仪，比重计
gravimetric analysis 重量分析法
gravitational attraction 万有引力
gravitational balancing machine 重力式平衡机
gravity 重力
gravity anomaly 重力异常
gravity casting 重力铸造
gravity casting machines 重力铸造机
gravity corer 重力式取样管
gravity correction 重力修正
gravity gradient survey 重力梯度测量
gravity gradient torque 重力梯度力矩
gravity gradiometer 重力梯度仪
gravity horizontal gradient survey 重力水平梯度测量
gravity load 重力荷载
gravity nut 重心铊
gravity platform 重力平台
gravity profile 重力剖面
gravity survey 重力测量
gravity type seawall 重力式海堤
gravity vertical gradient survey 重力垂向梯度测量
gravity water supply 天然水压供水，重力供水
gray cast iron 灰口铸铁，灰铸铁，灰口铁
gray pig iron 灰口铸铁，灰铸铁，灰口铁
graybody 灰体
grease 油脂，膏油，图形
grease chamber 油脂室
grease gun 油脂枪，注油枪
grease stains 油污

grease trap 除油器，油脂分离器
greasing 涂油脂
green design 绿色设计
Green function 格林函数
grey cast iron 灰铸铁，灰口铸铁
grid 网格，格子，栅，控制极，栅极
grid bias 栅偏压
grid bias supply 栅偏压电源
grid capacitance 栅极电容
grid control 栅极控制
grid lead wire 栅极引线
grid lines device 格线
grid nephoscope 栅状测云器
grid paper 网格纸，方格纸
grid resistance 栅极电阻
grid-controlled X-ray tube 栅控 X 射线管
grill 烤炉
grille 栅格，栏栅，铁花
grind 磨
grinder 磨床，磨削机，砂轮机
grinder bench 磨床工作台
grinders thread 螺纹磨床
grinders ultrasonic 超声波打磨机
grinding cracks 磨削裂纹
grinding defect 磨痕
grinding disc 研磨盘
grinding machine 磨床，研磨机，砂轮机，磨石机
grinding machines centerless 无心磨床
grinding machines cylindrical 外圆磨床
grinding machines universal 万能磨床
grinding paste 研磨膏
grinding stone 磨石
grinding tools 磨削工具
grinding wheel 砂轮
grinding wheel groove 砂轮越程槽
grinding wheel machine 砂轮机，砂轮锯
grinding wheels 磨轮
grip 夹头
grip length 握固长度
gripper 夹具
gripper feed 夹持进料
gripper feeder 夹紧传送
grit 砂砾，砂粒
grit maker 抽粒机

grit removal 除砂
grommet 孔环，索环，金属孔眼
groove 槽，压线，沟
groove cam 槽凸轮
groove punch 压线冲子
groove welding 坡口焊
grooved tile 槽纹瓷砖
grooving machine 刻槽机
gross error 过失误差
gross mass 总质量
gross weight 总重量
ground 地下，土地，地，接地，地线，接地装置
ground bearing pressure 地面承载力
ground bus inlet 接地总线入口
ground clearance 离地净高
ground condition 土地状况，地面条件
ground connection 接地，接地线
ground control 接地控制
ground electrochemical extractor 地电化学提取法仪器
ground electromagnetic instrument 地面电磁法仪器
ground fault 接地事故
ground floor level 地面层水平
ground gamma spectrometer 地面γ能谱仪
ground gravity survey 地面重力测量
ground instrument 地面仪器
ground investigation 探土，土地勘测
ground level contour 地面水平等高线
ground level roundabout 地面回旋处
ground line 地平线
ground noise 本底噪声
ground plane 地平面，水平投影
ground point 接地点
ground potential 大地电位
ground protection 接地保护
ground pulse electromagnetic instrument 地面脉冲电磁仪
ground receiving station 地面接收站
ground relay 接地保护继电器
ground resistance 接地电阻，地电阻
ground rod 接地棒
ground row light 地脚排灯
ground steel wire 磨光钢丝
ground strap 接地母线

ground survey method 地面测量法
ground switch 接地开关
ground system 接地系统，地线系统
ground terminal 接地端
ground transportation 地面运输
ground visibility 地面能见度
ground water 地下水
ground water level 地下水位
ground X-ray fluorimeter 地面X射线荧光仪
ground-bus 接地母线
grounded 接地的
grounded circuit 接地电路
grounded input 接地输入
grounded junction 接地端
grounded neutral 接地中点
grounded noise 接地噪声
grounded output 接地输出
grounded probe 接地探针
grounded-base transistor 共基极晶体管
grounded-collector transistor 共集电极晶体管
grounded-emitter transistor 共发射极晶体管
grounding 接地，接地装置，地线
grounding electrode 接地电极
groundwater drainage works 地下水排水工程
group amplifier 组合放大器
group delay equalizer 群延时均衡器
group detail display 分组详细显示画面
group display 分组显示画面，操作组画面
group graphic display 分组流程图显示画面
group of an instrument for explosive atmosphere 防爆仪表类别
group pulse generator 脉冲群发生器
group technology 成组技术
group trend display 组趋势画面
group work 成组工作
grout 灌浆，灰浆，水泥浆，浆液
grout pipe 灌浆管
grout sleeve 灌浆套筒
grout tube 灌浆管
grout valve 灌浆阀

grout vent pipe 灌浆排气管
grouting works 灌浆工程
GTO (gate turn-off thyristor) 可关断晶闸管
GTR (giant transistor) 电力晶体管
guarantor 保证人
guard 外罩，保护，防护，保护装置，防护装置
guard against damp 禁止潮湿
guard column 保护柱
guard net 保护网
guard rail 护栏，护轨
guard wire 保护线
guarded hot plate apparatus 防护热板法
guarded input 保护输入
guidance system 制导系统，导航系统，导向系统
guide bushing 引导衬套
guide pad 导料块
guide pile 导桩
guide pin 导销，导正销
guide plate 导板，定位板
guide post 引导柱
guide rail 导轨
guide ring 导环
guide sign 指示标志
guide wall 导墙
guide wheel 导向轮

guide wire 尺度索，准绳
guide 领路人，向导
gully 集水沟
gully trap 集水沟隔气弯管
gun metal 炮铜
gunk 料斗
gust effect of wind 阵风效应
gutter 雨水槽，边沟，砌沟，檐沟，锻模飞边槽
guy 拉索
guy wires 张索，长绳
gypsum 煅石膏
gypsum mold 石膏铸模
gypsum plaster 石膏灰泥，石膏抹面，石膏胶凝材料，粉饰用石膏粉
gyrator 回转器
gyratory crusher 回转压碎机
gyratory traffic 回旋交通
gyro balancing machine 陀螺测斜仪
gyro drift rate 陀螺漂移率
gyro-level 陀螺水平仪
gyromagnetic ratio 旋磁比
gyroscope 回转器，陀螺仪
gyroscopic mass flowmeter 陀螺型质量流量计
gyrostat 陀螺体
gyro-theodolite 陀螺经纬仪

H

hack saw 钢锯
hafnium (Hf) 铪
hair crack 发裂，毛细裂缝
hair hygrograph 毛发湿度计，毛发湿度记录仪
hair hygrometer 毛发湿度表
hair wire 游丝
hairpin turn 急转弯
hairspring 游丝，细弹簧
Haldane's apparatus 霍尔丹空气分析仪
half bridge converter 半桥电路
half bridge measurement 半桥测量
half cycle 半周
half duplex transmission 半双工传输
half period 半周
half 一半，一半的
half-bridge high temperature strain gauge 半桥路高温应变仪
half-bridge measurement 半桥测量
half-cell 半电池
half-life 半衰期
half-power point 半功率点
half-round iron wire 半圆铁线
half-round steel wire 半圆钢丝
half-shift register 半移位寄存器
half-subtracter 半减法器
half-wave 半波
half-wave plate 半波片
half-wave potential 半波电位
half-wave rectification 半波整流
Hall 霍尔
Hall conductivity detector 霍尔电导检测器
Hall constant 霍尔常数
Hall displacement transducer 霍尔式位移传感器
Hall effect 霍尔效应
Hall effect device 霍尔效应器件
Hall effect displacement transducer 霍尔效应位移传感器
Hall effect fluxmeter 霍尔效应磁通计
Hall effect linear detector 霍尔效应线性检测器
Hall effect magnetometer 霍尔效应磁力计
Hall effect multiplier 霍尔效应倍增器
Hall effect tachometric transducer 霍尔效应式转速传感器
Hall electrolytic conductivity detection 霍尔电解质电导率检测
Hall element 霍尔元件，霍尔传感器
Hall probe 霍尔探头
Hall type pressrue transducer 霍尔式压力传感器
halogen 卤，成盐元素
halogen counter 卤素计数器
halogen leak detector 卤素检漏仪
halogen light 卤素灯
halt instruction 停机指令
halve 二等分，平分
hammer 锤，榫机，铁锤，锤子
hammer man 锻工
hammer test 锤击试验
hammering method 敲击法
hammering type Brinell hardness tester 锤击式布氏硬度计
Hamon-Pair potentiometer 哈蒙-佩尔电位差计
hand anemometer 手持风速仪
hand brace 手摇钻
hand brake 手制动器
hand compass 手持罗盘
hand electric drill 手电钻
hand face shield 手握面罩
hand finishing 手工修润，手工精削
hand goniometer 手持测角器
hand hydrometer 手提式比重计
hand lamp 手提灯
hand of cutter 刀盘方向
hand of spiral 螺旋方向
hand operated valve 手动阀
hand operation 手动操作

hand pattern level　手持水准仪
hand press　手动冲床
hand printed character　手写字符
hand pump　手摇泵
hand rack pinion press　手动齿轮齿条式冲床
hand reset　手动复位
hand screw press　手动螺旋式冲床
hand signal　手动信号
hand spectrophotometer　手提式分光光度计
hand tool　手工具
hand truck　手推车
hand wheel　手轮
hand　手，指针
hand-and-feet contamination monitor　手、足放射性污染监测仪
hand-aspirated psychrometer　手动吸气式湿度计
hand-dug caisson　手挖沉箱，人工挖掘沉箱
hand-feed punch　手动输入穿孔机
hand-held digital thermometer　手提式数字温度计
hand-held grinder　手持研磨机
hand-held logic circuit probe　手提式逻辑电路故障探测器
hand-held multimeter-oscilloscope　手提式小型万用表-示波器两用仪
handle　手柄，门柄
handle mold　手持式模具
handle with care　小心搬运
handlevel　手持水准仪
handling　手动
hand-operated regulator　手动调节器
hand-operated valve　手动操作阀
hand-over　释放延迟
hand-rolling tester　手动滚动试验机
hands of worm　蜗杆旋向
handset　手持通话器
handshake　握手
handshaking　同步交换，信号交换
handwheel　手轮机构，手轮，驾驶盘
handy digital tachometer　手持式数字转速表
handy type thermometer　手提式温度计
hanger　吊架，吊

hanging clinometer　悬式倾斜计
hanging level　悬式水准仪
hanging mercury electrode　悬汞电极
hanging spring　吊丝
hanging theodolite　悬式经纬仪
Harber earth current ammeter　哈伯接地电流表
Harcourt photometer　哈考特光度计
hard alloy steel　超硬合金钢
hard bearing balancing machine　硬支承平衡机
hard constraint　硬约束
hard copy　硬拷贝
hard disk drive (HDD)　硬碟机，硬盘驱动器
hard drawn copper　冷拉铜
hard drawn wire　冷拉线，硬拉线
hard facing　熔接硬面法，加焊硬面法，表面硬化（淬火）
hard film　硬膜
hard finishing　硬齿面精加工
hard impervious material　坚硬不透水物料
hard seal butterfly valves　金属密封蝶阀
hard switching　硬转换
hardcore　硬底层，碎砖垫层
hard-drawn aluminium wire　硬铝线
hard-drawn copper strand wire　硬铜绞线
hardenability　硬化性
hardenability curve　硬化性曲线
hardened concrete　硬化混凝土
hardener　硬化剂
hardening　淬火，硬化
hardening agent　硬化剂
hardening and tempering　调质处理
hardening strength　硬化强度
Hardie spectrophotometer　哈迪分光光度计
hardness　硬度
hardness ratio factor　硬度比
hardness sensor　硬度传感器
hardness tester　硬度计
hardness transducer　硬度传感器
hardware　五金器件，硬件，硬体
hardware verification test system　硬

检验测试系统
harmonic 谐波，谐波的
harmonic amplifier 谐波放大器
harmonic analysis 谐波分析
harmonic analyzer 谐波分析器
harmonic balance analysis 谐波平衡分析
harmonic balance technique 谐波平衡，谐波平衡技术
harmonic balancer 谐波平衡器
harmonic component 谐波分量
harmonic content 谐波含量
harmonic content of AC power supply 交流电源的谐波含量
harmonic current 谐波电流
harmonic distortion 谐波畸变，谐波失真
harmonic drive 谐波驱动
harmonic driving 谐波传动
harmonic filter 谐波滤波器
harmonic function 谐波函数，调和函数
harmonic gear 谐波齿轮
harmonic generation 谐波发生
harmonic generator 谐波发生器
harmonic power 谐波功率
harmonic response 谐波响应
harmonic response characteristic 谐波响应特性
harmonic search 谐振追踪
harmonic sweep 谐振扫描
harmonic wave 谐波
HART（highway addressable remote transducer） 可导址远程传感器数据公路
HART protocol intelligent linear electric actuator HART协议直行程电动执行器
HART protocol intelligent pressure transmitter HART协议智能压力变送器
Hartley circuit 哈脱莱振荡电路
Hartley oscillator 哈脱莱振荡器
hashing 散列法
hatch 影线
hatch box 检查箱
hatchet stake 折铁砧

hatching 剖面线
hauling 搬运，拖运
havoc 严重破坏，损害
hazard assessment study 危险程度评估研究
hazard sign 危险标志
hazard warning lantern 危险警告灯
hazardous substances mark 危险品包装标志
haze 雾度
h-bomb 氢弹
HCI（human-computer interaction） 人机相互作用，人机交互
HCMM（heat-capacity mapping mission） 热能勘测任务
HCRF（hierarchical chromatographic response function） 等级色谱响应函数
HDD（hard disk drive） 硬碟机，硬盘驱动器
head amplifier 前置放大器
head of screwdriver 起子头
head pressure 水头压力
head space gas chromatography 顶空气相色谱法
header 头部，标题
header byte 首标字节
header tank 定压池，配水池
heading machine 掘进机，顶镦机，螺丝打头机，封头机
headroom 净空高度
headset 耳机
headstock 主轴箱
headway 前后两车时间间隔
hearing aid 助听器
heart sound sensor 心音传感器
heart sound transducer 心音传感器
heart-rate meter 心率计
heat 热量
heat balance 热平衡
heat conducting 热传导的，导热
heat conduction 热传导
heat conductivity 热导率，导热系数
heat conductor 热导体
heat convection 热对流
heat cycle test 热循环试验
heat detector 热探测器

heat dissipation 散热
heat efficiency 热效率
heat emission 散热，发热
heat exchanger 换热器，热交换器
heat flow 热量流
heat flow meter 热流计
heat flux meter 热流计，热通量计
heat flux sensor 热流传感器
heat flux transducer 热流传感器
heat insulated box 保温箱
heat integration 热量集成
heat loss 热损耗
heat of combustion 燃烧热
heat preserving furnaces 保温炉
heat process 加热过程
heat radiation 热辐射
heat resistance of asphalt 沥青的耐热度
heat sensing cable 热敏电缆
heat shield 隔热屏障
heat shrinkable tubing 遇热收缩软管
heat sink device 散热设备
heat test 耐热试验
heat transfer flowmeter 热传导式流量计
heat treated glass window 热处理玻璃观察窗
heat treatment 热处理
heat unit (HU) 热单位
heat value 热值
heat 热，加热
heat capacity mapping mission (HCMM) 热能勘测任务
heat capacity mapping satellite 热容量制图卫星
heated 加热的
heated and cooled enclosed location 升温和降温封闭场所
heated or cooled enclosed location 升温或降温封闭场所
heater 加热器，热水炉
heater band 加热片
heater characteristic of X-ray tube X 射线管灯丝特性
heater cooler 加热器冷却
heater wire 加热丝
heat-flow meter 热流计
heat-flux differential scanning calorimeter 热流型差示扫描量热仪
heating 加热，发热
heating appliance 电热器
heating blocks 试验管加热板
heating box 保温箱
heating boxes 加热室
heating by far infra-red radiation 远红外辐射加热
heating curve determination 升温曲线测定
heating curve determination apparatus 升温曲线测定仪
heating curves 加热曲线
heating element 加热器，加热元件
heating quantity of bomb cylinder 弹筒发热量
heating quantity of constant capacity on high position 恒容高位发热量
heating rate 升温速率
heating rate curves 加热速率曲线
heating system 暖气系统
heating treatment furnaces 熔热处理炉
heating unit 加热体，加热装置，暖气片，加热元件
heat-transfer coefficient 传热系数
heavy compaction plant 重型固土机
heavy condensation 高度冷凝
heavy current engineering 强电工程
heavy duty 重型
heavy duty electrical conductor 大功率导电体
heavy fuel oil 重燃料油
heavy gauge wire 粗导线
heavy hex nuts 六角重型螺帽
heavy iron 厚度层铁皮
heavy load 重载
heavy metal eliminator 重金属排液处理
heavy oil engine 重油引擎
heavy water 重水
heavy wire 粗钢丝
heavy-duty lathe 重型车床
hectare (ha) 公顷（面积单位）
HEED (high electron energy diffractometer) 高能电子衍射仪
heel 基跟，柱脚，轮齿大端
heigh strength nuts 高张力螺帽

163

height adjustment knob 调高旋钮
height gauge 高度计,测高规
height of capillary rise 毛细提升高度,毛细上升高度
height of tide 潮高
height series 高度系列
helical bellows 螺旋波纹管
helical bevel gear 螺旋锥齿轮
helical compression spring 螺旋压缩弹簧
helical duplex 双重螺旋法
helical gear 螺纹齿轮,螺旋齿轮,斜齿圆柱齿轮
helical line 螺旋线
helical motion 螺旋运动
helical pair 螺旋副
helical spring 螺旋弹簧
helical torsion spring 扭簧
helical-spur gear 斜齿圆柱齿轮
helicograph 螺旋规
helicopter 直升机
helicopter control 直升机控制
helicopter dynamics 直升机动力学
heliograph 日照计,日光计
helium (He) 氦
helium mass spectrometer leak detector 氦质谱检漏仪
helium survey 氦气测量
helix 螺旋
helix angle 螺旋角
helix angle at reference cylinder 分度圆柱螺旋角
Helmholtz coils 亥姆霍兹线圈
Helmholtz resonator 亥姆霍兹共振器
help 帮助文件
help display 帮助显示画面
hematite pig iron 低磷生铁
hematocyte counter 血球计数器,血细胞计数器
hemi-homolysis cleavage 半均裂
hemming 卷边加工
henry 亨,亨利(电感单位)
hermetically sealed enclosure 气密外壳
hermitic 密封的
hermitic motor 密封式电动机
herringbone gear 人字齿轮

hertz (Hz) 赫兹(频率单位)
hertz equation 赫兹公式
hertz oscillator 赫兹振荡器
heterodyne 外差法
heterodyne analyzer 外差式分析仪
heterodyne converter 外差变频器
heterodyne detector 外差检波器
heterodyne frequency 外差频率
heterodyne oscillator 外差振荡器
heterodyne receiver 外差接收机
heterogeneous membrane electrode 非均相膜电极
heterogeneous reactor 非均匀反应堆
heterolytic cleavage 异裂(非均裂)
heteronuclear lock signal 异核锁信号
heuristic 启发式的,试探法,直观推断
heuristic inference 启发式推理
heuristic programming 启发式编程,启发式规划
heuristic search 启发式搜索
hex cap nuts 六角盖头螺帽
hex flange nuts 六角轮缘螺帽
hex jam nuts 薄型螺帽
hex machine screw nut 机械螺丝用六角螺帽
hex nuts 普通六角螺帽
hex serrated nuts 六角锯齿螺帽
hex slotted nut 六角割沟螺帽
hexagon bolts with slot on head 六角头头部带槽螺栓
hexagon bolts with split pin hole on shank 六角头螺杆带孔螺栓
hexagon fit bolts 六角头铰制孔用螺栓
hexagon fit bolts with split pin hole on shank 六角头螺杆带孔铰制孔用螺栓
hexagon headed bolt 六角头螺栓
hexagon nut 六角螺帽
hexagonal steel wire 六角钢丝
HF-bus HF 总线
hidden feedback loop 隐含反馈回路
hidden oscillation 隐蔽振荡
hierarchical 分层的,分级的
hierarchical chart 层次结构图
hierarchical chromatographic response

function (HCRF) 等级色谱响应函数
hierarchical computer control system 分级计算机控制系统
hierarchical control 分级控制
hierarchical control system 分级控制系统
hierarchical decision making 分层决策
hierarchical design 层次设计
hierarchical intelligence 分级智能
hierarchical parameter estimation 递阶参数估计
hierarchical structure 递阶结构，分层结构
hierarchical system 分层系统，分级系统，递阶系统
hierarchically intelligent 分层智能，分级智能
hierarchically intelligent control 分层智能控制，分级智能控制
hierarchies 多层，分级
high accuracy 高精度
high alloy steel wire 高合金钢丝
high coolant temperature shutdowns 高水温故障停机
high current density 高电流密度
high definition television 高清晰度电视
high density 高密度
high density I/O subsystem 高速度I/O子系统，大容量I/O子系统
high density polyethylene 高密度聚乙烯
high direct voltage 直流高压
high dispersion diffraction image 高分散衍射象
high dynamic strain indicator 超动态应变仪
high electron energy diffractometer (HEED) 高能电子衍射仪
high elongation strain gauge 大应变应变计
high EMF potentiometer 高电热电位差计
high energy varistor 高能电压敏电阻器

high energy X-rays 高能X射线
high frequency dielectric splitter 高频介电分选仪
high frequency electrotome 高频电刀
high frequency fatigue testing machine 高频疲劳试验机
high frequency mobile X-rays machine 高频移动X射线机
high frequency varistor 高频电压敏电阻器
high frequency X-rays diagnostic machine 高频X射线诊断机
high impact polystyrene 高冲击聚苯乙烯
high impact polystyrene rigidity 高冲击性聚苯乙烯
high impedance 高阻抗
high level analog input (HLAI) 高电平模拟输入
high level process interface unit (HLPIU) 高电平过程接口单元
high limited value 高顶值
high limiting control 上限值控制
high low average algorithm 高低平均算法
high mast lighting 高桅（杆）照明
high pass filter 高通滤波器
high performance capillary electrophoresis (HPCE) 高效毛细管电泳法
high performance computing and communicating 高性能计算与通信
high performance computing and communication program 高性能计算机与通信规划
high performance liquid chromatograph 高效液相色谱仪
high performance liquid chromatography (HPLC) 高效液相色谱法
high performance liquid chromatography-mass spectrometry (HPLC-MS) 高效液相色谱-质谱联用
high performance scanning electron microscope 高性能扫描电子显微镜
high performance thin layer chromatography 高效薄层色谱法
high polymer microphone 高聚物传声器

high precision stereotaxic 高精度脑立体定向仪
high pressure 高压
high pressure burner 高压炉头
high pressure cabinet 高压柜
high pressure chamber 正压室
high pressure cleaner 高压清洁器
high pressure DTA unit 高压差式热分析仪
high pressure limiter 高压限压器
high pressure liquid chromatography (HPLC) 高压液相色谱
high pressure mercury lamp 高压汞灯，高压水银灯
high pressure pneumatic diaphragm 高压气动薄膜执行机构
high pressure sodium lamp 高压钠灯，高压钠光灯
high pressure water jet 高压喷水器，高压水力喷射器
high pressure water propagating cleaner 高压水推送清洗器
high resolution diffraction attachment 高分辨衍射附件
high resolution diffraction image 高分辨衍射像
high resolution electron microscope 高分辨电子显微镜
high resolution gas chromatography (HRGC) 高分辨率气相色谱法
high resolution visible (HRV) 高分辨率
high resolution visible image instrument 高分辨率图像仪
high Rf value 高比移值
high side 高压侧
high speed balancing 高速平衡
high speed balancing installation 高速平衡设备
high speed belt 高速带
high speed liquid chromatography 高速液相色谱法
high speed micro-centrifuges 微量高速离心机
high speed refrigerated centrifuge 高速冷冻离心机
high speed simulation 快速仿真

high speed tool steel 高速度工具钢
high temperature device 高温
high temperature electric furnaces 高温电阻炉
high temperature electric resistance tubular furnace 高温管式电阻炉
high temperature furnaces heating apparatus 高温炉加热设备
high temperature pressure power station gate valves 高温高压电站闸阀
high temperature steel 高温钢
high temperature strain gauge 高温应变计
high temperature test 高温试验
high temperature test chamber 高温试验箱
high temperature testing machine 高温试验机
high temperature thermistor 高温处理电热调节器
high tensile steel 高抗拉钢
high tensile steel tendon 高抗拉钢缆
high tension bolt 高抗拉螺栓
high tension loop 高压回路
high tension wire 高压电线，高强度铁线
high vacuum 高真空
high value standard resistor 高阻标准电阻器
high volt starter 高压起动器
high voltage 高压
high voltage and insulation technology 高电压与绝缘技术
high voltage bridge 高压电桥
high voltage cable trough 高压缆槽
high voltage electricity 高压电力
high voltage electron microscope 高压电子显微镜
high voltage IC 高压集成电路
high voltage injection 高压输入
high voltage insulating tank 高压绝缘箱
high voltage probe 高压探针
high voltage shunt reactor 高压并联电抗器
high voltage source 高压电源
high voltage stability 高压稳定度

high voltage terminal　高压端
high voltage tester　高压测试器
high definition　高精度
high-capacity motor　大功率电动机
high-capacity water power station　大容量水电站
high-efficiency　高效益，高效率
higher actual measuring range value　实际测量范围上的限值
higher harmonic　高次谐波
higher harmonic resonance　高次谐波共振
higher measuring range value　测量范围上限值
higher order　高次
higher pair　链系的线点对偶
higher-order differential equation　高阶微分方程
higher-order statistics　高阶统计表，样本函数统计量
higher-order system　高阶系统
highest face temperature　最高表面温度
high-frequency　高频，高频率
high-frequency choke coil　高频扼流圈
high-frequency communication　高频通信
high-frequency control　高频控制
high-frequency diffraction　高频衍射
high-frequency heating　高频加热
high-frequency instrument　高频仪表
high-frequency interference　高频干扰
high-frequency measuring instrument　高频测量仪表
high-frequency modulation　高频调制
high-frequency noise　高频噪声
high-frequency oscillator　高频振荡器
high-frequency performance　高频性能
high-frequency titration　高频滴定
high-frequency transformer　高频变压器
high-frequency wire communication　有线高频通信
high-gain　高增益
high-gain feedback　高增益反馈
high-impedance differential probe　高阻抗差式探头
high-level language　高级语言
high-limit adjustment　上限调整

high-low action　高低作用
high-low selector　高低选择器
high-low signal selector　高低信号选择器
high-low temperature chamber　温度交变试验箱
highly sensitive　高灵敏度的
high-performance　高性能的
high-performance process manager　高性能过程管理站
high-power station　大电厂
high-precision　高精度的
high-recovery valve　高压力恢复阀
high-resolution gamma detector　高分辨率γ探测器
high-resolution mass spectroscope　高分辨质谱仪器
high-resolution NMR spectrometer　高分辨核磁共振波谱仪
high-resolution NMR spectroscope　高分辨率核磁共振波谱仪
high-signal selector　高信号选择器
high-speed lathe　高速车床
high-speed plotter　高速绘图仪
high-speed printer　高速打印机
high-speed reader　高速阅读器
high-speed switch　高速开关
high-speed transmission　高速传递
high-temperature hook-up wire　高温安装线
high-temperature stability　高温稳定
high-temperature superconductor　超高温导体
high-tensile steel wire　高强度钢丝
high-tension line　高压线
high-tension motor　高压电动机
high-tension supply main　高压馈电干线
high-tension transformer　高压变压器
high-tension transmission line　高压输电电线
high-voltage　高电压
high-voltage cable　高压电缆
high-voltage DC amplifier　高压直流放大器
high-voltage fence　高压电网
high-voltage test　高压试验

highway 道路，公路，高速通路，总线，信息公路
highway addressable remote transducer (HART) 可导址远程传感器数据公路
highway communication processor 高速数据通道通信处理器
highway control state 高速数据通道控制状态
highway coupler module 高速数据通道耦合模块
highway driver 信息公路驱动器
highway frame 数据公路帧
highway gateway 高速数据通道接口
highway gateway library 高速数据通道接口字库
highway gateway status display 高速数据通道接口状态显示画面
highway interface module 高速数据通道接口模块
highway protocol 数据公路协议
highway status 高速数据通道状态
highway status display 高速数据通道状态显示画面
highway traffic director 高速通信指挥器，高速数据通道通信指挥器
highway unit 数据公路单元
Hilbert 希尔波特
Hilbert space 希尔波特空间
Hilbert transformer 希尔波特变换器
H-infinity H∞，H 无穷
H-infinity control H∞控制
H-infinity optimization H∞优化
hinge 铰，铰链，铰链，枢纽
hinge bar 铰链杆
hinge pin 铰链销
HIS (human interface station) 人机界面站
HIS setting window 人机界面站设定窗口
histogram 柱状图，频率曲线，频率分布
historian 历史文件
historical message reports 历史信息报告
historical trend 历史趋势
historical trend panel 历史趋势画面

history 历史
history module 历史模块
history module processing unit 历史模块处理单元
history module status display 历史模块状态显示画面
history process and system 过程和系统的历史数据
history processor 历史处理器
hit 命中
hit quality index (HQI) 命中质量
hitting 压缩
HLAI (high level analog input) 高电平模拟输入
HLPIU (high level process interface unit) 高电平过程接口单元
HMI (human machine interface) 人机接口，人机界面
hoarding 围板
hoarfrost point 结霜点
hob 滚刀
hobbing 滚齿，滚齿加工
hobbing cutter 齿轮滚刀
hod 灰砂斗，砂浆桶
hoist 起重机，吊重机
hoisting bearing 起重轴承
hoisting chain 起重链
hoisting hook 提升钩
hoisting motor 起重马达
hoisting rope 起重绳
hoisting stopper 起吊止挡
hoisting test 起吊试验
hoisting wire 起重钢索
hold 保持
hold element 保持元件
holder base plate 支座垫板
holding 保持，存储
holding action 保持作用
holding bolt 固定螺栓
holding circuit 保持电路，吸持电路
holding contact 保持触点，吸持触点
holding current 吸持电流
holding down bolt 地脚螺栓，定位螺栓，压紧螺栓
holding element 保持环节，保持元件
holding systems 支持系统

holding voltage 保持电压,吸持电压
holding wire 测试线
hole 空穴,孔
hole carrier 空穴载流子
hole current 空穴电流
hole density 孔密度
hole location 孔位
hole micrometer 测孔千分尺
hole mobility 空穴迁移率
hole pattern 孔图
hole-gauge 孔径规
hole-plate flowmeter 孔板流量计
hollow blow molding 中空吹出成形
hollow copper wire 空心铜线
hollow flank worm 圆弧圆柱蜗杆
hollow section steel fencing 空心钢栏
hollow wire 管状线
hollow-cathode atomizer 空心阴极原子化器
hollow-cathode lamp 空心阴极灯
holmium (Ho) 钬
holographic grating 全息光栅
holographic interferometer 全息干涉仪
holographic SDRS 全息地震仪
holography 全息照相
homeostasis 内稳态
homogeneity spoiling pulse 均匀性突变脉冲
homogeneous 均质,均匀的
homogeneous membrane electrode 均相膜电极
homogeneous quantities 同类量
homogeneous radiation thermometer 单色辐射温度计
homogeneous reactor 均匀反应堆
homomorphic model 同态系统
homonuclear lock signal 同核锁信号
homopolar 单极的,同极的
hone 磨刀石
honeycombing 蜂窝
Honeywell verification test system 霍尼韦尔验证测试系统
honing machines 搪磨机
hood 机罩
hook angle 断面前角
hook cavity 钩穴

hooks coupling 万向联轴器
hookup 接线图
hook-up wire 架空电缆
hoop 卡箍
hop 突跃
hopper 漏斗,料斗,装料斗,有倾卸斗的手推车
hopper feed 料斗送料
hopper scale 漏斗秤,自动戽斗定量秤,库秤
hopper weigher 料斗秤
horizontal alignment 水平线向,平面线形
horizontal and vertical machining centers 卧式及立式加工中心
horizontal area 水平面积
horizontal balancing machine 卧式平衡机
horizontal boring machine 卧式镗孔机,卧式镗床
horizontal clearance 水平净空
horizontal component 水平分量
horizontal curve 平面曲线,水平曲线,水平断面曲线
horizontal damper 水平减震器
horizontal decomposition 横向分解
horizontal diameter 横向直径
horizontal dimension 水平尺寸
horizontal displacement 水平位移
horizontal glazing bar 水平玻璃横格条
horizontal guide face 水平导面
horizontal length measuring machine 水平测长仪
horizontal limit 水平极限
horizontal machine 卧式电机
horizontal machine center 卧式加工制造中心
horizontal machining centers 卧式加工中心
horizontal milling machines 卧式铣床
horizontal offset 水平偏置
horizontal plan drawing 平面图则
horizontal pressurized steam sterilizer 普通卧式压力蒸汽灭菌器
horizontal restraint 横向约束
horizontal scanning 行扫描

horizontal sliding gate 横向滑动闸门
horizontal surface 水平表面
horizontal tail 水平翼
horizontal visibility 水平能见度
horizontal wire 横丝
hormonal control 内分泌控制
horn wire 喇叭线
horsepower 马力，功率
horse-shoe electromagnet 马蹄形电磁铁
horseshoe magnet 马蹄形磁铁
hose 软管，软喉，水龙带
hose armoring wire 铠装胶管（用）钢丝
hose coupler 软管接头
hose reel 软管卷盘
hose-proof machine 防水式电机
hospital information system（HIS） 医院信息系统
host computer 主计算机
host processor 主处理机
hot air sterilizer 热空气消毒箱
hot bath quenching 热浴淬火
hot bend test 热弯试验
hot chamber die casting 热室压铸
hot dipping 热浸镀
hot drawing wire 热拉钢丝
hot forging 热锻
hot junction 热端
hot line 电话热线
hot mark 热斑
hot plates 加热板
hot rolled steel bar 热轧钢筋条
hot runner 热浇道
hot sprue 热嘴
hot stamping 烫印
hot sync 热同步
hot water boiler 热水锅炉
hot water circulation pump 热水循环泵
hot wire 热电阻线，热线，热丝
hot wire anemometer 热线风速表
hot wire flow measuring device 热丝流量测量仪表
hot wire flow sensor 热线流量传感器
hot wire flow transducer 热线流量传感器
hot work die steel 热锻模用钢

hot 热的
hot circuit 通电线路
hot start 热态启动
hot-dip galvanized steel wire 热镀锌钢丝
hot-dip galvanizing 热浸镀锌
hot-runner mold 热流道模具
hot-wire instrument 热线式仪表
hot-wire respiratory flow sensor 热丝（式）呼吸流量传感器
hot-wire respiratory flow transducer 热丝（式）呼吸流量传感器
hot-wire turbulence meter 热线湍流计
hotwork 热加工
hourglass 沙漏，水漏
hour-glass 沙漏，水漏
hourly average display 小时平均画面
hourly report 时报
house automation 家庭自动化
house supply 厂用电
house telephone 内部电话
household appliances 家用电器
housing 外壳
housing limit temperature 外壳极限温度
HPCE（high performance capillary electrophoresis） 高效毛细管电泳法
HPLC（high performance liquid chromatography） 高效液相色谱法
HPLC（high pressure liquid chromatography） 高压液相色谱
HPLC-MS（high performance liquid chromatography-mass spectrometry） 高效液相色谱-质谱联用
HQI（hit quality index） 命中质量系数
HRGC（high resolution gas chromatography） 高分辨率气相色谱法
HRV（high resolution visible） 高分辨率
H-shaped iron 工字钢
HU（heat unit） 热单位
hub 仓库，集线器
Huffman code 霍夫曼编码
huge system 巨系统
hum 交流声
human brain 人脑
human error 人为误差，主观误差

human factor 人为因素
human factors engineering 人因工程学
human information processing 人类信息处理
human interface station (HIS) 人机界面站
human machine interaction 人机交互
human machine interface (HMI) 人机接口，人机界面
human perception 人类知觉
human reliability 人工可靠性
human resource management 人力资源管理
human supervisory control 人工监督控制
human system diagnosis and improvement 人体系统诊断与改善
human system interaction 人工系统交互
human system interface 人工系统接口
human-centered design 以人为中心设计
human-computer interaction (HCI) 人机相互作用，人机交互
human-machine system design 人机系统设计
humicap 湿敏电容器
humid air 湿空气
humid heat test 湿热试验
humidification 加湿
humidifier 喷雾机，加湿器
humidistat 恒湿箱
humidity 湿度
humidity constant state 恒湿状态
humidity controller 湿度控制器
humidity fluctuation 湿度波动度
humidity of the air 空气湿度
humidity range 湿度范围
humidity sensitive switch 湿敏开关
humidity sensor 湿度传感器
humidity transducer 湿度传感器
humidity uniformity 湿度均匀性
humidity-dependent resistor 湿敏电阻
hummer 蜂鸣器
hunting 摆动，猎振
HVDC 高压直流（电）
HVDC transmission lines 高压直流传输线
hybrid 混合
hybrid circuit 混合电路
hybrid computer 混合计算机
hybrid connection 混联
hybrid faceplate block 混合面板功能块
hybrid junction 混合连接
hybrid model 混杂模型
hybrid parameter 混合参数
hybrid simulation 混合仿真
hybrid system 混杂系统
hybrid vehicle 混合驱动汽车
hybridization affect 杂化影响
hybridization oven 分子杂交仪
hydracid 氢酸
hydrant 消防龙头
hydrant outlet 消防龙头出口
hydrate 水合物
hydrated lime 氢氧化钙，熟石灰
hydraulic 液压，水利的
hydraulic accumulator 水力蓄力器
hydraulic actuator 液动执行机构，液动执行器
hydraulic amplifier 液动放大器，液力放大器
hydraulic and hydro-power engineering 水利水电工程
hydraulic components 液压元件
hydraulic control 液动控制，液压控制
hydraulic couplers 液力耦合器
hydraulic cylinder 液压缸
hydraulic diameter 水力直径
hydraulic engineering 水利工程
hydraulic excavator 液压式挖土机
hydraulic fluid 液压油
hydraulic generator 水轮发电机
hydraulic handjack 油压板车
hydraulic jack 液压千斤顶
hydraulic machine 油压机
hydraulic mechanism 液压机构
hydraulic motor 液动马达，液力马达，液压马达
hydraulic natural frequency 液压固有频率
hydraulic oil 液压油
hydraulic platform 液压升降工作台
hydraulic power 液压动力，水力发电

hydraulic power supply 液压动力源
hydraulic power tools 液压动力工具
hydraulic power units 液压动力元件
hydraulic pressure 液压
hydraulic pump 液压泵
hydraulic ram cylinder 液压伸缩筒
hydraulic relay 液动继电器,液力继电器,液压继电器
hydraulic relay valves 液压继动阀
hydraulic relief valve 液压减压阀
hydraulic rotary cylinders 液压回转缸
hydraulic servo-motor 液压伺服电动机
hydraulic shock absorber 液压减震器
hydraulic steering system 液压转向系统
hydraulic step motor 液压步进马达
hydraulic stepless speed changes 液压无级变速
hydraulic structure engineering 水工结构工程
hydraulic testing machine 液压式试验机
hydraulic transmission control system 液压传动控制系统
hydraulic turbine 水轮机
hydraulic vibrator 液压振动器
hydraulic-formed bellows 液压波纹管
hydraulics and river dynamics 水力学及河流动力学
hydraumatic lubrication 液压润滑
hydrocarbon 碳氢化合物,烃
hydrochloric acid 盐酸
hydrodynamic drive 液力传动
hydrodynamic injection 动力进样
hydrodynamic noise 液体动力噪声
hydrodynamic volume 流体力学体积
hydroelectric 水电的
hydroelectric system 水力发电系统
hydrofoil shape vehicle 水翼型拖曳体
hydrogen (H) 氢
hydrogen bomb 氢弹
hydrogen chloride 氯化氢
hydrogen electrode 氢电极
hydrogen flame ionization detector 氢焰离子化检测器
hydrogen pressure gauge 氢压力表
hydrogen regulator 氢气减压阀

hydrogen sulfide 硫化氢
hydro-generator 水轮发电机
hydrographic survey 水文测量,河海测量
hydrographic winch 水文绞车
hydrology and water resources 水文学及水资源
hydrolysis 水解
hydrometer 比重计,流速表
hydrophobic interaction chromatography 疏水作用色谱法
hydrophone 水诊器,检漏器,水中听音器,水听器
hydrophone calibrator 水听器校准器
hydro-pneumatic system 水压气动系统
hydropower station 水电站
hydrostatic head 静压头
hydrostatic level 液体静力水准仪
hydrostatic pressure test 液体静水压试验
hydrostatic test 流体静力学试验,水压试验
hydrosulphuric acid 氢硫酸
hydrothermal power system 热电站系统
hydroxide 氢氧化物,羟化物
hygrograph 湿度计
hygrometer 湿度表,湿度计
hygrometer calibration chamber 湿度表检定箱
hygrothermograph 温湿计
hyperboloid gear 双曲面齿轮
hypercap diode 变容二极管
hyperchromic effect 增色效应(浓色效应)
hypercycle theory 超循环理论
hyper-high-frequency 超高频
hyperstability 超稳定性
hypertension 高血压,超高压
hyphenated techniques chromatography 色谱联用技术
hypochromic effect 减色效应,淡色效应
hypoid gear 准双曲面齿轮
hypoid offset 准双曲面齿轮偏置距
hypothesis 假设,假说
hypsometer 沸点测高表
hysteresis 磁滞,滞后,滞环,滞后

现象
hysteresis comparison 滞环比较方式
hysteresis curve 磁滞曲线，磁滞回线
hysteresis error 回差，滞环误差，滞后误差
hysteresis loop 磁滞回线，磁滞回路，磁滞损耗，滞环
hysteresis loss 磁滞损失
hysteresis motor 磁滞电动机，理想发动机，永动机
hysteretic 磁滞的
hysteretic error 回差

I

I beam 工字梁
I controller 积分控制器
I/O channel 输入/输出通道，I/O 通道
I/O link extender I/O 连接扩展器
I/O module (IOM) 输入/输出模块，输入/输出模件
I/O module nest 输入/输出模件箱
I/O report 输入/输出报告
I/O unit 输入/输出单元
I/O point 输入/输出点
I/O (input/output) 输入/输出
IAE (integral of the absolute error) 绝对偏差积分
IC (integrated circuit) 集成电路
ice pack 冰袋
ice point 冰点
ICMP (internet control message protocol) 控制信息协议
ICU monitor 重症监护仪
ideal 理想的
ideal condition 理想条件
ideal controller 理想控制器
ideal current source 理想电流源
ideal element 理想元件
ideal final value 理想终值
ideal gas temperature scale 理想气体温标
ideal gyrator 理想回转器
ideal low-pass filter 理想低通滤波器
ideal network 无损耗网络
ideal source 理想电源
ideal transformer 理想变压器
ideal value 理想值
ideal voltage amplifier 理想电压放大器
ideal voltage source 理想电压源
ideal wire 理想导线，尤勒卡导线，铜镍合金丝
idealized system 理想系统
identifiability 可辨识性，能辨识性
identification 辨识，标识
identification algorithm 辨识算法
identification mark 识别标志
identification test 鉴别试验，药物的鉴别试验
identifier 辨识器
identifying sheet list 标示单
IDF (intermediate data file) 中间数据文件
idle 无效的，无功的，不工作的，空转，空载的
idle component 无功分量
idle current 无功电流
idle gear 惰轮
idle running 空转
idle run-normal run selector switch 急速-快速运行选择键
idle speed control 空速控制
idle stage 空站
idler 空转轮
IDSS (intelligent decision support system) 智能决策支持系统
IEC (ion exchange chromatography) 离子交换色谱法
IEC (International Electrotechnical Commission) 国际电工委员会
IEEE (Institute of Electrical and Electronics Engineers) 电气与电子工程师学会
IF amplifier 中频放大器
If-Then operator If-Then 操作
ignition 点火，火，灼热
ignition electrode 点火电极
ignition wire 点火线
ignitor 点燃剂，起燃剂，点火剂，点火器
illuminance 照度，照明度
illuminance level 照明度
illuminance sensor 照度传感器
illuminance transducer 照度传感器
illuminate 照明
illuminated nameplate 光字排
illuminated sign 照明标志

illuminating 照明的
illuminating aperture angle 照明孔径角
illuminating brightness 照明亮度
illuminating system 照明系统
illumination 照明
illumination level 照明度
illumination meter 照度计
illuminator 照明器
illuminometer 照度计
ILS (ionization lose spectroscopy) 离子化损失谱法
image 图像，影像，映像
image aerial 电视天线
image amplification 图像放大
image analysis 图像分析，映像分析
image coding 图像编码
image compression 图像压缩
image converter 图像转换器，光电变换器
image distortion 图像失真，图像变形
image electro-magnetic shift 图像电磁位移
image enhancement 图像增强
image frequency 像频，视频
image frequency interference 像频干扰
image intensifier 图像放大器，图像增强器
image interpolation 图像插补
image matching 图像匹配
image method 镜像法
image modelling 图像模型化
image motion compensation 图像移动补偿
image pickup tube 摄像管
image plane 像平面
image point 像点
image processing 图像处理
image processor 图像处理器
image recognition 图像识别，图形识别
image reconstruction 图像重构，图形重构
image registration 图像配准，光栅重合
image restoration 图像修复，图像恢复
image rotation 像旋转
image segmentation 图像分割

image sensor 图像传感器
image smoothing 图像平滑
image transducer 图像传感器
imaginary 虚的，虚数的，假象的
imaginary axis 虚轴
imaginary component 虚部，无功部分
imaginary number 虚数
imaginary part 虚部
imaginary quantity 虚数
imaginary root 虚根
imaging system 成像系统
imbalance 不平衡
IMC (internal model control) 内模控制
imitation 模拟，仿造
imitator 模拟器
immediate change in feedback pressure 反馈压力瞬间变化值
immediate vicinity 紧邻
immersed tube 沉管
immersed tube tunnel 沉管隧道
immersion cooler 浸没式冷却器
immersion depth 浸入深度
immersion error 浸入误差
immersion reflectometer 浸入式反射计
immersion testing 浸渍试验，浸泡试验，浸没试验
immersion type probe 水浸探头
imminent danger 迫切危险
immobilized phase open tubular column 固定化相开管柱
impact 碰撞，脉冲，冲击
impact damper 缓冲器
impact energy 冲击能量
impact extrusion 冲击挤压加工
impact flowmeter 冲击式流量计
impact force 撞击力，冲击力
impact hammer 冲击锤体
impact load 冲击荷载
impact pendulum 冲击摆锤
impact specimen support 冲击试校支架
impact test 冲击试验
impact testing machine 冲击试验机
impact toughness 冲击韧性
impact velocity of the pendulum 摆锤的冲击速度
impacter 卧式锻造机，冲击器

impair 减损
IMPC (internal model predictive control) 内模预测控制
impedance 导纳,阻抗
impedance analysis 阻抗分析
impedance blood volume sensor 阻抗式血容量传感器
impedance blood volume transducer 阻抗式血容量传感器
impedance coil 阻抗线圈
impedance control 阻抗控制
impedance convertor 阻抗变换器
impedance head 阻抗头
impedance matching 阻抗匹配
impedance matrix 阻抗矩阵
impedance plane diagram 阻抗平面图
impedance respiratory frequency sensor 阻抗式呼吸频率传感器
impedance respiratory frequency transducer 阻抗式呼吸频率传感器
impedance transformer 阻抗变换器
impedance triangle 阻抗三角形
impedance value 阻抗值
impeller 旋桨,叶轮,推进器
impeller-turbine mass flowmeter 叶轮-涡轮式质量流量计
imperforate 无孔
impermeable 防渗
impervious 不透水
impervious material 不透水物料
implementation 实施,执行
implication operator 隐含算子
implicit system 隐式系统
import 进口,引入
important tag 重要工位号
imposed load 附加荷载,外加荷载
impregnated base strain gauge 浸胶基应变计
impregnated thin layer chromatography 浸渍薄层色谱法
impregnating insulation paper 浸渍绝缘纸
impulse 脉冲,冲击,冲量
impulse amplitude 脉冲幅度
impulse analyzer 脉冲分析仪
impulse circuit 脉冲电路
impulse condition 脉冲条件
impulse control 脉冲控制
impulse counter 脉冲计数器
impulse current generator 脉冲电流发生器
impulse distance meter 脉冲式测距仪
impulse function 冲击函数,脉冲函数
impulse generator 脉冲发生器
impulse input 脉冲输入
impulse pipe 脉冲(导压)管线
impulse precision sound level meter 脉冲精密声级计
impulse regulator 脉冲调节器
impulse relay 脉冲继电器
impulse response 冲激响应,脉冲响应
impulse response model 脉冲响应模型
impulse sequence 脉冲序列
impulse signal 脉冲信号
impulse wave 冲击波,脉冲波
impulse-forced response 脉冲响应
impulsive sound 脉冲声
impurity 杂质
in equilibrium 平衡状态
in isolation 绝缘
in parallel 并联
in situ quantitation 原位定量
in synchronism 同步
inaccuracy 不精确度
inblock cast 整体铸造
inboard rotor 内重心转子
inboard 内侧
incandescent lamp 白炽灯
incendivity 点燃性
inch 英寸
inching 寸动,点动,微调尺寸,缓动
incidence 入射
incidence matrix 关联矩阵
incident 入射的,事件,事故
incident wave 入射波
incinerator 垃圾焚化炉
inclination 坡度,斜度,倾角,倾斜度
inclination error 倾斜误差
incline impact test 斜面冲击试验
inclined plane 倾斜面
inclined strut 斜撑
inclined wire 斜网
inclinometer 倾角罗盘,倾斜计,倾

indicial

角计
inclusion 杂质
incombustible material 不燃材料
incoming 输入的，引入的
incoming circuit 输入电路
incoming current 输入电流
incoming degree 出度
incoming quality control（IQC） 进料品质管制人员
incompatibility principle 不相容原理
incomplete data 不完备数据，不完全数据
inconsequential 不合逻辑的
incorporated 合成一体的
increase 增加，增长
increase of area 断面增大率
increase of voltage 升压
increased safety 增安型
increased safety electrical apparatus 增安型电气设备
increased safety electrical instrument 增安型电动仪表
increasing engineering productivity component and module 增加工程能力的成分和模块
increment 增量
increment or decrement work 盈亏功
incremental control 增量控制
incremental integrator 增量积分器
incremental key 增加键
incremental motion control 增量运动控制
incremental motion control system 增量运动控制系统
incremental range 微调（增量）范围
incremental summer algorithm 增量加法器算法
incremental testing 增量测试
incrustation 水锈
incubator 培养箱
indefinite integral 不定积分
indentation 压痕
indented pre-stressed concrete steel wire 预应力混凝土结构用刻痕钢丝
indenting 压痕加工
independent conformity 独立一致性
independent linearity 独立线性度

independent source 独立源
indeterminate error 不可定误差
index 检索，索引，指示，指示器，指数
index gears 分度齿轮
index head 分度头
index mark 指数，索引
index method 检索方法
index of asymmetry 非对称性指数
index of merit 品质因数
index plate 分度盘
index profile 指示曲线
index register 变址寄存器
index tolerance 分度公差
index variation 分度变化量
India-rubber wire 橡胶绝缘线
indicate 指示，表示，表明
indicated angle 指示角度
indicated electrode 指示电极
indicated strain 指示应变
indicated value 指示值
indicating 指示，指示的，指示式的
indicating controller 指示控制器
indicating device 指示，指示机构，指示装置
indicating error 指示误差，示值误差
indicating instrument 指示仪表，指示仪器仪表
indicating lamp 指示灯
indicating pressure controller 指示压力调节器
indication 示值
indication error 指示误差
indication of a measuring instrument 测量仪表示值
indication point 指示点
indication range 示值范围
indicative mark 指示标志
indicator 指示器，显示器，指示仪
indicator anchorage 指示表支撑座
indicator constant 指示剂常数
indicator electrode 指示电极
indicator lamp 指示灯
indicator light 指示灯
indicator travel 指示仪表行程
indicial response 单位阶跃响应，指数响应

177

indirect 间接的
indirect acting instrument 间接作用仪表
indirect acting measuring instrument 间接作用式测量仪表
indirect acting recording instrument 间接作用记录仪
indirect control 间接控制
indirect current control 间接电流控制
indirect DC-DC converter 间接电流变换电路
indirect drive 间接传动
indirect measurement 间接测量
indirect method of measurement 间接测量法
indirect resistance heating 间接电阻加热
indirectly controlled system 间接被控系统
indirectly controlled variable 间接被控变量
indirectly heated type thermistor 旁热式热敏感电阻器，旁热式热敏电阻
indium (In) 铟
individual line 专线
individual nominal characteristic 单个名义特性
individual stop valve 独立截止阀
induce 感应
induced 感应，导致
induced current 感生电流
induced current image 感生电流像
induced current method 感应电流法
induced draft fan 抽风式风扇，抽风机
induced EMF 感应电动势
induced pulse transient system 感应脉冲瞬变系统
induced voltage 感应电压
induced wire 感应电路
inducer 电感器
inductance 感应，电感，感应系数
inductance box 电感箱
inductance coil 电感线圈
inductance pressure transducer 电感式压力传感器
inductance transducer 感应式传感器

induction 感应，激发
induction generator 感应发电机
induction hardening 感应硬化，感应淬火
induction instrument 感应式仪表，感应系仪表
induction light 感应光
induction logger 感应测井仪
induction machine 感应电机，感应式机械，感应式电机
induction meter 感应式电度表
induction motor 感应电动机，感应式电机
induction motor design 感应式电机设计
induction salinometer 感应式盐度计
induction voltage divider 感应分压器
induction motor 感应电动机
inductive character 电感性
inductive component 感性分量，无功分量
inductive control 感应控制
inductive coupled plasma emission spectrometer 电感耦合等离子体发射光谱仪
inductive displacement measuring instrument 电感式位移测量仪
inductive displacement transducer 电感式位移传感器
inductive EMF 感应电动势
inductive feedback 电感反馈
inductive force transducer 电感式力传感器，电感式位移传感器
inductive load 感性负载
inductive micrometer 电感测微计
inductive modeling method 归纳建模法
inductive pick-off 感应式检出器
inductive sensor 电感式传感器
inductive susceptance 感纳
inductive tensometer 电感式张力计
inductive transducer 电感式传感器
inductive reactance 感抗
inductometer 电感计
inductor 电感器，感应器
inductosyn 感应式传感器
inductosyn displacement transducer 感应同步式位移传感器

industrial 工业
industrial alcohol 工业酒精
industrial area 工业区
industrial automation 工业自动化
industrial bimetallic thermometer 工业双金属温度计
industrial catalysis 工业催化
industrial communication 工业通信
industrial computer 工业用计算机
industrial control 工业控制
industrial control system 工业控制系统
industrial data processing 工业数据处理
industrial distribution system 工业布线系统
industrial enclosure 工业机箱, 工业机柜
industrial environment evaluation 工业环境评估
industrial frequency 工业频率
industrial moulding design 工业造型设计
industrial organizations and management 工业组织与管理
industrial process 工业过程, 工业工序
industrial process measurement and control instrument 工业自动化仪表, 工业过程检测控制仪表
industrial production system 工业生产系统
industrial robot 工业机器人
industrial safety 工业安全
industrial total radiation pyrometer 工业辐射高温计
industry standard architecture (ISA) 工业标准架构
inelastic background 非弹性本底
inelastic scatter 非弹性散射过程
inert gas 惰性气体
inert gas ovens 惰性气体恒温器
inert metal indicated electrode 惰性金属指示电极
inertia 惯性, 惯量
inertia factor 惯量
inertia force 惯性力

inertia load 惯性负载
inertial 惯性的, 惯量的
inertial attitude sensor 惯性姿态敏感器
inertial coordinate system 惯性坐标系
inertial matrix 惯性矩阵
inertial measurement unit 惯性测量部件
inertial motion 隋走量, 惯性运动
inertial navigation 惯性导航
inertial platform 惯性平台
inertial reference unit 惯性参考部件
inertial sensor 惯性传感器
inertial wheel 惯性轮
inference 推理, 推断
inference engine 推理机, 推理控制
inference mechanism 推理机
inference model 推理模型
inference process 推理过程
inference strategy 推理策略
inferential control 推断控制
inferential flowmeter 推导式流量计
inferential type flowmeter 推导式流量计
infill 填料
infinite 无穷大
infinite dimensional system 无穷维系统
infinite voltage gain 无穷大电压增益
infinity control 无穷控制
inflammable 易燃, 易燃物
inflammable gas 易燃气体
inflatable dam 充气水闸
inflation 充气, 膨胀
inflator 充气机
inflection point 拐点
influence characteristic 影响特性
influence coefficient method 影响系数法
influence error 影响误差
influence of magnetic field 磁场影响
influence of ship-body 船体影响
influence of static pressure 静压影响
influence of vibration 振动影响
influence quantity 影响量
information 信息, 情报, 资料
information acquisition 信息采集
information analysis 信息分析, 资料

分析
information and communication engineering 信息与通信工程
information capacity 信息容量
information depth 信息深度，信息长度
information flow 信息流
information flow diagram 信息流图
information integration 信息集成
information network 信息网
information network interface 信息网络接口
information parameter 信息参数
information parameter of an electrical signal 电信号的信息参数
information pattern 信息模式
information processing 信息处理
information retrieval 信息检索，信息恢复
information structure 信息结构
information system 信息系统
information system for process control 过程控制信息系统
information technology 信息技术
information theory 信息论
information wire 信息线
information-carrying wire 信息传递线
infrared absorption spectra hygrometer 红外吸收光谱湿度计
infrared camera 红外照相机
infrared detector 红外检测器，红外探测器
infrared distance meter 红外测距仪
infra-red electronic distance meter 红外线电子测距仪
infrared gas analyzer 红外线气体分析器
infrared hygrometer 红外湿度表
infrared light sensor 红外光传感器
infrared light transducer 红外光传感器
infrared photoelectric switch 红外光电开关
infrared radiation 红外辐射，红外线，热辐射
infrared radiation detection apparatus 红外检测仪

infrared radiation thermometer 红外辐射温度计
infrared radiometer 红外辐射计
infrared radiometry 红外线探伤法
infrared ray (IR) 红外线
infrared remote sensing 红外遥感
infra-red security system 红外线保安系统
infrared spectrometry 红外光谱法
infrared spectrophotometer 红外分光光度计，红外分光光度法
infrared spectroscopy 红外吸收光谱法
infrared type gas analyzer 红外线气体分析器
ingot 铸锭
ingot blank 铸坯
ingot iron 锭铁，低碳钢
ingot mold 钢锭模
ingredient 配料，成分
ingress and egress 进出
ingress point 入口
inherent 固有的，本征的
inherent characteristic of a system 系统固有特性
inherent diaphragm pressure range 固有膜片压力范围
inherent feedback 固有反馈
inherent filtration 固有滤过当量
inherent flow characteristic 固有流量特性
inherent instability 固有不稳定性
inherent nonlinearity 固有非线性
inherent regulation 固有自动调整，固有调整
inherent stability 本征稳定，固有稳定性
inherent weakness failure 固有弱质失效
inherently 固有地
inheritance 继承，遗传
inhibit 禁止
inhibit wire 禁止线
inhibitor 阻抑剂
init 起始，初始化
initial characterization 初始特征化
initial condition 初始条件
initial contact 起始啮合点，初接触
initial deviation 初始偏差

initial load 初负荷
initial phase 初相位
initial phase angle 初相角
initial pitting 初期点蚀
initial position 初始位置
initial program loader (IPL) 初始程序的装入程序
initial pulse 初始脉冲
initial state 初始状态
initial temperature 起始温度，初始温度
initial unbalance 初始不平衡量
initial value 初始值
initial verification 首次检定
initial 初始的，最初的
initialization 初始化
initialization of control algorithm 控制算法初始化
initialize 初始化
initialize personality 初始化属性
initial-value theorem 初值定理
initiator 启动站，发起站
injection attitude 入轨姿势
injection cross-section 注入横截面
injection molding 喷射造型法
injection moulding 注射模
injection nozzle 注射管嘴，射出喷嘴
injection pipe 喷射管，注射管
injection plunger 注射柱塞，注射活塞
injection ram 注射活塞，压射柱塞
injection set 注射器
injector 注射器
injunction 禁止令，强制令
ink jet printer 喷墨印刷机
inkless recorder 无墨水式记录仪
inland lot 内地段
inlay 镶嵌
inlay casting 镶铸法
inlet 进口，入口
inlet pipe 入口管
inlet system 进样系统
inlet valve 进气阀，进给阀
in-line or crank-slider mechanism 对心曲柄滑块机构
in-line roller follower 对心滚子从动件

in-line slider-crank mechanism 对心曲柄滑块机构
in-line translating follower 对心直动从动件
inner 内部的
inner addendum 小端齿顶高
inner cone distance 小端锥距
inner dedendum 小端齿根高
inner guiding post 内导柱
inner hexagon screw 内六角螺钉
inner loop 内环
inner matrix 内部矩阵
inner package 内包装
inner parts inspect 内部检查
inner plate 内板
inner plunger 内柱塞
inner punch 内冲头
inner ring 内圈
inner slot width 小端槽宽
inner spiral angle 小端螺旋角
inner stripper 内脱料板
inner wire 内部钢丝，（钢丝蝇中的）内索
inoculated cast iron 孕育铸铁
inorganic chemistry 无机化学
inphase 同相的
in-phase 同相的
inphase component 同相分量
in-phase opposition 反相的
in-phase voltage 同相电压
in-plane bending vibration 面内弯曲振动
input 输入，输入端
input admittance 输入导纳
input and output with isolated common point 带有隔离公共点的输入和输出
input centralized system 集中输入系统
input circuit 输入电路
input compensation 输入补偿
input decentralized system 分散输入系统
input device 输入设备，输入装置
input dynamics 输入动态
input element 输入环节，输入元件
input equipment 输入设备
input estimation 输入估计

input impedance 输入阻抗
input impedance of the magnetizing winding 激磁绕组输入阻抗
input impedance of the measuring winding 测量绕组输入阻抗
input impulse 输入脉冲
input indicator 输入指示器
input indicator with deviation alarm 带偏差报警的输入指示器
input link 输入构件
input matrix 输入矩阵
input open check 输入开路检查
input parameter 输入参数
input prediction method 输入预估法
input quantity to frequency conversion type 输入量-频率转换型
input resistance 输入电阻
input signal 输入信号
input terminal 输入端
input unit 输入设备
input variable 输入变量
input vector 输入向量，输入矢量
input with isolated common point 公共端隔离输入
input/output (I/O) 输入/输出
input/output operation 输入/输出操作
input/output port 输入/输出端口
input-output analysis 投入产出分析
input-output channel 输入-输出通道
input-output device 输入-输出设备
input-output equipment 输入-输出装置
input-output interface 输入-输出接口
input-output linearization 输入-输出线性化
input-output model 输入-输出模型，投入产出模型
input-output processor 输入-输出处理器
input-output table 投入产出表
input-output unit 输入-输出单元，输入-输出设备
inrush current 涌流，浪涌电流
INS (ion neutralizing spectroscopy) 离子中和谱法
inscribed circle 内接圆面

inscription 注字（指铭牌），写上
insecticidal fluid 杀虫液体
insensitive 不灵敏的，低灵敏度的
insensitivity 不灵敏性
insert 入块（嵌入件），嵌件，插件，插入，镶件
insert core 放置入子，插入型芯
insert length 插入长度
insert pin 嵌件销
inserted blade cutter 镶片刀盘
insertion length 插入深度
insertion turbine meter 插入式涡轮流量计
insert-type 插入式
inset 插图，插页
inside blade 内切刀齿
inside calipers 内卡钳，内径尺
inside calliper 内径尺，内卡钳
inside coil 内插线圈
inside glazing 内镶玻璃
inside micrometer 内径千分尺
inside point diameter 内切刀尖直径
inside thread 内螺纹
inside 内侧，内部
in-situ colour meter 现场水色计
in-situ concrete strength 现场混凝土强度
in-situ cube strength 现场立方体强度
in-situ extraction sampler 现场萃取采水器
in-situ material testing 现场物料测试
in-situ moisture content 现场含水量
in-situ permeability 现场渗透度
insolation 日射，曝晒
inspection 检验，检查，视察，观察
inspection chamber 检查井
inspection manhole 检查井
inspection pit 检查井
inspection specification 成品检验规范
inspector 督察员，视察员，检查员，检测
instability 不稳定，不稳定性
install 安装
installation 安装
installation environment specification 安装环境说明书

installation of oceanographic survey 海洋调查装备
installation specification 安装规格
instantaneous 瞬间的
instantaneous availability 瞬时可用度
instantaneous center 瞬心
instantaneous center of velocity 速度瞬心
instantaneous contact pattern 瞬时接触斑点
instantaneous electric power 瞬时电功率
instantaneous flowmeter 瞬时流量计
instantaneous mechanical power 瞬时机械功率
instantaneous sound pressure 瞬时声压
instantaneous value 瞬时风速，瞬时值
instantaneous water heater 即热式热水炉
Institute of Electrical and Electronics Engineers (IEEE) 电气与电子工程师学会
instruction counter 指令计数器，指令地址寄存器
instruction level language 指令级语言
instruction manual 使用说明书
instruction register 指令寄存器
instruction set architecture (ISA) 指令集体系结构
instruction 指令，指导，指示，说明书，
instrument 仪表，仪器，器具
instrument accessory 仪器附件
instrument auto transformer 仪用自耦互感器
instrument block valve 仪表截止阀
instrument block valve with pressed coupling 压套式仪表截止阀
instrument box 仪表箱
instrument case 仪表箱
instrument communication network 仪表通信网络
instrument constant 仪表常数，仪表常数
instrument desk 仪表台
instrument error 仪表误差
instrument faceplate 仪表面板
instrument faceplate data entry 仪表面板数据输入
instrument flange 仪表法兰，仪表安装边缘
instrument for automatic sequence control 自动顺序控制仪表
instrument for nondestructive testing 无损检测仪
instrument front 仪表面板，仪表正面
instrument lead 仪表导线
instrument of ground electrochemical prospecting method 地电化学法找矿仪器
instrument of magneto telluric method 磁大地电流法仪器
instrument panel 仪表屏，仪表盘
instrument panel-board inside wiring diagram 仪表盘盘内接线图
instrument panel-board layout 仪表盘正面布置图
instrument pen 仪表记录笔
instrument protecting box 仪表保护箱
instrument rack 计测器支架，计测器框架
instrument science and technology 仪器科学与技术
instrument screen 仪器屏幕
instrument security factor 仪表的安全因数
instrument tag No. 仪表位号
instrument transformer 仪用互感器
instrument type and specification 设备型号规格
instrument with contacts 带触点仪表
instrument with electromagnetic screening 电磁屏蔽仪表
instrument with electrostatic screening 静电屏蔽仪表
instrument with locking device 带有锁定的仪表
instrument with optical index 光标式仪表
instrument with suppressed zero 压缩零位仪表
instrumental analysis 仪器分析，仪

器分析法
instrumental background 仪器本底
instrumental error 仪器误差
instrumentation 仪表化，使用仪表，仪表装置，仪表
instrumentation accuracy 仪表精度，仪器精度
instruments of the pneumatic aggregate 气动单元组合仪表
insufficient rigidity 强度不够
insufficient 不足的，不够的
insulate 绝缘，绝热，隔离
insulated aluminum wire 绝缘铝线
insulated cable 绝缘电缆
insulated ferrule 绝缘套圈
insulated gate bipolar transistor 绝缘栅双极型晶体管
insulated runner 绝缘浇道方式
insulated system 不接地系统
insulating barrier 绝缘障
insulating layer 绝缘层
insulating material 绝缘材料
insulating rod 绝缘棒
insulating shield 绝缘护片
insulating strength 绝缘强度
insulating test voltage 绝缘试验电压
insulating-resistance tester 绝缘电阻测试器
insulation 绝缘，隔离，绝热，保温
insulation class 绝缘等级
insulation displacement connection 绝缘层信移连接件
insulation fault detecting instrument 绝缘损坏检示仪表
insulation lagging 绝缘层
insulation material 绝缘物，绝缘材料
insulation resistance 绝缘电阻
insulation resistance meter 高阻表，兆欧表
insulation sleeve 绝缘护套
insulation spacer 绝缘垫片
insulation strength 绝缘强度
insulation test 绝缘试验
insulation voltage breakdown 绝缘击穿电压，绝缘强度
insulator 绝缘体，隔离器
insulin sensitivity 胰岛素敏感

intake fan 进气风扇
intake manifold 进气歧管
intake valve 进气阀
integer programming 整数规划
integer 整数
integral 积分
integral action 积分作用
integral action coefficient 积分作用，积分作用系数
integral action factor 积分作用因子
integral action limiter 积分作用限幅器，积分作用限制器
integral action rate 积分速率，积分作用率
integral action time 积分作用时间
integral action time constant 积分作用时间常数
integral control 积分控制
integral control action 积分控制作用
integral controller 积分控制器，I 控制器，积分调节器
integral cross section 截面积分
integral detector 积分检测器，积分型检测器
integral electric actuator 积分式电动执行机构
integral equation 积分方程
integral equation formulation 积分方程式
integral error criteria 积分偏差判据
integral feedback 积分反馈
integral flow orifice differential pressure transmitter 基地式差压流量变送器，内孔板式差压流量变送器
integral formulation 积分公式
integral of absolute value of error criterion 绝对误差积分准则
integral of squared error criterion 平方误差积分准则
integral of the absolute error (IAE) 绝对偏差积分
integral of the squared error (ISE) 偏差平方积分
integral of the time and absolute error (ITAE) 绝对值偏差乘时间积分
integral performance criterion 积分性能准则

integral performance index 积分性能指标
integral representation 积分表示法
integral time 积分时间
integral time constant 积分时间常数
integral windup 积分饱和
integrated 集成
integrated circuit (IC) 集成电路,积分电路
integrated circuit amplifier 集成电路放大器
integrated circuit antenna 集成电路天线,集成线路天线
integrated circuit injection logic 集成注射逻辑
integrated data processing 统一数据处理
integrated gate-commutated thyristor 集成门极换流晶闸管
integrated optics 集成光学
integrated plant control 工厂集成控制
integrated regulator 稳压集成电路
integrated sensor 集成传感器
integrated services digital network (ISDN) 综合服务数字网
integrated vehicle highway system (IVHS) 集成汽车高速公路系统
integrated voltage regulator 集成稳压器
integrating 积算,积分,集成
integrating actinometer 累计光能计
integrating amplifier 集成放大器,积分放大器
integrating circuit 积分电路
integrating conversion 积分转换
integrating element 集成元件,积分元件
integrating gyro 集成陀螺
integrating induced type polarization potentiometer 积分式激发电位仪
integrating instrument 积分仪,积算仪器,积分器
integrating measuring instrument 积分测量仪表
integrating meter 积算仪表,积分器
integrating process 过程集成
integrating recording instrument 积分式记录仪

integrating sphere 积分球
integration 积分,求积,积分下限
integration instrument 积算仪器
integration method 积算法
integration testing 集成测试,综合测试
integrator 积分器,求积器,计算运算器,积分仪,积分仪器
integrity 完整,完整性,整体性
Intel 英特尔公司
intellectual property laws 知识产权法
intellectualized simulation software 智能化仿真软件
intelligence 智力,智能
intelligent 智能的,智力的
intelligent automation series 智能自动化系列
intelligent computer 智能计算机
intelligent control 智能控制
intelligent control system 智能控制系统
intelligent controller 智能控制器
intelligent cruise control 智能巡航控制
intelligent decision support system (IDSS) 智能决策支持系统
intelligent design 智能化设计
intelligent instrument 智能仪表
intelligent instrumentation 智能仪表
intelligent keyboard system 智能键盘系统
intelligent knowledge-base 智能知识库
intelligent machine 智能机器
intelligent management 智能管理
intelligent management system 智能管理系统
intelligent manufacturing system 智能制造系统
intelligent power module 智能功率模块
intelligent robot 智能机器人
intelligent sensor 智能传感器
intelligent simulation 智能仿真
intelligent station 智能站
intelligent system 智能系统
intelligent system model 智能系统模型
intelligent temperature transmitter 智能温度变送器

intelligent terminal 智能终端
intelligent transmitter 智能变送器
intensifying screen 光增强屏
intensity 强度，亮度
intensity change 强度改变
intensity control 强度控制，亮度控制
intensity modulation 强度调制
intensity modulation method 强度调制方法
intensity noise 强烈噪声
intensity of absorption band 吸收带的强度
intensity of field 场强
intensity of light 光强
intensity of magnetic field 磁场强度
intensity of polarization 极化强度
intensity of rainfall 降雨强度
interacted system 互联系统，关联系统
interacting 相互作用，反应，感应
interacting control loop 相关联控制回路，耦合控制回路
interacting service station 交互服务站
interacting system 耦合系统
interaction 互联，关联，相互耦合，相互作用
interaction index 相关指数，相关指标
interaction mechanism 相互作用机理
interactive 人-机对话的，交互的，相互作用的
interactive approach 人-机对话方法
interactive drawing design 交互式制图设计
interactive frequency response synthesis 相关频率响应综合
interactive multimedia association 交互式多媒体协议
interactive prediction approach 互联预估法，关联预估法
interactive program 人机交互程序
interactive terminal 交互式终端
interactive vehicle control 人-机机车控制，人-机汽车控制
interactive vehicle dynamics 人-机机车动力学，人-机汽车动力学
interactor matrix 交叉矩阵
intercepting channel 截水沟

intercepting ditch 截水沟
intercepting drain 截流沟，截水沟
intercepting sewer 截流污水管
interchange 交汇处，转车处
interchangeability 互换性
interchangeable accessory 可互换附件
interchangeable gears 互换性齿轮
interchangeable terminal 可互换终端
intercommunication system 内部通信系统
interconnected system 互联系统，耦合系统
interconnection 互联，耦合，互联
interconnection matrix 连接矩阵
interconnection network 互联网络
interconnection technology 连接技术
interdigital transducer 叉指式换能器
interdisciplinary design 交叉设计，多学科设计
interface 接口，界面，分界面，干扰
interface echo 界面反射波
interface message processor 接口通信处理机
interface state 接口状态
interface state generation 接口状态发生
interface unit 接口单元
interference current 干扰电流
interference detection 干扰探测
interference error 干扰误差
interference filter 干扰滤光片，干涉滤光片
interference fit 过盈配合
interference level 干扰电平
interference point 干涉点
interference prevention 防止干扰，抗干扰
interference 干扰，干涉
interfering 干扰性的
interfering signal 干扰信号
interferometer 干涉仪
interim improvement works 临时改善工程
interim measure 临时措施
interior package 内包装
interleaved 交叉
interleaved memory 交叉存储器

interlock 联锁
interlocking 联锁
interlocking contact 联锁触点
interlocking device 联锁装置
interlocking relay 联锁继电器
intermediate 中间，中层，中级
intermediate data file (IDF) 中间数据文件
intermediate distribution frame 中间分配线架
intermediate file 中间文件
intermediate frequency 中频
intermediate frequency amplifier 中频放大器
intermediate frequency receiver 中频接收机
intermediate frequency transformer 中频变压器
intermediate lens 中间镜
intermediate zone 中心区
intermediate relay 中间继电器
intermittent DC non-capacitive arc 非电容间歇直流电弧
intermittent duty 间歇工作方式，间歇任务
intermittent gearing 不完全齿轮机构，间歇传动装置
intermittent index 间断分度
intermittent motion mechanism 间歇运动机构
intermittent signal 继续信号，脉动信号
intermittent spray 间断喷雾
intermodulation distortion meter 互调失真仪
internal 内部，内部的
internal area 内部面积
internal bevel gear 内锥齿轮，内斜齿轮
internal calibrator 内部校准器
internal combustion engine 内燃机
internal component 内部组件
internal conversion 内转换
internal cylindrical machine 内圆磨床
internal diameter 内径，内直径
internal disturbance 内扰
internal dynamics 内部状态

internal face 内面，内壁面
internal force 内力
internal friction 内摩擦
internal gear 定子齿轮，内齿轮
internal grinding 内圆磨削
internal lead 内引线
internal lock signal 内锁信号
internal model control (IMC) 内模控制
internal model predictive control (IMPC) 内模预测控制
internal model principle 内模原理
internal porosity 内部气孔
internal radius 内半径
internal reference electrode 内参比电极
internal reference sample 内参比试样
internal reflection element 内反射元件
internal reflection spectrometry 内反射光谱法
internal relief valve 内置泄压阀
internal resistance 内阻
internal setpoint 内部设定
internal standard 内标准
internal standard line 内标线
internal standard method 内标法
internal status switch 内部状态开关
internal stop 内挡块
internal surface 内表面，内层
internal threaded block valve 内螺纹截止阀
internal topology 内拓扑学
internal-convection sensitive element 内对流敏感元件
international 全世界的
International Electrotechnical Commission (IEC) 国际电工委员会
international practical temperature scale (IPTS) 国际实用温标
International Science Organization (ISO) 国际科学组织
international stability 全局稳定
international standard 国际标准器，国际标准
International Standardization Organization (ISO) 国际标准化组织
international survey 全局观测，全局测量

international system of units (SI) 国际单位制
International Telecommunication Union - Telecommunication Standardization Sector 国际电信联盟-电信标准部
International Telecommunications Union (ITU) 国际电信联盟
International Union of Pure and Applied Chemistry (IUPAC) 国际纯粹与应用化学联合会
international workman standard 工艺标准
internet control message protocol (ICMP) 控制信息协议
internuclear double resonance 核间双共振法
interoffice 局间的，（公司或其他组织的）各办公室间的
interoperable system protocol (ISP) 可互操作系统协议
interphase 相同的
interplanetary spacecraft 宇宙飞船
interpolating oscillator 内插振荡器
interpolation 插补，插入
interpolation approximation 内插近似
interpolation error 内插误差
interpolation algorithm 插入算法，内插算法
interpolator 校对机
interpretation tree 描述树，判断树
interpreter 解释程序器
interpretive language 解释语言
inter-program communication 程序间通信
interrupt 中断
interrupt mask 中断屏蔽
interrupt priority 中断优先权
interrupt priority system 中断优先权系统
interrupt source 中断源
interrupter 中断器
interruption of supply 供应中断
intersect 交叉
intersection 交汇点，交叉点，交叉路口
interspace of double glazing 双层玻璃空隙
interstage load resistance 级间负载电阻

interstitial hole 中间孔
interstitial volume 粒间体积
inter-system crossing 体系间跨越
inter-terminal connection 端子间连接
interval 间隔
interval timer 间隔时钟
intracranial pressure sensor 颅内压传感器
intracranial pressure transducer 颅内压传感器
intra-plant system 厂内系统
intrinsic 固有的，内在的，本征的
intrinsic astigmatism 固有像散
intrinsic bistability 固有双稳定性
intrinsic error 固有误差，基本误差
intrinsic mode 固有模式
intrinsic region transistor 本征区晶体管
intrinsic safety barrier 本质安全栅，本质栅
intrinsic semi-conductor 本征半导体
intrinsic viscosity 固有黏度
intrinsically safe electrical instrument and wiring 本质安全型电动仪表和接线
intrinsically safe equipment and wiring 本质安全设备和接线
intrinsically safe circuit 本质电路
intruder alarm 防盗报警
invalid event 无效事件
invalidation 失效
invariance 不变性
invariant 不变式，标量
invariant embedding principle 不变嵌入原理
invariant system 不变性系统
inventory control 库存控制
inventory management system 库存管理系统
inverse 反变换，反相的，反向的，相反的
inverse kinematics 反向运动学
inverse cam mechanism 凸轮倒置机构
inverse current 反向电流
inverse dynamic problem 逆动态问题
inverse dynamics control 逆动态控制
inverse gear ratio 反齿数比
inverse heating rate curves 逆加热速

率曲线
inverse kinematic problem　逆动力学问题
inverse Laplace transform　拉普拉斯反变换
inverse matrix　逆矩阵
inverse Monte Carlo　逆蒙特卡罗
inverse Nyquist array　逆奈奎斯特阵列，逆奈奎斯特数组
inverse Nyquist diagram　逆奈奎斯特图
inverse scattering　逆散射
inverse scattering problem　逆散射问题
inverse system　可逆系统，逆系统
inverse transfer　逆变换
inverse transfer function　逆变换函数，逆转换函数
inverse transfer locus　逆变换轨迹，逆转换轨迹
inverse transform　逆变换，逆转换，反变换
inverse z-transformation　反 z 变换
inversion　倒置，反向，求逆，逆变
invert　倒拱，倒置，内底
inverted bell manometer　钟罩式压力计
inverted microscope　倒置显微镜
inverted output　反相输出
inverted sequence　逆序
inverter　逆变器，反向变流器，变换器，反向器，换流器，非门
inverter drive　反向驱动
inverting amplifier　倒相放大器，反相放大器
inverting telescope　倒像望远镜
invertor　逆变器
investigation　勘测，调查，勘查，勘察
investment casting　熔模铸造
investment decision　投资决策
involute　渐开线，渐开线的
involute equation　渐开线方程
involute function　渐开线函数
involute gear　渐开线齿轮
involute helicoid　渐开螺旋面
involute interference point　渐开线干涉点
involute profile　渐开线齿廓
involute spiral angle　渐开线螺旋角
involute spline　渐开线花键

involute teeth　渐开线齿
involute worm　渐开线蜗杆
involve　包含
iodimetry　碘量法
iodine (I)　碘
IOM (I/O module)　输入/输出模块，输入/输出模件
ion　离子
ion abundance　离子丰度
ion accelerating voltage　离子加速电压
ion beam thinner　离子减薄机
ion bombardment secondary electron image　离子轰击二次电子像
ion carbonitriding　离子渗碳氮化
ion carburizing　离子渗碳处理
ion chromatography　离子色谱法，离子色谱仪
ion counter　离子计数器
ion detector　离子检测器
ion exchange capacity　离子交换容量
ion exchange chromatography (IEC)　离子交换色谱法
ion exchange thin layer chromatography　离子交换薄层色谱法
ion exchanger　离子交换剂
ion flow anemometer　离子流风速表
ion kinetic energy spectra　离子动能谱
ion lens　离子透镜
ion liquid chromatography　离子液相色谱法
ion mobility　离子淌度
ion mobility spectrometer　离子淌度光谱仪
ion neutralization　离子中和
ion neutralization spectrometer　离子中和谱仪
ion neutralizing spectroscopy (INS)　离子中和谱法
ion optics　离子光学
ion pair　离子对（离子缔合物）
ion pair chromatography　离子对色谱法
ion pair extraction method　离子对提取法
ion pair reagent　离子对试剂
ion plating　离子电镀
ion pump　离子泵
ion repeller　离子排斥极
ion scattering spectrometer　离子散射

谱仪
ion selective electrode 离子选择电极
ion sensor 离子传感器
ion source 离子源
ion strength 离子强度
ion suppression chromatography 离子抑制色谱法
ion transducer 离子传感器
ion transmission efficiency 离子传输效率
ion-activity meter 离子计
ion-exchange chromatography 离子排斥色谱法
ion-exchange electrokinetic chromatography 离子交换动电色谱
ion-exchange low pressure chromatography 离子交换低压色谱法
ionic 离子的
ionization 电离，离子化，电离作用
ionization by sputtering 溅射电离
ionization chamber 电离室
ionization efficiency curve 电离效率曲线
ionization lose spectroscopy (ILS) 离化损失谱法
ionization smoke detector 电离烟尘检测器
ionized layer 电离层
ionized stratum 电离层
ionizing sensor 电离式传感器
ionizing transducer 电离式传感器
ionizing voltage 电离电压
ionospheric recorder 电离层记录器
ion-pair 离子对
ion-scattering spectroscopy (ISS) 离子散射谱法
ion-scattering spectrum 离子散射谱
ion-selective electrode 离子选择电极
ion-selective electrode analysis 离子选择电极分析
ion-selective electrode gas sensor 离子选择电极式气体传感器
ion-selective electrode gas transducer 离子选择电极式气体传感器
ion-selective measuring system 离子选择测量系统
IPL (initial program loader) 初始程序的装入程序

IPTS (international practical temperature scale) 国际实用温标
IQC (incoming quality control) 进料品质管制人员
IR (infrared ray) 红外线
iridium (Ir) 铱
iron 铁
iron chain 铁索
iron core 铁芯
iron loss 铁损
iron ore 铁矿石
iron plate 铁板
iron wire 低碳钢丝
iron/copper-nickel thermocouple 铁-铜镍热电偶
iron-carbon equilibrium diagram 铁-碳合金相图
iron-core coil 铁芯线圈
iron-cored 铁芯的
ironing 引缩加工，变薄压延
iron-Kober reagent 铁-柯柏试剂
ironless 无铁芯的
iron-loss 铁损
iron-phenol reagent 铁-酚试剂
irradiance 辐射照射度，辐照度
irreversible 不可逆的
irreversible reaction 不可逆反应
irritant 刺激剂
irritant gas 刺激性毒气
irritating smoke 刺激性烟气
ISA 1932 nozzle ISA1932 喷嘴
ISA (industry standard architecture) 工业标准架构
ISA (instruction set architecture) 指令集体系结构
ISDN (integrated services digital network) 综合服务数字网
ISE (integral of the squared error) 偏差平方积分
ISFET 离子选择场效应管
ISFET ion sensor 场效应管离子传感器
ISFET ion transducer 场效应管离子传感器
ISM (interpretive structure modeling) 解释结构建模法
ISO (International Science Organiza-

tion) 国际科学组织
ISO (International Standardization Organization) 国际标准化组织
isobaric mass-change determination 等压质量变化测定
isobaric weight-change curve 等压重量变化曲线
isobaric weight-change determination 等压重量变化测定
isoelectric focusing 等电点聚焦
isohydric solvent 等水溶剂
isolated amplifier 隔离放大器
isolated analogue input 隔离模拟输入，隔离的模拟输入
isolated junction 绝缘端
isolated junction type sheathed thermocouple 绝缘型铠装热电偶
isolated network 隔离网络
isolated system 孤立系统
isolating link 隔离开关
isolating partition 隔离间壁，绝缘隔板
isolating plate 隔板
isolating transformer 隔离变压器
isolating valve 隔断阀，隔离阀
isolation 隔离，分离，绝缘，隔振
isolation pad 隔离盘
isolator 隔离，刀闸，绝缘器，隔离器
isomer 同质异能素，同分异构物
isotherm 等温线
isothermal annealing 等温退火
isothermal forging 恒温锻造
isothermal gas chromatography 等温层跟踪仪
isothermal point 恒温式热量计

isothermal weight-change curve 等温重量变化曲线
isothermal weight-change determination 等温重量变化测定
isotope 同位素
isotope dilution mass spectrometry 同位素稀释质谱法
isotope mass spectrometer 同位素质谱计
isotope peak 同位素峰
isotope ratio measurement 同位素丰度测定
isotope X-ray fluorescence spectrometer 同位素 X 荧光光谱仪
isotopic ion 同位素离子
isotropism 各向同性
ISP (interoperable system protocol) 可互操作系统协议
ISS (ion-scattering spectroscopy) 离子散射谱法
ITAE (integral of the time and absolute error) 绝对值偏差乘时间积分
item 项（目）
iteration chromatography 核对色谱法
iterative 迭代，重复
iterative coordination 迭代协调
iterative improvement 迭代矫正
iterative method 迭代法
ITU (International Telecommunications Union) 国际电信联盟
IUPAC (International Union of Pure and Applied Chemistry) 国际纯粹与应用化学联合会
IVHS (integrated vehicle highway system) 集成汽车高速公路系统

J

jack 千斤顶，起重器，升降机
jacket 夹套
jack-hammer 手持式风钻，气锤
jacking force 顶推力
jacking pile 顶压桩，压入桩
Jacobian 雅可比
Jacobian matrix 雅可比矩阵
jamproof 抗干扰的
jam-to-signal 噪声信号比
Japanese Industrial Standard (JIS) 日本工业标准
jaw 钳口
jerk 跃度
jerk diagram 跃度曲线，阶跃曲线
jerk sensor 加速度传感器
jerk transducer 加速度传感器
jet lubrication 喷射润滑
jet orifice separator 喷嘴分离器
jet propulsion 喷气推进
jet recorder 喷墨式记录仪，喷射式记录仪
jib 起重臂，吊机臂
jib crane 旋臂吊机，起重吊机，桅杆式起重机
jib type cranes 臂架型起重机
jig 冶具，钻模，机床夹具，模具
jig boring machine 冶具镗孔机，坐标镗床
jig grinding machine 冶具磨床，坐标磨床
jig welding 工模焊接
JIS (Japanese Industrial Standard) 日本工业规标准
JIT (just-in-time) 运行时编译执行的技术
jitter 振动，不稳定的信号
jitter tester 抖动检测器
job order 工作通知单，工作通知
job-lot control 分批控制
jogging track 缓跑径
joggling 摇动加工
joinery 细木工作

joint 接合，接合处，连接处
joint box 接线箱
joint plate 结合板
joint probability 联结概率
joint sealant 填缝料，夹口胶
joint sheet 结合垫片
joint trajectory 轨迹对接
jointed manipulator 关节型操作器
jointing compound 接合剂，密封剂
joist 格栅，托梁
Jordan block 约当块
Jordan canonical form 约当标准型，约当范式
Jordan normal form 约当标准型
Jordan sunshine recorder 暗筒日照计，乔丹日照计
Joule (J) 焦耳（能量、热量、功的单位）
Joule-Lenz's law 焦耳-楞次定律
Joule's law 焦耳定律
journal 日记，日志，学报，轴颈
journal axis 轴颈中心线
journal centre 轴颈中心
journal manager special event 特殊事件日记管理
joy stick 操纵杆
jump 跳刀
jump process 跳变过程
jumper 跳线，跨接，跨接片，跨接件，跨接线
jumper wire 跳线，（架空线路的）跨接线
junction 接合，路口，道路连接处，交界处，接合处，交汇处
junction box 接线盒，分线盒，接线箱，分线箱
junction capacitor 结电容器，接面电容器
junction capacity 路口容量
junction field effect transistor 结型场效应晶体管
junction line 中继线

junction panel 接线盘
junction transistor integrator 结型晶体管积分器
junction type varistor 结型压敏电阻器
junction type zinc oxide varistor 结型氧化锌电压敏电阻器
junction box 接线盒
just-in-time（JIT） 运行时编译执行的技术

K

Kaiser effect 凯撒效应
Kalman filter 卡尔曼滤波器
Kalman-Bucy filter 卡尔曼-布西滤波器
Karman swirlmeter 卡曼漩涡流量计
Karnaugh map 卡诺图
KBC (keyboard controller) 键盘控制器
KBMS (knowledge base management system) 知识库管理系统
KBS (knowledge based system) 知识库系统，基于知识的系统
keep 保持，维持
keep away from boiler 远离锅炉
keep cool 保持冷藏
keep dry 保持干燥
keep in cool place 保持冷藏
keep out of the direct sun 避免日光直射
Kelvin 开[尔文](温度单位)
Kelvin absolute electrometer 开尔文绝对静电计
Kelvin bridge 开尔文电桥
Kelvin bridge ohmmeter 开尔文电桥式欧姆表
Kelvin double bridge 开尔文双臂电桥
Kelvin temperature scale 开氏温标
kenotron 高压整流二极管，大型热阴极二极管
kerosene 煤油
Kew pattern barometer 寇乌气压表
key 键，按键，键销，钥匙，键槽
key clutch 键槽离合器
key diagram 索引图，要览图，解说图，原理图
key in 键盘输入
key level 键锁级别，功能键级别
key panel 键盘
key plan 索引平面，索引图，平面布置总图
key switch 键
keyboard 键盘
keyboard control key 键盘控制键
keyboard controller (KBC) 键盘控制器
keyboard processor 键盘处理器
keyboard receive 键盘接收
keyboard send 键盘发送
keyed amplifier 键控放大器
keyer 调制器，电键器
keylock 键锁
keypunch 键控穿孔，键控穿孔机
keystone distortion 梯形畸变
keyway 键槽
Kharitonov theorem 哈里多诺夫定理
kickback power supply 回扫脉冲电源
killer switch 断路器开关
kilo (K) 千
kilobyte 千字节（1024字节）
kiloton 千吨
kilovolt-ampere (KVA) 千伏安
kilovoltmeter 千伏表
kilowatt-hour meter 电度表
kilowatt-hour meter calibration equipment 电度表校验装置
kinematic 运动学
kinematic analysis 运动分析
kinematic chain 运动链
kinematic design 运动设计
kinematic design of mechanism 机构运动设计
kinematic inversion 反转法，机架变换，运动倒置
kinematic pair 运动副
kinematic precept design 运动方案设计
kinematic sketch 运动简图
kinematic sketch of mechanism 机构运动简图
kinematic synthesis 运动综合
kinematic viscosity 运动黏度
kinematic viscosity baths 黏度测定槽
kinematical seal 运动密封
kinematics 运动学
kinetic control system 反应动力学控制系统
kinetic energy coefficient 动能系数
king bolt 主栓，中枢销
king pin 中心销，主销，转向销，

大王销
king pin bush 主销衬套
kink 弯曲，缠绕
Kirchhoff's current law 基尔霍夫电流定律
Kirchhoff's law 基尔霍夫定律
Kirchhoff's voltage law 基尔霍夫电压定律
kit 工具箱
kitol 鳕醇
knack 技巧，窍门，诀窍
kneader 混合机
knee point voltage 拐点电压
knife edge contact width 刀子接触宽度
knife edge curvature radius 刀子曲率半径
knife switch 闸刀开关
knife-edge follower 尖底从动件
knife-edge pointer 刀形指针，刃形指针
knife-switch 闸刀
knives 刀子
knives linear 刀子联线
knob 按钮，调节器，球形把手
knock pin 顶出销
knockout 脱模
knockout bar 脱模杆，钎子
Knoop hardness number 努普硬度值
Knoop hardness penetrator 努普硬度

压头
knotted bar iron 竹节钢
know how 秘诀
knowledge 知识
knowledge acquisition 知识获取
knowledge aided protocol automation 知识辅助协议自动化
knowledge assimilation 知识同化
knowledge base 知识库
knowledge base management system (KBMS) 知识库管理系统
knowledge based system (KBS) 知识库系统，基于知识的系统
knowledge engineering 知识工程
knowledge inference 知识推理
knowledge model 知识模型
knowledge representation 知识表达
knowledge tool 知识工具
knowledge transfer 知识转换器
knowledge-based adaptive control 基于知识自适应控制
knowledge-based control 基于知识控制
Knudsen pipette 克努森移液管
knurled shank 滚花身，压花身
knurling 滚花，滚纹，切辊纹
konstantan 康铜
kovar wire 科伐丝
krypton (Kr) 氪

L

label 标号，标签，商标，标志
labelled molecule 标记分子
labelling 加标
lability 不稳定性
laboratory 实验室，实验室应用资料
laboratory application data 实验室应用资料
laboratory education 实验教学
laboratory furniture 实验台
laboratory glassware washers 实验室玻璃器皿清洗机
laboratory reliability test 实验室可靠性试验
laboratory salinometer 实验室盐度计
laboratory technique 实验技术
laboratory testing 实验室测试
labyrinth seal 迷宫式密封，曲径密封垫片
lacing wire 绑扎用铁丝，系束穿束线
lack of painting 烤漆不到位
lacquered wire 漆包线
ladder algorithm 梯形算法
ladder attenuator 梯形衰减器
ladder diagram 梯形图
ladder filter 梯形滤波器
ladder logic 梯形逻辑
ladder 梯形，梯子，阶梯
ladder logic diagram 逻辑梯形图
ladder-type filter 梯形滤波器
ladder-type network 梯形网络
lag 滞后
lag compensation 滞后补偿
lag compensator 滞后补偿器
lag element 滞后环节
lag module 滞后组件
lag network 滞后网络
lagging 滞后，防护套，保温包扎
lagging current 滞后电流
lagging phase 滞后相位
lagging power-factor 滞后功率因数
lag-lead compensation 滞后-超前补偿
lag-lead controller 滞后-超前控制器
lag-lead network 滞后-超前网络
Lagrange duality 拉格朗日对偶性
Lagrangian method 拉格朗日法
laid 敷设，铺砌
laitance 水泥浆
LAM (look-at-me) 请求注意
laminar flow 层流
laminar flowmeter 层流式流量计
laminate 碾压，层压板
laminate for additive process 加成法用层压板
laminated core 叠片铁芯
laminated glass window 层压玻璃观察窗
laminated magnet 积层磁铁
laminating method 被覆淋膜成形
lamination 层，叠片
lamp 灯，灯泡，灯管，光源
lamp dimming resistor 灯光调节电阻器
lamp flasher 闪光器
lamp holder 灯架
lamp pole 灯柱，灯杆
lamp synchroscope 同步指示灯
LAN (local area network) 局部区域网络，局域网，局域网络
lancing die 切口模
land 合模平坦面，模具直线刀面部
land area 合模面
land drainage 地面排水
land-based interceptor sewer 地面污水截流管
landfill 垃圾堆填区，堆填区
landform 地形，地貌
landing 楼梯平台，斜路平台，梯台
landing impact testing machine 落锤式冲击试验机
landing wires 降落张线
landless hole 无连接盘孔
landless via hole 无连接盘导通孔
landscape 园景，景观
landscaping works 环境美化工程，景观美化工程，美化景物工程

lane capacity 车道容量
lane direction control signal 车道方向指示信号
lantern 灯具
lanthanum (La) 镧
Laplace transform 拉氏变换，拉普拉斯变换
Laplace-Gauss distribution 拉普拉斯-高斯分布
Laplace's equation 拉普拉斯方程式
Laplace's transformation 拉普拉斯变换，拉氏变换
lapping 研磨，研磨修润，精磨，抛光
lapping machine 研磨机
lapping machines 精研机
lapping machines centerless 无心精研机
large 大的，大容量的
large AC motor 大的交流电机
large aperture interferometer 大孔径干涉仪
large capacity filter 大功率过滤器
large capacity generating set 大容量发电机组
large capacity pressure regulator 大功率调压器
large deviation 大偏差
large glass fiber reinforced plastic cooling towers 大型玻璃纤维增强塑料冷却塔
large scale compound integration 大规模混合集成电路
large scale hybrid integrated circuit 大规模混合集成电路
large scale integrated circuit 大规模集成电路
large scale integration (LSI) 大规模集成电路
large scale system 大系统
large scale system cybernetics 大系统控制论
large signal 大信号
large space structure 大空间结构
large volume sampler 大容量采水器
large-scale system 大规模系统
largest singular value 最大特征值

laser 激光，激光器
laser alignment instrument 激光导向仪
laser anemometer 激光风速计
laser atomizer 激光原子化器
laser beam ionization 激光电离
laser beam machining 雷射加工，激光加工
laser beam welding 雷射光焊接，激光焊
laser ceilometer 激光测云仪
laser cutting 激光切割
laser cutting for SMT stensil 激光钢板切割机
laser diode 激光二极管
laser distance meter 激光测距仪
laser ellipticity measuring instrument 激光椭圆度测量仪
laser fluorescence detector 激光荧光探测仪
laser gravimeter 激光重力仪
laser inside diameter measuring instrument 激光内径测量仪
laser interferometer 激光干涉仪
laser interferometry 激光干涉测量法
laser jet printer 激光喷射打印机
laser length measuring instrument 激光式测长仪
laser level 激光水准仪
laser microspectral analyzer 激光显微光谱分析仪
laser orientation instrument 激光指向仪
laser outside diameter measuring instrument 激光外径测量仪
laser plummet 激光垂准仪
laser printer 激光印刷机，激光打印机，雷射打表机
laser probe mass spectrometer 激光探针质谱计
laser scatterometer 激光散射计
laser sensor 激光传感器
laser spectrum radiator 激光光谱辐射计
laser tachometer 激光转速仪
laser theodolite 激光经纬仪
laser triggered switching 激光器触发开关
laser-induced fluorescence detection 激光诱导荧光检测

lashing wire 拉系索，系固钢丝
last amplifier 末级放大器，终端放大器
last 最后的
latch 锁扣，碰锁，弹簧锁
latch circuit 锁定电路
latching effect 擎住效应
latent heat 潜热
lateral dimension 横向尺寸
lateral force 横向力，侧向力
lateral inhibition network 侧抑制网络
lateral logger 侧向测井仪
lateral movement 横向移动，横向位移，侧向位移
lateral positioning groove 横向定位凹槽
lateral pressure 横向压力，侧向压力
lateral stability assessment 横向稳定性评估
lateral stopper 横向限动块
lateral support system 横向承托系统
lateral support works 横向承托工程
latest PC technology 最新 PC 技术
latex solution 乳胶液
lath 板条
lathe 车床
lathe bench 车床工作台
lathe cutting 车床车削
lathe tool 车刀
lattice 网络结构，晶格
lattice filter 网络滤波器
lattice type filter 桥形滤波器
launching erection 曳进架设法
launching girder 曳进吊梁机
launching nose 曳进导梁，导梁
law 定律
law of conservation of energy 能量守恒定律
law of electric network 电网络定律，基尔霍夫定律
law of electromagnetic induction 电磁感应定律，法拉第定律
law of electrostatic attraction 静电吸引定律，库仑定律
lawrencium (Lr) 铹
lay 敷设，铺设
lay out 布置
lay wire 模网

layout 布置图，布图设计，规划设计，设计，规划图
layout area 蓝图区，详细设计区
layout drawing 布置图
layout efficiency 布线完成率
layout of cam profile 凸轮廓线绘制
layout plan 发展蓝图，分布图，详细蓝图，布局平面图
LB (logic block) 逻辑块
LBS 镑
LCD thermometer 液晶显示数字温度计
LCD (liquid crystal display) 液晶显示屏
LCL (lower control limit) 控制下限
LC-MS (liquid chromatograph-mass spectrometer) 液相色谱-质谱联用仪
LC-MS (liquid chromatography-mass spectrometry) 液相色谱-质谱法
LCN (local control network) 局部控制网络
LCN subsystem test program LCN 子系统测试程序
LCNE (local control network extender) 局部控制网络扩展器
leachate 沥滤液
lead (Pb) 铅
lead 导程，螺纹导程，超前，导前，引导，导向，铅导线，引线，导线
lead angle 超前角，导程角，螺纹升角
lead angle at reference cylinder 分度圆柱导程角
lead cam 导程凸轮
lead capacitance 接线电容
lead compensation 超前补偿
lead compensator 超前补偿器
lead fuse wire 保险铅丝
lead glazing 铅条玻璃窗，铅釉
lead inductance 导线电感，引线电感
lead module 超前组件
lead network 超前网络，导向网络
lead oxide paint 氧化铅漆
lead pipe 铅管
lead rail 导轨
lead-covered wire 铅包线，铅皮线
leader busher 导销衬套
lead-in wire 引入线
leading 超前，导前

leading area 前伸峰
leading edge 前缘，前沿，上升沿
leading peak 前伸峰，前沿峰
leading-in cable 引入电缆
leading-out wire 引出线
lead-lag controller 超前-滞后控制器
leadless component 无引线元件
lead-sheathed wire 铅包线，铅皮线
leaf spring 钢板弹簧
leaf switch 片状开关
leak 漏电，渗漏，泄漏
leak detection 检漏，密闭性检查
leak detector 检漏器
leak pressure 泄漏压力
leak rate 渗漏率
leak sealing 封漏
leak sealing bag 封漏袋
leakage 渗漏，泄漏
leakage control 防漏
leakage current 泄漏电流，漏电流
leakage current reduction 降低泄漏电流
leakage current screen 泄漏电流屏蔽
leakage detector 检漏仪，检漏器，漏电指示器
leakage flux 漏磁通
leakage inductance 漏感，漏磁
leakage magnetic field 漏磁场
leakage magnetic field inspection 漏磁探伤法
leakage magnetic flow detector 漏磁探伤仪
leakage property 泄漏特性
leakage reactance 漏电抗，漏磁电抗
leakage relay 漏电继电器，接地继电器
leakage resistance 漏电阻
leakage 漏，泄漏，渗漏
lean manufacturing 精实生产
lean mix concrete 少灰混凝土
learning 学习
learning algorithm 学习算法
learning control 学习控制
learning system 学习系统
least square (LS) 最小二乘法
least-squares algorithm 最小二乘算法
least-squares approximation 最小二乘近似

least-squares estimation (LSE) 最小二乘估计
least-squares fitting (LSF) 最小二乘拟合
least-squares identification 最小二乘辨识
least-squares method 最小二乘方法
least-squares problem 最小二乘问题
leather chemistry and engineering 皮革化学与工程
LED display 发光二极管显示器
LEED (low electron energy diffractometer) 低能电子衍射仪
left fork 叉车
leg wire 脚线
legal metrology 法制计量学
legend 图例
length of a scale division 标度分格间距，分格（度）长度，分割间距
length of action 啮合长度
length of line of action 啮合线长度
length of magnetic path 磁路长度
length 长度
lengthening coil tuner 加长线圈调谐器
lengthwise mismatch 纵向失配
lengthwise sliding velocity 纵向滑动速度
lengthy bellows 长波纹管
lens micrometer 透镜测微器
Lenz's law 楞次定律
lethal 致死的
lethality 致死性
levee 河堤
level 水准仪，水准泡，级，电平，物位，能级，水平，水平线，液面
level area 平坦地方
level assembly 杠杆组件
level control 液面控制，电平调节
level controller 液位控制器，液位调节器
level gauge 水准仪，水平规，水平仪
level indicator 液位指示器
level instrument 位面计，水平仪
level of significance 显著性水平
level of smoke 烟浓度
level recorder 电平记录仪
level regulator 液位调节器

level sensor 物位传感器
level switch 电平开关
level tracer 电平图示仪
level transducer 物位传感器
level transmitter 能级传送器，液面传感器，物位变送器
level-braked motor 杠杆式制动电动机
leveling agent 匀染剂，均化剂
leveller 水平测量员，校平机
levelling course 整平层
levels of control 控制级别
lever arm 杠杆臂，活动臂
lever 杠杆，杠杆控制杆
lever-arm recording flowmeter 杠杆记录式流量计
lever-type vibrograph 杠杆式测振仪
liaison 联络单
library 程序库
library searching 谱库检索
libration damping 天平动阻尼
life 寿命，使用期
life cycle 生命周期
life factor 寿命系数，使用年限因素
life of X-ray tube X射线管寿命试验
life span 使用期限
lifetime 寿命，生命时间
lift 提，升，升距，垂直位移，升降机
lift car 升降机厢
lift check valves 升降式止回阀
lift control 电梯控制
lift controller 电梯控制器
lift of hydrofoil 水翼升力
lift pin 升降机顶升销
lift pit 升降机坑
lift shaft 升降机井
lift works 升降机工程
lift-drag of towed vehicle 拖曳体升阻比
lifter 斜顶，升降器
lifter guide pin 浮升导料销
lifter pin 顶料销，升降销
lifting appliance 起重机械，起重设备，吊具
lifting chain 起重链，吊链
lifting cylinder 起卸液压伸缩筒，起重缸，千斤顶

lifting electromagnet 起重电磁铁
lifting frame 吊架
lifting gear 起重装置，提升机构，升降装置
lifting hooks 起重吊钩
lifting jack 千斤顶，举重机
lifting pin 顶料销，顶升杆
lifting platform 升降平台，举重台
lifting rod 起重杆
lifting tackle 起重滑车，提升滑轮
ligand chromatography 配位体色谱法
ligand exchanger 配基交换剂
light activated switch 光敏开关，光致开关
light beam oscillograph 光线示波器
light button 光按钮
light chopper 光斩波器
light control junction 灯号控制路口
light dependent resistance 光敏电阻
light digital logging instrument 轻便数字测井仪
light emitting diode 发光二极管
light flux 光通量
light industry technology and engineering 轻工技术与工程
light intensity 光强
light modulation 光调制
light modulator 光调制器
light oil 轻油
light pen 光笔
light rail system 轻便铁路系统
light relay 光继电器
light scattering detector 光散射检测器
light sensor 光敏元件，光敏感器
light source 光源
light switch 照明开关，电灯开关
light transmittance 透光率
light triggered thyristor 光控晶闸管
light weight aggregate 轻骨料
light 光，光亮，点，点燃，照亮
light run 空转
light-beam galvanometer 光束检波计
lighting 照明
lighting cable 照明电缆
lighting mast 电灯杆
lighting-intensity meter 光强计
lightning 闪电，雷电

lightning arrester 避雷器,避雷装置
lightning conductor 接地极,避雷针
lightning protection 避雷保护
lightning rod 避雷针
lightning shielding 避雷
lightning arrestor 避雷器
light-regulator 照明调节器
light-sensitive cell 光敏电池
lightweight cover 轻型上盖
lightweight filler 轻填料
likelihood 似然
likelihood function 似然函数
lime 石灰
limit 极限,限制
limit alarm sensor 极限报警检测器
limit cycle 极限环,极限周期
limit equivalent conductance 极限当量电导
limit frequency 极限频率
limit gauge 极限量规
limit of concentration 浓度极限
limit of detection (LOD) 检测限
limit of flammability 易燃性极限
limit of inflammability 燃烧极限
limit of intrinsic error 基本误差限
limit of quantitation (LOQ) 定量极限
limit point width 极限刀顶距
limit pressure angle 极限压力角
limit resolution 极限分辨率
limit state 极限状态
limit switch 限位开关,极限开关,终端开关
limit theorem 终极定理
limit valves 限位阀
limitation of sensibility 灵敏度
limited 有限的
limited breathing enclosure 限制通气外壳
limited code 有限编码
limited current 极限电流
limited data 有限数据
limited gain amplifier 有限增益放大器
limited integrator 有限积分器
limiter 限位器,限幅器,限制器,限值器
limiter diode 限幅二极管

limiting 限制,有限,极限的
limiting amplifier 限幅放大器
limiting availability 极限可用度
limiting control 极限控制,限值控制
limiting control action 有限控制作用
limiting distribution 有限分布
limiting effective wavelength 极限有效波长
limiting operating condition 极限工作条件
limiting position 极限位置
limiting resister 限流电阻
limiting temperature 极限温度
limiting vacuum 极限真空度
limiting value for operation 工作极限值
limiting value for storage 贮存极限值
limiting value for transport 运输极限值
limits of error 误差极限
limits of operating error 工作误差极限
line 线,直线,线路
line balance converter 线性平衡变换器
line check 小检修
line commutation 电网换流
line conditioner 线性调节器
line coupler 线路耦合器
line current 线电流
line driver 线驱动器
line drop compensation 线路补偿器
line editor 行编辑程序,行编辑器
line equalizer 线路均衡器
line focus 线焦点
line frame 线路帧
line impedance 线路阻抗
line loss 线损
line of action 啮合线
line of centers 中心线
line of communication 通信线路
line of contact 接触线
line of magnetic force 磁力线
line of magnetization 磁化线,磁力线
line pair 线对
line printer 行式打印机,宽行打印机
line protocol 线路协议
line pulling 线拉伸
line radio 有线载波通信
line relay 线路继电器

line scanning　行扫描
line sensitivity　线灵敏度
line strain　线应变
line supervisor　生产线主管，线长
line switch　线路开关，导线机
line tester　线路试验器
line to line resolution　线分辨力
line trap　线路陷波器
line turnaround　线路换向
line voltage　线电压
line voltage regulation　电源电压调整率
linear　线性的，直线的
linear absorption coefficient　线性吸收系数
linear acceleration　线加速度
linear acceleration sensor　线加速度传感器
linear acceleration transducer　线加速度传感器
linear algebra　线性代数
linear amplification　线性放大
linear amplifier　线性放大器
linear analysis　线性分析
linear array　线性排列，直线天线阵
linear block code　线性分组码，线性区块码
linear code　线性编码
linear control　线性控制
linear control system　线性控制系统
linear control system theory　线性控制系统理论
linear conversion　线性转换
linear cutting　线切割
linear dependence　线性相关
linear detector　线性检波器
linear differential transformer　线性差分变换
linear dispersion　线色散
linear displacement　线位移
linear displacement grating　直线位移光栅
linear displacement transducer　线性位移传感器
linear distortion　线性失真
linear dual slope type　线性双斜型
linear electric actuator　电子直线加速器
linear element　线性元件

linear equation　线性方程
linear equations　线性方程组
linear estimation　线性估计
linear filter　线性滤波
linear in the parameter　参数线性
linear independence　线性独立
linear integrated optics　线性集成光学
linear lag　线性滞后
linear lead　线性超前
linear location　线定位
linear model　线性模型
linear modulation　线性调制
linear motion　直线运动
linear motion electric drive　直线运动电气传动
linear motion valve　直行程阀
linear motor　线性电机
linear multivariable system　线性多变量系统
linear network　线性网络
linear optimal　线性优化
linear optimal control system　线性最优控制系统
linear output feedback　线性输出反馈
linear phase　线性相角
linear potentiometer　线性电位计
linear power amplifier　线性功率放大器
linear prediction　线性预报，线性预测
linear programming　线性规划
linear quadratic regulator (LQR)　线性二次调节器
linear quadratic regulator problem　线性二次调节器问题
linear ramp type　线性斜坡型
linear range　线性范围
linear regression　线性迭代
linear regulator　线性调节器
linear relation　线性关系
linear resistance　线性电阻
linear scale　线性标度
linear sweep generator　线性扫描振荡器
linear system　线性系统
linear system simulation　线性系统仿真
linear theory　线性理论
linear thermistor　线性热敏电阻器

linear time-invariant control system 线性定常控制系统
linear transducer 线性转换器
linear variable reluctance transducer 线性可变磁阻传感器
linear velocity 线速度
linear velocity sensor 线速度传感器
linear velocity transducer 线速度传感器
linear viscous damping 线性黏性阻尼
linear viscous damping coefficient 线性黏性阻尼系数
linear zone 线性区
linear scale 线性标度
linearity 线性,线性度
linearity and range 线性与范围
linearity control 线性控制
linearity error 线性度误差
linearity of amplifier 线性放大器,垂直线性
linearity of time base 时基线性
linearizable system 线性化系统
linearization 线性化
linearization technique 线性化方法
linearized model 线性化模型
line-at-time printer 行式印刷机,一次一行印刷机,行式印刷机打印机
line-by-line scanning 面扫描
lineman's spectrometer 携带式光谱仪
linescan 线扫描曲线
line-segment function unit 折线函数运算器
line-stabilized oscillator 传输线稳频振荡器
line-to-line 线间的
line-to-neutral 线与中性点间的
linguistic 语言的
linguistic information 语言信息
linguistic support 语言支持
linguistic synthesis 语言综合
linguistic variable 语言变量
lining 衬砌,衬层,衬管,内衬
lining ball valves 衬里球阀
lining butterfly valves 衬里蝶阀
lining check valves 衬里止回阀
lining globe valves 衬里截止阀
lining t-cock valves 衬里三通旋塞阀
link 构件,连接,联结
link resistance 跨线电阻
linkage 键接,连杆,联动机构,耦合,连锁,磁链,连接
linkage editor 连锁编辑,链接编辑器
linked switch 联动开关
linked-scanning 联动扫描
linker 连接程序
linking 耦合,连锁
lintel beam 水平横楣梁
lip rubber seal 唇形橡胶密封
liquefaction 液化过程
liquefied petroleum gas(LPG) 石油气
liquefied petroleum gas cylinders 液化石油气钢瓶
liquid chiller 液体冷冻机
liquid chromatograph 液相色谱仪
liquid chromatograph-mass spectrometer (LC-MS) 液相色谱-质谱联用仪
liquid chromatography 液相色谱法
liquid chromatography-mass spectrometry (LC-MS) 液相色谱-质谱法
liquid column chromatography 柱液相色谱法
liquid column manometer 液柱压力计
liquid control valve 液体控制阀
liquid crystal display(LCD) 液晶显示器
liquid crystal thermometer 液晶温度计
liquid densitometer 液体密度计
liquid density sensor 液体密度传感器
liquid density transducer 液体密度传感器
liquid displacement system 液体置换系统
liquid displacement technique 液体置换法
liquid filled case 充液压力表
liquid filled thermal system 充液式感温系统
liquid honing 液体喷砂法
liquid indicator 液位计
liquid junction boundary 液接界面
liquid junction potential 液接电位
liquid level 液位
liquid level alarm 液位报警器

liquid level instrument of open vessel 敞口容器液位器
liquid level meter 液位计
liquid level system model 液面系统模型
liquid manometer 液体压力表
liquid penetrant examination 液体渗透探伤
liquid phase 液相
liquid phase loading 液相载荷量
liquid positive displacement flowmeter 容积式液体流量计
liquid pressure recovery factor 液体压力恢复系数
liquid receiver 集液器
liquid refraction meter 标尺折光仪
liquid scintillation counting 液滴闪烁计数
liquid seal 液封
liquid sealed drum gas flowmeter 液封转筒式气体流量计
liquid solid chromatography (LSC) 液固色谱法
liquid spring 液体弹簧
liquid surface acoustical holography 液面声全息
liquid thermometer 液体温度表
liquid thermostatic bath 液体恒温槽
liquid visual expansion coefficient 液体视膨胀
liquid-in-glass thermometer 玻璃温度计
liquid-liquid chromatography (LLC) 液液色谱法
liquid-liquid extraction 液-液提取法
liquid-solid adsorption chromatography 液固吸附色谱法，液固色谱法
liquid-solid chromatography 液固色谱法
liquid-solid extraction 液-固提取法
liquid-to-glass thermometer 玻璃管液柱式温度计
Lissajou's figures 利萨如图形
list 表，列表，目录
list a file 显示一个文件
list archive catalog 列出归档目录
list command file 列出命令文件清单
list directory 列出目录清单
list of builder definition items 定义项

目生成的列表
list of equipment for process control 自控设备表
list of function blocks 功能块的功能列表
list processing 表处理
listener 收听站
liter 公升
lithium (Li) 锂
lithium chloride humidity-dependent resistor 氯化锂湿敏电阻器
lithotrite 碎石机
litmus 石蕊
litmus paper 石蕊试纸
litre (L) 升（容积单位）
little 小的，少许，少的
litz wire 编织线，绞合线
live conductor 火线，带电导体
live crack 活裂缝
live load 活荷载，活载重
live time 活时间
live wire 载电线，火线，带电电线
live zone 活区
living system 生命系统
LLC (liquid-liquid chromatography) 液液色谱法
LLPIU(low level process interface unit) 低电平过程接口单元
LM(logic manager) 逻辑管理综合控制器
load 负载，加载，装载，荷载，负载量
load adjustment 负载调节
load admittance 负载导纳
load amplitude 负荷幅
load archive file 装载归档文件
load bearing 承载，承重
load bearing structure 承载构筑物，荷载支承结构
load break switch 负载断路开关
load carrying capacity 负荷能力，载重量
load cell 称重传感器，负载传感器
load characteristic 负荷特性，负载特性，负载特性曲线
load circuit 负载电路，负荷电路

load coil 感应加热线圈
load commutation 负载换流
load conditions 负荷状态
load control 负载控制
load curve 负荷曲线
load device parameter 加载装置参数
load diagram 负载图，负荷图
load dispatching 负载等，负载配电
load distribution 负荷分配
load divider 负载分配器
load equalization 负载平衡
load flow 负载流
load flow solution 负载流率解
load forecasting 载荷预测
load frequency control 负载频率控制
load impedance 负载阻抗
load line 负载线
load modelling 负载模型
load moment 负载转矩
load moment limiters 起重力矩限制器
load rating 额定载荷
load ratio voltage regulator 负载电压调制器
load regulation 负载调节，负载调整率
load regulator 负载调整器
load resistance 负载电阻
load response 负载响应，负荷响应
load scope 负载范围，装入范围
load sharing ratio 负荷分配比
load switch 负荷开关
load testing 负载测试，荷载测试，承重测试
load variable 负载变量，负荷变量
load variation 负载变动
loadamatic control 负载变化自动控制
load-deformation curve 载荷-变形曲线
load-deformation diagram 载荷-变形图
loaded line 加感线路
loaded waveguide 负载波导管
loaded-potentiometer function generator 负载电位器函数发生器
loader 铲泥车，供料器
loading 负荷量，承重量，荷载量，加载，配置

loading capacity 负载容量，负载容积
loading condition 荷载状况
loading error 负载误差
loading intensity 荷载强度
loading shoe mold 料套式模具
loading shovel 推土机
loading survey 载重勘测
loading system 加荷系统
load-saturation curve 负载饱和曲线
load-shedding equipment 电力平均分配装置
local 局域，就地，局部
local algorithm 逻辑算法
local area network(LAN) 局部区域网络，局域网，局域网络
local asymptotic stability 局部渐近稳定性
local automation 局部自动化
local batch operator station 就地批量操作站
local computer system 局域计算机系统
local control 局部控制，就地控制
local control network(LCN) 局部控制网络
local control network exerciser 局部控制网络练习程序
local control network extender (LCNE) 局部控制网络扩展器
local control network segment 局部控制网络段
local control panel 就地控制柜
local control unit 就地控制单元
local controllability 局部可控性
local controller 局部控制器
local data base 局部数据库
local diamagnetic shielding 局部抗磁屏蔽
local exchange 本地交换网
local feedback 局部反馈
local lightning counter 局地闪电计数器
local loop 局部回路
local magnetization 局部磁化
local manual 局部手动
local master reference ground 局部主

参考地
local mounting 就地安装
local operating networks (LON) 局域操作网络
local optimum 局部最优
local oscillations 本机振荡
local oscillator 本机振荡器
local panel 现场配电盘
local run-stop-remote starting selector switch 运行-停机-遥控启动选择键
local self-test 本机自测试,本机自检验
local set 本机设定,就地设定
local set point 本机设定点
local stability 局部稳定性
local station 局部操作站,本地电台,本地站
local structure 局部结构
local attendant 现场值班员
local repair 现场检修
localization of spot 斑点定位法
localized repair 局部修茸
localized seepage 局部渗漏
localized tooth contact 齿局部接触
local-mounted controller 基地式调节仪表
located block 定位块
located pin 定位销
locating bar 定位横栏
locating center punch 定位中心冲头
locating piece 定位块
locating pilot pin 定位导销
locating pin 定位销
locating plate 定位片
locating ring 定位环,定位圈
locating surface 定位表面
location 地点,处所,位置
location counter 位置计数器,指令计数器,地址寄存器
location lump 定位块
location of storage 存储单元
location pin 定位销
location plan 位置图
lock block 压块
lock in synchronism 牵入同步
lock nut 锁紧螺母

lock plate 锁模块
lock seaming 固定接合
lock signal 锁信号
lock washer 锁固垫圈
lock 闭锁,密封舱,固定
locked in oscillator 同步振荡器,锁定振荡器
locked-rotor 锁定转子
locked-rotor torque 堵转转矩,锁定转子转矩
locking 锁定,同步
locking block 定位块
locking coil 吸持线圈
locking device 关锁,锁定装置
locking gear 制动,锁闭齿轮,锁定装置
locking lever 栏板锁杆
locking plate 定位板
locking wire 锁线,锁紧用钢丝
locknut 自锁螺母,防松螺母
locksmith 钳工
lockup 锁定,锁定阈
lock-up relay 自保持继电器
lock-up valve 定位阀
locomotive 机车
locomotive diesel engine 机车柴油机
locomotive shed 机车库,机车棚
locus 轨迹
LOD (limit of detection) 检测限
log 报表,记录,原木
log console 记录控制台
log magnitude-frequency characteristics 对数幅频特性
log magnitude-phase diagram 对数幅相图
log phase-frequency characteristics 对数相频特性
logarithmic 对数(调节)
logarithmic A/D converter 对数式模/数转换器
logarithmic amplifier 对数放大器
logarithmic decrement 对数衰减,对数衰减率
logarithmic gain 对数增益
logarithmic magnitude 对数幅值
logarithmic plot 对数坐标图
logarithmic time 对数时间

logarithmic time dependence 对数时间相关
logger 测井仪器，记录器，拖车
logging 记录，存入
logging recorder 测井记录仪
logging tool 下井仪器
logic 逻辑
logic algebra 逻辑代数
logic analyser 逻辑分析器
logic application 逻辑应用
logic array 逻辑阵列，逻辑数组
logic block(LB) 逻辑块
logic circuit 逻辑电路
logic control 逻辑控制
logic control pressure 逻辑控制压力
logic controller 逻辑控制器
logic data point 逻辑数据点
logic design 逻辑设计
logic design automation 逻辑设计自动化
logic diagram 逻辑图，逻辑网
logic gate 逻辑闸，逻辑门
logic gate circuit 逻辑门电路
logic manager(LM) 逻辑管理综合控制器
logic minimization 逻辑最小化
logic mix 逻辑混合
logic operation 逻辑运算
logic point 逻辑点
logic simulation 逻辑模拟
logic state analyzer 逻辑分析仪
logic unit 逻辑单元
logical 逻辑的
logical address 逻辑地址
logical circuit 逻辑电路
logical control 逻辑控制
logical device identifier 逻辑设备标识符
logical diode circuit 二极管逻辑电路
logical model 逻辑模型
logical node 逻辑节点
logical operation 逻辑操作，逻辑运算
logical operation block 逻辑操作功能块
logical product 逻辑乘
logical ring 逻辑环

logical sum 逻辑和
logical unit 逻辑部件
logitron 磁性逻辑元件
logometer 流比计，比率表
LON(local operating networks) 局域操作网络
long distance telephone 长途电话
long distance transmission line 长距离输电线路
long distance water level recorder 遥测水位计
long nozzle 延长喷嘴方式
long period seismograph 长周期地震仪
long radius elbow 长径肘管，大半径弯头
long radius nozzle 长径喷嘴
long range coupling 远程耦合
long range shielding effect 远程屏蔽效应
long term drift 长期偏移
long term memory 长期记忆
long term running test 长期运转试验
long vehicle 拖板车
long wave 长波，长波的
long 长
longitudinal 纵向的
longitudinal axis 纵轴
longitudinal gradient 纵向坡度
longitudinal interference 纵向干扰
longitudinal magnetic flux 纵向磁通
longitudinal magnetization 纵向磁化
longitudinal magnetomotive 纵向磁动势
longitudinal section 纵截面，纵截面图
longitudinal wave 纵波
longitudinal wave probe 纵波探头
longitudinal wave technique 纵波法
longitudinal wind analysis 纵向风力分析
long-line effect 长线效应
long-radius nozzle 长径喷嘴
long-term memory 长期存储器
long-term trend function 长时间趋势功能
look-ahead adder 前式加法器
look-ahead buffer 先行缓冲器

look-at-me(LAM)　请求注意
look-at-me signal　请求注意信号
loom　织布机
loop　回路，循环，环，框，匝
loop adapter　回路适配器
loop cinerator　环形燃烧管灭菌器
loop circuit　环路，环形电路
loop class module　回路级模块
loop communication card for single loop controller　供单回路调节器用的回路通信卡
loop communication unit　回路通信单元
loop communications card　回路通信卡
loop connection definition　回路连接定义
loop connection status display panel　回路连接状态显示画面
loop control　环路控制
loop controller　闭回路控制器
loop current　环流，回路电流
loop display　回路显示
loop gain　回路增益，环路增益
loop gain characteristic　回路增益特性
loop manual　回路手动
loop method　回路电流法
loop phase angle　回路相角
loop phase characteristic　回路相位特性
loop road　旋路，环路
loop status　回路状态
loop test　环路测试
loop transfer　回路转换器
loop transfer function　回路传递函数
loop tune　回路整定
loop wire　环线，回线
loopback checking　回送检验
loop-couple-mode attenuator　环耦合型衰减器
loop-locked　闭环的
loops　多回路，线圈
loose bush　活动衬套
loose detail mold　活零件模具
loose mold　活动式模具
loose piece　松件
loose　松的，不牢固的
loosen　松开，松动
LOQ(limit of quantitation)　定量极限
Lorentz electron　洛仑兹电子
lorry crane　货车吊机

loss　损失，减少，损耗，亏损
loss angle　损耗角，衰减角
loss contact　接触不良
loss coupler　弱耦合器
loss minimization　损失极小化
loss of power　功率损耗
loss of synchronization　失去同步
loss of voltage　电压损失
loss tangent　损耗角正切
lossless　无损耗
lossless line　无损线路
lossy coaxial cable　有损耗同轴电缆
lost wax casting　脱蜡铸造
loudspeaker　扬声器，喇叭
louver window　百叶窗，气窗
louvering　百叶窗板加工
louvering die　百叶窗冲切模
low　低
low alloy steel wire　低合金钢钢丝
low alloy tool steel　特殊工具钢
low altitude sonde　低空探空仪
low constant temperature shaking baths　振荡式低温水槽
low cost automation　低成本自动化
low density polyethylene　低密度聚乙烯
low drive power　低驱动电源
low electron energy diffractometer(LEED)　低能电子衍射仪
low EMF potentiometer　低电位电位差时
low energy　低能
low energy process interface unit　低能过程接口单元
low frequency fatigue testing machine　低频疲劳试验机
low heat expansive cement　低热微膨胀水泥
low level analog input　低电平模拟量输入
low level process interface unit(LLPIU)　低电平过程接口单元
low limiting control　下限控制
low manganese casting steel　低锰铸钢
low noise valve　低噪声阀
low oil pressure shutdowns　低油压故障停机

low potential 低电位
low pressure 低压力
low pressure casting 低压铸造
low pressure chamber 负压室
low pressure galvanized pipe 低压镀锌焊接钢管
low speed balancing 低速平衡
low speed motor 低速电动机
low temperature annealing 低温退火
low temperature humidity chamber 低温恒湿箱
low temperature incubators 低温培养器
low temperature plasma sterilizers 低温等离子灭菌器
low temperature program type incubators 可程式低温培养器
low temperature stability incubators 低温稳定性培养器
low temperature steel 低温钢
low temperature strain gauge 低温应变计
low temperature test 低温试验
low temperature testing machine 低温试验机
low temperature thermistor 低温热敏电阻器
low tension 低电压
low tension loop 低压回路
low threshold 低门限，低阈值
low viscosity polymer resin 低黏滞性聚合树脂
low voltage 低压，低电压，低压的
low voltage terminal 低压端
low water 低潮
low-and-high-pass filter 高低通滤波器，带阻滤波器
lower actual measuring range value 实际测量范围下限值
lower control limit(LCL) 控制下限
lower die 下模
lower die base 下模座
lower explosion limit 爆炸下限
lower flammable limit 易燃下限
lower limit 下限，下限值
lower limit of flammability 易燃性下限
lower measuring range value 测量范围下限值
lower plate 下模板
lower range limit 范围下限，量程下限
lower range-value 低范围值
lower side-band 下边带
lower sliding plate 下滑块板
lower stripper 下脱料板
lower switching value 下切换值
lower wire 下网
lower 较低的，降低
lower-cut-off frequency 下限截止频率
low-frequency 低频，低频的
low-frequency amplification 低频放大
low-frequency amplifier 低频放大器
low-frequency differential magnetometer 低频差动磁强计
low-frequency dispersion 低频漂移
low-frequency filter 低频滤波器
low-frequency induction furnace 低频感应电炉，工频感应电炉
low-frequency intensity 低频强度
low-frequency noise 低频噪声
low-frequency scattering 低频扫描
low-frequency signal 低频信号
low-frequency stage 低频段
low-frequency transformer 低频变压器
low-half 下半
low-impedance 低阻抗
low-level language 低级语言
low-level logic circuit 低电平逻辑电路
low-level modulation 低电平调制
low-limit adjustment 下限调整
low-limiting control 下限值控制
low-noise 低噪声
low-noise and low drift amplifier 低噪声低漂移放大器
low-noise chanel 低噪声通道，低噪声信道
low-noise optimization 低噪声优化
low-pass 低通
low-pass filter 低通滤波器
low-potential DC potentiometer 低电位直流电位差计
low-recovery valve 低压力恢复阀

low-signal selector 低信号选择器
low-speed magneto synchronous motor 永磁式低速同步电动机
low-temperature test chamber 低温试验箱
low-temperature treatment 低温处理
low-tension distribution box 低压配电箱
low-tension fuse 低压熔断器
low-tension motor 低压电动机
low-tension network 低压电力网
low-tension switch 低压开关
low-tension transformer 低压变压器
low-threshold current 低门限电流
low-volt transformer 低压变压器
low-voltage bus bar 低压母线
low-voltage circuit breaker 低压断路器
low-voltage distribution system 低压配电系统
low-voltage equipment 低压设备
low-voltage line 低压线
low-voltage network 低压配电网
low-voltage protection 低压保护
low-voltage system 低压系统
LPG(liquefied petroleum gas) 石油气
LQG control 线性二次高斯控制
LQG control method 线性二次高斯控制方法
LQR control method 线性二次调节器控制方法
LQR(linear quadratic regulator) 线性二次调节器
LS(least square) 最小二乘法
LSC(liquid solid chromatography) 液固色谱法
LSE(least-squares estimation) 最小二乘估计
LSF(least-squares fitting) 最小二乘拟合
LSI(large scale integrated circuit) 大规模集成电路
LSI chip 大规模集成电路芯片
lub oil 润滑油
lub oil pump 润滑油泵

lubricant 润滑剂，润滑油
lubricant film 润滑膜
lubricating oil 润滑机油
lubrication 润滑，润滑作用
lubrication device 润滑装置
lubrication systems 润滑系统
lubricator 注油器
lubricators 注油机
lubricity 润滑性
lumen 流明(光通量单位)
lumeter 照度计
luminaire 发光体，照明设备
luminance 亮度
luminance sensor 亮度传感器
luminance transducer 亮度传感器
luminous flux 光通量
luminous intensity 发光强度
lumped 集中的，集总的
lumped capacitance 集中电容，集总电容
lumped constant element 集总常数元件
lumped inductance 集中电感，集总电感
lumped model 集中模型
lumped parameter 集总参数，集中参数
lumped parameter model 集总参数模型
lumped parameter system 集中参数系统
lumped resistance 集中电阻，集总电阻
lux(lx) 勒克斯(光照度单位)
Lyapunov 李雅普诺夫
Lyapunov equation 李雅普诺夫方程
Lyapunov function 李雅普诺夫函数
Lyapunov method 李雅普诺夫方法
Lyapunov stability 李雅普诺夫稳定性
Lyapunov theorem of asymptotic stability 李雅普诺夫渐近稳定性定理
Lysimeter 溶度计，测渗计，浓度计
Lysimeter experiment 管柱试验

M

MAC (model algorithm control) 模型算法控制
macadam 碎石
Mach number 马赫数
machine cast pig iron 可切削铸铁
machine center 机械中心
machine center to back 机床中心至工件安装基准面
machine check indicator 机器检查指示仪
machine code 机器代码
machine description format (MDF) 机器描述格式
machine intelligence 机器智能
machine language 机器语言
machine learning 机器学习
machine plane 机床切削平面
machine recognition 机器识别
machine room 机房
machine tool 机床
machine type rock ore densimeter 机械式岩矿密度仪
machine with inherent self-excitation 内在自励磁电机
machine with nature cooling 自然冷却式电机
machine 机器，机械
machine-oriented language 面向机器语言
machinery 机械，机器，机构，机械装置
machining 制造，机械加工，加工
machining allowance 机械加工余量
machining center 加工中心
macro 宏，巨大的
macro analysis 常量分析
macro assembly language 宏汇编语言
macro instruction 宏指令
magmeter 直读式频率计
magnalium 镁铝合金
magnesium (Mg) 镁
magnet 磁铁

magnet assembly 磁体
magnet band 磁带
magnet card 磁卡片
magnet core 磁芯
magnet damping 磁性阻尼
magnet dynamic instrument 磁式动态仪器
magnet meter 磁通计
magnet starter 磁力起动器
magnetic 磁的，磁性的
magnetic amplifier 磁放大器
magnetic anisotropy 磁各向异性
magnetic balance 磁秤
magnetic base 磁性座
magnetic bearing 磁轴承
magnetic blow-out 磁吹
magnetic blow-out circuit breaker 磁吹式灭弧断路器
magnetic break 磁力制动器
magnetic card 磁卡
magnetic circles 磁路
magnetic circuit 磁路
magnetic clutch 电磁离合器
magnetic conductance 磁导
magnetic conductivity 磁导率
magnetic contactor 磁接触器
magnetic control component 磁控元件
magnetic control relay 磁控继电器
magnetic co-operate globe valves 磁耦合截止阀
magnetic core 磁芯
magnetic crane 磁力吊机，磁力起重机
magnetic damper 磁性阻尼器
magnetic damping 磁阻尼
magnetic deflection 磁偏转
magnetic detector for lightning currents 闪电电流磁检示器
magnetic dipole excitation 磁偶激励
magnetic disc 磁盘
magnetic domain 磁畴，磁域，磁场范围

magnetic domain attachment 磁畴附件
magnetic doublet 磁偶极子
magnetic drum 磁鼓
magnetic dumping 磁卸载
magnetic element 磁性元件
magnetic equivalence 磁等价
magnetic field 磁场
magnetic field computation 磁场计算
magnetic field intensity 磁场强度
magnetic field interference 磁场干扰
magnetic field meter 磁场计
magnetic field strength 磁场强度
magnetic field strength sensor 磁场强度传感器
magnetic field strength transducer 磁场强度传感器
magnetic figure 磁力线图
magnetic flaw detection ink 磁悬液
magnetic flow measuring device 电磁流量测量仪表
magnetic flow transducer 磁性流量传感器
magnetic flow tube 磁通管
magnetic fluid bearing 磁流体轴承
magnetic fluid clutch 磁流体离合器
magnetic flux 磁通, 磁通量
magnetic flux sensor 磁通传感器
magnetic flux transducer 磁通传感器
magnetic flux-density 磁感应强度, 磁通密度
magnetic grating displacement transducer 磁栅式位移传感器
magnetic induction 磁感应
magnetic locator 磁定位器
magnetic logger 磁测井仪
magnetic material 磁性材料
magnetic modulator 磁调制器
magnetic motive force 磁动势
magnetic oxygen sensor 磁式氧传感器
magnetic oxygen transducer 磁式氧传感器
magnetic particle 磁粉
magnetic particle inspection 磁粉探伤, 磁粉探伤机, 磁粉检查
magnetic path 磁路
magnetic permeability 磁导率
magnetic pole 磁极

magnetic potentiometer 磁位计
magnetic power clutch 磁动力离合器
magnetic property 磁性
magnetic prospecting instrument 磁法勘探仪器
magnetic quantity sensor 磁学量传感器
magnetic quantity transducer 磁学量传感器
magnetic quantum number 磁量子数
magnetic recording channel 磁记录通道
magnetic recording wire 磁带
magnetic reluctance 磁阻
magnetic resistivity instrument 磁电阻率仪
magnetic resistor 磁敏电阻器
magnetic resonance microscopy 磁共振显微镜
magnetic response 磁响应
magnetic rotation comparison 磁旋比
magnetic saturated voltage stabilizer 磁饱和稳压器
magnetic saturation 磁饱和
magnetic scale width gauge 磁栅式宽度计
magnetic scale width meter 磁栅式宽度计
magnetic screen 磁屏蔽
magnetic sensor 磁传感器
magnetic separator 磁性分选仪
magnetic shield 磁屏蔽
magnetic shielding 磁屏蔽
magnetic starter 磁力启动器
magnetic stirrers 磁力搅拌器
magnetic storage 磁存储器
magnetic superlattice 磁超点阵
magnetic susceptibility logger 磁化率测井仪
magnetic suspension 磁悬浮
magnetic suspension type density sensor 磁悬式密度传感器
magnetic tape 磁带
magnetic tape drive 磁带驱动器
magnetic tape unit 磁带单元, 磁带装置
magnetic tools 磁性工具

magnetic transducer 磁传感器
magnetic valve 电磁阀
magnetic wind 磁风
magnetic wire 录音钢丝
magnetically coupled level detector 磁耦合液位检测器
magnetically hard material 永磁材料，硬磁材料
magnetically insulated gap 磁绝缘间隙
magnetically soft material 软磁材料
magnetic-coupled rotameter 磁耦合转子流量计
magnetic-induction flowmeter 磁感应流量计
magnetic-operated float switch 磁驱动浮子开关
magnetic-sector mass spectrometer 磁质谱仪
magnetism 磁，磁学
magnetism force pumps 磁力泵
magnetization 磁化
magnetization curve 磁化曲线
magnetization method 磁化方法
magnetization reversal 反向磁化
magnetize 磁化
magnetized ferrite 铁氧磁体
magnetizer 导磁体，磁化机
magnetizing 磁化，磁化的
magnetizing assembly 磁化装置
magnetizing coil 磁化线圈
magnetizing current 磁化电流，励磁电流
magnetizing reactance 磁化电抗
magnetizing time 磁化时间
magneto 磁发电机
magnetoconductivity 导磁性
magneto electric balance 电磁天平
magneto sensor 磁敏元件
magnetoelastic effect 压磁效应
magnetoelastic force transducer 磁弹性式力传感器
magnetoelastic rolling force measuring instrument 磁弹性式轧制力测量仪
magnetoelastic torque measuring instrument 磁弹性式转矩测量仪

magnetoelastic torque transducer 磁弹性式转矩传感器
magnetoelastic weighing cell 磁弹性式称重传感器
magnetoelectric instrument 磁电式仪表
magnetoelectric phase difference torque measuring instrument 磁电相位差式转矩测量仪
magnetoelectric phase difference torque transducer 磁电相位差式转矩传感器
magnetoelectric tachometer 磁电式转速表
magnetoelectric tachometric transducer 磁电式转速传感器
magnetoelectric velocity measuring instrument 磁电工速度测量仪
magnetoelectric velocity transducer 磁电式速度传感器
magnetoelectricity 电磁学
magnetometer 磁强计，磁力仪
magnetomotive force 磁通势，磁动势
magneto-optical effect magnetometer 磁光磁强计
magnetostriction 磁致伸缩
magnetostriction magnetometer 磁致伸缩磁力仪
magnetostriction testing meter 磁致伸缩测试仪
magnetostrictive transducer 磁致伸缩振动器
magnetreater 磁处理机
magnetrol 磁放大器
magnetron oscillator 磁控管振荡器
magnification 放大倍率
magnitude 幅值，大小，量级
magnitude contour 等高幅值
magnitude margin 幅值裕度，幅值裕量
magnitude scale factor 幅值比例尺
magnitude-frequency characteristic 幅频特性
magnitude-phase characteristic 幅相特性，幅相频率特性
magslip 无触点式自动同步机，旋转

变压器
main anti-syphonage pipe 总反虹吸管
main block 主体，主要部分
main circuit 主电路
main contact 主触点
main contactor 主接触器
main contract 主体合约
main control room 主控制室，主控室
main distribution frame (MDF) 主配线架
main drain 排水干管，总排水管
main feedback path 主反馈通路
main flux 主磁通
main frame 主机架
main manifold 主集流脉，主歧管
main memory 主存储器
main memory database system (MMDBS) 主存储器数据库系统
main movement 主运动
main pipe 排水干管，总排水管
main reinforcement 主钢筋
main shear wire 主剪力线
main span 主跨
main staircase 主楼梯
main storage 主存储器
main storey 主楼层
main switch 总开关
main switchboard 总配电板，总开关板
main transformer 主变压器
main valve 主阀，总阀门
main wall 主墙，承重墙
main waste pipe 主要废水管
main window 主窗口
main 主要的
main wire 电源线
mainboard 主机板，主板
maintenance 维修，保养，维护，检修，小修
maintenance device 维修仪器
maintenance engineer 维修工程师，维护工程师
maintenance log book 保养记录册，保养日志
maintenance manual 保养手册，检修手册
maintenance period 保养期
maintenance pit 维修坑
maintenance recommendation 维修建议
maintenance recommendation message 维护建议信息
maintenance service 保养维修
maintenance support software 维护支持软件
maintenance system 维修系统
maintenance test 维护试验
maintenance track 维修轨道
maintenance works 维修工程
major 主要的
major accident 严重事故
major defect 主要缺陷
major emergency 重大紧急事件
major loop 主回路
major repair 大修
major overhaul 大修
majority carrier 多数载流子
make contact 常开触点，动合触点
make good 修复
make sure 确定
making die 打印冲子
male die 公模，凸模
male plug 插头
male thread 外螺纹
malfunction 误动作，失灵，故障，故障出错，误动
malleable iron 可锻铸铁
malleablizing 可锻化退火，脱碳，可锻化处理
mallet 木槌
managed object 管理目标
management decision 管理决策
management information system (MIS) 管理信息系统
management level 管理级
management science 管理科学
management system 管理系统
management 管理，处理
manager 管理站
manager module test system 管理模块测试系统
Manchester encoding 曼彻斯特编码
mandatory standard 强制性标准

manganese(Mn) 锰
manganese iron 锰铁
manganese steel 锰钢
manganin wire 锰铜线
Manhattan distance 曼哈顿距离
manhole 沙井,检查井
manhole cover 井盖
manifold 歧管
manifold block 歧管挡块
manifold die 分歧管模具
manifold gauge 歧管仪表
manifolds 集合管
manipulate range 操纵范围
manipulated variable(MV) 操纵变量,输出值
manipulated variable high limit 操作量上限
manipulated variable low limit 操作量下限
manipulation 操纵,操作
manipulation task 操作任务
manipulator 机械手,操纵器,调制器
manipulator inertia matrix 操作器惯性矩阵
man-machine 人机
man-machine and environmental engineering 人机与环境工程
man-machine communication 人机通信
man-machine coordination 人机协调
man-machine interaction 人机交互
man-machine interface(MMI) 人机界面,人机接口
man-machine interface processor 人机接口处理器
man-machine system 人机系统
man-machine interaction 人机对话
manned submersible 载人潜水器
manoeuvrability 机动性,可移动,操纵的灵敏性
manometer 压力计,压力表,测压器
manometer with condensing tube 带冷凝管压力表
manometric thermometer with electric contacts 电接点压力温度计
manostat 稳压器,恒压器

manual 手动,手工的,手动的,指南
manual control 人工控制,手动控制
manual control panel 手动控制盘
manual control unit 手动操作器
manual data entry module 手动数据进入组件
manual data input programming 手动数据输入编程,人工数据输入编程
manual loader 手动操作器
manual loader with input indicator 带输入指示的手动操作器
manual mode 手动模式,人工模式
manual mode loop status switching key 手动方式回路状态转换键
manual mode operation 手控操作
manual oil pumps valves 手摇油泵阀
manual operating device 手动
manual operating mode 手动运转方式
manual operation 人工操作,手动操作
manual output card 手动输出卡
manual output station 手动输出操作站,手动输出操纵器
manual regulation 人工调节
manual scanning 手动扫查
manual set 手动设定
manual station 手动站,手动操作器
manual switch 手动开关
manual to automatic transfer 手动-自动的切换
manual with bias station 带偏置手动操作器
manual/automatic 手动/自动
manual/automatic station 手动/自动切换站
manufacture management 制造管理
manufacture procedure 制程,制造过程
manufacture quality assurance 制造品质保证
manufacturing 制造
manufacturing automation 制造自动化
manufacturing automation protocol (MAP) 制造自动化协议,生产自动化协议

manufacturing documentation 制造文件
manufacturing engineering 制造工程
manufacturing engineering of aerospace vehicle 航空宇航器制造工程
manufacturing execution systems 制造执行系统
manufacturing management 制造管理
manufacturing message service (MMS) 加工制造报文服务
manufacturing process 制造过程，制造工艺，生产过程
manufacturing specifications 制造规范
manufacturing system 制造系统
manufacturing systems and management 制造系统与管理
many degree of freedom 多自由度
many degree of freedom system 多自由度系统
MAP(manufacturing automation protocol) 制造自动化协议
mapper 测绘仪
mapping 页面寻址，面分布图
mapping analyzer 形态观察分析系统
mapping survey 地图制作测量
marble 大理石，云石
margin 裕度，利润，边缘
marginal 边界，临界，临界的
marginal distribution 边界分布
marginal stability 极限稳定性，临界稳定性
marine barometer 船用气压表
marine boiler 船用锅炉
marine borrow area 海洋采泥区，海洋采料区
marine deposit 海洋沉积土
marine diesel engine 船用柴油机
marine digital seismic apparatus 海洋数字地震仪
marine engine engineering 轮机工程
marine gravimeter 海洋重力仪
marine instrument 船用仪器仪表
marine optical pumping magnetometer 海洋光泵磁力仪
marine proton gradiometer 海洋质子梯度仪
marine proton magnetometer 海洋质子磁力仪
marine sand 海沙
marine seismic prospecting 海洋地震勘探
marine steam turbine 船用汽轮机
marine system 航海系统
marine vibrating-string gravimeter 海洋振弦重力仪
marine works 海洋工程
mark 标志，符号
mark of conformity 合格标志
marker 市价，标志，路标
marker post 标杆
marker sweep generator 扫频标志发生器
market and marketing 市场与行销
marking 标志，刻印加工
marking device 记录装置，压印器，压印机
marking gauge 划线规
marking iron 打号冲具，印记冲头
marking knife 划线刀
marking of an instrument for explosive atmosphere 防爆仪表标志
marking out 划线
marking terminal 标记型接线端子
marking tool 划线工具
Markov decision problem 马可夫决策问题
Markov decision process 马可夫决策过程
Markov model 马可夫模型
Markov parameter 马可夫参数
martempering 麻回火处理，分级淬火
Martens heat distortion temperature test 马顿斯耐热试验
martensite 马氏体
mask 防毒面具
masking tape 封口胶纸
Mason's gain formula 梅森增益公式
mass 质量
mass absorption coefficient 质量吸收系数
mass analyzed ion kinetic energy spectrometer(MIKES) 质量分析离子动能谱仪

mass analyzer 质量分析器
mass balance 质量平衡
mass balance equation 质量平衡式
mass centering 质量定心
mass centering machine 质量定心机
mass chromatography(MC) 质量色谱法
mass detection 质量检测
mass discrimination effect 质量歧视效应
mass dispersion 质量色散
mass flow computer 质量流量计算机
mass flow rate sensitive detector 质量流量敏感型检测器
mass flow transmitter 质量流量变送器
mass flowmeter 质量流量计
mass flow-rate 质量流量
mass fragmentography(MF) 质量碎片谱法
mass indicator 质量指示器
mass number 质量数
mass peak 质量峰
mass range 质量范围
mass scanning 质量扫描
mass selective detection 质量选择检测
mass spectrograph 质谱仪
mass spectrometer 质谱计
mass spectrometry(MS) 质谱学，质谱法，质谱分析法
mass spectrometry-mass spectrometry(MS-MS) 质谱-质谱法
mass spectroscope 质谱仪器，质谱仪
mass spectroscopy 质谱学，质谱法
mass spectrum 质谱
mass stability 质量稳定性
mass storage 大容量存储器，海量存储器
mass transit line 集体运输路线
mass transit system 集体运输系统
mass-radius product 质径积
mass-spectrometric method 质谱法
mass-spring system 质量弹簧系统
mass-to-charge ratio 质荷比
mast 柱杆，机柱(起重机)
master blade 标准刀齿
master control 主控，中心控制
master control board 主控板
master control panel 主控制屏
master control room 主控制室，中央控制室
master controller 主控制器，主令控制器
master drawing 布设总图
master file 主文卷，主文件
master gear 标准齿轮
master instruction 使用说明书
master plate 靠模样板
master program 主程序
master reference ground 主参考地
master signal 主信号
master station 主控站
master switch 总开关，主控接线机
master viscometer 标准黏度计
master 主要，控制者
master/slave discrimination 主从鉴别，主副鉴别
master-slave flip-flop 主从触发器
master-slave system 主-从系统
mastic 胶黏水泥
mastic sealant 胶泥封合剂
mat footing 席式基脚
mat foundation 席式地基，筏形基础
match index 匹配指数
match plate 分型板
matched die method 配合成形法，对模成形法
matched filter 匹配滤波器
matching 匹配，配合
matching circuit 匹配电路
matching criterion 匹配准则
matching of load 负载匹配
matching transformer 匹配变压器
material 材料，物料，原料
material balance 物料平衡
material balance control 物料平衡控制
material check list 物料检查表
material chemical analysis and mechanical capacity 材料化学成分和机械性能
material control 物料控制
material flows automation 物流自动化
material for engineering mold testing

工程试模材料
material measure 实体量器
material reject bill 退货单
material requirements planning (MRP) 物料需求计划，库存管理
material statistics sheet 物料统计明细表
material system 物料系统
material testing machine 材料试验机
material thickness 料片厚度
materials 物料
materials expenses 材料费
materials physics and chemistry 材料物理与化学
materials processing engineering 材料加工工程
materials science and engineering 材料科学与工程
mathematic model 数学模型
mathematical 数学的
mathematical programming 数学规划
mathematical similarity 数学相似
mathematical simulation 数字仿真
mathematical systems theory 数学系统理论
mating face 接触面，接合面
mating part 接合部件
matrass 卵形（蒸馏）瓶
matrice 矩阵，真值表
matrix 矩阵
matrix adder 矩阵式加法器
matrix algebra 矩阵代数
matrix amplifier 矩阵式放大器
matrix correction 基本修正
matrix determinant 确定矩阵
matrix effect 基体效应
matrix element 矩阵元素
matrix equation 矩阵方程
matrix formulation 矩阵公式
matrix inversion 矩阵求逆
matrix method 矩阵方法
matrix polynomial equation 矩阵多项式方程
matrix printer 点阵印刷机，点阵打印机
matrix Riccati equation 矩阵黎卡提方程
matrix triangularization 三角矩阵
matt finish 粗面，消光整饰
matte side 粗糙面
mattress netting steel wire 钢丝床用钢丝
mature 成熟的
max allowable continuous working current 最大允许连续工作电流
max allowable pressure 最高允许压力
max allowable temperature 最高允许温度
max deflection of linearity 最大线性偏转
max discharging capacity 最大排水量
max exclusion limit 排除极限
max loop resistance 最大回路电阻
max operating temperature 最高工作温度
max/min current and voltage 最高/低电流及电压
maximum acceleration 最大加速度
maximum allowable noise level 最高允许噪声声级
maximum allowed deviation 最大允许偏差
maximum amplitude 最大幅值
maximum capacity 最大称量
maximum controller gain 最大控制器增益
maximum cyclic load 最大循环负荷
maximum cyclic stress 最大循环应力
maximum difference work between plus and minus work 最大盈亏功
maximum displacement 最大位移
maximum entropy 极大熵
maximum excitation 最大激励
maximum floating voltage 最大浮置电压，最大空载电压
maximum flow-rate 最大流量
maximum likelihood 极大似然法
maximum likelihood estimation 最大似然估计
maximum likelihood estimator 极大似然估计器
maximum load 最大荷载
maximum load of the test 最大试验负荷

maximum load of the testing machine 最大负荷试验机
maximum operating pressure differential 最大工作压差
maximum operating water depth 最大工作（水）深度
maximum output 最大输出
maximum output inductance 最大输出电感
maximum output resistance 最大输出电阻
maximum overshoot 最大超调量
maximum penetration power 最大穿透力
maximum permitted load 最大容许负载量
maximum power consumption 最大功耗
maximum power supply voltage 最高电源电压
maximum principle 极大值原理，最大值原理
maximum profit programming 最大利润规划
maximum rated circumferential magnetizing current 额定周向磁化电流，最大周向磁化电流
maximum rated force under sinusoidal conditions 正弦态最大激振力
maximum relay 过载继电器，过电流继电器，过电压继电器
maximum revolutions of output shaft 输出轴最大转数
maximum rule 极大规则
maximum safe speed 最高速度
maximum safe working load 最大施工负载量
maximum scale value 标度终点值，最大标度
maximum sound pressure level of microphone 传声器最高声压级
maximum static pressure 最高静态压力
maximum strain 最大应变
maximum stress 最大应力
maximum temperature 最高温度
maximum temperature-rise 最高温升

maximum thermometer 最高温度计，最高温度表
maximum transverse load 最大横向负荷
maximum value of the frequency response 频率响应的最大值
maximum velocity 最大速度
maximum wind speed 最大风速
maximum working pressure 最大工作压力
maximum 极大，最大限度，最大值
Maxwell bridge 麦克斯韦电桥
Maxwell equation 麦克斯韦方程
MB(model base) 模型库
MB(megabyte) 兆字节
MBC(model-based control) 基于模型控制
MBMS(model base management system) 模型库管理系统
MBT (mechanical bathythermograph) 机械式深温计
MC(mass chromatography) 质量色谱法
MC(micellar chromatography) 胶束色谱法
MC(multifunction controller) 多功能控制器
MCA (microchannel architecture) 微通道结构
McLeod vacuum gauge 麦氏真空计
MDF(machine description format) 机器描述格式
MDF(main distribution frame) 主配线架
MDF database 机器描述格式数据库
mean 平均值，平均的，中央
mean addendum 中点齿顶高
mean cone distance 中点锥距
mean dedendum 中点齿根高
mean diameter 中径
mean diametral pitch 中点径节
mean dynamic pressure in a cross-section 横截面内的平均动压
mean effective wavelength 平均有效波长
mean error 平均误差
mean flow-rate 平均流量

mean height 平均高度
mean level 平均水平
mean life 平均寿命
mean linear velocity of mobile phase 流动相平均线速
mean load 平均负荷
mean measuring addendum 中点测量齿顶高
mean measuring depth 中点测量齿高
mean measuring thickness 中点测量厚度
mean normal base pitch 中点法向基节
mean normal diameter pitch 中点法向径节
mean normal module 中点法向模数
mean point 中点，平均点
mean radius 中点半径
mean screw diameter 平均中径
mean section 中点截面
mean slot width 中点齿槽宽
mean spiral angle 中点螺旋角
mean squared spectral density 均方谱密度
mean strain 平均应变
mean strength 平均强度
mean stress 平均应力
mean switching point 切换中值，切换中点
mean time 平均时间
mean time between failure(MTBF) 平均无故障时间
mean time between failures(MTBF) 平均故障间隔时间
mean time to failure(MTTF) 平均故障时间
mean time to repair(MTTR) 平均修理时间
mean value analysis 均值分析
mean value 平均值
means of access 进出途径，途径
mean-square error 均方误差
mean-square error criterion 均方误差准则
meantime auto-spectrometer 同时式自动光谱仪
measurand 被测量
measure target 被测目标

measure 量度，测量，尺寸
measured 测量的
measured accuracy 测量精度
measured deviation 测量偏差
measured feedback 测量反馈
measured object 被测对象
measured point 检测点
measured quantity 被测量
measured signal 被测信号
measured strength 量度所得强度
measured value 测定值，被测值，测量值
measured variable 被测变量
measurement 测量
measurement bridge 测量电桥
measurement condition 检测条件
measurement dynamics 测量动态
measurement error 测量误差
measurement hardware 测量硬件
measurement indicator 测量指示仪
measurement noise 测量噪声
measurement of directional response pattern 指向性响应图案测量
measurement of exciting force 激振力的测量
measurement of vibration quantity 振动量的测量
measurement procedure 测量步骤
measurement range 量程，测量范围
measurement signal 测量信号
measurement standard 计量标准器，测量标准
measurement time 测量时间
measuring 测量，测量的
measuring addendum 测量齿顶高
measuring amplifier 测量放大器
measuring and testing technologies and instruments 测试计量技术及仪器
measuring apparatus 测量装置
measuring bridge 测量电桥
measuring chain 测量链
measuring current transformer 测量用电流互感器
measuring device 测量装置
measuring distance 测量距离
measuring element 测量元件

measuring equipment 测量设备
measuring error 测量误差
measuring hole 测量孔
measuring indication system 测量指示
measuring installation 测量设备
measuring instrument 测量仪器,测量仪表,测量器具
measuring instrument with circuit control device 带有电路控制器件的测量仪表
measuring junction 测量端,测量端区
measuring microphone 测试传声器
measuring plane 测量平面
measuring point 测量点
measuring point for the humidity 湿度测定点
measuring point for the temperature 温度测定点
measuring potentiometer 测量电位差计
measuring section 测量段
measuring sequence 测量顺序
measuring slide wire 测量用滑触电阻线
measuring span 量程
measuring spark gap 测量球隙
measuring system 测量系统
measuring terminal 测量端
measuring time 测量时间
measuring tooth thickness 测量齿厚
measuring transducer 测量传感器
measuring transmitter 变送器,测量变送器
measuring unit 测量单位
measuring voltage transformer 测量用电压互感器
measuring wire 尺度索,准绳
measuring range 测量范围
measuring range higher limit 测量范围上限值
measuring range lower limit 测量范围下限值
MEC(most economic control) 最经济控制
MECC(micellar electrokinetic capillary chromatography) 胶束动电毛细管色谱法
mechanical accelerometer 机械加速度计
mechanical advantage 机械利益
mechanical bathythermograph (MBT) 机械式深温计
mechanical behavior 机械特性
mechanical brake 机械制动器
mechanical capacity 机械性能
mechanical coupler 机械套管
mechanical creation design 机械创新设计
mechanical design 机械设计
mechanical design and theory 机械设计与理论
mechanical drawing 机械制图
mechanical efficiency 机械效率
mechanical engineering 机械工程
mechanical hygrometer 机械湿度计
mechanical impedance 机械阻抗
mechanical linkage 机械联动装置
mechanical manipulator 机械操纵器
mechanical manufacture and automation 机械制造及其自动化
mechanical parts 机械零件
mechanical property 机械性能,力学性能
mechanical property tester 机械性能测定仪,力学性能测定仪
mechanical quantity 机械量
mechanical quantity sensor 机械量传感器
mechanical quantity transducer 机械量传感器
mechanical rectifier 机械式整流器
mechanical regulator 机械调节器
mechanical resonance 机械共振
mechanical resonance frequency of the moving element 运动部件机械共振频率
mechanical runout 机械脱出
mechanical sensor 力敏元件
mechanical shock 机械冲击
mechanical speed governors 机械调速
mechanical stepless speed changes 机械无级变速
mechanical stethoscope 机械金属探伤

器
mechanical strain 机械应变
mechanical strength 机械强度
mechanical stress 机械应力
mechanical structure type sensor 结构型传感器
mechanical structure type transducer 结构型传感器
mechanical system 机械系统
mechanical system design 机械系统设计
mechanical test 机械性能试验
mechanical testing machine 机械式试验机
mechanical top-loading balance 机械式上皿天平
mechanical ventilation 机械通风，机动通风
mechanical vibration meter 机械测振仪
mechanical vibrator 机械振动器
mechanical zero 机械零位，机械零点
mechanical zero adjuster 机械零位调节器
mechanical 机械的，力学的
mechanical-electrical integration 机电一体化
mechanical-electrical integration system design 机电一体化系统设计
mechanics 力学
mechanics of motor vehicle 汽车力学
mechanism and component parts 机构和零部件
mechanism in common use 常用机构
mechanism model 机理模型
mechanism with flexible elements 挠性机构
mechatronic engineering 机械电子工程
media access control 媒介存取控制，介质存取控制
medial 中间的，平均的
median 中值，中央
median barrier 中央路栏，路中护栏
median filter 中值滤波器
median frequency 中频
medical application 医学应用

medical catheter 医用导管
medical electronic linear accelerator 医用电子直线加速器
medical equipments urine analyzer 尿液分析仪
medical injection pump 医用灌注泵
medium temperature strain gauge 中温应变计
medium wave 中波
medium 中间的，中等的，介质，工质
medium-high frequency 中高频
medium-power distribution 中压配电
medium-wave band 中波段
meehanite cast iron 米汉纳铸钢，高强度铸铁
meehanite metal 米汉纳铁
meeting minutes 会议记录
megabyte (MB) 兆字节
megameter 兆欧表
megaton 百万吨级
megger 兆欧表，摇表，高阻表
megger test 绝缘测试
megger tester 绝缘测试器
MEKC (micellar electrokinetic chromatography) 胶束动电色谱
melamine formaldehyde resin 三聚氰胺甲醛树脂
melted quartz capacitor 熔融石英电容器
melting heat 熔解热
melting point 熔解点，熔点
melting point measuring instrument 熔点测定仪
melting point type disposable fever thermometer 熔点型消耗式温度计
membership 隶属度，相关
membership degree 相关度
membership function (MF) 隶属函数，相关函数
membrane 膜片
membrane differential pressure gauge 膜片式差压计
membrane switch 薄膜开关
membrane vacuum-gauge 膜片真空表
memory 存储，存储器
memory and I/O bridge controller 内

存和 I/O 桥控制器
memory application 存储器应用
memory bank 存储体
memory cell 存储单元
memory element 存储元件
memory interface 存储器接口
memory junction cell 存储连接单元
memory protect and watch variable test program 存储器保护和监测程序
memory protection 存储保护
memory register 存储寄存器
memory swap 交换记忆
memory transfer hub 内存转换中心
memory unit 存储部件
memoryless 无记忆
memoryless source 无记忆信源
Mendeleev weighing 门捷列夫称量法
mendelevium(Md) 钔
mental workload 脑力负荷
menu selection mode 选单选择式，菜单选择式
menu 菜单
MEO(most economic observing) 最经济观测
mercury(Hg) 汞
mercury barometer 水银气压表
mercury drop amplitude 汞滴振幅
mercury filled thermal system 充水银式感温系统
mercury lamp 水银灯具
mercury manometer 水银压力计
mercury motor meter 水银电机式仪表
mercury pool electrode 水银电极
mercury switch 水银开关
mercury thermometer 水银温度表
mercury-in-glass thermometer 玻璃水银温度计
mercury-wetted relay 汞浸继电器，水银继电器
mesa transistor 台面式晶体管
mesh 筛孔，网孔，筛目
mesh current 网孔电流
mesh point 啮合点
mesh sieve 网筛
mesh wire 金属网线
meshing 啮合

meson 介子
message 报文，信息
message capacity 信息容量
message mode 报文方式
message screen 信息屏幕
message summary display 信息总貌画面
message switching 报文交换，信息交换
messenger wire 悬缆线，承力吊索，吊线
metadyne 磁场放大机，旋转式磁场放大机，微场扩流发电机
metadyne generator 微场扩流发电机
meta-knowledge 元知识
metal 金属
metal arc welding 金属电弧焊
metal base copper-clad laminate 金属基覆铜层压板
metal base indicated electrode 金属基指示电极
metal base printed board 金属基印制板
metal braid 金属编织电缆
metal casing 金属壳，金属套管
metal chip 金属碎片
metal clad 金属壳
metal core copper-clad laminate 金属芯覆铜箔层压板
metal core printed board 金属芯印制板
metal cover 金属封盖
metal cutting 金属切削
metal disc 金属圆片
metal dust 金属粉末
metal elemental analysis 金属元素分析仪
metal filing 金属屑
metal gripping rim 金属夹边
metal halide lamp 金属卤化物灯
metal lath 金属拉网
metal pail 铁桶
metal plate 金属板
metal saw 金工锯
metal spike 金属钉
metal stud 金属嵌钉
metal U-tube manometer U形金属管

压力计
metal-ceramic X-ray tube 金属陶瓷X射线管
metal-clad bade material 覆金属箔基材
metal-insoluble salt indicated electrode 金属-难溶盐指示电极
metal-level knowledge 中位知识
metallic diaphragm 金属膜片
metallic gasket 金属垫片
metallic halide lamp 金属卤化物灯
metallic iron 精炼铁
metallic material testing machine 金属材料试验机
metallikon 金属喷镀法
metallizing 真空涂膜
metalloid 非金属
metallurgical automation 冶金自动化
metallurgical engineering 冶金工程
metallurgical microscopy 金相显微镜
metallurgy 冶金学
metal-oxide gas sensor 金属氧化物气体传感器
metal-oxide gas transducer 金属氧化物气体传感器
metal-oxide humidity sensor 金属氧化物湿度传感器
metal-oxide humidity transducer 金属氧化物湿度传感器
metal-oxide semiconductor (MOS) 金属氧化物半导体
metal-shielded wire 金属屏蔽线
metal-spring gravimeter 金属弹簧重力仪
metalster 金属膜电阻
metalwork 金属部分,金属工艺
metastable decomposition 亚稳分解
metastable ion 亚稳离子
metastable peak 亚稳峰
metastable scanning 亚稳扫描
meteorograph 气象计
meteorological data 气象数据
meteorological instrument 气象仪器
meteorological observation 气象观测
meteorological radar 气象雷达
meteorological rocket 气象火箭
meteorological satellite 气象卫星
meteorological tower 气象塔

meter 计,表
meter case 仪表外壳
meter dial 仪表度盘
meter electrodes 测量电极
meter flow-rate 仪表流量
meter needle type globe valves 仪表针形截止阀
meter resistance 仪表电阻
meter sensitivity 仪表灵敏度
meter terminal box 仪表端子盒
meter transformer 仪用互感器
meter with maximum demand indicator 最大需量电度表
metering panel 计量盘
metering pump 计量泵
meters for measuring amplitude by a reading microscope 读数显微镜测振幅法
methane 甲烷,甲烷沼气
methane content 沼气含量
methane detection 沼气检查
methane emission 沼气泄出
methane gas 瓦斯
methane ignition 甲烷着火
methane-air mixture 沼气-空气混合物
method 方法,方式,规律,程序
method of correction 校正方法
method of electro-mechanical analogy 机电模拟方法
method of electron diffraction 电子衍射法
method of field emission microscope 场发射显微镜法
method of field parameter measurement 场参数测量法
method of instability 不稳定法
method of least square 最小二乘法
method of measurement 测量方法,计量方法
method of spot parameter measurement 点参数测量法
method of substitution 替代方法
method of successive comparison 逐次比较法
method of supposition 叠加法
method of three key factors 三要素法
method of trial and error 试探法,逐

步逼近法
method of weighted 加权方法
method of weighted residual 残差加权法
method standard 方法标准
method of operation 运行方式
methodology 方法论
methods of representing sequences 描述顺序的方法
metre (m) 米(长度单位)
metric 量度,尺度
metric gears 公制齿轮
metrological performance 计量性能
metrology 计量学
MF (mass fragmentography) 质量碎片谱法
MF (membership function) 隶属函数,相关函数
mica capacitor 云母电容器
mica condenser 云母电容器
mica diaphragm 云母膜片
micellar chromatography (MC) 胶束色谱法
micellar electrokinetic capillary chromatography (MECC) 胶束动电毛细管色谱法
micellar electrokinetic chromatography (MEKC) 胶束动电色谱
micro analysis 微量分析
micro balance 微量天平
micro coulometric detector 微库仑检测器
micro ray 微波,微波光线
micro wire board 微线印制板
microammeter 微安计
microbarograph 微(气)压计
microbiological incubator 微生物培养箱
microbore column 微径柱
microbore packed column 微填充柱
microchannel architecture (MCA) 微通道结构
microcomputer 微型计算机,微机
microcomputer alternating current resistivity instrument 微机化交流电阻率仪
micro-computer field measuring system 微电脑野外检测系统

microcomputer induced polarization instrument 微机激电仪
microcomputer system 微机系统
microcomputer-based control 基于微机控制
microcomputer-based system 基于微机系统
micro-densitometer 微密度计
micro-economic model 微观经济模型
micro-economic system 微观经济系统
microelectronic device 微电子器件
microelectronics 微电子学
microelectronics and solid state electronics 微电子学与固体电子学
microemulsion electrokinetic chromatography 微滴乳状液动电色谱法
microfarad 微法[拉](电容单位)
microfilm 微缩影片
microhardness number 显微硬度值
micro-heat of adsorption detector 微量吸附热检测器
microinching equipment 微动装置
micrometer 千分尺,测微计
micrometer calipers 千分卡尺
micrometer checker 千分表检查仪
micron 微米
micro-packed column 微填充柱
microparticle column 微粒柱
microphone 传声器,麦克风,话筒
microphone calibration apparatus 传声器械校准仪
microphone protection grid 传声器保护罩
microphone response frequency 传声器共振频率
microphone stand 传声器架
microphone temperature coefficient 传声器温度系数
microphotometer 显微光度计
microplate spectrophotometer 微型分光光度计
microprocessor 微处理器,微处理机
microprocessor control 微处理器控制
microprocessor-based central processor unit 以微处理器为基础的中央处理单元
microscope 显微镜

microsoft-disk operating system 微波磁盘操作系统
microstrip 微波传输带
micro-structure 微观结构
microswitch 微动开关
microswitch transmitter 微动开关式变送器
microsystem 微系统
microwave 微波
microwave band 微波段
microwave detection apparatus 微波检测仪
microwave distance meter 微波测距仪
microwave distance method 微波探伤法
microwave Doppler meter 微波多普勒流量计
microwave filter 微波滤波器
microwave inductive plasma emission spectrometer 微波等离子体光谱仪
microwave network analyzer 微波网络分析仪
microwave plasma detector 微波等离子体检测器
microwave radar 微波雷达
microwave radiometer 微波辐射计
microwave remote sensing 微波遥感
micro-wave scatterometer 微波散射计
microwave spectrum 微波谱
microwave thickness meter 微波厚度计
microwave transmitter-receiver 微波收发两用机
microwave tube 微波管
mid infrared range (MIR) 中红外
mid infrared range remote sensing 中红外遥感
mid infra-red spectrum (MIRS) 中红外光谱
mid 中间的
middle 中间的
middle voltage 中压
mid-frequency band 中频带
mid-infrared absorption spectrum 中红外吸收光谱
mid-plane 中间平面
mid-point 中点
mid-span 中跨
mid-span joint 中跨接头

migration 迁移,移动,移植
MIKES(mass-analyzed ion kinetic energy spectrometer) 质量分析离子动能谱仪
mil 密耳(千分之一英寸)
mild drawn wire 软拉钢丝
mild steel barrier 软钢围栏
mild steel pipe 软钢管
mildew 发霉
military chemistry and pyrotechnics 军事化学与烟火技术
military-standard(MIL-STD) 军用标准
mill 锉,铣削,磨机
milled helicoids worm 锥面包络圆柱蜗杆
miller 铣床
milli(m) 毫(千分之一)
milliammeter 毫安表
milliampere 毫安(电流单位)
milliliter(ml) 毫升
millimetric wave magnetron 毫米波磁控管
milling cutter 铣刀
milling heads 铣头
milling machine 铣床
millivolt 毫伏(电压单位)
millivolt ammeter 毫伏安培计
millivoltmeter 毫伏计
milliwatt 毫瓦(功率单位)
MIL-STD(military-standard) 军用标准
mimic control panel 模拟控制板
MIMO control system 多输入多输出控制系统
min 轻微的
min load resistance 最小负载电阻
mine 矿场
mineral 矿物
mineral engineering 矿业工程
mineral oil 矿物油,石油
mineral processing engineering 矿物加工工程
mineral resource prospecting and exploration 矿产普查与勘探
miniaturization 小型化
minicomputer 小型计算机

mini-max technique 极小极大技术
minimization 最小化
minimum 最小，最小的，最小值
minimum current relay 低电流继电器
minimum distance 最短距离
minimum load of the testing machine 试验机最小负荷
minimum of objective function 最小目标函数
minimum output 最小输出
minimum phase system 最小相位系统
minimum power supply voltage 最低电源电压
minimum power voltage 最小电源电压
minimum principle 最小值原理，极小值原理
minimum rate of benefit 最低收益率
minimum redundancy 最小冗余
minimum reserve 最低储备
minimum scale value 标度始点值，最小标度
minimum temperature 最低温度
minimum thermometer 最低温度计，最低温度表
minimum scale value 标度始点值
minimum value 最小值
minimum variance control 最小偏差控制
minimum-energy problem 最小能量问题
minimum-fuel problem 最小燃料问题
minimum-time control 最小时间控制
minimum-time problem 最小时间问题
mining compass 矿山罗盘仪
mining engineering 采矿工程
minioscilloscope 小型示波器
miniwatt amplifier 小功率放大器
minor 轻微的
minor defect 次要缺陷
minor diameter 小直径，小径，内径
minor structural works 小型结构工程
minor works 小工程
minor loop 副回路，辅助回路，小回路
minor overhaul 小修
minority carrier 少数载流子

MIO (multiuser information outlet) 多用户信息插座
MIR (mid infrared range) 中红外
mirror image switch 镜像开关
mirror iron 镜铁
mirror telescope 反射望远镜
mirroring 镜像
MIRS (mid infra-red spectrum) 中红外光谱
MIS (management information system) 管理信息系统
mishandle 胡乱操作，误操纵
mismatch 失配，偏模
misread 错读
misreading 误读，错读
miss operation 误动作，误操作
missile 导弹
missile-target relative movement simulator 弹体-目标相对运动仿真器
missing part 漏件
mist eliminator 除雾器
miter gears 等齿数整角锥齿轮副
mitigation measure 减轻措施，舒缓措施
mix 混合物，组合，混合，配料
mix design 混合设计
mix proportion 混合用料比例，混配比例
mixed color 杂色
mixed column 混合柱
mixed flow pumps 混流泵
mixed indicator 混合指示剂
mixed lubrication 混合润滑
mixed sensitivity problem 混合灵敏度问题
mixer 搅拌器，混频器，混合器
mixing 混频，混合
mixing drum 搅拌滚筒，搅拌鼓
mixing length 混合长度
mixing machine 拌和机
mixing ratio 混合比
mixing temperature 混合温度
mixture 混合，混合物
MMDBS (main memory database system) 主存储器数据库系统
MMI (man-machine interface) 人机界面，人机接口

MMS (manufacturing message service) 加工制造报文服务
mobile crane 起重车，移动式起重机
mobile ct system 移动 ct 系统
mobile daughter card 移动式子卡
mobile phase 流动相
mobile phase front 流动相前沿
mobile robot 移动机器人
mobile weather station 流动气象站
mobile X-ray detection apparatus 移动式 X 射线探伤机
mobility 移动性，迁移率
mock up test 模型试验
modal 模式的，模态的
modal aggregation 模态集结
modal control 模态控制
modal coupler 模态耦合器
modal matrix 模态矩阵
modal transformation 模态变换
mode 方式，方法，模式
mode action 控制作用，控制方式
mode analysis 模式分析
mode of transport 运输模式
mode of vibration 振形，振动模态，振动形式
mode shape 振形
mode structure 模式结构
mode theory 模式理论
model 模型，型号，样机，典型
model accuracy 模型精确度
model algorithm control(MAC) 模型算法控制
model analysis 模型分析
model approximation 模型近似
model base(MB) 模型库
model base management system(MBMS) 模型库管理系统
model confidence 模型置信度
model coordination method 模型协调法
model decomposition 模型分解
model design 模型设计
model designation of centrifuges 离心机型号编制方法
model designation of filters 过滤机型号编制方法
model designation of separators 分离机型号编制方法
model evaluation 模型评价
model experiment 模型实验
model fidelity 模型逼真度
model following control system 模型跟踪控制系统
model following controller 模型跟踪控制器
model loading 模型装载
model management 模型管理
model modification 模型修改
model of strain gauge 应变计形式
model predictive heuristic control (MPHC) 模型预测启发控制
model reduction 模型降价
model reduction method 模型降价法
model reference 模型参考
model reference adaptive(MRA) 模型参考自适应
model reference adaptive control 模型参考自适应控制
model reference adaptive control system 模型参考自适应控制系统
model reference control 模型参考控制
model reference control system 模型参考控制系统
model simplification 模型简化
model state feedback(MSF) 模型状态反馈
model test 模型测试
model transformation 模型变换
model validation 模型确认
model variable 模型变量
model verification 模型验证
model-based control(MBC) 基于模型控制
model-based recognition 基于模型识别
model-following control 跟随模型控制
modeling 建模
modeling error 建模误差
model-predictive control(MPC) 模型预测控制
modem 调制解调器
moderator 减速剂，调节器
modern control theory 现代控制理论
modern polarography 近代极谱法

modifiability 可修改性
modification 改装，修订，更改
modification coefficient 变位系数
modified 修改的
modified contact ratio 修正总重合度
modified gear 变位齿轮
modified roll 滚修正比
modified sine acceleration motion 修正正弦加速度运动规律
modified trapezoidal acceleration motion 修正梯形加速度运动规律
modifier 改性剂
modify 修改，更改
modular design 模块化设计
modular keyboard 组合式键盘
modular mold 组合式模具
modular programming 模块化程序设计
modular system 模块式传动系统
modular type annunciator 组件式信号器
modular control 模块控制
modularity 组合性
modularization 模块化
modulated wave 已调波
modulating 调制的
modulating control system 模拟量控制系统
modulating action 调制作用
modulating amplifier 调制放大器
modulating valve 调节阀
modulating wave 调制波，调幅波
modulation 调制
modulation analysis 调制分析
modulation control 调制控制
modulation degree 调制度
modulation depth 调制深度
modulation distortion 调制失真
modulation eliminator 解调器
modulation regulation 调节
modulation sideband 调制边带
modulator 调节器，调制器，调幅器
modulator band filter 调幅器带通滤波器
module 模块，组件，模件
module summary display 模块总貌画面
modulus 模数

modulus of elasticity 弹性模量
moisture 水分，湿度，湿汽
moisture analyzer 湿度分析器
moisture content 含湿量，水汽含量
moisture of aerated concrete 加气混凝土含水率
moisture sensor 湿敏元件
moisture-proof storage 防湿保管库
mold 成型，模具
mold base 塑胶模座，模胚（架）
mold clamp 铸模紧固夹
mold components 模具零件
mold platen 模用板
molded circuit board 模塑电路板
molding 成型
molding factory 成型厂
molecular absorption spectrometry 分子吸收光谱法
molecular beam mass spectrometer 调制分子束质谱计
molecular fluorometry 分子荧光分析法
molecular ion 分子离子
molecular spectrum 分子光谱
molecular weight(MW) 分子量
molecule 分子，微粒
molecule ion 分子离子
molten metal 熔融金属，熔态金属，金属熔液
molten slag 熔渣
molybdenum(Mo) 钼
molybdenum high speed steel 钼系高速钢
molybdenum steel 钼钢
molybdenum wire 钼丝
moment 力矩，动量
moment method 力矩法
moment of couple 力偶矩
moment of flywheel 飞轮矩
moment of inertia 惯性力矩，转动惯量
moment of pendulum 摆锤力矩
moment of torque 扭矩
momentum 动量
momentum-type mass flowmeter 动量式质量流量计
monel 蒙乃尔铜-镍合金
mongline 单线

monitor 监听器，监视器，监控器，监控
monitor light 指示灯
monitored control system 监控系统
monitoring 监测，监视
monitoring element 监控环节，监控元件
monitoring feedback 监控反馈
monitoring hardware 监视硬件
monitoring loop 监控回路
monitoring program 监督程序
monitoring station 监测站
monitoring wire 监听线，监视线路
monochromatic radiation 单色辐射
monochromator 单色仪
monocrystalline wire 单晶丝
monometallic wire 单金属线
monopolar 单极
monopolar D/A converter 单极性数/模转换器
monopole mass spectrometer 单极质谱计
monorail 单轨铁路
monostabillity 单稳态
monostabillity multivibrator 单稳态多谐振荡器
monostable trigger element 单稳态触发元件
monotone system 单调系统
monotonicity 单调性，单一性
monovalent 单价的
Monte Carlo 蒙特卡罗
Monte Carlo calculation 蒙特卡罗计算
Monte Carlo method 蒙特卡罗法（统计检验法）
Monte Carlo simulation 蒙特卡罗仿真
month 月
monthly load curve 月负荷曲线
monthly load factor 月负荷率
monthly report 月报
Morse taper gauge 莫氏锥度量规
mortar 沙浆，砂浆
mortise 榫眼，卯孔
mortise chisel 榫凿
mortise gauge 双线规，榫规
MOS (metal-oxide semiconductor) 金属氧化物半导体
MOSFET 金属氧化物半导体场效应晶体管
most economic control（MEC） 最经济控制
most economic observing（MEO） 最经济观测
mother board 母板
motion 移动，运动
motion control 运动控制
motion curves 运动曲线
motion estimation 移动估计
motion parameter 移动参数
motion space 可动空间
motive power 驱动功率，动力
motor 马达，电动机，发动机
motor alternator 电动交流发电机
motor and soft starters 电机及软起动器
motor control 电动机控制，发动机控制
motor controller 电机控制器
motor cycle 摩托车
motor element 电动机元件
motor generator 电动发电机
motor management systems 电机管理系统
motor pattern 发动机模式
motor rotor 电动机转子
motor starter 马达起动器，电动机起动器
motor torque 电机扭矩
motor unit 发动机单元
motor vehicle 汽车
motor vehicle for agricultural use 农业用汽车
motor 马达
motor winding 电动机组绕组
motor-drive 电动机拖动
motor-driven 电动机拖动的
motor-operated valve 电机控制阀
mottle 斑点
mould 铸模，模子
mould clamping force 锁模力
mould growth test 长霉试验
mould release agent 脱模剂
moulding 铸模成形，倒模成形，模塑，铸造
mountain barometer 高山气压表

mounting bracket 托架
mounting of I/O units 输入/输出单元的装置
mounting position 安装位置
mounting strain error 安装应变误差
mouse 鼠标
mouthpiece 接口管
movable cross-beam 移动横梁
movable frame 活动构架
move 移动
movement 移动，运动
movement joint 移动接缝，伸缩缝
moving 动的，活动的，移动
moving armature 动衔铁
moving average model 移动平均模型，滑动平均模型
moving average unit 移动平均运算器
moving band interface 传送带接口
moving bolster 活动工作台，移动式工作台
moving bolster plate 可动侧模板
moving coil 动圈，可动线圈
moving coil ammeter 动圈式安培计，磁电式安培计
moving coil galvanometer 动圈式检流计，磁电式检流计
moving coil instrument 动圈式仪表，磁电式仪表
moving coil oscillograph 动圈式示波器
moving coil type temperature indicating controller 动圈式温度指示调节仪
moving contact 动触点
moving core type relay 活动铁芯式继电器
moving element 运动部件，可动部分
moving field 移动磁场，动磁场
moving index measuring instrument 固定标度测量仪表
moving iron 动铁式，转动铁芯
moving iron ammeter 铁式安培计，电磁式安培计
moving link 运动构件
moving magnet instrument 动磁式仪表
moving object 移动目标
moving part 活动部分
moving point device 移点器
moving scale measuring instrument 活动标度测量仪表
moving table 滑台
moving-average filter 移动平均滤波器
moving-coil instrument 动圈式仪表
moving-coil microphone 动圈传声器
moving-conductor microphone 电动传声器
moving-iron instrument 动铁式仪表，电磁式仪表
moving-iron logometer 动铁式比率计，电磁式比率计
moving-magnet galvanometer 动磁系振动子
moving-magnet instrument 动磁式仪表
moving-scale instrument 动标度尺式仪表
MPC(model-predictive control) 模型预测控制
MPHC(model predictive heuristic control) 模型预测启发控制
MRA(model reference adaptive) 模型参考自适应
MRP(material requirements planning) 物料需求计划
MS(mass spectrometry) 质谱学，质谱法，质谱分析法
MSF-(model state feedback) 模型状态反馈
MS-MS scanning 质谱-质谱法扫描
MS-MS(mass spectrometry-mass spectrometry) 质谱-质谱法
MSS(multispectral scanner) 多光谱扫描仪
MTBF(mean time between failure) 平均无故障时间
MTBF(mean time between failures) 平均故障间隔时间
MTTF(mean time to failure) 平均故障时间
MTTR(mean time to repair) 平均修理时间
mud 泥浆
mud flow meter 泥浆流量计
mud hydrometer 泥浆比重计
mud logger 泥浆电阻仪
mud lubrification meter 泥浆润滑性测定仪

mud resistance meter　泥浆电阻仪
mud sand content meter　泥浆含砂量测定仪
mud water loss meter　泥浆失水量测定仪
mud wave　泥浆波
muff coupling　套接
Muffle Furnaces　马弗炉
muffler　灭声器,消声器
multi band seismograph　多频带地震仪
multi collectors mass spectrometer　多接收器质谱计
multi crystal thermistor　多晶热敏电阻器
multi-access system　多存取系统
multi-action controller　多作用控制器
multi-attributive utility function　多属性效用函数
multi-axial strain gauge　多轴应变计
multibus　多总线
multi-cache　多高速缓冲寄存器
multi-cavity mold　多模穴模具
multichannel　多通道的,多信道的
multichannel amplifier　多路放大器,多通道放大器
multichannel analyzer　多道分析器
multi-channel controller　多通道控制器
multichannel cross correlation　多通道互相关
multi-channel logging truck　多线式自动测井仪,测井站
multi-channel photo-recorder　多线照相记录仪
multi-channel pulse height analyzer　多道脉冲高度分析仪
multichannel system　多路制,多路系统
multichannel telephony　多路电话
multichannel transmission　多路传输
multi-channel X-ray spectrometer　多道X射线光谱仪
multi-chip　多芯片
multi-circle control　多周波控制
multi-circuit switch　多路转换开关
multi-colour radiation thermometer　多色辐射温度计
multi-colour thermometry　多色测温法
multi-computer system　多计算机系统
multi-conductor　多导体
multi-conductor system　多导体系统
multi-conductor transmission line　多导体传输线
multi-contact relay　多触点继电器
multicore cable　多芯电缆
multicore type current transformer　多铁芯型电流互感器
multicoupler　多路耦合器
multi-criteria　多重判据
multi-criteria decision making　多目标规划
multi-criteria decision methods　多准则决策分析
multidimensional　多维
multidimensional digital filter　多维数字滤波器
multidimensional gas chromatograph　多维气相色谱仪
multidimensional gas chromatography　多维气相色谱法
multidimensional system　多维系统
multi-electrode　多电极
multi-element control system　多冲量控制系统
multi-emitter transistor　多发射极晶体管
multifrequency channel ground detector　多频道地电仪
multifrequency generator　多频发电机,多频振荡器
multifrequency system　多频制,多频系统
multifunction controller(MC)　多功能控制器
multi-function graphics　多功能流程图
multi-function measuring instrument　多功能测量仪表
multi-function sensor　多功能传感器
multifunction subsystem　多功能子系统
multi-function transducer　多功能传感器
multifunctional　多功能的
multi-functional dual-pen strip chart re-

corder 多功能双笔长图记录仪
multigrid control 多栅控制
multi-idler belt conveyor scale 多托辊电子皮带秤
multi-input multi-output (MIMO) 多输入多输出
multi-input multi-output control system 多输入多输出控制系统
multilayer coil 多层绕组
multilayer control 多层控制
multi-layer printed board 多层印制板
multi-layer printed circuit board 多层印制电路板
multilayer system 多层系统
multi-layered film substrate 多层膜基板
multilevel 多级,多层
multilevel code 多级编码
multilevel computer control system 多级计算机控制系统
multilevel control 多级控制
multilevel controller 多级控制器
multilevel coordination 多级协调
multilevel decision 多级决策
multilevel hierarchical structure 多级递阶结构
multi-level logical circuit 多级逻辑电路
multilevel process 多级过程
multilevel structure 多级结构
multilevel system 多级系统
multilink 多链路
multiloop 多回路的,多匝的
multiloop control 多回路控制
multiloop control strategy 多回路控制策略
multiloop control system 多回路控制系统
multiloop controller 多回路控制器
multiloop system 多路系统
multi-machine 多机
multimedia 多媒体
multimedia show 多媒体展示
multimeter 万用表,多用仪表,万用电表
multi-microprocessor test program 多微处理机测试程序
multi-objective decision 多目标决策
multi-objective optimization 多目标优化
multi-order lag 多阶滞后
multi-parameter monitor 多参数监护仪
multipath 多路的
multi-path diagonal-beam ultrasonic flowmeter 多声道斜束式超声流量计
multipath transmission 多路输送
multiphase 多相的
multiphase circuit 多相电路
multiphase current 多相电流
multiphase generator 多相发电机
multiphase motor 多相电动机
multi-plane balancing 多面平衡
multiple analog input 多点模拟输入
multiple analog input-output card 多路模拟量输入-输出卡
multiple channel 多路通道
multiple channel recorder 多通道记录仪
multiple contact input 多点触点输入
multiple contact output 多点触点输出
multiple development 多次展开
multiple echo method 多次反射法
multiple interferometer 复式干涉仪
multiple ion detection 多离子检测
multiple keys 花键
multiple loop system 多路系统
multiple of a unit of measurement 测量的倍数单位
multiple range 多量程的
multiple scale 多量程,多标度
multiple scattering event 多重散射过程
multiple socket outlet 复式插座
multiple step plug 多级芯
multiple target tracking 多目标跟踪
multiple telephony 多路电话通信
multiple tide staff 群验潮杆,水尺组
multiple unit control 多元控制
multiple way switch 多路开关
multiple-cage rotor 多笼型转子
multiple-channel indicator 多点指示仪
multiple-channel recorder 多点记录仪
multiple-core cable 多芯电缆
multiple-criterion 多判据

multiple-criterion optimization 多判据优化
multiple-input multiple-output (MIMO) 多输入多输出
multiple-input multiple-output system 多输入多输出系统
multiple-jet water meter 多注束水表
multiple-loop control system 多回路控制系统
multiple-motor drive 多机传动
multiple-pen recorder 多笔记录仪
multiple-pointer indicator 多针指示仪
multiple-pole switch 多极开关
multiple-project configuration 多计划结构
multiple-sensor cross correlation 多传感器互相关
multiple-speed floating action 多速无定位作用
multiple-speed floating controller 多速无定位控制器
multiple-speed motor 多速电动机
multiple-stage amplification 多级放大
multiple-stage amplifier 多级放大器
multiplex link 复用链路
multiplex telephony 多路电话
multiplex transmission 多路传输
multiplexed module 多路模块
multiplexer 多路转换器，多路器，多路转接器
multiplexer card 多路器卡
multiplexing 多路复用
multiplicator 倍增器
multiplier 乘法器，倍频器，倍增器，扩量程器
multiplier divider 乘除器
multiplying factor 放大系数，乘数
multipoint analog control I/O module 多点模拟量控制输入/输出模块
multipoint connection 多点连接
multipoint indicating 多点指示
multipoint network 多点网络
multi-point recorder 多点记录仪
multipoint recording 多点记录
multipoint status input card 多点状态量输入卡
multipoint status output card 多点状态量输出卡
multipoint switch 多点开关
multipoint ultrasonic viscometer 多点超声黏度计
multipole switch 多极开关
multi-port 多端口
multi-port network 多端口网络
multi-position controller 多位式控制器，多位式调节器，多位置控制器
multiprocessing 多道处理，多处理（技术）
multiprocessing system 多重处理系统
multiprocessor 多处理器
multi-processor 多功能处理机
multiprocessor system 多重过程，多重系统
multiprogramming 多道程序，多道程序设计
multi-projecting plotter 多位投影测图仪
multi-purpose 多用途
multipurpose pliers 万能手钳
multirange 多量程的，多波段的
multirange instrument 多量程仪表
multi-range measuring instrument 多范围测量仪表
multi-rate 多速率
multi-rate meter 复费率电度表
multi-row bearing 多列轴承
multiscale analysis 多尺度细化分析
multi-scale instrument 多标度尺仪表
multi-scale measuring instrument 多标度尺测量仪表
multisegment model 多段模型
multi-sensor integration 多传感器集成
multispectral camera 多光谱照相机
multispectral scanner (MSS) 多光谱扫描仪
multi-speed 多速
multi-speed controller 多速控制器
multi-speed floating action 多速度漂移作用
multi-speed motor 多速电动机
multi-stage accelerating electron gun 多极加速电子枪
multistage amplifier 多级放大器
multistage decision process 多段决策

过程
multistage flash distillation method for desalination 多级闪急蒸馏淡化法
multistate logic 多态逻辑
multi-step action 多位作用
multi-step avalanche chamber 多步离子雪崩腔
multstep control 多步控制
multi-step controller 多步控制器,多位控制器
multistep prediction 多步预测
multistratum control 多段控制
multistratum hierarchical control 多段递阶控制
multistratum system 多段系统
multiterminal 多端的
multiterminal network 多端网络
multi-tester 万用表
multi-turn electric actuator 多转电动执行机构
multiuser information outlet(MIO) 多用户信息插座
multi-user simulation 多用户仿真
multivalued mapping 多值匹配
multivariable control 多变量控制
multi-variable control system 多变量控制系统
multivariable feedback control 多变量反馈控制
multivariable feedback system 多变量反馈系统

multivariable polynomial 多变量多项式
multivariable quality control 多变量质量控制
multivariable system 多变量系统
multi-version software 多版本软件
multivibrator 多谐振荡器
multi-vibrator 多谐振荡器
multi-voltmeter 多量程电压表
multiway 多路的,多向的
multi-window display 多窗口显示
multi-window mode 多窗口方式
multiwire 多股的,多线的
multi-wiring printed board 多重布线印制板
municipal engineering 市政工程
municipal power supply system 公用电力系统
mushroom cloud 蘑菇云
mutual 相互的
mutual admittance 互导纳
mutual conductance 互电导
mutual flux 互感磁通,交互磁通
mutual impedance 互阻抗,互感,互感系数
mutual interference 互干扰
mutual-inductor 互感
MV(manipulated variable) 操纵变量,输出值
MW(molecular weight) 分子量
myoelectric control 肌电控制

N

nail 钉，钉子
nail iron 制钉铁
naked flame 明火
name of parts 零件名称
nameplate 铭牌
NAND 与非
NAND element 与非元件，与非单元
NAND gate 与非门
NAND operation 与非算子
narrow band 窄频带
narrow band amplifier 窄频道放大器
narrow V belt 窄 V 带
narrow-band crystal filter 窄带晶体滤波器
narrow-band filter 窄带滤波器
narrow-band frequency modulation 窄带调制
narrow-band system 窄带制
narrow-band transmission 窄带传输，窄带发送
narrow-tow-wide-heel 小端窄大端宽接触
Nash game 纳什游戏，博弈
Nash optimality 纳什最优化
national electrical code 国家电气规范
national standard 国家标准
native driver 内置驱动器
natural 自然的，固有的，本征的
natural characteristic 固有特性，自然特性
natural color television 彩色电视机
natural cooling 自然冷却
natural frequency 固有频率，自然频率
natural function generator 自然函数发生器
natural gas 天然气
natural language 自然语言
natural language generation 自然语言生成
natural lighting 天然照明
natural line width 自然行宽度

natural oscillations 固有振荡，自然振荡
natural quartz 天然石英
natural undamped frequency 自然无阻尼频率
natural ventilation 天然通风
naught line 零线
naval architecture and ocean engineering 船舶与海洋工程
navigation 导航
navigation channel 航道
navigation clearance 航道净空
navigation, guidance and control 导航、制导与控制
navigation span 通航宽度
navigation system 导航系统
navigation waterway 航道
navigator window 导航窗口
NCF working file NCF 工作文件
NCF(network configuration file) 网络组态文件
near infra-red spectrum (NIRS) 近红外光谱
nearest-neighbor 最近邻
necessity measure 必然性测度
neck 颈，管颈
necking 缩颈，颈缩加工
need 需求
needle 指针，针形的
needle bearing 滚针轴承
needle destroyer 针头销毁器
needle indicator 指针式指示器
needle pointer 指针
needle roller 滚针
needle roller bearing 滚针轴承
needle thermocouple 指针式热电偶
negative 负的
negative charge 负电荷
negative electron 负电子
negative feedback 负反馈
negative feedback amplifier 负反馈放大器

negative feedback stabilized electrometer 负反馈稳定静电计
negative impedance 负阻抗
negative impedance converter 负阻抗变换器
negative ion 负离子
negative ion generator 负离子发生器
negative peak 反峰
negative polarity 负极性
negative pole 负极
negative resistance 负电阻
negative sequence 负序,逆序
negative sequence impedance 负序阻抗
negative suppression 负迁移
negative transconductance 反向跨导
negative wire 电源负极引线
negative pressure 负压
negligence 疏忽
negligible 可以忽略的
neighbourhood 邻近
neodymium(Nd) 钕
neon(Ne) 氖
neon light 氖灯,霓虹灯
neon oscillator 氖管振荡器
neon tester 试电表
neoprene 合成橡胶
neptunium(Np) 镎
nest 组件箱,机箱
nest assembly 组件箱,组件箱装置
nest common card 电源卡,控制箱公用卡
nest memory card 控制箱存储器卡
nest processor card 控制箱微处理器卡
nested diaphragm type pressure sensor 波纹膜片式压力传感器
net 线网
net control system 网络监控系统
net list 网络表
net weight 净重
netting wire 网钢丝
network 网络,电力网,广播网
network analyser 网络分析仪
network analysis 网络分析
network buffer 网络缓冲器
network configuration file(NCF) 网络组态文件
network control 网络控制

network fault 网络故障
network function 网络函数
network interface module(NIM) 网络接口模块
network layer 网络层
network layout 网络布置
network loss 网络损耗
network management system(NMS) 网络管理系统
network number 网络数量
network observability 网络可观察能力
network reliability 网络可实现性
network topology 网络拓扑
network version 网络版
network voltage 电力网电压
neural 神经的
neural activity 神经激活
neural assembly 神经集合
neural control 神经控制
neural dynamics 神经动力学
neural net 神经网
neural network 神经网络
neural network computer 神经网络计算机
neural network model 神经网络模型
neuron 神经元
neutral 中性点,中性线,中性的,中和的
neutral axis 中性轴,中和轴
neutral conductor 中线,中性线,中性导体
neutral current 中线电流
neutral earthing 中性接地
neutral grounding 中性接地
neutral steer 中性操作
neutral wire 中线
neutral zone 中间带,中间区,中性区
neutral zone control 中和区控制,中间带调节,中间带控制
neutral line 中性线
neutral point 中性点
neutralizing resistance 中和电阻
neutralizing tank 中和池
neutralizing wire 中和线
neutron 中子

neutron beam collimator 中子束瞄准仪
neutron flux 中子通量
new version 新版
newton(N) 牛顿(力的单位)
next batch set value 下批批量设定值
Nichols 尼柯尔斯
Nichols chart 尼柯尔斯图表
Nichols diagram 尼柯尔斯图
nichrome resistance wire 镍铬电阻线
nichrome wire 镍铬合金线,镍铬电热丝
nick 缺口,割痕
nickel(Ni) 镍
nickel cast iron 不锈镍铸铁
nickel chrome wire 镍铬丝
nickel chromium steel 镍铬钢
nickel plated steel wire 镀镍钢丝
nickel silver 镍银
nickel steel wire 镍钢丝
nickel white iron 镍白口铁
nickel-molybdenum thermocouple 镍-钼热电偶
NIM(network interface module) 网络接口模块
niobium(Nb) 铌
nipper 钳子,镊子,拔钉钳
nipple 螺纹接套,喷嘴
NIRS(near infra-red spectrum) 近红外光谱
nitric acid 硝酸
nitric oxide 一氧化氮
nitriding 渗氮法
nitrocarburizing 软氮化
nitrocellulose 硝酸纤维素
nitrogen(N) 氮
nitrogen cylinder 氮气钢瓶
nitrogen oxide 氧化氮
nitrogen phosphorous detection 氮磷检测
nitrosation reaction 亚硝基化反应
nitrosation titration 亚硝基化滴定法
nixie decoder 数码管译码器
NMR equipment 核磁共振装置
NMR spectroscopy 核磁共振波谱法
NMR spectrum 核磁共振波谱
NMR two dimensional 二维核磁共振
NMR(nuclear magnetic resonance) 核磁共振
NMS(network management system) 网络管理系统
no hooks 请勿用钩
no load 空载
no touch relay 无触点继电器
nobelium(No) 锘
nodal plane 结面
node 节点,结点
node assignment 节点安排
node bus 节点总线
node interface unit 节点接口单元
node number 节点数量
node pair 节点对
node pair number 节点对数
nodular cast iron 球墨铸铁
noise 噪声,噪音
noise abatement measure 减低噪声措施
noise absorption factor 噪声吸收系数
noise analysis 噪声分析
noise barrier 隔音屏障,隔声板
noise characteristic 噪声特性
noise characterization 噪声特性化
noise control 噪声控制
noise current 噪声电流
noise enclosure 隔音盖罩
noise factor 噪声系数
noise filter 噪声滤波器
noise interference 噪声干扰
noise level 噪声声级,噪声级别
noise margins 噪声容限,杂讯容限,噪声安全系数
noise meter 噪声计
noise of motor vehicle 汽车噪声
noise pollution 噪声污染
noise power spectrum 噪声功率谱
noise sensitive receiver 对噪声感应强的地方
noise suppressing case 灭音套
noise suppression 噪声抑制
noise suppressor 噪声抑制器
noise voltage 噪声电压
noise remover 消声器,消音器,噪声抑制器
noise-free 无噪声的
noisy channel 有噪信道
noisy image 有噪映像

noisy speech 有噪语言
no-load and shot-circuit method 空载和短路法
no-load characteristic 空载特性
no-load current 空载电流
no-load curve 空载曲线
no-load loss 空载损耗
no-load operation 空载运行
no-load speed 空载转速
no-load starting 空载启动
no-load state 空载状态
no-load test 空载试验
no-load voltage 空载电压
nominal 额定的，标称的
nominal bore 公称通径
nominal capacity 额定容量
nominal cross sectional area 标称横截面面积
nominal current 额定电流
nominal diameter 公称直径
nominal frequency 额定频率
nominal load 额定负载
nominal moment 公称力矩
nominal parameter 额定参数
nominal power 额定功率
nominal pressure 公称压力
nominal stress 名义应力，公称应力
nominal torque 额定转矩
nominal value 额定值，标定值，公称值
nominal voltage 额定电压，标称电压
nominal power 额定功率
nomogram 诺模图，列线图
non-absorbent material 非吸收性物料
nonaqueous titrations 非水滴定法
non-circular gear 非圆齿轮
non-circular pad 非圆形盘
noncoherent system 非单调关联系统
non-combustibility test 不可燃性测试
non-combustible material 不可燃物料
non-condensable gas 不可凝气体
non-conductive pattern 非导电图形
non-contact displacement meter 非接触式位移计
non-contact seal 非接触式密封
non-contact sensor 非接触式传感器
non-contacting pickup 无触点传感器

non-contacting switch 无触点开关
noncooperative game 非合作博弈
non-crystalline electrodes 非晶体电极
non-destructive test 非破坏性测试
nondestructive testing system 非破坏性测试系统
non-electric 非电的
nonequilibrium state 非平衡态
nonferrous metal 有色金属
non-ferrous metallurgy 有色金属冶金
non-Gaussian process 非高斯过程
non-generated gear 非展成大轮
non-good parts 不良品
non-ground neutral system 不接地中线制
non-homing switch 不归位机键
non-inductive 无感的
non-inductive capacitor 无感电容器
non-inductive circuit 无感电路
non-inductive resistance 无感电阻
non-interacting control 无互作用控制，非相关控制
non-interacting control system 无互作用控制系统
non-interacting process 非相关过程
non-interchangeable accessory 专用附件
non-interference 无相互干扰
non-isolated analog input 非隔离模拟输入
nonius 游标卡尺
nonlinear 非线性的
nonlinear analysis 非线性分析
nonlinear capacitance 非线性电容
nonlinear characteristic 非线性特性
nonlinear circuit 非线性电路
nonlinear control 非线性控制
nonlinear control system 非线性控制系统
nonlinear controller 非线性控制器，非线性调节器
nonlinear conversion 非线性转换
nonlinear distortion 非线性失真
nonlinear element 非线性环节，非线性元件
nonlinear equation 非线性方程
nonlinear external cavity 非线性外部

空穴
nonlinear filter 非线性滤波器
nonlinear gain 非线性增益
nonlinear interface 非线性接口
nonlinear mirror 非线性镜像，非线性映射
nonlinear model 非线性模型
nonlinear optical interaction 非线性光学相关
nonlinear optimization 非线性优化
nonlinear Poisson equation 非线性泊松方程
nonlinear potentiometer 非线性电位差计
nonlinear programming 非线性规划
nonlinear reactance amplifier 非线性电抗放大器
nonlinear refraction 非线性折射
nonlinear refractive index 非线性折射率
nonlinear regression 非线性迭代
nonlinear resistance 非线性电阻
nonlinear system 非线性系统
nonlinear theory 非线性理论
nonlinear scale 非线性标度
non-linearity 非线性
nonlinearity 非线性
nonmagnetic 无磁性的
non-methane hydrocarbons 无甲烷碳氢化合物
non-minimum phase 非最小相位
non-minimum-phase response 非最小相位响应
non-minimum-phase system 非最小相位系统
nonmonotonic logic 非单调逻辑
non-organic soil 无机泥土
non-orthogonal problem 非正交问题
non-parametric 非参数
non-parametric identification 非参数辨识
non-parametric regression 非参数回归
nonparametric training 非参数训练
non-periodic 非周期的
non-polar liquid 非极化液体
non-retentive material 软磁性材料
non-return valve 单向阀，止逆阀，单向活门
non-reversible 不可逆的
nonreversible electric drive 不可逆电气传动
non-salient pole 隐极
non-salient pole synchronous generator 隐极同步发电机
non-shrinking concrete 不收缩混凝土
nonsingular perturbation 非奇异摄动
non-sinusoidal 非正弦的
non-sinusoidal current 非正弦电流
non-sinusoidal curve 非正弦曲线
non-sinusoidal voltage 非正弦电压
non-sinusoidal wave 非正弦波
non-spark impulse type alarm bell 无火花冲击式电铃
non-stabilizable system 非可稳定化系统
nonstandard gear 非标准齿轮
non-stationary 非稳态
non-stationary learning characteristic 非稳态学习特性
non-stationary random process 非平稳随机过程
non-stationary signal 非稳态信号
non-stationary system 非稳态系统
non-steady 不稳定的
non-structural crack 非结构性裂缝
non-structural works 非结构工程
non-symmetrical network 不对称四端网络
non-uniform 不均匀的
non-uniform magnetic field 不均匀磁场
NOR 或非，或非元件
NOR circuit 或非电路
NOR element 或非元件
NOR gate 或非门
NOR operation 或非算子
normal 正常，标准，法向，法线，法面，正常的，常规的
normal atmosphere 标准大气压
normal backlash 法向侧隙
normal backlash tolerance 法向侧隙公差
normal base pitch 法向基节
normal bend 标准弯管

normal chordal addendum 法向弦齿高
normal chordal thickness 法向弦齿厚
normal circular pitch 法向周节，法面齿距
normal circular thickness 法向弧齿厚
normal concrete small hollow block 普通混凝土小型空心砌块
normal contact ratio 法向重合度
normal control wire 定位控制线
normal current 正常电流
normal diametral pitch 法向径节
normal direction 法线方向
normal distribution 常态分布，常态分配
normal electrode 标准电极
normal force 法向力
normal illumination 正常照明
normal indication wire 定位表示线
normal load 垂直载荷，法向载荷，正常负载
normal module 法面模数
normal parameters 法面参数
normal phase 正相
normal phase liquid chromatography 正相液相色谱法
normal pitch 法向齿距
normal plane 法面，法向平面
normal pressure angle 法面压力角
normal running 机组运行正常
normal section 法向截面
normal stress 正应力，法向应力
normal thickness taper 正常齿厚缩
normal tilt 法向刀倾
normal tooth profile 法向齿廓
normal wear 正常磨损
normal closed contact 常闭触点
normal mode 正常方式
normal mode attribute 正常状态特性
normal mode interference 串模干扰，常态干扰
normal mode rejection 串模抑制，常态抑制
normal mode voltage 串模电压
normal operating condition 正常工作条件
normal operation 正常运行，正常操作
normal output 正常输出

normal value 正常值
normal value status 正常值状态
normalization 归一化
normalization method 归一法
normalized 归一化的，标准化的
normalized device 标准化设备坐标
normalized response 标准响应
normalizing 正火，正常化
normally closed 常闭
normally closed contact 常闭触点
normally open 常开
normally open contact 常开触点
normal-phase HPLC 正相 HPLC
norms 定额，规范
no-roll roughing 无滚动粗切
Norton's theorem 诺顿定理
noscapine 诺司卡品
nose angle 刀角
nose of tool 刀尖
not applicable 不适用
NOT circuit 非电路
NOT element 非门元件
NOT gate 非门
not good 不良
notch 刻痕，槽口，缺口
notch effect 切口效果
notch filter 陷波滤波器
notching 冲口加工
notching press 冲缺口压力机
notes 说明
no-voltage protection 失压保护
no-voltage relay 无压继电器
noxious gas 有害气体
nozzle 喷嘴，管嘴，喷射器
N-P-N transistor NPN 晶体管
N-type semiconductor N 型半导体
nuclear 原子能的，细胞核的，中心的，原子核的
nuclear energy science and engineering 核能科学与工程
nuclear fuel cycle and materials 核燃料循环与材料
nuclear magnetic resonance(NMR) 核磁共振
nuclear magnetic resonance flowmeter 核磁共振流量计
nuclear magnetic resonance spectrometer

核磁共振分光计
nuclear physics 核物理
nuclear plant 核工厂
nuclear power 核电
nuclear power plant 核电站
nuclear power station 核电站
nuclear radiation level meter 核辐射物位计
nuclear reactor 核反应堆，核反应器
nuclear science and technology 核科学与技术
nuclear technology and applications 核技术及应用
nuclear tests 核实验
nucleon 核子
nucleus 核
null 清零
number of phases 相数
number of poles 极数
number of revolutions 转数
number of starts 启动次数
number of separate loci 根轨迹的条数
number of teeth 齿数
number of threads 蜗杆头数
number of waves 波数
number or logic slot 逻辑槽数量
number system 数字系统
numerator 分子，计算器

numeric control 数字控制
numeric point 数值点
numeric variable 数值变量
numerical 数字的
numerical algorithm 数值算法
numerical analysis 数值分析
numerical control 数字控制
numerical control system 数控系统
numerical method 数值方法
numerical simulation 数值仿真
numerical solution 数值解
numerical value 数值
nut 螺母，螺帽
nutation sensor 章动敏感器
nutsert 螺母
nylon 尼龙塑料
nylon ferrule 尼龙套圈
nylon filter 尼龙过滤器
nylon insert lock nuts 尼龙嵌入防松螺帽
nylon insulator 尼龙绝缘块
Nyquist 奈奎斯特
Nyquist diagram 奈奎斯特图
Nyquist filter 奈奎斯特滤波器
Nyquist plot 奈奎斯特图
Nyquist stability criterion 奈奎斯特稳定判据
Nyquist's criterion 奈奎斯特判据

O

OAM (operation, administration and maintenance) 操作、管理和维护
object 目标
object code 目的码
object modelling technique 目标模型技术
object recognition 目标辨认
object to be measured and controlled 测量和控制对象
objective 目标
objective function 目标函数
objective noise-meter 绝对噪声计
object-oriented programming 面向对象的程序设计
oblong pad 长方形焊盘
oblong steel wire 矩形钢丝
observability 可观性,可观测性
observability index 可观测指数
observable 可观的
observable canonical form 可观测规范型
observer 观测器
obstacle 障碍
obstacle avoidance 避障碍,排除障碍
obstacle detection 障碍探测
obstacle detector 障碍探测器
obstruction 堵塞物,堵塞,障碍,障碍物
obstruction detection system 障碍探测系统
obstructionless digital flowmeter 无阻塞式数字流量计
obtuse angle 钝角
octamonic amplifier 倍频放大器,八倍频放大器
octave filter 倍频程滤波器
odd electron 含奇数个电子
odd item 零头
odometer 里程表
odor 气味,臭味
odour control equipment 辟除气味的控制设备
off 断开,关

off centre 偏心,偏离中心
off line 离线
off position 断路位置
off process 离线
off shore 离岸
off unit 未定义单元
off-diagonal terms 分离对角线条件
off-gas pump 抽气泵
office automation 办公自动化
off-line control 离线控制
off-line diagnostic 离线诊断,脱机诊断
off-line equipment 脱机设备,离线设备
off-line measurement 离线测量
off-line programming 离线编程
off-normal lower 下限越界
off-normal upper 上限越界
off-process environment 离线环境
off-process test(OPT) 离线测试
offset 偏置位,偏差,偏置,偏置式,偏置距
offset circle 偏距圆
offset coefficient 静差系数,偏移系数
offset current 补偿电流
offset distance 偏距
offset knife-edge follower 偏置尖底从动件
offset land 偏置连接盘
offset roller follower 偏置滚子从动件
offset slider-crank mechanism 偏置曲柄滑块机构
offset voltage 补偿电压,偏移电压,残余电压
offset waveguide 偏移波导管
off-tune 失调的
ohm 欧姆(电阻单位)
ohmer 欧姆计
ohmic 电阻性的
ohmic contact 欧姆接触
ohmic drop 电阻性压降

ohmic loss	电阻损耗
ohmmeter	电阻表
Ohm's law	欧姆定律
oil and natural gas engineering	石油与天然气工程
oil baths	油槽
oil bearing	含油轴承
oil bottle	油杯
oil can	油壶
oil chamber	储油器，储油室
oil circuit breaker	油压断路器
oil consumption	耗油量
oil consumption factor	耗油量系数
oil cooling system	油冷却系统
oil damper	油阻尼器
oil depot	油库
oil drain hose	排油软管
oil filler pipe	加油管，注油管
oil filter	滤油器
oil filter canister	滤油罐
oil interceptor	集油器，油污截流井
oil manometer	油压计
oil nozzle	燃油喷嘴
oil pressure gauge	油压表
oil pump	油泵
oil quenching	油淬化，油淬火
oil reclaimer	集油器
oil return tube	回油管
oil seal	油封
oil spillage	燃油溢出
oil stains	油污
oil storage installation	贮油装置
oil sump	油槽，油底壳
oil switch	油开关
oil tank volume	油箱容积
oil temper wire	油回火钢丝
oil well cement	油井水泥
oil breaker	油断路器
oil gun	油枪
oil level	机油平面
oil-break fuse	油熔断器
oiler	注油器
oi-gasl field development engineering	油气田开发工程
oil-gas storage and transportation engineering	油气储运工程
oil-gas well engineering	油气井工程
oil-immersed key switch	油浸琴键
oil-immersed multi-point switch	油浸式多点切换
oil-immersed power transformer	油浸电力变压器
oil-immersed regulating transformer	油浸调整变压器
oil-immersed starter	油浸启动器
oil-immersed transformer	油浸变压器
oilless	缺油的
oily ditch seal	油沟密封
oldham coupling	滑块联轴器，十字滑块联轴器，欧氏联轴节
Oldham coupling	十字头联轴节，欧氏联轴节
omni-bearing convertor	全向方位变换器
OMS (outage management system)	停电管理系统
on deck	置于甲板
on line	在线的，联机的
on low-tension side	低压侧的
on position	开位置
on-and-off switch	通断开关
ondograph	高频示波器
one machine-infinity bus system	单机无穷大系统
one piece casting	整体铸件
one stroke	一行程
one-dimensional search	一维搜索
one-disk watt-hour meter	单转盘式电度表
one-path ultrasonic flowmeter	单通道超声流量计
one-pen recorder	单笔记录仪
one-phase	单相的
one-phase short-circuit	单相短路
one-shot battery	一次电池
one-shot execution function	单拍摄执行功能
one-shot multivibrator	单稳线路
one-stage amplifier	单级放大器
one-touch multi-window display function	单触多窗口显示功能
on-hand inventory	现有库存
on-line assistance	在线帮助

online 联机的，在线的
online closed loop 在线闭环回路
online control 在线控制
online diagnosis 在线诊断，联机诊断
online documentation 在线文件，在线纪录
online equipment 联机设备
online estimation 在线估计
online maintenance function 在线维护功能
online manual 在线手动
online operation 在线操作
online optimization 在线优化
online reconfiguration 在线重组态
online security analysis 在线安全分析
online tuning 在线整定
on-load 带负载的
on-load operation 负载操作
on-load running test 负载运行测试
on-net design 网上设计
on-off 开关
on-off action 通断作用，开关作用
on-off control 开关控制，双位控制，通断控制
on-off controller 开关控制器，通断控制器，位式调节器
on-off element 开关元件
on-off servo mechanism 开关伺服机构，开关随动机构
on-off time proportional controller 开关时间比例控制器
on-process analysis(OPA) 在线分析
on-process tests 在线测试
on-site 实地，就地，现
on-site inspection 实地调查，现调查
on-site test 实地测试，现测试
OPA(on-process analysis) 在线分析
opacimeter 烟度计
opaquer 遮光剂
open air 露天地方
open area 露天场地
open balcony 无遮蔽露台
open chain mechanism 开链机构
open interface 开放界面
open kinematic chain 开式链
open loop 开环
open loop control 开环控制

open loop control system 开环控制系统
open loop electromagnetic flowmeter 开路电磁流量计
open loop frequency response 开环频率响应
open loop gain 开环增益
open loop pole 开环极点
open loop transfer function 开环传递函数
open structure matrix 开环结构矩阵
open system interaction(OSI) 开放系统互联
open tubular column(OTC) 空心柱，开管柱
open type diaphragm-box 敞口式膜盒
open type induction motor 开启式感应电动机
open wire 裸线，明线，架空线
open circuit 开路
open-belt drive 开口皮带传动
open-circuit jack 开路插座
open-circuit parameter 开路参数
open-circuit position 开路状态
open-circuit test 开路试验
open-circuit voltage 开路电压
open-hearth iron 平炉生铁
opening 出入口，缺口，开口，门口，排料逃孔
opening switch 打开开关
open-loop control system 开环控制系统
open-loop 开环
open-phase protection 断相保护装置
open-phase relay 断相继电器
operameter 运数计，转速计，运转计
operand register 操作数寄存器
operate 操作，运作
operater seat 司机座椅
operating 运行的，操作的
operating circuit 操作电路
operating coil 工作线圈，动作线圈
operating condition 运行状态，工作条件，工况
operating console 操作控制台
operating cost 营运成本，运作费用
operating current 工作电流，操作电流
operating influence 工作条件影响

operating load　工作负荷
operating objective　操作目标
operating pressure　工作压力
operating pressure angle　工作压力角
operating range　工作范围,运行范围
operating screen mode　操作屏幕方式
operating system　操作系统,开放系统
operating temperatures　工作温度
operating torque　工作扭矩
operating voltage　操作电压,工作电压
operating panel　操作盘
operation, administration and maintenance(OAM)　操作、管理和维护
operation and monitoring function　操作和监视功能
operation circuit　运算电路
operation control device　操纵及控制装置
operation decoder　操作译码器
operation group　操作组
operation instruction　操作指示
operation keyboard　操作员键盘
operation manual　操作手册
operation mark　操作标记
operation mechanism　工作机构
operation procedure　操作程序
operation record　操作记录
operation register　操作寄存器
operation test　运行试验
operation　操作,运营,工作,运行,控制
operational amplifier　运算放大器
operational calculus　算子演算,运算微积分
operational characteristic　操作特性
operational impedance　运算阻抗
operational logbook　操作日志
operational research model　运筹学模型
operational vehicle　工作车辆
operational log　运行记录
operations research　运筹学,作业研究
operative limit　极限工作条件,运行极限
operator　营运者,操作员,经营者
operator console　操作台,操作员控制台
operator control panel　操作员控制键盘
operator guide message panel　操作指导信息画面
operator guide window　操作指导窗口
operator interface station　操作员接口站
operator keyboard　操作员键盘
operator panel display block　操作画面显示块
operator personality　操作员属性
operator station　操作站,操作员站,操作员操作台
operator station definition　操作站定义
operator station status display panel　操作站状态显示画面
operator utility menu panel　操作应用菜单画面
operator keyboard　操作员键盘
operators console　操作台,操作站,操作员控制台
operators request control panel　操作员请求控制台
opposing　反向的,相反的,反作用的
opposing voltage　反向电压
opposite in phase　反相
OPT(off-process test)　离线测试
optic fiber tachometer　光纤式转速表
optical　光学的
optical amplifier　光电放大器,光学放大器
optical average power meter　光平均功率计
optical band gap　光禁带
optical birefringence　光折光,光学双折射
optical bistability　光学双稳定性
optical character recognition　光学字符辨识
optical communication　光学通信
optical component　光学元件
optical constant　光学常数
optical data storage　光学数据存储
optical directional coupler　光学定向耦合器

optical disk　光盘
optical engineering　光学工程
optical feedback　光学反馈
optical fiber　光纤，光学纤维，光导纤维
optical fiber acoustic sensor　光纤声传感器
optical fibre network　光纤网络
optical field　光场
optical film thickness meter　光学薄膜厚度计
optical filter gas analyzer　滤波式气体分析仪
optical flat　光学平晶
optical flow　光通量
optical flowmeter　光学流量计
optical focus switch　光学聚焦转换开关
optical implementation　光实现
optical isolator　光频隔离器
optical mask　滤光片，光学掩模
optical microscope　光学显微镜
optical modulation　光调制
optical modulator　光调制器
optical nonlinearity　光非线性
optical parallel　光学平行
optical parametric oscillator　光参量振荡器
optical polarization bistability　光极化双稳定性
optical property　光学性质
optical property tester　光学性能测定仪
optical pulse　光脉冲
optical pyrometer　光学高温计，光学温度计
optical receiver　光接收器
optical response　光响应
optical sensor　光传感器
optical solution　光溶解
optical spectrograph　光学摄谱仪
optical spectroscopy　分光光度计
optical spectrum instrument　光谱仪器
optical stochastic control　光随机控制
optical storage device　光存储装置
optical switch　光开关
optical time division multiplex(OTDM)　光时分复用
optical to electrical converter　光电转换器
optical transducer　光传感器
optical transmission　光传输
optical-control　光学控制
optimal　优化的，最佳的，最理想的
optimal control　优化控制，最优控制
optimal design　优化设计
optimal device　最优装置
optimal estimation　优化估计
optimal experiment design　优化试验设计
optimal filtering　优化滤波
optimal load flow　优化负载流
optimal power flow　优化功率流
optimal problem　最优化问题
optimal regulator　优化调节器
optimal rejection　优化抑制
optimal replacement unit　最优更换单元
optimal search technique　寻优技术
optimal system　优化系统
optimal trajectory　优化轨迹
optimality　最优性
optimization　最优化，最佳选择
optimization design　优化设计
optimization in controller tuning　控制器优化整定
optimization in process operation　过程优化操作
optimized　最佳的
optimizing control　最佳化控制
optimum　最佳的，最佳条件
optimum control　优化控制
optimum filter　最佳滤波器
optimum operating point　优化操作点
optimum replacement unit　最佳可换单元，最佳可换装置
optimum steady state control　优化稳态控制
option　选择
option packages for communicating with subsystem　和子系统通信的可选软件包
option switch　选择开关

optional 可选的
optional device 选用设备
optoelectronic cell 光电池,光电管
optoelectronic device 光电器件,光电设备
optoelectronic pulse amplifier 光电子脉冲放大器
optoelectronics 光电子学
optron 光导发光元件
OR 或
OR circuit 或电路
OR gate 或门
orbit gyrocompass 轨道陀螺罗盘
orbit perturbation 轨道摄动
orbit 轨道,轨迹
orbital electron 轨道电子
orbital rendezvous 轨道交会
order parameter 有序参数
order wire 传号线,联络线,记录线,挂号线
ordinary differential equation 常微分方程
ordinary gear train 定轴齿轮系,普通齿轮系
ordinary Portland cement 普通硅酸盐水泥
ordinary terminal 普通型接线端子
ordinate 纵坐标
organic acid 有机酸
organic chemistry 有机化学
organic quantitative analyzer 有机定量分析仪
organic solvent 有机溶剂
organization and management 组织与管理
organizational factor 编排因子,有机因子
orientation control 定向控制
orientation device 定向装置
orientation 方位,定向,定位
oriented film 定向薄膜
orifice and plug-type variable area flowmeter 孔-塞式可变面积流量计
orifice fitting 安装孔板
orifice flowmeter 孔板流量计
orifice plate 孔板,挡板
orifice plate flowmeter 孔板流量计

orifice plate with circular pressure-equalizing 圆形均压环取压的孔板
origin 原点,起点
origin of coordinate 坐标原点
original design 原始设计
original equipment manufacture 原设备制造
original mechanism 原始机构
original plan 原来图则,原始图则
original 初始的,原始的
OR-operation 或运算
oscillating 振荡的
oscillating bar 摆杆
oscillating circuit 振荡电路
oscillating coil 振荡线圈
oscillating follower 摆动从动件
oscillating guide-bar mechanism 摆动导杆机构
oscillating period 振荡周期
oscillating piston type flowmeter 摆动活塞式流量计
oscillation 振荡
oscillation crystal 振荡晶体
oscillation frequency 振荡频率
oscillator 振荡器
oscillator doubler 振荡倍频器
oscillator quartz 振荡器石英
oscillatory 振荡的
oscillatory discharge 振荡放电
oscillistor 半导体振荡器
oscillograph 示波器
oscillometer 示波器
oscilloscope 示波器
oscilloscope tube 示波管
oscillosynchroscope 同步示波器
OSI(open system interaction) 开放系统互联
osmium(Os) 锇
osmotic pressure meters 浸透压测定表
OTC(open tubular column) 空心柱,开管柱
OTDM(optical time division multiplex) 光时分复用
other common fault alarm display and input 其他故障显示及输入
out amplifier 输出放大器

out of balance 不平衡的
out of phase 不同相
out of repair 失修
out of roughness 真圆度
out of step 失步
out switch 输出开关
outage management system（OMS） 停电管理系统
outage 中断供应，运行中断，停止，停电
outboard 外侧的
outdoor 户外的，露天的
outdoor equipment 户外设备
outer 外部的
outer addendum 大端齿顶高
outer cone distance 外锥距
outer dedendum 大端齿根高
outer diameter 外径
outer edge 外缘
outer guiding post 外导柱
outer loop 外环
outer ring 外圈
outer shield 外护罩
outer slot width 大端槽宽
outer spiral angle 大端螺旋角
outer stripper 外脱料板
outlet 出口，插座
outlet pipe 去水管，出口管
outlet pipework 出水管
outlet valve 排出阀
outlets for maintenance 维护出口
out-of-balance load 不平衡负载
out-of-plane bending vibration 面外弯曲振动
out-of-step protection 失步保护
out-phasing modulation 反相调制
out-phasing modulation system 反相调制系统
output air pressure gauge 风压表
output amplifier 输出放大器
output axis 输出轴
output brush 输出电刷
output capacity selection 输出容量选择
output circuit 输出电路
output compensation 输出补偿
output device 输出装置
output equations 输出方程

output error identification 输出误差辨识
output feedback 输出反馈
output filter 输出滤波器
output high limit in percent 以百分比表示的输出高限
output impedance 输出阻抗
output injection 输出注射
output limit 输出限，输出极限
output link 输出构件
output load resistance 输出负载电阻
output low limit in percent 以百分比表示的输出低限
output matrix 输出矩阵
output mechanism 输出机构
output network 输出网络
output open alarm check 输出开路报警检查
output power 输出功率
output prediction method 输出预估法
output printer 输出打印机
output regulation 输出调整
output regulator 输出调节器
output resistance 输出电阻
output reverse action 输出反方向作用
output shaft 输出轴
output signal 输出信号
output signal processing 输出信号处理
output signal type 输出信号类型
output stage 输出级
output station 输出操作站
output torque 输出力矩
output tracking 输出跟踪
output transfer function 输出传递函数
output transformer 输出变压器
output value 输出值
output variable 输出量，输出变量
output vector 输出向量
output velocity check 输出变化率检查
output voltage 输出电压
output winding 输出绕组
output work 输出功
output 产量，产品，输出
outrigger 起重臂，悬臂梁，斜撑
outside blade 外切刃点，外切刀齿

outside diameter 外径，大端直径
outside glazing 外装玻璃法
outside indicator 外指示剂
outside interference 外部干扰
outside micrometer 外径千分尺
outside point diameter 外切刀尖直径
outside radius 齿顶圆半径
outside surface 外表面
outside thread 外螺纹
oval gear flowmeter 椭圆齿轮流量计
oval wheel flow measuring device 椭圆齿轮流量测量仪表
oval wheel flowmeter 椭圆齿轮流量计
oval wire 椭圆钢丝
ovality 椭圆度
oven 烤箱
oven-dried density 烘干密度
over pressure impact from either direction 单门过载冲击
over current 过流
over loading 过载
over voltage 过压
overageing 过老化
overall design 总体设计
overall dimensions 整体尺寸，外形尺寸
overall error 总误差
overall progress 整体进度
overall stability 整体稳定性，总体稳定性
overall thermal transfer value 总热传送值
overcurrent 过电流
overcurrent protection 过电流保护，过电流保护装置，过载保护
overcurrent relay 过电流继电器
overdamped process 过阻尼过程
overdamped response 过阻尼响应
overdamping 过阻尼
overdriven amplifier 过激励放大器
over-excitation 过励磁，过激励
overflow 溢出，溢流，液泛
overflow capacity 溢流容量
overflow pipe 溢流管
overhauling 检修
overhead crane 高架起重机，桥式起重机

overhead earth wire 架空地线
overhead ground wire 架空地线，架空避雷线
overhead line 架空电缆，架空线路
overhead telecommunication line 架空电信线路
overhead telephone line 架空电话线
overhead travelling crane 移动式高架起重机
overhead wire 高架电线
overhead 顶部
overheat 过热，使过热
overheat protective relay 过热保护继电器
overheated 过热的
overheating 过热
overlap 互搭，重叠，叠加
overlap contact ratio 纵向重合度
overlapping 交接，重叠，网纹
overlapping decomposition 交叠分解
overlaying 堆焊
overload 超荷载，超重，超载，负荷过重，超负荷，过载，过载的
overload breakage 超负荷破裂
overload capacity 过载容量
overload characteristic 过载特性
overload factor 过载系数，超负荷系数
overload limit 过载限，过负荷极限
overload limit of short duration 短时过范围极限
overload protection device 防超载装置，过载保护装置
overload relay 过载继电器
overload test 过载试验
overload protection 过载保护
overloading 过载
overly 过度地
overrange 过范围
overrange limit of short duration 短时过范围极限
override control 超驰控制
override initialization 超驰预置
override selector algorithm 超驰选择算法
overshoot 超调，超调量，过冲
overspeed shutdowns 超速故障停机

oversteer 过渡转弯，过渡转向
overstress 超限应力
overtension 过电压
overturning 倾覆
overturning moment 倾覆力矩
overview 总貌
overview display 总貌画面
overview object 总貌目的
overview of functions 功能总貌
overview of security functions 安全功能总貌
overview panel 总貌画面
overview panel definition 总貌画面定义
overvoltage 超压，过电压
overvoltage fuse 过电压熔断器
overvoltage protective device 过电压保护装置
Owen bridge 欧文电桥
owning node 专用节点
oxidant 氧化性介质
oxidation 氧化
oxidation-reduction potential electrode assembly 氧化-还原电位电极装置
oxidation-reduction potential transmitter 氧化-还原电位变送器
oxidation-reduction titration 氧化还原滴定法
oxide 氧化物
oxide-film capacitor 氧化膜电容器，电解电容器
oxidization 氧化
oxy-acetylene welding 气焊，氧气乙炔焰焊接
oxygen (O) 氧
oxygen flask combustion method 氧瓶燃烧法
oxygen pressure gauge 氧气压力表
oxygen-deficiency safety device 缺氧安全装置
ozone resistance test 抗臭氧试验

P

P controller 比例控制器
PA(polyamide) 聚酰胺,尼龙
package specification 包装规范
package 包装,封装
packaged transistor 密封式晶体管,封装晶体管
packaging 包装,打包
packaging option 组件选择,软件选择
packaging tool 打包机
packed bed reactor 填充床反应器
packed capillary column 填充毛细管柱
packed column 填充柱
packed group 组合组
packer 打包机
packet 小包,信息包
packet instrument 袖珍仪表
packing 包装,填料函
packing list 装箱单
packing materials 包装材料,包装物
packing piece 垫片
padding 填充物
padding block 垫块
paddle 桨叶,搅棒
paddle type flow switch 闸板式流量控制开关
Pade approximation 帕德近似,培德近似
page printer 页式打印机,页式印刷机
page selector alarm panel 分页选择报警盘
pager 寻呼机,呼机
paint 油漆
paint coating 油漆层
painted steel 涂漆钢
painting factory 烤漆厂
painting make-up 补漆
paired cable 双线电缆,成对电缆
paired ion chromatography 离子对色谱法
paired mounting 成对安装

pairs 双(对等)
palladium (Pd) 钯
pallet 栈板,垫盘,托盘
pallet fork lift truck 托盘叉式起重车
pamphlet 小册子
pan head rivet 截锥铆钉
panel 屏,盘,面板,安装板,安装盘,安装屏,配电盘
panel board 镶块
panel comment 画面注释
panel connector 面板插座
panel meter 配电板式仪表,面板式仪表
panel mounting 配电盘装配,面板装配,盘装
panel name 画面名称
panel set 面板设定
panel switching 面板开关,控制盘开关
panel-board cut-out 仪表盘开孔
panel-type instrument 配电盘式仪表
panic barrier 紧急栏障
panoramic amplifier 全景放大器
pantometer 经纬测角仪,万测仪
paper chromatography 纸色谱分析法
paper industry 造纸工业
paper tape 纸带
paper tape punch 纸带穿孔机
paper tape reader 纸带阅读机
paper tape unit 磁带机,纸带机,纸带穿孔读出装置
parabolic 抛物线的
parabolic antenna 抛物面天线
parabolic input 抛物线输入
parabolic motion 抛物线运动
paraffin 石蜡
paraffin copper wire 浸蜡铜线
paraffin wire 浸蜡线
parallax error 视差
parallel 平行,并联,并联的,并行,并行的

parallel admittance 并联导纳
parallel algorithm 并行算法
parallel block 平行垫块
parallel branch 并联支路
parallel capacitance 并联电容
parallel circuit 并联电路，平行电路
parallel combined mechanism 并联组合机构
parallel computation 平行计算，并行运算
parallel computer 并行计算机
parallel connection 并联
parallel control device 并联控制装置
parallel feedback 并联反馈
parallel feedback integrator 并联反馈积分器
parallel helical gears 平面轴斜齿轮
parallel highway 并行信息公路
parallel impedance 并联阻抗
parallel input card 并行输入插件板
parallel input power module (PIPM) 并联输入电源组件
parallel key 普通平键，平面键
parallel mechanism 并联机构
parallel memory 并行存储器
parallel network 并行网络
parallel operation 并联运行
parallel operator keyboard (POK) 并联操作键盘
parallel output card 并行输出插件板
parallel processing 并行处理
parallel processor 并行处理器
parallel program 并行程序
parallel register 并行寄存器
parallel resistance 并联电阻
parallel resonance 并联谐振
parallel running 并联运行
parallel slide valves 浆液阀
parallel storage 并行存储器
parallel transducer 并行传感器
parallel winding 并联绕组
paralleling 并联，并行
parallelism 平行度
parallel search storage 并行检索存储器
parallel-serial 并-串联，混联
paramagnetic body 顺磁体

paramagnetic material 顺磁性物质，顺磁性材料
paramagnetic oxygen analyzer 磁氧分析器
paramagnetic substance 顺磁物质
paramagnetism 顺磁性
parameter 参数，参量
parameter design 参数设计
parameter entry display 参数输入显示，参数输入显示画面
parameter estimation 参数估计
parameter identification 参数辨识
parameter list 参数清单
parameter optimization 参数优化
parameter values 参数值
parameterization design 参数化设计
parameters 参数
parametric 参数的
parametric amplifier 参数放大器
parametric excitation 参量激励
parametric predictive control (PPC) 参数预测控制
parametric resonance 参数共振
parametric variation 参数波动
parametrization 参数化
paraphase amplifier 倒相放大器
parasitic disturbance 寄生干扰
parasitic element 寄生元件，寄生组件，无源元件，二次辐射体
parasitic feedback 寄生反馈
parasitic oscillation 寄生振荡
parasitics 寄生现象
parent ion 母离子
pareto diagram 排列图
Pareto optimality 帕累托最优
paring line 分模线
parser 分析程序
part 零件，部分
part drawing 零件图
part number 料号，零件号
partial 部分的，局部的
partial balance of shaking force 惯性力部分平衡
partial closure 局部封闭
partial collapse 局部坍塌
partial differential 偏微分
partial differential equation 偏微分

方程
partial expansion 部分展开
partial failure 局部故障，局部失灵
partial fraction expansion 部分分式展开
partial least square (PLS) 部分最小二乘
partial least squares method 偏最小二乘法
partial pressure analyzer 分压分析仪
partial radiation pyrometer 部分辐射温度计
partial response channel 部分响应信道
particle 粒子，微粒，颗粒
particle counters 尘埃计数器
particle detector 粒子探测器
particle size analysis 粒径分析
particle size analyzer 粒度分析仪
particle size measurement 颗粒尺寸测量
particular specification 特别规格
partition chromatography 分配色谱法
partition coefficient 分配系数
parts 部件，零件
parts drawing 零件图
parts list 零件表，零部件清单
pascal(Pa) 帕［斯卡］（压力单位）
Pascal preprocessor Pascal 预处理器
pass guide 穴型导板
passage quality control 段检人员
pass-band 通带
pass-band limiting switch 通带宽带限制开关
pass-band width 通带宽度
passenger capacity 载客量，客运量
passenger hoist 载客升降机
passer-by 行人
passimeter 自动卖票器，旋转栅门，步数计，内径指示规
passive 无源，无源的，被动的
passive attitude stabilization 被动姿态稳定
passive block 无源组件
passive circuit elements 无源电路元件
passive compensation 无源补偿
passive component 无源元件
passive constraint 消极约束，虚约束
passive constraint 虚约束

passive degree of freedom 局部自由度，无效自由度
passive element 无源元件
passive filter 无源滤波器
passive four-terminal network 无源四端网络
passive linear two-port network 无源线性二端口网络
passive network 无源网络
passive ranging 无源测距
passive suspension 无源悬挂
password 口令
patch repair 小块修补
patching 修补
patent 专利
patent glazing 专利装玻璃配件，无油灰镶玻璃法
patera 接线盒，插座
path 径
path generation 轨迹生成
path generator 轨迹发生器
path of action 啮合点轨迹
path of contact 接触迹
path of motion 动作路径
path planning 路径计划，路径规划
path repeatability 路径可重复性
pattern 图案，花样，图形，模式
pattern alarm 模式报警
pattern generation 模型产生
pattern identification 模式辨识
pattern primitive 模式基元
pattern recognition (PR) 模式识别
pattern recognition and intelligent systems 模式识别与智能系统
pawl 棘爪
pay-off function 罚函数
PBX(private branch exchange) 程控用户交换机，专用分组交换机
PC(personal computer) 个人计算机
PC(polycarbonate) 聚碳酸酯
PCA (principal component analysis) 主成分分析，主元分析
PCB(printed circuit board) 印刷电路板
PCBA(printed circuit board assembly) 印刷电路板装配
PCI (peripheral component interconnect) 互联外围设备

PCR (principal component regression) 主成分回归法
PD controller 比例微分控制器
PD (principal datum) 主水平基准, 主基准面
PD (proportional-derivative) 比例微分
PDA (personal digital assistant) 个人数字助理
PDM (pulse duration modulation) 脉宽调制
PE (polyethylene) 聚乙烯, 软胶
peak 峰值, 波峰
peak amplitude ratio 峰值比
peak base 峰底
peak current 峰值电流
peak height 峰高
peak hours 繁忙时间, 繁忙时段
peak limiter 峰位限制器, 限峰器
peak load 最大负载, 高峰负荷, 峰值负荷, 尖峰负载
peak meter 峰值电度表
peak period 繁忙时间, 繁忙时段
peak reading diode voltmeter 峰值二极管电压表
peak time 峰值时间
peak torque 峰值扭矩
peak traffic flow 最高交通流量, 高峰交通流量
peak value 峰值
peak voltage 峰值电压
peak voltmeter 峰值电压表
peak width 峰宽
peak width at half height 半高峰宽
peaker 微分电路, 脉冲削波电路
peak-to-peak value 峰-峰值
pearlitic iron 珠光体铸铁
pedestal generator 基准电压发生器
pedestrain signal 行人过路灯
peeling 剥离
peeling test 剥离试验
peer-to-peer 平分, 同等, 点对点
peer-to-peer communication 等权通信
pellet 粒料
pellicular packing 薄壳型填充剂
pen 笔, 记录笔
pen arm assembly 笔杆组件
pen tension 记录笔张力

penalty 罚则
penalty function method 罚函数, 补偿函数
pendulum type skid resistance tester 摆摆式防滑试验仪
pendulum viscometer 摆锤黏度计
penetrameter 透度计, 透光计
penetration 渗入
penetration of moisture 水分渗入
penetration test 渗透试验, 透入度试验
penstock 水闸
pen-writing recording instrument 笔式记录器
peptide synthesizer 多肽合成仪
percent of accuracy 准确度
percent per million (ppm) 百万分之一
percentage 百分数, 百分比
percentage of current load 负载电流百分比显示
perceptron 感知器
percolating 渗流, 渗透
perfect control 完美控制
perfect 完全的, 理想的
perforated pipe 穿孔管
perform 预先形成, 预制, 预成型坯, 粗加工的成品
performance 完成, 执行, 表现, 性能, 动作性能
performance analysis 性能分析
performance characteristic 性能特征, 工作特性
performance criteria 性能判据
performance drive 效能驱动
performance evaluation 性能评估
performance function 性能函数
performance index 性能指标
performance limit 性能极限
performance measure 性能度量
performance monitoring 性能监视
performance torque 性能扭矩
perimeter 周界, 周界长度
period 周期, 时期
period of oscillation 振荡周期
periodate 高碘酸盐
periodic 周期的, 周期性的, 循环的
periodic current 周期电流

periodic damping 周期阻尼
periodic duty 周期工作制
periodic function 周期函数
periodic load 周期性负载
periodic motion 周期运动
periodic pulse train 周期脉冲列
periodic replacement 周期替换
periodic speed fluctuation 周期性速度波动
periodic structure 周期结构
periodic wave 周期波
periodic inspection 定期检查
periodicity 周期性
peripheral component interconnect (PCI) 互联外围设备
peripheral equipment 外围设备,外部设备
peripheral 周围的,外围设备,周边的
perishable goods 易腐物品
peristaltic pump 蠕动泵
permalloy 坡莫合金
permalloy film 坡莫合金膜
permanent connection 固定连接
permanent deformation 永久变形
permanent magnet 永久磁铁
permanent magnet motor 永磁电动机
permanent magnet open magnetic resonance system 永磁开放式磁共振系统
permanent magnet undulator 永磁波动器
permanent magnetism 永磁,永久磁性
permanent partition 永久间隔
permanent 永久的,持久的
permanent-magnet moving-coil ammeter 永磁动圈式电流表
permeability 磁导率,导磁性,渗透性
permeability coefficient 渗透系数,渗滤系数,透气系数,磁导系数
permeability limit 渗透极限
permeability test 渗透试验,透水试验,透气性试验
permeance 磁导,导磁性;浸透,透过
permeation coefficient 渗透系数

permissible limit 容许极限
permissible stress 许用应力
permit 允许,牌照,许可证,准许证,准许
permit API interrupt 允许的API中断状态
permitted plot ratio 准许地积比率
permitted site coverage 准许上盖面积
permittivity 介电常数
permutation 重排
permutation algorithm 重排算法
permutation and combination 排列组合
perpendicular load 法向负载,垂直负载
personal computer (PC) 个人计算机
personal computer enclosure 个人计算机外设
personal computer serial interface 个人计算机串行接口
personal digital assistant (PDA) 个人数字助理
personal workstation on node bus 节点总线上的个人工作站
personality 属性,特性
perspective drawing 透视图
perspex 有机玻璃,防风玻璃,透明塑胶
pertain 合适
perturbation 扰动
perturbation analysis 扰动分析
perturbation theory 干扰理论,摄动理论
perturbed coefficient 扰动系数,干扰系数
per-unit value 标幺值
pessimistic value 悲观值
Petri-net Petri网
petrochemical plant 石油化工装置,石油化学工厂
petrol 汽油
petrol engine 汽油发动机,柴油引擎
petrol intercepting trap 石油截流隔
petroleum industry 石化工业
petroleum jelly 凡士林
petroleum products 石油产品

petroleum refinery 炼油厂
PFA(pulverized fuel ash) 粉煤灰，飞灰，煤灰
PFC(predictive functional control) 预测函数控制
pH control pH 控制
pH control system pH 控制系统，pH 调节系统
pH controller 酸碱控制器
pH detector 酸碱探测器
pH electrode assembly pH 电极装置
pH indicator pH 值指示剂，氢离子（浓度的）负指数指示剂
pH measurement system pH 测量系统
pH meter 酸碱度表，pH 计
pH transmitter pH 变送器
pH value 酸碱度值
phantom antenna 仿真天线
pharmaceutical equipments 制药设备
pharmacokinetic data 药物动力学数据
phase 阶段，状态，方面，相
phase contour 相位轮廓线，等相线
phase demodulation 相位解调，鉴相
phase advance 相位超前
phase advance controller 超前相位控制器
phase advance network 超前相位网络
phase advancer 进相器，进相机，相位移前器
phase angle 相角，相位角
phase angle of unbalance 不平衡相位
phase balance 相平衡
phase boundary potential 相界电位
phase calibration 相位校正，相位角校正
phase centre 相位中心
phase change 相位变化
phase changer 换相器
phase characteristic 相位特性
phase comparator 相位比较器
phase compensation 相位补偿
phase conjugation 相位连接，相位连接耦合
phase control 相位控制
phase correction 相位校正
phase corrector 相位校正器

phase crossover frequency 相位交越频率
phase current 相电流
phase diagram 相图
phase difference 相位差
phase difference indicator 相位差指示器
phase difference instrument 相位差仪
phase discriminator 鉴相器
phase displacement 相位移
phase distortion 相位失真
phase epitaxy 相位外延
phase frame analysis 相位系统分析
phase inversing amplifier 倒相放大器
phase inversion 倒相
phase inverter 倒相器，相位逆变器
phase lag 相位滞后
phase lead 相位超前
phase line 相线
phase load 相负载
phase locking 锁相
phase locus 相轨迹
phase margin 相补角，相位容限，允许相位失真
phase meter 相位计，功率因数表
phase modifier 调相器
phase modulation 调相，相位调制
phase noise 相噪声
phase perturbation technique 相位干扰技术
phase plane 相位平面，相平面
phase regulator 调相机
phase response 相位响应
phase reversal 反相
phase reversing transformer 倒相变压器
phase rotation 相序，相位旋转
phase sensitive rectifier 相敏整流器
phase sequence 相序
phase sequence relay 相序继电器
phase shift 相移
phase shifter 移相器
phase space 相空间
phase stability 相位稳定度，相位稳定性
phase system 相位系统
phase system identification 相位系统辨识

phase trajectory 相轨迹
phase transition 相暂态，相瞬态，相变
phase voltage 相电压
phase wire 相线
phase not together 缺相，失相
phase sequence 相序
phase voltage 相电压
phase-angle difference 相角差
phased array 相控阵，相位排列
phase-failure relay 断相继电器
phase-failure protection 断相保护
phase-frequency characteristic 相频特性
phase-in 同步
phase-lag control 相位滞后控制
phase-lag controller 相位滞后控制器
phase-lead controller 相位超前控制器
phase-locked array 锁相阵列
phase-locked loop 锁相回路
phase-only modulation 纯相位调制
phase-rotation relay 相序继电器
phasing back 反相
phasing switch 调相开关
phasometer 相位计
phasor 相量
phasor diagram 相量图
phasor sum 相量和
phenolic cellulose paper copper-clad laminates 酚醛纸质覆铜箔板
phenolic resin 酚醛树脂
phenyl propanol 苯丙醇
phosphate 皮膜化成，磷酸盐
phosphating 磷酸盐皮膜处理，磷化，磷酸盐化
phosphor-bronze wire 磷青铜线
phosphorous bronze 磷青铜
phosphorus(P) 磷
phosphorus sulfur flame ionization detection 磷硫火焰离子化检测
photo electric control 光度感应控制
photo electric switch 光电开关
photo parametric amplifier 光点参数放大器
photo telegraph 传真电极
photocell 光电池，光电管，光电流，感光器
photocell amplifier 光电管放大器
photoconductive cell 光敏电阻
photocoupler 光电耦合器，光隔离器
photodiode 光电二极管
photodiode array 二极管阵列
photodiode array detector (DAD) 光电二极管检测器，光二极管阵列检测器
photoeffect 光电效应
photoelectric 光电的
photoelectric cell 光电元件，光电池，光电管
photoelectric effect 光电效应
photoelectric multiplier 光电倍增器
photoelectric relay 光电继电器
photoelectric sensor 光电遥感器，光电传感器
photoelectric tachometric transducer 光电式转速传感器
photoelectric transducer 光电变换器
photoelectric transformer 光电变换器
photoelectric tube 光电管
photoelectricity 光电
photo-electromagnetic effect 光电磁效应
photoemission 光电发射
photo-emissive cell 光电发射管
photo-FET 光控场效应管
photo-flash-lamp 闪光灯
photoformer 光电波形发生器，光函数发生器，光电管振荡器
photogenerator 光电信号发生器
photogrammetry and remote sensing 摄影测量与遥感
photoimpact 光控脉冲，光冲量，光电脉冲
photomagnetic 光磁的
photomodulator 光调制器
photomultiplier 光电倍增器，光电倍增管
photomultiplier cell 光电倍增晶体管
photomultiplier tube 光电倍增管
photon 光子
photoresistance 光敏电阻
photoresistance relay 光敏电阻继电器
photosensitive 光敏的
photosensitive resin 感光性树脂

photosensitive resistance 光敏电阻
photoswitch 光电开关,光控继电器
phototransistor 光电晶体管
phototriode 光电三极管
phototube 光电管
phototube control 光电管控制
photounit 光电元件
photovalve 光电元件,光电管,光发射元件
photovaristor 光敏变阻器
physical analysis 物理分析
physical chemistry of metallurgy 冶金物理化学
physical constraint 实际限制
physical design 物理设计,结构设计,实体设计
physical dimension 实际尺寸,物理维度,外形尺寸,体积
physical electronics 物理电子学
physical inventory 盘点数量
physical layer 物理层
physical layout 实际布置图
physical measurement 实际丈量,物理测量
physical model 物理模型,实物模型,实体模型
physical node 物理节点
physical properties library 物理程序库
physical property 物理特性
physical property analysis 物性分析
physical realizability 物理可行性
physical symbol system 物理符号系统
physical vapor deposition 物理气相沉积
physicochemical analysis 物理化学分析
physics 物理学
physiological model 生理模型
physiology 生理学
PI controller (proportional-integral controller) 比例积分控制器,比例积分调节器,PI 控制器
PI film (polyimide film) 聚酰亚胺薄膜
piano wire 钢琴丝
pick 选择项
pick off 摘取
pickled steel wire 酸洗钢丝,酸浸钢线

pickling 泡浸,酸洗,浸洗,浸蚀
pick-up wire 吸起线
picture 图像
picture editor 画面编辑程序
picture element 像素
picture processing 图像处理
PID (proportion integration differentiation) 比例积分微分
PID control 比例积分微分控制
PID controller PID 控制器,PID 调节器,比例积分微分控制器,比例积分微分调节器
PID controller with batch switch 带批量开关的 PID 控制器
PID extend with EXACT self-tuning algorithm 带 EXACT 自整定算法的扩展 PID
PID with external reset feedback algorithm 用外部重置反馈的 PID 算法
PID with feedforward algorithm 用前馈的 PID 算法
piecewise linear 分段线性化
piecewise linear analysis 分段线性化分析
piecewise linear controller 分段线性化控制器
pierce 冲孔,剪内边
pierce die 冲孔模
piercing 冲孔加工
piezocrystal 压电晶体
piezo-electric control 压电控制
piezoelectric crystal 压电晶体
piezoelectric force transducer 压电式测力传感器
piezoelectric oscillator 晶体振荡器,压电振荡器
piezoelectric pressure sensor 压电压力敏感元件
piezoelectric quartz 压电石英
piezoelectric stabilizer 压电晶体稳频器
piezoelectric type vibration gauge 压电式振动仪
piezoelectric vibrator 石英振动片,压电振动片
piezometer 测压计,压强计

piezometric level 测压管水位
piezo-quartz 压电石英,压电晶体
piezoresistance accelerometer 压电电阻加速度计
piezoresistive effect 压阻效应
piezo-resonator 压电谐振器
pig iron 铸造生铁
pillow ball bearing 滑枕式滚珠轴承
pilot burner 导燃器,引燃器,长明灯
pilot channel 导频信道,领示通信电路,控制电路
pilot controller 辅助控制器,导频控制器,领示控制器
pilot frequency oscillator 导频振荡器
pilot hole 导孔,导向钻孔,定位孔
pilot light 指示灯
pilot motor 辅助电动机,伺服电动机
pilot project 试验计划
pilot relay 辅助信号继电器,控制继电器
pilot run 初步操作试验
pilot scheme 试验计划
pilot signal 导频信号,控制信号
pilot tube 指示灯,风速指示器
pilot tunnel 隧道导洞
pilot wire 引示线,电缆附线
pilot 导向的,辅助的,控制的
pin 销,镶针
pin bearing 轴承销
pin bush 销套
pin gate 针尖浇口,针点浇口
pin nose 销尖
pinboard 插接板
pincers 铁钳,钉钳
pincers tongs 夹钳
pinch roll 导正滚轮
pinchers 钳子
pinhole test 针孔试验机
pinion 小轮,小齿轮
pinion and rack 齿轮齿条,齿轮齿条机构
pinion axial displacement 小轮轴向位移
pinion cone 小轮锥距
pinion cutter 齿轮插刀

pinion front bearing 小轮前端轴承
pinion head bearing 小轮止端前轴承
pinion offset 小轮偏置距
pinion rear bearing 小轮后端后轴承
pinion rougher 小轮粗切机
pinion unit 齿轮传动系
pinion wire 小齿轮线坯
pin-point gate 细水口,针点式浇口
pintle valve 针形阀,销形阀,配流轴,分流器,轴针式喷油器
PIO (process input output) 过程输入输出
pipe barrel 管筒
pipe bender 弯管机
pipe bending machine 弯管机
pipe culvert 管渠,管状暗渠,管状排水渠
pipe filter 管式过滤器
pipe fittings 管配件
pipe graphite 管状电极
pipe jacking method 顶管法
pipe joint 管接头,管节,喉管连接处
pipe nuts 管用螺帽
pipe plug 喉塞,管堵头
pipe stanchion 管支柱
pipe syphon 冷凝弯(圈)
pipe works 管道工程
pipe 管,管道
piped gas supply system 喉管式气体供应系统
pipe-in-pipe method 套管法
pipeline 管道,管线
pipeline inspection gauge 检管器
pipeline reserve 管道预留地
pipelining processing 管道敷设过程
pipette 吸液管
piping 喉管,管路,接管,管道系统
piping centrifugal pumps 管道离心泵
piping diagram on the back of instrument panel-board 仪表盘盘后接管图
piping pumps 管道泵
piping safety valves 管道安全阀
PIPM (parallel input power module) 并联输入电源组件

Pirani vacuum gauge 皮拉尼真空计
piston 活塞
piston actuator 活塞执行机构，气动执行机构
piston rod 活塞连杆
piston type pressure gauge 活塞式压力计
piston valve 活塞阀
pit 坑穴，井
pitch 间距，周节，节距，齿距，斜度，齿节
pitch angle 节锥角
pitch apex 节锥顶
pitch apex to crown 节锥顶至轮冠
pitch circle 齿节圆，节距圆
pitch cone 节锥，节圆锥
pitch cone angle 节圆锥角
pitch curve 节面曲线，理论廓线，凸轮理论廓线
pitch diameter 节径，节圆直径
pitch element 节面母线
pitch line 节线
pitch of thread 螺距
pitch plane 节面
pitch point 节点
pitch radius 节圆半径
pitch surfaces 节曲面
pitch tolerance 齿距公差
pitch variation 齿距变化量
pitch-line chuck 节圆夹具
pitch-line runout 节线跳动
Pitot static tube 毕托静压管，皮托静压管
Pitot tube 毕托管，皮托管
pitting 点蚀
pivot 中支枢，枢轴，转轴
pixel 像素
placement 布局
plain bearing 平面轴承
plain cut-out 保险丝，熔丝
plain die 简易模
plain structure 平纹组织
plain tubular railing 光面管状栏杆
plan 图则，图，蓝图，计划，纲领，平面
planar cam 平面凸轮
planar cam mechanism 平面凸轮机构
planar chromatography 平面色谱法
planar kinematic pair 平面运动副
planar linkage 平面连杆机构
planar mechanism 平面机构
planar pair 平面副
plane 刨，平刨；平面
plane chromatography 平板色谱法
plane circuit 平面电路
plane grinding 平面磨削
plane iron 刨刀，刨铁
plane of action 啮合平面
plane of rotation 旋转平面
plane position indicator 平面位置显示器
plane strain 倒角应力
plane wave 平面波
planer 龙门刨床
planet gear 行星轮
planetary electron 轨道电子
planetary gear train 行星轮系
planetary speed changing devices 行星轮变速器
planetary transmission 行星齿轮变速箱
planimeter 测面器，求积仪
planing 龙门刨削
planing generator 展成法刨齿机
planing machine 刨床
planishing 打平，精轧
Plank constant 普朗克常数
planning process 计划流程
planometer 测平器
plant 装备，工厂，机器，设备，工场，厂房，车间
plant control 设备控制
plant control system 工厂控制系统
plant information network 工厂信息网络
plant network 厂用电力网
plant network module 工厂网络模块
plant overview display 工厂总貌显示画面
plant room 机房
plant floor data 工厂底层数据
plasma 血浆
plasma nitriding 离子氮化

plaster 批荡,灰浆,灰泥
plaster board 灰泥板
plaster mold 石膏模
plastering 抹灰,批荡
plastic 塑胶,塑料,塑料的
plastic basket 胶筐
plastic deformation 塑性变形
plastic distortion 塑性变形
plastic flow 塑性流动
plastic limit 塑限
plastic parts 塑料件,塑料零件
plastic settlement 塑性沉降
plastic tube 塑胶管
plastic wash bottle 洗瓶
plastic yield test 塑料变形测试
plasticine 橡皮泥
plastic-insulated wire 塑料绝缘线
plasticity coefficient 塑性系数
plasticizer 可塑剂
plasticizers 增塑剂
plastics industry 塑料工业
plastics 塑料
plate 电镀,衬板,压条
plate anchor 锚定板
plate clutch 碟式离合器,盘式离合器,圆片离合器
plate efficiency 板极效率,阳极效率
plate glazing 平板压光
plate glazing calender 平板抛光机,平板压光机
plate index 分度盘
plate mark 模板印痕
plate orifice with handle 带柄平孔板
plate pinch 压板
plate wire 镀线
plated 电镀的
plated through hole 镀通孔
plated wire 镀磁线,磁膜线
platform balance 托盘天平,台秤
platinum (Pt) 铂
platinum resistance bulb 测温铂电阻器
platinum resistance temperature detector 铂热电阻温度检查器
platinum resistance thermometer 铂热电阻温度计
platinum resistance thermometer with movable flange 可动法兰普通式铂热电阻温度计
platinum wire 铂丝,白金丝
platinum-rhodium platinum thermocouple 铂铑-铂热电偶
playback robot 示教再现式机器人
PLC (programmable logic controller) 可编程逻辑控制器,可编程序逻辑控制器,可编程序控制器,可编程控制器
PLC process control PLC过程控制
pliability 柔性,可挠性
pliable conduit 可弯曲导管
pliers 铗钳,钳子,老虎钳
plinth 基座,基脚,柱脚
plotter 绘图机,绘图仪
plotting device 绘图仪,曲线绘制仪
PLS (partial least square) 部分最小二乘
plug 塞,塞子,栓,插头,柱塞头,丝堵
plug braking 反接制动
plug connector block 插塞式连接器,插头,插塞接头
plug flow model 柱塞流模型
plug mill 自动轧管机
plug receptacle 插座
plug socket 插座,插孔,电源插座,塞孔
plug tee 带丝堵三通
plug valve 旋塞阀,塞嘴阀
plug welding 塞焊,塞孔熔接
plug wire 插线,火星塞高压线
plug-and-socket 插头插座
plugboard 插接板,插件
plugging 反向制动
plug-in amplifier 插入式放大器
plug-in board 插件
plug-in unit 插入单元,插入式部件,插入式分歧器
plumb bob 铅锤
plumb line 铅垂线
plunger 柱塞,压料柱塞
plunger globe valves 柱塞截止阀
plunger valves 柱塞阀,柱塞式阀,活塞式阀
plutonium (Pu) 钚

plywood 夹板,胶合板
plywood board 夹板
PM box datapoint PM 箱数据点
PM (post meridiem) 下午
PM (process manager) 过程管理机,过程管理站,工程管理综合控制器
PMC (product material control) 生产和物料控制
PMDP (process module data point) 过程模块数据点
PMM (process manager module) 过程管理机模块,过程管理控制模块
PMMA (polymethyl methacrylate) 聚甲基丙烯酸甲酯,有机玻璃
PMR (proton magnetic resonance) 质子核磁共振
pneumatic 气动,气动的
pneumatic absolute pressure transmitter 气动绝对压力变送器
pneumatic actuator 气动执行机构,气动执行器,气动促动器
pneumatic adder-subtractor 气动加减器
pneumatic amplifier 气动放大器
pneumatic analog station 气动模拟操作器
pneumatic brake system 气动制动系统
pneumatic butterfly valve 气动蝶阀
pneumatic by-pass remote control station 气动旁路遥控板
pneumatic by-pass station 气动副线板
pneumatic cascade indicating and recording 气动串级指示记录调节器
pneumatic colour-strip indicator 气动色带指示仪
pneumatic control 气动控制
pneumatic control and safety shutoff valve 气动调节及安全停止阀
pneumatic control hydraulic pump 气动控制液压泵
pneumatic control valve 气动控制阀,气动调节阀
pneumatic control valve with additional metal bellows seal 具有附加金属膜盒密封的气动调节阀

pneumatic delivery capability 供气能力
pneumatic diaphragm control valve 气动薄膜调节阀
pneumatic differential pressure transmitter 气动差压变送器
pneumatic double-pen recorder 气动二针记录仪
pneumatic exhaust capability 气动排气能力
pneumatic flanged differential pressure transmitter 气动法兰式差压变送器
pneumatic force-balance transducer 气动力平衡式传感器
pneumatic indicating and recorder controller 气动指示记录调节器
pneumatic indicating controller 气动指示调节器
pneumatic integrating counter 气动积算器,气动积分器
pneumatic integrator 气动积算器,气动积分器
pneumatic level controller transmitter 气动液位调节变送器
pneumatic limit operator 气动限幅器
pneumatic limit switch 气动限位开关
pneumatic loading diaphragm actuator 气动薄膜执行机构
pneumatic lock 气动夹紧
pneumatic long-stroke actuating mechanism 气动长行程执行机构
pneumatic mechanism 气动机构
pneumatic or hydraulic control 气动/液动系统
pneumatic piston cut-off valve 气动活塞切断阀
pneumatic positioner 气动阀门定位器
pneumatic pressure 气压
pneumatic pressure gauge 气动压力表
pneumatic pressure transmitter 气动压力变送器
pneumatic pump 气动泵
pneumatic relay 气动继电器,气动放大器
pneumatic remote control station 气动

遥控板
pneumatic remote transmitting rotameter 气远传转子流量计
pneumatic rotary actuator 气动旋转执行器
pneumatic screw driver 气动起子
pneumatic self-locking valve 气动自锁阀
pneumatic signal 气动信号
pneumatic single-pen recorder 气动一针记录仪,气动单针记录仪
pneumatic strip-type indicator 气动条形指示仪
pneumatic summing unit 气动加法器
pneumatic system 气动系统
pneumatic temperature transmitter 气动温度变送器
pneumatic to current converter 气电转换器
pneumatic transmitter 气动变送器
pneumatic tube 气动管
pneumatic tyred roller 气胎压路机
pneumatic valve positioner 气动阀门定位器,气动阀位控制器
pneumatic-electrical converter 气-电转换器
pock 麻点
pock mark 痘斑
point diameter 刀尖直径
point drift 点漂
point execution state 点执行状态
point form 点形式
point load 集中载重,集中荷载
point of contact 接触点
point process 点处理
point radius 刀尖半径
point scheduling 点调度
point width 刀顶距
point width taper 刀顶距收缩
point 点
pointer 指针
pointer instrument 指针式仪表
pointing system 定向系统
point's IOP type IOP 点的类型
point-to-point control 点位控制,点对点控制
point-to-point interface (PPI) 点对点接口
poison 有毒物品
poisonous exhaust composition 排气有害成分
poisonous substance 有毒物质
Poisson 泊松
Poisson process 泊松过程
POK (parallel operator keyboard) 并联操作键盘
poker vibrator 插入式振捣器
polar coordinate manipulator 球坐标操作器
polar coordinate type potentiometer 极坐标式电位差计
polar plot 极坐标图,极点绘图,极坐标作图
polar robot 极坐标型机器人
polarity 极性
polarization 极化
polarization analysis 极化分析
polarization dependence 极化关系
polarograph 极谱仪
polarography 极谱法
pole 极点,磁极,极,电杆,机,柱
pole assignment 极点配置
pole coil 磁极线圈,电极线圈
pole core 磁极铁芯
pole line 架空线路
pole mounting 柱上安装
pole placement 极点配置
pole zero assignment 零极点配置
pole-mounted transformer 杆上变压器
poles and zeros 极点和零点
poles definition 极点定义
pole-zero cancellation 零极点相消
polish 磨光
polisher 磨光器
polishing 抛光,磨光
polishing processing 表面处理
polluting substance 污染性物质
pollution 污染
polonium (Po) 钋
poly V-belt 多楔带
polyacrylamide gel electrophoresis 聚丙烯酰胺凝胶电泳
polyacrylic acid 聚丙烯酸

polyamide (PA) 聚酰胺，尼龙
polyamide epoxy paint 聚酰胺环氧漆
polyamide fiber 聚酰胺纤维
polycarbonate (PC) 聚碳酸酯
polyester 聚酯，聚酯薄膜
polyester resin 聚酯树脂，水晶胶
polyester woven glass fabric copper-clad laminates 聚酯玻璃布覆铜箔板
polyethylene (PE) 聚乙烯，软胶
polyethylene insulated cable 聚乙烯绝缘电缆
polyethylene insulated wire 聚乙烯绝缘线
polyethylene lining 聚乙烯内搪层
polyethyleneglycol 聚乙二醇
polyfunctional epoxy resin 多官能环氧树脂
polygon 多面体
polygonal 多边形的
polyimide film 聚酰亚胺薄膜
polyimide resin 聚酰亚胺树脂
polyimide woven glass fabric copper-clad laminates 聚酰亚胺玻璃布覆铜箔板
polymer 聚合物
polymerization 聚合作用
polymerization resin 聚合树脂
polymethyl methacrylate (PMMA) 聚甲基丙烯酸甲酯，有机玻璃
polynomial 多项式，多个多项式
polynomial functional basis 多项函数基
polynomial input 多项式输入
polynomial method 多项式方法
polynomial model 多项式模型
polynomial motion 多项式运动规律
polynomial transform 多项式转换
polyoxyethylene 聚氧化乙烯
polyphase 多相，多相的
polyphase asynchronous motor 多相异步电动机
polyphase current 多相电流
polyphase generator 多相发电机
polyphase induction motor 多相感应电动机
polyphase network 多相网络
polyphase rectifier 多相整流器
polyphase synchronous generator 多相同步发电机
polyphase system 多相系统，多相制
polyphenylene oxide 聚苯醚
polypropylene (PP) 聚丙烯，百折胶
polystyrene (PS) 聚苯乙烯，硬胶
polytetrafluoroethylene (PTFE) 聚四氟乙烯
polythene 聚乙烯
polyurethane (PU) 聚氨酯，聚亚胺酯
polyvinyl acetate 聚醋酸乙烯
polyvinyl alcohol (PVAC) 聚乙烯醇
polyvinyl butyral 聚乙烯醇缩丁醛
polyvinyl chloride (PVC) 聚氯乙烯，PVC 胶
polyvinyl chloride drop wire 聚氯乙烯用户引入线
polyvinyl chloride plastic sheets for waterproofing 聚氯乙烯防水卷材
polyvinyl fluoride 聚氟乙烯
polyvinylidene chloride 聚偏二氯乙烯
pontil 铁棒
Pontryagin's maximum principle 庞德里亚金极大值原理
pooled standard deviation 合并标准偏差，组合标准差
poor incoming part 事件不良
poor processing 制程不良
poor staking 铆合不良
Popov criterion 波波夫判据
porcelain 陶瓷，瓷器
porcelain insulator 瓷绝缘体，瓷隔电子
porosity 气孔，孔隙率
porous drain 多孔排水管道
porous mold 通气性模具
porous packing 多孔型填充剂
porous-layer open tubular column 多孔层开管柱
portable 手提式，轻便的，可移动的，携带方式
portable DC potentiometer 携带式直流电位差计
portable distribution board 轻便式配电屏

portable driller 手提钻孔机
portable fire extinguisher 手提灭火筒
portable instrument 便携式仪表
portable mold 手提式模具
portable pneumatic calibrator 便携式气动校验仪
portable saturated standard cell 便携式饱和标准电池
portable single-arm DC bridge 携带式直流单臂电桥
portable type-B ultrasonic 便携式B超
portal 隧道门，桥门，入口
portal frame 桥门构架，龙门架
Portland cement 波特兰水泥，普通水泥，硅酸盐水泥
Portland fly-ash cement 粉煤灰硅酸盐水泥
pose 位姿
pose overshoot 位姿过调量
posed problem 提出问题
position 位置，位移，职务
position accuracy 位置精度
position and orientation 位姿，定位与定向
position control 位置控制
position error 位置偏差，位置误差
position estimation 位置估计
position feedback 位置反馈，定位反馈
position finder 测位器
position location 定位
position measuring instrument 位置测量仪
position proportional algorithm 定位比例算法
position scale 位置刻度
position transmitter 位置变送器
position velocity 位置速度
positioner 定位器，位置控制器
positioning 位置控制，定位
positioning control system 定位控制系统
positioning relay 位置继电器
positioning system 定位系统
position-sensitive 位置灵敏度
position-sensitive photomultiplier 定位光电倍增器
positive 正的
positive carrier 空穴，正电荷载流子
positive column 正电柱，阳极区，阳辉区
positive displacement 容积式，实际位移
positive displacement flow measuring device 容积式流量测量仪表
positive displacement pump 正排量泵，容积泵，排代泵
positive feedback 正反馈
positive feedback control 正反馈控制
positive hole 空穴，正电荷载流子
positive ion 阳离子，正离子
positive mold 全压式模具，阳模，不溢式压缩模
positive phase-sequence 正相序
positive pole 正极，阳极
positive sequence impedance 正序阻抗
positive transfer 正迁移
positive wire 正极导线
positive-negative action 正负作用
positive-negative three step action 正负三位作用
positron 正电子，阳电子
post 接线柱
post amplifier 后置放大器
post cure 后固化
post design processing 设计后处理
post meridiem (PM) 下午
post store 后存
post-column reaction 后柱反应
post-column reactor 后柱反应器
posteriori estimate 后验估计
post-tensioned concrete beam 后张混凝土梁
post-tensioned prestressing 后张预应力
post-tensioned tendon 后张钢筋束
post-tensioning 后张法，后加拉力
potable water 可饮用的水，食水
potable water pump 食水泵，饮用水泵
potassium (K) 钾
potassium bromate method 溴酸钾法
potassium carbonate 碳酸钾
potassium dichromate method 重铬酸

钾法
potassium permanganate method 高锰酸钾法
potential 电位，电势，电压
potential difference 电位差，电势差
potential distribution 电位分布，电势分布
potential divider 分压器
potential drop 电位降，电压降，电势降
potential energy 势能
potential rise 电位升，电压升，电势升
potential transformer 电压互感器，变压器
potentially hazardous installation 潜在危险安装
potentiometer 电位计，电位差计，分压器，电位器，电势计
potentiometer pick off 电位发送器
potentiometric displacement transducer 电位器式位移传感器
potentiometric recorder 电位差计式记录仪
potentiometric titration 电位滴定法
potentiometry 电势测定法，电位测定法
potted component 密封元件
pouring process 浇注法
powder forming 粉末成形
powder metal forging 粉末锻造
powder metallurgy 粉末合金
power 功率，动力，电力，电力的，动力的
power amplification 功率放大
power amplifier 功率放大器
power angle 负载角，功率角
power assisted control 功率辅助控制
power balance 功率平衡，动力平衡
power button 电源按键，电源开关
power cable 电力电缆
power cable termination 电缆终端，电缆末端，电线封端
power capacitor 电力电容器
power circuit 电源线路，电力网
power control 功率控制，电源控制，动力控制

power density spectrum 功率密度频谱
power deposition 功率沉积
power deposition characterization 功率沉积特性
power device 功率器件，功率元件，电源设备，用电装置
power dissipation 功耗，功率损耗
power distribution 功率分布，配电
power distribution circuit 功率分布电路
power distribution component 电源分配组件
power distribution panel 电源分配箱，配电盘
power distribution unit 电源分配单元
power divider 功率分配器
power drive 电力传动，功率激励
power economy 电能经济，动力经济
power electronics 电力电子学，动力电子设备
power electronics and power drives 电力电子与电力传动
power engineering and engineering thermophysics 动力工程及工程热物理
power factor 功率因数，功率系数
power failure 停电
power flow 功率流，电力潮流
power frequency 电源频率，工业频率，工频
power fuse 电力熔断器
power generation 发电，发电量，发电机设备
power house 动力室，发电站
power industry 电力工业
power input 电源输入，功率输入
power law description 动力律描述
power line 电力线，电源线，输电线
power loss 功率损耗，功率损失
power machinery and engineering 动力机械及工程
power management 电源管理，功率管理，动力管理
power meter 功率表，瓦特计
power network 电力网，电力网络
power of attorney 授权书
power outage protection 断电保护

power output 功率输出，电源输出
power plant 电厂，发电厂
power rammer 动力夯
power rate 功率比
power reactor 原子动力反应堆，动力反应堆，动力堆
power regulator 功率调节器
power screw 螺旋传动
power socket 电源插座
power source 电源，主电源，动力源
power spectra 功率谱
power spectral density 功率谱密度
power spectrum 功率谱
power spectrum density 功率谱密度
power station 发电站，发电厂
power station control 电站控制
power sub-station 电力支站，配电站
power supply 电源，供电，供电器
power supply card 电源卡
power supply common 电源公共端
power supply device 电源装置
power supply switch 电源开关
power supply system 电源系统，供电系统
power supply voltage 供电电压
power system 电力系统，电网，动力系统
power system and its automation 电力系统及其自动化
power system automation 电力系统自动化
power system control 动力系统控制
power system stabilizator 电力系统稳定器
power system stabilizer 动力系统稳定器
power system voltage 动力系统电压
power transfer 能量输送，动力分配装置
power transformer 电源变压器，电力变压器，功率变换器
power transmission 电力传输，动力传输
power transmission system 输电系统
power triangle 功率三角形
power voltage 电源电压
power winding 功率绕组

power wire 电源线
power-angle 功角，功率角
power-factor capacitor 提高功率因数用电容
power-factor control 功率因数控制，功率因数调节
power-factor improvement 功率因数改善
power-factor meter 功率因数表
powerful 功能强大
powerful control and communication function 强大的控制与通信功能
powerful operation and monitoring function 强大的操作与管理功能
power-up 加电，上电
power-up diagnostic 加电诊断
PP（polypropylene） 聚丙烯，百折胶
PPC（parametric predictive control） 参数预测控制
PPI（point-to-point interface） 点对点接口
PPM（percent per million） 百万分之一
PR（pattern recognition） 模式识别
practical current source 实际电流源
practical voltage source 实际电压源
praseodymium（Pr） 镨
preamplifier 前置放大器，预放大器
pre-annealing 预备退火
precast 预制
precast concrete beam 预制混凝土梁
precast concrete segment 预制混凝土砌块
precautionary measure 预防措施
precipitation forms 沉淀形式
precipitation hardening 析出硬化
precipitation statics 雨滴静电干扰
precipitation titration 沉淀滴定法
precipitator 聚尘器
precision 精密度，精度，准确度，精确度
precision attenuator 精密衰减器
precision constant temperature ovens 精密恒温器
precision forging 精密锻造
precision instrument 精密仪表
precision instrument and machinery

精密仪器及机械
precision low constant temperature water baths 精密低温恒温水槽
precision measurement 精密测量
precombustion chamber 预燃室
predetermining impulse counter 预置脉冲计数器
predicate logic 谓词逻辑
prediction 预测，推断
prediction error method 预测误差方法
prediction interval 预测区间
prediction method 预测方法，推断方法
prediction problem 推断问题
predictive control 预测控制，预估计理论
predictive functional control (PFC) 预测函数控制
predominant axis 供设计优化坐标轴，主导轴
pre-excitation 预激励
prefabricated parts 预组构件，预制构件
prefabricated water tank 预制式水箱
preferential 优先的，优先权
preferred access device 优先存取设备
preferred device 优先设备
prefetch 预取
prefilter 预滤器，前置滤波器
pre-fixed finishing date 预定完成日
preheat 预热
preheater 预热器
preheating 预热
preliminary mold design 初步模图设计
preliminary works 初步工程
preliminary 准备工作
preload 预紧力
preloading 预加荷载
pre-operational check 用前检查，操作前检查
pre-operational examination 用前检查，操作前检查
preparative liquid chromatograph 制备液相色谱仪
preparative liquid chromatography 制备液相色谱法
preparative thin layer chromatography 制备薄层色谱法
prepared by 制表
prepolymer 预聚物
prepreg 预浸材料，预浸料坯
preprocessing 预处理
preprocessor 预处理器
press forging 冲锻，压锻
press quenching 加压硬化
press specification 冲床规格
press 压，按，压力
press-button switch 按钮开关
pressfit 压入
pressure 压力
pressure adjusting screw 调压螺钉
pressure and flow controller 压力流量控制器，压力流量调节器
pressure and temperature controller 压力温度控制器，压力温度调节器
pressure angle 压力角
pressure angle of base circle 基圆压力角
pressure angle of involute 渐开线压力角
pressure bulb 压力泡
pressure compensator 压力补偿器
pressure control 压力控制
pressure control loop 压力控制回路
pressure die 压紧模
pressure difference transmitter 压差变送器
pressure differential flowmeter 压差式流量计
pressure element 压力元件，测压元件
pressure gauge 压力计，压力表
pressure gauge with electric contact 电接点压力表
pressure governor 调压器
pressure grouting 压力灌浆
pressure instrument 压力仪表，压力计
pressure level measuring device 差压液位测量仪表
pressure loss 压力损失
pressure lubrication 压力润滑

pressure measurement 压力测量
pressure measuring instrument 压力测量仪表
pressure pipe 进水管，加压管
pressure plate 压板
pressure port 压力端口
pressure pump 压力泵
pressure range 压力范围
pressure recorder 压力记录仪
pressure reducing valve 减压阀
pressure regulator 调压器，压力调节器
pressure relief valve 安全阀，卸压阀
pressure seal 加压密封
pressure switch 压力开关
pressure tap 测压孔，压力计接口，取压分接管
pressure tapping point 取压点
pressure transducer 压力传感器
pressure transmitter 压力变送器，压力传送器
pressure valve 施压阀，压力控制阀，压力增压阀
pressure vessel 压力容器
pressure volume relationship 压力真空关系
pressure welding 压焊
pressure 压力
pressure-compensated flowmeter 压力补偿式流量计
pressure-sensitive diaphragm 压感膜片
pressure-sensitive element 敏感元件，灵敏部分，感受元件
pressure-sensitive probe 压力敏感电极
pressurized air 压缩空气
pressurized electrical instrument 过压型电动仪表
pressurized enclosure 过压外壳
pressurized water reactor 压水堆，压力水冷反应堆
prestressed concrete beam 预应力混凝土梁
prestressed element 预应力构件
prestressing 预（加）应力
prevention 预防，防止
preventive maintenance 防止污染，预防性维修，预防性保养
primary 主的，初级的，一次的
primary cell 原电池
primary control module 主控制模块
primary controller 主控制器，主调节器
primary device of a differential pressure device 差压装置的一次元件
primary frequency zone 主频区
primary gate 主级门路
primary instrument 一次仪表
primary loop 主控制回路
primary measuring element 一次测量元件
primary reference electrode 一级参比电极
primary regulation 主调节
primary site 基本点
primary test 初步试验
primary treatment of sewage 初级污水处理
primary winding 一次绕组，原绕组
primary wire 原电路，初级电路
prime amplifier 前置放大器，前级放大器
prime motor 原动机
prime mover 原动力
prime mover torque 原动机扭矩
principal component analysis (PCA) 主成分分析
principal component regression (PCR) 主成分回归法
principal datum (PD) 主水平基准，主基准面
principle of conservation of energy 能量守恒原理
principle of duality 对偶原理，二元原理
principle of electric engineering 电工原理
principle of measurement 测量原理
principle of reciprocity 互易原理
principle of superposition 叠加原理
principle of the argument 幅角原理
principle 原理，原则
principles of conservation 变换原理
print a file 打印一个文件

print from off-line archive 打印离线档案
printed 印刷,复制,打印
printed antenna 印刷天线
printed board 印制板
printed board assembly 印制板装配
printed board assembly drawing 印制板组装图
printed circuit 印刷电路
printed circuit antenna 印刷电路天线
printed circuit board (PCB) 印刷电路板,印制板电路
printed circuit board assembly (PCBA) 印刷电路板装配
printed component 印制元件
printed contact 印制接点
printed dipole 印刷偶极
printed wire layout 印制线路布设
printed wired assembly 印制线路装配
printed wiring 印制线路
printed wiring board (PWB) 印制线路板
printer 打印机
printing 印制,印刷
printing industry 印刷工业
printout of individual items 单项输出
priori estimate 先验估计
priority 优先级,优先权
private automatic branch exchange 程控数字自动交换机
private branch exchange (PBX) 程控用户交换机,专用分组交换机
private communication interface 专用通信接口
private wire 私人专用电报线,私人专用电话线
probabilistic 概率的,随机的
probabilistic data association 伴随随机数据
probabilistic load flow 随机负载流
probabilistic logic 概率逻辑,随机逻辑
probabilistic model 概率模式
probabilistic risk assessment 危险概率评估
probabilistic simulation 随机仿真

probabilities integration 概率积分
probability 概率
probability density function 概率密度函数
probability distribution function 概率分布函数
probe 探头,探针,探测器,测件
probe amplifier 探头放大器
probe lead 探头引线
probe sensor 探头传感器
probe type magnetic flowmeter 探头式磁流量计
probing 探测
problem solver 问题求解器
problem-oriented language 面向问题语言
procedure 步骤
procedure-oriented 面向过程
process 过程,处理,加工,工艺,方法
process alarm 过程报警
process alarm function 过程报警功能
process alarm message 过程报警信息
process alarm window 过程报警窗口
process annealing 中间退火,临界温度以下退火
process area 过程区域
process arithmetic and logical unit 处理算术和逻辑单元
process automation 过程自动化
process computer 过程计算机,控制计算机
process control 过程控制
process control computer 过程控制计算机
process control system 过程控制系统
process correlation 过程相关
process data 过程数据
process design 工艺流程设计,工艺过程设计
process dynamics 工艺动态
process engineer 工艺工程师,程序工程师,过程工程师
process engineers console 过程控制工程师控制台
process equipment 工艺设备
process fluid manipulated 被调介质

271

process fluid measured 被测介质
process gain matrix 过程增益矩阵
process identification 过程辨识
process input output (PIO) 过程输入输出
process instrumentation 过程仪表
process instrumentation and analytics 过程仪表及分析仪器
process interface 过程接口
process interface unit 过程接口单元
process load 过程负荷
process management and control 过程管理和控制
process manager test executive 过程管理机测试执行
process manager (PM) 过程管理机,过程管理站,工程管理综合控制器
process manager module (PMM) 过程管理机模件,过程管理控制模块
process measurement 过程测量
process model 过程模型,进程模型,程序模型
process modeling and optimization 过程模型化和优化
process module 过程模件,过程模块
process module data point (PMDP) 过程模块数据点
process module point 过程模件点
process network interface 过程网络接口
process network modem 过程网络调制解调器
process of self-excitation 自励过程
process operator 过程操作员
process optimizer 过程优化器
process parameter estimation 过程参数估计
process pressure 过程压力
process reaction curve 过程反应曲线
process reaction curve method 过程反应曲线法
process report menu panel 过程报告菜单画面
process reports 过程报告
process simulator 过程仿真
process special 特殊处理
process temperature 过程温度

process timer 过程定时器
process tomography (PT) 过程层析成像,过程断层摄影
process unit 过程单元
process variable (PV) 过程变量
process variable high limit 过程变量上限
process variable in percent 百分比的过程变量
process variable low limit 过程变量下限
process-control language 过程控制语言
processing cycle 处理周期
processing technique 生产工艺,制造工艺
processing unit 处理单元,处理部件
processor 处理器,处理程序,加工者
processor array 处理器阵列
processor gateway 处理器接口
processor system 处理器系统
process-oriented sequential control 过程顺序控制
process-oriented simulation 面向过程的仿真
product material control (PMC) 生产和物料控制
product integrator 乘积积分器
production rule 产生式规则
production system 制造系统
productive resistance 生产阻力
productivity 产率
products industry 制造工业
profile 轮廓,外形,概况,外观,剖面图
profile angle 齿廓角
profile bridge 齿廓桥形接触
profile chart 剖视图
profile contact ratio 齿廓重合度
profile die 轮廓模
profile grinding machine 投影磨床,仿形磨床
profile mismatch 齿廓啮合失配
profile projector 轮廓光学投影仪
profile radius of curvature 齿廓曲率半径
profiled iron 型钢

profiled sheet iron 成型薄钢板
profiler display function 剖面显示功能
profit function 利益函数
profitability 收益
program 程序，计划，项目
program assembler 程序汇编
program composition 程序设计
program control 程序控制
program control device 程序控制装置
program controller 程序控制，程序控制器
program controlling element 程序控制单元
program diagnostic 程序诊断
program documentation 程序文件编制，程序记录，程序文本
program order address 程序指令地址
program register 程序寄存器
program set station 程序设定操作器
program set unit 程序设定器
program store 程序存储
programmable computing 可编程运算器
programmable controller 可编程控制器，可编程序控制器
programmable function key assignment 可编程功能键分配
programmable gain amplifier 可编程增益放大器
programmable indicating controller 可编程指示调节器
programmable indicating controller with pulse width output 可编程脉冲宽度输出指示调节器
programmable logic block 可编程逻辑块
programmable logic controller (PLC) 可编程序控制器，可编程序逻辑控制器，可编程控制器，可编程逻辑控制器
programmable logic controller gateway 可编程逻辑控制器接口
programmable read only memory (PROM) 可编程只读存储器
programmable read-only memory 可编程的只读存储器

programmable 可设计的，可编程的，可编程序
programme 计划，方案，纲领
programmed control 编程控制，程控
programmed controller 程序控制器
programmed flow 程序流速
programmed pressure 程序压力
programmed solvent 程序溶剂
programmers console 程序员控制台
programming 程序设计，编程序
programming approach 编程方法
programming control panel 程序控制盘
programming environment 编程环境，程序设计条件
programming language 编程语言，程序设计语言
programming support 编程支持，程序设计支持
programming system 程式设计系统，程序系统
programming theory 编程理论
programs management 程式管理
progress chart 进度表
progressive 连续送料
progressive bending 连续弯曲加工
progressive blanking 连续下料加工
progressive die 顺序模，连续模
progressive drawing 连续引伸加工
progressive error 累进误差
progressive forming 连续成形加工
prohibit 禁止
prohibited zone permit 禁区许可证
project agreement 工程项目协议
project data 工程资料
project engineering 设计工程，工程项目
project limit 工程界限
projectile 抛射体
projection 凸出，投影，突出物，伸出物
projection grinder 投影磨床
PROM (programmable read only memory) 可编程只读存储器
promenade 散步广场，散步长廊
promethium (Pm) 钷

proof pressure 耐压
propagation 传播，繁殖
propagation of error 误差传递
propane gas cutting 丙烷气切割
propeller flowmeter 螺旋桨式流量计
propeller-type current meter 叶轮流速计
properties of direct tension 抗拉性能
properties of impact resistance 抗冲击性能
proportion 比例
proportion band of a controller 控制器的比例带，调节器的比例带
proportional 比例
proportional action 比例作用
proportional action coefficient 比例作用系数
proportional action factor 比例作用因子
proportional amplifier 比例放大器
proportional band 比例度，比例带，比例范围
proportional control 比例控制，比例控制器
proportional control action 比例调节作用
proportional control factor 比例控制因子
proportional controller 比例控制器
proportional counter 比例计数器
proportional gain 比例增益，比例放大率
proportional plus derivative action 比例微分作用，PD作用
proportional plus derivative control action 比例微分调节作用
proportional plus derivative controller 比例微分控制器，比例微分调节器
proportional plus integral action 比例积分作用，PI作用
proportional plus integral control 比例积分控制
proportional plus integral control action 比例积分调节作用
proportional plus integral controller 比例积分控制器，PI控制器
proportional plus integral plus derivative action 比例积分微分作用，PID作用
proportional plus integral plus derivative control action 比例积分微分调节作用
proportional plus integral plus derivative controller 比例积分微分调节器
proportional speed floating controller 比例速度浮动控制器
proportional time controller 比例时间控制器
proportional-derivative (PD) 比例微分
proportional-derivative controller 比例微分控制器
proportional-integral (PI) 比例积分
proportional-integral controller 比例积分控制器，比例积分调节器，PI控制器
proportional-integral-derivative (PID) 比例积分微分
proportional-integral-derivative controller 比例积分微分控制器，PID控制器
proportioner 比例调节器
proposal improvement 提案改善
proprietary product 专卖物料，专利物料
propulsion control 推进控制
protactinium (Pa) 镤
protecting hood 防护罩
protection 防护，保护，保护开关
protection box 保护箱
protective barrier 防护栏障，防护栏杆
protective coating 保护涂料
protective cover 保护罩
protective device 防护装置
protective earthing 保护接地
protective goggle 护目镜
protective guard 护罩
protective relay 保护继电器
protective system 保护系统
protector 保险器
protocol 规程，协议
protocol engineering 协议工程
proton 质子
proton balance equation 质子平衡式

proton magnetic resonance (PMR) 质子核磁共振
proton magnetic resonance spectrum 质子核磁共振谱
protonic solvent 质子溶剂
prototype 模型，标准，雏形，原型，首版
prototype mold 雏形试验模具
prototyping 圆形，样机
protrude 突出
proximity effect 邻近效应，接近效应，近距离效应
PS (polystyrene) 聚苯乙烯，硬胶
pseudo random sequence 伪随机序列
pseudo-rate-increment control 伪速率增量控制
PT (process tomography) 过程层析成像，过程断层摄影
PTFE (polytetrafluoroethylene) 聚四氟乙烯
PU (polyurethane) 聚氨酯，聚亚胺酯
public lighting 街道照明设施
public telecommunication service 公众电信服务
public-address system 扩音系统
puddle 胶泥
pull box 穿线盒
pull wire 拉线，牵引线
puller 拉出器，拔取器
pulley 滑轮，滑车
pulley block 滑轮组，滑车组
pulling 拉伸
pull-out test 拔拉测试
pulp and paper engineering 制浆造纸工程
pulp industry 纸浆工业
pulsating 脉动的
pulsating current 脉动电流
pulsating stepless speed changes 脉动无级变速
pulsating voltage 脉动电压
pulsation 脉动
pulse 脉冲，脉动
pulse accumulation value 脉冲累加值
pulse amplifier 脉冲放大器
pulse circuit 脉冲电路
pulse controlled 脉冲控制的

pulse distributor 脉冲分配器
pulse duration 脉冲持续时间
pulse duration control 脉冲宽度控制
pulse duration controller 脉宽控制器
pulse duration modulation (PDM) 脉宽调制
pulse encoder 脉冲编码器
pulse form 脉冲波形
pulse frequency control 脉冲频率控制
pulse frequency modulation control system 脉冲调频控制系统
pulse input analog output card 脉冲输入模拟量输出卡
pulse input module 脉冲输入模件
pulse keyer 脉冲键控器
pulse manipulator 脉冲控制器
pulse modulated 脉冲调制的
pulse modulation 脉冲调制
pulse modulation method 脉冲调制法
pulse modulator 脉冲调制器
pulse motor 脉冲电动机
pulse multiple input card 脉冲多路输入卡
pulse position modulation 脉冲位置调制
pulse radiation 脉冲发射
pulse rate indicator 脉冲速率指示仪
pulse response 脉冲响应
pulse shape 脉冲波形，脉冲形式
pulse shape synthesis 脉冲形状合成
pulse signal 脉冲信号
pulse spike 脉冲尖峰
pulse synchroscope 脉冲同步示波器
pulse train 脉冲序列，脉冲群，一串脉冲
pulse transfer function 脉冲传递函数
pulse transformer 脉冲变压器
pulse width 脉冲宽度，脉宽
pulse width modulated 调制的脉冲宽度
pulse width modulation control system 脉冲调宽控制系统
pulsed-amperometric detection 脉冲电流检测
pulse-length 脉冲宽度
pulse-width keyer 脉宽键控器

pulse-width modulation (PWM) 脉冲宽度调制,脉宽调制
pulverized fuel ash (PFA) 粉煤灰,飞灰,煤灰
pump 抽水机,泵
pump foundation 泵基座
pump house 泵房,抽水房
pump impeller 抽水机叶轮,泵叶轮
pump motor 泵马达,泵电动机
pump riser 泵水竖管
pump shaft 泵转轴
pump shaft sleeve 泵转轴套筒
pump body 泵体
pumping main 泵水干管
pumping station 抽水站,泵房,泵站
punch 冲床,冲头
punch card 穿孔卡
punch mark 冲压记号
punch pad 上垫板
punch press 冲床
punch riveting 冲压铆合
punch tape 穿孔带
punched 穿孔
punched hole 冲孔
puncher 推杆,冲孔器
punching die 落料模
punching machine 冲床,冲压机,冲孔机
pupin cable 加感电缆
pure aluminium 纯铝
pure copper 纯铜
pure nickel electrode 纯镍熔接条,纯镍焊条
pure time delay 纯滞后
purge directory 清除目录
purge valve 放气阀
purge 净化,吹扫
purging 清除
purlin 桁条
purpose-made bend 特制弯管
push bar 推杆
push button 按钮
push button input 按钮输入
push button input card 按钮输入卡
push button station 按钮式控制站
push button switch 按钮,按钮开关
push button traffic light 按钮式交通灯
push 推
push and pull switch 推拉开关
pushdown automaton 下推自动机
pusher feed 推杆式送料
push-fit spigot and socket joint 推进式套筒接头
push-pull 推挽的
push-pull amplifier 推挽式放大器
push-pull circuit 推挽电路
push-pull output amplifier 推挽输出放大器
putty 油灰
putty glazing 油灰镶玻璃法
puttyless glazing 无油灰镶玻璃法
PV auto value PV 自动值
PV flag PV 标志
PV flag parameter PV 标识参数
PV high alarm trip point PV 值高报警点
PV high high alarm trip point PV 值高高报警点
PV high range in engineering units PV 高量程
PV in percent 以百分比表示的 PV 参数
PV low alarm trip point PV 值低报警点
PV low low alarm trip point PV 值低低报警点
PV low range in engineering units PV 低量程
PV negative rate of change trip point PV 行程的负向变化率
PV normal state PV 正常状态
PV positive rate of change trip point PV 行程的正向变化率
PV raw value PV 原始值
PV source PV 来源
PV source selection 过程变量源选择
PV (process variable) 过程变量
PVAC (polyvinyl alcohol) 聚乙烯醇
PVC insulated copper soft wire 铜芯聚氯乙烯绝缘软线
PVC insulated copper wire 铜芯聚氯乙烯绝缘线
PVC insulated shielded soft wire 聚氯

乙烯绝缘屏蔽软线
PVC sheathed rubber insulated control cable 橡皮绝缘聚氯乙烯护套控制电缆
PVC (polyvinyl chloride) 聚氯乙烯
PVC-sheathed 塑料护套
PVC-insulated control cable 塑料绝缘控制电缆
PWB (printed wiring board) 印制线路板
PWM (pulse-width modulation) 脉宽调制
PWM inverter 脉宽调制逆变器
pylon 桥塔,指示塔,高压线铁塔
pyrocondensation power cable terminal 热缩式电力电缆终端头
pyroelectric thermodetector 热电式温度检测器
pyrometer 高温计
pyrometric cone 温度锥

Q

Q factor 品质因数
QA (quality assurance) 质量保证,品质保证
QC (quality control) 质量管理,品质管制
QCC (quality control circle) 品质圈
QDMC (quadratic dynamic matrix control) 二次动态矩阵控制
QIT (quality improvement team) 品质改善团队
quadrant 象限
quadratic 二次型
quadratic control 二次型控制
quadratic dynamic matrix control (QDMC) 二次动态矩阵控制
quadratic interpolation 二次插值,二次插值法
quadratic lag 平方滞后
quadratic optimal regulator 二次型优化调节器
quadratic performance criteria 二次性能指标,二次性能判据
quadratic performance index 二次型性能指标
quadratic potentiometer 正交电位差计
quadratic programming 二次规划
quadratic stability 二次型稳定
quadrature 求积,速度
quadrature axis 直角坐标轴
quadrature detection 积分检波
quadrature mirror filter 积分镜像滤波器
quadripole 四端网络
quadrupler 四倍器,四频器
quadrupole mass spectrometer 四极杆质谱仪,四极质谱仪
quadrupole type mass analyzer 四极质量分析仪
qualified person 合格人员
qualified products 良品
qualitative 定性的

qualitative analysis 定性分析
qualitative control 定性控制
qualitative physical model 定性物理模型
qualitative simulation 定性仿真
quality 质量,品质
quality ameliorate notice 品质改善活动
quality analysis 质量分析
quality assurance (QA) 质量保证,品质保证
quality assurance scheme 品质保证计划
quality control (QC) 质量管理,品质管制
quality control circle (QCC) 品质圈
quality engineering 品质工程
quality estimation 质量估计
quality factor 品质因数,品质因子,质量指标
quality improvement 品质改善
quality improvement team (QIT) 品质改善团队
quality logic test 质量逻辑测试
quality management techniques and practice 品质管理技术和实践
quality measurement system 质量检测系统
quality of material 物料质量
quality of work life 工件生命质量
quality policy 目标方针
quality system 质量体系,品质系统
quality target 品质目标
quality tracking 质量跟踪
quantitative analysis 定量分析
quantitative electron probe 定量电子探头
quantity 量,分量,数量
quantity of electricity 电量
quantization 量化
quantization error 量化误差
quantization noise 量化噪声

quantization signal 量化信号
quantized 量化的
quantized noise 量化噪声
quantized signal 量化信号
quantized state 量化状态
quantizer 量化器
quantizer design 量化器设计
quantizing encoder 量化编码器
quantum number 量子数
quarry 石矿场,采石场
quarry tile 缸砖
quarter circle orifice plate 1/4 圆孔板
quartet 四重峰
quartz 石英,水晶
quartz compensating element 石英补偿元件
quartz crystal 石英晶体
quartz crystal filter 石英晶体滤波器
quartz crystal oscillator 石英晶体振荡器
quartz elastic element 石英弹性元件
quartz piezoelectric accelerometer 石英晶体压电式加速度计
quartz sand 石英砂
quasilinear characteristics 准线性特性
quasi-optical component 准光学元件
quaternion feedback 四元反馈
quench 淬火
quench ageing 淬火老化
quench hardening 淬火硬化
quenching 淬火
quenching crack 淬火裂痕
quenching cracks 淬火裂纹
quenching die 淬火压模
quenching distortion 淬火变形
quenching press 淬火压床,加压淬火机,成形淬火机

quenching stress 淬火应力
queue 排队
queue railing 轮候处铁栏
queue support database 队列支撑数据库
queueing theory 排队论,大系统服务理论
queuing 排队
queuing network model 排队网络模型
queuing theory 排队论,排队理论
quick die change system 快速换模系统
quick draining valves 快速排污阀
quick switch 快速开关
quick-acting regulator 速动调节器
quick-acting relay 速动继电器
quick-action 速动,快作用
quick-action fuse 速动熔断器
quick-action switch 速断开关
quick-break switch 速断开关
quick-closing valve 速闭阀
quick-drying paint 快干油漆,快干漆
quick-open type 快开式
quick-opening valve 快开阀
quick-release coupling 速脱联轴节,速脱离合器
quick-return characteristics 急回特性
quick-return mechanism 急回机构
quick-return motion 急回运动
quick-setting cement 快凝水泥
quick-setting level 速测水准仪
quiescent point 静态工作点
quiescent state 静态
quiet circuit 无噪声电路
quintant 五分仪
quintet 五重峰
quintuple harmonic 五次谐波

R

raceway 滚道
rack 齿条，机架，机柜，上料
rack cutter 齿条插刀
rack gear 齿条传动
rack mounted instrument 架装仪表
rack mounting 架装
rack earth 机壳接地
rack-mount 机架安装
radar 雷达
radar control 雷达控制
radar display 雷达显示器
radar guidance system 雷达制导系统
radar network 雷达网
radial 径向的，径向刀位
radial base function network 径向基函数网络
radial basis function (RBF) 径向基函数
radial bearing 向心轴承
radial contact bearing 径向接触轴承
radial development 径向展开
radial direction 径向
radial internal clearance 径向游隙
radial load 径向载荷，径向负荷
radial load factor 径向载荷因素
radial locating surface 径向定位表面
radial plane 径向平面
radial planimetric plotter 径向平面绘图仪，辐射绘图仪
radial pump 径向泵
radial rake angle 径向前角
radial reciprocating follower 对心移动从动件
radial roller follower 对心滚子从动件
radial translating follower 对心直动从动件
radian (rad) 弧度
radiant energy 辐射能
radiation 照射，辐射
radiation ammeter 辐射安培计
radiation and environmental protection 辐射防护及环境保护
radiation level 辐射量
radiation power 辐射功率
radiation pyrometer 辐射高温计
radiation thermometer 辐射高温计
radiation type switch 辐射式开关
radiation wall 防火墙
radiation-balance type pyrometer 辐射平衡式高温计
radiation-energy thermometer 辐射温度计
radiation-pressure power meter 辐射压力功率表
radiator 散热器，辐射器，冰箱
radiator fin 散热片
radical 基
radio 无线电
radio control 无线电控制
radio direction finding 无线电测向
radio distance 无线电测距
radio engineering 无线电技术，无线电工程
radio facsimile 无线电传真
radio frequency ammeter 射频电流计
radio frequency amplification 射频放大
radio frequency amplifier 射频放大器
radio frequency interference 射频干扰，无线电频率干扰
radio frequency micro-potentiometer 射频微电位器
radio frequency pulse generator 射频脉冲发生器
radio frequency sensor 射频敏感器
radio frequency transformer 射频变压器
radio immuno assay 放射免疫测定
radio influence voltage 高频干扰电压
radio instrument 无线电测量仪器
radio interference 无线电干扰
radio receptor assay 放射受体测定
radio wire 绞合天线

radioactive cloud 放射云
radioactive elements 放射性元素
radioactive fallout 放射性沉降物
radioactive substance 放射性物质
radioactivity 放射性
radioactivity detection 放射性检测
radiocast 无线电广播
radioisotope 放射性同位素
radiological data 辐射数据
radiological hazard 辐射危害
radiology 放射学
radiometer 辐射计,放射计
radiophone 无线电话
radiotelephony 无线电话
radiotherapeutic equipment 放射疗法设备
radiotherapy 放射疗法
radio-wave frequency 射频
radium (Ra) 镭
radius 半径
radius of base circle 基圆半径
radius of curvature 曲率半径
radius of roller 滚子半径
radon (Rn) 氡,镭射气氡
rain-proof instrument 防雨式仪表
raise 升高
raised planter 高身花槽
rake angle 前角
rake face 前刀面
rake moment 制动力矩
RAM (random access memory) 随机存取存储器
Raman scattering light 拉曼散射光
Raman spectrophotometer 拉曼分光光度计
ramp 斜坡,滑行台,斜路,坡道
ramp delay circuit 斜坡延时电路
ramp error constant 斜坡误差常数
ramp function 斜坡函数
ramp function response 斜坡函数响应
ramp input 斜坡输入
ramp response 斜坡响应
ramp response time 斜坡响应时间
random 随机,随机的
random access 随机存取
random access device 随机存取装置

random access memory (RAM) 随机存取存储器
random access storage 随机存取存储器
random disturbance 随机扰动
random drift 随机漂移
random error 随机误差
random field 随机领域
random function 随机函数
random input 随机输入
random inspection 随机检查,随机检验
random media 随机介质
random noise 随机噪声
random number 随机数
random number generator 随机数字发生器,随机数生成程序
random perturbation 随机扰动
random process 随机过程,随机程序
random search 随机搜寻
random signal 随机信号
random telegraph noise 随机电报噪声
random uncertainty 随机不确定度
random variable 随机变量
randomizer 随机函数发生器
random-wound 散绕
range 范围,量程,值域,区域,距离
range data 数据范围
range finder 测距仪
range of disturbance 扰动幅度
range of regulation 调节范围
range of set value 设定值范围
range sensor 距离传感器
range splitting 区间分化
rangeability 上限值范围,可调范围
range-sweep generator 距离扫描发生器
rank 排列
rapid deterioration 迅速变质
rapid hardening 快硬,速凝
rapid hardening cement 快硬水泥
rapid harding Portland cement 快硬硅酸盐水泥
rapid programming 快速编程
rapid transit system 快速运输系统
rapid-scan monochromator 快速扫描单色仪

rapping rod 起模杆
rasp 粗锉，木锉
ratchet 棘轮
ratchet mechanism 棘轮机构
rate 比率，微分
rate action 比率作用
rate constant 比率常数
rate control 速率控制，微分控制
rate control action 比率调节作用，微分调节作用
rate feedback 速率反馈
rate gain 微分增益，比率增益
rate integrating gyro 速率积分陀螺
rate of change limiting control 变化率限值控制
rate of sampling 取样比率
rate time 微分时间，预调时间
rate 速度，速率
rated 额定的，标称的，比率的
rated capacity 额定容量，设计效率，额定功率
rated current 额定电流，规定电流
rated efficiency 额定效率
rated load 额定负载
rated output 额定输出，额定产量，额定功率
rated power 额定功率，额度输出
rated power factor 额定功率因数
rated speed 额定速度，额定转速
rated torque 额定转矩，额度扭矩
rated value 额定值
rated voltage 额定电压
rate-of-change filter 变速率滤波器
rating 额定功率，定额，评级
rating life 额定寿命
ratio 比例，比，比率，变比
ratio control 比例控制，比值控制
ratio control algorithm 比例控制算法
ratio control roughing 变滚比粗切
ratio control system 比值控制系统，比率调节系统
ratio controller 比率调节器，比率控制器
ratio gain 比率增益
ratio of winding 匝数比
ratio rate 脉冲比率
ratio set 比值设定

ratio set unit 比率设定器
ratio station 比值操作器
ratio transmitter 比值变送器
rational matrix 有理矩阵
rattler test 磨耗试验
raw data 原始数据
raw materials 原料
raw sewage 未经处理的污水，未经净化的污水
raw water 生水，未经净化水
ray 光线，射线
RBF (radial basis function) 径向基函数
RC oscillator RC振荡器，阻容振荡器
RC (reserved controller) 备用控制器
RCD (reserved controller director) 备用控制器指挥器
RCM (redundancy control module) 冗余化控制模块
reach ability 能达性
reactance 电抗
reactance amplifier 电抗耦合放大器
reaction coupling 电抗耦合
reaction curve 反应曲线
reaction kinetics 反应动力学
reaction motor 反应式电动机
reaction wheel control 反作用轮控制
reactive 无功的，电抗性的
reactive circuit 电抗电路
reactive component 电抗成分，虚数部分，无功部分
reactive current 无功电流，电抗性电流
reactive current component 无功电流分量
reactive current generator 无功电流发生器
reactive element 无功元件，电抗元件
reactive factor 无功功率因数
reactive in respect to 相对呈感性
reactive iron 电抗铁（附加在变压器或电抗器中，以加大电抗）
reactive kilovolt-ampere-hour meter 无功电度表
reactive load 无功负载，电抗负载

reactive loss　无功损耗
reactive power　无功功率
reactive power compensation　无功补偿
reactive power factor　无功功率因数
reactive voltage component　无功电压分量
reactor　电抗器
reactor control　反应器控制
reactor starting　电抗器启动
read　读，读出
read frequency input　读频输入
read in　读入
read only memory (ROM)　只读存储器
read write check indicator　读写检验指示器
read out　读出，结果传达
readily combustible　随时可燃烧
readily removable cover　易于移走的封盖
reading error　读数误差
reading　读数
readout unit　读数装置
ready-mix plant　混凝土厂房
reagent　试剂
reagent bottles　试剂瓶
real alarm　实时报警
real axis　实轴
real component　实时部分
real part　实部，实数部分
real time journal　实时日志
real time telemetry　实时遥测
real time trend　实时趋势
real power　有效功率
real time　实时
realignment　重新定线
realizability　可实现性，能实现性
realization　实现，实践计划
realization theory　实践论
real-time　实时，实时的
real-time AI　实时人工智能
real-time clock　实时时钟
real-time communication　实时通信
real-time computer　实时计算机
real-time computer system　实时计算机系统
real-time digital correlator　实时数字相关器

real-time expert system　实时专家系统
real-time harmonic analyzer　实时谐波分析仪
real-time language　实时语言
real-time network operating system　实时网络操作系统
real-time operating system　实时操作系统
real-time optimizer　实时优化程序
real-time system　实时系统
real-time task　实时任务
reamer　铰刀，扩孔钻
reaming　铰孔加工
rear　后端，后方，背部
rear plate　后板
reassembling　重新装配
rebate of glazing　镶嵌玻璃槽口
rebate plane　边刨，槽口刨，子口刨
rebuild　重建
recalibrate　重新校准
recalibration　重新校准
recasting　重铸
receive　接收，领取
receive tank　回收箱，接收箱
receiver　接收机，收音机，接收器
receiver element　接收机单元
receive-side　受端
receiving　接收的
receiving channel　接收通道
receiving circuit　接收电路
receiving selsyn　自动同步接收机，自整角接收机
receiving wave range　接收波段
receptacle　容器，盛载器，插座
reception　接收
reception tank　接收池
receptor　接收器
recess　凹位，凹槽，凹进处
recess action　啃出
recharge　再充电
recipe　诀窍，配方，制法，方法，处方
recipe data　诀窍数据
recipe detail display　配方细目显示画面
reciprocating compressor　往复式压缩

机，活塞式压缩机
reciprocating follower　移动从动件
reciprocating motion　往复移动
reciprocating piston air compressors　往复活塞空气压缩机
reciprocating piston flowmeter　往复活塞流量计
reciprocating pump　往复泵，往复式泵
reciprocating screw　往复螺杆
reciprocating seal　往复式密封，往复运动唇形密封圈
reciprocity theorem　互易定理，倒易理论
recirculating pump　循环泵
recirculation　再循环
recirculation damper　再循环阻尼器
reclosing　重合闸
recognition　识别
recognized standard　认可标准
recognizer　识别器
recommended spare parts list　推荐的备件清单
recommission　再度投入运作，重新校验
reconciliation of process data　过程数据调理
recondition　修复，修整
reconditioning　再调质
reconnection　重新接驳
reconstruction　重新建造
reconstruction of continuous signal　连续信号重组
reconstructive chromatogram　重建色谱图
record　记录
recorder　记录仪，录音机，记录装置，记录器
recording　记录，录音，唱片
recording alarmer　记录式报警器
recording ammeter　记录式电流表
recording analyzer　记录分析仪
recording channel　录音通道
recording code　记录编码
recording device　记录装置
recording head　记录头，录音磁头
recording instrument　记录仪表，自动记录仪表
recording media　记账依据，记录工具
recording media noise　录音介质噪声
recording medium　记录介质，记录纸，记录装置
recording pen　记录笔
recording performance　记录性能
recording property technology　记录技术
recording voltmeter　记录式电压表
recovery　补偿，还原，回复，恢复，再生
recovery circuit　补偿电路
recovery tank　回水箱
recovery time　补偿时间，恢复时间
recovery vehicle　维修车辆
recruitment　补充，复原
recrystallization　再结晶
rectangle　方形，矩形
rectangle filling　矩形填充
rectangular　矩形的
rectangular aluminum wire　矩形铝线
rectangular handrail　矩形扶手
rectangular robot　直角坐标型机器人
rectangular wave　矩形波，方形波
rectangular wave transform　方波变换
rectangular waveguide　矩形波导
rectification　修整，纠正，整流，检波，调整
rectification circuit　整流电路，检波电路
rectified　已整流的，已检波的
rectifier　整流器，检波器，整流管，矫正器
rectifier doubler　倍压整流器
rectifier element　整流器元件
rectifier filter　整流滤波器，平滑滤波器
rectifier instrument　整流式仪表
rectifier photocell　整流光电管
rectifier stage　整流级
rectifier transformer　整流器用变压器
rectifier type instrument　整流式仪表
rectiformer　整流变压器
rectifying　整流，求曲线长

rectifying action 整流作用
recursive 递推，递归
recursive algorithm 递推算法，递归算法
recursive control algorithm 递推控制算法
recursive digital filter 递推数字滤波
recursive estimation 递推估计
recursive filter 递推滤波器
recursive least square (RLS) 递推最小二乘，递归最小二乘
recycling elution 再循环洗脱，循环洗脱
red lead 红丹
red oxide primer 红氧化铁底漆
red shift 红移
redevelopment 重建，重新发展
reduced order observer 降阶观测器
reduced-order model 降阶模型
reducer 还原剂，异径管
reducing valve 减压阀
reductant 还原性介质
reduction 减少，简化，还原，缩小，降低
reduction factor 减缩因数，换算系数
reduction gear 减速齿轮，减速装置
reduction gearbox 减速箱
reduction ratio 减速比
redundancy 冗余，冗余化
redundancy control 冗余控制
redundancy control module (RCM) 冗余化控制模块
redundancy reduction 冗余简约
redundant 冗余，多余
redundant constraint 虚约束
redundant control module 冗余控制模件
redundant degree of freedom 冗余自由度
redundant information 冗余信息
redundant manipulator 冗余操纵器
reel stretch 卷圆压平
reel-stretch punch 卷圆压平冲子
re-energization 重供电
re-entrant mold 凹入模，倒角式模具
reentry control 再入控制
reference 参考，基准，依据，坐标，

参数，基准端，查询
reference accuracy 基准精确度
reference adaptive control 参考自适应控制
reference architecture 参考结构
reference circle 分度圆
reference cone 分度圆锥
reference direction 参考方向
reference electrode 参比电极
reference generator 参考信号发生器
reference input 参考输入
reference input element 参考输入环节，参考输入元件，基准输入元件
reference input signal 参考输入信号
reference input variable 参考输入变量
reference instrument 标准仪表，校正用仪表
reference junction 基准结
reference junction compensator 基准结补偿器，冷端补偿器
reference level 参考电平，基准电平，基准液面
reference line 分度线
reference mark 基准记号
reference material 参比物
reference meter 基准尺
reference node 参考节点
reference operating condition 参考工作条件，参比工作条件
reference performance 参考性能，参比性能
reference phasor 参考向量
reference point 参考点，基准点
reference signal 参考信号，参比信号，标准信号
reference standard 参考标准，参考标准器
reference trajectory 参考轨迹
reference value 参考值
reference variable 基准变量，参考变量
reference voltage 参考电压，基准电压
reference winding 参考线圈，参考绕组
reference performance characteristic 参比性能特性
reference-input variable 参考输入变量
reference-value scale of a quantity 量

的基准值标度，量的参考值标度
reference-value standard　参考值标准器
re-fill　重新注满
refilling　回填，再填
reflector　反射器，反光罩
reflux valve　回流阀
reflux　倒流，回流
refract　折射，曲折
refracted　折射的
refraction　折射
refractive index detection　示差折光检测，折光率检测
refractive index detector（RID）　示差折光检测器，折光率检测器
refractory lining　耐火内衬
refrigerant　致冷剂，制冷剂
refrigerant circuit　致冷剂回路，制冷剂回路
refrigeration and cryogenic engineering　制冷及低温工程
refrigeration compressor　致冷压缩器，制冷压缩机
refrigeration plant　冷藏设备，冷凝装置
regeneration　再生
regenerative amplifier　再生式放大器
regenerative braking　回馈制动，再生制动
regenerative feedback　再生反馈，正反馈，正回授
regenerative receiver　再生式收音机
region　区域
region filling　填充区
regional network　区域电力网，区域网络
regional planning model　区域规划模型
regional power station　区域发电厂
register　登记册，记录器，寄存器
register allocation　寄存器配置
regression　回归
regression algorithm system　回归算法系统
regression analysis　回归分析
regression estimate　回归估计
regression relationship　回归关系
regrinding　再次研磨

regularization　正则化
regulate　调节，控制，校正，管理
regulated　已调节的，已调整的
regulated power supply　稳压电源，稳定电源
regulated value　调节值
regulated variable　调整变量
regulating　调节，控制，调节的，控制的
regulating amplifier　调节放大器
regulating apparatus　调节器，调节装置
regulating characteristic　调节特性
regulating circuit　调节电路
regulating device　调节设备
regulating element　调节元件，调节环节
regulating energy　调节能量
regulating loop　调节回路
regulating mechanism　调节机构
regulating motor　调节电动机
regulating range　调节范围，控制范围
regulating relay　调节继电器
regulating resistance　调节电阻
regulating spring　调节弹簧
regulating system　调节系统，自动控制系统
regulating transformer　调节变压器
regulating unit　调节单元
regulating valve　调节阀，溢流阀，控制阀
regulating voltage　调整电压，调节电压
regulation　调节，控制，调整率，规程，整顿，规则
regulation of line voltage　线电压调整率
regulation of output　输出调整
regulation PV　调节PV点
regulation PV data point or algorithm　常规PV数据点或算法
regulation PV point　调节PV点
regulation tube　稳压管
regulator　调节器，自力式控制器，自力式调节器
regulator control　调节器控制
regulator governor　调速器
regulator theory　调节器理论

regulators for gas cylinders used in welding cutting and allied processes 焊接、切割及类似工艺用气瓶减压器
regulatory control 调节控制点，常规控制
regulatory control data point 调节控制数据点
regulatory control data point or algorithm 常规控制数据点或算法
regulatory control function block 调节控制功能块
regulatory control point 调节控制点
regulatory sign 管制标志
rehabilitation 修复
reinforced concrete 钢筋混凝土
reinforced concrete cover 钢筋混凝土封盖
reinforced excitation 强行励磁
reinforcement 钢筋，加固，加强
reinforcement bar 钢筋条
reinforcement of weld 加强焊接
reinforcement wire 预应力钢丝
reinforcing agent 增强剂
reinforcing material 增强材料
reinforcing wire 钢筋丝，增力丝，加固镍
reinstatement 回复原状，恢复原貌
reinstatement of supply 恢复供应
reinstatement works 修复工程
reject 拒收
rejection 拒绝，丢弃
rejective amplifier 抑制放大器
related display panel 相关显示画面
related properties 有关特性
relational algebra 关系代数
relational database 关系数据库
relational expression 相关表达式
relational data 关系型数据
relationship 关系
relative 相关的
relative abundance 相对丰度，相对强度
relative average deviation 相对平均偏差
relative damping 相对阻尼
relative density 相对密度
relative displacement 相对位移，相对排量

relative error 相对误差
relative gain 相对增益
relative gain array（RGA） 相对增益阵
relative gap 相对间隙
relative humidity 相对湿度
relative humidity sensor 相对湿度传感器，相对湿敏组件
relative motion 相对运动
relative movement 相对运动
relative radius of curvature 相对曲率半径
relative retention value 相对保留值
relative Rf value 相对比移值
relative stability 相对稳定性
relative standard deviation（RSD） 相对标准差
relative value 相对值
relative velocity 相对速度
relativistic 相对论
relaxation 松弛
relaxation analysis 松弛分析
relaxation circuit 张弛电路，弛张电路
relaxation frequency 弛豫频率
relaxation mechanism 弛豫历程
relaxation oscillation 弛张振荡，弛豫振荡
relaxation oscillation frequency 衰减振荡频率
relaxation oscillator 弛张振荡器
relay 继电器
relay characteristic 继电器特性
relay control 继电器控制
relay I/O module 继电器输入/输出模件
relay line 中继线
relay logic card 继电器逻辑卡
relay protection 继电保护装置
relay selsyn motor 中继自整角机
relay-contactor control 继电器-接触器控制
relaying protection 继电保护
release agent 脱模剂
release nozzle 泄放喷嘴
release spring 回位弹簧
release valve 泄气阀，放气阀
release wire 释放线
release 释放

releasing 释放，解除，松释动作
releasing of relay 继电器释放
relevant experience 有关经验
reliability 可靠性，可靠的
reliability analysis 可靠性分析
reliability design 可靠性设计
reliability engineering 可靠度工程
reliability evaluation 可靠性评估
reliability test system 可靠性测试系统
reliability theory 可靠性理论
reliable 可靠
relief sewer 溢流污水渠
relief valve 溢流阀，安全阀
relief 去载，卸载，释放，解除
relieving arch 减压拱
relieving beam 减压梁
relocation 迁移
reluctance 磁阻
reluctance generator 磁阻发电机
reluctance motor 磁阻电动机
remaining slag 剩余熔渣
remaining time 保留时间
remaining works 余下工程
remanence 剩磁，剩磁感应强度
remanence magnetism 剩磁
remark 备注
remedial measure 补救措施
remedial works 补救工程
remedy 补救
remote 远距离的，模糊的
remote control 遥控，远距离控制
remote controlling 遥控
remote display protocol 远程显示协议，远程显示规程
remote manipulator 遥控操作器，远程控制器
remote manual loader 远方手动操作器
remote measuring float level meter 浮标式遥控测液位计
remote notify 远程通告
remote regulating 遥调，远距离调节
remote set 远方设定
remote set controller 远方设定控制器，远方设定调节器
remote set point adjuster 远程设定点调整器

remote terminal panel 远程端子板
remote variable 遥控变量，遥测变量
remote-control apparatus 遥控设备
remote-control equipment 遥控设备
remote-control system 遥控系统
remote-controlled 遥控的
remote-controller 遥控器
remote-indicating instrument 遥测指示仪表
remote-indicating rotameter 遥示转子流量计
remoulding 重塑，改造
removable balanced mixer 可卸式平衡混频仪
removable media 可移动介质，盒式盘或软盘
removal 移去，删除
remove 除去，拆卸
rename a file 重新命名一个文件
rendering technique 演绎技术（设计）
renewable energy system 可再生能源系统
renewal 更新，续期，更换
renovate 翻新，整修
repaint 重新漆
repair 修理，修葺，修补
repair paint 补用漆
repairer 修理工，检修工
repeatability 重复性
repeatability error 重复性误差
repeatability of measure 测量的重复性
repeated fluctuating load 交变载荷
repeated load 交变载荷，重复载荷
repeated stress 交变应力
repeater 中继器，中间继电器，中间继动器
repeating alarm 重复报警
repetitive shock 重复冲击
replace 重新，启动，更换，替换
replacement parts 备件，替换零件
report 报告，报表，记录
report archiving 报告存档
report scheduler 报告调度程序
reporting functions 报告功能
represent 表示，代表

representative alarm unit 代表报警器
reproducibility 可再生性，可重复性，再现性
reproducibility of measurement 测量的再现性
reproducible 可再生的
repulsion motor 推斥电动机
repulsion type meter 推斥式仪表
request 请求
request letter 请求函
rerouting 重布
rescue equipment 拯救设备
research direction 研究方向
reserve 专用范围，保留地，保护区，备用，预备，储备
reserve controller directory 保留控制器目录
reserve equipment 备用设备
reserve supply 备用电源
reserve parts 备件
reserved controller (RC) 备用控制器
reserved controller director (RCD) 备用控制器指挥器
reserved entity 专用实体
reserved 备用的
reservoir 储气缸，水库，储水池
reset 复归，返回，复位，重置
reset action 重定作用
reset button 复原按钮，重置键
reset control action 积分调节作用
reset controller 积分调节器，积分控制器
reset key 复位键，回原键
reset time 积分时间，回复时间，再调时间，重定时间
reset value 重新设定值
reshaping 整形
residual deflection 残余偏转
residual strength 剩余强度
residual stress 残余应力
residue 滤渣，残渣，残留物，余数，剩余
residue feedback 残差反馈
residue number system 余数系统
residue on ignition 炽灼残渣
resilience 回弹性
resilient mounting 弹性机垫，柔性支撑
resilient pad 弹性垫片
resin 树脂
resin coated copper foil 涂胶脂铜箔，涂树脂铜箔，背胶铜箔
resin emulsion paint 树脂涂料，树脂漆
resin injection 树脂射出法
resin insulator 树脂绝缘器
resin mortar 树脂胶浆
resin streak 树脂流纹
resin wear 树脂脱落
resinoid bond 树脂胶合
resinoid grinding wheel 半树脂型砂轮
resist 阻抗，抗蚀护膜
resistance 电阻，阻尼，阻力，抵抗力，阻抗，阻抗力
resistance box 电阻箱
resistance bulb 测温电阻器
resistance coefficient 阻抗系数，电阻系数
resistance coupling 电阻耦合
resistance furnace 电阻炉
resistance loss 电阻损耗
resistance range 电阻范围，电阻量程
resistance temperature bulb 电阻温度球
resistance temperature detector 电阻温度检查器
resistance thermometer 热电阻，电阻温度计
resistance thermometer assembly 热电阻组件
resistance thermometer sensor 热电阻传感器，热电阻
resistance to electronic transmitter 带阻尼的电子变送器
resistance to pneumatic transmitter 带阻尼的气动变送器
resistance type potential divider 电阻分压器
resistance 电阻，阻抗
resistance-tuned oscillator 电阻调谐振荡器
resistive 有抵抗力的
resistivity 安定性，电阻率

resistivity test 电阻测试
resistor 电阻,电阻器
resolution 分辨率,分辨力,分辨能力,决定分解,决议分解,分离度
resolution in analog-to-digital 模-数转换分辨率
resolution principle 归结原理
resolved 分解,分析,决心
resolved gain measurement 分解增益测量
resolvent matrix 求解矩阵
resonance 谐振,共振
resonance amplifier 调谐放大器
resonance circuit 谐振电路
resonance curve 谐振曲线
resonance frequency 谐振频率
resonance frequency wavemeter 谐振波长计
resonance state 谐振状态
resonant 谐振的
resonant circuit 谐振电路
resonant frequency 共振频率,谐振频率
resonant type instrument 谐振式仪表
resonanting 谐振的,调谐
resonate 共振
resource allocation 资源分配
resource plan 资源计划
resource usage status display 资源利用状态显示
respirable suspended particulate 可吸入的悬浮颗粒
respirator 呼吸机
response 响应,反应
response curve 响应曲线,应答曲线
response function 响应函数
response measurement 响应测量
response time 响应时间
responsible department 负责单位
rest position 止动位置
restart 再启动,重新启动
restart option 重新启动选项
restoration works 修复工程
restore 修复
restore archive file 恢复归档文件
restore archived definition 恢复归档定义

restore event data 恢复事件数据
restore event descriptor 恢复事件描述
restore saved archive 恢复已存档案
restricted instruction set 约束指令集
restricted manual access 限制手动存取
restricted tender 局限性投标
restrictor valves 过流阀,节流阀
restriking 二次精冲加工,整形锻压
result 结果
result of measurement 测量结果
resultant bending moment 合成弯矩
resultant force 合力,总反力
resultant moment of force 合力矩
resultant moment of inertia 惯性主矩
resultant movement of cutting 合成切削运动
resultant movement of feed 合成进给运动
resultant vector of inertia 惯性合成向量
retained austenite 残留奥氏体,残留沃斯田铁
retainer pin 开口销,固定销,嵌件销
retainer plate 托料板
retaining circuit 保持电路
retardation 减速
retardation network 迟延网络
retardation of phase 相位滞后
retarder 缓凝剂,减速剂
retarding action 延迟动作
retarding torque 制动转矩
retention money 保证金
retention time 保留时间
retention volume 保留体积
retentive ferromagnetic material 顽磁性铁磁材料
retentive material 硬磁材料
retentivity 剩磁,保特性
retighten 重新紧固
retract 可伸缩的,缩回
retrieve data 回线数据
retroactive 再生的,反馈的
retroactive amplification 再生放大,反馈放大
return 回程,反射,回复,返回
return air inlet 回风进口,回气进口

return difference 回差
return difference matrix 回差矩阵
return difference ratio 回差比
return period 重现期
return pin 回位销,反顶针
return ratio matrix 回比矩阵
return signal 反射信号,返回信号
return spring 回位弹簧
return transfer function 返回传递函数
return oil 回油
reuse engineering data 重新使用工程数据
reverberation 回响
reversal of phase 倒相
reversal of pole 极性变换
reverse 反向的,反的,倒的,倒换
reverse acting actuator 反作用执行机构
reverse acting controller 反作用控制器,反作用调节器
reverse action 反作用
reverse angle 倒角,反角度
reverse control wire 反位控制线
reverse current circuit breaker 反流断路器
reverse current cut-out 反向电流自动断路器
reverse rotation 反转
reversed bias 反偏压,反向偏压,逆向偏压
reversed blanking 反转下料
reversed feedback amplifier 负反馈放大器
reversed phase 反相
reversed phase liquid chromatography 反相液相色谱法
reversed phase thin layer plate 反相薄层板
reverser 反向器,换向开关
reversibility 可逆性
reversible 可逆的,可反向的
reversible controller 双向控制器
reversible electric drive 可逆电气传动
reversible motor 可逆电动机
reversible system 可逆系统
reversible water meter 可逆式水表

reversing 反向
reversing controller 可逆控制器
reversing motor 可逆电动机,双向电动机,双向旋转电动机
reversing starter 可逆启动器
reversing switch 反向开关,换向开关
reversion 反向,倒换,回复
revetment 护岸墙,护墙,护坡
revex 直齿锥齿轮粗拉法
revision 版本
revolute joint 转动关节
revolute pair 转动副
revolute robot 关节型机器人
revolution 动圈,回转体
revolution indicator 转数指示器,转速计
revolution per minute 每分钟转数
revolution speed transducer 转速传感器
revolution switch 旋转开关
revolutions per minute 每分钟转数,转/分
revolutions per second 转/秒
revolving 旋转的
revolving door 旋转门
revolving magnetic field 旋转磁场
revolving rheostat 旋转变阻器
revolving shaft 转轴
rewinding 重新绕线
rewriting rule 重写规则
Reyleith scattering light 瑞利散射光
Reynolds's equation 雷诺方程
RF generator 射频发生器
RF millivoltmeter 射频毫伏计
Rf value 比移值
RGA (relative gain array) 相对增益阵
rhenium (Re) 铼
rheochord 滑线变阻器
rheology 流变学
rheometer 电流计,血流速度计,流变仪
rheostat 变阻器
rheostatic braking 电阻制动,变阻器制动
rheostatic control 变阻调整,变阻控制
rhodium (Rh) 铑
rhythm 韵律,节奏

rib working 肋部加工
ribbon cable 带状电缆
ribbon iron 带钢［铁］
ribbon punch 压筋冲子
Riccati 黎卡提
Riccati equation 黎卡提微分方程
RID (refractive index detector) 示差折光检测器
right triangle 直角三角形
right 右
right-hand rule 右手定则
right-hand screw rule 右手螺旋定则
right-left bearing indicator 左右偏位指示器
rigid bearing 刚性轴承
rigid body 刚体
rigid conduits 刚性导管
rigid coupling 刚性联轴器
rigid double-sided printed board 刚性双面印制板
rigid frame 刚性构架
rigid impulse 刚性冲击
rigid joint 刚性接缝
rigid matrix electrode 刚性基质电极
rigid metallic pipework 硬金属制的喉管
rigid multilayer printed board 刚性多层印制板
rigid plastic 硬性塑胶
rigid polyvinyl chloride pipe 硬质聚氯乙烯管
rigid printed board 刚性印制板
rigid rotor 刚性转子
rigid shock 刚性冲击
rigid single-sided printed board 刚性单面印制板
rigid spacecraft dynamics 刚性航天动力学
rigid steel conduit 钢制电缆管
rigid-flex double-sided printed board 刚性双面印制板
rigid-flex multilayer printed board 刚性多层印制板
rigid-flex printed board 刚性印制板
rigidity 刚度，硬度
rigidity criterion 刚度准则
rim 周围，缘

ring 圈，环
ring balance manometer 环称压力计，环平衡式压力计
ring circuit 环形电路
ring current 环流
ring gate 环形浇口
ring gauge 环规
ring gear 内齿圈，大轮，环形齿轮
ring magnetic circuit 环形磁路
ring size effect 环大小效应
ring wire 塞环线
rinse 水洗
RIO bus RIO 总线
RIO bus arrangement RIO 总线装配
ripple 脉动，波纹
ripple factor 脉动系数，波纹系数
rippling 振纹
rise 升程，上升
rise time 上升时间
riser 立管，竖管，梯级高度，踢脚，竖板
rising main 上行水管，上行电缆，泵送干管
rising stem 明杆，升杆
risk 危险
rivet 铆钉，窝钉
rivet gun 拉钉枪，铆钉枪
riveting die 铆合模
riveting machine 铆钉机
RL bus RL 总线
RLS (recursive least square) 递推最小二乘，递归最小二乘
Rm value 保留常数值
RNA 核糖核酸
road lighting lantern 路灯具
road lighting pillar box 路灯电箱
robot 机器人，机械手
robot arm 机械臂
robot calibration 机械手校准
robot control 机器人控制
robot dynamics 机器人动力学
robot kinematics 机器人运动学
robot navigation 机器人导航
robot programming 机器人编程
robot programming language 机器人编程语言，机器人程序设计语言
robot vision 机器人视觉，电脑视觉

robotic manipulator 机器人操纵器
robotics 机器人学
robust 鲁棒
robust control 鲁棒控制,强健控制
robust design 稳健设计
robust estimation 鲁棒估计,抗差估计,稳健估计
robust estimator 鲁棒估计器
robust performance 鲁棒性能
robust stability 鲁棒稳定
robust system 鲁棒系统
robust transmission 鲁棒变换
robustness 鲁棒性,耐用性
rock 岩石
rock anchor 石锚
rock bolt 石栓,岩层锚杆
rocker 摇杆
rocker arm 摇臂,摇杆
rocker switch 摇臂开关,摇杆式开关,提杆开关,跷板开关
rockfill 碎石,填石
rocking die forging 摇动锻造
Rockwell apparatus 洛氏硬度计
Rockwell hardness 洛氏硬度
Rockwell hardness test 洛氏硬度试验
rod 棒
rodding 以棒条通渠
rodding eye 通管孔,通渠孔
roentgen 伦琴
roll bending 滚筒弯曲加工
roll centering 滚动定心
roll chart 卷纸,卷筒记录纸
roll chart drive unit 卷纸机构
roll finishing 滚压加工
roll forging 轧锻
roll forming machine 辊轧成形,辊轧成形机
roll gap measuring instrument 辊缝测量仪
roll material 卷料
roll release 脱辊
roll surface temperature element 辊面温度元件
rolled annealed copper foil 压延退火铜箔
rolled copper foil 压延铜箔
rolled hardcore 经碾压碎石底层

rolled steel 盘条,铁丝盘辊轧钢丝,卷铁丝,轧制钢
rolled steel channel 轧制钢槽
rolled surface 轧制表面
rolled wire 盘条,铁丝盘辊轧钢丝,卷铁丝
roller 滚轴,滚子,滚筒,辊子,压路机
roller bearing 滚子轴承
roller chain 滚子链,滚柱链条
roller clutch 滚柱式单向超越离合器
roller conveyor 滚柱式输送机
roller electrode 滚子电极,滚轮式电极,辊式电极
roller feed 辊式送料
roller follower 滚子从动件
roller member pivot 辊构件枢轴
roller shutter 卷闸
rolling 滚压,辊压,滚轧,压延加工
rolling bearing 滚动轴承
rolling bearing identification code 滚动轴承代号
rolling element 滚动体
rolling friction 滚动摩擦
rolling velocity 滚动速度
ROM (read only memory) 只读存储器
roof cover 上盖
roof load 屋顶荷载
roof tank 天台贮水箱
root 根
root angle 安装角,齿根圆锥角,伞齿轮底角
root angle tilt 齿根角倾斜
root apex 根锥顶
root apex beyond crossing point 根锥顶至相错点的距离
root apex to back 根锥顶至安装基准面距离
root circle 齿根圆
root cone 根锥
root diameter 齿根直径
root line 齿根线
root loci 根轨迹
root locus 根轨迹
root locus diagram 根轨迹图
root locus method 根轨迹法
root locus segments on the real axis 实

轴上的根轨迹段
root mean square 均方根
root mean square value 均方根值
root method 根轨迹法
root radius 根部半径,齿根半径
root run 根部焊道,焊根焊道
root sensitivity 根灵敏度
root surface 齿根曲面
rope 缆绳,绳索
rope anchorage 缆索紧固锚
rope clamp 绳夹,索夹
rope drum groover 缆索卷筒槽
rope guide 导绳器
rope round pulley 绕绳滑轮
rope stranded wire 多股绞合线
rope suspended current meter 悬索式流速计
rope thimble 绳端套环,绳索套环
Rossi-Peakes flow test 罗西-皮克斯流动试验
rotameter 转子流量计,转式测速仪
rotary bender 卷弯成形机,回转弯曲机
rotary DC resistance box 旋转式直流电阻箱
rotary displacement type flowmeter 旋转式容积流量计
rotary drum 滚筒
rotary drum screen 滚筒筛滤器
rotary eccentric plug valve 偏心旋转阀
rotary evaporators 旋转式汽化器
rotary forging 回转锻造
rotary motion 旋转运动
rotary motion valve 角行程阀
rotary phase converter 旋转式相位变换器,旋转式变相机
rotary pump 旋转泵,回转泵,旋式泵
rotary surface 旋转面
rotary switch 旋转开关
rotary vane flowmeter 旋转叶片式流量计
rotary vane type vacuum pump 旋片式真空泵
rotary vane water meter 翼轮式水表
rotary water meter 旋转式水表

rotary switch 转换开关
rotating commutator 旋转整流子,旋转换向器
rotating disk 旋转磁盘,旋转碟片
rotating machine 旋转电机
rotating magnetic field 旋转磁场
rotating phasor 旋转相量
rotating seal 旋转式密封,回转式密封装置
rotating speed revolution 转速
rotating transformer 旋转变压器
rotating vane type cold water-meter 旋翼式冷水水表
rotating vector 旋转矢量
rotating 旋转
rotating-coil variometer 动圈式变感器
rotation 旋转,转动,回转
rotational molding 旋转模塑,旋转成型
rotor 转子
rotor circuit 转子电路
rotor coil 转子线圈
rotor core 转子铁芯,转子铁心
rotor frequency 转子频率
rotor generator 转子发动机
rotor induction motor 转子感应电动机
rotor leakage reactance 转子漏磁电抗
rotor loss 转子损耗
rotor resistance 转子电阻
rotor starter 转子启动器
rotor winding 转子绕组
rotor with several masses 多质量转子
rough 毛坯
rough machining 粗切削,粗加工
rough sand 粗砂
rougher 粗切机,粗轧机座
roughing 粗加工
roughing forge 粗锻
roughness coefficient 粗糙系数,糙度系数
round belt 圆带
round belt drive 圆带传动
round core wire 圆形心线
round nuts 环形螺帽,圆螺母
round pad 圆形盘
round punch 圆冲子
round steel 圆钢

rounding 圆形加工
rounding chamfer 倒角
round-off errors 舍入误差，进位错误（指计算机将十进制数和二进制数互转时产生的错误）
round-off noise 舍入噪声
Routh approximation method 劳斯近似判据
Routh stability criterion 劳斯稳定判据
Routh tabulation 劳斯表
Routh-Hurwitz criterion 劳斯-赫尔维茨判据
routine 例行的，日常的
routine inspection 日常检查，日常检测
routine maintenance 日常维护
routing 路由选择，工艺路线，选择途径
routing algorithm 程序算法
routing problem 路径问题
rovings 无捻粗纱
RSD (relative standard deviation) 相对标准差
RSR 弧齿锥齿轮条形刀齿铣刀盘
RTD input signals 热电阻输入信号
rubber anti-vibration mounting 橡胶避震垫
rubber boot 橡胶支座
rubber bump stop 橡胶缓冲器
rubber gasket 橡胶垫圈
rubber graphite board 橡胶石墨板
rubber grommet 橡胶孔环，橡胶索环
rubber hose 橡胶软管
rubber insulated low voltage lacquer wire 橡皮绝缘低压腊克线
rubber molding 橡胶成形
rubber moulding 胶模
rubber seal ring 橡胶密封环
rubber sealing washer 橡胶密封垫圈
rubber sheathed wire 橡胶外包线
rubber sleeve 橡胶管套
rubber spring 橡胶弹簧
rubber strip 橡胶条
rubber tube 橡胶喉管
rubber 橡胶

rubidium (Rb) 铷
rule 规则
rule-base 规则库
rule-based system 法则系统，规则系统
rule-based control 基于规则控制
run on generic PC 在一般个人计算机上运行
run program 运行程序，执行程序
run 运行
run back 返回
Runge-Kutta integration method 龙格-库塔积分方法
runner 流道
runner assembly 转动组件
runner balance 流道平衡
runner ejector set 流道顶出器
runner lock pin 流道拉销
runner plate 流道模块
runner stripper plate 流道脱料板
runner system 流道系统
runnerless mould 无流道冷料模具，无浇道模
running noise 运行噪声
running online manual viewer 运行在线手动观察窗
running rpm 运行转速
running speed 运行速度
running test 额定负荷试验，运行测试
running time 运行时间
running torque 旋转力矩
runout 径向跳动
runout tolerance 径向跳动公差，偏转公差
run-time system 转动时间系统
rupture pressure 破坏压力
rural distribution wire 农村电话线
rush current 冲击电流
rush current state 冲击电流状态
rust 腐蚀，生锈，铁锈
rust prevention 防锈处理
rust stain 锈渍
rustless iron 不锈钢
ruthenium (Ru) 钌

safe

S

safe 安全的,稳定的,可靠的
safe operation for safety shutdown 安全停车操作
safe voltage 安全电压
safeguard 防护
safety 安全,稳定,可靠
safety actuator 安全执行器
safety alarm 安全报警器
safety analysis 安全分析
safety appliance 防护用具,安全装置
safety control 安全控制,安全保障,安全管制
safety curtain 防护幕
safety device 安全装置,保护装置,保险装置,安全防护装置,过载安全装置
safety goggle 护目镜
safety hardware 安全硬件
safety independent check 安全独立检查
safety interlock system 保险联锁系统,安全联锁系统
safety lamp 安全灯
safety lighting 安全照明
safety of motor vehicle 安全汽车
safety plan 安全工作计划
safety rules for lifting appliances 起重机械安全规程
safety valve 安全阀,安全活门
safety visor 护目面罩
safety wire 保险丝,熔丝
sag 垂度
sagging 松垂
sales and service quality assurance 销售及服务品质保证
salient poles 凸极
saline water 盐水,海水
salt 盐
salt bath quenching 盐浴淬火
salt glazing 盐釉
salt water 盐水,海水,咸水
salt water flushing system 海水冲水系统
salt water pump 海水抽水机,海水泵
salt water pumping station 海水抽水站
salt water resistant material 抗咸水物料
samarium (Sm) 钐
sample 抽样,采样,样品,样本,取样,举例
sample and hold 采样与保持
sample data control system 采样数据控制系统
sample flip flop 取样触发器
sample handling 样品处理
sample handling system 样气处理系统,凝水分离器
sample hold device 采样保持器
sample order 指样订货,试购
sample probe 试样探针,取样探头
sample probe for dilution air 稀释空气取样探头
sample set 样本集合
sample size 抽样检验样本大小
sample space 样本空间,试样空间,取样空间
sample time clock 采样时钟
sample-and-hold amplifier 取样及保持放大器
sampled 采样,取样
sampled data 采样数据
sampled signal 采样信号
sampled-data control 采样数据控制
sampled-data control system 采样控制系统
sampled-data system 数据采样系统
sampler 采样器
sampling 采样,抽样,取样,连续选择,脉冲调制
sampling action 采样作用
sampling control 采样控制
sampling control system 采样控制系统
sampling controller 采样控制器
sampling disturbance 取样扰动

296

sampling element 采样元件，采样环节
sampling frequency 采样频率，取样频率，抽样频率，量化频率
sampling inspection 抽样检查
sampling interval 采样间隔
sampling oscilloscope 采样示波器
sampling period 采样周期
sampling PI controller 采样PI调节器
sampling point 采样点，取样点
sampling pulse 抽样脉冲，选通脉冲
sampling rate 采样速率，取样率
sampling scope 取样显示器
sampling system 采样系统
sampling type turbine flowmeter 取样用涡轮流量计
sand 沙
sand blast 喷砂处理
sand blasting 喷沙，喷沙打磨法
SAP (sideband address port) 边带寻址端口
satellite 卫星
satellite artificial 人造卫星
satellite commumication 卫星通信
satellite control 卫星控制
satellite control application 卫星控制应用
satellite order wire 卫星联络线
saturate 饱和
saturated 饱和的
saturated density 饱和密度
saturation 饱和，饱和度，饱和作用
saturation characteristics 饱和特性
saturation control 饱和控制
saturation current 饱和电流
saturation effect 饱和效应
saturation limiter 饱和限幅器
saturation of controller 控制器饱和
saturation power 饱和功率
saturation range 饱和区
saturation state 饱和状态
save 保存
save archive data 存储归档数据
save archives to tape 存储档案到磁带
save event data 存储事件数据
save event descriptors 存储事件描述
saving to other media 存储到其他媒体
saw blade 锯片
saw dust 锯屑
saw turn-off diode 锯齿波截止二极管
saw 锯
sawing 锯削
sawing machine 机械锯床，电锯
saw-tooth oscillator 锯齿波振荡器
saw-tooth wave 锯齿波
saw-tooth wave generator 锯齿波发生器
SBA (side band addressing) 边带寻址
SC (signal common) 信号公共端，信号公共点
SCADA (supervisory control and data acquisition) 数据采集与监控系统，监测控制和数据采集
scalar 标量，常系数
scalar Lyapunov function 标量李雅普诺夫函数
scale 标度，刻度，标度尺，比例尺
scale disk 刻度盘
scale division 分格，刻度
scale factor 标度因数
scale high value 标度上限值
scale interval 分格值，刻度值，刻度间隔
scale low value 标度下限值
scale mark base 标度基线
scale numbering 标度数字
scale range 标度范围，刻度范围
scale spacing 分格间距
scale value 标度值
scale division 标度分格
scale length 标度长度
scale mark 标度标记
scale spacing 标度分格间距
scaling constant 标定常数
scaling factor 比例因子
scan filling 扫描填充
scan 扫描
scandium (Sc) 钪
scan-fold chart 折纸机构，折叠式记录纸
scan-fold chart drive unit 折纸机构驱动装置
scanister 扫描装置，扫描仪

scanner 扫描器
scanning 扫描
scanning circuit 扫描电路
scanning monitor 扫描监测器
scanning probe microscope 扫描探针电子显微镜
scanning probe microscopy 扫描探针显微镜
scanning switch 换向开关，按序切换开关
SCARA (selective compliance assembly robot arm) 驯服的装配机器人臂
scatter 扩散器，散射器
scatter diagram 散布图
scattered data 分数数据
scattered light 分散光
scattering 消散，散射
scattering holographic interferometer 散射全息干涉仪
scattering light 散射光
scattering parameter 耗散参数
scattering problem 耗散问题
schematic 示意图
schematic diagram 原理图，示意图
schematic display 图形画面
scheme 计划，方案
scheme of external cable and wire connections 电缆电线外部连接系统图
scheme of external pneumatic piping connections 气动管线外部连接系统图
scheme of power supply 供电原理图
Schering bridge 西林电桥
Schottky barrier diode 肖特基势垒二极管
Schottky effect 肖特基效应
scissoring vibration 剪式振动
sclerometer 硬度计
scoop conveyor 戽式输送机
scoring 胶合，刻痕，划痕，凹槽
scoring index 胶合指数
scotch 止动块，制动棒，切口，浅刻痕
scrap cutter 废料切刀
scrap iron 铁屑，废铁
scrap jam 废料阻塞
scrap material 废料

scrap press 废料冲床，废料打包压力机，废料压块压力机
scrap wire 废线
scraper 三角刮刀，刮刀，刮土刀
scratch 刮痕，划痕，抓
scratch hardness 划痕硬度
screeding 沙浆底层，抹平层
screen 遮板，屏障，筛，网罩，屏幕
screen wire 网线
screened cable 屏蔽电缆
screening 屏蔽
screening effect 屏蔽效应
screening wire 屏蔽线
screw 螺旋，螺钉，螺栓，螺杆，螺丝，旋转
screw base fuse 螺旋式熔断器
screw driver 起子，螺丝刀，螺钉起子，螺钉旋具
screw efficiency 螺纹效率
screw joint 螺纹套管接头，螺旋接合
screw mechanism 螺旋机构
screw nut 螺母
screw pile 螺旋桩
screw plug 头塞，螺旋塞，安全塞
screw pumps 螺杆泵
screw rod 丝杠
screw tap 螺旋式水龙头
screw socket 螺口插座
screwdriver holder 起子插座
screws 螺钉
scriber 划针，划线器
scribing 划线
scripting 脚本
scroll down 下卷
scroll up 上卷
scum gate valves 排渣闸阀
s-domain s域
SDS (sodium dodecyl sulfate) 十二烷基磺酸钠
SE (system environment) 系统环境
sea water booster pump 海水增压泵
sea water screen 海水隔滤网
seabed 海床
seal 阀座，密封面，封闭，密封，围封，封口，封料，封孔

seal belt 密封带
seal chamber 密封室,隔离容器
seal gum 密封胶
seal washer 密封垫圈
seal wire 铅封丝
sealant 密封剂
sealing 密封件
sealing arrangement 密封装置
sealing bar 密封棒
sealing compound 电缆膏,封口膏,密封剂,油灰,腻子
sealing fitting 密封接头,密封配件
sealing liquid 隔离液
sealing pot 密封罐
sealing ring 密封圈,密封环
sealing rubber strip 密封胶条
sealing strip 封密条,密封片
sealing tape 密封带
sealing washer 密封垫圈
sealing wax 火漆
sealing wire 铅封丝
seam 缝,裂痕,焊缝
seam welding 缝焊
seaming 接合,折弯重叠加工
seamless 无缝
seamless forging 无缝锻造
seamless steel gas cylinders 钢质无缝气瓶
search 搜索,寻找,查找
search engine 搜索引擎
search method 搜索方法
searching system 检索系统
seating 基座
seawall 海堤,防波堤
seawall opening 海堤通孔
seawall trench 海堤壕沟
seawater 海水
seawater pump 海水泵
seawater pumping station intake 海水抽水站入口
second central moment 二阶中心矩
second derivative action 二阶微分作用
second order determinant 二阶行列式
second order spectrum 二级光谱,二级图谱
second stage annealing 第二段退火

second 秒,第二
secondary access road 支路
secondary air 二次空气
secondary breakdown tester 二次击穿测试仪
secondary control loop 次级控制回路
secondary controller 副控制器,副调节器
secondary gate 次级门路
secondary instrument 二次仪表
secondary loop 二次回路,次级回路
secondary measurement 二次测量
secondary meter 分表,二次测量仪表
secondary reference point 次级基准点,次级参考点
secondary sequence program 次顺序程序
secondary set 副环设定
secondary standard 次级标准器
secondary storage 二级存储器,辅助存储器
secondary treatment 二级处理
secondary treatment of sewage 二级污水处理
secondary voltage 次级电压
secondary winding 次级绕组,复卷绕组
second-order lag 二阶滞后
second-order plus time delay model 二阶加纯滞后模型
second-order system 二阶系统
section 截面,剖面,组,部分,区段
section analysis 截面分析
sectional area 截面面积
sectional axonometric drawing 立体剖面图
sectional die 拼合模,对合模具
sectional elevation 立剖面图,截视立面图,剖视图
sector magnetic field mass spectrometer 扇形磁场质谱计
secular distortion 经年变形
secular equation 特征方程,久期方程
securing dowel 紧固定位销
securing hook 紧固钩

securing nut 紧固螺帽
securing plate 紧固板
securing ring 紧固环
securing rod 紧固连杆
securing screw 紧固螺钉
securing strap 紧固索带
security functions 安全功能
security gate 保安闸，防盗闸
sediment 沉积物
seepage force 渗流力
Seger cone 西格示温熔锥，测温锥
segment 段，一段，扇形齿
segment mold 组合模
segmental orifice plate 圆缺孔板
segmental-blade cutter 大轮精切刀
seismic angular motion sensor 地震角运动传感器
seismic coefficient 地震系数
seismic effect 地震效应
seismic load 地震载重，地震荷载
select switch 选择开关
select switch terminal 选择开关接线端子，选择开关接线板
select 选择
selected ion monitoring 选择性离子监测
selection of sampling period 采样周期选择
selection wire 选择线
selective 选择的，选择性的
selective circuit 选择电路
selective compliance assembly robot arm (SCARA) 驯服的装配机器人臂
selective hardening 局部淬火，选择硬化
selective ion monitoring 选择性离子监测
selective ionization gauge 选择性电离真空计
selective network 选择性网络
selectivity characteristic 选择性，选择度，选择特性
selectivity curve 选择曲线
selector 选择器
selenium (Se) 硒
self 自己，自身
self biased amplifier 自偏差放大器

self braking 自制动
self diagnostic 自诊断
self documentation 同一文件
self excited 自励
self oscillation 自振荡
self phase 自相位
self tapping screw 自攻螺丝
self-adapting algorithm 自适应算法
self-adaptive control 自适应控制
self-adaptive controller 自整定自适应控制器
self-adjusting system 自调整系统
self-adjustment 自动调节
self-aligning ball bearing 调心球轴承
self-aligning bearing 调心轴承
self-aligning roller bearing 自位滚柱轴承
self-aligning structure 自均衡结构
self-balancing electrometer 自平衡静电计
self-balancing indicator with rotating scale 旋转刻度自动平衡指示仪
self-balancing instrument 自平衡仪表
self-balancing slide wire potentiometer 自平衡滑线电位差计
self-balancing strip indicator 条形自动平衡显示仪
self-bias resistor 自偏置电阻
self-check key 自校开关
self-compensation 自补偿
self-control 自动控制，自动调整
self-correction 自校正
self-damping 自阻尼
self-excitation 自励，自激
self-excitation process 自励过程
self-excitation winding 自激励绕组，自励（磁）绕组
self-excited 自励的，自激的
self-excited machine 自励电机
self-excited oscillation 自激励振荡，自激振荡，自振荡
self-excited oscillator 自激振荡器
self-exciting 自励的
self-extinguishing material 自动灭火材料
self-feedback amplifier 自动反馈放大器，自反馈放大器

self-glazing 自动研光
self-heating 自动加热
self-hold 自保持,自锁
self-induced EMF 自感电动势
self-inductance 自感,自感应
self-inductor 自感体,自感应线圈
self-learning 自学习
self-locking 自锁,自同步
self-locking nuts 自动防松螺帽
self-operated control 自操作控制
self-operated controller 自力式控制器
self-optimizing control 自优化控制
self-optimizing system 自优化系统
self-organizing strategy 自组织策略
self-organizing system 自组织系统
self-perpetuating 自保持
self-phase modulation 自相位调制
self-recorder 自动记录器
self-recording 自动记录的
self-rectifying 自整流
self-regulating 自动调节的
self-regulating process 自调制过程
self-regulation 自调节,自调整
self-reproducing automata 自再生自动化,自繁殖系统
self-running 自运行
self-starting 自启动
self-supply power plant 自备电厂,自备动力厂
self-tuning control 自整定控制,自校正控制
self-tuning regulator 自校正调节器
self-verifying 自动检测
self-weight of structure 结构自重
selsyn 自整角机
selsyn generator 自整角发送机
selsyn motor 自整角接收机
selsyn system 自整角机系统
semantic network 语义网络,语义信息网络
semi 半
semi-active 半活性的,半衰期
semi-active damper 半衰期阻尼器
semi-active suspension 半作用悬挂
semiautomatic 半自动的
semi-automatic biochemical analyzer 半自动生化分析仪
semi-automatic controller 半自动控制器
semi-automatic hob grinder 半自动滚刀磨床
semi-automatic operation 半自动操作
semi-automation 半自动化
semi-batch operation 半间歇操作
semiconductor 半导体
semiconductor amplifier 半导体放大器
semiconductor component 半导体元件
semiconductor device 半导体器件
semiconductor diode 半导体二极管
semiconductor element 半导体元件
semiconductor integrated circuit 半导体集成电路
semiconductor pressure sensor 半导体压力传感器
semiconductor pressure transducer 半导体压力传感器
semiconductor production management 半导体生产管理
semiconductor strain gauge 扩散型半导体应变计
semi-continuous operation 半连续操作
semi-destructive testing 半破坏性测试
semi-digital indication 半数字示值
semi-digital read out 半数字示值
semi-empirical model 半经验模型
semi-finished product 半成品
semi-graphic panel 半图形板,半图解式面板
semihard-drawn aluminium wire 半硬铝线
semi-Markov process 半马尔可夫过程
semi-mechanized 半机械化
semi-physical simulation 半实物仿真
semi-positive mold 半全压式模具
semirigid PVC plastics floor tiles 半硬质聚氯乙烯块状塑料地板砖
semi-shearing 半剪
semiwave 半波
sender 发射机,发送器,引向器,电键
sending 发送,发射

sending message 发送信息
sense line 读出线
sense wire 读出线
sensing capillary 传感毛细管
sensing diaphragm 传感膜片
sensing element 敏感元件，传感元件
sensing element 敏感元件，检出器
sensing element elevation 敏感元件高度
sensing probe 传感探头，感测探针，敏感元件
sensitive 灵敏，敏感，灵敏器
sensitivity 灵敏度，灵敏性，灵敏度特性
sensitivity analysis 灵敏度分析
sensitivity control 灵敏度控制
sensitivity controller 灵敏度控制器，灵敏度调节器
sensitivity function 灵敏度函数
sensitivity-time control 灵敏度时间控制
sensor 敏感器，传感器，检出器，敏感元件
sensor failure 传感器失效
sensor fusion 传感器融合
sensor system 传感器系统
sensors 传感器
sensory control 传感控制，感觉控制
separate 分离
separate galvanometer 分装式检流计
separate grounding 单独接地
separated layer 分层
separated time 分线
separately excited 他励，他励的
separately excited generator 他励发电机，他激式发电机
separately excited motor 他励电动机，外激式电动机
separately excited multivibrator 他励多谐振荡器
separating factor 分离系数
separating force 分离力
separation 分离，分离间隙
separation number 分离数
separator 分离器，分隔器，分隔物
sephadex 葡聚糖
sequence 顺序，序列

sequence auxiliary block 顺序辅助功能块
sequence control 程序控制
sequence control function 顺序控制功能
sequence control function block 顺序控制功能块
sequence control system 顺序（程序）控制系统
sequence controller 顺序控制器
sequence debug display 顺序调试画面
sequence element 顺序元素
sequence element reference panel 顺序元素参考画面
sequence element search panel 顺序元素搜索画面
sequence estimation 序列估计
sequence execution mode 顺序执行方式
sequence execution state 顺序执行状态
sequence faceplate block 顺序面板功能块
sequence in time 时序
sequence message output 顺序信息输出
sequence of operation 操作程序，作业程序
sequence oriented procedural language (SOPL) 顺序定向的过程语言
sequence processing 程序处理
sequence program 顺序程序
sequence slot 顺序槽路
sequence slot size 程序槽规模
sequence table 顺序表
sequence table block 顺序表功能块
sequence table reference 顺序表参考
sequence table status 顺序表状态
sequence table status report 顺序表状态报告
sequence table window 顺序表窗口
sequence test 程序测试，联锁顺序试验
sequence tester 程序测试器
sequencers and synthesizers for DNA and protein DNA及蛋白质的测序和合成仪

sequencing and scheduling 排序与排程
sequential 连续
sequential computer 顺序计算机
sequential control 连续控制，顺序控制，时序控制
sequential control algorithm 连续控制算法
sequential decomposition 顺序分解
sequential injection analysis 顺序注射分析法
sequential least squares estimation 序贯最小二乘估计
sequential machine 时序机
sequential phase control 顺序相位控制
sequential programme 顺序程序
sequential switching 连续开关
sequential transducer 顺序传感器
sequentially-laminated multilayer 顺序层压多层印制板
serial 串行的，串联的，系列的
serial computer 串行计算机
serial device interface 串行设备接口
serial element 串联元件
serial link 串联接驳
serial logic 串行逻辑
serial mode 串行方式
serial operation 串行操作
serial register 串行寄存器
serial storage 串行存储器
series 串联，连续，串行，串励，系列
series combined mechanism 串联式组合机构
series compensation 串联补偿
series connection 串联
series excited 串励
series feedback 串联反馈
series inductance 串联感应
series limiter 串联限幅器
series machine 串励电机
series mode interference 串模干扰
series mode rejection 串模抑制
series mode rejection ratio 串模抑制比
series mode signal 串模信号
series mode voltage 串模电压
series motor 串激电机

series negative feedback 串联负反馈
series resistance 串联电阻
series resonance 串联谐振，电压谐振
series seam welding 串联缝熔接
series transductor 串联饱和电抗器
series-compensated amplifier 串联补偿放大器
series-in-series-out register 串入串出寄存器
series-opposing connection 反向串联
series-parallel 串并联
series-parallel circuit 串并联电路
series-parallel control 串并联控制
serration spline 三角形花键
service 维修，保养，服务，伺服，维护
service ability 工作能力
service life 使用期限，服务期限，使用寿命
service regulator 工作调节器
service wire 引入线
serviceable condition 可用状况，营运条件
servo 伺服，伺服系统，伺服机构
servo control 伺服控制，随动控制
servo drive 伺服传动
servo hydraulics 伺服水力学
servo problem 伺服问题
servo system 伺服系统，随动系统
servo-actuated control 伺服控制
servoamplifier 伺服放大器
servo-balance type tank gauge 伺服平衡式液位计
servo-control 伺服控制，随动控制
servomechanism 伺服机构，自动驾驶装置，自动控制装置，跟踪器
servomechanism control 伺服机械控制
servomotor 伺服电机，伺服电动机，伺服马达
servomotor actuator 伺服电机执行机构，伺服电机执行器
servo-operated potentiometer 伺服驱动电位差计
servo-selsyn system 伺服自整角机系统
set 设定

set archive pointers 设置归档指针
set default directory 设置缺省目录
set point 设定值，给定值，设定点
set point change 设定值变化
set point control 设定点控制
set point generator 定值器
set point in percent 百分比的给定值
set point trajectory 设定轨迹
set point unit and measurement receive assembly 给定与测量机构
set screw 固定螺丝
set top box (STB) 机顶盒
set value 给定值，给定点，参比变量
setback 向后退入
setback line 退入界
set-in 补充切入，进刀
set-reset operation 设定重定操作
setting 设定
setting out 放样，开线，测定，定线
setting time 建立时间，置位时间
settle 沉淀
settlement 沉降
settling time 校正时间，过渡时间，调节时间
set-value 设定值
sewage disposal 污水处理
sewage pipe 污水管
sewage pump 污水泵
sewage pump house 污水泵房
sewage sump 污水坑，集污槽
sewage treatment 污水处理
sewage treatment plant 污水处理厂
sewer 污水管，污水渠
sewerage reticulation system 网状污水渠系统
sextant 六分仪
SFC block function SFC 块功能
SFC window SFC 窗口
SFC (supercritical-fluid chromatography) 超临界流体色谱法
shaded-pole motor 罩极式电动机
shading effect 遮光效果
shadowless lamp 无影灯
shaft 竖井，轴，通风井，手柄，矿井
shaft angle 轴角，轴转角
shaft collar 轴环

shaft encoder 轴解码器，计数鼓
shaft end ring 轴端挡圈
shaft seal 轴封
shaft shoulder 轴肩
shake 摇动，振动
shakeproof 防振
shaker 摇床
shakers 振荡器
shaking couple 振动力矩
shaking incubators 振荡培养器
shank 柄，柄部，模柄
Shannon's sampling theorem 香农采样定理
shape 形状，轮廓
shape description 形状描述
shape discrimination 形状区别
shape iron 型钢
shaped reflector 物形反光镜
shaper 整形器，脉冲整形器，牛头刨床，造型者，塑造者
shaping 整形，成形，成形加工
shaping circuit 整形电路，折线电路
shaping filter 整形滤波器
shaping machine 牛头刨床，成形机
shaping network 整形网络
share memory architecture (SMA) 共享内存结构
share 共享，分配
shared time control 分时控制
sharp bend 急弯
sharp edge 锐边，锐角部
shaving 缺口修整加工
shear 剪切
shear angle 剪角
shear key 抗剪键，剪力榫，剪力键
shear reinforcement 剪力钢筋，剪切钢筋
shear strength 切变强度
shear wall 剪力墙，抗震墙，耐震壁，风力墙
shearing 剪断，切断加工
shearing die 剪边模
shearing force 剪力
shearing stress 剪应力
shearing test 抗剪试验，剪切试验
shears 剪子

sheath 护套,护层,保护管(指热电阻、热电偶)
sheathed thermocouple 铠装热电偶
sheathed wire 铠装线,金属护皮电线
sheave 滑车轮
shedder 脱模
sheer leg 起重机支架
sheet 表格,纸张,塑胶片
sheet iron 铁片
sheet loader 薄板装料机
sheet metal parts 冲件,板料冲压
sheet moulding compounds for general purposes 通用型片状模塑料
sheet pile 板桩
sheet pile anchorage 板桩锚,板桩锚定
sheet pile cofferdam 板桩围堰
sheet piled wall 板桩墙
sheet resistance 薄膜电阻,表面电阻
sheet steel 钢片
sheet stock 片料,薄钢板
sheet wire 扁线
shelf 托架
shelf mounted instrument 盘装仪表
shell 套管,壳
shell casting 壳模铸造
shell of expert system 专家系统的壳
shell test pressure 壳体试验压力
shelter 遮蔽处,庇护所,收容所,避雨亭
sheltered area 掩蔽区
sheltered location 掩蔽场所
shield 防护罩,遮护,屏蔽层
shield assembly 屏蔽装置
shielded cable 屏蔽电缆,屏蔽线
shielded wire 屏蔽线,隔离线
shielding 屏蔽
shielding action 屏蔽作用
shielding constant 屏蔽常数
shielding wire 屏蔽线,遮蔽线
shift 偏移,替换
shift pulse 移位脉冲
shift pulse driver 移位脉冲驱动器
shift register 移位寄存器
shift roster 轮班
shifting function 移位操作

shim 薄垫片,填隙片,分隔片
shim member 补偿元件
shim plate 垫板
shiny side 光面
ship control 船控制,舰控制
shipment 出货
shock eliminator 消震器
shock line 模口挤痕;激波线
shock load 冲击荷载
shock testing systems 冲击试验系统
shock wave 冲击波
shock 冲击,震动,使受电击
shock-absorber 缓冲
shockproof device 防振装置
shoot 流道
Shore hardness 肖氏硬度,回跳硬度
Shore hardness tester 肖氏硬度计
shore up 承托
shoring 横撑板,撑柱
short 短路,使短路
short circuit 短路,漏电
short iron 脆性铁
short shot 充填不足
short switch 短路开关
short time horizon coordination 短时程协调
short trouble 短路故障
short wave 短波
short wave band 短波波段
short-circuit 短路
short-circuit characteristic 短路特性
short-circuit loss 短路损耗
short-circuit parameter 短路参数
short-circuit test 短路试验
short-circuit voltage 短路电压
short-circuited 短路的
short-circuited resistance 短路电阻
short-circuited winding 短路绕组
short-circuiting ring 短路环
short-term memory 短期存储器
short-time duty 短时工作制
short-time load 短时负荷
short-time rating 短时出力,短时额定出力
shot 注射,单行程工作
shot blast 喷丸处理
shot blasting 珠粒喷击清理

shot cycle　射出循环，压射周期
shot peening　珠击法，喷丸强化，喷丸硬化
shotcrete　喷射混凝土，喷射水泥沙浆
shoulder bolt　肩部螺丝
shoulder peak　肩峰
show batch queue　显示批任务单
show logical parameter　显示逻辑参数
show own process parameter　显示过程本身的参数
show print queue　显示打印任务单
show system　显示系统清单
show　展览，显示，指示
shrink fit　热压配合
shrinkage　收缩，收缩量
shrinkage coefficient　收缩系数
shrinkage fit　收缩配合
shrinkage hole　缩孔
shroud　保护罩
shunt　并励，分路，分流器，分路器
shunt capacitor　并联电容器，分路电容器，旁路电容器
shunt circuit　并联电路，分流电路
shunt compensation　旁路补偿器，并联补偿
shunt connection　并联，旁路连接
shunt displacement current　旁路位移电流
shunt excited　并励
shunt feedback　并联反馈
shunt field　并励磁场
shunt impedance　并联阻抗
shunt pattern flowmeter　分流式流量计
shunt reactor　并联电抗器，分路电抗器，分馏力，分流扼流圈
shunt winding　并励绕组
shunted capacitor　分路电容器
shunted instrument　带分流器的仪表
shunted meter　带分流器的电流表
shunt-excited machine　并励电机
shunting resistance　并联电阻，分流电阻
shunt-opposed limiter　并联反接式限制器
shut down　关闭，停车，停机

shut height　闭合高度
shut height of a die　架模高度
shut off　断开，关闭
shut off switch　断路开关
shut off valve　关闭阀，截流阀
shut　关上
shutdown　停止，停机
shutoff valve　截止阀，断流阀
shuttering　模板
SI（International System of Units）　国际单位制
siccative　干燥剂
side band　边带
side band addressing (SBA)　边带寻址
side cut　切边
side elevation　侧视图
side frequency　边频，旁频
side gate　侧浇口
side mounting　侧面安装
side movement　侧向位移
side pipe　旁管
side planer　边刨床
side rake angle　侧前角
side road　旁路
side span　旁跨，边跨
side stretch　侧冲压平
side　侧
sideband address port (SAP)　边带寻址端口
side-mounted float switch　侧置式浮子开关
siemens　西门子（电导单位）
sifter　筛子，滤波器
sight line　视线，视准线，照准线
sighting wire　照准丝
sign　标志，告示牌，招牌，标记，注册
sign detection　标记检测
sign gantry　标志架
sign mount　标志托架
signal　讯号，交通灯，信号，符号
signal amplifier　信号放大器，小信号放大器
signal amplitude sequencing　信号幅度排序
signal analysis　信号分析
signal and information processing　信

号与信息处理

signal aspect 信号式样,信号形态
signal cable termination 信号电缆端头
signal characterizer 信号表征器
signal characterizer card 函数卡
signal circuit 信号电路,信号线路
signal common (SC) 信号公共端,信号公共点
signal conditioning 信号调理,信号调节,信号波形加工
signal conditioning amplifier 信号调节放大器
signal converter 信号转换器
signal correlation 信号相关
signal delay 信号延迟
signal detection 信号检测
signal detection and estimation 信号检测和估计
signal distortion 信号失真,信号畸变
signal distribution component 信号分配组件
signal duration 信号间隔
signal flow diagram 信号流图
signal flow graphs 信号流图
signal generator 信号发生器,讯号产生机
signal gong 讯号钟,信号钟
signal impulse 信号脉冲
signal indicator 信号指示器,信号表示器
signal isolation 信号隔离
signal level 信号电平,信号级
signal light 信号灯,讯号灯
signal line 信号线
signal noise ratio (SNR) 信噪比
signal pressure 信号压力
signal processing 信号处理
signal processing algorithm 信号处理算法
signal processor 信号处理器,讯息处理者
signal reconstruction 信号重构
signal relay 信号继电器
signal screen 信号屏
signal selector 信号选择器
signal shot noise 散粒噪声
signal source 信号源
signal state code 信号状态编码
signal synthesis 信号合成
signal to noise ratio (SNR) 信噪比,信号噪声比
signal transmission 信号传输,信号发送
signal transmitter 信号发送器
signal voltage 信号电压
signal lamp 信号灯,讯号灯
signal-controlled pedestrian crossing 交通灯控制行人过路处
signaling 信号
signature 特征,标记
signature analysis 特征分析
signature register 标记寄存器
signboard 招牌
significant figure 有效数字
silane 硅烷
silence 消音,消声,沉寂
silencer 灭声器,灭音器,消声器
silent chain 齿形链,无声链
silent zone 静寂地带
silica 硅石,二氧化硅
silica gel 硅胶
silicate compound plaster for thermal insulation 硅酸盐复合绝热涂料
silicon (Si) 硅
silicon diode 硅二极管,硅晶体二极管
silicon iron 硅钢,硅铁
silicon on sapphire 蓝宝石硅片
silicon photodiode 硅光电二极管
silicon rubber 聚硅酮橡胶,硅橡胶
silicon steel sheet 硅钢板
silicon transistor 硅晶体管
silicon unijunction transistor 硅单结晶体管
silicone paint 硅树脂油漆
silicone resin 硅树脂
silicone sealant for building 硅酮建筑密封膏
silk-and-cotton-covered insulated wire 丝棉包线
sill 窗台,门槛
sill of window 窗台

silt trap 淤泥收集器
silver (Ag) 银
silver jacketed wire 镀银线
silver silver-chloride electrode 银-氯化银电极
similarity 相似性
similarity transformation 相似变换,相似转换
SIMM (single in-line memory modules) 单边接触模组
simple germ test 微生物简单测试仪
simple harmonic motion 简谐运动
simple network management protocol (SNMP) 简单网络管理协议
simulated interrupt 模拟中断
simulated positioner 模拟定位机
simulation 仿真,模拟,模拟分析
simulation analysis 仿真分析
simulation block diagram 仿真框图
simulation data 仿真数据,模拟数据
simulation experiment 仿真实验,模拟实验
simulation languages 仿真语言
simulation velocity 仿真速度
simulative generator 模拟发生器
simulator 仿真器,模拟设备,模拟器
simultaneous computer 同步操作计算机
simultaneous stabilization 同时稳定
sine 正弦
sine bar 正弦量规
sine generator 正弦发生器
sine voltage 正弦电压
sine wave output 正弦波输出
sine-cosine potentiometer 正、余弦电位计
sine-forced response 正弦响应
sine-wave 正弦波
sine-wave generator 正弦波发生器,正弦波发电机
single 单
single autosyn indicator 单指针自同步指示器
single axle table 单轴转台
single band 单波段,单频带
single board computer 单板微型计算机

single busbar 单母线
single cable 单芯电缆
single cavity mold 单腔模具
single cycle 单循环法
single degree of freedom gyro 单自由度陀螺
single ended input 单端输入
single ended output 单端输出
single -frequency gas laser 单频气体激光器
single in-line memory modules (SIMM) 单边接触模组
single input single output (SISO) 单输入单输出
single input single output system 单输入单输出系统
single level process 单级过程
single loop controller 单回路调节器
single loop indicating controller 单回路指示调节器
single loop programmable controller 单回路可编程序调节器
single mode 单模式
single mode operation 单模式操作
single opening exhaust valves 单口排气阀
single phase 单相
single pole double throw 单刀双掷
single pole single throw 单刀单掷
single roll 单滚动
single row bearing 单列轴承
single selector terminal 信号选择单元型端子
single setting 单面调整法
single side 单面
single side band (SSB) 单边带
single side-band amplitude modulation 单边带调幅
single side-band circuit 单边带电路
single span 单跨
single squirrel cage 单鼠笼的
single stage pump 单级泵
single strategy controller Ⅱ (SSCⅡ) Ⅱ型单回路控制器
single strategy controller port (SSCP) 单回路控制器接口
single track railway 单轨铁路

single universal joint 单万向联轴节
single value nonlinearity 单值非线性
single 单个的，个体的
single blade switch 单刀开关
single-arm and double-arm DC bridge 单双臂两用直流电桥
single-arm and double-arm double-purpose DC bridge 单双臂两用直流电桥
single-cell box culvert 单孔盒形暗渠
single-channel control 单通道控制，单路控制
single-channel ultrasonic flowmeter 单通道超声流量计
single-chip microcomputer 单片微型计算机
single-circuit 单回路的
single-direction thrust bearing 单向推力轴承
single-ended termination 单一终端终止
single-pen strip chart recorder 单笔长图记录仪
single-phase bus 单相母线
single-phase circuit 单相电路
single-phase current 单相电流
single-phase induction motor 单相感应电动机
single-phase motor 单相电动机
single-phase regulating transformer 单相调整变压器
single-phase socket 单相插座
single-phase source 单相电源
single-ported globe valve 单孔球阀，单座球阀
single-range 单量程的，单波段的
single-range instrument 单量程仪表
single-sided copper-clad laminate 单面覆铜箔层压板
single-sided printed board 单面印制板
single-speed floating action 单速无定位作用
single-speed floating controller 单速无定位控制器
single-stage 单级的
single-stage modulation 单级调制
single-tag display block 信号-工位号显示块

single-throw switch 单掷开关
singular 奇异
singular attractor 奇异吸引子
singular control 奇异控制
singular perturbation 奇异扰动，奇异摄动
singular perturbation method 奇异扰动方法
singular point 奇异点
singular position 奇异位置
singular system 奇异系统
singular value 奇异值，特征值
singular value decomposition 奇异分解
singularity 奇点，奇异值，特性
sink well 掘井
sinking 凹陷，碟形凹陷法
sinter forging 烧结锻造
sintered plate 烧结板
sintering of sand 铸砂烧贴，烧结料
sinusoid 正弦形的，正弦波信号
sinusoidal 正弦波的，正弦的
sinusoidal current 正弦电流
sinusoidal density wave 正弦磁密度
sinusoidal oscillator 正弦振荡器
sinusoidal response 正弦响应
sinusoidal signal 正弦信号
sinusoidal steady state 正弦稳态
sinusoidal time function 正弦时间函数
siphon 虹吸，虹吸管
SISO（single input single output） 单输入单输出
site 地盘，地点，现场
site boundary 地界范围，地盘界线
site coverage 上盖面积
six sides forging 六面锻造
six-bar linkage 六杆机构
size 尺寸，大小，浸润剂
size content 浸润剂含量
size exclusion chromatography 尺寸排阻色谱法
size factor 尺寸因素
size marking 尺寸标注
skeleton 骨架
sketch 图纸，草图
skew bevel gears 歪轮，斜伞齿轮
skid resistance 防滑

skid resistance test	防滑试验	sliding dowel block	滑块固定块
skid resistant surfacing	防滑面层	sliding force	滑动力
skill	技巧	sliding friction	滑动摩擦
skill-based production	基于技巧生产	sliding mode	滑模
skill-based system	基于技巧系统	sliding mode control (SMC)	滑模控制
skin effect	集肤效应,趋肤效应	sliding rack	滑料架
skin friction	表面摩擦	sliding ratio	滑动率
skin inclusion	表皮折叠	sliding surface	滑动表面
skinned wire	裸线	sliding velocity	滑动速度
skip crane	吊斗起重机	sling hygrometer	摆动湿度计,悬挂式湿度计
skip hoist	吊斗吊重机	slip agent	光滑剂
skip index	跳齿分度	slip form	滑模,活动板模
skip welding process	跳焊法	slip joint	滑配接头
skip	空指令,跳跃	slip ratio	转差率,滑移比率,滑率,滑差系数
skirting	墙脚线	slip road	连接路
skiving	表面研磨	slip speed	转差率
slab	平板	slip torque	滑移扭矩
slag	熔渣,结垢	slipped screwhead	螺丝滑头
slave	从属装置	slippery screw head	螺丝滑头
slave operation	从动运行	slipping	打滑
slave station	从属站	slip-ring motor	滑环电动机
slaved system	受役系统	slit gate	缝隙浇口
sleeve	套管,套袖	slitting	切缝量,割缝加工
sleeve wire	套线,塞套引线	slope	斜率,斜坡,坡度
slewing crane	旋臂起重机,回转式吊机	slot	槽路,槽,插槽
slide	滑动,行位(滑块)	slot line	槽线
slide balancer	滑动平衡器	slot width	齿缝宽度
slide caliper	游标卡尺	slotting machine	插床
slide coupler	滑动耦合器	slotting tool	切槽刀
slide gauge	滑尺,游标卡尺	slot-width taper	槽宽收缩
slide rail	滑轨	slow processor	慢速处理器
slide rheostat	滑线电阻,滑触变阻器	slow subsystem	慢变子系统
slide switch	滑动开关	slow-burning wire	慢燃线,耐火绝缘
slide valve	滑阀	slower-than-real-time simulation	欠实时仿真
slide vane flowmeter	划片式流量计	slow-speed motor	低速电动机
slide wire	滑线,滑触电阻线	slug hole	逃料孔,废料孔
slider	滑块	sluice gate	水闸门
slider-crank mechanism	曲柄滑块机构	sluice valve	水闸,闸式阀
slide-roll ratio	单位滑滚比,比滑	slush molding	沥铸成形法
sliding	滑动	SMA (share memory architecture)	共享内存结构
sliding bearing	滑动轴承,滑动支承	small computer system interface	小型计算机系统接口
sliding block	滑块		
sliding coefficient	滑动系数		
sliding curve	滑动曲线		

small ordinary relay　小型通用继电器
small signal mode　小信号模式
small strip-type indicator　小条型指示仪
small strip-type self-balancing indicator　小条型自动平衡指示仪
small-gauge wire　细钢丝
smart field communicator　智能现场通信器
smart field instrumentation　智能化现场仪表
smart instrument　智能仪表
smart power application　智能电源应用
smart transmitter interface　智能变送器接口
smart transmitter interface module　智能变送器接口模件
smart valve　智能阀
SMC（sliding mode control）　滑模控制
Smith predictor technique　史密斯预估器技术
Smith's method　史密斯方法
smls　无缝
smog　光化学烟雾
smog absorber　烟雾吸收器
smoke density sensor　烟度传感器
smoke detector　感烟式探测器，烟雾报警器
smoke test　冒烟测试，通烟试验，用烟试验
smoke　烟，冒烟
smooth　平滑的，光滑的
smoother　平滑器，滤波器，稳定器
smoothing　平滑，滤波
smoothing capacitor　滤波电容器，平流电容器
smoothing choke　滤波扼流器，平流扼流圈
smoothing circuit　滤波电路，平流电路
smoothing filter　平滑滤波器
smoothly surfaced steel wire　光面钢丝
smoothness criterion　平滑判据
snake wire　绳索包皮线
snap gauge　卡规，外径规
snapshot　瞬时值
snips　铁剪

SNMP（simple network management protocol）　简单网络管理协议
snowman pad　雪人盘
SNR（signal noise ratio）　信噪比
SNR（signal to noise ratio）　信噪比，信号噪声比
soakaway　渗水坑，渗水井
socket　卡套，插座，套，外接头，承口
socket wrench　管钳子，套筒扳手
socket-and-spigot joint　承插接头，承窝接合
socketed pipe　套接管
soda　苏打
sodium（Na）　钠
sodium carbonate　碳酸钠
sodium dodecyl sulfate（SDS）　十二烷基磺酸钠
sodium nitrite method　亚硝酸钠法
sodium wire　钠线
soffit formwork　底模板
soft and free expansion sheet making plant　软板材及自由发泡板机组
soft key　软键，自定义功能键，软键盘
soft measurement　软测量
soft sensor　软传感器
soft shock　柔性冲击
soft-annealed wire　软金属线
software　软件
software architecture　软件体系结构，软件架构，软件构架
software engineering　软件工程
software environment　软件环境
software metrics　软件标准，软件度量
software performance　软件性能
software productivity　软件产率
software project management　软件项目管理
software psychology　软件心理学
software reliability　软件可实现性，软件可靠性
software safety　软件安全性
software specification　软件规格
software tool　软件工具
soil　土壤，污物，便溺污水

311

soil nail 泥钉
soil stabilization works 土壤稳定工程，加固工程
soild wire 单线，实线，单股线
solar 太阳
solar array 太阳能电池阵，太阳电池板
solar array pointing control 太阳能板指向控制
solar battery 太阳能电池
solar cell 太阳能电池
solar energy 太阳能
solar module 太阳能电池组件
solar power plant 太阳能发电厂
solder 焊料，焊锡
solder side 焊接面，焊接板面，焊锡面
solder-covered wire 锡包线
solenoid 电磁线圈，筒形线圈，螺线管
solenoid actuator 电磁执行机构，螺线管执行机构
solenoid valve 螺线管电磁阀，电磁阀
solicited message structure 请求信息结构
solid conductor 实心导线
solid cutter 整体刀盘
solid flowmeter 固体流量计
solid forging die 整体锻模
solid lubricant 固体润滑剂
solid mechanics 固体力学
solid model 实体模型
solid phase extraction 固相萃取
solid raft 实体筏基
solid state 固态，固体
solid state cell 固态电池
solid state component 固态元件
solid state laser 固态激光器，固体激光器
solid state logic card 固态逻辑组件
solid state micrologic element 固体微型逻辑元件
solid state phase changes 固态相变
solid state photosensor 固体光敏元件，固态光电传感器
solid 固体的，坚固的，固体的，刚性

solid platform 稳固平台
solidifying point 凝固点
solid-stem liquid in glass thermometer 棒式玻璃温度计
solute property detector 溶质性能检测器
solution 溶解
solution treatment 固溶退火，溶体处理
solvent 溶剂
solvent extraction 溶剂萃取法
solvent recovery unit 溶剂回收单元
sonar manometer 声呐压力计
sonic flowmeter 声学流量计
sonic nozzle 音速喷嘴，声速喷嘴
sonic Venturi-nozzle 音速文丘里喷嘴
soot 碳烟，煤烟，烟灰
SOPL (sequence oriented procedural language) 顺序定向的过程语言
sorted unloading 分级卸载
sorter 分类机
sound intensity 声强度
sound level meter 声级计
sound spectrograph 声谱仪
sound test 噪声试验
sounding device 回声探测装置
sounding wire 测深绳
sound-level meter 声级计
sour condition 腐蚀性环境
source 源，电源，源点，信号源，源极
source code 源码
source device parameter 源装置参数
source follower circuit 源极输出电路
source impedance 电源阻抗
source language 源语言
source of power 电源，能源
source program 源程序
space 空间
space charge 空间电荷
space frame 空间构架
space robot 太空机器人
space structure interaction 空间结构相互作用
space vehicle 空间运载工具
spacecraft autonomy 空间飞船自主性
spacer 垫片，钢筋定位物，隔离物

spacer arm effect 间隔臂效应
spacer block 间隔块
spacer ring 间隔环
space-width taper 齿距收缩
spacing 间距
spacing tolerance 齿距公差
spacing variation 齿距变动量
span 量程，跨距，跨度
span adjustment 量程调整器
span error 量程误差
span shift 量程迁移
span tolerance 量程允差
span wire 悬索
spanner 扳手
spare battery 备用电池
spare capacity 备用容量
spare dies 模具备品
spare line 备用线路
spare molds location 模具备品仓
spare motor 备用电动机
spare parts 备件，备品
spark 火花
spark advance control 提前点火控制
spark discharge 火花放电
spark gap 火花间隙
spark lighter 火花点火器
spatial cam 空间凸轮机构
spatial kinematic chain 空间运动链
spatial kinematic pair 空间运动副
spatial linkage 空间连杆机构
spatial mechanism 空间传动机构
spatial waveform 空间波形
SPC (statistical process control) 统计过程控制
speaker 扬声器，话筒
spear head 刨尖头
special 专用的，特殊的，特别的，专门的
special cross section steel wire 特殊断面钢丝
special data point 应用数据点
special instruction 特殊指令，专用指令
special kinematic chain 特殊运动链
special nuts 其他特殊螺帽
special purpose computer 专用计算机，特殊用途计算机

special semaphore command 特殊命令信号
special shape punch 异形冲子
special size 专用尺寸
special test unit 专用测试仪表
special work request 特殊工作需求
special tool 专用工具
specialized standard 专用标准
specific conductance 电导率
specific gravity 比重
specific gravity hydrometer 液体比重计
specific heat capacity 比热容
specific impulse 比冲量
specific inductive capacity 电容率，介电常数
specific resistance 比电阻，电阻系数
specified characteristic curve 在规定特性曲线
specify type of printing 打印的指定类型
specimen 样品，试件
spectra 光谱
spectral 光谱的
spectral analysis 光谱分析，频谱分析
spectral characteristic 频谱特性
spectral correlation 光谱相关，频谱相关
spectral density 频谱密度
spectral density function 频谱密度函数
spectral estimation 频谱估计
spectral filter 分光滤光器，滤光片，光谱滤器
spectral matching 光谱匹配法
spectral pyrometer 光谱高温计
spectro-chemical analysis apparatus 光谱化学分析仪器
spectrograph 摄谱仪，光谱仪
spectrometer 分光计
spectrophotometer 分光光度计
spectroscopic analysis 光谱分析法，分光镜分析法
spectroscopy 光谱学
spectrum 频谱，光谱
spectrum analyzer 频谱分析仪
spectrum generator 频谱发生器

speech 语音	spheroidal graphite cast iron 球墨铸铁
speech amplifier 音频放大器	spheroidizing 球化处理
speech analysis 语音分析	spiegel iron 镜铁
speech control 语音控制	spigot 塞子，插口
speed change 变速	spike 峰值，尖峰，尖峰信号，道钉，大钉
speed changer 变速器	spike pulse 尖峰脉冲，窄脉冲
speed control 速度控制，速度控制器	spike voltage 峰值电压
speed control servo-motor 速度伺服电动机	spillage 溢出物
speed control system 调速系统，速度控制系统，变速系统	spillway 溢水道，溢洪口
speed fluctuation 速度波动，速率变动，速度忽高忽低	spin 反旋
	spin axis 自旋轴
speed gears 变速齿轮	spin forming machine 旋压成形机
speed governing 调速，速度调节	spindle 针，主轴，心轴，转轴
speed governor 调速器，限速器	spindle rotation angle 主轴旋转角
speed hump 缓冲路拱	spinner 自旋体
speed indicator 速度计，示速器	spinner magnetometer 旋转磁场计，旋转式磁力仪，旋转磁强计
speed limit 时速限制	
speed measurement 速度测量	spiral angle 螺旋角
speed of action 动作速度	spiral bevel gear 弧齿锥齿轮，螺旋伞齿轮
speed probe 速度探测器，车速探示器	
	spiral flow test 螺旋流动试验
speed range 速度范围	spiral reinforcement 螺旋钢筋
speed ratio 速度比	spiral seal 螺旋密封
speed reducer 减速器	spiral spring 螺旋弹簧
speed regulation 速度调节	spirit level 水平尺，气泡水准仪
speed regulator 调整器	splash lubrication 飞溅润滑
speed relay 限速继电器	splice 拼接，镶接
speed transmitter 速度变送器	splice bar 拼接板，鱼尾板
speed variator 变速器	splice joint 拼合接头
speed 速度	spline 键，花键，连接，塞缝片，槽栓
speed-changing gear boxes 齿轮变速箱	
	spline broach 方栓槽拉刀
speedometer 速度表	splines 样条，仿样
speed-torque characteristic 速度转矩特性	split 分割，拼合，裂缝
	split field motor 分割磁场电机
speed-torque curve 转速力矩特性曲线	split forging die 拼合锻模
	split injection 分流进样
sphere-pin pair 球销副	split mold 拼合铸模，可拆模，组合模
spheric pair 球面副	
spherical bearing 球形支承	spoke wire 辐条钢丝，辐线
spherical involute 球面渐开线	sponge iron 海绵铁
spherical involute teeth 球面渐开线齿	sponge rubber 海绵橡胶
	spontaneous combustion 自燃
spherical motion 球面运动	spontaneous ignition 自燃
spherical roller 球面滚子	spot 斑点

spot facing machining 孔加工
spot welding 点焊,点焊接
spotlight 射灯,聚光灯
spotting 合模
spout 喷水孔
spray dryer 喷雾干燥器,有机溶剂喷雾干燥器
spray dryer for product line 生产线喷雾干燥器
spray glazing 喷釉法
spray gun 喷枪
spray plotter 喷墨式彩色绘图机
spray nozzle 喷嘴
sprayed concrete 喷射混凝土
spraying glazing 喷釉
spraying nozzle 喷嘴
spraying painting 喷漆
spreader 涂布器
spreadsheet 电子制表软件,电子数据表,试算表
spring ammeter 弹簧式电流表
spring box 弹簧箱
spring compressed length 弹弓压缩量
spring constant 弹簧常数
spring nuts 弹簧螺帽
spring rod 弹弓柱,弹簧杆
spring steel 弹簧钢
spring test 弹簧片
spring 弹簧
spring clutch 弹簧离合器
spring-tube manometer 弹簧管式压力计
sprinkler 喷洒器
sprinkler system 自动喷水灭火系统,自动洒水装置
sprocket 链轮,链轮齿
sprocket gear 链轮
sprocket-wheel 链轮
sprue 注入口,铸口,溶渣
sprue bushing 唧嘴,浇口套,注口衬套
sprue bushing guide 注道导套
sprue diameter 唧嘴口径
sprue gate 射料浇口,直浇口
sprue lock bushing 注道定位衬套
spur gear 直齿圆柱齿轮,正齿轮
spurious error 疏忽误差

spurious oscillation 寄生振荡
SQC (statistical quality control) 统计质量控制,统计品质管制
square 直角尺
square bar iron 方钢,方铁
square derivative algorithm 导数平方程序
square edged thin orifice plate 直角边薄孔板
square key 方键
square nut 方螺帽
square pad 方形盘
square root 方根
square root extractor 开方器
square sleeker 方形镘刀
square thread 方螺纹
square threaded form 矩形螺纹
square wave 方波
square wave signal generator 矩形波信号发生器
square wave-form oscillator 方波振荡器,矩形波发生器
square-jaw positive-contact clutch 矩形牙嵌式离合器
squaring circuit 方波整形电路,矩形波整形电路
squaring circuits 方形电路
squeeze casting 高压铸造
squeezing die 挤压模
squirrel cage 鼠笼
squirrel cage induction motor 笼式感应电动机
SSB (single side band) 单边带
SSC Ⅱ (single strategy controller Ⅱ) Ⅱ型数字式单回路控制器
SSCP (single strategy controller port) 单回路控制器接口
stability 稳定,稳定性,稳定度
stability analyses 稳定性分析
stability condition 稳定条件,稳定状态
stability criterion 稳定性判据
stability domain 稳定域
stability factor 稳定因数
stability limit 稳定极限
stability of a linear control system 线性控制系统的稳定性

stability of a sampled-data system 数据采样系统的稳定性
stability of numerical methods 数值方法稳定性
stability property 稳定性性质
stability range 稳定范围
stability robustness 鲁棒稳定性
stability test 稳定性测试
stabilization 稳定，稳定化，镇定
stabilization method 镇定方法
stabilization network 稳定网络
stabilized power supply 稳定电源
stabilizer 稳定剂，稳定器
stabilizing 稳定化，安定
stabilizing agent 稳定剂
stabilizing controller 稳定化控制器
stabilizing feedback 稳定反馈
stabilizing feedforward 稳定前馈
stabilizing moment 稳定力矩
stabilizing network 稳定网络
stabilizing transformer 稳定变压器
stable 稳定，稳定的，稳态
stable equilibrium 稳定平衡
stable state 稳定状态
stable switch 稳定开关
stable system 稳定系统
stack feeder 堆叠拨送料机
stack mounting 组合安装
Stackelberg decision theory 施塔克尔贝格决策理论
stadia wire 准距线视距丝
stage die 工程模
stagger tuning amplifier 参差调谐放大器
stagger wire 交错线
stain proofing 防锈处理
stainless acid-resisting steel 不锈耐酸钢
stainless and graphite 不锈钢和石墨
stainless iron 不锈钢
stainless mild steel plate 不锈软钢板
stainless steel 不锈钢
stainless steel flange nuts 不锈钢突缘螺帽
stainless steel flange nylon insert lock nuts 不锈钢六角轮缘尼龙嵌入螺帽
stainless steel heavy hex nuts 不锈钢六角重型螺帽
stainless steel hex nuts 不锈钢六角螺帽
stainless steel nuts 不锈钢螺帽
stainless steel nylon insert lock nuts 不锈钢尼龙嵌入螺帽
stamp letter 冲字，打印字模
stamp mark 冲记号，刻印
stamped printed wiring board 模压印制线路板
stamped punch 字模冲子
stamping 冲压，压模，锻压加工
stamping factory 冲压厂
stamping press 冲压
stamping-missing 漏冲
stanchion 支柱
standalone 单独的，单机
stand-alone instrument 独立仪器
standard 标准，标准的，标准化
standard pitch circle 分度圆
standard atmospheric conditions 标准大气状况
standard bus 标准总线
standard calomel electrode 标准甘汞电极
standard cell 标准电池
standard component 标准件
standard depthwise taper 标准深锥度
standard deviation 标准偏差（标准差）
standard electrode 标准电极
standard engineering functions 标准的工程功能
standard FCS 标准的现场控制站
standard frequency 标准频率
standard gas cylinder 标准气钢瓶
standard gear 标准齿轮
standard hydrogen electrode 标准氢电极
standard inductance 标准电感
standard instrument 标准仪表
standard mercury thermometer 标准水银温度计
standard operation and monitoring window 标准的操作和监视窗口
standard operation procedure 制造作

业规范
standard orifice 标准孔口
standard parts 标准件
standard pitch cone 分度圆锥
standard pitch line 分度线
standard platinum resistance thermometer 标准铂电阻温度计
standard pressure gauge 标准压力表
standard quality control 标准质量控制
standard sand for cement strength test 水泥强度试验用标准砂
standard signal 标准信号
standard signal generator 标准信号发生器
standard spur gear 标准直齿轮
standard state 标准状态
standard taper 正常收缩
standard thermocouple 标准热电偶
standard thickness 正常齿厚收缩
standard time 标准时间
standard vacuum gauge 标准真空表
standardization 标准化
standby generator 后备发电机
standby register 备用寄存器
standby water pump 后备水泵
standby 备用的
standing manual station 便携式手动操作器
standing wave 驻波
staple U U形钉
star connection 星形连接
star network 星形网络
star-delta connection 星形-三角形连接，Y-Δ连接
star-delta starter 星形-三角形启动器，Y-Δ启动器
start button 启动按钮
start 启动，开始
start up 启动
starter 启动器
starting button 启动按钮
starting compensator 启动补偿器
starting contactor 启动接触器
starting current 启动电流
starting equipment 启动装置
starting period 启动时间，启动周期

starting resistance 启动电阻
starting rheostat 启动变阻器
starting torque 启动力矩，起动力矩，起动扭矩
starting winding 启动绕组
starting conditions 启动条件
start-up 启动，开车
startup mode 启动方式
start-up sequence 启动程序
state 状态
state assignment 状态赋值
state diagram 状态图
state equation model 状态方程模型
state equations 状态方程
state estimation 状态估计
state feedback 状态反馈
state monitoring 状态监控
state observer 状态观测器
state scintillation detector 状态变化检测器
state sequence estimation 状态序列估计
state space 状态空间
state space description 状态空间描述
state trajectory 状态空间轨迹
state transition equation 状态转移方程
state transition matrix 状态转移矩阵
state variable 状态变量
state variable method 状态变量法
state variables 状态变量
state vector 状态矢量，状态向量
statement 报告，命题，语句，说明
state-space formula 状态空间公式
state-space method 状态空间方法
state-space model realization 状态空间模型实现
state-space realization 状态空间实现
static 静力学，静态，静态的，静止的
static accuracy 静态精度
static balance 静平衡
static characteristics curve 静态特性曲线
static charge measuring instrument 静电荷测量仪
static controller 静态控制器
static decoupling 静态解耦
static electrification 静态充电

static equivalent axial load 轴向当量静载荷
static equivalent radial load 径向当量静载荷
static error 静态误差
static force 静力
static friction 静摩擦，库仑摩擦
static gain 静态增益
static induction 静电感应
static induction transistor 静态感应半导体
static load 静载重，静荷载
static measurement 静态测量
static model 静态模型
static multivibrator 自励多谐振荡器
static precipitator 静电除尘器
static pressure 静压
static RAM 静态随机存取存储器
static regulation 静态调节
static seal 静密封
static sensitivity 静态灵敏度
static stability 静态稳定度
static storage 静态存储器
static wire 导电丝
station control nest 站控制箱
station default access level 站默认存取级别
station 站，发电厂，地点，操作器，操作站
stationary liquid 固定液
stationary phase 固定相
stationary random process 平稳随机过程
statistic pattern recognition 统计模式识别
statistical 统计的
statistical analysis 统计分析
statistical design 统计设计
statistical detection theory 统计决策理论
statistical inference 统计推理
statistical method 统计方法
statistical process control（SPC） 统计过程控制
statistical quality control（SQC） 统计质量控制
statistics 统计，统计资料

stator 定子
stator circuit 定子电路
stator core 定子铁芯
stator leakage reactance 定子漏磁电抗
stator winding 定子绕组
stator coil 定子线圈
stator core 定子铁芯
status contact output connection 状态触点输出连接
status input-output card 状态输入/输出卡
status output 状态输出
status report 状态报告
status 状态
status display 状态显示
stay wire 系紧线，拉线
STB（set top box） 机顶盒
steadiness 稳定性
steady 稳定的，不变的
steady direct current 恒稳直流电
steady motion period 稳定运转阶段
steady state 稳态，稳定状态
steady state deviation 静态偏差
steady state error coefficient 稳态误差系数
steady-state availability 稳态有效性
steady-state condition 稳态条件
steady-state deviation 稳态偏差
steady-state error 稳态误差
steady-state power condition 稳态电源条件
steady-state response 稳态响应
steady-state stability 稳态稳定性
steady-state value 稳态值
steam 蒸汽
steam condensate 蒸汽凝结水
steam boiler 蒸汽锅炉
steam converting valve 蒸汽转换阀
steam flow meter 蒸汽流量计
steam generator 蒸汽发生器，蒸汽锅炉
steam plant 蒸汽工厂
steam trace 加热蒸汽管道
steam trap 疏水器
steam turbine 汽轮机
steam-cured concrete 蒸汽养护混凝土

steam-heated exchanger	蒸汽热交换器
steel	钢，钢铁
steel casting iron	碳素钢铸件
steel channel	槽钢
steel fabric	钢筋网
steel guy wire	钢拉线
steel industry	钢铁工业
steel manufacture	钢铁制造
steel pile	钢桩
steel pipe	钢管
steel plate	钢板
steel reinforced aluminium wire	钢芯铝线
steel reinforcement	钢筋
steel structure	钢结构
steel structure bracket tubing	钢结构支架配管
steel wire	钢丝
steel wire netting	钢丝网
steel wire rope	钢丝绳
steel wool	钢丝绒
steel-cored aluminium wire	钢芯铝线
steel-cored copper wire	钢芯铜线
steepest descent	最速下降，急速下降
steering wheel	舵轮，驾驶盘
stem mut	杆螺母
stem seal	填料
stencil pen	铁笔
step	步，位，级，步进，阶跃，步幅
step action	阶跃作用
step by step control	步进控制
step change method	阶跃法
step control	分级控制，分步控制
step down transformer	降压变压器
step error constant	阶跃误差常数
step execution	执行步骤
step function	阶跃函数
step function response	阶跃响应
step generator	阶梯信号发生器
step input	阶跃输入
step motor	步进电机
step pulley	塔轮
step response	阶跃响应
step response model	阶跃响应模型
step response time	阶跃响应时间
step-by-step control	步进控制
step-by-step motor	步进式电动机
step-down transformer	降压变压器
step-function input signal	阶跃函数输入信号
stepless action	无级作用
stepless speed changes devices	无级变速
stepper	台阶器，步进器
stepper motor	步进电机
stepping	步进
stepping action	步进作用
stepping controller	步进控制器
stepping motor	步进电机
stepping relay	步进继电器
stepping switch	步进开关
step-up transformer	升压变压器
stepwise development	分步展开
stepwise refinement	逐步精化
stere vision	立体视角
steric exclusion chromatography	空间排阻色谱法
sterilization and disinfection equipment	消毒灭菌设备
sterilizer	清毒器，杀菌器
stethoscope	听筒，听诊器
STF (strong tracking filter)	强跟踪滤波
stickiness	黏性
sticking	附着
stiffener	加劲杆，硬化剂，增强板
stiffener material	增强板材
stiffening	固化
stiffness	刚度，刚性，劲性，劲度
stiffness coefficient	刚度系数，劲度系数
stimulus	激发
stirred-tank heater	带搅拌储槽加热器
stirrer	搅拌器
stirring rod	搅拌棒
stitch marker	划线轮
stitch welding	针角焊接
stochastic	随机的
stochastic approximation	随机近似
stochastic automation	随机自动化
stochastic complexity	随机复杂性
stochastic control	随机控制
stochastic control system	随机控制系统
stochastic finite automaton	随机有限自

动机
stochastic input 随机输入
stochastic jump process 随机跳变过程
stochastic modelling 随机模型
stochastic parameter 随机参数
stochastic process 随机过程
stochastic programming 随机编程
stochastic property 随机特性
stochastic realization 随机实现
stochastic system 随机系统
stochastic theory 随机理论
stochastic variable 随机变量
stock locater block 定位块
stockpiling area 贮料区
stoichiometric point 化学计量点
Stokes shift 斯托克斯位移
stone ware pipe 粗陶管
stop band filter 阻带滤波器
stop button 停止按钮
stop cock 龙头，停止旋塞
stop collar 限动环
stop end 封端
stop pin 定位销，止动销
stop plate 挡板
stop ring 止动环
stop screw 防松螺钉
stop valves 截止阀
stop watch 秒表
stop 停止
stopcock 活塞，开关
stopper 阻挡器，定位停止销
stopping 停车，制动
stopping phase 停车阶段
storage 存储，存储器
storage area 存放区
storage battery 蓄电池
storage buffer 存储器缓冲器
storage capacity 贮存量
storage device 存储设备，存储装置
storage element 存储元件
storage environment 存储环境
storage oscilloscope 存储示波器
storage program computer 存储程序计算机
storage register 存储寄存器
storage tank 贮水箱
store 存储

storer 存储器
stormwater drain 雨水渠
stormwater drainage system 雨水疏导系统
stormwater main drain 雨水排放主渠
stormwater outfall 雨水渠排水口
stormwater overflow chamber 雨水溢流室
straddle cutter 跨式铣刀
straddle mounting 跨装
straight amplification 直接放大
straight bevel gear 直齿锥齿轮
straight edge 直尺
straight external threaded block valve 直形外螺纹截止阀
straight grooving iron 直槽刨刀
straight pin 圆柱销
straight shaft 直轴
straight side frame 冲床侧板
straight sided normal worm 法向直廓蜗杆
straight toothed spur gear 直齿圆柱齿轮
straightener 整直器，调直器
straightening annealing 矫直退火
straightness 直度
straight-through current transformer 穿心式电流互感器
strain 应变
strain ageing 应变时效
strain capacity 应变量
strain gauge 应变仪
strain gauge load cell 应变式称重传感器
strain measuring instrument 应变仪
strainer 隔滤器，过滤器，滤网
strain-gauge calibration device 应变仪，校准器
strand 线，串
strand wire 绞合线，多股线，绳索
stranded cable 绞合电缆
stranded galvanized steel wire 镀锌钢绞线
strap 皮带
strap wire 带状电线
strategic function 策略函数
stray 杂散

stray capacitance 杂散电容
stray capacity 杂散电容，寄生电容
stray current 杂散电流
stray field 杂散场，漏磁场
stray light 杂散光
stray loss 杂散损耗
stray magnetic field 杂散电磁场
stray voltage 杂散电压
streak 条状痕，条痕，纹理
stream channel 河道，河槽
streaming tape drive 数据流式磁带驱动器
streamline wire 流线型张线
street lighting 街道照明，街道照明设备
strength 强度，力
strength factor 强度系数
strength test 强度测试
strength testing pressure 强度试验压力
strength training 力量训练
strengthen 加固
strengthening works 加固工程
stress 应力，压力
stress amplitude 应力幅，振幅
stress analysis 应力分析
stress concentration 应力集中
stress concentration factor 应力集中系数
stress crack 应力龟裂，应力裂纹
stress diagram 应力图
stress relieving annealing 应力消除退火
stress-strain diagram 应力-应变图
stretch corner 伸拉角
stretch form die 拉伸成形模
stretch zone 伸拉中心
stretcher leveler 拉伸矫直机
stretching vibration 伸缩振动
stretching 拉伸，伸长
striking 拆除
string point 字符串点
string potentiometer 弦丝电位计
strip-chart self-balancing recorder 长图自动平衡记录仪
stripper bolt 脱料螺栓
stripper bushing 脱模衬套

stripper pad 脱料背板
stripper plate 推板，剥料板，脱料板
stripping 剥离工具
stripping method 带状法，截片法
stripping pressure 弹出压力
stroboscope 闪频观测仪
stroke 冲程，行程
stroke end block 行程止销
strong anion exchange 强阴离子交换
strong band 强带
strong cation exchange 强阳离子交换
strong tracking filter (STF) 强跟踪滤波
strongly coupled system 强耦合系统
strontium (Sr) 锶
structural 结构的
structural analysis 结构分析
structural appraisal 结构勘测评估，结构检定
structural behaviour 结构性能
structural condition survey 结构状况勘测
structural constraint 结构约束
structural design 结构设计
structural element 结构构件，结构元件
structural engineering 结构工程
structural formula 分子式
structural frame 结构构架
structural hardware 构件
structural improvement works 结构改善工程
structural integrity 结构完整程度
structural investigation 结构勘查
structural iron 结构钢
structural member 构件
structural optimization 结构优化
structural parameter 结构参数
structural property 结构性能
structural relaxation 结构松弛
structural silicone sealants for building 建筑用硅酮结构密封胶
structural skeleton 结构骨架
structural stability 结构稳定性
structural stability analysis 结构稳定性分析

structural steel　结构钢
structural steel member　结构钢构件
structural strength　结构强度
structural survey　结构勘测
structural use　结构用途
structural works　结构工程
structure　结构，建筑物，构筑物
structure system　结构系统
structure test　结构试验
structured　结构的
structured analysis　结构分析
structured programming　结构化程序设计
structured singular value　结构奇异值
strut　支撑，支柱
stub teeth　短齿
studs　双头螺柱
stylus　触针
sub multiple of a unit　分数单位
sub multiplexer　分多路转换器，分多路转换箱
subassembly　分部件，分组件
subcarrier generator　副载波发生器
subdivide　细分
subjective probability　主观频率
sub-line　支线
submarine　深陷式
submarine gate　潜入式浇口
submarine outfall　海底沟管出口处，海底排放管，海底排污管
sub-mechanism　子机构
submerged discharge valve　淹没式泄水阀
submerged motor pumps　潜水电泵，排污泵
submersible sewage pump　可沉浸的污水泵
submit job　提交任务
suboptimal　次优
suboptimal control　次优控制
suboptimal system　次优系统
suboptimality　次优性
subprogram　子程序
subsequent verification　后续检定
subsidence ratio　递减比，衰减比
subsidiary loop　子回路
subsidiary wire　辅线

subspace　子空间
subspace method　子空间方法
substandard instrument　副标准仪表
substation transformer　配电变压器
substitution　取代，置换
substrate　基底
substructure　下层结构，底层结构
sub-surface building works　地下建筑工程
subsurface initiated fatigue breakage　金属表面斜层初始疲劳破裂
subsynchronous　次同步
subsynchronous oscillations　次同步振荡
subsynchronous resonance　次同步共振
subsystem　子系统
subsystem communication function　子系统通信功能
subsystem integration function　子系统综合功能
subtracter　减法器
subtraction　减法运算
subway　隧道，地道，地下电缆管道，地下铁路
subzero　深冷处理
subzero treatment　生冷处理
suction　吸力
suction fan　排气通风机，抽风机
suction main　吸水干管，吸入总管
suction strainer　吸水口滤网，吸入管过滤器
suction valve　吸水阀
sugar engineering　制糖工程
suitable surrounding　使用环境
sulfate resistance Portland cement　抗硫酸盐硅酸盐水泥
sullage water　淤泥水
sulphated ash　硫酸灰分
sulphate　硫酸酯，硫酸盐
sulphur (S)　硫，硫磺，硫黄
sulphuric acid　硫酸
sum velocity　总速度
summary of machine settings　机床调整卡
summation　总和
summator　求和器
summed current　总电流
summer　加法器

summing 相加，求积
summing amplifier 加法放大器，求和放大器
summing circuit 加法电路
summing element 相加环节，相加元件
summing point 相加点
summing unit 相加单元
sump 集水坑，机油箱
sump lubrication 油槽润滑
sump pit 集水坑，排液槽
sump pump 集水坑泵，油池泵
sun gear 太阳轮，中心齿轮
sunk key 嵌入键，暗键
supercomputer 巨型计算机，超级计算机
superconducting level indicator 超导液位指示器
superconductive 超电导的，超传导现象的
superconductive wire 超导体丝
superconductor 超导体
supercooling 过冷
supercritical fluid extraction 超临界流体萃取
supercritical-fluid chromatography (SFC) 超临界流体色谱法
supercurrent 超导电流
superelevation 超高
superficial area 表面面积
superficial Rockwell hardness tester 洛氏表面硬度计
super-grade thermostat with oil bath 超级恒温油浴
super-grade thermostat with water bath 超级恒温水浴
superheater 过热器
superhet 超外差，超外差收音机
superheterodyne 超外差，超外差收音机
superposition 叠加，重叠
superposition method 叠加法
supersonic generator 超声波发生器
supertension power network 超高压电力网
supervise 监督，监管，督导
supervised training 监督学习

supervision 监督
supervision plan 监工计划书
supervisor 监督人
supervisory computer control 管理计算机控制，监督计算机控制
supervisory computer control system 计算机监控系统
supervisory control 管理控制，监督控制
supervisory control and data acquisition (SCADA) 数据采集与监控系统，监测控制和数据采集
supervisory control computer 管理控制计算机
supervisory information system 监控信息系统
supervisory level network 管理级网络
supervisory relay 监督继电器
supervisory sequence program 监控顺序程序
supervoltage 超高压
supervoltage transmission power line 超高压输电线
supply 供电，供给，电源
supply circuit 供电电路
supply fan 进气扇
supply fault recovery circuit 电源故障恢复电路
supply frequency 电源频率
supply meter 电度表
supply network 供电网
supply pipe 供水管
supply pressure 供气压力
supply regulator 电源调节器
supply source 电源
supply voltage 电源电压，供电电压
support 载体，支座，支承，支持，支撑，承托，承托物
support pillar 撑头，支撑支柱
support pin 支撑销
support plate 载板，托板
supported hole 支撑孔
supporting facilities 辅助设备，辅助设施
supporting frame 支承构架，承重构架
supporting member 支承构件

supporting plate	托板
supporting wire	吊线,支撑线
suppression	正迁移,压缩,抑制
suppression column	抑制柱
suppression ratio	正迁移比,抑制率
surface	表面,表层
surface abrasion test	表面磨耗试验
surface analyzer	表面分析仪
surface asperities	表面粗糙度
surface coefficient of heat transfer	表面传热系数,表面换热系数,表面散热系数
surface conduit	明敷导管,明敷线管
surface contacting thermometer	表面接触式温度计
surface course	表面层
surface deformation	表面变形
surface durability	表面耐久度
surface fatigue	表面疲劳
surface finish	表面抛光
surface friction	表面摩擦
surface gauge	平面规,划针盘
surface gradient	表面坡度
surface grinder	平面磨床
surface hardening	表面硬化处理
surface initiated fatigue breakage	表面初始疲劳破裂
surface ionization detection	表面离子化检测
surface laminar circuit	表面层合电路板
surface mounting	盘面安装,表面安装
surface of action	啮合面
surface of revolution	回转面
surface processing	表面处理
surface resistance	表面电阻
surface resistance thermometer	表面热电阻
surface roughness	表面光洁度,表面粗糙度
surface run-off	地面径流
surface science	表面科学
surface tension	表面张力
surface texture	表面纹理
surface thermocouple	表面热电偶
surface transfer impedance	表面传输阻抗
surface treatment	表面处理
surface water channel	地面水渠道,排水明渠
surface weather station	地面气象站
surface wiring	明线
surface-charge effect	表面电荷效应
surge	电涌,涌波
surge pressure	冲击压力,峰值压力
surge suppressor	电涌抑制器
surge voltage	冲击电压,浪涌电压
surging	波动
survey	勘测,测量,调查
survey plan	测量图
surveying and mapping	测绘科学与技术
surveyor	测量员
susceptance	电纳
susceptibility	敏感性
susceptible	易受影响的
suspend to disk	磁盘唤醒
suspend to ram	内存唤醒
suspended ceiling	垂吊式天花板
suspended truss	悬挂式桁架
suspension	悬架
suspension bridge	吊桥,悬索桥
suspension wire	吊线
sustained oscillation	自持振荡,等幅振荡
swaging	挤锻,挤锻压加工,型锻
swan neck fire hydrant	鹅颈消防栓
sweep frequency	扫描频率
sweep template	造模刮板
sweeper	清扫器
sweeping mold	平刮铸模
swing	摇摆
swing check valves	旋启式止回阀
swing die	振动模具
swing pinion cone	摆动小轮节锥法
swinging base	回转底座
swirl-meter	漩涡流量计
swirlmeter	旋进流量计,涡流式流量计
switch	开关,切换,电闸
switch algorithm	开关算法
switch closure	开关闭合信号
switch desk	开关台,控制台
switch instrument	开关仪表
switch instrument block	开关仪表功

能块
switch off 断开，切断
switch on 合闸，接通
switch position label 开关位置标志
switch position selector 开关位置选择器
switch room 转换室
switch station 开关站
switch terminal 切换开关型端子
switch unit 开关单元
switch blade 闸刀
switchboard 配电盘，电键板
switchbox 开关柜，配电箱
switched 换向，切换
switched capacitor 换向电容器
switched capacitor filter 换向电容滤波器
switched reluctance motor 换向磁阻电机
switching 配电，整流，开关切换
switching algebra 开关代数
switching algorithm 开关算法
switching characteristic 开关特性
switching function 开关函数
switching knob 合闸旋钮，切换按钮
switching law 开关定律，换路定律
switching mechanism 切换机构
switching network 开关网络
switching point 开关点，转接点，切换点
switching power supply 电源箱
switching pulse generator 开关脉冲发生器
switching rectifier 开关整流器
switching surface 开关表面
switching theory 开关理论
switching time 开关时间
switching value 开关值，切换值
switching variable 开关变量
switching off 断开
switching on 接通
switching push button 按钮
switch-over valve 切换阀
swivel 刀转，转环
swivel angle 刀转角
swivel joint 旋转接合，回转接头
symbol 符号，记号
symbol of a unit of measurement 测量

单位符号
symbolic method 符号法
symbolic processing 符号处理
symmetrical component 对称分量
symmetrical deformation vibration 对称变形振动
symmetrical loading 对称荷载
symmetrical network 对称网络
symmetrical phase control 对称相位控制
symmetrical rack 对称齿条
symmetrical rack proportions 对称齿条比例
symmetrical stretching vibration 对称伸缩振动
symmetrical three-phase circuit 对称三相电路
symmetrical two-port network 对称二端口网络
symmetry 对称
symmetry circulating stress 对称循环应力
symmetry factors 对称因子
synaptic plasticity 突触可塑性
synchro control receiver 同步控制接收器
synchro control transformer 同步控制变压器
synchro control transmitter 同步控制发射机
synchro drive 同步传动
synchro error 同步误差
synchro indicator 同步指示器
synchro motor 同步电动机
synchro resolver 同步分解器
synchro switch 同步开关
synchro torque receiver 同步力矩接收器
synchro torque transmitter 同步力矩变送器
synchro 同步，同步机
synchrocyclotron 同步回旋加速器
synchronization 同步，同时性
synchronization control 同步控制
synchronized operation 同步运行
synchronizer 同步器，同步装置
synchronous 同步的
synchronous analysis 同步分析

synchronous architecture 同步建筑
synchronous belt 同步带，同步驱动皮带
synchronous belt drive 同步带传动
synchronous compensator 同步补偿机，同步自耦变压器
synchronous condenser 同步调相机
synchronous data flow 同步数据流
synchronous diagnosis 同步诊断
synchronous documentation 同步文件
synchronous failure 同步故障
synchronous failure and recovery 同步故障与修复
synchronous generator 同步发电机
synchronous identification 同步辨识
synchronous integration 同步积分
synchronous machine 同步机器
synchronous matrix 同步矩阵
synchronous model 同步模型
synchronous motor 同步电动机
synchronous motors system 同步电机系统
synchronous noise 同步噪声
synchronous operation 同步运行
synchronous order 同步阶次
synchronous order reduction 同步降阶
synchronous reactance 同步电抗
synchronous regulator 同步校准器
synchronous reliability 同步可靠性
synchronous security 同步保护
synchronous sensitivity 同步灵敏度
synchronous speed 同步转速
synchronous state estimation 同步状态估计
synchronous synthesis 同步结合
synchronous theory 同步理论
synchronous transfer function 同步传递函数
synchrotron 同步加速器
synergetics 协同学
synthesis 合成
synthesis of mechanism 机构综合
synthesized signal generator 合成信号发生器
synthetic 合成物
synthetic fibre 合成纤维
syren 报警器
syringe pump 注射泵，注射器泵
syringe 注油器，注射器
system 系统，制，方式
system alarm 系统报警
system alarm message panel 系统报警信息画面
system alarm notification 系统报警通知
system alarm window 系统报警窗口
system analysis and design in large scale 大型系统分析与设计
system assessment 系统评价
system builder menu 系统生成菜单
system concept 系统概念
system configuration 系统构成
system control 系统控制，系统调节
system database 系统数据库
system design 系统设计
system engineering 系统工程
system environment (SE) 系统环境
system equipment 系统设备
system error handler 系统误差处理
system function key name 系统功能键名
system gain 系统增益
system homomorphism 系统同态
system integrity module 系统完整性组件，系统完整性模件
system isomorphism 系统同构
system maintenance 系统维护
system maintenance control center 系统维护控制中心
system maintenance panel 系统维护画面
system menu display 系统菜单画面
system message window 系统信息窗口
system methodology 系统方法论
system mounting drawing No. 系统安装图号
system of quantities 量制
system of units of measurement 测量单位制
system overview 系统总貌
system performance evaluation 系统性能评估
system quality assurance engineering 系统品质保证工程
system save 系统存储

system scale 系统等级
system specification 系统规格说明
system stability 系统稳定性
system state executive 系统状态执行程序
system status display 系统状态画面
system status overview window 系统状态总貌窗口
system transfer function 系统传递函数
system utility 系统应用
system view 系统观察
system voltage 系统电压,电网电压
systematic deviation 系统偏差
systematic error 系统误差
systematic uncertainty 系统不确定度
systematology 系统学
systems engineering 系统工程
systems simulation 系统仿真

T

T connection　T形连接
tab gate　搭接浇口，辅助浇口
tab　表格
table origin　表格原点
table profile projector　台式轮廓投影仪
tacho-generator　测速传感器，测速发电机
tachogenerator　测速发电机
tachometer　转速，转速计，转速表
tachometer generator　测速发电机，转速计数传感器
tachometer type flowmeter　转速表式流量计
tack coat　冷油
tag memory　标记存储器
tag wire　终端线
tagged molecule　标记分子
tail gas　废气，尾气
tailing area　拖尾峰
tailing factor　拖尾因子
take out device　取料装置
take-off amplifier　选送放大器
taker　取料机
talc　滑石，云母
talker　扬声器
talk-listen button　通话按钮
tamper　夯土机，捣固机
tamper blade　捣固掌
tandem　串联
tangency　接触
tangent　切线，正切
tangent galvanometer　正切检流计，正切电流计
tangent mechanism　正切机构
tangent plane　切平面
tangent point　切向点
tangential force　切向力
tangential load　切向负荷
tank　箱，油缸罐，槽
tank gauge　液位计
tank level meter　油箱液位计
tank lorry ball valves　槽车球阀

tantalum (Ta)　钽
tantalum wire　钽丝
tap　分接头，攻螺纹，攻丝
tap casting　顶注
tap position　档位
tape　磁带
tape drive　磁带驱动器，纸袋驱动器，纸袋驱动装置
tape feed　磁带馈送
taper　锥度
taper key　斜键，钩头楔键
taper pin　锥形销
taper turning　锥度车削
tapered nut　锥形螺母
tapered roller　圆锥滚子
tapered roller bearing　圆锥滚子轴承
tapered transformer　锥形变压器
tapping　开孔
target　目标，指标
target computer　目标程序计算机
target control　目标控制
target flow measuring device　靶式流量测量仪表
target flow transmitter　靶式流量变送器
target flowmeter　靶式流量计
target position indicator　目标位置指示器
target test function　对象测试功能
target tracking　目标跟踪
target tracking filter　目标跟踪滤波器
target-type flowmeter　靶式流量计
tasking program　任务程序
Taylor control language　泰勒控制语言
Taylor ladder logic　泰勒梯形逻辑
Taylor remote I/O　泰勒远程I/O
Taylor series approximation　泰勒级数近似
TCD (thermal conductivity detector)　热传导检测器
TEA (thermal energy analyzer)　热能

分析仪
teaching programming 示教编程
teak 柚木
teardrop pad 泪滴盘
technetium (Tc) 锝
technical and economic evaluation 技术经济评价
technical assistance center 技术援助中心
technical conditions 技术条件
technical information 技术资料
technical requirements 技术要求
technical specification 技术规格
technical specifications 技术条件
technical specifications for centrifugal pumps 离心泵技术条件
technique process 技术过程
technique system 技术系统
technological design 工艺设计
technological forecasting 技术预报
technological support network 技术支持网络
technology 技术
technology of metals 金属工艺学
technology transfer 技术转换
teflon insulated wire 聚四氟乙烯绝缘电线
telecommunication 电信,远程通信,无线电通信
telecontrol 遥控
telegraph wire 电报线
telemanipulation 遥控操纵
telemechanical apparatus 运动装置
telemechanics 远动学,遥控动力学
telemechanism 遥控机构
telemeter 测距器
telemetering 遥测
telemetering eddy current detector 遥测涡流探伤仪
telemetering equipment 遥测装置
telemetering system 遥测系统
telemetering system of frequency division type 频分遥测系统
telemetry 遥测,遥控技术
teleological system 目的系统
teleology 目的论
telephone 电话,打电话

telephone call wire 电话挂号线
telephone line 电话线
telephone network 电话网络
telerobotics 遥控机器人技术
telescope 望远镜
telescope detector 望远镜探测器
telescopic gauge 伸缩性量规
teleswitch 遥控开关
television 电视
television antenna 电视天线
television equipment 电视设备
television scanning generator 电视扫描发生器
television system 电视系统
television tower 电视塔
television transmitter 电视发射机
televisor 电视接收机
tell-tale 驾驶动作分析仪
tellurium (Te) 碲
temper brittleness 回火脆性
temper colour 回火颜色
temperature 温度
temperature calculation 温度计算
temperature chamber 恒温箱
temperature coefficient 温度系数
temperature compensated electrometer 温度补偿静电计
temperature compensating device 温度补偿装置
temperature control 温度控制
temperature control loop 温度控制回路
temperature control relay 温度控制继电器
temperature controller 温度调节仪表
temperature converter 温度变换器
temperature distributions 温度分布
temperature error 温度误差
temperature factor 温度系数
temperature humidity infrared radiometer 温湿度红外辐射计
temperature indicators 温度指示
temperature measurement 温度测量
temperature measuring instrument 温度测量仪表
temperature meter 温度计
temperature profile 温度轨线

temperature recorder 温度记录仪
temperature rising 温升
temperature scale 温标
temperature sensor 温度传感器
temperature stability 温度稳定度
temperature transducer 温度传感器
temperature transmitter 温度变送器
temperature uniformity 温度均匀性
temperature variation 温度变化
temperature compensation 温度补偿
tempering 回火
tempering crack 回火裂痕
template base 模板库
templet directory 模型目录
templet display 模型画面,模型显示
temporal 暂时的
temporal logic 暂态逻辑
temporal reasoning 暂态推理
temporary base strain gauge 临时基底应变计
temporary construction 临时结构
tender 投标,标书
tender assessment 投标评估,评审标书,标书评估
tender selection criteria 选标准则
tenon 榫头,凸榫
tenon saw 榫锯,夹背锯
tensile force 拉力
tensile gauge 张力计,拉力计
tensile impact test 拉伸冲击试验
tensile reinforcement 抗拉钢筋
tensile strength 抗拉强度
tensile strength of asphalt 沥青的拉力
tensile stress 拉应力
tensile testing machine 拉力试验机
tensile 拉力的,张力的
tensiometer 张力计
tension 张紧力,拉力,张力,压力,拉紧
tension assembly 拉条组件
tension force 张力
tension pulley 张紧轮
tension test 张力试验
terbium (Tb) 铽
terminal 终点,终端,终端机,接线端,电路接头

terminal block 端子板,接线端子板
terminal box 端子箱
terminal cable 终端电缆
terminal clearance hole 端接全隙孔
terminal connection 端子连接
terminal control 终点控制,终端控制
terminal generation 试验发生
terminal indicator 终端指示器
terminal length 试验长度
terminal panel (TP) 终端面板
terminal reliability 终点可实现性
terminal voltage 端电压,终点电压
terminal with adjustable resistance 可调电阻接线端子
terminal with fuse 熔断器接线端子
terminal with switch 接线端子
terminal-based conformity 端基一致性
terminal-based linearity 端基线性度
terminal-type digital I/O modules 端子型数字量输入/输出模件
termination 终端适配器,终止
terminator 终结器
terminology of centrifuge and filter 离心机和过滤机名词术语
terminology standard 术语标准
terms 项,期限,术语
terms and definitions of water-turbine pumps 水轮泵名词术语及定义
ternary logic 三重逻辑
terrace 平台,露台,台阶
terraced platform 梯状平台
terraced slope 梯状斜坡
terrain 地形
terrain vehicle mechanics 地面车辆力学
terrestrial magnetic field 地磁场
terrestrial radiation 地球辐射
tesla 特［斯拉］(磁通量密度单位)
test 试验,测试,检测
test board 测试板,实验盘
test box 测试箱
test coil 试验线圈
test data 试验数据
test data adequacy 试验数据充分度
test equipment 测试设备,实验设备

test for nominal samples 标样试验
test frequency 激励频率，试验频率
test function 测试功能
test load 测试荷载
test mass 试验质量
test method for hardness of fiber reinforced plastics by means of a Barcol impressor 纤维增强塑料巴氏硬度试验方法
test method for specular gloss 镜面光泽度试验方法
test method for thermal conductivity of glassfiber reinforced plastics 玻璃钢导热试验方法
test methods 试验方法
test methods for compressive strength 压缩强度试验方法
test methods for lightweight aggregates 轻骨料试验方法
test methods for wall bricks 砌墙砖试验方法
test methods of aerodynamic performance for fans 通风机空气动力性能试验方法
test of axial compressive strength 轴心抗压强度试验
test of bending strength 抗折强度试验
test of compressive strength 抗压强度试验
test of tensile splitting strength 劈裂抗拉强度试验
test operator control station 测试操作员控制站
test piece 试件
test point 测试点
test procedure standard 试验程序标准
test pump 试验泵
test result 试验结果
test run 测试运行
test sample 测试样本
test signal 测试信号
test solution 试验溶液
test space 试验空间
test strip 测试条
test surface 探伤面
test table 工作台
test tube 试管

test value 测量值
testability 可测试性
tester 检测者，校验器，检测器
testing bench 试台
testing device 试验装置
testing facility 试验设备
testing machine 试验机
testing methods of water-turbine pump 水轮泵试验方法
testing of motor vehicle 汽车试验
testing of refrigerating systems 制冷装置试验
testing stand for pneumatic instruments 气动仪表校验台
testing system flexibility 试验系统的柔度
testing voltage 试验电压
testorder 试验顺序
tetracyclines 四环素类
text editor 文本编辑程序
text input port 文本输入口
textile chemistry and dyeing and finishing engineering 纺织化学与染整工程
textile engineering 纺织工程
textile glass 纺织玻璃纤维
textile science and engineering 纺织科学与工程
texture 构造，纹理，咬花
thallium (Tl) 铊
the compressive strength 抗压强度
thebaine 蒂巴因
thematic mapper 专题制图仪
theodolite 经纬仪
theorem 定理，原理，法则
theorem proving 定理证明
theoretical 理论的
theoretical line of action 理论啮合线
theoretical models 理论模型
theoretical slope factor 理论斜率因数
theory 理论，学说
theory and new technology of electrical engineering 电工理论与新技术
theory of constitution 组成原理
theory of machines and mechanisms 机械原理
therapeutic drug monitoring 治疗药

物浓度监测
therapy model 治疗模型
thermal 热的,热量的,由热驱动的
thermal analysis 热分析,热学分析
thermal analysis curve 热分析曲线
thermal analysis instrument 热分析仪器
thermal analysis range 热分析范围
thermal analyzer 热分析仪
thermal bulb 测温包,测温筒
thermal capacitance calorimeter 热容式热量计
thermal capacity 热容量
thermal chemical gas analyzer 热化学式气体分析器
thermal conductivity 导热性,热导率
thermal conductivity cell 热导池
thermal conductivity detector (TCD) 热传导检测器
thermal conductivity gas analyzer 热导式气体分析器
thermal conductivity gas sensor 热导式气体传感器
thermal conductivity gas transducer 热导式气体传感器
thermal conductivity humidity sensor 热导式湿度传感器
thermal conductivity humidity transducer 热导式湿度传感器
thermal conductivity meter 热导率计
thermal conductivity of mixture gas 混合气体热导率
thermal conductivity superconductor 超导热体
thermal cone 温度锥
thermal convertor 热电变换器
thermal cycle 热循环
thermal degradation 热降解,热老化
thermal detector 感温探测器
thermal diffusivity 热扩散率,热扩散系数
thermal dilatometer 热膨胀仪
thermal electricity 热电
thermal electromotive force 热电动势,温差电动势
thermal energy analyzer (TEA) 热能分析仪

thermal equilibrium 热平衡
thermal error 温度误差
thermal expansion 热膨胀
thermal expansion coefficient 热膨胀系数
thermal fatigue testing machine 热疲劳试验机
thermal flowmeter 热式流量计
thermal hysteresis 热滞现象
thermal infrared range remote sensing 热红外遥感范围
thermal instrument 热系仪表,电热仪表
thermal insulation 隔热
thermal ion detector 热离子检测器
thermal ionization mass spectrometer 热电离质谱计
thermal magnetic oxygen analyzer 热磁式氧分析器
thermal mass flowmeter 热式质量流量计
thermal neutron 热中子
thermal nitridation 热渗氮
thermal noise 热噪声
thermal output 热输出
thermal output coefficient 热输出系数
thermal physical property tester 热物理性能测定仪
thermal power engineering 热能工程
thermal printer 热敏印刷机,热感式打印机
thermal property 热特性,热性质
thermal radiation 热辐射
thermal radiator 热辐射体
thermal reactor 热反应堆
thermal recorder 热式记录仪
thermal refining 调质处理,热精炼
thermal relay 热继电器
thermal response time 热响应时间
thermal sensitivity drift 热灵敏度漂移
thermal shock 热冲击
thermal shock test 热冲击试验
thermal shock test chamber 热冲击试验箱
thermal stability 热稳定性

thermal strain 热应变
thermal stress 热应力
thermal time-delay relay 热效式延时继电器
thermal titration 热滴定,热滴定法
thermal transmittance value 传热值
thermal wave electron image 热波电子像
thermal wave electron microscope 热波电子显微镜
thermal zero drift 热零点漂移
thermal conduction 热传导
thermal couple 热电偶
thermal relay 热继电器
thermal-conductivity hydrogen analyzer 热磁式氢气分析器
thermionic 热电子的
thermistor 热敏电阻,电热调节器
thermistor bolometer 热敏电阻辐射测量仪
thermistor chain 热敏电阻测温链
thermistor flowmeter 热敏电阻流量计
thermo hydrometer 温差式比重计
thermo spray 热喷雾
thermo system 热分析系统
thermobalance 热天平
thermocouple 热电偶,温差电偶
thermocouple assembly 热电偶组件
thermocouple burnout 热电偶烧断
thermocouple circuit 热电偶电路
thermocouple element 热电偶元件
thermocouple input module 热电偶输入模件
thermocouple instrument 热电偶仪表,热偶式仪表
thermocouples 热电偶,探针
thermodynamic scale (of temperature) 热力学温标
thermoelectric effect 热电效应,温差电效应
thermoelectric pyrometer 热电高温计
thermoelectric wire 热电导线
thermoelement 热电偶,温差电元件
thermogravimetric analysis 热重分析法
thermometer 温度计
thermometer well 温度计套管
thermonuclear 热核的

thermo-operated steam trap 热动式疏水器
thermoplastic 热塑性,热塑性塑料
thermoplastic material 热塑材料
thermoplastic resin 热塑性树脂
thermoplastic road marking paint 热塑路标漆
thermoplastic-covered wire 塑料绝缘电线,聚氯乙烯绝缘线
thermoregulator 温度调节器,调温器
thermo-relay 热电偶继电器,温差电偶继电器
thermo-sensitive element 热敏元件
thermosetting 热固性
thermosetting plastic 塑胶
thermosetting resin 热固性树脂
thermostat 自动调温器,稳定调节,恒温器
thermowell 热电偶套管
thero expansion 热膨胀
Thevenin's theorem 戴维南定理
thick film 厚膜
thick film amplifier 厚膜放大器
thick film circuit 厚膜电路
thick film trimming potentiometer 厚膜微调电位器
thicked 加厚
thickness 厚度
thickness gauge 厚薄规,厚度计
thickness meter 厚度计
thickness on pitch circle 节圆齿厚
thickness taper 齿厚收缩
thimble 套管,套圈
thin copper foil 薄铜箔
thin film 薄膜
thin film chromatography 薄膜色谱法
thin film Hall effect magnetometer 薄膜霍尔效应磁强计
thin film hybrid circuit 薄膜混合电路
thin gate valves 薄型闸阀
thin laminate 薄层压板
thin layer chromatography (TLC) 薄层色谱法,薄层层析法
thin layer plate 薄层板
thin layer rod chromatography 薄层棒色谱法
thin layer scanner 薄层扫描仪

thin-film notch filter 薄膜陷波滤波器
thinner 稀释剂
third harmonic 三次谐波，第三谐波
thorium (Th) 钍
thoroughfare 大道，通路
thread 螺纹
thread count 织物经纬密度，线程数
thread cutting 螺纹切削
thread pitch 螺距
thread processing 螺纹加工
threaded bolt 螺纹栓
threading 穿线
three phase fault 三相故障
three plates mold 三片式模具
three start screw 三条螺纹
three state controller 三位控制器
three valve bypass manifold 三阀组
three-axis attitude stabilization 三轴姿态稳定
three-column transformer 三绕组变压器
three-conductor power cable 三线电源电缆
three-digit DC digital voltmeter 三位数字显示直流电压表
three-dimensional cam 三维凸轮
three-mode controller 三作用控制器
three-phase 三相
three-phase AC 三相交流电
three-phase asynchronous motor 三相异步电动机
three-phase circuit 三相电路
three-phase four-wire system 三相四线制
three-phase four-wire watt-hour meter 三相四线电度表
three-phase generator 三相发电机
three-phase half wave rectifier circuit 三相半波整流电路
three-phase load 三相负载
three-phase motor 三相电动机
three-phase power network 三相电力网
three-phase power transmission 三相电力传输
three-phase squirrel cage motor 三相笼式电动机
three-phase synchronous generator 三相同步发电机
three-phase transformer 三相变压器
three-phases system 三相制，三相系
three-point chiral recognition model 三点手性识别模式
three-point plug 三点插头
three-position controller 三位式调节器，三位式控制器
three-position on/off controller 三位式开关调节器
three-step control 三位控制，三位调节
three-step controller 三位控制器，三位调节器
three-term 三项
three-term action 三作用
three-term control 三作用控制
three-term controller 三作用控制器，三作用调节器
three-way conduit fitting with cover 电气带盖三通
three-way solenoid valve 三通电磁阀
three-way valve 三通阀
three-wire system 三线制
threshold 阈值，门限，临界值，阈限，界限，门槛
threshold current 门限电流
threshold decomposition 门限消失
threshold element 门限元件
threshold function 阈值函数
threshold level 门槛值，阈电平
threshold logic 阈值逻辑
threshold of resolution 阈值分辨率
threshold selection 阈值选择
threshold value 阈值
threshold voltage 门限电压
throttle 节流阀，风门
throttle valve 节流阀
throttling device with drilled holes 法兰上钻孔取压的节流装置
throttling type instrument block valve 节流式仪表截止阀
through crack 贯通裂缝
through traffic 直通交通

through-hole form 通孔形式
throughput 生产能力
thrust ball bearing 推力球轴承
thrust bearing 推力轴承
thrust pin 推力销
thrust vector control system 推力矢量控制系统
thrust 冲击,推力
thruster 推力器
thulium (Tm) 铥(稀土金属元素)
thumb 大拇指
thumb screw 蝶形螺钉,大头螺丝
thymol blue 百里酚蓝
thyristor 晶闸管,闸流管
tick-mark farside 反面压印
tick-mark nearside 正面压印
tidal power station 潮汐发电站
tidal range 潮差
tie 系杆,系条
tie band 捆扎用带材
tie bar 拉杆
tie rod 拉杆,系杆
tier 层级
tight 紧密的,紧固的
tighten 扣紧,变紧
tightening wire 拉线
tight-side 紧边,紧固件
tile 瓷砖,瓦
tilt 刀倾,倾斜
tilt angle 刀倾角,倾角
tilting manometer 倾斜式压力计
tilting 摆动
timber partition 木料间隔
timber pile 木桩
timber yard 木料堆置场
time 时间
time base sweep multivibrator 时基扫描多谐振荡器
time constant 时间常数
time control 时间控制
time cycle controller 时间周期控制器
time delay 纯滞后,时间延迟,时延,延迟时间
time delay compensation 时间滞后补偿
time delay device 延时装置
time delay estimation 时延估计
time delay model 时间延迟模型
time delay relay 延时继电器,时间继电器
time delay spread 时延扩张
time domain analysis 时域分析
time element 限时元件,延时元件
time equipment 定时装置
time invariant 时不变的
time invariant system 时不变系统
time lag 时滞,时间滞后
time lag relay 延时继电器
time limit 时限
time meter 计时表
time of concentration 集流时间
time of response 响应时间
time Petri-net 时域Petri网
time programme 时间程序
time programmed control 时序控制
time proportioning control 时间比例控制
time proportioning on/off controller 时间比例开关调节器
time quenching 时间淬火
time relay 时间继电器
time response 时间响应,时域响应
time reversal 时域逆变器
time schedule control 时域规划控制
time schedule controller 时域规划控制器,时序控制器
time setting 时间整定
time setting dialog 时间设定对话
time setting range 时间整定范围
time shared control 分时控制,时域共享控制
time signal 时域信号
time synchronization 时域同步
time system 时域系统
time variable 计时变量
time varying system 时变系统
time stamp resolution 时间标签分辨率
timed pulse 计时脉冲,同步脉冲
time-domain 时域
time-domain analysis 时域分析
time-domain calculation 时域计算
time-domain correlation 时域相关
time-domain method 时域方法
time-domain reflectometer 时域反射仪,时域反射计

time-domain response 时域响应
time-domain responses for discrete time model 离散时间模型时域响应
time-domain spectroscopy 时域分光光度计
time-domain synthesis 时域综合
time-frequency 时域频率
time-frequency localization 时域频率定位
time-frequency representation 时域频率表示
time-integral criteria 时间积分判据
time-invariant 时域不变式
time-invariant plant 时域不变对象
time-invariant system 时域不变系统，定常系统
time-invariant systems 时不变系统
time-lag relay 延时继电
time-limit protection 时限保护
time-optimal control 时间优化控制
time-oriented sequential control 时间顺序控制
time-phase 时间相位
timer 计时器，定时器
timers point 时钟点
time-series analysis 时间串联分析
time-sharing control 分时控制
time-sharing program 时域共用程序
time-sharing system 时域共用系统
time-slot assignment 时域通道分配
time-varying parameter 时变参数
time-varying plant 时变对象
time-varying system (TVS) 时变系统
timing 定时，计时器定时器
timing analysis 定时分析
timing circuit 计时电路
timing jitter 定时振动
timing pulse 定时脉冲
timing ramp amplifier 定时斜坡放大器
timing recovery 定时复位
timing relay 时间继电器，时限继电器
timing simulation 时序模拟
tin 锡，白铁，马口铁
tin plated steel sheet 镀锡铁板
tin solder 焊锡
tinned iron 镀锡铁板，镀锡铁皮
tinned wire 镀锡铜线
tinning 镀锡
tinplate 马口铁，洋铁
tiny drag slow shut check valves 微阻缓闭止回阀
tip radius 齿顶圆角半径
tipper 倾卸斗车，运泥车
titanium (Ti) 钛
titanium wire 钛线，钛丝
titration 滴定
titrator 滴定仪
titrimetry analysis 滴定分析法
T-junction 三通
TLC (thin layer chromatography) 薄层色谱法
toaling hole 定位孔
toggle 触发器
toggle mechanism 肘形机构，弯头接合
toggle press 肘杆式压力机
toggle switch 触发控制，钮子开关
toggle type mould clamping system 肘杆式锁模装置
token pass 令牌传送
token passing 令牌传递方式
token ring 令牌环
token 标志
token-ring protocol 令牌规程
tolerance 冗余度，偏差容限，公差
tolerant 冗余的
toluene 甲苯
tong-test ammeter 钳式电流表
tonne (t) 吨（质量单位）
tool 工具，刀具
tool advance 刀具进刀
tool edge radius 刀刃圆角半径
tool for lathe 车刀
tool point angle 刀刃角
tool point width 刀顶距
tool post 刀架
tool withdrawal groove 退刀槽
tool box 工具箱
tooth angle 齿角
tooth bearing 齿支撑面，轮齿接触面
tooth contact analysis 轮齿接触分析
tooth contact pattern 轮齿接触斑点
tooth curve 齿廓曲线
tooth horizontal 齿水平面

tooth layout 轮齿剖面图
tooth number 齿数
tooth plane iron 梳形刨刀
tooth profile 齿形，齿廓
tooth ratchet mechanism 齿式棘轮机构
tooth space 齿槽
tooth spacing testing 齿距检查仪
tooth spiral 齿螺旋线
tooth surface 轮齿表面
tooth taper 轮齿收缩
tooth thickness 齿厚
tooth trace 齿线
tooth vertical 齿垂直面
toothed belt idler plate 皮带惰轮锯齿板
tooth-mesh frequency 齿啮合频率
tooth-to-tooth composite variation 一齿度量中心距变量
top 齿顶
top block 上垫脚
top gate 顶注浇口
top plate 上托板，顶板
top rail 顶栏杆，上横梁
top relief angle 顶刃后角
top slope angle 刀齿顶刃倾角
top storey 顶层
top wire 上网，面网
top-down method 自上向下方法
top-down testing 自上而下测试
topographic survey 地形测量
topography 地形，地势，地志
topological structure 拓扑结构
topology graph 拓扑图
torch-flame cut 火焰切割
toroid helicoids worm 环面蜗杆
torpedo spreader 鱼雷形分流板
torque 力矩，转矩，扭矩，转矩
torque amplifier 力矩放大器
torque at starting 启动转矩
torque constant 扭矩常数
torque control 力矩控制
torque motor 力矩电动机
torque rod 扭矩杆，反作用杆
torque synchro 力矩同步
torque variator 转矩变换器
torque wrench 扭力扳手
torquemeter 转矩计

torsion 扭力，扭曲
torsion load 扭转载荷
torsion stress 扭转应力
torsion test 扭曲试验
torsion wire 扭力丝
torsional force 扭力
torsional strength 扭转强度
torsional vibration damper 扭振阻尼器
torsional vibration frequency 扭振频率
tosecan 划线盘
total 合计，总计
total carbon analysis 总碳分析
total collapse 整体坍塌
total composite tolerance 总综合公差
total composite variation 总度量中心距变量
total contact ratio 总重合度，总接触比
total current 总电流
total distributed control 集散控制
total error 总误差
total index tolerance 总分度公差
total index variation 总分度变动量
total load 总荷载
total losses 总损耗
total number of turns 总匝数
total power 总功率
total pressure Pitot tube 总压毕托管
total productive maintenance (TPM) 全面生产维护，全面生产保养
total quality control (TQC) 全面质量管理
total quality management (TQM) 全面品质管理
total resistance 总电阻
total settlement 总沉降
total voltage 总电压
total wt. 总重量
totalization 整体化
totalizer 加法计算器
totalizing measuring instrument 累计仪表
totally integrated power 全集成能源管理
touch screen 触摸屏，触屏
tough bronze 韧青铜
tough pitch copper 韧铜，紫铜
toughness 韧性

tower 塔
tower crane 塔式吊机，塔式起重机
towing cable 牵索
towing force 牵力
town gas 民用煤气，民用燃气
tox machine 自铆机
TP（terminal panel） 终端面板
TPM（total productive maintenance） 全面生产维护，全面生产保养
TQC（total quality control） 全面质量管理
TQM（total quality management） 全面品质管理
traceability and audit records 可追踪的审计数据
tracer 描图员
tracer element 示踪原子
tracing 描图，描摹图，摹绘图
tracing pipe 伴热管线
track 轨道
track circuit wire 轨道电路连接线
trackball 球标，轨迹球，跟踪球
tracker wire 跟踪线
tracking 跟踪
tracking application 跟踪应用
tracking characteristic 跟踪特性
tracking error 跟踪误差
tracking system 跟踪系统
traction 牵引，牵引力
traction assistance 辅助索引
traction control 索引控制
traction yaw control 索引偏航控制
tractive 牵引的，曳引的
tractive effort torque 牵引力扭矩
tractor 牵引机，拖拉机
trade effluent 工业污水，工业废水
trade test 行业技能测验
trade-off analysis 权衡分析
traffic actuated signal 车辆触发交通灯号
traffic capacity 交通容量，容车量，运输能力
traffic control 交通控制
traffic control and surveillance system 交通管制及监视系统
traffic control centre 交通控制中心
traffic corridor 交通走廊
traffic information engineering & control 交通信息工程及控制
traffic light 交通灯
traffic light signal 交通灯号
traffic noise 交通噪声
traffic sign 交通标志
traffic signal 交通信号（灯）
traffic system 运输系统
traffic volume 行车量，交通容量
trailer 拖车，拖板车
train control 火车控制
training 训练
trajectory 轨迹，弹道
trajectory planning 轨迹计划
tramway 电车轨道
transaction 会刊
transceiver 收发报机，收发器
transcoder 译码器
transconductance 跨导，互导
transcriber 转录器
transducer 变换器，转换器，换能器
transductor 饱和电抗器
transductor element 饱和电抗器元件
transfer 传递，传送，转换
transfer device 转换装置
transfer element 传递环节，传递元件
transfer feed 连续自动送料
transfer function 传递函数，转移函数
transfer function in the frequency domain 频率特性函数
transfer function matrix 传递函数矩阵
transfer station 转运站
transfer switch 转换开关
transferred 已转运
transform 变换
transformation 变换，交换，转换
transformation grammar 转换文法
transformation matrix 转换矩阵
transformation ratio 变比，变压系数
transformation ratio of voltage transformer 变压器的变压比
transformed value of a measured quantity 被测量的变换值
transformer 变压器
transformer amplifier 变压器耦合放大器
transformer capacity 变压器容量

transformer coupling 变压器耦合
transformer oil 变压器油
transformer room 电力变压房,配电站
transformer station 电力变压站,变电所
transformer substation 变电站
transient 瞬态,暂态,瞬变,瞬变的,过渡的,瞬时
transient analysis 暂态分析
transient current 暂态电流,过渡电流
transient deviation 瞬态偏差
transient electrical discharges 暂态放电
transient energy transfer 瞬态能量传递
transient error 暂态误差
transient grating 瞬态光栅
transient load 瞬时荷载
transient oscillation 暂态振荡
transient overshoot 过冲,瞬态超调
transient overvoltage 瞬态过电压
transient power disturbance 瞬态电源扰动
transient process 过渡过程
transient radiation 瞬态辐射
transient response 暂态响应,瞬态响应
transient scattering 瞬态散射
transient signal 暂态信号
transient solution 暂态解
transient stability 瞬态稳定性,暂态稳定
transient stability analysis 暂态稳定性分析
transient stability assessment 暂态稳定性评估
transient state 瞬时状态
transient state component 暂态分量
transient state travelling wave 暂态行波
transient system deviation 瞬时系统误差
transient torque 瞬时力矩
transient vibration 瞬时振动
transient voltage 暂态电压,过渡电压
transistor 晶体管,半导体管
transistor amplifier 晶体管放大器
transistor circuit 晶体管电路

transistor multistage amplifier 晶体管多级放大器
transistor parameter tester 晶体管参数测试仪
transistorized millivoltmeter 晶体管毫伏表
transistorized surface film thickness meter 晶体管化表面膜厚计
transistorized voltage stabilizer 晶体管稳压电源
transistor-transistor logic 晶体管-晶体管逻辑
transition 转变
transition point 转变点
transition diagram 推移图,转移图
transition matrix 转移矩阵,跃迁矩阵
transition mode 转移模式
transition noise 转移噪声
transition system 转移系统
transition time 转移时间
transmissible pressure gauge 电远传压力表
transmission 输送,传递,发射,传动,传输
transmission angle 传动角
transmission characteristic 传输特性
transmission electron microscope 透射式电子显微镜
transmission gear 变速箱,传动齿轮
transmission line 传输线,输电线
transmission line malfunction 输电线路故障
transmission line matrix 传递矩阵
transmission line resonator 反射谐振器
transmission network 传输网络,输电网
transmission of electricity 输送电力
transmission rack 输送架
transmission range 传输范围
transmission ratio 传动比
transmission shaft 传动轴
transmission system 传输系统
transmission type electron microscope 透射式电子显微镜
transmission zero 传输零点

transmit 传递
transmitter 变送器,发送器,发射机
transmitting 发送,发射,传动,传递
transmitting selsyn 自整角发送机
transmitting unit 变送单元
transom window 顶窗,气窗
transparency 透明性
transparent lacquer 透明漆
transponder 转发器
transport 运输
transport delay 输送延迟,运输滞后
transport demand 运输需求
transport interchange 交通交汇处
transport layer 传递层,传输层
transport property 传输特性,迁移特性
transportation 运输
transportation and storage condition 运输和储存条件
transportation control 运输控制
transportation facilities 运输设施
transportation lag 传输滞后
transportation planning and management 交通运输规划与管理
transshipment depot 中转仓库
transtat 可调变压器
transversal filter 横向滤波器,截断滤波器
transverse circular pitch 端面齿距
transverse circular thickness 端面弧齿厚
transverse contact ratio 端面重合度
transverse diametral pitch 端面径节
transverse interference 横向干扰
transverse module 端面模数
transverse parameters 端面参数
transverse plane 端面
transverse pressure angle 端面压力角
transverse space-width taper 端面槽宽收缩
transverse thickness taper 端面齿厚收缩
transverse tooth profile 端面齿廓
transverse wind 横向风力
trap 隔气弯管,存水弯管
trap circuit 陷波电路
trap outlet 隔气弯管去水口
trapezoidal channel 梯形渠,梯形河槽

trapezoidal steel wire 梯形钢丝
trapezoidal wave 梯形波
trapped gully 装有隔气弯管的集水沟
trapper 陷波器
trapping 隔气
travel 过程、运转、进行、移动
traveling wave protection 行波保护
travelling 旅行,移动
travelling crane 移动式起重机,行车
travelling derrick crane 移动式转臂起重机
travelling standard 移动标准器
travelling wave 行波
travelling wave amplifier 行波放大器
travelling wave modulator 行波调制器
travelling wave relay 行波继电器
travelling wave signal 行波信号
travelling-wave transponder 行波应答器
traverse station 导线测量站
traversing crane 桥式吊车
tread 楼梯踏板
treatment 处理
tree 树
tree search 树形检索
tree structure 树形结构
trembler bell 电铃
tremie 混凝土导管
trench 线坑,坑道
trench excavator 挖坑机
trend 趋势,倾向
trend analysis 趋势分析
trend function 趋势功能
trend instrument 趋势仪表
trend panel 趋势画面
trend point panel 点画面
trend point window 趋势点窗口
trend record definition 趋势记录定义
trend recorder 趋势记录仪
trend window 趋势窗口
trial and error tuning 试差整定
trial hole 试孔
trial run 试车,操作测试
triangle-shape grade of membership function 三角形隶属度函数
triangular file 三角锉
triangular resonator 三角波谐振器
triangular symbol 三角符号

tribology 润滑与磨损学，摩擦学
tribology design 摩擦学设计
trickle charge 涓流充电
trigger 触发器
trigger circuit 触发电路，同步起动电路
trigger comparator 触发比较器
trigger element 启动元件
triggering level 触发电平
triggering signal 触发信号
trigger-operated tension meter 触发传动式张力计
trigonometric transformations 三角变换
trim 切边，修边，剪外边
trimming 去毛边，整缘加工
trimming die 切边模
trimming punch 切边冲头
trinitrotoluene 三硝基甲苯，烈性炸药
triode 三极管
triphase 三相
triple modulation telemetering system 三重调制遥测系统
triple point of water 水的三相点
triple valves 三通阀
tripler 三倍器
tripod 三脚架
trochometer 里程计，车程计
trolley 台车，手推车
trolley bus 无轨电车
trolley wire 架空线，接触导线
trouble shooting 故障分析，故障查找
trouble spot 故障点
trouble 事故，故障，干扰
trowel 灰匙，泥铲，泥刀
truck crane 起重汽车，卡车起重机
true position 位置度
true power consumption 有效功率消耗
true value 真值
true 真实的
truncated Venturi tube 截尾文丘里管
trunk road 干路
trunk route 干线，主航线
trunk sewer 污水干管
truth table 真值表
tube 筒管，管子，电子管
tube voltmeter 电子管电压表
tubing gauge 油管压力表

tubular compass 管式罗盘
tubular diesel pile hammer 筒式柴油打桩锤
tubular fluorescent lamp 管状荧光灯
tubular handrail 管状扶手
tubular railing 管状栏杆
tunable filter 可调谐滤波器
tunable preamplifier 可调前置放大器
tune 调节
tuned power amplifier 调整电源放大器
tuner 调谐器
tungsten (W) 钨
tungsten filament lamp 钨丝灯
tungsten wire 钨丝
tuning 整定，调谐
tuning characteristic 整定特性
tuning control systems 控制系统整定
tuning panel 调整画面
tuning parameter printout 调整参数打印输出
tuning parameter save function 调整参数存储功能
tuning trend 调整趋势
tuning-fork level switch 音叉式液位控制开关
tunnel 隧道
tunnel cable 隧道电缆
tunnel junction 隧道连接
tunnel junction receiver 隧道连接器
tunnel lining 隧道衬砌
tunnel portal 隧道口
tunnel ventilation fan 隧道通风风扇
tunnelling works 隧道工程，开挖隧道工程
turbidity 混浊度
turbidity meter 浊度计
turbine 涡轮机
turbine flow measuring device 涡轮流量测量仪表
turbine flow transmitter 蜗轮流量变送器
turbine flowmeter 涡轮流量计
turbine generator 涡轮发电机
turbine meter 涡轮流量计
turbine pump 涡轮式泵
turbine supervisory instrumentation 汽轮机监测仪表

turbine type flow transducer 涡轮式流量传感器
turbine type inferential flowmeter 涡轮式间接流量计
turbo-generator 涡轮发电机
turbo-generator set 汽轮发电机组
turbulence 湍流
turbulent convection 湍流对流
Turing machine 图灵机，图灵计算机
turn off 断开
turn on 接通
turn 转动，转向
turning 车削
turn-key contract 全包合约
turnkey system 可以立即启用的电脑系统
turns ratio 匝数比
turret 小塔
turret lathe 六角车床，转塔车床
turret punch press 转塔冲床
turret vertical milling machines 六角立式铣床
TVS（time-varying system） 时变系统
tweezers 镊子，小钳子
twin pipe 双管式管道
twin thermocouple 双支热电偶
twin track railway 双轨铁路
twin wire 双股线
twin-turbine mass flowmeter 双涡轮质量流量计
twist 扭曲，扭转
twist drill 麻花钻
twisted wire 绞合线，双绞线
twisting apparatus 测扭仪
twisting force 扭力
two core cable 双芯电缆
two phase 两相，两相的
two plate 两极式
two position power actuated valve 二位式调节阀
two speed clutch 双速离合器
two-channel carrier amplifier 双通道载波放大器
two-dimensional cam 两维凸轮，二维凸轮
two-dimensional systems 二维系统
two-phase balancing motor 双相平衡电动机
two-phase circuit 两相电路
two-phase induction motor 两相感应电动机
two-phase motor 两相电动机
two-phase servomotor 两相伺服电动机
two-phase wattmeter 两相瓦特计
two-port network 二端口网络
two-position on/off controller 二位式开关调节器
two-position controller 二位式调节器，二位式控制器
two-position relay 双位继电器
two-pressure humidity apparatus 双压力湿度计
two-speed controller 双速控制器
two-state device 双稳态装置
two-step action 两位作用
two-step controller 二位控制器，二位调节器，两级控制器
two-term 二项
two-term action 二项作用
two-term control 二项作用控制
two-term controller 二项作用控制器
two-terminal element 二端元件
two-terminal network 二端网络
two-time scale system 双时标系统
two-way 双向的
two-way conduit fitting with cover 电气带盖二通
two-way configuration 二线制
two-way drive 双向传动
two-way switch 双向开关
two-way valve 二通阀
two-winding transformer 双绕组变压器
two-wire pressure transmitter 双线压力传感器
two-wire transmitter input module 二线制变送器输入模块
type 形态，型式
type and specification 型号及规格
type number 型数
type of control 控制方式，调节方式
type selection 选型
types of sequence control 顺序控制的类型
typical section 典型横切面

U

U bolt　U形螺栓
U nuts　U形螺帽
UAC（uninterrupted automatic control）无中断自动控制
Ubbelohde viscometer　乌氏黏度计
U-bolt　U形螺栓
UCN（universal control networks）　万能控制网络
UHF insertion-type voltmeter　超高频插入式电压表
ULL（up line loading）　向上加载
ultimate bearing capacity　极限承载力
ultimate bearing failure　极限承载故障
ultimate gain method　极限增益法
ultimate limit state design　极限状态设计
ultimate load　极限载重，极限荷载
ultimate strength　极限强度
ultimate stress　极限应力
ultimately controlled variable　最终被控变量
ultra low frequency signal generator　超低频信号发生器
ultra thin laminate　超薄型层压板
ultra-efficient　超高效率
ultra-high frequency　超高频
ultrahigh purity filter　超滤器
ultra-high temperature electric furnaces　超高温电气炉
ultra-high vacuum flange-dimensions　超高真空法兰尺寸
ultra-high vacuum flange-types　超高真空法兰结构型式
ultra-high voltage　超高压
ultra-high-frequency wave　超短波
ultra-high-speed internal grinder　超速内圆磨床
ultra-low frequency　超低频
ultra-low noise preamplifier　超低噪声前置放大器
ultra-low temperature freezer　超低温冰箱
ultra-low-frequency phase meter　超低频相位计
ultra-pure water purifier system　超纯水制造系统
ultra-pure water purifiers　超纯水制造仪
ultrashort wave　超短波
ultrasonator　超声波发生器
ultrasonic cable leak detector　超声电缆检漏仪
ultrasonic cell disruptor　超声破碎仪
ultrasonic cleaners　超声波清洗机
ultrasonic color Doppler diagnostic system　彩色超声多普勒诊断系统
ultrasonic crack detection　超声波裂缝检测
ultrasonic Doppler flowmeter　超声多普勒流量计
ultrasonic electrode cleaner　超声波电极清洗器
ultrasonic flaw detector　超声波探伤仪
ultrasonic flow measuring device　超声流量测量仪表
ultrasonic flowmeter　超声流量计
ultrasonic generator　超声波发生器
ultrasonic level measuring device　超声物位测量仪表
ultrasonic level sensor　超声波液位传感器
ultrasonic machining　超声波加工
ultrasonic obstacle detector　超声波障碍探测器
ultrasonic pipet washers　超声波试管清洗机
ultrasonic sensor　超声传感器
ultrasonic testing　超声波测试
ultrasonic transducer　超声波传感器
ultrasonic wave　超声波
ultrasound transducer　超声传感器
ultraviolet detection　紫外光检测
ultraviolet photomultiplier　紫外光电倍增管

unadjustable speed electric drive 非调速电气传动
unbalance 不平衡
unbalanced attenuator 不平衡衰减器
unbalanced bridge 不平衡电桥
unbalanced circuit 不平衡电路
unbalanced load 不平衡负载
unbalanced three phase load 不平衡三相负载
unbiased estimate 无偏估计
unbiased estimation 无偏估计
unbolt 取下螺栓，打开
uncertain 不确定的
uncertain dynamic system 不确定动态系统
uncertain linear system 不确定线性系统
uncertain polynomial 不确定多项式
uncertain value status 不定值状态
uncertainty 不确定性
uncertainty of measurement 测量的不确定度
unclad laminate surface 层压板面
uncoiler 闭卷送料机
uncoiler and straightener 整平机
uncontrollable 不可控性
uncorrected result 未修正结果
undamped frequency 非衰减频率，无阻尼频率
under annealing 不完全退火
under cut 凹割，咬边
under damping 欠阻尼，周期阻尼
under water pumps 液下泵
undercoat 涂底层，涂底漆
under-current relay 低电流继电器
undercut 低切，根切
underdamped 欠阻尼
underdamped response 欠阻尼响应
underdamped system 欠阻尼系统
underdamping 欠阻尼
underground 地面之下
underground cable 地下电缆，管道电缆
underground chamber 地下池
underground fuel tank 地下油缸
underground space 地下空间
underload relay 欠载继电器

underpass 地下通道
undershoot 负脉冲信号
under-voltage protection 欠电压保护装置
undervoltage 欠压，电压低
underwater acoustics engineering 水声工程
underwater lighting 水底照明
underwater test 水下试验
undeveloped settings 试切前调整
undulated sheet iron 陨铁，瓦楞铁[钢]皮
undulation 波动，起伏
unfold install 明装
uniaxial load 单轴荷载
unicontrol 单向控制，单向调整
unidirectional coupler 单向耦合器
unidirectional current 单方向电流
unidirectional element 单向元件
unidirectional pulse train 单向脉冲列
unified system diagnostic manager（USDM） 统一系统监测管理器
uniform 均匀的，一致的
uniform electric field 均匀电路
uniform magnetic field 均匀磁场
uniform motion 等速运动规律
uniform roll 匀速滚动
uniform velocity method 等速法
uniform velocity tester 匀速试验机
uniformity 均匀性，一致性
uniformly asymptotic stability 一致渐近稳定性
uniformly distributed load 均布荷载
unilateral 单向性
unilateral switch 单向开关
uninterrupted automatic control（UAC） 无中断自动控制
uninterrupted duty 不间断工作制，长期工作制
uninterrupted power supply 不间断供电
uninterrupted power supply（UPS） 不间断电源，不停电电源
uniphase amplifier 单相放大器
unipolar 单极性
unipolar charge injection 单极电荷注射

unipolar converter 单极变换器
unipolar injection 单极注射
uniprocessor version 单机版
unique action 唯一作用
uniqueness 唯一性
unit 单元，单位，部件
unit action potential 单元动作电位
unit alarm summary 单元报警总貌
unit assignment display 单元分配画面
unit circle 单位圆
unit commitment problem 单元约定问题
unit configuration 单元构成
unit data link terminal 单元数据连接型端子
unit feedback 单位反馈
unit impulse 单位脉冲
unit impulse function 单位脉冲函数
unit impulse response 单位脉冲响应
unit impulse signal 单位脉冲信号
unit instrument 单元仪表
unit load 单位负荷
unit mold 单元式模具
unit names configuration 操作单元名称组态
unit operation 单元操作
unit operation block 单元操作块
unit price 单价
unit status overview display panel 单元状态总貌显示画面
unit step 单位阶跃
unit step function 单位阶跃函数
unit step response 单位阶跃响应
unit step signal 单位阶跃信号
unit supervisory function 单元管理功能
unit switch 单元开关
unit tend display 单元趋势画面，单元趋势显示
unit testing 单元测试
unit vector 单位矢量
unit weight 单位重量
unit 单元，机组，单位
unit-ramp signal 单位斜坡信号
units 单位
unit-step response 单位阶跃响应
universal 通用的
universal asynchronous transmitter 通用异步发送器
universal ball joint 万向球节
universal beam 通用钢梁
universal bridge 万用电桥
universal calibration function or curve 校准函数或校准曲线
universal comparator 通用比较仪
universal control 通用控制
universal control network 通用控制网络
universal control network interface 通用控制网接口
universal control network modem 通用控制网调制解调器
universal control networks (UCN) 万能控制网络
universal coupling 万向联轴器，万向节
universal coupling shaft 万向联结轴
universal data compression 通用数据压缩
universal fitting 通用穿线盒
universal joint 万向接头
universal measuring instrument 通用测量仪表
universal meter 万用表
universal milling machines 万能铣床
universal mold 通用模具
universal motor 交直流两用电动机，通用电动机
universal personality 万能属性
universal pliers 万能手钳
universal ramp generator 通用斜坡发生器
universal serial bus (USB) 通用串行总线
universal station 通用操作站，多功能操作站，万能操作站
universal synchroscope 通用同步示波器
universal test ammeter 通用测试电流表
universal tool grinding machine 万能工具磨床
universal work station 通用工作站，万能工作站
universe 宇宙，整体
unleaded petrol 不含铅汽油，无铅汽油
unlined galvanized iron pipe 无内搪

层镀锌铁管
unload 卸载
unload archive 卸下档案
unload audit trail 卸去尾部检查
unloader 卸料机,卸载器,卸货机
unlock 打开,解锁,释放
unmonitored control 非监控
unmonitored control system 非监控系统
unobservable 不可观测
unpacking procedure 开放程序
unplasticized polyvinyl chloride (UPVC) 非塑化聚氯乙烯
unplasticized polyvinyl chloride fitting for water supply 给水用硬聚氯乙烯管件
unprotected 未保护的,无屏蔽的
unreliable machine 不可靠机器
unsaturated polyester 不饱和聚酯
unscreened cable 非屏蔽电缆
unscrewing mold 退扣式模具
unstable 不稳定的
unstable closed-loop system 不稳定闭环系统
unstable controller 不稳定控制器
unstable open-loop process 不稳定开环过程
unstable system 不稳定系统
unsteady state 不稳定状态
unsteady-state operation 不稳定状态操作
unsupported adhesive film 无支撑胶黏剂膜
unsupported hole 非支撑孔
unsymmetrical 不对称的
unsymmetrical load 不对称负载
untight 松动的
unworked casting 未加工铸件
up line loading (ULL) 向上加载
up to grade 合格
up 向上
update 更新,修改,校正
upender 翻料机,翻转装置
up-escalator 上行电动扶梯
upgrade 升级,提高,改进
up-half 上部,上半
uphill casting 底铸,底注

up-hill lane 上坡车道
uplift force 上升力,浮力
upper atmosphere 上层大气,高空大气
upper catchment area 上段集水区
upper control limit 控制上限
upper die base 上模座
upper harmonic 高次谐波
upper holder block 上压块
upper level problem 上级问题
upper level processor 上位处理机,上位处理器
upper limit 上限,上限值
upper mid plate 上中间板
upper padding plate blank 上垫板
upper plate 上模板
upper platform 上层平台
upper range-limit 量程上限
upper range-value 量程上限值
upper storey 上层
upper supporting blank 上承板
upper switching value 上切换值
upper wire 上层网
upper 上部
uppermost storey 最高楼层
upright cycle 立式健身车
upright wall 直立墙
UPS (uninterrupted power supply) 不间断电源
upsetting 锻粗加工
upstream 上游
UPVC (unplasticized polyvinyl chloride) 非塑化聚氯乙烯
upward 向上
uranium (U) 铀
urine analyzer 尿液分析仪
urine sediments analyzer 尿沉渣分析器
urology 泌尿学
usage 用途
USB (universal serial bus) 通用串行总线
USDM (unified system diagnostic manager) 统一系统监测管理器
useful 有效的,有用的
useful power 有功功率,有用功率
useful resistance 有益阻力
useless resistance 有害阻力

user 用户，使用人
user assignable key 用户可定义的键
user average 用户平均值
user calculation function class module 用户计算功能级模块
user group 用户组
user interface 用户接口
user name 用户名称
user permission 用户许可
user screen 用户屏幕
user volume 用户卷

user adjustment 用户调整
using table of contents and index 目录和索引的应用表
utility 有用，公用，实用工具
utility bench 三角椅
utility function 公用函数，应用功能
utility program 常用程序
utilization rate 使用率
U-tube density transmitter U形管密度传感器
U-tube manometer U形管压力计

V

V belt 三角皮带，V形皮带，V带
V net V网
V net cable V网电缆
V net coupler installation V网耦合器安装
V thread screw 三角形螺纹
vaccum cleaner 吸尘器
vacuometer 真空计
vacuum air pump 真空泵
vacuum carburizing 真空渗碳处理
vacuum contacting dilatometer 真空接触式膨胀计
vacuum cryostat 真空低温保持器
vacuum drying ovens 真空干燥箱
vacuum flanges 真空法兰
vacuum flask 冷藏瓶
vacuum gauge 真空计
vacuum hardening 真空淬火
vacuum heat treatment 真空热处理
vacuum nitriding 真空氮化
vacuum photodiode 真空光电二极管
vacuum pump 真空泵
vacuum relief valve 真空卸压阀，真空解除阀
vacuum technology 真空技术
vacuum technology screwed type quick release flange 拧紧型真空快卸法兰
vacuum thermionic detector 真空热离子检测计
vacuum tube 真空管，电子管
vacuum type pneumatic micrometer 真空式气动测微计
vacuum 真空，空间，真空吸尘器
vacuum pump 真空泵
valid 有效的，正当的
validation 确认，生效
validity 有效性
valley load 低谷负荷
value 值，价值
value of quantity 量值
value of scale division 分格值，格值
value status 值状态

value of scale division 标度分格值
valve 阀，调节阀，控制阀
valve chamber 阀室
valve coefficient 阀系数
valve core 阀芯
valve gate 阀门浇口
valve injection 阀进样
valve monitoring block 阀门监视功能块
valve pit 阀井
valve positioner 阀门定位器
valve relay 阀电子管继电器，阀继电器
valve spindle 阀轴
valve stem 阀杆
valve tube 阀电子管
valve disk 阀芯
valve seat 阀座
valves 阀
vanadium (V) 钒
Vandermonde matrix 范德蒙矩阵
vane 轮叶，叶片
vane flowmeter 叶片式流量计
vane pump 叶轮泵
vapor pressure thermal system 蒸汽压力式感温系统
vaporization 汽化作用，蒸发
vapour fractometer 气相分离计
var 无功伏安
varactor 变抗器，变容二极管
varactor diode 变容二极管
variability 可变性
variable 量，变量，可变的，变化的
variable aperture type flowmeter 可变口径式流量计
variable area flow measuring device 变面积式流量测量仪表
variable area flowmeter 可变面积量计，变截面流量计
variable capacitor 可变电容器，变容器
variable capacity 可变电容
variable condenser 可变电容器
variable delay multivibrator 可变延迟

多谐振荡器
variable delivery pump　变量输送泵
variable duration impulse system　脉冲调制系统
variable gain　可变增益
variable gain amplifier　可变增益放大器
variable head flowmeter　可变压差流量计
variable holographic interferometer　可变全息干涉仪
variable impedance power meter　可变阻抗功率表
variable inductance　可变电感
variable orifice flow indicator　可变孔径流量计
variable pairing　变量配对
variable quantity　变量,可变参量
variable reluctance type pressure transducer　变磁阻式压力传感器
variable resistance　可变电阻
variable resistance type flowmeter　变阻式流量计
variable structure　可变结构
variable structure control　变结构控制
variable valve timing control　可变阀门时间控制
variable voltage control　变压控制,变压调节
variable-length code　可变长度编码
variable-speed drive　无级变速传动
variable-speed motor　变速电动机
variable-structure control　可变结构控制
variable-structure system　可变结构系统
variable-value control　跟踪控制,变值控制
variance　方差,差异
variance matrix　方差矩阵
variation　变化,变异
variation in load　负载变化
variation in voltage　电压波动
variational analysis　变分分析
variator　变化器,变速器
varistor　变阻器,压敏电阻
varistor rectifier　变阻整流器
varnish　罩光漆,绝缘漆,清漆
varying-voltage control　变压控制
varying-voltage generator　变压发电机

V-belt　V带,三角皮带,三角带
VCCS (voltage-controlled current source)　电压控制电流源
VCVS (voltage-controlled voltage source)　电压控制电压源
VDU (visual display unit)　显示器,视频显示装置
vectogram　矢量图
vector　矢量,向量
vector diagram　矢量图,向量图
vector equation　向(相)量方程,矢量方程
vector Lyapunov function　向量李雅普诺夫函数
vector method　矢量法
vector quantization　矢量量化
vector quantizer　矢量量化器
vector sum　矢量和
vehicle　速度,速率,机车
vehicle access road　车辆通道
vehicle aerodynamics　机车空气动力学
vehicle barrier　车辆栏障
vehicle control　速度控制
vehicle dynamics　机车动力学
vehicle engineering　车辆工程
vehicle error　速度误差
vehicle feedback　速度反馈
vehicle measurement　速度测量
vehicle operation engineering　载运工具运用工程
vehicle overshoot　速度超调
vehicle parapet　车辆护栏
vehicle saturation　速度饱和
vehicle simulator　机车仿真器
vehicle suspension　悬浮机车
vehicle traffic　车辆交通
vehicular access　车路,车辆通道
velocimeter　测速仪,速度计
velocity　速度
velocity alarm check　变化率报警检查
velocity check　速度检查
velocity control system　速度控制系统
velocity diagram　速度曲线
velocity error coefficient　速度误差系数
velocity error constant　速度误差常数
velocity flowmeter　流速型流量计
velocity integrator　速度积分器

velocity limit	变化速率限值
velocity limiter	变化率限幅值，速度限制器
velocity limiting control	速度限值控制
velocity transducer	速度传感器
velodyne	调速发电机，测速发电机
vent	通风口，通气孔
vent duct	通风管道
vent pipe	排气管
vent plug	通风螺塞，通风阀塞
vent screw	排气螺钉
venter	排气风扇
ventilating pipe	通风管，通气管
ventilating system	通风系统
ventilation fan	通风扇，通风机
ventilation opening	通风口
ventilation shaft	通风塔
ventilation vent	通风
ventilator	通风器，通风机，风扇，通风设备
venturi	文丘里管，文氏管
Venturi air gauge	文丘里管，空气压力计
Venturi flowmeter	文丘里管流量计
Venturi nozzle	文丘里喷嘴
Venturi tube	文丘里管，文丘里流量计
verification	检验，核对
vernier	游标
vernier caliper	游标卡尺
vernier depth gauge	深度游标卡尺
vernier micrometer	游标千分尺
vernier scale	游标尺，游标度盘
versatile U-steel	万用槽钢
vertical	垂直的
vertical and horizontal milling machines	立式及卧式铣床
vertical clearance	竖向净空，垂直净空
vertical curve	竖向曲线
vertical decomposition	纵向分解
vertical direction	垂直方向
vertical displacement	垂直位移
vertical distance	垂直距离
vertical down pipe	直立式落水管
vertical factor	垂直系数
vertical force	垂直力
vertical formwork	垂直模板
vertical galvanometer	直立式检流计
vertical imposed load	垂直外加荷载
vertical lathe	立式车床
vertical lift check valves	立式止回阀
vertical load	垂直荷载
vertical machine center	立式加工制造中心
vertical machining centers	立式加工中心
vertical milling machines	立式铣床
vertical offset	垂直偏置距
vertical panel	竖直面板
vertical plane	垂直面
vertical situation indicator	垂直状态指示器
vertical-tube manometer	立管式压力计
vessel	容器
VHF modulation signal generator	甚高频调制信号发生器
via-in-pad	在连接盘中导通孔
vibrating load	振动荷载
vibrating of output	输出振荡
vibrating tamper	震动式捣固机
vibrating wire force transducer	振弦式测力传感器
vibrating-reed micro quantity electrometer	振簧式微静电计
vibration	振动
vibration damper	减震器
vibration feeder	振动送料机
vibration galvanometer	振动式检流计
vibration measurement	振动测量
vibration meter	振动测量计
vibration test systems	振动试验系统
vibration testing method and limiting levels	振动试验方法和限值
vibrational severity	振动烈度
vibration-proof mercury-in-glass thermometer	防振式汞柱温度计
vibrator	振捣器，振动器
vibratory compaction	震动压实
vibratory pile hammer	振动桩锤
vibratory roller	震动压路机
vibro viscometer	振动式黏度计
vibrometer	振动计
vice	台钳，虎钳，老虎钳
Vickers hardness test	维氏硬度试验

VID (voltage identification definition) 电压识别认证
video amplifier 视频放大器
video conferphone system 会议电视系统
video display monitor 视频监视器
video frequency 视频
video interphone 可视对讲门铃
video module 视频组件,视频模块
video pair cable 视频双芯电缆
video second detector 视频第二检波器
video signal 视频信号
video voltage amplifier 视频电压放大器
videocorder 录像机
video-on-demand (VOD) 视频点播
Vierendeel truss 空腹桁架
view 观察,视图
view angle 视角
vinyl tapped steel sheet 塑胶覆面钢板
virtual instrument 虚拟仪器
virtual number of teeth 当量齿数
virtual pitch radius 当量节圆半径
virtual reality 虚拟现实
virtual reality design 虚拟现实设计
virtual reality technology 虚拟现实技术
virtual test function 虚拟测试功能
viscometer 黏度计
viscosimeter 黏度计
viscosity 黏度,黏性
viscous 黏性的,黏的
viscous damping 黏滞阻尼,黏性阻尼
viscous damping coefficient 黏性阻尼系数
viscous force 黏性力,黏力
viscous friction 黏性摩擦,黏滞摩擦
viscous meter 黏度计
vise 老虎钳
visibility 能见度,可见性
visible detection 可见光检测
vision and colors 视觉与色彩
vision 视觉,视力,显示
visor 护目镜
visual 目视的,可见的
visual display unit (VDU) 显示器,视频显示装置

visual inspection 目视检查,检视,外观检查
visual motion 目视移动
visual pattern recognition 目视模式识别
visualization 显谱,显影
visualizer 观察仪
vitamin 维生素
vitreous tile 釉瓷瓦,玻璃瓦
vitrified 陶瓷的
vitrified-clay pipe 缸瓦管,陶土管
VLSI 超大规模集成电路
VOD (video-on-demand) 视频点播
voice 声音,音频的
voice amplifier 音频放大器
voice interference analysis system 语音干扰分析系统
voice modulation 音频调制
voice-activated control unit 声敏控制装置
voice-frequency band 音频带
void 孔隙,空隙,空间
void content 空隙度,孔隙量,孔隙容积,孔率,空洞率
voided slab 空心桥板,空心板
volatile 挥发物
volatility 挥发性
volatilization method 挥发法
volcanic rock 火山岩
volt 伏特,伏(电压单位)
volt ampere 伏安(功率单位)
volt ohmmeter 伏欧表
voltage 电压
voltage across the terminals 端电压
voltage adjuster 电压调整器
voltage amplification 电压放大
voltage amplification factor 电压放大系数
voltage amplifier 电压放大器
voltage balancer 均压器
voltage between lines 线间电压
voltage between phases 相间电压
voltage by one phase 单相电压
voltage by three phase 三相电压
voltage changer 变压器
voltage characteristic 电压特性
voltage collapse 电压损失

voltage comparator 电压比较器
voltage control 电压控制
voltage control system 电压控制系统
voltage corrector 电压校正器
voltage dependent resistor 压敏电阻
voltage distribution 电压分布
voltage divider 分压器
voltage doubler 倍压器
voltage drop 电压降
voltage follower 电压跟随器
voltage grade 电压等级
voltage identification definition (VID) 电压识别认证
voltage indicator 电压指示器
voltage input module 电压输入模件
voltage inverter switch 电压逆变开关
voltage limiter 限压器
voltage meter 电压表,伏特计
voltage multiple input card 多路电压输入卡
voltage output 电压输出
voltage ratio 电压比
voltage reference 基准电压
voltage regulating transformer 调压变压器,调压器
voltage regulation 电压调整
voltage regulation factor 电压调整率
voltage regulator 稳压器,调压器
voltage regulator module 电压调整模块
voltage relay 电压继电器
voltage resonance 电压谐振,串联谐振
voltage source 电压源
voltage source inverter 电压源型逆变器
voltage stability 电压稳定度,电压稳定性
voltage stabilized source 稳压电源
voltage stabilizer 电压稳定器,稳压器
voltage stabilizing tube 稳压管
voltage standard 标准电压
voltage transformer 电压互感器
voltage tripler 三倍倍压器
voltage 电压,伏特数
voltage-controlled current source (VCCS) 电压控制电流源
voltage-controlled voltage source (VCVS) 电压控制电压源
voltage-current characteristic 伏安特性
voltage-doubler rectifier 倍压整流器
voltage-regulator tube 稳压管
voltaic wire 导线
voltameter 电量计,伏特计
voltammetry 伏安法
volt-ampere characteristics 伏安特性
voltmeter 伏特计,电压计
volume 容积,体积,音量
volume control 音量控制
volume density of aerated concrete 加气混凝土体积密度
volume flowmeter 容积流量计
volume marker 体积标记器
volumetric analysis 容量分析法
volumetric flasks 比重瓶
volumetric precipitation method 容量滴定法
vortex 漩涡
vortex analog speed sensor 涡流模拟速度传感器
vortex flowmeter 涡流流量计
vortex precession flowmeter 旋进流量计
V-shaped pad V形盘
V-tool V形刀具
vulcanized rubber 硫化橡胶

W

wafer check valves 对夹式止回阀
wafer prober 晶片检测计
wafer-type-resistance thermometer 膜片式电阻温度计
waffle die flattening 压纹校平
waffle iron 对开式铁芯
wagging vibration 面外摇摆振动
wait state 等待状态
waiting-time 等待时间
walkie-talkie 步话机
wall coated open tubular column 壁涂开管柱
wall effect 管壁效应
wall-coated open tubular column 壁涂开管柱
warehouse 仓库
warehouse automation 仓储自动化
warhead 弹头
warm forging 温锻
warm-up period 预热时间,预动作时间
warning apparatus 报警信号器
warning circuit 警报电路
warning device 报警器
warning lantern 警告灯
warning sign 警告标志
warning signal 危险信号,警报信号
warning 报警
warp yarn 经纱
warpage 翘曲,热曲线
warpage test 翘曲试验
warping 扭曲,变形
wash 洗
washer 垫圈,介子,衬垫
washers 洗净器
waste fitments 废水设备
waste iron 废铁
waste pipe 废水管
waste treatment 废物处理
waste valves 排污箱,排污阀
waste water treatment 废水处理系统
watch 手表,注视

watchdog 看门狗
watchdog timer 看门狗定时器
water 水
water absorption of aerated concrete 加气混凝土吸水率
water absorption of asphalt 沥青的吸水度
water absorption test 吸水测试
water absorptivity 吸水性
water booster pump 升压水泵
water bowser 水车
water carrying services 输水设施
water cement ratio 水灰比
water content 含水率
water current meter 水流速计
water displacement meter 容积式水表
water flowmeter 水表,水流量计
water gauge 水标尺,量水表
water hammer 水锤
water impermeability of asphalt 沥青的防水度
water jet 喷水器
water jet pump 喷水泵
water leakage alarm meter 漏水报警器
water level 水平面,水位
water level meter 水位计
water load power meter 水负载功率计
water main 总水管
water main diversion works 输水管改移工程
water manometer 水柱压力计
water meter 水表
water of crystallization 结晶水
water of imbibition 吸入水
water pipe 水管
water pollution 水污染
water pressure 水压力
water pump 水泵
water purifier 净水器
water purifiers system 大容量纯水制造系统
water quality tester strips 简易水质

检查试验纸
water quenching 水淬火
water seal 水封
water seal gate valves 水封闸阀
water spots 水渍
water spray system 喷水系统
water sprinkler system 洒水系统
water storage tank 贮水箱
water supply 供水
water supply point 供水点
water tank 水箱，水缸
water test kits 水质分析仪
water tight concrete 不透水混凝土
water tower 水塔
water trap 聚水器
water treatment plant 净水厂，滤水厂
water vapour 水蒸气
waterproof 防水
waterproof cement 防水水泥
waterproof dial gauge 防水千分表
water-proof flexible conduit 防水挠性管
waterproof instrument 防水式仪表
waterproof membrane 防水膜
water-proof packing gland 防水密封接头
water-proof packing gland for connecting pipe 接管式防水密封接头
waterproof socket 防水插座
water-sealed rotary gas meter 水封式旋转气体流量计
water-temperature gauge 水温表
watertight 水密
watertight seal 不透水密封垫
watt (W) 瓦特（功率单位）
watt loss 功率损耗
wattage 瓦特数，瓦数
wattful-power 有功功率
watt-hour meter 瓦时计，电度表，有功电能表
wattless component watt-hour meter 无功电度表
wattless power meter 无功功率表
wattmeter 瓦特计，功率表
wattmeter calibrator 功率表校验器
wave 波

wave analyzer 波形分析仪
wave band 波段，频带
wave changing switch 波段转换开关
wave converter 波形变换器
wave equation 波动方程
wave filter 滤波器
wave generator 波发生器
wave guide 波导
wave length 波长
wave modulated oscilloscope 波调制示波器
wave train 波列
waveform 波形
waveform distortion 波形失真
waveform monitor 波形监视器
waveguide 波导，波导管
waveguide branching filter 波导管分支滤波器
waveguide flow calorimeter 波导流体式热量计
waveguide tube 波导管
wavelength deviation sensing detector 波长偏差传感检测器
wavelength meter 波长计
wavelength spectrometer 波长分光计
wavelet transform 小波变换
wavelet transformation 小波变换
wave-range switch 波段开关
wayleave 通行权
weak anion exchange 弱阴离子交换
weak band 弱带
weak cation exchange 弱阳离子交换
weak interaction of loops 回路弱相关
weapon systems and utilization engineering 武器系统与运用工程
wear 磨损
wear and tear 消耗，磨损
wear pad 垫磨片
wear plate 垫磨板
wear resistance 耐磨性，耐磨度
wear resistant 耐磨性
wear ring 磨耗环
wearing course 磨耗层
weather glass 气压计，晴雨表
weather indicator 气象指示器，天气指示器
weather resistance 抗风化

weather 天气
weathering 风化
weatherproof 不受天气影响,防风雨
weave structure 织物组织
weber 韦伯(磁通量单位)
wedge 楔形
wedge cam 移动凸轮,楔形凸轮
wedge disc 闸板
wedge gate valves 楔式闸阀
wedge spectrometer 楔形光谱仪
wedge wear plate 耐磨板
weed killer spraying 喷射除草剂
weekly delivery requirement 周出货需求
weep hole 泪孔,疏水孔,泄水孔
weft yarn 纬纱
weft-wise 纬向
Weibull distribution 韦布尔分布
weighbridge 桥秤,地秤
weighing cell 称重传感器
weighing forms 称量形式
weight of measurement 测量的权
weight per epoxy equivalent 环氧当量
weight sets 权重集
weight thermometer 恒重温度计
weight training 负荷训练
weight 重量
weighted mean 加权平均,加权平均值
weighted moving average 加权移动平均
weight-flow meter 重量流量计
weighting efficient 有效加权
weighting factor 权因子
weighting function 加权函数
weighting method 加权法
weir 堰,坝
weir meter 溢流水位计,堰顶水位计
weir plate 堰板,挡水板
weir type flow measuring device 堰式流量测量仪表
Weissenberg effect 威森伯格效应
weld 焊,焊接
weld bead 焊缝
weld flow mark 焊接流痕
weld flush 焊缝凸起
weld gauge 焊缝量规
weld line 焊接纹

weld mark 焊痕,焊接痕
weld nuts 焊接螺帽
weld penetration 焊透深度
weld wire 铜包钢丝
weld zone 焊接区
welded joint 焊接接点,焊缝
welded steel gas cylinders 钢质焊接气瓶
welder 电焊机,焊工,焊机
welding 焊接,烧焊
welding arc 焊弧
welding bead 焊缝,焊珠
welding direction 焊接方向
welding distortion 焊接变形
welding electrode 焊条,焊接电极
welding flux 焊剂
welding ground 电熔接地
welding interval 焊接周期
welding line 熔合痕,焊缝
welding machine 焊接机
welding mark 熔合痕,焊接痕迹
welding powder 焊粉,焊药
welding rod 焊枝,焊条,焊棒
welding slag 焊渣
welding stress 熔接应力
welding torch 熔接气炬
welding wire 焊条,焊丝
well 水井,井
well type 蓄料井
well water pump 井水泵
wet and dry bulb thermometer 干湿球温度计
wet leg 隔离管
wet station 沾湿台
wet strength retention 湿强度保留率
wet test meter 测湿计
wet type gas flowmeter 湿式气体流量计
wetted parts 接触液体的部件
Wheatstone bridge 惠斯通电桥,单臂电桥
Wheatstone bridge recorder 惠斯通电桥记录仪
wheel 车轮
wheel axle 轮轴
wheel load 车轮载重,车轮荷载
wheel nuts 轮壳螺帽

wheel slip torque 车轮打滑扭矩
wheel 轮,车轮,轮子,转盘,旋转
white cast iron 白口铸件
white iron 白心铁,白铸铁
white noise 白噪声
white noise test set 白噪声测定仪
white smoke 白烟
white 白色
whiteness 白度
whitening 白化
Whittaker-Shannon sampling theorem 惠特克-香农采样定理
whole blood 全血
whole depth 全齿高
whole set of distribution box 成套配电箱
wide angle 广角
wide angle total radiation pyrometer 广角全辐射高温计
wide area network 宽带网
wide band amplifier 宽带放大器
wide band differential DC amplifier 宽带差动直流放大器
wide meter 宽量程仪表
wide range 宽量程
wide range AC millivoltmeter 宽量程交流毫伏表
wide range meter 宽量程仪表
wide range orifice meter 宽压插孔板流量计
wide range temperature controller 宽范围温度控制器
wide scale self-balancing recorder 宽标尺自动平衡记录仪
widening 扩展,加宽
width 宽
width of flat-face 从动件平底宽度
width series 宽度系列
Wien bridge 文氏电桥,维恩电桥
Wiener filter 维纳滤波器
Wiener filtering 维纳滤波
winch 绞车,绞盘,卷扬机
wind 风
wind force 风力
wind gauge 风速表
wind load 风荷载
wind pressure 风压

wind shutter 防风板
wind speed 风速
wind stress 风应力
wind suction 风吸力
wind tunnel testing 风洞测试,风力模拟试验
windbox 风箱
wind-driven generator 风动发电机
winding 绕组,线圈
winding loss 绕组损耗,铜损耗
winding temperature indicator 绕组温度指示器
winding wire 绕组线
windmill 风车
window catch 窗扣,窗钩
window sizing 窗口尺寸
window to the process 过程窗口
windup 饱和
wing nuts 蝶形螺帽
wire 电线,金属丝,接线,导线,金属线
wire brush 钢丝刷
wire cutters 剪线钳
wire EDM 线割
wire fuse 熔丝
wire guard 钢丝护网
wire mesh 金属丝网,钢丝网
wire netting 导线网,铁丝网,金属丝网
wire resistance strain gauge 电阻丝应变仪
wire rope 钢丝绳
wire rope grips 钢丝绳夹
wire soft shaft 钢丝软轴
wire spring 圆线弹簧
wire stripper 电线退皮钳,剥线钳
wire-cutting 线切割
wired edge 卷刃
wired program computer 插线程序计算机
wireless 无线的
wireless link 无线电线路
wireless remote control 无线电遥控
wire-space detector 接线检测仪
wiring 敷设电线,接线,抽线加工
wiring diagram 线路图
wiring diagram on the back of instru-

ment panel-board 仪表盘盘后接线图
wiring layout 线路配置图
wiring press 嵌线卷边机
with hand wheel radiator 带手轮散热片
within-day precision 日内精密度
within-run precision 批内精密度
witnessed inspections 现场检测
wobble pump for pressure testing 手摇试压泵
wobbulator 摆频振荡器
wolfram-rhenium thermocouple 钨-铼热电偶
woodruff key 半圆键
work 功，工件，工作
work cell 工作间
work in progress product 在制品
work order 工令，工作令，工作通知单
work schedule 工作进度表，施工进度表
work sheet 工作单，工艺卡，订货联系单
work station 工作站
work station for computer aided design 计算机辅助设计工作站
workhardening 加工硬化
workhead 工件头座
workholding equipment 工件夹具
working allowance 加工余量
working cycle diagram 工作循环图
working depth 工作齿高
working drawing 施工详图，施工图则
working gauge 工作量规
working life 工作寿命，使用寿命
working load 使用荷载，施工荷载，工作荷载
working order 工作状态
working platform 工作平台
working pressure angle 啮合角
working space 工作空间
working stress 工作应力，工作允许应力
working temperature 工作温度
workpiece 工件
works order 施工通知，施工令
works programme 工程进度，工程程序
worksheet display 工作单画面
workshop 车间，工场
workstation 工作站
workstation processor 工作站处理机
world class 一流的
worm 蜗轮，蜗杆
worm and worm gear 蜗杆蜗轮机构
worm cam interval mechanism 蜗杆形凸轮步进机构
worm gear 蜗轮
worm gearing 蜗杆传动机构
worm drive 蜗杆驱动
wortle plate 拉丝模板
wound rotor 线绕式转子
woven wire 铁丝网
w-plane w平面
wrench 扳手，扳钳，螺旋扳手，拧
wrinkle 皱纹
write 写
write circuit 写入电路
wrong 错误的，失常的
wrought iron 熟铁，锻铁
wye Y形接法
wye-delta Y-△连接

X

X axis amplifier　X轴信号放大器，水平信号放大器
X-band micro-spin spectrometer　X频带微自旋波谱仪
X-band microwave interferometer　X波段微波干涉仪
xenon (Xe)　氙
xor　异或
xor gate　异或门
X-ray　X射线，X光
X-ray detection apparatus　X射线探伤仪
X-ray emission spectrometer　X射线发射光谱仪
X-ray examination　X射线检查
X-ray gas density gauge　X射线气体密度计
X-ray generator　X射线发生器
X-ray goniometer　X射线测角仪
X-ray interference　X射线干扰
X-ray spectrometry　X射线度谱术
X-ray warning device　X射线报警器
X-Y oscilloscope　X-Y示波器
X-Y plotter　X-Y绘图仪
X-Y recorder　X-Y记录器

Y

Y axis amplifier Y轴线放大器，Y轴信号放大器，垂直信号放大器
Y-△ starter Y-△启动器
yarn 纱线
Y-connection Y形连接，星形连接
yearly load curve 年负荷曲线
yield factor 合格率
yield management for semiconductor products 半导体制造用检查
yoke 偏转线圈
Yokogawa computer 横河公司计算机系统

Z

z transfer function　z传递函数
z transformation　z变换
Z-axis amplifier　Z轴信号放大器
Zeeman laser　塞曼效应激光仪
Zener diode　齐纳二极管,稳压二极管
Zener diode regulator　齐纳二极管校准器,稳压二极管校准器
zero　零,零点
zero adjuster　零位调整器
zero adjustment　零位调整器,零位调整
zero bit　零位
zero control　零位调节
zero correction　零点校正
zero drift　零漂,零漂移
zero elevation　零点提升
zero error　零点误差,始点误差,零位误差
zero frequency　零频率
zero frequency gain　零频增益
zero indicator　零位指示器
zero input response　零输入响应
zero of a measuring instrument　测量仪表的零位
zero offset　零静差
zero phase current relay　零相电流继电器
zero phase-sequence relay　零相序继电器
zero scale mark　零标度标记,零标度线
zero sequence　零序
zero sequence impedance　零序阻抗
zero set　调零
zero shift　零点迁移,零点漂移
zero state response　零状态响应
zero suppression　零点压缩,零点正迁移
zero time reference　零时间调节
zero tolerance　零点允差
zero value　零值
zero of a measuring instrument　测量仪器仪表的零位
zero-address　零地址
zero-based conformity　零基一致性
zero-based linearity　零基线性度
zero-center instrument　中心零位式仪表
zero-cross-level detector　零交叉电平检测器
zero-drift　零漂
zero-order hold　零阶保持器,零阶保持
zero-range trigger generator　零距触发脉冲发生器
zero-resistance ammeter　零电阻电流表
zero-sequence component　零序分量
zero-sequence symmetrical component　零序对称分量
zero-sequence system　零序系统
zero-setting phase shifter　调零移相器
zero-state response　零状态响应
Ziegler-Nichols method　齐格勒-尼科尔斯方法
Ziegler-Nichols setting　齐格勒-尼科尔斯整定
zone　区带
zone tailing　区带脱尾
zone-position indicator　区位指示器
zoom microscope　变焦距显微镜

汉英部分

仓储英文

A

埃克曼流速仪　Ekman current meter
阿贝测长仪　Abe metro-scope
安培　ampere
安培定律　Ampere's law
安培环路定律　Ampere circuit law
安培计　ammeter
安培右手螺旋定则　Ampere's right-handed screw rule
安全　safety
安全报警器　safety alarm
安全灯　safety lamp
安全电压　safe voltage
安全阀　emergency valve, safety valve
安全分析　safety analysis
安全改进维护　corrective maintenance
安全控制　safety control
安全联锁系统　safety interlock system
安全气囊顶杆　air-cushion eject-rod
安全塞　screw plug
安全停车操作　safe operation for safety shutdown
安全硬件　safety hardware
安全照明　safety lighting
安全执行器　safety actuator
安全装置　safety appliance, safety device
安匝　ampere turn
安匝数　ampere-turns, number of ampere turns
安装　install
安装规格　installation specification
安装环境说明书　installation environment specification
安装角　root angle
安装孔板　orifice fitting
安装位置　mounting position
安装应变误差　mounting strain error

按键　key
按键开关　key switch
按钮　button
按钮开关　button switch, press-button switch
按钮式交通灯　push button traffic light
按钮式控制站　push button station
按钮输入　push button input
按钮输入卡　push button input card
按序切换开关　scanning switch
暗槽　covered channel
暗场电子像　dark field electron image
暗钉　dowel
暗管敷设　concealed piping
暗键　sunk key
暗筒日照计　Jordan sunshine recorder
暗销杆　dowel bar
暗噪声　dark noise
暗装　concealed installation
凹割　under cut
凹角　dish angle
凹口剪床　gap shear
凹面　concave side, concavity
凹面方钢　fluted bar iron
凹面光栅　concave grating
凹面铣刀　concave cutter
凹模　female die
凹模固定板　die holder
凹入模　re-entrant mold
凹位　recess
凹陷　sinking
凹形　concave
螯合物　chelate
奥氏体　austenite
奥氏体回火法　austempering
奥氏体钢　austenitic steel
奥氏体铸铁　austenitic cast iron

B

巴特沃斯滤波器　Butterworth filter
扒钉　cramp iron
拔拉测试　pull-out test
拔丝机　drawing machines
钯　palladium (Pd)
靶式流量变送器　target flow transmitter
靶式流量测量仪表　target flow measuring device
靶式流量计　target flowmeter, target-type flowmeter
白炽灯　incandescent lamp
白度　whiteness
白化　whitening
白口铸件　white cast iron
白色　white
白心铁　white iron
白噪声　white noise
白噪声测定仪　white noise test set
百分比　percentage
百分之一　centi
百万分之一　percent per million (PPM)
摆锤的冲击速度　impact velocity of the pendulum
摆锤空击　free swing of pendulum
摆锤力矩　moment of pendulum
摆锤黏度计　pendulum viscometer
摆动　tilting
摆动从动件　oscillating follower
摆动从动件凸轮机构　cam with oscillating follower
摆动导杆机构　oscillating guide-bar mechanism
摆动活塞式流量计　oscillating piston type flowmeter
摆动湿度计　sling hygrometer
摆动小轮节锥法　swing pinion cone
摆动周期　deflection period
摆杆　oscillating bar
摆频振荡器　wobbulator
摆线齿轮　cycloidal gear
摆线齿形　cycloidal tooth profile
摆线的　cycloidal
摆线轮廓　cycloidal profile
摆线运动规律　cycloidal motion
摆线针轮　cycloidal-pin wheel
摆线质谱计　cycloidal mass spectrometer
摆轴　axis of rotation
拜拉姆风速表　Byram anemometer
扳手　spanner, wrench
斑点　mottle, spot
斑点定位法　localization of spot
搬运　hauling
板　board
板边插头　edge-board contact
板簧　flat leaf spring
板极效率　plate efficiency
板料冲压　sheet metal parts
板落锤　board drop hammer
板条　lath
板牙绞手　die holder
板桩　sheet pile
板桩锚　sheet pile anchorage
板桩墙　sheet piled wall
板桩围堰　sheet pile cofferdam
版本　revision
版权　copyright
办公自动化　office automation
半　semi
半波　half-wave, semiwave
半波电位　half-wave potential
半波片　half-wave plate
半波整流　half-wave rectification
半成品　semi-finished product
半导体　semiconductor
半导体二极管　semiconductor diode
半导体放大器　semiconductor amplifier
半导体集成电路　semiconductor integrated circuit
半导体器件　semiconductor device
半导体生产管理　semiconductor production management

半导体压力传感器 semiconductor pressure sensor
半导体元件 semiconductor component, semiconductor element
半导体振荡器 oscillistor
半导体制造用检查 yield management for semiconductor products
半电池 half-cell
半高峰宽 peak width at half height
半功率点 half-power point
半活性的 semi-active
半机械化 semi-mechanized
半间歇操作 semi-batch operation
半减法器 half-subtracter
半剪 semi-shearing
半经验模型 semi-empirical model
半径 radius
半均裂 hemi-homolysis cleavage
半连续操作 semi-continuous operation
半马尔可夫过程 semi-Markov process
半破坏性测试 semi-destructive testing
半桥测量 half bridge measurement, half-bridge measurement
半桥电路 half bridge converter
半桥路高温应变仪 half-bridge high temperature strain gauge
半全压式模具 semi-positive mold
半实物仿真 semi-physical simulation
半树脂型砂轮 resinoid grinding wheel
半数字示值 semi-digital indication, semi-digital read out
半数字显示模拟测量仪表 analogue measuring instrument with semi-digital presentation
半衰期 half-life
半衰期阻尼器 semi-active damper
半双工传输 half duplex transmission
半图形板 semi-graphic panel
半旋转式容积式流量计 displacement flowmeter of semi-rotary pattern
半移位寄存器 half-shift register
半硬铝线 semihard-drawn aluminium wire
半硬质聚氯乙烯块状塑料地板砖 semi-rigid PVC plastics floor tiles
半圆钢丝 half-round steel wire

半圆键 woodruff key
半圆铁线 half-round iron wire
半圆凿 gouge
半周 half cycle, half period
半自动操作 semi-automatic operation
半自动的 semiautomatic
半自动滚刀磨床 semi-automatic hob grinder
半自动化 semi-automation
半自动控制器 semi-automatic controller
半自动生化分析仪 semi-automatic biochemical analyzer
半自动样品制备系统 advanced automated sample processor (AASP)
半作用悬挂 semi-active suspension
伴热管线 tracing pipe
伴随变量 adjoint variable
伴随矩阵 companion matrix
伴随算子 adjoint operator
伴随随机数据 probabilistic data association
拌和机 mixing machine
邦德防蚀钢板 bonderized steel sheet
帮助 help
帮助显示画面 help display
绑扎用铁丝 lacing wire
棒 rod
棒式玻璃温度计 solid-stem liquid in glass thermometer
棒图 bar graph
棒图指示器 bar graph indicator
棒形电流互感器 bar primary bushing type current transformer
镑 LBS
包铝钢丝 aluminium clad wire
包络 envelope
包络线 envelop curve
包铜钢线 copper-clad covered steel wire
包铜铝导体 copper-clad aluminium conductor
包围 enclose
包线 covered wire
包装 package, packing
包装材料 packing materials
包装规范 package specification

饱和的 saturated
饱和电抗器 transductor
饱和电抗器元件 transductor element
饱和电流 saturation current
饱和功率 saturation power
饱和控制 saturation control
饱和密度 saturated density
饱和区 saturation range
饱和曲线 saturation curve
饱和特性 saturation characteristics
饱和限幅器 saturation limiter
饱和效应 effects of saturation, saturation effect
饱和状态 saturation state
保持触点 holding contact
保持电路 holding circuit, retaining circuit
保持电压 holding voltage
保持干燥 keep dry
保持环节 holding element
保持冷藏 keep cool, keep in cool place
保持元件 hold element
保持作用 holding action
保存 save
保护继电器 protective relay
保护接地 protective earthing
保护输入 guarded input
保护涂料 protective coating
保护网 guard net
保护系统 protective system
保护线 guard wire
保护箱 protection box
保护罩 protective cover, shroud
保护柱 guard column
保护装置 fender system
保角变换 conformal transformation
保角映射 conformal mapping
保留 reserve
保留常数值 Rm value
保留控制器目录 reserve controller directory
保留时间 remaining time, retention time
保留体积 retention volume
保守系 conservative system
保温炉 heat preserving furnaces
保温器 attemperator
保温箱 heating box, heat insulated box
保险盒 fuse holder
保险连杆 fusible link
保险器 protector
保险铅丝 lead fuse wire
保险丝 plain cut-out, safety wire
保形变换 conformal transformation
保形映射技术 conformal mapping technique
保养期 maintenance period
保养日志 maintenance log book
保养手册 maintenance manual
保养维修 maintenance service
保真度 fidelity
保证金 retention money
保证人 guarantor
堡垒墙 concrete wall
报告 report
报告存档 report archiving
报告调度程序 report scheduler
报告功能 reporting functions
报警 warning
报警处理 alarm process
报警单元 alarm unit
报警灯显示画面 alarm annunciator display
报警汇总画面 alarm summery panel
报警记录程序包 alarm logger package
报警控制台 alarm console
报警类型 alarm type
报警器 announciator, syren, warning device
报警器信息 annunciator message
报警切除 alarm cutout
报警系统 alarm system
报警信号 alarm signal
报警信号灯 alarm lamp
报警信号继电器 alarm relay
报警信号开关 alarm switch
报警信号器 alarm enunciator, warning apparatus
报警信号装置 alarm device
报警优先度 alarm priority
报警指示器 alarm indicator

报警状态　alarm status
报文　message
报文方式　message mode
报文交换　message switching
曝光量　exposure
曝光曲线图　exposure chart
曝光时间　exposure time
曝气池　aeration tank
爆裂　bursting
爆燃过程　deflagration
爆丝　exploding wire
爆炸　blasting, burst, explosion
爆炸声源　explosive sound source
爆炸下限　lower explosion limit
杯模式流动度试验　cup flow test
杯突试验机　cupping testing machine
悲观值　pessimistic value
贝克曼梁试验　Benkelman beam test
贝克曼温度计　Beckman differential thermometer
贝克钳位电路　Baker clamping circuit
贝塞尔函数　Bessel function
贝森测云器　Besson nephoscope
贝氏体　bainite
贝叶斯分类器　Bayes classifier
备份　backup
备份软盘　backup diskette
备份文件　backup file
备件　replacement parts, reserve parts, spare parts
备用按钮　emergency button
备用的　reserved, standby
备用电池　spare battery
备用电动机　spare motor
备用电缆　emergency cable
备用电源　reserve supply
备用寄存器　standby register
备用控制器　backup controller, reserved controller (RC)
备用控制器指挥器　reserved controller director (RCD)
备用容量　spare capacity
备用设备　reserve equipment
备用系统　backup system
备用线路　spare line
备注　remark
背板　backing plate, backplane

背对背安装　back-to-back arrangement
背胶铜箔　resin coated copper foil
背角距　back angle distance
背景　background
背景板　background plate
背景颜色　background color
背景噪声　background noise
背靠背的　back-to-back
背面接线开关　back connected switch
背散射电子像　backscattered electron image
背砂　backing sand
背压　back pressure
背压阀　counterbalance valve
背锥　back cone
背锥角　back angle
背锥距　back cone distance
背锥母线　back cone element
钡　barium
倍频程滤波器　octave filter
倍频放大器　octamonic amplifier
倍频器　frequency doubler
倍压器　voltage doubler
倍压整流器　rectifier doubler, voltage-doubler rectifier
倍增器　multiplicator, multiplier
被测变量　measured variable
被测对象　measured object
被测介质　process fluid measured
被测量　measured quantity
被测量的变换值　transformed value of a measured quantity
被测目标　measure target
被测信号　measured signal
被测元件　component under test
被测值　measured value
被动姿态稳定　passive attitude stabilization
被覆板　clad sheet
被覆淋膜成形　laminating method
被覆熔接　clad weld
被覆线　field wire
被调介质　process fluid manipulated
本底噪声　ground noise
本底质谱　background mass spectrum
本地电台　local station

本地交换网　local exchange
本地扰动　disturbance localization
本机设定　local set
本机设定点　local set point
本机振荡　local oscillations
本机振荡器　local oscillator
本机自测试　local self-test
本生灯　Bunsen burner
本体　body
本征半导体　intrinsic semi-conductor
本征的　inherent, intrinsic
本征函数　eigenfunction
本征结构测定　eigen structure assignment
本征模式分析　eigenmode analysis
本征区晶体管　intrinsic region transistor
本征稳定　inherent stability
本质安全栅　intrinsic safety barrier
本质安全设备和接线　intrinsically safe equipment and wiring
本质安全型电动仪表和接线　intrinsically safe electrical instrument and wiring
本质电路　intrinsically safe circuit
苯　benzene
苯丙醇　phenyl propanol
苯甲酸及其钠盐　benzoic acid and sodium benzoate
苯扎溴铵　benzalkonium bromide
泵　pump
泵电动机　pump motor
泵房　pump house, pumping station
泵基座　pump foundation
泵马达　pump motor
泵水干管　pumping main
泵水竖管　pump riser
泵体　pump body
泵叶轮　pump impeller
泵转轴　pump shaft
泵转轴套筒　pump shaft sleeve
比测仪　comparator
比冲量　specific impulse
比电阻　specific resistance
比尔定律　Beer law
比滑　slide-roll ratio
比较　compare

比较标准器　comparison standard
比较测量法　comparison method of measurement
比较的　comparative
比较电桥　comparison bridge
比较读出　comparative read-out
比较法　comparison method
比较法标定　comparison method of calibration
比较法校准　comparison calibration
比较和交换　compare and swap
比较逻辑　compare logic
比较同步　file compare
比较文件　compare file difference
比较线圈　comparator coil
比较元件　comparing element
比较值　comparison value
比例　proportion, proportional
比例尺范围　proportional band
比例带　proportional band
比例放大器　proportional amplifier
比例积分　proportional-integral
比例积分控制　proportional plus integral control
比例积分控制器　proportional-integral (PI) controller, proportional plus integral controller, PI controller
比例积分调节作用　proportional plus integral control action
比例积分微分　proportional-integral-derivative (PID)
比例积分微分控制　PID control
比例积分微分控制器　PID controller, proportional-integral-derivative controller
比例积分微分调节器　proportional plus integral plus derivative controller
比例积分微分调节作用　proportional plus integral plus derivative control action
比例积分微分作用　proportional plus integral plus derivative action
比例积分作用　proportional plus integral action
比例计数器　proportional counter
比例控制　proportional control
比例控制器　P controller, propor-

tional controller
比例控制算法　ratio control algorithm
比例控制因子　proportional control factor
比例时间控制器　proportional time controller
比例速度浮动控制器　proportional speed floating controller
比例调节器　proportioner
比例调节作用　proportional control action
比例微分　proportional-derivative (PD)
比例微分控制器　PD controller, proportional-derivative controller, proportional plus derivative controller
比例微分调节作用　proportional plus derivative control action
比例微分作用　proportional plus derivative action
比例因子　scaling factor
比例增益　proportional gain
比例作用　proportional action
比例作用系数　proportional action coefficient
比例作用因子　proportional action factor
比率　rate, ratio
比率表　logometer
比率常数　rate constant
比率控制器　ratio controller
比率控制系统　ratio control system
比率设定器　ratio set unit
比率调节作用　rate control action
比率增益　rate gain, ratio gain
比率作用　rate action
比热容　specific heat capacity
比色计　colorimeter
比色器　color comparator
比声计　acoustical meter
比特　bit
比移值　Rf value
比值变送器　ratio transmitter
比值操作器　ratio station
比值控制　ratio control
比值设定　ratio set
比重　specific gravity
比重计　hydrometer
比重瓶　volumetric flasks

笔杆组件　pen arm assembly
笔式记录器　pen-writing recording instrument
闭端　closed end
闭合　close-up
闭合触点　closing contact
闭合磁路　closed magnetic circuit
闭合的　closed
闭合高度　shut height
闭合排队网络　closed queuing network
闭合铁心　closed core
闭环　closed loop, closed-loop
闭环辨识　closed loop identification
闭环传递函数　closed loop transfer function, closed-loop transfer function
闭环的　loop-locked
闭环极点　closed loop pole
闭环控制　closed loop control
闭环控制电路　control circuit closed-loop
闭环控制器　closed loop controller
闭环控制系统　closed loop control system
闭环零点　closed loop zero
闭环频率响应　closed-loop frequency response
闭环稳定　closed loop stabilization
闭环系统　closed loop system
闭环相角　closed loop phase angle
闭环增益　closed loop gain
闭回路控制器　loop controller
闭卷送料机　uncoiler
闭口铁心　closed core
闭口铁芯变压器　closed core transformer
闭链机构　closed chain mechanism
闭路　closed circuit
闭路电视　closed circuit television
闭路设计原理　closed-loop design principle
闭路式轨道电路　closed type track circuit
闭路天线　closed antenna
闭模高度　die height
闭塞　blocking

闭式铁芯　closed core
闭式运动链　closed kinematic chain
闭锁　lock
闭锁层　blocking layer
毕托管　Pitot tube
毕托静压管　Pitot static tube
铋　bismuth (Bi)
铋铸模　bismuth mold
壁涂开管柱　wall coated open tubular colum, wall-coated open tubular column
避雷　lightning shielding
避雷保护　lightning protection
避雷导线　lightning conductor
避雷器　arrester, lightning arrester, lightning arrestor
避雷针　diverter, lightning conductor, lightning rod
避免日光直射　keep out of the direct sun
避障碍　obstacle avoidance
臂架型起重机　jib type cranes
边带　side band
边带寻址　side band addressing (SBA)
边带寻址端口　sideband address port (SAP)
边界层　boundary layer
边界层流型质量流量计　boundary-layer type mass flowmeter
边界分布　marginal distribution
边界搜索　boundary detection
边界条件　boundary condition
边界值分析　boundary value analysis
边刨床　side planer
边频　side frequency
边沿触发的触发器　edge triggered flip-flop
边缘查找器　edge finder
边缘处理　edging
边缘连接　edge joint
边缘连接器　edge connector
边缘效应　edge effect
边值积分公式　boundary integral formulation
边值问题　boundary-value problem
边值原理方法　boundary element method

编程方法　programming approach
编程环境　programming environment
编程理论　programming theory
编辑　edit
编辑命令文件　edit command file
编辑器　editor
编辑一个文件　edit a file
编码　encode, encoding
编码地址　coded address
编码度盘　coded circle
编码方案　coding scheme
编码器　coder, encoder
编码调制　coded modulation
编目数据集　cataloged data set
编目系统　cataloging system
编排因子　organizational factor
编线　braided wire
编译　compile
编译程序　compiler
编译程序的生成程序　compiler generator
编译程序的诊断程序　compiler diagnostics
编译和连接一个程序　compile and link a program
编译器优化　compiler optimization
扁锉　flat file
扁钢　flat steel
扁钢丝　flat wire
扁母线　flat bus bar
扁平电缆　flat cable
扁线　sheet wire
便携电源适配器　desk top
便携式 B 超　portable type-B ultrasonic
便携式饱和标准电池　portable saturated standard cell
便携式气动校验仪　portable pneumatic calibrator
便携式手动操作器　standing manual station
便携式仪表　portable instrument
变比　transformation ratio
变齿轮　change wheel
变磁阻式压力传感器　variable reluctance type pressure transducer
变电站　electric substation, transformer

substation
变分分析　variational analysis
变滚比粗切　ratio control roughing
变化　variation
变化检查器　change detect
变化率报警检查　velocity alarm check
变化率限值控制　rate of change limiting control
变化速率限值　velocity limit
变换器　changer, transducer
变换因数　conversion factor
变换原理　principles of conservation
变极电动机　change-pole motor
变焦距显微镜　zoom microscope
变结构控制　variable structure control
变截面流量计　variable area flowmeter
变抗器　varactor
变量　variable
变量配比　variable pairing
变量输送泵　variable delivery pump
变流器　converter, convertor
变面积式流量测量仪表　variable area flow measuring device
变频机　frequency convertor
变频器　frequency changer, frequency converters, frequency transformer
变频调速　frequency control of motor speed
变容二极管　hypercap diode, varactor, varactor diode
变容器　adjustable capacitor, adjustable condenser, variable capacitor, variable condenser
变色　discolouration, discoloration
变色范围　colour change interval
变送单元　transmitting unit
变送器　transmitter
变速　speed change
变速齿轮　change gear, speed gears, transmission gear, change wheel
变速电动机　adjustable-speed motor, variable-speed motor
变速率滤波器　rate-of-change filter
变速器　speed changer, variator
变速系统　speed control system
变速箱　gear shift housing
变速箱体　gearbox casing

变位齿轮　modified gear
变位系数　modification coefficient
变形　deformation
变形特征　deformation characteristic
变形振动　formation vibration
变压发电机　varying-voltage generator
变压控制　varying-voltage control
变压器　potential transformer, transformer, voltage changer
变压器的变压比　transformation ratio of voltage transformer
变压器耦合　transformer coupling
变压器耦合放大器　transformer amplifier
变压器容量　transformer capacity
变压器油　transformer oil
变压调节　variable voltage control
变异系数　coefficient of variation
变值控制　variable-value control
变址寄存器　B-register, index register
变阻器　rheostat, varistor
变阻器制动　rheostatic braking
变阻式流量计　variable resistance type flowmeter
变阻整流器　varistor rectifier
辨别溶剂　differentiating solvent
辨别效应　differentiating effect
辨识　identification
辨识器　identifier
辨识算法　identification algorithm
标称的　nominal, rated
标称电压　nominal voltage
标称横截面面积　nominal cross sectional area
标尺折光仪　liquid refraction meter
标定常数　scaling constant
标定值　nominal value
标定转子　calibration rotor
标度　scale
标度标记　scale mark
标度长度　scale length
标度范围　scale range
标度分格　scale division
标度分格间距　length of a scale division, scale spacing
标度分格值　value of scale division

标度基线　scale mark base
标度盘　graduated dial
标度上限值　scale high value
标度始点值　minimum scale value
标度数字　scale numbering
标度下限值　scale low value
标度因数　scale factor
标度值　scale value
标度终点值　maximum scale value
标杆　marker post
标记　event marker, flag
标记存储器　tag memory
标记分子　labelled molecule, tagged molecule
标记寄存器　signature register
标记检测　sign detection
标记型接线端子　marking terminal
标距长度　gauge length
标量　scalar
标量李雅普诺夫函数　scalar Lyapunov function
标幺值　per-unit value
标示单　identifying sheet list
标样试验　test for nominal samples
标印装置　marking device
标志　mark
标志变量　flag variable
标志架　sign gantry
标志托架　sign mount
标准　prototype
标准铂电阻温度计　standard platinum resistance thermometer
标准成分　standard component
标准齿轮　master gear, standard gear
标准大气压　normal atmosphere
标准大气状况　standard atmospheric conditions
标准刀齿　master blade
标准的　standard
标准的操作和监视窗口　standard operation and monitoring window
标准的工程功能　standard engineering functions
标准的现场控制站　standard FCS
标准电池　standard cell
标准电感　standard inductance
标准电极　normal electrode, standard electrode
标准电压　voltage standard
标准甘汞电极　standard calomel electrode
标准函数　criterion function
标准化　standardization
标准化的　normalized
标准化设备坐标　normalized device
标准件　standard parts
标准孔口　standard orifice
标准偏差　standard deviation
标准黏度计　master viscometer
标准频率　standard frequency
标准气钢瓶　standard gas cylinder
标准器　etalon
标准氢电极　standard hydrogen electrode
标准热电偶　standard thermocouple
标准深锥度　standard depthwise taper
标准时间　standard time
标准试块　calibration block
标准水银温度计　standard mercury thermometer
标准弯管　normal bend
标准显示画面　display standard
标准响应　normalized response
标准信号　standard signal
标准信号发生器　standard signal generator
标准压力表　calibrator for pressure gauges, standard pressure gauge
标准仪表　standard instrument
标准真空表　standard vacuum gauge
标准直齿轮　standard spur gear
标准质量控制　standard quality control
标准状态　standard state
标准总线　standard bus
表　meter
表层温度表　bucket thermometer
表处理　list processing
表达式　expression
表格　tab
表格原点　table origin
表观 pH 值　apparent pH
表观功率　apparent power
表观温度　apparent temperature

表面　surface
表面安装　surface mounting
表面变形　surface deformation
表面层　surface course
表面层合电路板　surface laminar circuit
表面重修　resurfacing
表面初始疲劳破裂　surface initiated fatigue breakage
表面处理　polishing processing, surface processing, surface treatment
表面传热系数　surface coefficient of heat transfer
表面传输阻抗　surface transfer impedance
表面粗糙度　surface roughness
表面电荷效应　surface-charge effect
表面电阻　sheet resistance, surface resistance
表面分析仪　surface analyzer
表面光洁度　surface roughness
表面接触式温度计　surface contacting thermometer
表面科学　surface science
表面宽度　face width
表面离子化检测　surface ionization detection
表面面积　superficial area
表面摩擦　skin friction, surface friction
表面磨耗试验　surface abrasion test
表面耐久度　surface durability
表面抛光　surface finish
表面疲劳　surface fatigue
表面坡度　surface gradient
表面热电偶　surface thermocouple
表面热电阻　surface resistance thermometer
表面散热系数　surface coefficient of heat transfer
表面纹理　surface texture
表面研磨　skiving
表面硬化　hard facing
表面硬化处理　surface hardening
表面张力　surface tension
表面中部波皱　center buckle
表面总面积　aggregate superficial area

表盘式指示仪表　dial instrument
表皮折叠　skin inclusion
表压　gauge pressure
表压传感器　gauge pressure sensor, gauge pressure transducer
冰袋　ice pack
冰点　ice point
冰片　borneol
兵器发射理论与技术　armament launch theory and technology
兵器科学与技术　armament science and technology
丙类电路　class C circuit
丙酮　acetone
丙烷气切割　propane gas cutting
丙烯胶片　acrylic sheet
丙烯酸的　acrylic
丙烯酸片　acrylic sheet
丙烯酸树脂　acrylic resin
柄　shank
并-串联　parallel-serial
并励　shunt excited
并励磁场　shunt field
并励电机　shunt-excited machine
并励绕组　shunt winding
并联　connecting in parallel, in parallel, parallel connection, paralleling, shunt connection
并联补偿　shunt compensation
并联操作键盘　parallel operator keyboard (POK)
并联触点　collateral contact
并联导纳　parallel admittance
并联电抗器　shunt reactor
并联电路　parallel circuit, shunt circuit
并联电容　parallel capacitance
并联电容器　shunt capacitor
并联电阻　parallel resistance, shunting resistance
并联反接式限制器　shunt-opposed limiter
并联反馈　parallel feedback, shunt feedback
并联反馈积分器　parallel feedback integrator
并联机构　parallel mechanism

并联控制装置　parallel control device
并联绕组　parallel winding
并联式组合　combination in parallel
并联输入电源组件　parallel input power module（PIPM）
并联谐振　parallel resonance, antiresonance
并联谐振频率　antiresonance frequency
并联运行　parallel operation, parallel running
并联支路　parallel branch
并联阻抗　parallel impedance, shunt impedance
并联组合机构　parallel combined mechanism
并列的　apposable
并行　concurrency
并行程序　concurrent program, parallel program
并行处理　parallel processing
并行处理器　parallel processor
并行传感器　parallel transducer
并行存储器　parallel memory, parallel storage
并行的　concurrent
并行工程　concurrent engineering
并行计算机　parallel computer
并行寄存器　parallel register
并行检索存储器　parallel search storage
并行结构　concurrent architecture
并行控制　concurrency control
并行连接　connected parallel
并行连接计算机　connected parallel computer
并行设计　concurred design
并行输出插件板　parallel output card
并行输入插件板　parallel input card
并行搜索　concurrent search
并行算法　parallel algorithm
并行网络　parallel network
并行系统　concurrent system
并行信息公路　parallel highway
并行运算　parallel computation
拨号　dial up
拨号终端　dial terminal
波　wave

波长　wave length
波长分光计　wavelength spectrometer
波长计　band meter, wavelength meter
波长精确度　accuracy of the wavelength
波长偏差传感检测器　wavelength deviation sensing detector
波导　wave guide, waveguide
波导管　waveguide tube
波导管分支滤波器　waveguide branching filter
波导流体式热量计　waveguide flow calorimeter
波登管　Bourdon tube
波登管压力传感器　Burdon pressure sensor
波动　undulation
波动方程　wave equation
波动功率　fluctuating power
波段　wave band
波段开关　wave-range switch
波段调整开关　change-tune switch
波段选择器　band selector
波段转换开关　band switch, wave changing switch
波发生器　wave generator
波峰因数　crest factor
波高校正系数　calibration coefficient of wave height
波列　wave train
波数　number of waves
波特　baud
波特兰水泥　Portland cement
波特率　baud rate
波特率产生器　baud rate generator
波调制示波器　wave modulated oscilloscope
波纹　ripple
波纹管　bellows
波纹管阀　bellows valve
波纹管密封型上盖　bellows seal bonnet
波纹管式　bellow type
波纹管压力表　bellows pressure gauge
波纹膜片式压力传感器　nested diaphragm type pressure sensor
波纹系数　ripple factor

波纹研磨机	corrugation grinding machine
波形	waveform
波形变换器	wave converter
波形垫	corrugation pad
波形分析仪	wave analyzer
波形监视器	waveform monitor
波形失真	waveform distortion
波状钢	corrugated iron
玻棒	glass rod
玻耳兹曼机	Boltzman machine
玻璃布	glass fabric
玻璃槽口	glazing rebate
玻璃窗装配行业	glazing industry
玻璃瓷	glazed porcelain
玻璃电极	glass circle
玻璃钉	glazing brad, glazing sprig
玻璃钢导热试验方法	test method for thermal conductivity of glassfiber reinforced plastics
玻璃钢管	glass reinforced plastic pipe
玻璃格条	glazing bar
玻璃管式比重计	glass hydrometer
玻璃管式转子流量计	glass-tube rotameter
玻璃管液面计	glass-tube level gauge
玻璃管液柱式温度计	liquid-to-glass thermometer
玻璃活塞式压力表	glass piston pressure gauge
玻璃卡子	glazing clip
玻璃漏斗	glass funnel
玻璃密封条	glazing tape
玻璃泡	glass bulb
玻璃器皿	glassware
玻璃嵌板	glass panel
玻璃强化胶管	glass reinforced plastic pipe
玻璃水银温度计	mercury-in-glass thermometer
玻璃丝编织线	fiber-glass braided wire
玻璃温度计	liquid-in-glass thermometer
玻璃纤维	fibre glass, glass fiber
玻璃纤维垫	glass mats
玻璃纤维绝缘器	glass fibre insulator
玻璃纤维增强聚酯波纹板	glass fiber reinforced polyester corrugated panels
玻璃纤维增强塑料	glass fiber reinforced plastics
玻璃珠	glass beading
玻璃转子流量计	glass type rotameter
玻璃装配	glazing
剥孔机	broaching machine
剥离	peeling
剥离工具	stripping
剥离试验	peeling test
剥线钳	wire stripper
伯德图	Bode diagram
泊松	Poisson
泊松过程	Poisson process
铂	platinum (Pt)
铂铑-铂热电偶	platinum-rhodium platinum thermocouple
铂热电阻温度计	platinum resistance thermometer
铂热电阻温度检查器	platinum resistance temperature detector
铂丝	platinum wire
博弈论	game theory
箔剖面轮廓	foil profile
箔式应变计	foil strain gauge
薄板装料机	sheet loader
薄层板	thin layer plate
薄层棒色谱法	thin layer rod chromatography
薄层层析法	thin layer chromatography (TLC)
薄层扫描仪	thin layer scanner
薄层色谱法	thin layer chromatography (TLC)
薄层压板	thin laminate
薄垫片	shim
薄壳型填充剂	pellicular packing
薄膜	thin film
薄膜吹制法	film blowing
薄膜电阻	sheet resistance
薄膜混合电路	thin film hybrid circuit
薄膜霍尔效应磁强计	thin film Hall effect magnetometer
薄膜浇口	film gate
薄膜开关	membrane switch
薄膜色谱法	thin film chromatography

薄膜式测辐射热计　film type bolometer
薄膜陷波滤波器　thin-film notch filter
薄膜样品　film sample
薄膜执行机构　diaphragm actuator
薄铜板　copper sheet
薄铜箔　thin copper foil
薄型螺帽　hex jam nuts
薄型闸阀　thin gate valves
补偿　compensate, compensation
补偿波纹管　compensating bellows
补偿导线　compensatory leads, extension wire
补偿电流　offset current
补偿电路　recovery circuit
补偿电压　offset voltage
补偿电阻　compensating resistance
补偿定理　compensation theorem
补偿度　degree of compensation
补偿阀　compensating valve
补偿法　compensation method
补偿函数　penalty function method
补偿环节　compensating components, compensating element
补偿计　compensating gauge
补偿密度测井仪　compensation density logger
补偿器　compensator
补偿绕组　backing coil, compensating winding, compensation coil
补偿热电偶　dummy thermocouple
补偿设定　compensation set
补偿声波测井仪　borehole compensated sonic logger
补偿时间　recovery time
补偿式航电仪　compensation type airborne electromagnetic instrument
补偿式气体流量计　compensation-type gas flowmeter
补偿输出信号　compensation output signal
补偿输出信号作用　compensation output action
补偿网络　compensating network
补偿微压计　compensated micromanometer
补偿温度　compensated temperatures
补偿线圈　backing coil, compensating winding, compensation coil
补偿型延长导线　compensating extension lead
补偿元件　compensating components, compensating element
补偿装置　compensation equipment
补充　recruitment
补充过滤器　afterfilter
补充切入　set-in
补充设备　complement
补充修正　additional correction
补救　remedy
补救措施　remedial measure
补救工程　remedial works
补料过程优化　fed-batch processes optimization
补料间歇操作　fed-batch operation
补码　complement, complementary code
补漆　painting make-up
补气阀　compensating valve
补水　body wetting before glazing
补用漆　repair paint
不饱和聚酯　unsaturated polyester
不变量　invariant
不变嵌入原理　invariant embedding principle
不变性　invariance
不变性系统　invariant system
不等臂误差　arm error
不定积分　indefinite integral
不定值状态　uncertain value status
不对称　dissymmetry
不对称变形振动　asymmetrical deformation vibration
不对称的　asymmetric, unsymmetrical
不对称电位　asymmetry potential
不对称多相电流　asymmetrical polyphase current
不对称负载　unsymmetrical load
不对称伸缩振动　asymmetrical stretching vibration
不对称四端网络　non-symmetrical network
不工作齿侧　coast side
不够的　insufficient
不归位机键　non-homing switch
不规则辉光放电　abnormal glow

不合逻辑的　inconsequential
不间断电源　uninterrupted power supply（UPS）
不间断工作制　uninterrupted duty
不间断供电　uninterrupted power supply
不接触的　contactless
不接地系统　insulated system
不接地中线制　non-ground neutral system
不精确的　coarse
不精确度　inaccuracy
不均匀磁场　non-uniform magnetic field
不均匀的　non-uniform
不可定误差　indeterminate error
不可观性　unobservable
不可靠机器　unreliable machine
不可控性　uncontrollable
不可逆的　irreversible, non-reversible
不可逆电气传动　nonreversible electric drive
不可逆反应　irreversible reaction
不可凝气体　non-condensable gas
不可燃物料　non-combustible material
不可燃性测试　non-combustibility test
不连续波　full wave discontinuity
不连续的　discontinuous
不连续控制系统　discontinuous control system
不连续性　discontinuity
不良　not good
不良标签　defective product label
不良品　defective product, non-good parts
不良品箱　defective product box
不良数据　bad data
不良数据辨识　bad data identification
不灵敏的　insensitive
不灵敏性　insensitivity
不平衡　imbalance, unbalance
不平衡的　out of balance
不平衡电路　unbalanced circuit
不平衡电桥　unbalanced bridge
不平衡负载　out-of-balance load, unbalanced load
不平衡量　amount of unbalance
不平衡量指示器　amount of unbalance indicator
不平衡三相负载　unbalanced three phase load
不平衡衰减器　unbalanced attenuator
不平衡相位　phase angle of unbalance
不确定的　uncertain
不确定动态系统　uncertain dynamic system
不确定多项式　uncertain polynomial
不确定线性系统　uncertain linear system
不确定性　uncertainty
不燃材料　incombustible material
不适用　not applicable
不收缩混凝土　non-shrinking concrete
不通电试验　cold test
不同位号的点　alias point
不同相　out of phase
不透水　impervious
不透水混凝土　water tight concrete
不透水密封垫　watertight seal
不透水物料　impervious material
不完全齿轮机构　intermittent gearing
不完全数据　incomplete data
不完全退火　under annealing
不稳定闭环系统　unstable closed-loop system
不稳定的　non-steady, unstable
不稳定法　method of instability
不稳定开环过程　unstable open-loop process
不稳定控制器　unstable controller
不稳定连接　connective instability
不稳定系统　unstable system
不稳定性　instability, lability
不稳定状态　unsteady state
不稳定状态操作　unsteady-state operation
不相容原理　incompatibility principle
不锈钢　rustless iron, stainless iron, stainless steel
不锈钢和石墨　stainless and graphite
不锈钢六角轮缘尼龙嵌入螺帽　stainless steel flange nylon insert lock nuts
不锈钢六角螺帽　stainless steel hex nuts
不锈钢六角重型螺帽　stainless steel

heavy hex nuts
不锈钢螺帽　stainless steel nuts
不锈钢尼龙嵌入螺帽　stainless steel nylon insert lock nuts
不锈钢突缘螺帽　stainless steel flange nuts
不锈耐酸钢　stainless acid-resisting steel
不锈镍铸铁　nickel cast iron
不锈软钢板　stainless mild steel plate
不溢式压缩模　positive mold
不用冶炼的生铁　all-mine pig iron
不载电导线　dead wire
布尔报警　Boolean alarm
布尔代数　Boolean algebra
布尔函数　Boolean function
布尔逻辑　Boolean logic
布尔运算　Boolean operation
布局　placement
布局平面图　layout plan
布拉格方程　Bragg's equation
布拉格衍射声成像　acoustical imaging by Bragg diffraction
布朗运动　Brownian motion
布劳恩管　Braun-tube
布设总图　master drawing
布氏硬度　Brinell hardness
布氏硬度计　Brinell hardness tester
布氏硬度试验　Brinell hardness test
布氏硬度压头　Brinell hardness penetrator
布氏硬度值　Brinell hardness number
布线完成率　layout efficiency
布置图　layout drawing
步话机　walkie-talkie
步进　stepping
步进电机　step motor, stepper motor, stepping motor
步进继电器　stepping relay
步进开关　stepping switch
步进控制　step by step control, step-by-step control
步进控制器　stepping controller
步进式电动机　step-by-step motor
步进作用　stepping action
步骤　procedure
钚　plutonium (Pu)
部分分式展开　partial fraction expansion
部分辐射温度计　partial radiation pyrometer
部分收集器　fraction collector
部分响应信道　partial response channel
部分谐波　fractional harmonic
部分展开　partial expansion
部分最小二乘　partial least square (PLS)

C

擦光机　glazing calender
材料费　materials expenses
材料化学成分和机械性能　material chemical analysis and mechanical capacity
材料加工工程　materials processing engineering
材料科学与工程　materials science and engineering
材料试验机　material testing machine
材料物理与化学　materials physics and chemistry
采矿工程　mining engineering
采样　sampled, sampling
采样 PI 调节器　sampling PI controller
采样保持器　sample hold device
采样点　sampling point
采样环节　sampling element
采样间隔　sampling interval
采样控制　sampling control
采样控制器　sampling controller
采样控制系统　sampled-data control system, sampling control system
采样频率　sampling frequency
采样器　sampler
采样时钟　sample time clock
采样示波器　sampling oscilloscope
采样数据　sampled data
采样数据控制　sampled-data control
采样数据控制系统　sample data control system
采样速率　sampling rate
采样系统　sampling system
采样信号　sampled signal
采样与保持　sample and hold
采样元件　sampling element
采样周期　sampling period
采样周期选择　selection of sampling period
采样作用　sampling action
彩色编码器　colour coder
彩色超声多普勒诊断系统　ultrasonic color Doppler diagnostic system
彩色电视　colour television
彩色电视机　colour television receiver
彩色电视显像管　chromatron
彩色对比度　colour contrast
彩色解调器　colour demodulator
彩色喷墨式打印机　color ink-jet printer
彩色平衡　colour balance
彩色摄像管　colour pick-up tube
彩色视频复印机　color video copier
彩色显示器　color display
彩色显像管　chromoscope, colour kinescope, colour picture tube
彩色信号解调器　chrominance demodulator
彩色信号调制器　chrominance modulator
彩色信号载波　chrominance signal carrier
彩色硬拷贝器　color hard copy unit
菜单　menu
菜单选择式　menu selection mode
参比电极　reference electrode
参比端恒温法　constant-temperature method of reference junction
参比物　reference material
参比性能特性　reference performance characteristic
参考标准　reference standard
参考点　reference point
参考电平　reference level
参考电压　reference voltage
参考方向　reference direction
参考工作条件　reference operating condition
参考轨迹　reference trajectory
参考节点　reference node
参考结构　reference architecture
参考输入　reference input
参考输入变量　reference input varia-

ble，reference-input variable
参考输入环节　reference input element
参考输入信号　reference input signal
参考线圈　reference winding
参考向量　reference phasor
参考信号　reference signal
参考信号发生器　reference generator
参考性能　reference performance
参考值　reference value
参考值标准器　reference-value standard
参考轴　axis of reference
参考自适应控制　reference adaptive control
参量激励　parametric excitation
参数　parameter
参数辨识　parameter identification
参数波动　parametric variation
参数的　parametric
参数放大器　parametric amplifier
参数共振　parametric resonance
参数估计　parameter estimation
参数化　parametrization
参数化设计　parameterization design
参数清单　parameter list
参数设计　parameter design
参数输入显示　parameter entry display
参数线性　linear in the parameter
参数优化　parameter optimization
参数预测控制　parametric predictive control（PPC）
参数值　parameter values
残差反馈　residue feedback
残差加权法　method of weighted residual
残留奥氏体　retained austenite
残留沃斯田铁　retained austenite
残余电压　offset voltage
残余偏转　residual deflection
残余应力　residual stress
残渣　residue
仓储自动化　warehouse automation
仓库　depot，hub，warehouse
操纵　manipulation
操纵变量　manipulated variable（MV）
操纵范围　manipulate range

操纵杆　joy stick
操纵及控制装置　operation control device
操纵器　manipulator
操纵室　control cabinet
操纵手柄　control handle
操纵索　control cable
操纵台　console, control console
操纵台监视器　console monitor
操纵台显示器　console display
操纵台信号处理机　console message processor
操纵台状态显示画面　console status display
操纵台子系统　console subsystem
操纵线　control wire
操纵装置　control gear
操作　operate，operation
操作、管理和维护　operation, administration and maintenance（OAM）
操作标记　operation mark
操作单元名称组态　unit names configuration
操作电路　operating circuit
操作电压　operating voltage
操作规程　operation procedure
操作和监视功能　operation and monitoring function
操作和监视功能的组态　configuration of operation and monitoring functions
操作画面显示块　operator panel display block
操作级别　function access level
操作记录　operation record
操作寄存器　operation register
操作开关　console switch
操作控制台　operating console
操作量上限　manipulated variable high limit
操作量下限　manipulated variable low limit
操作目标　operating objective
操作盘　operating panel
操作屏幕方式　operating screen mode
操作器惯性矩阵　manipulator inertia matrix

操作任务	manipulation task
操作日志	operational logbook
操作手册	operation manual
操作数寄存器	operand register
操作顺序	sequence of operation
操作特性	operational characteristic
操作图	application drawing
操作系统	operating system
操作译码器	operation decoder
操作应用菜单画面	operator utility menu panel
操作员操作台	operator station
操作员键盘	operation keyboard, operator keyboard
操作员接口站	operator interface station
操作员控制键盘	operator control panel
操作员控制台	operator console
操作员请求控制台	operators request control panel
操作员站	operator station
操作员属性	operator personality
操作站定义	operator station definition
操作站状态显示画面	operator station status display panel
操作指导窗口	operator guide window
操作指导信息画面	operator guide message panel
操作指示	operation instruction
操作组	operation group
操作组画面	group display
糙度系数	roughness coefficient
槽	groove
槽车球阀	tank lorry ball valves
槽盖	channel cover
槽钢	box iron, channel iron, steel channel
槽口刨	rebate plane
槽宽收缩	slot-width taper
槽轮	geneva wheel
槽轮机构	geneva mechanism
槽数	geneva numerate
槽凸轮	groove cam
槽纹瓷砖	grooved tile
槽线	slot line
侧壁皱纹	body wrinkle
侧边	side
侧冲压平	side stretch
侧浇口	side gate
侧面安装	side mounting
侧前角	side rake angle
侧视图	side elevation
侧隙变量	backlash variation
侧隙变量公差	backlash variation tolerance
侧隙公差	backlash tolerance
侧限应力	confining stress
侧向测井仪	lateral logger
侧向力	lateral force
侧向位移	side movement
侧向压力	lateral pressure
侧抑制网络	lateral inhibition network
侧翼	flank
侧置式浮子开关	side-mounted float switch
测地学测量	geodetic survey
测定	setting out
测定对映体	enantiomer
测氡仪	emanometer
测风塔	anemometer tower
测高规	height gauge
测高仪	altitude meter
测厚仪	thickness gauge
测绘科学与技术	surveying and mapping
测绘仪	mapper
测井记录仪	logging recorder
测径	caliper measure
测径规	caliber gauge
测距器	telemeter
测距仪	distance gauge, distance meter, range finder
测孔千分尺	hole micrometer
测力计	dynamometer
测力系统	dynamometric system
测量	measuring
测量变送器	measuring transmitter
测量标准	measurement standard
测量步骤	measurement procedure
测量齿顶高	measuring addendum
测量齿厚	measuring tooth thickness
测量传感器	measuring transducer

测量单位　coherent of measurement, measuring unit
测量单位符号　symbol of a unit of measurement
测量单位制　coherent system of measurement, system of units of measurement
测量的　measured
测量的倍数单位　multiple of a unit of measurement
测量的不确定度　uncertainty of measurement
测量的重复性　repeatability of measure
测量的导出单位　derived unit of measurement
测量的权　weight of measurement
测量的再现性　reproducibility of measurement
测量点　measuring point
测量电极　meter electrodes
测量电桥　measurement bridge, measuring bridge
测量电位差计　measuring potentiometer
测量电阻用的直流电桥　DC bridge for measuring resistance
测量动态　measurement dynamics
测量端　measuring junction, measuring terminal
测量段　measuring section
测量反馈　measured feedback
测量范围　measuring range
测量范围上限值　higher measuring range value, measuring range higher limit
测量范围下限值　lower measuring range value, measuring range lower limit
测量方法　method of measurement
测量放大器　measuring amplifier
测量和控制对象　object to be measured and controlled
测量结果　result of measurement
测量精度　accuracy of measurement, measured accuracy
测量距离　measuring distance
测量孔　measuring hole
测量链　measuring chain

测量偏差　measured deviation
测量平面　measuring plane
测量球隙　measuring spark gap
测量绕组输入阻抗　input impedance of the measuring winding
测量设备　measuring equipment, measuring installation
测量时间　measurement time, measuring time
测量顺序　measuring sequence
测量图　survey plan
测量误差　error of measurement, measurement error, measuring error
测量系统　measuring system
测量信号　measurement signal
测量仪表的常数　constant of a measuring instrument
测量仪表的零位　zero of a measuring instrument
测量仪表的误差曲线　error curve of a measuring instrument
测量仪表的修正值曲线　correction curve of a measuring instrument
测量仪表动态　dynamics of measuring instrument
测量仪表示值　indication of a measuring instrument
测量仪器　measuring instrument
测量仪器仪表的零位　zero of a measuring instrument
测量硬件　measurement hardware
测量用电流互感器　measuring current transformer
测量用电压互感器　measuring voltage transformer
测量用滑触电阻线　measuring slide wire
测量元件　measuring element
测量员　surveyor
测量原理　principle of measurement
测量噪声　measurement noise
测量值　measured value
测量指示　measuring indication system
测量指示仪　measurement indicator
测量装置　measuring apparatus, measuring device
测流标杆　current pole

测面器	planimeter
测扭仪	twisting apparatus
测频计	frequency meter
测平器	planometer
测深规	depth gauge
测深绳	sounding wire
测渗计	lysimeter
测湿计	wet test meter
测试板	test board
测试操作员控制站	test operator control station
测试传声器	measuring microphone
测试点	test point
测试功能	test function
测试荷载	test load
测试计量技术及仪器	measuring and testing technologies and instruments
测试频率	test frequency
测试设备	test equipment
测试条	test strip
测试线	holding wire
测试箱	test box
测试信号	test signal
测试样本	test sample
测试运行	test run
测速传感器	tachogenerator, tachometer generator
测速发电机	tacho-generator, tachometer generator
测速仪	velocimeter
测位器	position finder
测温铂电阻器	platinum resistance bulb
测温电桥	bridge for measuring temperature
测温电阻器	resistance bulb
测温筒	thermal bulb
测温锥	seger cone
测隙规	feeler gauge
测斜仪	clinometer
测压管水位	piezometric level
测压计	piezometer
测压孔	pressure tap
测压元件	pressure element
测云雷达	cloud detection radar
策略函数	strategic function
参差调谐放大器	stagger tuning amplifier

层	lamination
层次分析法	analytic hierarchy process (AHP)
层次结构图	hierarchical chart
层次设计	hierarchical design
层级	tier
层级结构	hierarchical structure
层间全内导通多层印制板	any layer inner via hole multilayer printed board
层理	bedding
层流	laminar flow
层流式流量计	laminar flowmeter
层压板面	unclad laminate surface
层压玻璃观察窗	laminated glass window
叉车	forklift, left fork
叉积	cross product
叉式起重车	fork-lift truck
叉指式换能器	interdigital transducer
插槽	slot
插齿	gear shaping
插齿机	gear shaper
插床	slotting machine
插件	plugboard, plug-in board
插件箱式仪器	card cage instrumentation
插接板	pinboard, plugboard
插入	interpolation
插入长度	insert length
插入单元	plug-in unit
插入深度	insertion length
插入式	insert-type
插入式测量仪表	bayonet gauge
插入式放大器	plug-in amplifier
插入式涡轮流量计	insertion turbine meter
插入式振捣器	poker vibrator
插入算法	interpolation algorithm
插塞式连接器	plug connector block
插头	male plug, plug connector block
插头插座	plug-and-socket
插头钳	connector plug
插图	inset
插线	plug wire
插线程序计算机	wired program computer

插页　inset
插座　female plug, patera, plug receptacle, plug socket, socket
查核点　check point
差错率指标　error rate performance
差动比较器　differential comparator
差动变压器式位移传感器　differential transformer displacement transducer
差动变压器式压力传感器　differential transformer pressure transducer
差动测量仪表　differential measuring instrument
差动传感器　differential transducer
差动磁场转子　differential field rotor
差动的　differential
差动电路　differential circuit
差动读出　differential read-out
差动活塞　differential piston
差动继电器　balanced relay, differential relay
差动检流计　difference galvanometer, differential galvanometer
差动接法　differential connection
差动轮系　differential gear train
差动器侧面伞齿轮　crown gear
差动前置放大器　differential preamplifier
差动热电偶式电压表　differential thermocouple voltmeter
差动输入电阻　differential input resistance
差动调节器　differential regulator
差动线圈　differential coil
差动元件　differentiating element
差动作用　differential action, differentiating action
差额博弈　differential game
差分方程　difference equation
差分方程模型　difference equation model
差分放大乘法器　differential amplifier multiplier
差分放大器　differential amplifier
差分分析　difference analysis
差分几何方法　differential geometric method
差分检测　differential detection
差分曼彻斯特编码　differential Manchester encoding
差分热电偶　differential thermocouple
差分输入　differential input
差分增益　differential gain
差复励　differential compound
差复励电动机　differential compound motor
差接变压器　differential transformer
差励　differential excitation
差励发电机　differential-excited generator
差拍测量法　beat method of measurement
差拍法　beat method
差频　beat frequency
差频振荡器　beat frequency oscillator
差频指示器　beat indicator
差热分析范围　DTA range
差热分析仪　differential thermal analyzer
差示热分析　differential thermal analysis (DTA)
差示热分析仪　differential thermal analysis meter
差示热膨胀法　differential dilatometry
差示热曲线　differential thermal curve
差示扫描量热法　differential scanning calorimetry (DSC)
差示色谱法　differential chromatography
差示温度滴定法　differential thermometric titration
差式吸引光谱　difference absorption spectrum
差速齿轮　differential gear
差隙　differential gap
差隙控制　differential gap control
差隙控制器　differential gap controller
差压　differential pressure
差压比　differential pressure ratio
差压传感器　differential pressure sensor, differential pressure transducer
差压流量传感器　differential pressure flow sensor, differential pressure flow transducer
差压流量计　differential pressure flowmeter

差压式流量计　differential pressure type flowmeter
差压式物位传感器　differential pressure level sensor，differential pressure level transducer
差压压力表　differential pressure gauge
差压液位测量仪表　pressure level measuring device
差压液位计　differential pressure level meter
差压装置　differential pressure device
差压装置的一次元件　primary device of a differential pressure device
差异　variance
差异量　difference quantity
差异移动　differential movement
拆除　striking
拆模　dismantle the die
拆散　disassembly
拆卸　demolition, remove
拆卸工程　demolition works
柴油　diesel fuel, diesel oil
柴油电力机车头　diesel-electric locomotive
柴油发电机　diesel generator
柴油机　diesel engine
柴油机车　diesel locomotive
柴油液动机车头　diesel-hydraulic locomotive
柴油引擎　petrol engine
掺和剂　admixture
掺杂剂　dope additive
产率　productivity
产生　generation
产生机制　generation mechanism
产生式规则　production rule
产形轮压力角　generating pressure angle
铲车　fork-lift truck
铲泥车　loader
阐释指引　explanatory guide
颤动　chattering, dithering
长　long
长波　long wave
长波纹管　lengthy bellows
长波形可移动输入红外线　amplitude shift keyed infra-red (ASKIR)
长度　length
长度测量工具　dimensional measuring instrument
长方形焊盘　oblong pad
长径喷嘴　long radius nozzle, long-radius nozzle
长径肘管　long radius elbow
长距离输电线路　long distance transmission line
长明灯　pilot burner
长期存储器　long-term memory
长期工作制　uninterrupted duty
长期记忆　long term memory
长期偏移　long term drift
长期运转试验　long term running test
长时间趋势功能　long-term trend function
长图自动平衡记录仪　strip-chart self-balancing recorder
长途电话　long distance telephone
长网线　fourdrinier wire
长线效应　long-line effect
长周期地震仪　long period seismograph
常闭　normally closed
常闭触点　break contact, normal closed contact, normally closed contact
常规PV数据点或算法　regulation PV data point or algorithm
常规操作　conventional operation
常规控制　conventional control, regulatory control
常规控制数据点或算法　regulatory control data point or algorithm
常开　normally open
常开触点　make contact, normally open contact
常量分析　macro analysis
常数　const, constant
常速试验机　constant velocity testing machine
常态分布　normal distribution
常态干扰　normal mode interference
常态抑制　normal mode rejection
常微分方程　ordinary differential equation
常温　atmospheric temperature

385

常温凝固的 air set
常温试验 cold test
常温应变计 general purpose strain gauge
常温自硬铸模 air set mold
常压电离 atmospheric pressure ionization (API)
常压蒸馏 atmospheric distillation
常用程序 utility program
常用机构 conventional mechanism, mechanism in common use
常用设备 common equipment
厂内系统 intra-plant system
厂用电 house supply
厂用电力网 plant network
场参数测量法 method of field parameter measurement
场发射电子枪 field emission gun
场发射电子像 field emission electron image
场发射显微镜 field emission microscope (FEM)
场发射显微镜法 method of field emission microscope
场解吸法 field desorption (FD)
场阑 field stop
场离子发射显微镜 field ion emission microscope
场频锁 field-frequency lock
场强 intensity of field
场强计 field intensity meter, field strength meter
场绕组 field wire
场扫描 field sweeping
场效应 field effect
场效应管 field effect transistor (FET)
场效应管离子传感器 ISFET ion sensor, ISFET ion transducer
场效应管气体传感器 FET gas sensor, FET gas transducer
场致电离 field ionization (FI)
场致发射 field emission
敞口容器液位器 liquid level instrument of open vessel
敞口式膜盒 open type diaphragm-box
超薄型层压板 ultra thin laminate

超驰控制 override control
超驰选择算法 override selector algorithm
超驰预置 override initialization
超纯水制造 ultra-pure water purifiers
超纯水制造系统 ultra-pure water purifier system
超大规模集成电路 VLSI
超导磁力梯度仪 gradient superconducting magnetometer
超导电流 supercurrent
超导分量磁力仪 component superconducting magnetometer
超导热体 thermal conductivity superconductor
超导体 superconductor
超导体丝 superconductive wire
超导液位指示器 superconducting level indicator
超低频 ultra-low frequency
超低频相位计 ultra-low-frequency phase meter
超低频信号发生器 ultra low frequency signal generator
超低温冰箱 ultra-low temperature freezer
超低噪声前置放大器 ultra-low noise preamplifier
超电导的 superconductive
超动态应变仪 high dynamic strain indicator
超短波 ultra-high-frequency wave, ultrashort wave
超负荷 overload
超负荷破裂 overload breakage
超负荷系数 overload factor
超高 superelevation
超高频 hyper-high-frequency, ultra-high frequency
超高频插入式电压表 UHF insertion-type voltmeter
超高速内圆磨床 ultra-high-speed internal grinder
超高温导体 high-temperature superconductor
超高温电气炉 ultra-high temperature

electric furnaces
超高效率 ultra-efficient
超高压 extra high voltage, extra-high voltage, supervoltage, ultra-high voltage
超高压电力网 supertension power network
超高压输电线 supervoltage transmission power line
超高真空法兰尺寸 ultra-high vacuum flange-dimensions
超高真空法兰结构型式 ultra-high vacuum flange-types
超过 exceed, excess
超级恒温水浴 super-grade thermostat with water bath
超级恒温油浴 super-grade thermostat with oil bath
超级计算机 supercomputer
超净工作台 bechtop
超量电度表 excess energy meter
超临界流体萃取 supercritical fluid extraction
超临界流体色谱法 supercritical-fluid chromatography (SFC)
超滤器 ultrahigh purity filter
超前 advancing, leading
超前补偿 lead compensation
超前补偿器 lead compensator
超前角 angle of advance, lead angle
超前网络 lead network
超前相位控制器 phase advance controller
超前相位网络 phase advance network
超前-滞后控制器 lead-lag controller
超前组件 lead module
超声波 ultrasonic wave
超声波测试 ultrasonic testing
超声波传感器 ultrasonic transducer
超声波打磨机 grinders ultrasonic
超声波电极清洗器 ultrasonic electrode cleaner
超声波发生器 supersonic generator, ultrasonic generator
超声波加工 ultrasonic machining
超声波裂缝检测 ultrasonic crack detection
超声波清洗机 ultrasonic cleaners
超声波试管清洗机 ultrasonic pipet washers
超声波探伤仪 ultrasonic flaw detector
超声波液位传感器 ultrasonic level sensor
超声波障碍探测器 ultrasonic obstacle detector
超声传感器 ultrasonic sensor, ultrasonic transducer
超声电缆检漏仪 ultrasonic cable leak detector
超声电视测井仪 borehole acoustic television logger
超声多普勒流量计 ultrasonic Doppler flowmeter
超声流量计 ultrasonic flow measuring device, ultrasonic flowmeter
超声破碎仪 ultrasonic cell disruptor
超声物位测量仪表 ultrasonic level measuring device
超实时仿真 faster-than-real-time simulation
超速故障停机 overspeed shutdowns
超调量 overshoot
超外差 superhet
超外差收音机 superhet, superheterodyne
超稳定性 hyperstability
超限应力 overstress
超循环理论 hypercycle theory
超压 overvoltage
超硬合金钢 hard alloy steel
潮差 tidal range
潮高 height of tide
潮湿 dampness
潮汐发电站 tidal power station
车程计 trochometer
车床 lathe
车床车削 lathe cutting
车床工作台 lathe bench
车刀 lathe tool, tool for lathe
车道 driveway
车道方向指示信号 lane direction control signal

车道容量　lane capacity
车房　garage
车间　workshop
车辆触发交通灯号　traffic actuated signal
车辆工程　vehicle engineering
车辆护栏　vehicle parapet
车辆交通　vehicle traffic
车辆栏障　vehicle barrier
车辆调光器　dimmer
车辆通道　vehicle access road
车轮　wheel
车轮车床　car wheel lathe
车轮打滑扭矩　wheel slip torque
车轮载重　wheel load
车身底盘　chasis
车厢控制器　controller vehicle
车厢门触点　car-door electric contact
车削　turning
车轴　axle
尘埃计数器　particle counters
尘量分析仪　dust analyzer
沉淀　settle
沉淀滴定法　precipitation titration
沉淀形式　precipitation forms
沉管　immersed tube
沉管隧道　immersed tube tunnel
沉积　consolidation
沉积物　sediment
沉降差　differential settlement
沉沙池　desilting sand pit, desilting sand trap
衬垫　gasket
衬里蝶阀　lining butterfly valves
衬里截止阀　lining globe valves
衬里球阀　lining ball valves
衬里三通旋塞阀　lining t-cock valves
衬里止回阀　lining check valves
衬套　bushing block
称量形式　weighing forms
称重传感器　load cell
撑杆　brace
撑压加工　bulging
成对安装　paired mounting
成分　composition, constituent
成分传感器　composition sensor
成分偏差变送器　composition deviation transmitter
成品　finished products
成品检验规范　inspection specification
成熟的　mature
成套配电箱　whole set of distribution box
成像系统　imaging system
成形　shaping
成形淬火机　quenching press
成形刀　form tool
成形机　shaping machine
成形加工　forming
成形磨床　form grinding machine
成型　molding
成型薄钢板　profiled sheet iron
成型厂　molding factory
成型模　forming die
成型刨齿机　gear planer
成圆机　roll forming machine
成组工作　group work
成组技术　group technology
承插接头　socket-and-spigot joint
承担　commitment
承压接缝　compression joint
承载构筑物　load bearing structure
承载力　bearing capacity
承载能力　carrying capacity
承重　load bearing
承重构架　supporting frame
承重能力　bearing force
乘除器　multiplier divider
乘法器　multiplier
乘积积分器　product integrator
乘数　multiplying factor
程度　degree
程控数字自动交换机　private automatic branch exchange
程控用户交换机　private branch exchange（PBX）
程式管理　programs management
程式设计系统　programming system
程序　program
程序编译器　compiler
程序槽规模　sequence slot size
程序测试　sequence test
程序测试器　sequence tester

程序处理　sequence processing
程序存储　program store
程序工程师　process engineer
程序汇编　program assembler
程序寄存器　program register
程序间通信　inter-program communication
程序控制　program control, programmed control, sequence control
程序控制单元　program controlling element
程序控制盘　programming control panel
程序控制器　program controller, programmed controller
程序控制装置　program control device
程序库　library
程序流速　programmed flow
程序溶剂　programmed solvent
程序设定操作器　program set station
程序设定器　program set unit
程序设计　program composition, programming
程序设计条件　programming environment
程序设计语言　programming language
程序设计支持　programming support
程序算法　routing algorithm
程序文本　program documentation
程序文件编制　program documentation
程序压力　programmed pressure
程序员控制台　programmers console
程序诊断　program diagnostic
程序指令地址　program order address
秤杆　beam
弛豫历程　relaxation mechanism
弛豫频率　relaxation frequency
弛张振荡　relaxation oscillation
弛张振荡器　relaxation oscillator
迟到反射波　delayed echo
迟缩剂　retarder
迟延网络　retardation network
持久强度试验机　creep rupture strength testing machine
持续负载　continuous load
持续过范围限　continuous overload limit, continuous overrange limit
持续时间　duration

尺寸　size
尺寸标注　size marking
尺寸测量　dimension survey
尺寸的　dimensional
尺寸公差　dimensional tolerance
尺寸排阻色谱法　size exclusion chromatography
尺寸系列　dimension series
尺寸因素　size factor
尺度　dimension
尺度传感器　dimension sensor, dimension transducer
尺度索　guide wire, measuring wire
齿槽　tooth space
齿槽底面　bottom land
齿长重合度　face contact ratio
齿垂直面　tooth vertical
齿顶　top
齿顶高　addendum
齿顶圆　addendum circle
齿顶圆半径　outside radius
齿顶圆角半径　tip radius
齿顶圆锥母线　face cone element
齿缝宽度　slot width
齿根半径　root radius
齿根高　dedendum
齿根过渡曲线　fillet curve
齿根角　dedendum angle
齿根角倾斜　root angle tilt
齿根曲面　root surface
齿根线　root line
齿根圆　dedendum circle, root circle
齿根圆角半径　fillet radius
齿根圆锥角　root angle
齿根直径　root diameter
齿冠　crown
齿厚　tooth thickness
齿厚收缩　thickness taper
齿厚游标卡尺　gear tooth vernier gauge
齿花螺帽　clinch nuts
齿角　tooth angle
齿节圆　pitch circle
齿局部接触　localized tooth contact
齿距　circular pitch
齿距变动量　pitch variation, spacing variation
齿距公差　pitch tolerance, spacing tol-

erance
齿距检查仪　tooth spacing testing
齿距收缩　space-width taper
齿廓　tooth profile
齿廓重合度　profile contact ratio
齿廓角　profile angle
齿廓啮合失配　profile mismatch
齿廓桥形接触　profile bridge
齿廓曲率半径　profile radius of curvature
齿廓曲线　tooth curve
齿廓样板　former
齿轮　gear, gear wheel
齿轮泵　gear pump
齿轮变速箱　speed-changing gear boxes
齿轮测量线　gear measuring wire
齿轮插刀　pinion cutter
齿轮齿条　pinion and rack
齿轮齿条机构　pinion and rack
齿轮传动链　gear train
齿轮传动系　pinion unit
齿轮粗切机床　gear rougher
齿轮的接触斑点　contact pattern
齿轮滚刀　hobbing cutter
齿轮加工　gear machining
齿轮加工机床　gear cutting machines
齿轮加工调整卡　gear manufacturing summary
齿轮减速箱　gear reducer
齿轮联轴器　gear coupling
齿轮啮合节点　pitch point
齿轮倾斜　gear tipping
齿轮系分度　geared index
齿轮箱　gearbox, gear box
齿轮中心　gear center
齿轮轴线　gear axis
齿轮轴向平面　gear axial plane
齿轮轴向位移　gear axial displacement
齿轮组　gears
齿轮组合　gear combination
齿螺旋线　tooth spiral
齿面宽　face width
齿面塌陷　case crushing
齿面研磨磨损　abrasive tooth wear
齿啮合频率　tooth-mesh frequency

齿式棘轮机构　tooth ratchet mechanism
齿数　number of teeth, tooth number
齿数比　gear ratio
齿水平面　tooth horizontal
齿条　rack
齿条插刀　rack cutter
齿条传动　rack gear
齿线　tooth trace
炽灼残渣　residue on ignition
充电　charge up
充电电流　charging current
充电电压　charging voltage
充电接线图　charging wiring diagram
充电开关　charger switch
充电器　charger
充电设备　charging equipment
充电时间　charging period
充电时间常数　electric charge time constant
充分的　fully
充分混杂湍流（紊流）　fully rough turbulent flow
充灌式感温系统　filled thermal system
充气　aeration
充气二极管　gas diode
充气光电管报警器　gas cell alarm
充气机　inflator
充气膜盒　gas-filled bellows
充气热系统　gas filled thermal system
充气水闸　inflatable dam
充气温度计　gas-filled thermometer
充水银式感温系统　mercury filled thermal system
充填不足　short shot
充液式感温系统　liquid filled thermal system
充液压力表　liquid filled case
冲裁模　blanking die, cutting die
冲厕喉管　flushing pipe
冲程　stroke
冲床　punch press
冲床侧板　straight side frame
冲床规格　press specification
冲淡　dilute

冲锻　press forging
冲锻法　drop forging
冲击　impact
冲击摆锤　impact pendulum
冲击波　blast wave, shock wave
冲击锤体　impact hammer
冲击电流　rush current
冲击电流计　ballistic galvanometer
冲击电流状态　rush current state
冲击电压　surge voltage
冲击函数　impulse function
冲击荷载　impact load, shock load
冲击挤压加工　impact extrusion
冲击力　impact force
冲击能量　impact energy
冲击韧性　impact toughness
冲击式流量计　impact flowmeter
冲击试校支架　impact specimen support
冲击试验　impact test
冲击试验机　impact testing machine
冲击试验系统　shock testing systems
冲击压力　surge pressure
冲积物　alluvial deposit
冲激响应　impulse response
冲记号　stamp mark
冲件　sheet metal parts
冲角　angle of attach
冲孔　punched hole
冲孔加工　piercing
冲孔模　pierce die
冲孔器　puncher
冲口加工　notching
冲模垫　die pad
冲模母模　die button
冲缺口压力机　notching press
冲水泵　flush water pump
冲水增压泵　flush water booster pump
冲突　collision
冲洗　flush
冲洗水　flushing water
冲压　stamping
冲压闭合高度　die height
冲压厂　stamping factory
冲压机　punching machine, stamping press
冲压记号　punch mark
冲压铆合　punch riveting
冲缘加工　burring
冲字　stamp letter
重布　rerouting
重氮化滴定法　diazotization titration
重氮化反应　diazotization reaction
重叠　overlap
重定作用　reset action
重复报警　repeating alarm
重复冲击　repetitive shock
重复性　repeatability
重复性误差　repeatability error
重复载荷　repeated load
重供电　re-energization
重合点　coincident points
重合度　contact ratio
重合闸　reclosing
重建　rebuild
重建色谱图　reconstructive chromatogram
重排　permutation
重排算法　permutation algorithm
重塑　remoulding
重现期　return period
重新定线　realignment
重新建立　reconstitute
重新建造　reconstruction
重新校验　recommission
重新校准　recalibrate
重新接驳　reconnection
重新紧固　retighten
重新命名一个文件　rename a file
重新漆　repaint
重新启动　restart
重新启动选择　restart option
重新绕线　rewinding
重新设定值　reset value
重新使用工程数据　reuse engineering data
重新注满　re-fill
重新装配　reassembling
重置键　reset button
重铸　recasting
抽点检验　check pointing
抽风机　exhauster, suction fan, air exhaust fan
抽风式风扇　induced draft fan

抽空　evacuation
抽粒机　grit maker
抽气泵　off-gas pump
抽气扇　extract fan
抽气式热电偶　gas pump thermocouple
抽湿机　dehumidifier
抽水厂　pumping plant
抽水机　pump
抽水机叶轮　pump impeller
抽水站　pumping station
抽象系统　abstract system
抽样　sample
抽样检查　sampling inspection
抽样检验样本大小　sample size
抽样脉冲　sampling pulse
畴壁　domain wall
出错率　error rate
出错维修　corrective maintenance
出度　incoming degree
出风口　air outlet
出货　shipment
出口　exit, export, outlet
出口包装　export package
出口开关　exit switch
出口路线　exit route
出入口　opening
出射角　angle of departure
出水管　outlet pipework
出现电热谱仪　appearance potential spectrometer (APS)
出现电位　appearance potential
初步操作试验　pilot run
初步工程　preliminary works
初步开发样机　conceptual development model
初步模图设计　preliminary mold design
初步试验　primary test
初负荷　initial load
初级污水处理　primary treatment of sewage
初期点蚀　initial pitting
初熔铁　fresh iron
初始不平衡量　initial unbalance
初始程序的装入程序　initial program loader (IPL)
初始　initial, original

初始化　initialize
初始化属性　initialize personality
初始脉冲　initial pulse
初始偏差　initial deviation
初始特征化　initial characterization
初始条件　initial condition
初始位置　initial position
初始值　initial value
初始状态　initial state
初相角　epoch angle, initial phase angle
初相位　initial phase
初学者通用符号指令码　beginner all-purpose symbolic instruction code
初样　engineering model
初值定理　initial-value theorem
除尘布袋　bag filter
除尘器　dust catcher
除法　division
除法电路　dividing circuit
除法器　divider
除氯　chloride extraction
除气　degassing
除砂　grit removal
除湿　dehumidification
除湿器　dehumidifier
除雾器　mist eliminator
除锈　derust
除锈机　derusting machine
除油器　grease trap
雏形锻模　blocker
雏形试验模具　prototype mold
储气缸　air reservoir
储油器　oil chamber
处理　treatment
处理单元　processing unit
处理品　disposed goods, disposed products
处理器　processor
处理器接口　processor gateway
处理器系统　processor system
处理器阵列　processor array
处理算术和逻辑单元　process arithmetic and logical unit
处理周期　processing cycle
处置　disposal
触点　contact
触点闭合　closing of contact

触点电流	contact current
触点断开　breaking of contact
触点输出　contact output
触点输入　contact input
触电　electric failure, electrocution
触发比较器　trigger comparator
触发传动式张力计　trigger-operated tension meter
触发电路　trigger circuit
触发电平　triggering level
触发电压　gate voltage
触发控制　toggle switch
触发器　flip-flop, toggle, trigger
触发器电路　flip-flop circuit
触发信号　triggering signal
触摸屏　touch screen
触头　contact terminal
触针　stylus
穿孔　punched
穿孔带　punch tape
穿孔管　perforated pipe
穿孔卡　punch card
穿透深度　depth of penetration
穿线　threading
穿线盒　pull box
穿心式电流互感器　straight-through current transformer
穿越频率　cross over frequency
传播　propagation
传导　conduct
传导单元　conductivity cell
传导电流　conduction current
传导角　conduction angle
传导类电法仪器　conducting electrical instrument
传递　transfer, transmit
传递层　transport layer
传递函数　transfer function
传递函数矩阵　transfer function matrix
传递环节　transfer element
传递矩阵　transmission line matrix
传递途径　pipeline
传动比　transmission ratio
传动齿轮　drive gear
传动角　transmission angle
传动链　drive chain
传动螺杆　drive screw

传动马达　drive motor
传动皮带　drive belt
传动桥试验机　axle testing machine
传动系统　driven system
传动销　drive pin
传动小齿轮　drive pinion
传动轴　drive axle, transmission shaft
传动轴承　drive bearing
传动装置　driving device
传感控制　sensory control
传感毛细管　sensing capillary
传感膜片　sensing diaphragm
传感器　sensor, transducer
传感器融合　sensor fusion
传感器失效　sensor failure
传感器系统　sensor system
传感探头　sensing probe
传感元件　sensing element
传号线　order wire
传热系数　heat-transfer coefficient
传热值　thermal transmittance value
传声器保护罩　microphone protection grid
传声器动态范围　dynamic range of microphone
传声器共振频率　microphone response frequency
传声器架　microphone stand
传声器扩散场灵敏度　diffuse-field sensitivity of microphone
传声器扩散场响应　diffuse-field response of microphone
传声器频率响应　frequency response of microphone
传声器温度系数　microphone temperature coefficient
传声器械校准仪　microphone calibration apparatus
传声器指向性频率响应　directional frequency response of microphone
传声器指向性图案　directional pattern of microphone
传声器指向性指数　directivity index of microphone
传声器自由声场灵敏度　free-field sensitivity of microphone
传声器自由声场频率响应　free-field

frequency response of microphone
传声器最高声压级 maximum sound pressure level of microphone
传输范围 transmission range
传输零点 transmission zero
传输特性 transmission characteristic, transport property
传输网络 transmission network
传输系统 transmission system
传输线 transmission line
传输线的通话能力 traffic capacity
传输线稳频振荡器 line-stabilized oscillator
传输滞后 transportation lag
传送带接口 moving band interface
传送机 conveyor
传真 facsimile
传真电极 facsimile telegraph, photo telegraph
传真式地震仪 facsimile seismograph
传纸机构 chart driving mechanism
船舶与海洋工程 naval architecture and ocean engineering
船控制 ship control
船体影响 influence of ship-body
船坞 dock
船用柴油机 marine diesel engine
船用锅炉 marine boiler
船用气压表 marine barometer
船用汽轮机 marine steam turbine
船用仪器仪表 marine instrument
串 bunch
串并联 series-parallel
串并联电路 series-parallel circuit
串并联控制 series-parallel control
串话 crosstalk
串激电机 series motor
串级 cascade
串级电动机 concatenated motor
串级放大器 cascade amplifier
串级激励 cascade exciter
串级记录调节器 cascade recording controller
串级控制 cascade control
串级控制系统 cascade control system
串级控制组件框 cascade control module

串级设定 cascade set
串级设定点 cascade setpoint
串级调节器 cascade controller
串级调速 cascade speed control
串级系统 cascade system
串级限幅器 cascade amplitude limiter
串励 series excited
串励电机 series machine
串联 connecting in series, series, series connection, tandem
串联饱和电抗器 series transductor
串联补偿 series compensation
串联补偿放大器 series-compensated amplifier
串联的 connected in series
串联电阻 series resistance
串联反馈 series feedback
串联缝熔接 series seam welding
串联负反馈 series negative feedback
串联感应 series inductance
串联校正网络 cascade compensation network
串联接驳 serial link
串联连接 cascade connection
串联式组合 combination in series
串联式组合机构 series combined mechanism
串联限幅器 series limiter
串联谐振 series resonance
串联元件 serial element
串模电压 normal mode voltage, series mode voltage
串模干扰 differential mode interference, series mode interference
串模信号 series mode signal
串模抑制 series mode rejection
串模抑制比 series mode rejection ratio
串入串出寄存器 series-in-series-out register
串行操作 serial operation
串行存储器 serial storage
串行打印机 character-at-time printer
串行的 serial
串行方式 serial mode
串行计算机 serial computer
串行寄存器 serial register
串行逻辑 serial logic

串行设备接口　serial device interface
窗口尺寸　window sizing
窗扣　window catch
窗台　sill of window
床身式铣床　bed type milling machines
创新设计　creation design
吹　blow
吹倒　blow down
吹风机　blower
吹膜　blow moulding
吹气管　bubble-tube
吹气式压力计　bubble-tube pressure sensing device
吹釉　glazing by sufflation
垂吊式天花板　suspended ceiling
垂度　sag
垂度规　dip gauge
垂直　vertical factor
垂直的　vertical
垂直方向　vertical direction
垂直负载　perpendicular load
垂直荷载　vertical load
垂直距离　vertical distance
垂直力　vertical force
垂直面　vertical plane
垂直模板　vertical formwork
垂直偏置距　vertical offset
垂直外加荷载　vertical imposed load
垂直位移　vertical displacement
垂直信号放大器　Y axis amplifier
垂直载荷　normal load
垂直状态指示器　vertical situation indicator
锤锻法　drop forging
锤击式布氏硬度计　hammering type Brinell hardness tester
锤击试验　hammer test
锤尖法　drawing-down
锤子　hammer
纯铝　pure aluminium
纯镍熔接条　pure nickel electrode
纯铜　pure copper
纯相位调制　phase-only modulation
纯滞后　pure time delay
纯滞后补偿　dead time-delay compensation, dead-time compensation
唇形橡胶密封　lip rubber seal

瓷隔电子　porcelain insulator
瓷绝缘体　porcelain insulator
瓷器　china, porcelain
瓷砖　ceramic tile
磁　magnetism
磁饱和　magnetic saturation
磁饱和稳压器　magnetic saturated voltage stabilizer
磁测井仪　magnetic logger
磁场　magnetic field
磁场变阻器　field rheostat
磁场范围　magnetic domain
磁场放大机　metadyne
磁场干扰　magnetic field interference
磁场计　magnetic field meter
磁场计算　magnetic field computation
磁场控制器　field controller
磁场强度　intensity of magnetic field, magnetic field intensity, magnetic field strength
磁场强度传感器　magnetic field strength sensor, magnetic field strength transducer
磁场失效保护装置　field-failure protection
磁场影响　influence of magnetic field
磁超点阵　magnetic superlattice
磁秤　magnetic balance
磁畴　magnetic domain
磁畴附件　magnetic domain attachment
磁处理机　magnetreater
磁传感器　magnetic sensor, magnetic transducer
磁吹　magnetic blow-out
磁吹式灭弧断路器　magnetic blow-out circuit breaker
磁存储器　magnetic storage
磁大地电流法仪器　instrument of magneto telluric method
磁带　magnet band, magnetic recording wire, magnetic tape, tape
磁带单元　magnetic tape unit
磁带馈送　tape feed
磁带驱动器　magnetic tape drive, tape drive
磁导　magnetic conductance, permeance

磁导率	magnetic conductivity, magnetic permeability, permeability
磁导系数	permeability coefficient
磁道柱面地址	cylinder address
磁等价	magnetic equivalence
磁电工速度测量仪	magnetoelectric velocity measuring instrument
磁电式安培计	moving coil ammeter
磁电式检流计	moving coil galvanometer
磁电式速度传感器	magnetoelectric velocity transducer
磁电式仪表	magnetoelectric instrument
磁电式转速表	magnetoelectric tachometer
磁电式转速传感器	magnetoelectric tachometric transducer
磁电相位差式转矩测量仪	magnetoelectric phase difference torque measuring instrument
磁电相位差式转矩传感器	magnetoelectric phase difference torque transducer
磁电阻率仪	magnetic resistivity instrument
磁碟片	floppy
磁定位器	magnetic locator
磁动力离合器	magnetic power clutch
磁动势	magnetic motive force
磁发电机	magneto
磁法勘探仪器	magnetic prospecting instrument
磁放大器	magnetic amplifier
磁粉	magnetic particle
磁粉探伤	magnetic particle inspection
磁风	magnetic wind
磁感应	magnetic induction
磁感应流量计	magnetic-induction flowmeter
磁感应强度	magnetic flux-density
磁各向异性	magnetic anisotropy
磁共振显微镜	magnetic resonance microscopy
磁鼓	magnetic drum
磁鼓存储器	drum memory
磁光磁强计	magneto-optical effect magnetometer
磁化	magnetize, magnetization, magnetizing
磁化饱和	saturation
磁化电抗	magnetizing reactance
磁化电流	magnetizing current
磁化方法	magnetization method
磁化机	magnetizer
磁化率测井仪	magnetic susceptibility logger
磁化曲线	B-H curve, curve of magnetization, magnetization curve
磁化时间	magnetizing time
磁化线圈	magnetizing coil
磁化循环	cycle of magnetization
磁化装置	magnetizing assembly
磁极	magnetic pole
磁极铁芯	pole core
磁极线圈	pole coil
磁记录通道	magnetic recording channel
磁接触器	magnetic contactor
磁绝缘间隙	magnetically insulated gap
磁卡	magnetic card
磁卡片	magnet card
磁控管振荡器	magnetron oscillator
磁控继电器	magnetic control relay
磁控元件	magnetic control component
磁力泵	magnetism force pumps
磁力搅拌器	magnetic stirrers
磁力启动器	magnet starter, magnetic starter
磁力起重机	magnetic crane
磁力线	line of magnetic force, line of magnetization
磁力线图	magnetic figure
磁力仪	magnetometer
磁力制动器	magnetic break
磁链	flux linkage
磁量子数	magnetic quantum number
磁流体离合器	magnetic fluid clutch
磁流体轴承	magnetic fluid bearing
磁路	magnetic circles, magnetic circuit, magnetic path
磁路长度	length of magnetic path
磁敏电阻器	magnetic resistor
磁敏元件	magneto sensor
磁膜线	plated wire

磁偶激励　magnetic dipole excitation
磁偶极子　magnetic doublet
磁耦合截止阀　magnetic co-operate globe valves
磁耦合液位检测器　magnetically coupled level detector
磁耦合转子流量计　magnetic-coupled rotameter
磁盘　disk, diskette, magnetic disc
磁盘存储器　disk memory
磁盘唤醒　suspend to disk
磁盘驱动器　disc drive, disk drive
磁盘容量　disk capacity
磁盘文件索引　disk file index
磁盘组　disk pack
磁偏转　magnetic deflection
磁屏蔽　magnetic screen, magnetic shield, magnetic shielding
磁漆　enamel paint
磁强计　magnetometer
磁驱动浮子开关　magnetic-operated float switch
磁栅式宽度计　magnetic scale width gauge, magnetic scale width meter
磁栅式位移传感器　magnetic grating displacement transducer
磁式动态仪器　magnet dynamic instrument
磁式氧传感器　magnetic oxygen sensor, magnetic oxygen transducer
磁弹性式称重传感器　magnetoelastic weighing cell
磁弹性式力传感器　magnetoelastic force transducer
磁弹性式轧制力测量仪　magnetoelastic rolling force measuring instrument
磁弹性式转矩测量仪　magnetoelastic torque measuring instrument
磁弹性式转矩传感器　magnetoelastic torque transducer
磁调制器　magnetic modulator
磁体　magnet assembly
磁铁　magnet
磁通表　fluxmeter, flux meter
磁通表校验仪　fluxmeter calibrator
磁通常数　flux constant

磁通传感器　magnetic flux sensor, magnetic flux transducer
磁通分布　flux distribution
磁通管　magnetic flow tube
磁通计　magnet meter
磁通量　magnetic flux
磁通量控制器　flux guide
磁通门磁力梯度仪　gradient flux-gate magnetometer
磁通门磁力仪　fluxgate magnetometer
磁通门罗盘　fluxgate compass
磁通密度　magnetic flux-density
磁通势　magnetomotive force
磁位计　magnetic potentiometer
磁响应　magnetic response
磁卸载　magnetic dumping
磁芯　core, magnetic core, magnet core
磁芯晶体管逻辑　core transistor logic
磁芯模件测试系统　core module test system
磁芯损耗　core loss
磁芯存储器驱动器　core memory driver
磁性　magnetic property
磁性材料　magnetic material
磁性的　magnetic
磁性分选仪　magnetic separator
磁性工具　magnetic tools
磁性流量传感器　magnetic flow transducer
磁性逻辑元件　logitron
磁性元件　magnetic element
磁性阻尼　magnet damping
磁性阻尼器　magnetic damper
磁性座　magnetic base
磁悬浮　magnetic suspension
磁悬式密度传感器　magnetic suspension type density sensor
磁悬液　magnetic flaw detection ink
磁旋比　magnetic rotation comparison
磁学量传感器　magnetic quantity sensor, magnetic quantity transducer
磁氧分析器　paramagnetic oxygen analyzer
磁质谱仪　magnetic-sector mass spectrometer

磁致伸缩　magnetostriction
磁致伸缩测试仪　magnetostriction testing meter
磁致伸缩磁力仪　magnetostriction magnetometer
磁致伸缩振动器　magnetostrictive transducer
磁滞　hysteresis
磁滞的　hysteretic
磁滞电动机　hysteresis motor
磁滞回线　B-H loop, hysteresis curve, hysteresis loop
磁滞损失　hysteresis loss
磁滞误差　hysteresis error
磁轴承　magnetic bearing
磁阻　magnetic reluctance, reluctance
磁阻电动机　reluctance motor
磁阻发电机　reluctance generator
磁阻尼　magnetic damping
次级标准器　secondary standard
次级参考点　secondary reference point
次级电压　secondary voltage
次级回路　secondary loop
次级基准点　secondary reference point
次级控制回路　secondary control loop
次级门路　secondary gate
次级绕组　secondary winding
次顺序程序　secondary sequence program
次同步　subsynchronous
次同步共振　subsynchronous resonance
次同步振荡　subsynchronous oscillations
次要缺陷　minor defect
次优　suboptimal
次优控制　suboptimal control
次优系统　suboptimal system
次优性　suboptimality
刺激　excite
刺激剂　irritant
刺激性毒气　irritant gas
刺激性烟气　irritating smoke
从动带轮　driven pulley
从动件　driven link follower
从动件平底宽度　width of flat-face
从动件停歇　follower dwell
从动件运动规律　follower motion

从动轮　driven gear
从动运行　slave operation
从动针　follow-up pointer
从属站　slave station
从属装置　slave
粗拔钢丝　coarse wire
粗糙-精细　coarse-fine
粗糙-精细继电器　coarse-fine relay
粗糙-精细开关　coarse-fine switch
粗糙-精细控制系统　coarse-fine control system
粗糙面　matte side
粗糙系数　roughness coefficient
粗锉　rasp
粗导线　heavy gauge wire
粗锻　roughing forge
粗钢丝　heavy wire
粗集料　coarse aggregate
粗加工　roughing
粗-精控制　coarse-fine control
粗面　matt finish
粗切机　rougher
粗切削　rough machining
粗砂　rough sand
粗筛　coarse screening
粗陶管　stone ware pipe
粗线　bold line
粗牙槽刨刀　coarse indented cut plating iron
粗牙螺纹　coarse thread
粗真空　coarse vacuum
促动器　actuator
醋酸丁酸纤维素　cellulose acetate butyrate
醋酸绝缘线　acetate wire
醋酸纤维素　cellulose acetate
醋酸盐　acetate
催化板材　catalyzed board coated catalyzed laminate
催化反应器　catalytic reactor
催化分析器　catalytic analyzer
催化剂　catalyst
催化色谱法　catalytic chromatography
催化式气体传感器　catalytic gas transducer
催化元件　catalysis element
催化作用　catalysis, catalytic action

脆性材料　brittleness material
脆性铁　short iron
淬火　quench, quenching
淬火变形　quenching distortion
淬火老化　quench ageing
淬火裂痕　quenching crack
淬火裂纹　quenching cracks
淬火深度　depth of hardening
淬火压床　quenching press
淬火压模　quenching die
淬火应力　quenching stress
淬火硬化　quench hardening
淬硬深度　depth of hardening
存储　store
存储保护　memory protection
存储部件　memory unit
存储程序计算机　storage program computer
存储单元　location of storage, memory cell
存储档案到磁带　save archives to tape
存储到其他媒体　saving to other media
存储归档数据　save archive data
存储环境　storage environment
存储寄存器　memory register, storage register
存储连接单元　memory junction cell
存储器　memory, storage, storer
存储器保护和监测程序　memory protect and watch variable test program
存储器缓冲器　storage buffer
存储器接口　memory interface
存储器应用　memory application
存储设备　storage device
存储示波器　storage oscilloscope
存储事件描述　save event descriptors
存储事件数据　save event data
存储体　memory bank
存储元件　memory element, storage element
存放区　storage area
存取级别　access level
存取控制　access control
存取控制寄存器　access control register
存取控制字段　access control field
存取码　access code
存取权限　access authority
存取时间　access time
寸动　inching
锉　mill
锉削加工　filing
锉屑　file dust
错读　misread
错误的　wrong
错误检测及校正　error checking and correcting, error detection and correction

D

搭接浇口　tab gate
达林控制器　Dahlin's controller
达林算法　Dahlin's algorithm
打包带褶皱　fold of packaging belt
打包机　packaging tool, packer
打点时间　dotting time
打点式长图记录仪　dotting strip chart recorder
打号冲具　marking iron
打滑　slipping
打浇口　degate
打开　unbolt
打开开关　opening switch
打平　planishing
打入桩　driven pile
打印冲子　making die
打印的指定类型　specify type of printing
打印机　printer, stamping press
打印离线档案　print from off-line archive
打印一个文件　print a file
打印字模　stamp letter
打字副本　carbon copy
大半径弯头　long radius elbow
大尺寸螺帽　big size nuts
大地测量学与测量工程　geodesy and survey engineering
大地测量仪器　geodetic instrument
大地电位　ground potential
大地电阻　earth resistance
大电厂　high-power station
大端槽宽　outer slot width
大端齿顶高　outer addendum
大端齿根高　outer dedendum
大端螺旋角　outer spiral angle
大功率导电体　heavy duty electrical conductor
大功率电动机　high-capacity motor
大功率过滤器　large capacity filter
大功率调压器　large capacity pressure regulator

大规模混合集成电路　large scale compound integration, large scale hybrid integrated circuit
大规模集成电路　large scale integrated circuit (LSI)
大规模集成电路芯片　LSI chip
大规模系统　large-scale system
大节距　coarse pitch
大空间结构　large space structure
大孔径干涉仪　large aperture interferometer
大理石　marble
大梁　girder
大轮　gear member
大轮精切刀　segmental-blade cutter
大轮锥距　gear cone
大螺距　coarse pitch
大拇指　thumb
大偏差　large deviation
大气不透明度　atmospheric opacity
大气电　atmospheric electricity
大气干扰　atmospherics
大气温度　atmospheric temperature
大气向下辐射　downward terrestrial radiation
大气压电离　atmospheric pressure ionization (API)
大气压离子化　atmospheric pressure ionization (API)
大气压力　atmospheric pressure
大容量采水器　large volume sampler
大容量纯水制造系统　water purifiers system
大容量存储器　bulk storage memory, mass storage
大容量的　large
大容量发电机组　large capacity generating set
大容量水电站　high-capacity water power station
大头螺丝　thumb screw

大系统　large scale system
大系统服务理论　queueing theory
大系统控制论　large scale system cybernetics
大小相等且方向相反　equal and opposite in direction
大信号　large signal
大型玻璃纤维增强塑料冷却塔　large glass fiber reinforced plastic cooling towers
大型超声波清洗机　aqueous ultrasonic cleaning systems
大型热阴极二极管　kenotron
大型系统分析与设计　system analysis and design in large scale
大修　major overhaul, major repair
大应变应变计　high elongation strain gauge
代表　represent
代表报警器　representative alarm unit
代码　code
代码输出　code output
代码输入　code input
代码透明的数据通信　code-transparent data communication
代码转换器　code converter
代数的　algebraic
代数方法　algebraic approach
代数黎卡提方程　algebraic Riccati equation
代数系统理论　algebraic systems theory
代数选择　algebraic selection
带　band
带柄平孔板　plate orifice with handle
带触点仪表　instrument with contacts
带电部分　energized part
带电导体　live conductor
带电电线　live wire
带电体　charged body
带电指示器　charge indicator
带分流器的电流表　shunted instrument, shunted meter
带负载的　on-load
带钢　ribbon iron
带环线　eyelet wire
带夹　band clamp

带搅拌储槽加热器　stirred-tank heater
带卷升降运输机　coil car
带宽　band width
带宽测量　bandwidth measurement
带宽测量仪　bandwidth meter
带宽电脉冲　bandwidth electrical pulse
带宽分配　band width allocation
带宽分析仪　band analyzer
带宽极小化问题　bandwidth minimization problem
带宽同轴探头　bandwidth coaxial probe
带宽延迟扫描示波器　bandwidth delayed sweep oscilloscope
带冷凝管压力表　manometer with condensing tube
带轮　belt pulley
带批量开关的 PID 控制器　PID controller with batch switch
带偏差报警的输入指示器　input indicator with deviation alarm
带偏置手动操作器　manual with bias station
带偏置自动操作器　automatic with bias station
带屏 γ 取样辐射仪　gamma sampling radiometer of banded screen
带式打磨机　belt sander
带式输送机　belt conveyor
带式制动器　band brake
带手轮散热片　with hand wheel radiator
带输入指示的常数设定器　data set unit with input indicator
带输入指示的手动操作器　manual loader with input indicator
带丝堵三通　plug tee
带通放大器　band pass amplifier
带通滤波器　band pass filter
带有电路控制器件的测量仪表　measuring instrument with circuit control device
带有隔离公共点的输入和输出　input and output with isolated common point
带有锁定的仪表　instrument with locking device

带制动的继电器　biased relay
带状电缆　ribbon cable, strap wire
带状法　stripping method
带阻滤波器　band elimination, band rejection filter
带阻尼的电子变送器　resistance to electronic transmitter
带阻尼的气动变送器　resistance to pneumatic transmitter
怠速-快速运行选择键　idle run-normal run selector switch
戴利检测器　Daly detector
戴维南定理　Thevenin's theorem
丹斯风速计　Dines anemometer
单板微型计算机　single board computer
单笔长图记录仪　single-pen strip chart recorder
单笔记录仪　one-pen recorder
单臂电桥　wheatstone bridge
单边带　single side band (SSB)
单边带电路　single side-band circuit
单边带调幅　single side-band amplitude modulation
单边接触模组　single in-line memory modules (SIMM)
单波段　single band
单车径　cycle track
单触多窗口显示功能　one-touch multi-window display function
单刀单掷　single pole single throw
单刀开关　single blade switch
单刀双掷　single pole double throw
单调关联系统　coherent system
单调系统　monotone system
单调性　monotonicity
单独接地　separate grounding
单端输出　single ended output
单端输入　single ended input
单方向电流　unidirectional current
单个　single
单个名义特性　individual nominal characteristic
单股线　soild wire
单轨铁路　monorail, single track railway
单滚动　single roll

单回路的　single-circuit
单回路可编程序调节器　single loop programmable controller
单回路控制器接口　single strategy controller port (SSCP)
单回路调节器　single loop controller
单回路指示调节器　single loop indicating controller
单机　standalone
单机版　uniprocessor version
单机无穷大系统　one machine-infinity bus system
单级泵　single stage pump
单级的　single-stage
单级放大器　one-stage amplifier
单级过程　single level process
单级调制　single-stage modulation
单极变换器　unipolar converter
单极的　homopolar, monopolar
单极电荷注射　unipolar charge injection
单极性数/模转换器　monopolar D/A converter
单极质谱计　monopole mass spectrometer
单极注射　unipolar injection
单价　unit price
单价的　monovalent
单金属线　monometallic wire
单晶丝　monocrystalline wire
单孔盒形暗渠　single-cell box culvert
单孔球阀　single-ported globe valve
单口排气阀　single opening exhaust valves
单跨　single span
单量程的　single-range
单量程仪表　single-range instrument
单列轴承　single row bearing
单门过载冲击　over pressure impact from either direction
单面　single side
单面覆铜箔层压板　single-sided copper-clad laminate
单面调整法　single setting
单面印制板　single-sided printed board
单模式　single mode
单模式操作　single mode operation

单母线　single busbar
单拍摄执行功能　one-shot execution function
单片微处理器　chip microprocessor
单片微型计算机　single-chip microcomputer
单频气体激光器　single-frequency gas laser
单腔模具　single cavity mold
单色辐射　monochromatic radiation
单色辐射温度计　homogeneous radiation thermometer
单色仪　monochromator
单输入单输出　single input single output（SISO）
单输入单输出系统　single input single output system
单鼠笼　single squirrel cage
单双臂两用直流电桥　single-arm and double-arm DC bridge, single-arm and double-arm double-purpose DC bridge
单丝　filament
单速无定位控制器　single-speed floating controller
单速无定位作用　single-speed floating action
单通道超声流量计　one-path ultrasonic flowmeter, single-channel ultrasonic flowmeter
单通道控制　single-channel control
单万向联轴节　single universal joint
单位　units
单位反馈　unit feedback
单位负荷　unit load
单位滑滚比　slide-roll ratio
单位阶跃　unit step
单位阶跃函数　unit step function
单位阶跃响应　indicial response, unit step response, unit-step response
单位阶跃信号　unit step signal
单位脉冲　unit impulse
单位脉冲函数　unit impulse function
单位脉冲响应　unit impulse response
单位脉冲信号　unit impulse signal
单位面积质量的测定　determination of mass per unit area

单位矢量　unit vector
单位斜坡信号　unit-ramp signal
单位圆　unit circle
单位重量　unit weight
单稳态　monostabillity
单稳态触发元件　monostable trigger element
单稳态多谐振荡器　monostabillity multivibrator
单稳线路　one-shot multivibrator
单线　mongline, soild wire
单相　single phase
单相插座　single-phase socket
单相的　one-phase
单相电动机　single-phase motor
单相电流　single-phase current
单相电路　single-phase circuit
单相电压　voltage by one phase
单相电源　single-phase source
单相短路　one-phase short-circuit
单相放大器　uniphase amplifier
单相感应电动机　single-phase induction motor
单相母线　single-phase bus
单向阀　non-return valve
单向开关　unilateral switch
单向控制　unicontrol
单向脉冲列　unidirectional pulse train
单向耦合器　unidirectional coupler
单向推力轴承　single-direction thrust bearing
单相调整变压器　single-phase regulating transformer
单向性　unilateral
单向元件　unidirectional element
单项输出　printout of individual items
单芯电缆　single conductor cable, single core cable
单循环法　single cycle
单一的　single
单一终端终止　single-ended termination
单元　unit
单元报警总貌　unit alarm summary
单元操作　unit operation
单元操作块　unit operation block
单元测试　unit testing

单元动作电位　unit action potential
单元分配画面　unit assignment display
单元构成　unit configuration
单元管理功能　unit supervisory function
单元开关　unit switch
单元逻辑　cellular logic
单元趋势画面　unit tend display
单元神经网络　cellular neural network
单元式模具　unit mold
单元数据连接型端子　unit data link terminal
单元误差　elemental error
单元仪表　unit instrument
单元约定问题　unit commitment problem
单元阵列处理机　cellular array processor
单元状态总貌显示画面　unit status overview display panel
单元自动化　cellular automation
单值非线性　single value nonlinearity
单指针自同步指示器　single autosyn indicator
单掷开关　single-throw switch
单轴荷载　uniaxial load
单轴转台　single axle table
单转盘式电度表　one-disk watt-hour meter
单自由度陀螺　single degree of freedom gyro
弹道　trajectory
弹弓压缩长　spring compressed length
弹弓柱　spring rod
弹体-目标相对运动仿真器　missile-target relative movement simulator
弹筒发热量　heating quantity of bomb cylinder
弹头　warhead
弹头托盘　bomb head tray
弹药筒　cartridge
弹药推动的工具　cartridge operated tool
淡水　fresh water
淡水泵　fresh water pump
淡水冷却水塔　fresh water cooling tower

氮　nitrogen（N）
氮化硼立方晶　borazon
氮磷检测　nitrogen phosphorous detection
氮气钢瓶　nitrogen cylinder
当量齿数　equivalent number of teeth, equivalent teeth number, virtual number of teeth
当量齿条　equivalent rack
当量传动比　equivalent gear ratio
当量电导　equivalent conductance
当量节圆半径　equivalent pitch radius, virtual pitch radius
当量摩擦系数　equivalent coefficient of friction
当量载荷　equivalent load
当量直齿轮　equivalent spur gear
当前文件　current file
当前行　current line
当前值缓冲区　current value buffer
挡板　flapper, stop plate
挡板变换器　baffle-plate converter
挡块　check block
挡土构筑物　earth-retaining structure
挡土墙　earth-retaining wall, retaining wall
档案重访模块　archive replay module
档位　tap position
刀齿齿廓角　blade angle
刀齿顶刃倾角　top slope angle
刀顶距　cutter point width, point width, tool point width
刀顶距收缩　point width taper
刀顶宽　blade point width
刀号　cutter number
刀架　tool post
刀尖　nose of tool
刀尖半径　cutter point radius, point radius
刀尖寿命　blade life
刀尖凸角代号　blade letter
刀尖圆角半径　blade edge radius, edge radius
刀尖直径　cutter point diameter, point diameter
刀角　nose angle
刀具　cutter

刀具进刀　tool advance
刀盘的轴向位置　cutter axial
刀盘方向　hand of cutter
刀盘体　cutter head
刀盘直径　cutter diameter
刀盘轴线　cutter axis
刀盘轴向平面　cutter axial plane
刀盘主轴　cutter spindle
刀盘主轴转角　cutter spindle rotation angle
刀片　blades
刀倾　tilt
刀倾角　tilt angle
刀刃角　tool point angle
刀刃圆角半径　cutter edge radius, tool edge radius
刀形指针　knife-edge pointer
刀轴　arbor
刀转　swivel
刀转角　swivel angle
刀子　knives
刀子接触宽度　knife edge contact width
刀子联线　knives linear
刀子曲率半径　knife edge curvature radius
氘　deuterium
氘核　deuteron
导程角　lead angle
导程凸轮　lead cam
导出量　derived quantity
导出树　derivation tree
导磁体　magnetizer
导磁性　permeability, permeance, magnetoconductivity
导弹　missile
导电　conduction
导电箔　conductive foil
导电部分　conductive part
导电的　conductive
导电计　conductivity meter
导电胶印制板　electroconductive paste printed board
导电介质　conducting medium
导电空穴　conduction holes
导电铝线　electrical aluminium wire
导电丝　static wire
导电通路　electrically conductive path
导电图形　conductive pattern
导管　conduit
导管引入　conduit entry
导轨　guide rail, lead rail
导航　navigation
导航、制导与控制　navigation, guidance and control
导航窗口　navigator window
导航系统　navigation system
导环　guide ring
导孔　pilot hole
导料块　guide pad
导流　diversion kerb
导流板　guide plate
导率　conductance
导纳　admittance
导纳测量仪　admittance measuring instrument
导纳继电器　conductance relay
导频继电器　pilot relay
导频控制器　pilot controller
导频信道　pilot channel
导频信号　pilot signal
导频振荡器　pilot frequency oscillator
导前角　angle of advance
导墙　guide wall
导热　heat conducting
导热计　conductivity meter
导热系数　heat conductivity, thermal conductivity
导热性　thermal conductivity
导绳器　rope guide
导数差示热分析　derivative differential thermal analysis
导数差示热曲线　derivative differential thermal curve
导数绝对值程序　absolute derivative algorithm
导数膨胀法　derivative dilatometry
导数平方程序　square derivative algorithm
导数热重法　derivative thermogravimetry
导数热重曲线　derivative thermogravimetric curve
导数吸收光谱　derivative absorption spectrum

导套孔	dowel hole
导体	conductor
导体层	conductor layer
导体痕迹线	conductor trace line
导体元件	conductor element
导通角	conduction angle
导销	guide pin
导销衬套	leader busher
导线	conducting wire, voltaic wire
导线测量站	traverse station
导线电感	lead inductance
导线电阻	conductor resistance
导线截面	conductor cross-section
导线距离	conductor spacing
导线宽度	conductor width
导线面	conductor side
导线网	wire netting
导向键	feather key
导向轮	guide wheel
导向网络	lead network
导行线	channelizing line
导正滚轮	pinch roll
导桩	guide pile
倒角	chamfering, rounding chamfer
倒角机	chamfering machine
倒角应力	plane strain
倒流	reflux
倒相	phase inversion, reversal of phase
倒相变压器	phase reversing transformer
倒相放大器	amplifier inverter, inverting amplifier, paraphase amplifier, phase inversing amplifier
倒相器	phase inverter
倒像望远镜	inverting telescope
倒易理论	reciprocity theorem
倒置	invert, inversion
倒置显微镜	inverted microscope
捣固掌	tamper blade
道路	highway
锝	technetium（Tc）
灯	lamp
灯光调节电阻器	lamp dimming resistor
灯号控制路口	light control junction
灯架	lamp holder
灯具	lantern

灯丝像	filament image
灯柱	lamp pole
等百分比	equal percent
等百分比阀	equal-percentage valve
等齿顶高齿	equal addendum teeth
等齿数整角锥齿轮副	miter gears
等待时间	waiting-time
等待状态	wait state
等电点聚焦	isoelectric focusing
等电位	equipotential
等电位点	equipotential point
等电位连接线	bonding wire
等电位屏蔽	equivalent potential screen
等电位区域	equipotential zone
等电位线	equipotential line
等分刻度	equally divided scale
等幅振荡	sustained oscillation
等高幅值	magnitude contour
等高套筒	equal-height sleeves
等高线	contour line
等高线图	contour map
等级	class
等级强度	grade strength
等级色谱响应函数	hierarchical chromatographic response function（HCRF）
等加等减速运动规律	constant acceleration and deceleration motion
等价类划分	equivalence partitioning
等精密度	equal precision measurement
等径凸轮	conjugate yoke radial cam
等宽凸轮	constant-breadth cam
等权通信	peer-to-peer communication
等水溶剂	isohydric solvent
等速法	uniform velocity method
等速运动	constant velocity motion
等速运动规律	uniform motion
等同变换	co-ordinate transformation
等同时间	co-ordinate time
等温层跟踪仪	isothermal gas chromatography
等温淬火	austempering
等温退火	isothermal annealing
等温线	isotherm
等温重量变化测定	isothermal weight-change determination

等温重量变化曲线　isothermal weight-change curve
等效 T 形电路　equivalent T circuit
等效电感　equivalent inductance
等效电路　equivalent circuit
等效电容　equivalent capacity
等效电阻　equivalent resistance
等效动力学模型　dynamically equivalent model
等效发电机　equivalent generator
等效构件　equivalent link
等效交流电阻　equivalent AC resistance
等效决策　decision feedback equalization
等效均匀粗糙度　equivalent uniform roughness
等效空气容积　equivalent air volume
等效力　equivalent force
等效力矩　equivalent moment of force
等效输入阻抗　equivalent input impedance
等效于　equivalent
等效载重　equivalent load
等效增益　equivalent gain
等效正弦波　equivalent sine wave
等效质量　equivalent mass
等效转动惯量　equivalent moment of inertia
等压线　constant pressure line
等压质量变化测定　isobaric mass-change determination
等压重量变化测定　isobaric weight-change determination
等压重量变化曲线　isobaric weight-change curve
等值波阻抗　equivalent value wave impedance
等值线图　contour map
低　low
低潮　low water
低成本自动化　low cost automation
低电流继电器　minimum current relay, under-current relay
低电平过程接口单元　low level process interface unit (LLPIU)
低电平逻辑电路　low-level logic circuit
低电平模拟量输入　low level analog input
低电平调制　low-level modulation
低电位　low potential
低电位电位差计　low EMF potentiometer
低电位直流电位差计　low-potential DC potentiometer
低电压　low tension
低范围值　lower range-value
低谷负荷　valley load
低合金钢钢丝　low alloy steel wire
低级语言　low-level language
低空探空仪　low altitude sonde
低磷生铁　hematite pig iron
低灵敏度的　insensitive
低门限　low threshold
低门限电流　low-threshold current
低锰铸钢　low manganese casting steel
低密度聚乙烯　low density polyethylene
低能　low energy
低能电子衍射仪　low electron energy diffractometer (LEED)
低能过程接口单元　low energy process interface unit
低黏滞性聚合树脂　low viscosity polymer resin
低频　low-frequency
低频变压器　low-frequency transformer
低频差动磁强计　low-frequency differential magnetometer
低频段　low-frequency stage
低频放大　low-frequency amplification
低频放大器　low-frequency amplifier
低频感应电炉　low-frequency induction furnace
低频滤波器　low-frequency filter
低频疲劳试验机　low frequency fatigue testing machine
低频漂移　low-frequency dispersion
低频强度　low-frequency intensity
低频扫描　low-frequency scattering
低频信号　low-frequency signal
低频噪声　low-frequency noise
低切　undercut

低驱动电源　low drive power
低热微膨胀水泥　low heat expansive cement
低速电动机　low speed motor, slow-speed motor
低速平衡　low speed balancing
低碳钢　ingot iron
低碳钢丝　iron wire
低通　low-pass
低通滤波器　low-pass filter
低温处理　low-temperature treatment
低温脆性　cold shortness
低温等离子灭菌器　low temperature plasma sterilizers
低温电流标准器　cryogenic current standard
低温钢　low temperature steel
低温恒湿箱　low temperature humidity chamber
低温恒温器　cryostat
低温培养器　low temperature incubators
低温器件　cryogenic device
低温热敏电阻器　low temperature thermistor
低温试验　low temperature test
低温试验机　low temperature testing machine
低温试验箱　low-temperature test chamber
低温室　constant low temperature facilities
低温退火　low temperature annealing
低温稳定性培养器　low temperature stability incubators
低温应变计　low temperature strain gauge
低污染核弹　clean bomb
低限报警　alarm low
低信号选择器　low-signal selector
低压　low voltage
低压保护　low-voltage protection
低压变压器　low-tension transformer, low-volt transformer
低压侧的　on low-tension side
低压电动机　low-tension motor
低压电力网　low-tension network

低压镀锌焊接钢管　low pressure galvanized pipe
低压端　low voltage terminal
低压断路器　low-voltage circuit breaker
低压封装成型　encapsulation molding
低压回路　low tension loop
低压开关　low-tension switch
低压力　low pressure
低压力恢复阀　low-recovery valve
低压母线　low-voltage bus bar
低压配电网　low-voltage network
低压配电系统　low-voltage distribution system
低压配电箱　low-tension distribution box
低压熔断器　low-tension fuse
低压设备　low-voltage equipment
低压系统　low-voltage system
低压线　low-voltage line
低压铸造　low pressure casting
低油压故障停机　low oil pressure shutdowns
低阈值　low threshold
低噪声　low-noise
低噪声低漂移放大器　low-noise and low drift amplifier
低噪声阀　low noise valve
低噪声通道　low-noise chanel
低噪声优化　low-noise optimization
低阻抗　low-impedance
滴定　titration
滴定分析法　titrimetry analysis
滴定管　burette
滴定仪　titrator
滴汞电极　dropping mercury electrode
滴谱仪　drop size meter
镝　dysprosium（Dy）
笛卡儿　Cartesian
笛卡儿算子　Cartesian manipulator
笛卡儿坐标　Cartesian coordinates
底部　bottom
底部隆起　bottom heave
底部密封胶　base sealing
底层　bottom layer
底浇铸模　bottom pour mold
底面　bottom surface
底面反射波　bottom echo

底模板　soffit formwork
底片观察用光源　film viewer
底色　background color
底铁　floor iron
底铸　uphill casting
底座支架　base support
抵消　balance out, cancellation
地表面辐射　emittance of the earth's surface
地层倾角测井仪　dip logger
地磁　geomagnetism
地磁场　geomagnetic field, terrestrial magnetic field
地磁的　geomagnetic
地磁动电计　geomagnetic electrokinetograph (GEK)
地磁力矩　geomagnetic torque
地道　subway
地电化学法找矿仪器　instrument of ground electrochemical prospecting method
地电化学提取法仪器　ground electrochemical extractor
地电流　earth current
地电位　earth potential
地电阻　ground resistance
地脚螺母　anchor nuts
地脚螺栓　anchor bolt
地脚排灯　ground row light
地界标志　landmark
地库　basement
地面 X 射线荧光仪　ground X-ray fluorimeter
地面 γ 能谱仪　ground gamma spectrometer
地面测量法　ground survey method
地面层水平　ground floor level
地面车辆力学　terrain vehicle mechanics
地面承载力　ground bearing pressure
地面电磁法仪器　ground electromagnetic instrument
地面回旋处　ground level roundabout
地面接收站　ground receiving station
地面径流　surface run-off
地面脉冲电磁仪　ground pulse electromagnetic instrument
地面能见度　ground visibility
地面排水　land drainage
地面气象站　surface weather station
地面水平等高线　ground level contour
地面水渠道　surface water channel
地面污水截流管　land-based interceptor sewer
地面仪器　ground instrument
地面运输　ground transportation
地面之下　underground
地面终端设备　ground terminal
地面重力测量　ground gravity survey
地盘　site
地盘界线　site boundary
地平线　ground line
地球辐射　terrestrial radiation
地球同步气象卫星　geostationary meteorological satellite
地球资源技术卫星　earth resource technology satellite (ERTS)
地势　topography
地图制图学与地理信息工程　cartography and geographic information engineering
地图制作测量　mapping survey
地下池　underground chamber
地下电缆　buried cable, underground cable
地下建筑工程　sub-surface building works
地下空间　underground space
地下水　ground water
地下水排水工程　groundwater drainage works
地下水位　ground water level
地下铁路　subway
地下通道　underpass
地下油缸　underground fuel tank
地线　earth
地线系统　ground system
地形　landform, terrain
地形测量　topographic survey
地压　earth pressure
地震荷载　earthquake loading
地震角运动传感器　seismic angular motion sensor

地震系数　seismic coefficient
地震效应　seismic effect
地震载重　seismic load
地址　address
地址比较器　address comparator
地址变换组态　address map configuration
地址代码　address code
地址读出线　address read wire
地址缓冲器　address buffer
地址寄存器　address register
地址解码器　address decoder
地址空间　address space
地址线　address wire
地址总线　address bus
地质　geology
地质工程　geological engineering
地质立体量测仪　geological stereometer
地质罗盘仪　geologic compass
地质状况　geological condition
地质资源与地质工程　geological resources and geological engineering
递归　recursive
递归最小二乘　recursive least square (RLS)
递减比　subsidence ratio
递阶参数估计　hierarchical parameter estimation
递阶系统　hierarchical system
递推估计　recursive estimation
递推控制算法　recursive control algorithm
递推滤波器　recursive filter
递推数字滤波　recursive digital filter
递推算法　recursive algorithm
第二段退火　second stage annealing
第三谐波　third harmonic
第四代语言　fourth-generation language
第一段退火　first stage annealing
第一机理模型　first principles model
蒂巴因　thebaine
碲　tellurium (Te)
颠簸试验　bump test
颠倒温度表检定设备　cailbration equipment of reversing thermometers

巅值电压　crest voltage
典型横切面　typical section
点　dot, point
点参数测量法　method of spot parameter measurement
点处理　point process
点调度　point scheduling
点对点接口　point-to-point interface (PPI)
点对点控制　point-to-point control
点焊　spot welding
点划线　chain dotted line
点画面　trend point panel
点火　ignition
点火电极　ignition electrode
点火线　ignition wire
点漂　point drift
点燃剂　ignitor
点燃性　incendivity
点蚀　pitting
点-条信号发生器　dot-bar generator
点形式　point form
点阵　dot matrix
点阵打印机　dot matrix printer, dot printer
点执行状态　point execution state
碘　iodine (I)
碘量法　iodimetry
电伴随加热　electric heat tracing
电报线　telegraph wire
电表箱　electricity meter box
电冰箱　electric refrigerator
电波　beam
电测法　electrical measurement method
电测井仪器　electric logger
电测温度表　electric thermometer, electrical thermometer
电测仪表　electric instrument
电插头　electrical plug
电厂　power plant
电场　electric field
电场传感器　electric field sensor
电场发射　field emission
电场控制仪　electric field controller
电场强度　electric field intensity
电场强度传感器　electric field strength sensor, electric field strength transducer

电场强度计　electric field meter
电场效应　field effect
电车　electric car
电车轨道　tramway
电池　cell
电池备用单元　battery backup unit (BBU)
电池常数　cell constant
电池充电器　battery charger
电池电解液　battery electrolyte
电池酸位　battery acid level
电畴　electric domain
电触头　electrical contact
电传导　electrical conduction
电传风向风速仪　electrical wind vane and anemometer
电吹风　electric blower
电磁　galvanomagnetism
电磁泵　electromagnetic pump
电磁波　electromagnetic wave
电磁波测距仪　electromagnetic distance meter
电磁波传播测井仪　electromagnetic wave propagation logging instrument
电磁波谱　electromagnetic spectrum
电磁场　electromagnetic field
电磁场问题　electromagnetic field problem
电磁场与微波技术　electromagnetic field and microwave technology
电磁单元　electromagnetic unit
电磁的　electromagnetic
电磁发射　electromagnetic transmission
电磁阀　magnetic valve, solenoid valve
电磁法　electromagnetic methods
电磁法仪器　electromagnetic method instrument
电磁辐射　electromagnetic radiation
电磁干扰　electro magnetic interference (EMI), electromagnetic interference
电磁感应　electromagnetic induction
电磁感应定律　law of electromagnetic induction
电磁供电接收机　battery operated receiver

电磁惯性　electromagnetic inertia
电磁计数器　electromagnetic counter
电磁继电器　electromagnetic type relay
电磁开关　electromagnetic switch
电磁控制单元　electromagnetic control unit (ECU)
电磁离合器　electromagnetic clutch, electromagnetic induction
电磁力　electromagnetic force
电磁流量测量仪表　magnetic flow measuring device
电磁流量传感器　electromagnetic flow sensor, electromagnetic flow transducer
电磁流量计　electromagnetic flowmeter
电磁脉冲　electromagnetic pulse
电磁密度计　electromagnetic densitometer
电磁模式　electromagnetic mode
电磁能　electromagnetic energy
电磁偏转对中系统　electromagnetic deflector alignment system
电磁屏蔽　electromagnetic screen
电磁屏蔽仪表　instrument with electromagnetic screening
电磁枪　electromagnetic gun
电磁散射　electromagnetic scattering
电磁散射问题　electromagnetic scattering problem
电磁式安培计　moving iron ammeter
电磁式比率计　moving-iron logometer
电磁式传感器　electromagnetic sensor, electromagnetic transducer
电磁式仪表　electromagnetic instrument, moving-iron instrument
电磁瞬态过程　electromagnetic transient
电磁天平　magneto electric balance
电磁铁　electromagnet
电磁铁起重机　electromagnetic crane
电磁铁芯　electromagnetic core
电磁透镜　electromagnetic lens
电磁系统　electromagnetic system
电磁效应　electromagnetic induction
电磁信号　electromagnetic signal
电磁信件　electromagnetic mail

电磁学　magnetoelectricity
电磁液体阻尼振动子　electromagnet fluid damping galvanometer
电磁应用　electromagnetic application
电磁元件　electromagnetic element
电磁闸　electromagnetic brake
电磁振动器　electromagnetic vibrator
电磁执行机构　solenoid actuator
电磁制动　electromagnetic braking
电磁制动器　electromagnetic brake
电磁转矩　electromagnetic torque
电磁装置　electromagnetic device
电磁阻尼器　electromagnetic damper
电磁阻尼振动子　electromagnet damping galvanometer
电导　conductance, electrical conduction
电导测定法　conductometry
电导滴定　conductometric titration
电导分析法　conductometric analysis, conductometry, conductimetry
电导计　conductometer
电导检测　conductivity detection
电导检测器　electrical conductivity detector
电导率　electric conductivity, electrical conductivity, specific conductance, conductivity
电导率变送器　conductivity transmitter
电导率测试　electrical conductivity test
电导率传感器　conductivity sensor
电导率记录仪　conductivity recorder
电导率仪器　conductivity instrument
电导耦合　conductive coupling
电导式分析器　conductometric analyzer
电导式气体传感器　conductive gas sensor, conductive gas transducer
电导式物位传感器　conductive level sensor, conductive level transducer
电导物位测量仪表　electrical conductance level measuring device
电导仪　electric conductivity meter
电的　electrical
电灯　electric lamp
电灯杆　lighting mast
电灯泡　bulb

电动操作台　electric operating station
电动差压传感器　electric differential pressure transmitter
电动抽水机　electric pump
电动传动控制设备　electric drive control gear
电动传声器　dynamic microphone, moving-conductor microphone
电动刀具　electric power tools
电动发电机　dynamotor, motor generator
电动扶梯　escalator
电动机　electric motor, electromotor
电动机控制　motor control
电动机起动器　motor starter
电动机拖动　electric motor drive, motor-drive
电动机拖动的　motor-driven
电动机械的　electromechanical
电动机元件　motor element
电动机转子　motor rotor
电动机组绕组　motor winding
电动交流发电机　motor alternator
电动截止阀　electric actuated stop valves
电动进样　electrokinetic injection
电动控制　electric control
电动力　electrodynamic force
电动力的　electrodynamic
电动螺丝刀　electric screw driver, automatic screwdriver
电动平行式双闸板闸阀　electric double disk parallel gate valves
电动起重机　electric hoist
电动起子　electric screw driver, automatic screwdriver
电动气动定位器　electropneumatic positioner, electro-pneumatic positioner
电动气动断路器　electro-pneumatic breaker
电动气动阀门　electro-pneumatic valve
电动气动制动器　electro-pneumatic brake
电动汽车　electromobile
电动式电桥　electrodynamic bridge
电动势　electromotive force (EMF)
电动势-电流变送器　EMF to current transmitter

电动势-气动变送器　EMF to pneumatic transmitter
电动调节阀　electrical control valve
电动挖掘机　electric excavator
电动系电度表　electrodynamic meter
电动系仪表　electrodynamic instrument
电动楔式闸阀　electric actuated wedge gate valves
电动-液压执行机构　electro-hydraulic actuator
电动振动器　electrodynamic vibrator
电动执行机构　electric actuator
电动指示仪　electronic indicator
电度表　electric meter, kilowatt-hour meter, supply meter, energy meter, watt-hour meter
电度表校验装置　kilowatt-hour meter calibration equipment
电镀　electroplating
电镀的　plated
电镀锌钢丝　electro-galvanized steel wire
电反馈　electrical feedback
电费　electric charge
电费率　electric rate
电感　inductance
电感测微计　inductive micrometer
电感反馈　inductive feedback
电感计　inductometer
电感耦合等离子体发射光谱仪　inductive coupled plasma emission spectrometer
电感器　inducer, inductor
电感式传感器　inductive sensor, inductive transducer
电感式力传感器　inductive force transducer
电感式位移测量仪　inductive displacement measuring instrument
电感式位移传感器　inductive displacement transducer, inductive force transducer
电感式压力传感器　inductance pressure transducer
电感式张力计　inductive tensometer
电感线圈　inductance coil
电感箱　inductance box

电感性　inductive character
电工　electrician
电工材料　electric material
电工测量　electrotechnical measurement
电工测量仪器仪表　electrical measuring instrument
电工技术　electrotechnics, electrotechnology
电工理论与新技术　theory and new technology of electrical engineering
电工手册　electrical engineering handbook
电工学　electrotechnics, electrotechnology
电工原理　principle of electric engineering
电功率　electric power
电功率表　electric dynamometer
电焊　electric soldering
电焊机　electric welding set
电焊条　electrode
电荷　charge, electric charge
电荷放大器　charge amplifier
电荷分布　distribution of electric charges
电荷灵敏度　charge sensitivity
电荷密度　charge density, electric-charge density
电荷耦合器件　charge-coupled device
电荷耦合摄像机　CCD camera
电荷耦合元件　charge coupled device (CCD)
电荷平衡　charge balance
电荷平衡式　charge balance equation
电荷守恒　charge conservation
电荷中和　charge neutralization
电荷转移　charge transfer
电弧　arc
电弧触点　arc contact
电弧电阻　arc resistance
电弧放电　arc discharge
电弧焊接　arc welding
电葫芦　electric hoist
电化腐蚀　galvanic corrosion
电化学分析　electrochemical analysis
电化学腐蚀　electrochemical corrosion

电化学加工　electro-chemical machining
电化学检测　electrochemical detection
电化学检测器　electrochemical detector
电化学式传感器　electrochemical sensor, electrochemical transducer
电化学式分析器　electrochemical analyzer
电话　telephone
电话挂号线　telephone call wire
电话热线　hot line
电话网络　telephone network
电话线　telephone line
电活动　electrical activity
电火花　electrical sparkle
电火花加工　electric spark machining
电火花线切割加工　electrical discharge wire-cutting
电击　electrical shock
电击穿　electrical breakdown
电机　electric machine
电机放大机　amplidyne
电机工程　electrical engineering
电机管理系统　motor management systems
电机活动盖板　cover plate
电机及软起动器　motor and soft starters
电机控制阀　motor-operated valve
电机控制器　motor controller
电机扭矩　motor torque
电机与电器　electric machines and electric apparatus
电机执行机构　control motor actuator
电激发　electrical stimulation
电极　electrode
电极电位　electrode potential
电极夹座　graphite holder
电极式盐度计　electrode type salinometer
电极缩小余量　graphite contraction allowance
电极信号　electrode signal
电键　keyer
电角　electrical angle
电角度　electric angle, electric degree

电接点　electric contact set
电接点玻璃温度计　electric contact liquid-in-glass thermometer
电接点水银气压表　contact mercury barometer
电接点弹簧压力表　Bourdon-tube manometer with electric contacts
电接点压力表　pressure gauge with electric contact
电接点压力温度计　manometric thermometer with electric contacts
电接风速计　contact anemometer
电解　electrolysis
电解池　electrolytic cell
电解淬火　electrolytic hardening
电解的　electrolytic
电解电容器　electrolytic capacitor, electrolytic condenser
电解法　electrolytic analysis method
电解抛光机　electropolisher
电解湿度计　electrolytic hygrometer
电解式湿度传感器　electrolysis humidity sensor, electrolysis humidity transducer
电解铁　electrolytic iron
电解铜箔　electrodeposited copper foil
电解研磨　electrolytic grinding
电解质　electrolyte
电解质分析仪　electrolytic analyzer
电解作用　electrolytic action
电介质　dielectric
电介质线　dielectric wire
电锯　sawing machine
电抗　reactance
电抗成分　reactive component
电抗电路　reactive circuit
电抗负载　reactive load
电抗耦合　reaction coupling
电抗耦合放大器　reactance amplifier
电抗器　reactor
电抗器启动　reactor starting
电抗铁　reactive iron
电抗性的　reactive
电抗性电流　reactive current
电抗元件　reactive element
电可编程只读存储器　EPROM
电缆　cable, electric cable

电　dian

电缆安装要求　cabling requirements
电缆包皮层　cable sheath
电缆槽　cable channel, cable trench
电缆电线外部连接系统图　scheme of external cable and wire connections
电缆断线事故　cable break
电缆敷设车　cable laying wagon
电缆干线　cable trunk
电缆膏　sealing compound
电缆故障定位器　cable fault locator
电缆管　cable conduit
电缆管道　culvert
电缆汇线槽　cable duct
电缆接头　cable joint, cable terminal
电缆接头盒　cable-joint box
电缆铠装　cable armor
电缆坑　cable trough
电缆路线　cable route
电缆路由选择　cable routing
电缆密封套　cable gland
电缆耦合器　cable coupler
电缆牵引线　drawing-in wire
电缆桥架　cable bridge
电缆沙井　cable draw pit
电缆式电流互感器　cable type current transformer
电缆隧道　cable tunnel
电缆损伤　cable fault
电缆损失　cable loss
电缆托架　cable bracket
电缆悬挂线　cable suspension wire
电缆引线　cable lead
电缆噪声　cable noise
电缆张力传感器　cable-tension transducer
电缆之间的距离　distance between cable
电缆纸　cable paper
电缆终端　power cable termination
电烙铁　electric soldering iron
电离　ionization
电离层　ionized layer, ionized stratum
电离层记录器　ionospheric recorder
电离常数　dissipation constant
电离电压　ionizing voltage
电离式传感器　ionizing sensor, ionizing transducer
电离室　ionization chamber

电离效率曲线　ionization efficiency curve
电离烟尘检测器　ionization smoke detector
电力　electrical energy
电力变压房　transformer room
电力变压器　power transformer
电力变压站　transformer station
电力潮流　power flow
电力传动　power drive
电力传输　power transmission
电力电缆　power cable
电力电瓶双电源机车　battery/electric locomotive
电力电容器　power capacitor
电力电子学　power electronics
电力电子与电力传动　power electronics and power drives
电力动态　electrical behaviour
电力断路接触器　electrical cut-off switch
电力扼流控制　electric throttle control
电力分配　electric power distribution
电力负载　electrical load
电力干线　feeder cable
电力工程技术员　electrical technician
电力工业　electric power industry, power industry
电力供应　electric power supply
电力机车　battery electric locomotive, electric vehicle
电力经济　electric power economy
电力晶体管　giant transistor (GTR)
电力平均分配装置　load-shedding equipment
电力熔断器　power fuse
电力输送　electric power transmission
电力特性　electrical characteristic
电力拖动控制　electric drive control
电力网　electric network, power circuit, power network
电力网电压　network voltage
电力网络　electrical network
电力网系统　electric power pool
电力系统　electric power system, electrical power system, power sys-

415

tem
电力系统及其自动化 power system and its automation
电力系统稳定器 power system stabilizator
电力系统自动化 power system automation
电力线 electric line of force, power line
电力性质 electrical property
电力支站 power sub-station
电力中断 electricity interruption
电炼生铁 electric pig iron
电量 quantity of electricity
电量传感器 electric quantity sensor, electric quantity transducer
电量滴定 coulometric titration
电量分析 coulometric analysis, coulometry
电量计 coulometer, voltameter
电量式分析器 coulometric analyzer
电铃 electric bell, trembler bell
电铃按钮 bell button
电铃指示装置 annunciator
电零位 electrical zero
电零位调节器 electrical zero adjuster
电流 electric current
电流饱和 current saturation
电流保护装置 current protection
电流倍增型晶体管 current multiplication type transistor
电流比 current ratio
电流比较器 current comparators
电流表 ammeter, current meter, electric current meter
电流常数 current constant
电流传感器 electric current sensor, electric current transducer
电流的电子主控元件 electronic controlling element for current
电流滴定法 amperometric titration
电流电压特性 current voltage characteristic
电流反馈 current feedback
电流范围 current range
电流方向 current direction
电流放大器 current amplifier

电流分布 current distribution
电流负载 current loading
电流过载 current overload
电流互感器 current transformer
电流互感器的变压系数 actual transformation ratio of current transformer
电流计 galvanometer, rheometer
电流计记录式强震仪 galvanometer record type strong-motion instrument
电流继电器 current relay
电流接触器 contactor
电流可逆斩波电路 current reversible chopper
电流控制电流源 current-controlled current source (CCCS)
电流控制电压源 current-controlled voltage source (CCVS)
电流灵敏度 current sensitivity
电流路径 current path
电流密度 current density
电流耦合 current coupling
电流匹配互感器 current matching transformer
电流平衡 current balance
电流强度 current intensity
电流容许量 current carrying capacity
电流三角形 current triangle
电流输出 current output
电流输出模件 current output module
电流损耗 current decay
电流损耗调节器 current loss regulator
电流损失 current loss
电流调节器 current regulator
电流调制 current modulation
电流稳定度 current stability
电流稳定器 current stabilizer
电流误差 current error
电流相位 current phase
电流消耗 current consumption
电流谐振 antiresonance
电流谐振频率 antiresonance frequency
电流型逆变电路 current source type inverter
电流需求量 current demand
电流源 current source
电流增益 current gain

电流整定值　current setting
电流指示器　current indicator
电流中心　current center
电流阻尼器　current damper
电炉　electric furnace
电滤波器　electric filter
电路　circuit, circuitry, electric circuit
电路板测试仪　board tester
电路变换　circuit transformation
电路参数　circuit parameters
电路插件板　circuit card
电路的连接　connections of circuits
电路方程式　circuit equation
电路仿真　circuit simulation
电路分析　circuit analysis
电路分析测试仪　circuit analyzer and tester
电路计算　circuit calculation
电路开关网络　circuit switched network
电路理论　circuit theory
电路模型　circuit model
电路设计　circuit design
电路损耗　circuit loss
电路图　circuit diagram
电路系统　circuitry
电路效率　circuit efficiency
电路性能　circuit performance
电路与系统　circuits and systems
电路元件　circuit components, circuit element
电路原理图解　circuit schematic diagram
电路制动器　circuit brake
电路阻抗　circuit impedance
电脉冲　electric pulse, electrical pulse
电纳　susceptance
电脑化混凝土立方块压力试验机　computerized automatic concrete cube crushing machine
电脑控制　control computer
电脑视觉　robot vision
电脑数控电火花机　CNC electric discharge machines
电脑数控电火花线切削机　CNC EDM wire-cutting machines

电脑数控雕刻机　CNC engraving machines
电脑数控机床配件　CNC machine tool fittings
电脑数控剪切机　CNC shearing machines
电脑数控镗床　CNC boring machines
电脑数控铣床　CNC milling machines
电脑数控线切削机　CNC wire-cutting machines
电脑数控压弯机　CNC bending presses
电脑数控钻孔机　CNC boring machines
电脑数值控制　computerized numerical control (CNC)
电能　electric energy, electrical energy
电能成本　cost of electric energy
电能储存　accumulation of electric energy
电能经济　power economy
电能转换器　energy converter
电暖气　electric heater
电偶　galvanic couple
电偶极子　electric doublet
电喷射　electrospray
电平记录仪　level recorder
电平开关　level switch
电平调节　level control
电平图示仪　level tracer
电瓶电压表　battery voltage meter
电气安装技术　electrical installation
电气布线　electrical wiring
电气传动　electric drive
电气带盖二通　two-way conduit fitting with cover
电气带盖三通　three-way conduit fitting with cover
电气干燥　electric drying
电气隔离　electrical isolation
电气工程　electrical engineering, electrical works
电气惯性　electric inertia
电气化　electrification
电气化铁路　electrified railroad
电气机车　electric locomotive
电气基础设施　electrical infrastruc-

ture
电气连接 electrical connection
电气连锁 electrical interlock
电气联动 electric shaft
电气列车 electric train
电气设备 electric apparatus, electrical apparatus, electrical device, electrical equipment
电气湿度计 electrical hygrometer
电气试验 electric test
电气伺服控制 electro-servo control
电气随动控制 electro-servo control
电气调节器 electric regulator
电气铁道 electric railway
电气信号 electric signalling
电气与电子工程师学会 Institute of Electrical and Electronics Engineers (IEEE)
电气照明 electric lighting
电气制动 electric braking
电气中心 electrical center
电-气转换器 electric pneumatic converter, electric-pneumatic transducer
电气装置 electric installation
电气自动化 electric automatization
电器 electric appliance, electrical appliance
电器产品 electrical products
电桥比 bridge arm ratio
电桥臂 bridge arm
电桥法 bridge method
电桥校准 bridge calibration
电桥平衡 bridge balance
电桥平衡范围 bridge's balance range
电桥式频率计 bridge-type frequency meter
电热的 electrothermal
电热防火闸 electro-thermal damper
电热干燥器强制空气循环 electric drying oven force-air circulation
电热干燥箱 electrically heated drying cabinet
电热鼓风干燥箱 draught drying cabinet, electric drying oven with forced convection
电热器 electric heater, heating appliance

电热式仪表 electrothermic instrument
电热式蒸汽锅炉 electric steam boiler
电热调节器 thermistor
电容 capacitance
电容补偿 capacitive compensation
电容测量仪表 capacitance measurement instrument
电容触发传感器 capacitor activated transducer
电容传声器 condenser microphone
电容的 capacitive
电容电动机 condenser motor
电容电流 capacitive current
电容电桥 capacitance bridge
电容反馈 capacitive feedback
电容分压器 capacitive divider
电容负载 capacitive load, capacitively loaded
电容话筒 capacitor microphone
电容计 capacitance meter
电容继电器 capacitance relay
电容量分析法 electro-volumetric analysis
电容滤波器 capacitor filter
电容率 specific inductive capacity
电容耦合 capacitance coupling, capacitive coupling
电容平衡 capacitance balance
电容启动 capacitor start
电容启动电动机 capacitor motor
电容器 capacitor, condenser, electric condenser
电容器存储器 capacitor storage
电容器励磁 condenser excitation
电容器耦合 condenser coupling
电容器组 condenser bank
电容湿度计 capacitance hygrometer
电容式测微计 capacitance micrometer
电容式辐射热测定器 capacitive bolometer
电容式膜片压力计 capacitance diaphragm manometer
电容式位移传感器 capacitive displacement transducer
电容式压力传感器 capacitance type

pressure sensor, capacitance pressure transducer
电容物位测量仪表　electrical capacitance level measuring device
电容效应　capacitance effect
电容阻抗　condensance
电熔焊　electro-fusion welding
电熔接地　welding ground
电熔套管　electro-fusion coupler
电渗　electroosmosis
电渗析淡化法　electrodialysis method for desalination
电升降机　electric elevator, electric lift
电声互易定理　electroacoustical reciprocity theorem
电声换能器　electroacoustic transducer
电石气　acetylene
电蚀　galvanic corrosion
电势　electric potential, potential
电势差　potential difference
电势分布　potential distribution
电视　television
电视发射机　television transmitter
电视接收机　televisor
电视扫描发生器　television scanning generator
电视设备　television equipment
电视塔　television tower
电视天线　image aerial, television antenna
电视系统　television system
电枢　armature
电枢磁场　armature field
电枢电流　armature current
电枢电路　armature circuit
电枢电阻　armature resistance
电枢反应　armature reaction
电枢漏磁电感　armature leakage inductance
电枢漏磁电抗　armature leakage reactance
电枢绕组　armature winding
电枢铁芯　armature core
电枢铜损　armature ohmic loss
电枢外电路　external armature circuit
电枢线圈　armature coil
电刷　brush

电刷触点　brush contact
电刷火花　brush spark
电梯　electric elevator, electric lift, elevator
电梯控制　lift control
电梯控制器　lift controller
电通量　electric flux
电通密度　electric flux density
电网换流　line commutation
电网络定律　law of electric network
电位　electric potential, potential
电位测定法　potentiometry
电位差　electric potential difference, potential difference
电位差计式记录仪　potentiometric recorder
电位滴定法　potentiometric titration
电位发送器　potentiometer pick off
电位分布　potential distribution
电位计　potentiometer
电位降　potential drop
电位器式位移传感器　potentiometric displacement transducer
电位升　potential rise
电位移　electric displacement, electric flux density
电位移通量　electrostatic flux
电涡流厚度计　eddy current thickness meter, eddy-current thickness meter
电线　electric wire, wire
电线退皮钳　wire stripper
电效率　electrical efficiency
电信　electric communication, telecommunication
电信号　electrical signal
电信号的信息参数　information parameter of an electrical signal
电性能测定仪　electrical property tester
电压　electric voltage, voltage
电压、电流互感器　combined voltage current transformer
电压比　voltage ratio
电压比较器　voltage comparator
电压表　voltage meter
电压波动　variation in voltage
电压等级　voltage grade

电压放大　voltage amplification
电压放大器　voltage amplifier
电压放大系数　voltage amplification factor
电压分布　voltage distribution
电压跟随器　voltage follower
电压互感器　potential transformer, voltage transformer
电压计　voltmeter
电压继电器　voltage relay
电压降　fall of potential, potential drop, voltage drop
电压校正器　voltage corrector
电压控制　voltage control
电压控制电流源　voltage-controlled current source (VCCS)
电压控制电压源　voltage-controlled voltage source (VCVS)
电压控制系统　voltage control system
电压逆变开关　voltage inverter switch
电压平衡　balance of voltage
电压升　potential rise
电压识别认证　voltage identification definition (VID)
电压输出　voltage output
电压输入模块　voltage input module
电压损失　loss of voltage, voltage collapse
电压特性　voltage characteristic
电压调整　voltage regulation
电压调整率　voltage regulation factor
电压调整模块　voltage regulator module
电压调整器　voltage adjuster
电压稳定度　voltage stability
电压稳定器　voltage stabilizer
电压谐振　voltage resonance
电压源　voltage source
电压源型逆变器　voltage source inverter
电压指示器　voltage indicator
电液控制　electrohydraulic control
电-液控制　electric-hydraulic control
电液伺服阀　electrohydraulic servo valve
电液伺服疲劳试验机　electro-hydraulic servocontrolled fatigue testing machine
电-液系统　electro-hydraulic system
电-液转换器　electric hydraulic converter
电泳　electrophoresis
电泳系统　electrophoresis system
电泳仪　electrophoresis meter
电涌　surge
电涌抑制器　surge suppressor
电源　power source, power supply, supply source
电源按键　power button
电源变压器　power transformer
电源插座　electrical socket outlet, power socket
电源电压　power voltage, supply voltage
电源电压调整率　line voltage regulation
电源分配单元　power distribution unit
电源分配箱　power distribution panel
电源分配组件　power distribution component
电源负极引线　negative wire
电源公共端　power supply common
电源供电器　power supply
电源故障恢复电路　supply fault recovery circuit
电源管理　power management
电源卡　power supply card
电源开关　power supply switch
电源控制　power control
电源频率　supply frequency
电源设备　power device
电源输出　power output
电源输入　power input
电源调节器　supply regulator
电源系统　power supply system
电源线　main wire, power wire
电源线路　power circuit
电源箱　switching power supply
电源装置　power supply device
电源阻抗　source impedance
电远传压力表　transmissible pressure gauge
电晕　corona
电晕放电　corona discharge
电晕损失　corona loss

电晕效应 corona effect
电熨斗 electric iron
电站控制 power station control
电致伸缩 electrostriction
电致伸缩继电器 capadyne
电重量分析 electro-gravimetric analysis
电重量分析法 electrogravimetry, electro-gravimetric analysis
电轴 electric axis
电铸成形模 electroformed mold
电灼式印刷机 electrosensitive printer
电子 electron
电子变送器 electronic transmitter
电子波长 electron wave length
电子捕获检测 electron capture detection
电子捕获检测器 electron capture detector（ECD）
电子采样开关 electronic sampling switch
电子测距光学经纬仪 electronic distance-meter theodolite
电子测量仪表 electronic measuring instrument
电子测倾器 electronic tilt sensor
电子层 electronic shell
电子秤 electronic weigher
电子磁通表 electronic fluxmeter
电子的 electronic
电子吊秤 electronic hanging scale, electronic hoist scale
电子发射 electron emission
电子阀 electronic valve
电子分析仪 electronic analyzer
电子伏特 electron volt
电子管 electronic tube
电子管电压表 tube voltmeter
电子管光电探测器 electron-tube photodetector
电子管间歇振荡器 blocking tube oscillator
电子管密封玻璃管制造机 glazing mill
电子管式放大器 electronic amplifier
电子光学 electron optics
电子轨道衡 electronic railway scale

电子轰击解吸 electron impact desorption（EID）
电子轰击离子化 electron impact ionization
电子轰击离子源 electron impact ion source
电子回旋加速器 betatron
电子积分式磁通表 electronic integrating fluxmeter
电子计数秤 electronic counting scale
电子计数器 electronic counter
电子继电器 electron relay, electronic relay
电子加速器 electron accelerator
电子交流稳压器 electronic AC voltage stabilizer
电子经纬仪 electronic theodolite
电子开关 electronic switch
电子科学与技术 electronics science and technology
电子可擦只读存储器 electrically-erasable ROM
电子空穴 electron hole
电子-空穴对 electron-hole pairs
电子控制 electronic control
电子控制的 electronically-controlled
电子控制发射 electronically-controlled transmission
电子控制器 electronic controller
电子理论 electron theory
电子料斗秤 electronic hopper scale
电子零位平衡记录仪 electronic null-balance recorder
电子模块 electronics module（EM）
电子模拟乘法器 electronic analog multiplier
电子模拟仿真设备 electronic analog and simulation equipment
电子模/数转换器 electronic analogue-to-digital converter
电子能量分析器 electron-energy analyzer
电子能量损失谱学 electron energy loss spectroscopy（EELS）
电子能量损失谱仪 energy loss of electron spectrometer
电子能谱仪 electron spectrometer

电子浓度 electron concentration
电子配料秤 electronic batching scale
电子碰撞 electron impact
电子皮带秤 electronic belt conveyor scale
电子平板仪 electronic plane table equipment
电子平台秤 electronic platform scale
电子汽车秤 electronic trunk scale
电子器件 electronic device
电子迁移率检测器 electron mobility detector
电子枪 electron gun, electron-gun
电子枪对中调节 electron gun alignment adjustment
电子枪交叉点 crossover of electron gun
电子枪阴极 cathode of electron gun
电子设计自动化 electric design automation (EDA)
电子射线 electron rays
电子深度温度计 electronic bathythermograph (EBT)
电子式试验机 electronic testing machine
电子式岩矿密度仪 electron type rock ore densimeter
电子束 bundle of electrons, electron beam, electron rays
电子束曝光机 electron beam exposure apparatus
电子束加工机 electron beam processing machine
电子束偏转器 beam deflector
电子束扫描声全息 acoustical holography by electron-beam scanning
电子束探测器 beam finder
电子束原子化器 electron-beam atomizer
电子数据表 spreadsheet
电子数据处理中心 electronic data processing centre
电子水准仪 electronic level
电子顺磁共振 electron paramagnetic resonance (EPR)
电子顺磁共振波谱仪 electron paramagnetic resonance spectrometer
电子速测仪 electronic tacheometer
电子探针 electron probe
电子探针 X 射线微分析仪 electron probe X-ray microanalyzer
电子探针的电子操纵台 electron operation desk of EPMA
电子探针的电子光学系统 electron optical system of EPMA
电子探针微分析 electron probe micro-analysis (EPMA)
电子天平 electronic balance
电子通道花样 electron channelling pattern
电子透镜 electron lens
电子微探针 electron microprobe
电子稳压器 electronic regulator
电子显微镜 electro microscopy, electron microscope
电子像增强器 electron image intensifier
电子学的 electronics
电子雪崩 electron avalanche
电子衍射法 electron-diffraction method (EED), method of electron diffraction
电子衍射谱仪 electron diffractometer
电子衍射像 electron diffraction image
电子仪表 electronic instrument
电子音乐工业 electrical-music industry
电子邮件 electrical-mail, e-mail
电子诱导解吸 electron induced desorption (EID)
电子云 cloud of electrons, electron cloud
电子噪声 electron noise
电子直线加速器 linear electric actuator
电子制表软件 spreadsheet
电子装置 electron device
电子自动补偿仪 electronic automatic compensator
电子自动测井仪 auto-compensation logging instrument
电子自动平衡仪 electronic automatic balancer
电子自动平衡仪表 electronic self-balance instrument

电阻　electric resistance, resistance, resistor
电阻表　ohmmeter
电阻测试　resistivity test
电阻范围　resistance range
电阻分压器　resistance type potential divider
电阻抗　electrical impedance
电阻炉　resistance furnace
电阻耦合　resistance coupling
电阻器控制　rheostatic control
电阻丝应变仪　wire resistance strain gauge, electric-resistance wire strain gauge
电阻损耗　ohmic loss, resistance loss
电阻调谐振荡器　resistance-tuned oscillator
电阻温度计　resistance thermometer
电阻温度检查器　resistance temperature detector
电阻系数　resistance coefficient, specific resistance
电阻箱　resistance box
电阻箱电桥　box bridge
电阻性　ohmic
电阻性压降　ohmic drop
电阻制动　rheostatic braking
电钻　electric drill
垫板　bench hook, check plate, shim plate
垫承物料　cushioning material
垫块　padding block
垫磨板　wear plate
垫磨片　wear pad
垫片　packing piece, spacer
垫片密封　gasket seal
垫圈　carrier ring, gasket, washer
雕刻机　engraving machine
吊车　crane
吊车秤　crane weigher
吊斗　bucket
吊斗吊重机　skip hoist
吊斗起重机　skip crane
吊杆　boom
吊架　lifting frame
吊桥　suspension bridge
吊丝　hanging spring
吊索　catenary wire
吊线　supporting wire, suspension wire
调度集中电码线　centralized traffic control code wire
调度站　control station
调度中心　control center
调用错误　call error
调用辅助程序　debugging aid
跌落试验　drop test
迭代　iterative
迭代法　iterative method
迭代矫正　iterative improvement
迭代协调　iterative coordination
叠加　superposition
叠加法　addition method, method of supposition, superposition method
叠加原理　principle of superposition
叠片铁芯　laminated core
碟式离合器　plate clutch
碟式制动器　disc brake
碟形弹簧　belleville spring
蝶点　set value
蝶阀　butterfly valve, butterfly gate
蝶式交汇处　cloverleaf interchange
蝶式调谐器　butterfly tuner
蝶形螺钉　thumb screw
蝶形旋阀　butterfly cock
蝶形闸门　butterfly gate
蝶形螺帽　wing nuts
丁钠橡胶　buna-n rubber
丁烷　butane
丁字头　cross head
钉子　nail
顶板　top plate
顶部　overhead
顶层　top storey
顶出板　ejector plate
顶出衬垫　ejection pad
顶出衬套　ejector sleeve
顶出导销　ejector guide pin
顶出导销衬套　ejector leader busher
顶出垫　ejector pad
顶出阀　ejector valve
顶出杆　ejector rod
顶出销　knock pin, ejector pin

顶出针　eject pin, ejector pin
顶窗　transom window
顶镦机　heading machine
顶盖末端　capping ends
顶管法　pipe jacking method
顶空气相色谱法　head space gas chromatography
顶栏杆　top rail
顶料销　lifter pin, lifting pin
顶刃后角　top relief angle
顶推力　jacking force
顶隙　bottom clearance
顶压桩　jacking pile
顶注　tap casting
顶注浇口　top gate
顶锥　face cone
顶锥顶　face apex
顶锥顶至相错点距离　face apex beyond crossing point
顶锥角　face angle
顶锥角距　face angle distance
定槽水银气压表　compensated scale barometer
定常系统　time-invariant system
定带局域网　broadband LAN
定点法标定　fixed points method of calibration
定点炉　furnace for reproduction of fixed points
定积分　definite integral
定界符　delimiter bit
定界字节　delimiter byte
定理　theorem
定理证明　theorem proving
定量电子探头　quantitative electron probe
定量分析　quantitative analysis
定量极限　limit of quantitation (LOQ)
定律　law
定期检查　periodic inspection
定圈式指示器　fixed-coil indicator
定时　timing
定时分析　timing analysis
定时复位　timing recovery
定时继电器　definite time relay
定时脉冲　timing pulse
定时器　timer
定时斜坡放大器　timing ramp amplifier
定时振动　timing jitter
定时装置　time equipment
定顺序　definite sequence
定铁芯　fixed core
定位　allocation, position location
定位板　locking plate
定位比例算法　position proportional algorithm
定位表面　locating surface
定位表示线　normal indication wire
定位导销　locating pilot pin
定位阀　lock-up valve
定位反馈　position feedback
定位光电倍增器　position-sensitive photomultiplier
定位横栏　locating bar
定位环　locating ring
定位孔　pilot hole
定位控制系统　positioning control system
定位控制线　normal control wire
定位块　located block, locating piece, location lump, locking block, stock locater block
定位螺栓　holding down bolt
定位片　locating plate
定位器　positioner
定位系统　positioning system
定位销　located pin, locating pin, stop pin
定位中心冲头　locating center punch
定向薄膜　oriented film
定向道路交汇处　directional interchange
定向发射　directional radiation
定向接收　directional reception
定向控制　orientation control
定向系统　pointing system
定向线圈　directional coil
定向装置　orientation device
定性的　qualitative
定性仿真　qualitative simulation
定性分析　qualitative analysis
定性控制　qualitative control
定性物理模型　qualitative physical model
定压池　header tank
定义固定点　defining fixed point

定义项目生成的列表 list of builder definition items
定义域 definitional domain
定值控制 fixed set point control
定值器 set point generator
定值调节 control with fixed set point
定中心 centering
定轴齿轮系 gear train with fixed axes, ordinary gear train
定子 stator
定子齿轮 internal gear
定子电路 stator circuit
定子漏磁电抗 stator leakage reactance
定子绕组 stator winding
定子铁芯 stator core
定子线圈 stator coil
定阻输出式分压箱 fixed resistance output type volt ratio box
定阻输入式分压箱 fixed resistance input type volt ratio box
锭铁 ingot iron
铥 thulium (Tm)
氡 radon (Rn)
动标度尺式仪表 moving-scale instrument
动槽气压表 adjustable cistern barometer
动触点 moving contact
动磁式仪表 moving magnet instrument
动磁系振动子 moving-magnet galvanometer
动电阻应变仪 dynamic resistance strain gauge
动断触点 break contact, normal closed contact, normally closed contact
动负荷 dynamic load
动刚度 dynamic stiffness
动刚度比 dynamic stiffness ratio
动合触点 make contact, normally open contact
动力 motive power
动力测试 dynamic test
动力传输 power transmission
动力电子设备 power electronics
动力堆 power reactor

动力反应堆 power reactor
动力分配装置 power transfer
动力工程及工程热物理 power engineering and engineering thermophysics
动力管理 power management
动力夯 power rammer
动力机械及工程 power machinery and engineering
动力计 dynamometer
动力进样 hydrodynamic injection
动力经济 power economy
动力控制 power control
动力律描述 power law description
动力模型 dynamic model
动力黏度 dynamic viscosity
动力平衡 power balance
动力润滑 dynamic lubrication
动力室 power house
动力系统电压 power system voltage
动力系统控制 power system control
动力系统稳定器 power system stabilizer
动力效应 dynamic effect
动力学 dynamics
动力学特性 dynamic behaviour
动力制动器 dynamic brake
动量 momentum
动量式质量流量计 momentum-type mass flowmeter
动能 dynamic energy
动能系数 kinetic energy coefficient
动平衡 dynamic balance
动圈 dynamic coil, moving coil
动圈传声器 moving-coil microphone
动圈式 moving coil
动圈式安培计 moving coil ammeter
动圈式变感器 rotating-coil variometer
动圈式检流计 moving coil galvanometer
动圈式示波器 moving coil oscillograph
动圈式温度指示调节仪 moving coil type temperature indicating controller
动圈式仪表 moving coil instrument, moving-coil instrument

动水槽　dynamic water tank
动态　dynamic state
动态标准应变　dynamic standard strain device
动态补偿　dynamic compensation
动态补偿器　dynamic compensator
动态不平衡　dynamic unbalance
动态参数　dynamic parameter
动态测量　dynamic measurement
动态称重法　dynamic weighing method
动态窗口设定　dynamic windows set
动态磁化曲线　dynamic magnetization curve
动态存储分配　dynamic storage allocation
动态存储器　dynamic memory
动态存储器分配　dynamic memory allocation
动态电感　dynamic inductance
动态电容　dynamic capacity
动态电阻　dynamic resistance
动态二次离子质谱法　dynamic SIMS
动态反应　dynamic reaction
动态范围　dynamic range
动态方程　dynamic equations
动态分辨力　dynamic resolution
动态分析设计　dynamic analysis design
动态规划　dynamic programming
动态降解　dynamic degradation
动态校准器　dynamic calibrator
动态解耦　dynamic decoupling
动态矩阵　dynamic matrix
动态矩阵控制　dynamic matrix control (DMC)
动态累计误差　dynamic accumulation error
动态灵敏度　dynamic sensitivity
动态模型　dynamic model, dynamic modelling
动态挠性板　dynamic flex board
动态偏差　dynamic deviation
动态偏置控制　dynamic bias control
动态平衡机　dynamic balancing machine
动态曲线　dynamic curve

动态热机械分析　dynamic thermomechanical analysis
动态热机械分析仪　dynamic thermomechanical analysis apparatus
动态容积测量法　dynamic gauging
动态输出反馈　dynamic output feedback
动态输入-输出模型　dynamic input-output model
动态双面平衡机　dynamic two-plane balancing machine
动态随机存储器　dynamic RAM, dynamic random access memory (DRAM)
动态特性　dynamic characteristic
动态特性校准仪　dynamic characteristic calibrater
动态吻合性　dynamic exactness
动态稳定度　dynamic stability
动态误差　dynamic error
动态误差系数　dynamic error coefficient
动态系统　dynamic system
动态显示图像　dynamic display image
动态响应　dynamic response
动态信道分配　dynamic channel assignment
动态性能分析　dynamic performance analysis
动态性质　dynamic property
动态压力传感器　dynamic pressure sensor, dynamic pressure transducer
动态应变仪　dynamic strain indicator
动态优化模型　dynamic optimization model
动态元件　dynamic element
动态运行　dynamic-state operation
动态噪声抑制器　dynamic noise suppresser
动态增益　dynamic gain
动态指标　dynamic specification
动态质谱仪器　dynamic mass spectrometer instruments
动铁式　moving iron
动铁式比率计　moving-iron logometer
动铁式仪表　moving-iron instrument
动衔铁　moving armature

动压	dynamic pressure
动应变	dynamic strain
动作	action
动作电流	operating current
动作路径	path of motion
动作偏差信号	actuating error signal
动作速度	speed of action
动作线圈	operating coil
冻干机	freeze drying equipment
冻结干燥器	freeze dryers
洞穴	cavern
斗式输送机	bucket conveyor
抖动检测器	jitter tester
读	read
读出线	sense line, sense wire
读码器	code reader
读频输入	read frequency input
读入	read in
读数	reading
读数误差	reading error
读数显微镜测振幅法	meters for measuring amplitude by a reading microscope
读数装置	readout unit
读写检验指示器	read write check indicator
独立截止阀	individual stop valve
独立式	free standing
独立线性度	independent linearity
独立一致性	independent conformity
独立仪器	stand-alone instrument
独立源	independent source
堵缝	caulk
堵塞物	obstruction
堵转转矩	locked-rotor torque
度	degree
度量图	dimensioned plan
度盘几何中心	geometric centre of the dial
镀磁线	plated wire
镀黄铜钢丝	brass-plated steel wire
镀金导线	goldclad wire
镀金科伐线	gold-plated kovar wire
镀镍钢丝	nickel plated steel wire
镀通孔	plated through hole
镀锡	tinning
镀锡铁板	tin plated steel sheet, tinned iron
镀锡铜线	tinned wire
镀线	plate wire
镀锌	galvanization, galvanize
镀锌的	galvanized
镀锌钢	galvanized steel
镀锌钢板	galvanized steel sheet
镀锌钢绞线	galvanized stranded wire, stranded galvanized steel wire
镀锌金属	galvanized metal
镀锌软钢板	galvanized mild steel plate
镀锌铁管	galvanized iron pipe, galvanized iron tube
镀锌铁皮	galvanized sheet iron
镀锌铁丝	galvanized iron wire
镀银线	silver jacketed wire
端点	end-points
端电压	voltage across the terminals, terminal voltage
端盖	end cover
端环	end ring
端基线性度	terminal-based linearity
端基一致性	terminal-based conformity
端接接头	edge joint
端接全隙孔	terminal clearance hole
端跨	end span
端面	transverse plane
端面参数	transverse parameters
端面槽宽收缩	transverse space-width taper
端面齿厚收缩	transverse thickness taper
端面齿距	transverse circular pitch
端面齿廓	transverse tooth profile
端面重合度	transverse contact ratio
端面弧齿厚	transverse circular thickness
端面晶粒	end grain
端面径节	transverse diametral pitch
端面距尺寸	face to face dimension
端面模数	transverse module
端面压力角	transverse pressure angle
端视图	end view
端支承	end bearing
端子板	terminal block
端子间连接	inter-terminal connection
端子型数字量输入/输出模件	terminal-type digital I/O modules

| 短波 | short wave
| 短波波段 | short wave band
| 短齿 | stub teeth
| 短路 | short circuit, short-circuit
| 短路参数 | short-circuit parameter
| 短路的 | short-circuited
| 短路电压 | short-circuit voltage
| 短路电阻 | short-circuited resistance
| 短路故障 | short trouble
| 短路环 | short-circuiting ring
| 短路开关 | short switch
| 短路绕组 | short-circuited winding
| 短路试验 | short-circuit test
| 短路损耗 | short-circuit loss
| 短路特性 | short-circuit characteristic
| 短期存储器 | short-term memory
| 短时程协同 | short time horizon coordination
| 短时负载 | short-time load
| 短时工作制 | short-time duty
| 短时过范围极限 | overload limit of short duration, overrange limit of short duration
| 段 | segment
| 段检人员 | passage quality control
| 断点 | breakpoint
| 断点控制 | breaking point control
| 断电保护 | power outage protection
| 断电器 | cut-out
| 断经 | end missing
| 断开 | breaking, switch off, switching off, turn off
| 断开电键 | cut off key
| 断裂 | fracture
| 断裂长度 | breaking length
| 断裂韧性 | fracture toughness
| 断流阀 | shutoff valve, stop valves
| 断路 | break, circuit break
| 断路开关 | breaker switch, shut off switch
| 断路器 | breaker, circuit breaker
| 断路器柜 | circuit breaker cabinet
| 断路器开关 | killer switch
| 断路位置 | off position
| 断面前角 | hook angle
| 断面增大率 | increase of area
| 断相保护 | phase-failure protection
| 断相保护装置 | open-phase protection
| 断相继电器 | open-phase relay
| 断续非稳态多谐振荡器 | chopping astable multivibrator
| 断续开关 | break-and-make switch
| 断续线记录仪 | dotted line recorder
| 断续作用 | discontinuous action
| 断源位置 | failure valve position
| 煅石膏 | gypsum
| 锻粗加工 | upsetting
| 锻工 | hammer man
| 锻铝 | forging aluminium
| 锻模 | forging dies
| 锻铁 | dug iron, wrought iron
| 锻造 | forge
| 锻造机 | forging machine
| 锻造模用钢 | forging die steel
| 堆叠拨送料机 | stack feeder
| 堆焊 | overlaying
| 队列支撑数据库 | queue support database
| 对比 | contrast
| 对比定律 | contrast law
| 对比灵敏度 | contrast sensibility
| 对策论 | game theory
| 对策树 | game tree
| 对称 | symmetry
| 对称变形振动 | symmetrical deformation vibration
| 对称齿条 | symmetrical rack
| 对称齿条比例 | symmetrical rack proportions
| 对称二端口网络 | symmetrical two-port network
| 对称分量 | symmetrical component
| 对称负载 | balanced load
| 对称荷载 | symmetrical loading
| 对称三线制 | balanced three-wire system
| 对称三相电路 | symmetrical three-phase circuit
| 对称三相系统 | balanced three-phase system
| 对称伸缩振动 | symmetrical stretching vibration
| 对称输出 | balance output
| 对称网络 | symmetrical network

对称系统的分量　component of a symmetrical system
对称相位控制　symmetrical phase control
对称循环应力　symmetry circulating stress
对称因子　symmetry factors
对地电容　earth capacity
对焊　butt welding, butt-welding
对焊连接阀　buttwelding valves
对合变换　convolution transform (CT)
对夹式阀门　clamp valves
对夹式止回阀　wafer check valves
对角电桥　diagonal bridge
对角电压　diagonal voltage
对角线　diagonal line
对角优势　diagonal dominance
对角主导矩阵　diagonally dominant matrix
对接合　butt joint
对接角榫　butt mitred joint
对开式铁芯　waffle iron
对流　contra-flow, convection
对偶原理　dual principle, principle of duality
对数的　logarithmic
对数放大器　logarithmic amplifier
对数幅频特性　log magnitude-frequency characteristics
对数幅相图　log magnitude-phase diagram
对数幅值　logarithmic magnitude
对数时间　logarithmic time
对数时间相关　logarithmic time dependence
对数式模/数转换器　logarithmic A/D converter
对数衰减　logarithmic decrement
对数相频特性　log phase-frequency characteristics
对数增益　logarithmic gain
对数坐标图　logarithmic plot
对头熔接　butt fusion welding
对象测试功能　target test function
对心滚子从动件　in-line roller follower, radial roller follower
对心曲柄滑块机构　in-line or crank-slider mechanism, in-line slider-crank mechanism
对心移动从动件　radial reciprocating follower
对心直动从动件　in-line translating follower, radial translating follower
吨（质量单位）　tonne (t)
钝化作用　deactivation
钝角　obtuse angle
多版本软件　multi-version software
多笔记录仪　multiple-pen recorder
多边形的　polygonal
多变量多项式　multivariable polynomial
多变量反馈控制　multivariable feedback control
多变量反馈系统　multivariable feedback system
多变量控制　multivariable control
多变量控制系统　multi-variable control system
多变量系统　multivariable system
多变量质量控制　multivariable quality control
多标度尺测量仪表　multi-scale measuring instrument
多标度尺仪表　multi-scale instrument
多波段的　multirange
多步控制　multistep control
多步控制器　multi-step controller
多步离子雪崩腔　multi-step avalanche chamber
多步预测　multistep prediction
多参数监护仪　multi-parameter monitor
多层控制　multilayer control
多层膜基板　multi-layered film substrate
多层绕组　multilayer coil
多层系统　multilayer system
多层印制板　multi-layer printed board
多层印制电路板　multi-layer printed circuit board
多尺度细化分析　multiscale analysis
多冲量控制系统　multi-element control system
多重布线印制板　multi-wiring printed board
多重处理系统　multiprocessing system

多重过程	multiprocessor system
多重判据	multi-criteria
多重散射过程	multiple scattering event
多处理（技术）	multiprocessing
多处理器	multiprocessor
多触点继电器	multi-contact relay
多传感器互相关	multiple-sensor cross correlation
多传感器集成	multi-sensor integration
多窗口方式	multi-window mode
多窗口显示	multi-window display
多次反射法	multiple echo method
多次展开	multiple development
多存取系统	multi-access system
多导体	multi-conductor
多导体传输线	multi-conductor transmission line
多导体系统	multi-conductor system
多道 X 射线光谱仪	multi-channel X-ray spectrometer
多道程序设计	multiprogramming
多道分析器	multichannel analyzer
多道脉冲高度分析仪	multi-channel pulse height analyzer
多点超声黏度计	multipoint ultrasonic viscometer
多点触点输出	multiple contact output
多点触点输入	multiple contact input
多点记录	multipoint recording
多点记录仪	multiple-channel recorder; multi-point recorder
多点开关	multipoint switch
多点连接	multipoint connection
多点模拟量控制输入/输出模块	multipoint analog control I/O module
多点模拟输入	multiple analog input
多点网络	multipoint network
多点指示	multipoint indicating
多点指示仪	multiple-channel indicator
多点状态量输出卡	multipoint status output card
多点状态量输入卡	multipoint status input card

多电极	multi-electrode
多端的	multiterminal
多端口	multi-port
多端口网络	multi-port network
多端网络	multiterminal network
多段递阶控制	multistratum hierarchical control
多段决策过程	multistage decision process
多段控制	multistratum control
多段模型	multisegment model
多段系统	multistratum system
多发射极晶体管	multi-emitter transistor
多范围测量仪表	multi-range measuring instrument
多高速缓冲寄存器	multi-cache
多工器	multiplexer
多功能测量仪表	multi-function measuring instrument
多功能处理机	multi-processor
多功能传感器	multi-function sensor, multi-function transducer
多功能的	multifunctional
多功能控制器	multifunction controller (MC)
多功能流程图	multi-function graphics
多功能双笔长图记录仪	multi-functional dual-pen strip chart recorder
多功能子系统	multifunction subsystem
多股的	multiwire
多股绞合线	rope stranded wire
多股线	strand wire
多官能环氧树脂	polyfunctional epoxy resin
多光谱扫描仪	multispectral scanner (MSS)
多光谱照相机	multispectral camera
多回路	loops
多回路的	multiloop
多回路控制	multiloop control
多回路控制策略	multiloop control strategy
多回路控制器	multiloop controller
多回路控制系统	multiloop control

多 duo

system, multiple-loop control system
多机　multi-machine
多机传动　multiple-motor drive
多级　multilevel
多级编码　multilevel code
多级递阶结构　multilevel hierarchical structure
多级放大　multiple-stage amplification
多级放大器　multistage amplifier, multiple-stage amplifier
多级过程　multilevel process
多级计算机控制系统　multilevel computer control system
多级结构　multilevel structure
多级决策　multilevel decision
多级控制　multilevel control
多级控制器　multilevel controller
多级逻辑电路　multi-level logical circuit
多级闪急蒸馏淡化法　multistage flash distillation method for desalination
多级系统　multilevel system
多级协调　multilevel coordination
多级芯　multiple step plug
多极加速电子枪　multi-stage accelerating electron gun
多极开关　multiple-pole switch, multipole switch
多计划结构　multiple-project configuration
多计算机系统　multi-computer system
多阶滞后　multi-order lag
多接收器质谱计　multi collectors mass spectrometer
多晶热敏电阻器　multi crystal thermistor
多孔层开管柱　porous-layer open tubular column
多孔排水管道　porous drain
多孔型填充剂　porous packing
多离子检测　multiple ion detection
多链路　multilink
多量程　multiple scale
多量程的　multiple range, multirange
多量程电压表　multi-voltmeter
多量程仪表　multirange instrument

多列轴承　multi-row bearing
多笼型转子　multiple-cage rotor
多路传输　multichannel transmission, multiplex transmission
多路的　multipath, multiway
多路电话　multichannel telephony, multiplex telephony
多路电话通信　multiple telephony
多路电压输入卡　voltage multiple input card
多路阀　gauge multiport valve
多路放大器　multichannel amplifier
多路复用　multiplexing
多路开关　multiple way switch
多路模件　multiplexed module
多路模拟量输入-输出卡　multiple analog input-output card
多路耦合器　multicoupler
多路器　multiplexer
多路器卡　multiplexer card
多路输送　multipath transmission
多路通道　multiple channel
多路系统　multichannel system, multiloop system, multiple loop system
多路转换开关　multi-circuit switch
多路转接器　multiplexer
多媒体　multimedia
多媒体展示　multimedia show
多面平衡　multi-plane balancing
多面体　polygon
多模穴模具　multi-cavity mold
多目标跟踪　multiple target tracking
多目标规划　multi-criteria decision making
多目标决策　multi-objective decision
多目标优化　multi-objective optimization
多判据　multiple-criterion
多判据优化　multiple-criterion optimization
多频带地震仪　multi band seismograph
多频道地电仪　multifrequency channel ground detector
多频发电机　multifrequency generator
多频系统　multifrequency system
多频振荡器　multifrequency generator
多普勒海流计　Doppler current meter

多普勒雷达　Doppler radar
多普勒流量计　Doppler flowmeter
多普勒频率跟踪仪　Doppler frequency tracker
多普勒声呐　Doppler sonar
多普勒声呐导航器　Doppler sonar navigator
多普勒效应　Doppler effect
多腔铸型　family mold
多色测温法　multi-colour thermometry
多色辐射温度计　multi-colour radiation thermometer
多栅控制　multigrid control
多声道斜束式超声流量计　multi-path diagonal-beam ultrasonic flowmeter
多输入多输出　multi-input multi-output（MIMO），multiple-input multiple-output（MIMO）
多输入多输出控制系统　MIMO control system，multi-input multi-output control system
多输入多输出系统　multiple-input multiple-output system
多数载流子　majority carrier
多速　multi-speed
多速电动机　multiple-speed motor，multi-speed motor
多速度漂移作用　multi-speed floating action
多速控制器　multi-speed controller
多速率　multi-rate
多速无定位控制器　multiple-speed floating controller
多速无定位作用　multiple-speed floating action
多态逻辑　multistate logic
多肽合成仪　peptide synthesizer
多探测器　detectors
多铁芯型电流互感器　multicore type current transformer
多通带滤波器　comb filter
多通道的　multichannel
多通道放大器　multichannel amplifier
多通道互相关　multichannel cross correlation
多通道记录仪　multiple channel recorder
多通道控制器　multi-channel controller
多托辊电子皮带秤　multi-idler belt conveyor scale
多微处理机测试程序　multi-microprocessor test program
多维　multidimensional
多维气相色谱法　multidimensional gas chromatography
多维气相色谱仪　multidimensional gas chromatograph
多维数字滤波器　multidimensional digital filter
多维系统　multidimensional system
多位式控制器　multi-position controller
多位投影测图仪　multi-projecting plotter
多位作用　multi-step action
多线式自动测井仪　multi-channel logging truck
多线照相记录仪　multi-channel photo-recorder
多相的　multiphase，polyphase
多相电动机　multiphase motor
多相电流　multiphase current，polyphase current
多相电路　multiphase circuit
多相电路的连接　connections of polyphase circuits
多相发电机　multiphase generator，polyphase generator
多相感应电动机　polyphase induction motor
多相流　flow heterogeneity
多相同步发电机　polyphase synchronous generator
多相网络　polyphase network
多相系统　polyphase system
多相异步电动机　polyphase asynchronous motor
多相整流器　polyphase rectifier
多项函数基　polynomial functional basis
多项式　polynomial
多项式方法　polynomial method
多项式模型　polynomial model
多项式输入　polynomial input

多项式运动规律	polynomial motion
多项式转换	polynomial transform
多楔带	poly V-belt
多谐振荡器	multivibrator, multi-vibrator
多芯电缆	multicore cable, multiple-core cable
多芯片	multi-chip
多样性	diversity
多义线	poly-line
多用户仿真	multi-user simulation
多用户信息插座	multiuser information outlet (MIO)
多用途	multi-purpose
多余金属	excess metal
多元控制	multiple unit control
多匝	multiloop
多针指示仪	multiple-pointer indicator
多值匹配	multivalued mapping
多质量转子	rotor with several masses
多周波控制	multi-circle control
多轴应变计	multi-axial strain gauge
多轴钻床	drilling machines multi-spindle
多属性效用函数	multi-attributive utility function
多注束水表	multiple-jet water meter
多转电动执行机构	multi-turn electric actuator
多准则决策分析	multi-criteria decision methods
多自由度	many degree of freedom
多自由度系统	many degree of freedom system
多总线	multibus
多作用控制器	multi-action controller
舵轮	steering wheel
惰轮	idle gear
惰性金属指示电极	inert metal indicated electrode
惰性气体	inert gas
惰性气体恒温器	inert gas ovens

E

俄歇电子光谱学　Auger electron spectroscopy (AES)
俄歇电子能谱仪　Auger electron spectrometer
俄歇电子像　Auger electron image
锇　osmium (Os)
鹅颈管　gooseneck
鹅颈消防栓　swan neck fire hydrant
额定参数　nominal parameter
额定的　nominal, rated
额定电流　current rating, nominal current, rated current
额定电压　nominal voltage, rated voltage
额定负荷试验　running test
额定负载　nominal load, rated load
额定功率　nominal power, rated power, rated capacity
额定功率因数　rated power factor
额定击穿电压　breakdown voltage rating
额定频率　full speed, nominal frequency
额定容量　nominal capacity, rated capacity
额定寿命　rating life
额定输出　rated output
额定效率　rated efficiency
额定载荷　load rating
额定值　rated value
额定周向磁化电流　maximum rated circumferential magnetizing current
额定转矩　nominal torque, rated torque
额定转速　rated speed
额度扭矩　rated torque
额外横向力　additional horizontal force
扼流变压器　choke transformer
扼流滤波器　choke filter
扼流线圈　choke turn, choke winding
恩氏黏度　Engler viscosity
耳机　earphone, headset
耳塞　ear plug
铒　erbium (Er)

二重精炼铁　double refined iron
二次测量　secondary measurement
二次插值法　quadratic interpolation
二次动态矩阵控制　quadratic dynamic matrix control (QDMC)
二次反射法　double bounce technique
二次辐射体　parasitic element
二次规划　quadratic programming
二次过滤器　afterfilter
二次回路　secondary loop
二次击穿测试仪　secondary breakdown tester
二次精冲加工　restriking
二次空气　secondary air
二次冷却器　aftercooler
二次型　quadratic
二次型控制　quadratic control
二次型稳定　quadratic stability
二次型性能指标　quadratic performance index
二次型优化调节器　quadratic optimal regulator
二次性能指标　quadratic performance criteria
二次仪表　secondary instrument
二等标准测力机　grade II standard load calibrating machine
二等分　halve
二端口网络　two-port network
二端网络　two-terminal network
二端元件　two-terminal element
二级处理　secondary treatment
二级存储器　secondary storage
二级光谱　second order spectrum
二级图谱　second order spectrum
二级污水处理　secondary treatment of sewage
二极管　diode
二极管电容器　diode capacity meter
二极管放大器　diode amplifier
二极管晶体管逻辑电路　diode-transistor logic

二极管逻辑电路　logical diode circuit
二极管钳位　diode clamping
二极管限幅器　diode limiter
二极管削波仪　diode clipper
二极管阵列　photodiode array
二价的　bivalent
二阶加纯滞后模型　second-order plus time delay model
二阶微分作用　second derivative action
二阶系统　second-order system
二阶行列式　second order determinant
二阶滞后　second-order lag
二阶中心矩　second central moment
二进位　binary bit
二进制编码的十进制　binary coded decimal
二进制的　binary
二进制卡片　binary card
二进制控制　binary control
二进制数　binary digit
二进制数字　binary digital
二进制系统　binary system
二进制信号　binary signal
二进制元件　binary element
二进制运算　binary arithmetic
二硫化碳　carbon disulfide
二茂铁　dicyclopentadienyl iron

二-十进制编码　binary coded decimal (BCD)
二态的　binary
二通　cock
二通阀　two-way valve
二维核磁共振　NMR two dimensional
二维系统　two-dimensional systems
二位控制器　two-step controller
二位式开关调节器　two-position on/off controller
二位式调节阀　two position power actuated valve
二位式调节器　two-position controller
二线制　two-way configuration
二线制变送器输入模件　two-wire transmitter input module
二相位键控　binary phase shift keying（BPSK）
二项　two-term
二项作用　two-term action
二项作用控制　two-term control
二项作用控制器　two-term controller
二氧化硅　silica
二元原理　principle of duality
二值存储元件　binary storage element
二组分电磁流量计　component electromagnetic flowmeter

F

发电 electric power generation, power generation
发电厂 generating plant
发电机 dynamo, dynamotor, electric generator, electrical machine, generator
发电机保护 generator protection
发电机磁场控制 generator field control
发电机导线 generator lead wire
发电机电路 generator circuit
发电机电压 generator voltage
发电机控制面板 generator control panel
发电机励磁 generator excitation
发电机励磁机 generator exciter
发电机励磁绕组 generator exciting winding
发电机容量 generator capacity
发电机设备 power generation
发电机线接头 generator terminal
发电机运行 generator operation
发电机组 generator group
发电量 generated energy, power generation
发电设备 generating equipment
发电站 power house, power station
发动机 motor
发动机单元 motor unit
发动机功率计 energy dynamometer
发动机控制 motor control
发动机模式 motor pattern
发动机模型 energy modelling
发动机扭矩 engine torque
发动机系统 energy system
发动机效率 energy efficiency
发光二极管 light emitting diode
发光二极管显示器 LED display
发光漆 glazing paint
发光强度 luminous intensity
发光体 luminaire
发酵 fermentation
发酵工程 fermentation engineering
发酵罐 fermenter
发酵过程 fermentation processes
发裂 hair crack
发霉 mildew
发泡剂 foaming agent
发热 heat emission
发热管 cartridge heater
发射 emission, sending
发射X射线谱法 emission X-ray spectrometry
发射电子显微镜 emission electron microscope
发射光谱 emission spectrum
发射机 transmitter
发射极 emitter
发射极电流 emitter current
发射结 emitter junction
发射率 emissivity
发生 generate, generating
发生面 generating plane
发生器 generator
发生线 generating line
发送信息 sending message
罚函数 penalty function
罚则 penalty
阀 valve
阀电子管 valve tube
阀电子管继电器 valve relay
阀盖 bonnet
阀杆 valve stem
阀继电器 valve relay
阀进样 valve injection
阀井 valve pit
阀控传动机构 control valve actuator
阀门 valves
阀门定位器 valve positioner
阀门监视功能块 valve monitoring block
阀门浇口 valve gate
阀室 valve chamber
阀系数 valve coefficient
阀芯 valve core, valve disk

阀轴	valve spindle
阀座	valve seat
法［拉］（电容单位）	farad
法拉第电磁感应定律	Faraday electromagnetic induction law
法拉第定律	Faraday's law
法拉第笼	Faraday cage
法拉第效应	Faraday effect
法兰	flange
法兰截止阀	flange globe valves
法兰连接端	flanged ends
法兰膜片隔离式压力表	flanged diaphragm sealed manometer
法兰球阀	flange ball valves
法兰取压口	flange pressure tappings
法兰上钻孔取压的节流装置	throttling device with drilled holes
法兰闸阀	flange gate valves
法面	normal plane
法面参数	normal parameters
法面齿距	normal circular pitch
法面模数	normal module
法面压力角	normal pressure angle
法线方向	normal direction
法向侧隙	normal backlash
法向侧隙公差	normal backlash tolerance
法向齿距	normal pitch
法向齿廓	normal tooth profile
法向重合度	normal contact ratio
法向刀倾	normal tilt
法向负载	perpendicular load
法向弧齿厚	normal circular thickness
法向基节	normal base pitch
法向截面	normal section
法向径节	normal diametral pitch
法向力	normal force
法向平面	normal plane
法向弦齿高	normal chordal addendum
法向弦齿厚	normal chordal thickness
法向应力	normal stress
法向直廓蜗杆	straight sided normal worm
法向周节	normal circular pitch
法因曼测云器	Fineman nephoscope
法则系统	rule-based system
法制计量学	legal metrology
翻料机	upender
翻新	renovate
翻转装置	upender
凡士林	petroleum jelly
钒	vanadium (V)
繁忙时段	peak hours, peak period
反 z 变换	inverse z-transformation
反变换	inverse transform
反波管	backward wave tube
反常辉光放电	abnormal glow
反齿数比	inverse gear ratio
反冲	back flushing
反冲洗	back wash
反磁材料	diamagnetic material
反磁的	diamagnetic
反磁线圈	bucking coil
反磁效应	diamagnetic effect
反磁性	diamagnetism
反萃取	counter extraction
反电动势	counter electromotive force
反电动势继电器	counter electromotive force relay
反电压	counter voltage
反峰	negative peak
反共振	antiresonance
反共振频率	antiresonance frequency
反光警戒线	reflectorised warning line
反光罩	reflector
反虹吸管	anti-syphonage pipe
反激电流	flyback converter
反角度	reverse angle
反接制动	plug braking
反馈	back feed, feedback
反馈变量	feedback variable
反馈变压器	flyback transformer
反馈波纹管	feedback bellows
反馈补偿	feedback compensation
反馈电路	feedback circuit
反馈电压	feedback voltage
反馈放大器	feedback amplifier
反馈回路	feedback loop
反馈回路动态	dynamics of feedback loop
反馈激光器	feedback laser
反馈校正	correcting feedback
反馈校正器	feedback modifier

反馈控制　feedback control
反馈控制方法　feedback method
反馈控制器　feedback controller
反馈控制器设计　feedback controller design
反馈控制设计　design of feedback control
反馈控制系统　feedback control systems
反馈控制仪表　feed control instrument
反馈控制仪表显示块　feed control instrument display block
反馈力矩　feedback moment
反馈能力　feedback capacity
反馈绕组　feedback winding
反馈式组合　feedback combining
反馈通道　feedback channel, feedback path
反馈稳定　feedback stabilization
反馈系数　feedback factor
反馈系统　feedback system
反馈线圈　feedback coil
反馈线性化　feedback linearization
反馈信号　feedback signal
反馈压力瞬间变化值　immediate change in feedback pressure
反馈元件　feedback component, feedback element
反馈增益　feedback gain
反馈振荡器　back-coupled generator
反粒子　antiparticle
反流断路器　reverse current circuit breaker
反面压印　tick-mark farside
反偏压　reversed bias
反射波　echo
反射波高度　echo height
反射计　anacamptometer
反射器　reflector
反射望远镜　mirror telescope
反射谐振器　transmission line resonator
反射信号　return signal
反生线　back born wire
反调制　demodulation
反凸模　counter punch
反位控制线　reverse control wire
反物质　antimatter

反相　opposite in phase, phase reversal, phasing back, reversed phase
反相薄层板　reversed phase thin layer plate
反相的　in-phase opposition
反相输出　inverted output
反相液相色谱法　reversed phase liquid chromatography
反向　reversing, reversion
反向差分　backward difference
反向传播　back propagation (BP)
反向传播算法　backpropagation algorithm
反向串联　series-opposing connection
反向磁化　magnetization reversal
反向电离层探测器　back ionospheric sounder
反向电流　inverse current
反向电流自动断路器　reverse current cut-out
反向电压　opposing voltage
反向电阻计　back-resistance
反向二极管　backward diode
反向跟踪　backtracking
反向开关　reversing switch
反向跨导　negative transconductance
反向耦合器　backward coupler
反向器　reverser
反向驱动　inverter drive
反相调制　out-phasing modulation
反相调制系统　out-phasing modulation system
反向信道　backward channel
反向运动学　backward kinematics, inverse kinematics
反向制动　plugging
反旋　spin
反压阀　back-pressure valve
反应动力学　reaction kinetics
反应动力学控制系统　kinetic control system
反应器控制　reactor control
反应曲线　reaction curve
反应式电动机　reaction motor
反褶积　deconvolution
反正切　arctan
反质子　antiproton

反转	reverse rotation
反转法	kinematic inversion
反转下料	reversed blanking
反作用	reverse action
反作用杆	torque rod
反作用控制器	reverse acting controller
反作用轮控制	reaction wheel control
反作用执行机构	reverse acting actuator
返回	run back
返回传递函数	return transfer function
泛光灯	floodlight
泛光灯柱	floodlight mast
范成法	generating cutting, generation method
范德蒙矩阵	Vandermonde matrix
范围下限	lower range limit
方案设计	concept design
方波	square wave
方波变换	rectangular wave transform
方波振荡器	square wave-form oscillator
方波整形电路	squaring circuit
方差	variance
方差分析	analysis of variance
方差矩阵	variance matrix
方程式	equation
方程组	equations set
方法	method
方法标准	method standard
方法论	methodology
方钢	square bar iron
方根	square root
方键	square key
方框图	block diagram
方螺帽	square nut
方螺纹	square thread
方栓槽拉刀	spline broach
方铁	square bar iron
方位标示仪	direction mark meter
方位对准	bearing alignment
方向变化	change of direction
方向聚焦	direction focusing
方向指示板	direction arrow plate
方形波	rectangular wave
方形电路	squaring circuits
方形镗刀	square sleeker
方形盘	square pad
方形桩	barrette
钫	francium (Fr)
防饱和	anti-windup
防爆插座	explosion-proof socket
防爆冷藏柜	explosion proof freezer and refrigerator
防爆区域	explosion-proof area
防爆式电机	fire proof machine
防爆泄压设施	explosion relief measure
防爆型的	explosion proof
防爆型电磁阀	explosion-proof solenoid valve
防爆型电动机	explosion proof motor
防爆型电动执行机构	explosion-proof electric actuator
防爆型电机	explosion proof electric machine
防爆型仪表	explosion proof instrument
防爆型仪器仪表	explosion-proof instrument
防爆仪表标志	marking of an instrument for explosive atmosphere
防爆仪表类别	group of an instrument for explosive atmosphere
防剥剂	anti-stripping agent
防超载装置	overload protection device
防潮	damp proofing
防潮的	damp-proof
防潮绝缘	damp-proof insulation
防尘包装	dustproof packaging
防尘封	dust seal
防尘式仪器仪表	dustproof instrument
防尘型电磁阀	dust-proof solenoid valve
防打滑控制	anti-wheelspin control
防盗报警	intruder alarm
防盗玻璃窗	anti-burglary glazing, attack resisting glazing
防盗警钟系统	burglar alarm system
防盗自动警铃	burglar alarm
防毒面具	mask
防风板	wind shutter

防风玻璃	perspex
防风雨	weatherproof
防腐蚀	corrosion preventative
防腐式仪器仪表	corrosion-proof instrument
防复位终止	anti reset windup
防护	safeguard, protection
防护等级	grade of protection
防护栏障	protective barrier
防护幕	safety curtain
防护热板法	guarded hot plate apparatus
防护用具	safety appliance
防护罩	protecting hood, shield
防护装置	protective device
防滑	anti-slip, skid resistance
防滑钢砂	anti-skid dressing
防滑控制	antiskid control
防滑面层	friction course, skid resistant surfacing
防滑试验	skid resistance test
防滑物料	anti-skid material
防滑装置	antiskid
防火的	fireproof
防火电气配件	flame-proof electrical fittings
防火间隔	fire barriers
防火墙	fire wall, radiation wall
防火设备	fire protection equipment
防火外壳	flameproof enclosure
防火性能分级	classification for fire retardancy
防火障板	fire resisting shield
防静电腕带	electrostatic discharge
防空化阀	anti-cavitation valve
防空转	anti-slip
防雷用电压敏电阻器	arrester varistor
防裂片	cooling fin
防漏	leakage control
防窃报警器	burglar alarm
防倾侧滚轮	anti-tip roller
防渗	impermeable
防升滚轮	anti-lift roller
防湿保管库	moisture-proof storage
防蚀铝线	alumite wire
防蚀漆	anti-corrosion paint
防蚀物料	corrosion-resistant material
防水	waterproof
防水插座	waterproof socket
防水密封接头	water-proof packing gland
防水膜	waterproof membrane
防水挠性管	water-proof flexible conduit
防水千分表	waterproof dial gauge
防水式电机	hose-proof machine
防水式仪表	waterproof instrument
防水水泥	waterproof cement
防松螺钉	stop screw
防锁相制动系统	antilock braking system
防污染装置	anti-contamination device
防锈处理	rust prevention, stain proofing
防锈漆	anti-rust paint
防雨式仪表	rain-proof instrument
防振	shakeproof
防振式汞柱温度计	vibration-proof mercury-in-glass thermometer
防振装置	shockproof device
防震	anti-vibration mounting
防震垫	anti-vibration pad
防止干扰	interference prevention
防止侵蚀工程	erosion protection works
防皱压板	blank holder
防撞护栏	crash barrier
防撞系统	fender system
防撞桩	fender pile
防自旋调节	anti-spin regulation
仿生学	bionics
仿形车床	copy lathe
仿形法	form cutting
仿形机床	contouring machine
仿形磨床	profile grinding machine
仿形铣床	copy milling machine, duplicating milling machines
仿造	imitation
仿真	simulation
仿真分析	simulation analysis
仿真负载	artificial load
仿真框图	simulation block diagram
仿真器	simulator

仿真实验　simulation experiment
仿真实验分析　analysis of simulation experiment
仿真实验设计　design of simulation experiment
仿真数据　simulation data
仿真速度　simulation velocity
仿真天线　artificial antenna（AA），dummy antenna，phantom antenna
仿真语言　simulation languages
访问环境　access environment
访问码　access code
纺织玻璃纤维　textile glass
纺织工程　textile engineering
纺织化学与染整工程　textile chemistry and dyeing and finishing engineering
纺织科学与工程　textile science and engineering
放大　amplification，amplify
放大倍率　magnification
放大环节　amplifying element
放大级　amplifier stage
放大器　amplifier
放大器变压器　amplifier transformer
放大器系统　amplifier system
放大系数　coefficient of amplification，gain constant，multiplying factor
放电　electric discharge
放电灯　discharge lamp
放电电路　discharge circuit
放电过程　discharge process
放电机　electric discharge machine（EDM）
放电加工　electric discharge machining，electric spark machining
放电能量　discharge energy
放电容量　discharge capacity
放电时间常数　electric discharge time constant
放电特性曲线　discharge characteristic
放电系数　discharge coefficient
放电状态　discharge condition
放气　air bleeding
放气阀　purge valve
放气螺钉　bleed screw

放气嘴　bleed nipple
放热反应　exothermic reaction
放热峰　exothermic peak
放热式焊接　exothermic welding
放射计　radiometer
放射量测定器　dosemeter
放射疗法　radiotherapy
放射疗法设备　radiotherapeutic equipment
放射免疫测定　radio immuno assay
放射热分析仪　emanation thermal analysis apparatus
放射受体测定　radio receptor assay
放射性　radioactivity
放射性沉降物　radioactive fallout
放射性检测　radioactivity detection
放射性热分析　emanation thermal analysis
放射性同位素　radioisotope
放射性物质　radioactive substance
放射性元素　radioactive elements
放射学　radiology
放射云　radioactive cloud
放射自显影法　autoradiography
放水龙头　draw-off tap
放水旋塞　drain cock
放置入子　insert core
飞弧　flashover
飞溅润滑　splash lubrication
飞轮　flywheel
飞轮矩　moment of flywheel
飞轮制动器　flywheel brake
飞行控制　flight control
飞行器操纵　aircraft operation
飞行器控制　aircraft control
飞行器设计　flight vehicle design
非标准齿轮　nonstandard gear
非参数　non-parametric
非参数辨识　non-parametric identification
非参数回归　non-parametric regression
非参数训练　nonparametric training
非单调关联系统　noncoherent system
非单调逻辑　nonmonotonic logic
非导电图形　non-conductive pattern
非电的　non-electric

非电路　NOT circuit
非电容间歇直流电弧　intermittent DC non-capacitive arc
非对称的　asymmetrical
非对称性指数　index of asymmetry
非对映异构体　diastereoisomer
非高斯过程　non-Gaussian process
非隔离模拟输入　non-isolated analog input
非合作博弈　noncooperative game
非极化液体　non-polar liquid
非监控　unmonitored control
非监控系统　unmonitored control system
非接触式传感器　non-contact sensor
非接触式密封　non-contact seal
非接触式位移计　non-contact displacement meter
非结构工程　non-structural works
非结构性裂缝　non-structural crack
非金属　metalloid
非晶态聚合物　amorphous polymer
非晶体电极　non-crystalline electrodes
非均相膜电极　heterogeneous membrane electrode
非均匀反应堆　heterogeneous reactor
非可稳定化系统　non-stabilizable system
非连续过程　discontinuous process
非连续控制　discontinuous control
非连续作用　discontinuous action
非门　NOT gate
非门元件　NOT element
非平衡态　nonequilibrium state
非平稳随机过程　non-stationary random process
非屏蔽电缆　unscreened cable
非破坏性测试　non-destructive test
非破坏性测试系统　nondestructive testing system
非奇异摄动　nonsingular perturbation
非衰减频率　undamped frequency
非水滴定法　nonaqueous titrations
非塑化聚氯乙烯　unplasticized polyvinyl chloride（UPVC）
非弹性本底　inelastic background
非弹性散射过程　inelastic scatter

非调速电气传动　unadjustable speed electric drive
非稳态　non-stationary
非稳态系统　non-stationary system
非稳态信号　non-stationary signal
非稳态学习特性　non-stationary learning characteristic
非吸收性物料　non-absorbent material
非线性　nonlinearity, non-linearity
非线性标度　nonlinear scale
非线性泊松方程　nonlinear Poisson equation
非线性的　nonlinear
非线性电抗放大器　nonlinear reactance amplifier
非线性电路　nonlinear circuit
非线性电容　nonlinear capacitance
非线性电位差计　nonlinear potentiometer
非线性电阻　nonlinear resistance
非线性迭代　nonlinear regression
非线性方程　nonlinear equation
非线性分析　nonlinear analysis
非线性光学相关　nonlinear optical interaction
非线性规划　nonlinear programming
非线性环节　nonlinear element
非线性接口　nonlinear interface
非线性控制　control nonlinearity, nonlinear control
非线性控制器　nonlinear controller
非线性控制系统　nonlinear control system
非线性理论　nonlinear theory
非线性滤波器　nonlinear filter
非线性模型　nonlinear model
非线性失真　nonlinear distortion
非线性特性　nonlinear characteristic
非线性外部空穴　nonlinear external cavity
非线性系统　nonlinear system
非线性映射　nonlinear mirror
非线性优化　nonlinear optimization
非线性增益　nonlinear gain
非线性折射　nonlinear refraction
非线性折射率　nonlinear refractive in-

dex
非线性转换　nonlinear conversion
非相关过程　non-interacting process
非相关控制　non-interacting control
非圆齿轮　non-circular gear
非圆形盘　non-circular pad
非展成大轮　non-generated gear
非振荡控制算法　deadbeat control algorithm
非正交问题　non-orthogonal problem
非正弦波　non-sinusoidal wave
非正弦的　non-sinusoidal
非正弦电流　non-sinusoidal current
非正弦电压　non-sinusoidal voltage
非正弦交流电　distorted alternating current
非正弦曲线　non-sinusoidal curve
非支撑孔　unsupported hole
非周期的　non-periodic
非周期分解　aperiodic decomposition
非周期衰减　aperiodic damping
非周期性速度波动　aperiodic speed fluctuation
非周期振动　aperiodic vibration
非周期阻尼　aperiodic damping
非最小相位　non-minimum phase
非最小相位系统　non-minimum-phase system
非最小相位响应　non-minimum-phase response
菲涅耳衍射条纹　Fresnel diffraction string
废料　scrap material
废料冲床　scrap press
废料打包压力机　scrap press
废料切刀　scrap cutter
废料压块压力机　scrap press
废料阻塞　scrap jam
废气　exhaust gas, tail gas
废气脱硫　flue gas desulphurization
废气循环　exhaust gas recirculation
废水处理系统　waste water treatment
废水管　waste pipe
废水设备　waste fitments
废铁　waste iron
废物处理　waste treatment
废线　scrap wire

沸点　boiling point
沸点测高表　hypsometer
费拉里感应测试仪器　Ferraris instrument
费用矩阵　cost matrix
镄　fermium (Fm)
分贝（声音强度单位）　decibel (dB)
分贝衰减　dB-loss
分辨率　resolution
分表　secondary meter
分布　distribute
分布参数　distributed parameter
分布参数电路　circuit with distribution parameters, distributed circuit
分布参数控制系统　distributed parameter control system
分布参数系统　distributed parameter system
分布电感　distributed inductance
分布电容　distributed capacitance
分布负载　distributed load
分布绕组　distribution winding
分布式处理单元　distributed processing unit
分布式放大器　distributed amplifier
分布式计算机控制数字地震仪　distributed computer-control SDAS
分布式控制系统　distributed control system (DCS)
分布式模型　distributed model
分布式人工智能　distributed artificial intelligence
分布式数据库　distributed database
分布式通信体系　distributed communication architecture
分布式通信网络　distributed network
分布式遥测型数字地震仪　distributed telemetry SDAS
分布式智能　distributed intelligence
分布式自动化系统　decentralized automation
分布温度　distribution temperature
分布系数　distribution coefficient
分布自动化　distribution automation
分步控制　step control
分步展开　stepwise development
分部件　subassembly

分层　separated layer
分层决策　hierarchical decision making
分叉　bifurcate
分担交通量的道路　relief road
分度　graduation
分度变化量　index variation
分度齿轮　index gears
分度公差　index tolerance
分度盘　index plate, plate index
分度曲线　pitch curve
分度圈　graduated circle
分度头　index head
分度线　reference line, standard pitch line
分度圆　reference circle, standard pitch circle
分度圆柱导程角　lead angle at reference cylinder
分度圆柱螺旋角　helix angle at reference cylinder
分度圆锥　reference cone, standard pitch cone
分度指示器　division indicator
分度锥　pitch cone
分段线性化　piecewise linear
分段线性化分析　piecewise linear analysis
分段线性化控制器　piecewise linear controller
分多路转换器　sub multiplexer
分割　split
分割磁场电机　split field motor
分格　scale division
分格间距　scale spacing
分格式办公室　cellular office
分格值　scale interval, value of scale division
分隔墙　compartment wall
分隔室　compartment
分光光度计　optical spectroscopy, spectrophotometer
分光计　spectrometer
分光晶体　dispersive crystal
分光镜分析法　spectroscopic analysis
分光滤光器　spectral filter
分级淬火　martempering

分级的　hierarchical
分级计算机控制系统　hierarchical computer control system
分级控制　hierarchical control, step control
分级控制系统　hierarchical control system
分级系统　hierarchical system
分级卸载　sorted unloading
分级智能　hierarchical intelligence
分级智能控制　hierarchically intelligent control
分解电压　decomposition voltage
分解定理　decomposition theorem
分解方法　decomposition method
分解集结法　decomposition-aggregation approach
分解增益测量　resolved gain measurement
分界面　boundary surface
分界线　boundary
分类　classification
分类机　sorter
分类器　classifier
分类数据　categorical data
分离　separate, separation
分离点　breakaway point
分离对角线条件　off-diagonal terms
分离机型号编制方法　model designation of separators
分离角　angle of departure
分离接线块　discrete wire block
分离力　separating force
分离器　separator
分离数　separation number
分离系数　separating factor
分量质子磁力仪　component proton magnetometer
分流电路　divided circuit, shunt circuit
分流电阻　parallel resistance, shunting resistance
分流扼流圈　shunt reactor
分流进样　split injection
分流器　current divider, diverter, pintle valve
分流式流量计　shunt pattern flowme-

ter
分馏　fractionation
分路电抗器　shunt reactor
分路电容器　shunted capacitor
分路开关　branch switch
分马力电动机　fractional-horsepower motor
分模线　paring line
分母　denominator
分配　allocate, dissipation
分配比　distribution ratio
分配器　distributor
分配色谱法　partition chromatography
分配系数　partition coefficient
分批进样　batch inlet
分批控制　job-lot control
分批配料装置　batching plant
分频　frequency division
分频器　frequency demultiplier, frequency divider
分歧管模具　manifold die
分区流程图显示　area graphic display
分区趋势画面　area trend display
分区显示　area display
分区状态画面　area status display
分散　decentralized
分散读出系统　distribution readout system
分散光　scattered light
分散化　decentralization
分散剂　dispersant
分散控制　decentralized control, distributed control, distribution control
分散控制系统　decentralized control system
分散鲁棒控制　decentralized robust control
分散模型　decentralized model
分散式反馈　distributed feedback
分散式仿真　distributed simulation
分散式非线性环节　distributed non-linear element
分散式检测　distributed detection
分散输入系统　input decentralized system
分散随机控制　decentralized stochastic control

分散网络　distribution network
分散型计算机控制系统　distributed computer control system
分散型通信体系　distributed communication architecture
分散性　decentrality
分散性强化复合材料　disperse reinforcement
分时控制　time-sharing control
分数　fraction
分数单位　sub multiple of a unit
分数数据　scattered data
分析　analyse, analysis
分析程序　parser
分析法　analytical method
分析管　analyzer tube
分析化学　analytical chemistry
分析间隙　analytical gap
分析力学　analyse mechanics
分析天平　analytical balance
分析线　analytical line
分析型电子显微镜　analytical electron microscope
分析仪　analyser, analyzer
分析仪器　analytical instrument
分析预估器　analytical predictor (AP)
分析质量管理　analytical quality control (AQC)
分线　separated time
分线盒　branch box, cable box
分型板　match plate
分压电路　bleeder circuit
分压分析仪　partial pressure analyzer
分压计　potentiometer
分压器　voltage divider, potential divider
分页选择报警盘　page selector alarm panel
分圆锥　pitch cone
分支电缆　branching cable
分支电路　branch circuit
分装式检流计　separate galvanometer
分子　molecule
分子光谱　molecular spectrum
分子离子　molecular ion, molecule ion
分子量　molecular weight (MW)

分子式　structural formula
分子吸收光谱法　molecular absorption spectrometry
分子荧光分析法　molecular fluorometry
分子杂交仪　hybridization oven
分组画面　control group panel
分组画面定义　control group panel definition
分组流程图显示画面　group graphic display
分组显示画面　group display
分组详细显示画面　group detail display
酚磺乙胺　etamsylate
酚醛树脂　phenolic resin
酚醛纸质覆铜箔板　phenolic cellulose paper copper-clad laminates
粉煤灰　pulverized fuel ash (PFA)
粉煤灰硅酸盐水泥　Portland fly-ash cement
粉末成形　powder forming
粉末锻造　powder metal forging
粉末合金　powder metallurgy
粉末取样　dust sampling
粉末压出成形　compacting molding
粉碎　crash
粉碎器　cutting mills
丰度灵敏度　abundance sensitivity
风　wind
风车　windmill
风挡　air damper
风道　air passage, duct
风动发电机　wind-driven generator
风洞测试　wind tunnel testing
风干清漆　airy dry varnish
风荷载　wind load
风化　weathering
风冷式冷冻机　air cooled chiller
风冷系统　air cooling system
风力　wind force
风力墙　shear wall
风帘风扇　air curtain fan
风量计　air flowmeter
风门　throttle
风扇　fan
风扇电动机　fan motor

风扇式空气加热器　fan heater
风扇通风槽　fan ducting
风速　wind speed
风速表　anemometer, wind gauge
风速计　anemograph
风速指示器　pilot tube
风吸力　wind suction
风箱　windbox
风向标的动力偏幅　dynamic vane bias
风向袋　air sleeve
风压　wind pressure
风压表　blast pressure gauge, output air pressure gauge
风应力　wind stress
封板　blanking plate
封闭　closure
封闭式插接　close type socket joint
封闭式电动机　enclosed motor
封闭式继电器　enclosed relay
封闭式开关　enclosed switch
封闭式渗漏测定计　confined lysimeter
封闭系统　closed system
封端　end capping, stop end
封口　seal
封口膏　sealing compound
封口胶纸　masking tape
封漏　leak sealing
封漏袋　leak sealing bag
封密条　sealing strip
封头机　heading machine
封箱渗碳　box carburizing
峰底　peak base
峰-峰值　peak-to-peak value
峰高　peak height
峰宽　peak width
峰位限制器　peak limiter
峰值　peak, peak value
峰值比　peak amplitude ratio
峰值电度表　peak meter
峰值电流　peak current
峰值电压　crest voltage, peak voltage, spike voltage
峰值电压表　peak voltmeter
峰值二极管电压表　peak reading diode voltmeter
峰值伏特计　crest voltmeter

峰值计	crest meter
峰值扭矩	peak torque
峰值时间	peak time
峰值系数	crest factor
峰值压力	surge pressure
蜂鸣器	buzzer, buzzle, hummer
蜂窝	honeycombing
蜂窝数字包数据	cellular digital packet data (CDPD)
缝焊	seam welding
缝隙浇口	slit gate
伏安（功率单位）	volt-ampere
伏安表	galvano-voltmeter
伏安法	voltammetry
伏安特性	voltage-current characteristic, volt-ampere characteristics
伏欧表	volt ohmmeter
伏特	volt
伏特计	voltage meter, voltameter
伏特数	voltage
俘获	capture
氟	fluorine (F)
氟化乙丙烯	fluorinated ethylene propylene (FEP)
氟化乙烯离聚物	fluoro-ethylene polymer (FEP)
氟利昂	freon
氟塑料	fluorine plastic
氟橡胶	fluorine rubber, fluorous rubber
浮标	buoy, float
浮标和缆索式物位测量仪表	float and cable level measuring device
浮标式遥控测液位计	remote measuring float level meter
浮标体	buoy float
浮标运动监测	buoy motion package
浮标站	buoy station
浮标阵	buoy array
浮充电	float-charge
浮点	floating point
浮点计算	floating point computation
浮点控制作用	control floating action
浮雕（图案）	emboss
浮动	floating
浮动冲头	floating punch
浮动开关	float switch

浮花压制加工	embossing
浮环	floating disc
浮空输出	floating output
浮空输入	floating input
浮力	buoyant force
浮力沉筒式液位变送器	buoyancy displacer level transmitter
浮力式液位变送器	buoyancy type level measuring transmitter
浮力修正	buoyancy correction
浮力液位测量仪表	buoyancy level measuring device
浮球	floating ball
浮球阀	float operated valve
浮球式疏水阀	free float type steam trap
浮球式液位记录仪	float-actuated recording liquid-level instrument
浮球调节阀	float adjusting valve
浮升导料销	lifter guide pin
浮式起重机	floating cranes
浮置输入	floating input
浮子流量计	float flowmeter
浮子面积式流量计	float-area-type flowmeter
浮子气压计	float barograph
浮子式差压记录仪	float-type differential pressure recorder
浮子式验潮仪	float tide gauge
浮子型液位调节	float level regulator
浮子液位测量	float level measuring device
符号	symbol
符号处理	symbolic processing
符号法	symbolic method
符合性测试	conformance test
幅度介电感应测井仪	dielectric amplitude induction logging instrument
幅度控制	amplitude control
幅度调制	amplitude modulation (AM)
幅角原理	principle of the argument
幅频特性	amplitude-frequency characteristic, magnitude-frequency characteristic
幅相特性	magnitude-phase characteristic
幅相误差	amplitude-phase error

幅移键控　amplitude shift keying（ASK）
幅值比　amplitude ratio
幅值比例尺　magnitude scale factor
幅值轨迹　amplitude locus
幅值量化误差　amplitude quantization error
幅值失真　amplitude distortion
幅值调制　amplitude modulation（AM）
幅值响应　amplitude response
幅值裕度　magnitude margin
福丁气压表　Fortin barometer
福斯特电桥　Forster bridge
辐射　radiation
辐射安培计　radiation ammeter
辐射防护及环境保护　radiation and environmental protection
辐射高温计　radiation pyrometer, radiation thermometer
辐射功率　radiation power
辐射计　radiometer
辐射量　radiation level
辐射能　radiant energy
辐射平衡式高温计　radiation-balance type pyrometer
辐射热计　bolometer
辐射式开关　radiation type switch
辐射数据　radiological data
辐射通量　flux of radiation
辐射危害　radiological hazard
辐射温度计　radiation-energy thermometer
辐射压力功率表　radiation-pressure power meter
辐射照射度　irradiance
辐条钢丝　spoke wire
辅件　auxiliary attachment
辅线　subsidiary wire
辅因子　cofactor
辅助操纵板　auxiliary control panel
辅助触点　auxiliary contact
辅助窗口　auxiliary window
辅助存储器　auxiliary storage, secondary storage
辅助单元　auxiliary unit
辅助电动机　auxiliary motor, pilot motor

辅助电源　auxiliary power supply
辅助端　auxiliary terminal
辅助多项式　auxiliary polynomial
辅助阀　auxiliary valves
辅助反馈　complementary feedback
辅助方程　auxiliary equation
辅助给水泵　auxiliary feedwater pump
辅助给水箱　auxiliary feedwater tank
辅助功能　auxiliary function
辅助回路　auxiliary loop
辅助计算功能块　auxiliary calculation blocks
辅助寄存器　background register
辅助浇口　dozzle, tab gate
辅助开关　auxiliary switch
辅助控制器　pilot controller
辅助控制器总线　auxiliary controller bus（ACB）
辅助控制台　auxiliary console
辅助气体　auxiliary gas
辅助绕组　auxiliary winding
辅助设备　auxiliary equipment, supporting facilities
辅助设计　aided design
辅助时间　accessory time
辅助输出信号　auxiliary output signal
辅助水泵　auxiliary water pump
辅助伺服电动机　auxiliary servomotor
辅助索引　traction assistance
辅助系统　auxiliary system
辅助仪表　accessory instrument
辅助装置　auxiliary device
腐蚀　corrosion
腐蚀破裂　corrosion failure
腐蚀试验机　corrosion testing machine
腐蚀性大气试验　corrosive atmospheres test
腐蚀性环境　sour condition
腐蚀性空气　corrosive air
腐蚀性磨损　corrosive wear
腐蚀性物质　corrosive substance
腐蚀性烟气　corrosive fume
腐蚀性盐　corrosive salt
腐蚀液　corrosive liquid
负　negative
负电荷　negative charge

负电子 negative electron
负电阻 negative resistance
负反馈 degenerative feedback, negative feedback
负反馈放大器 degeneration feedback amplifier, degenerative feedback amplifier, negative feedback amplifier, reversed feedback amplifier
负反馈稳定静电计 negative feedback stabilized electrometer
负荷保持时间 duration of load
负荷分配 load distribution
负荷分配比 load sharing ratio
负荷幅 load amplitude
负荷开关 load switch
负荷量 loading
负荷能力 load carrying capacity
负荷曲线 load curve
负荷容量 loading capacity
负荷训练 weight training
负荷状态 load conditions
负畸变 barrel distortion
负极 cathode, negative pole
负极性 negative polarity
负离子 negative ion
负离子发生器 negative ion generator
负脉冲信号 undershoot
负迁移 negative suppression
负序阻抗 negative sequence impedance
负压 negative pressure
负压室 low pressure chamber
负载 load
负载饱和曲线 load-saturation curve
负载比 duty ratio
负载变动 load variation
负载变化 variation in load
负载变化自动控制 loadamatic control
负载变量 load variable
负载波导管 loaded waveguide
负载波动 fluctuation of load
负载操作 on-load operation
负载测试 load testing
负载传感器 load cell
负载导纳 load admittance
负载电流百分比显示 percentage of current load
负载电路 load circuit
负载电位器函数发生器 loaded-potentiometer function generator
负载电压调制器 load ratio voltage regulator
负载电阻 load resistance
负载断路开关 load break switch
负载范围 load scope
负载分配器 load divider
负载换流 load commutation
负载角 power angle
负载控制 load control
负载流 load flow
负载流率解 load flow solution
负载模型 load modelling
负载配电 load dispatching
负载匹配 matching of load
负载频率控制 load frequency control
负载平衡 load equalization
负载特性交叠 crossover of load characteristic
负载特性线 load characteristic
负载调节 load regulation
负载调整率 load regulation
负载调整器 load regulator
负载图 load diagram
负载误差 loading error
负载线 load line
负载响应 load response
负载运行测试 on-load running test
负载转矩 load moment
负载阻抗 load impedance
负责单位 responsible department
负重 freeweight
负阻抗 negative impedance
负阻抗变换器 negative impedance converter
负阻效应 dynatron effect
附加的白高斯噪声 additive white Gaussian noise (AWGN)
附加公式 complementary formulation
附加荷载 imposed load
附加局部控制网络模块 additional LCN module
附加气 complementary gas
附加网络 additional network

附加系统　addition system
附件　accessory, attachment, enclosure
附力　adhesive force
附属设施　ancillary facilities
附属物　appurtenance
附属仪表　accessory instrument
附属硬件　accessory hardware
附着　sticking
复s平面　complex s plane
复z平面　complex z plane
复板模　duplicated cavity plate
复变量　complex variable
复电阻率仪　complex radiation
复费率电度表　multi-rate meter
复共轭　complex conjugate
复合pH电极　combination pH electrode
复合半导体　compound semiconductor
复合材料　cab composite material
复合层压板　composite laminate
复合成形　compound molding
复合的　composite
复合点　composite point
复合电极　combination electrode
复合电缆　composite cable
复合钢板　composite steel plate
复合继电器　compound relay
复合夹层结构　composite sandwich construction
复合铰链　compound hinge
复合金属材料　composite metallic material
复合控制　compound control
复合控制器　compound controller
复合控制系统　composite control systems, compound control system
复合块　composite block
复合轮系　compound gear train
复合模　compound die, gang dies
复合模具　composite dies
复合平带　compound flat belt
复合式数字电压表　combined type digital voltmeter
复合式组合　compound combining
复合同步信号　composite synchronizing signal

复合同步信号发生器　composite sync generator
复合系统　compound system
复合信号　composite signal
复合氧化物系气敏元件　compound oxide series gas sensor
复合应力　combined stress
复合柱　combined column
复合作用　composite action
复激绕组的　compound-wound
复卷绕组　secondary winding
复励　compound excitation
复励电动机　compound motor
复励发电机　compound dynamo, compound generator
复励绕组　compound winding
复平面　complex plane
复式插座　multiple socket outlet
复式反馈　compounding feedback
复式干涉仪　multiple interferometer
复式计算机系统　dual computer system
复式流量计　compound flowmeter
复式螺旋机构　compound screw mechanism
复式前馈　compounding feedforward
复式水表　combination water meter
复数　complex number
复数导纳　complex admittance
复数功率　complex power
复数形式　complex form
复数阻抗　complex impedance
复位　reset
复位杆　early return bar
复位键　reset key
复用链路　multiplex link
复原按钮　reset button
复杂机构　complex mechanism
复杂扰动　complex perturbation
复杂系统　complex system
复制　copy
复制一个文件　copy a file
副本　backup copy
副标准仪表　substandard instrument
副阀　auxiliary valves
副环设定　secondary set
副控制器　secondary controller

副载波发生器 subcarrier generator
傅里叶 Fourier
傅里叶变换 Fourier transform
傅里叶变换红外光谱法 Fourier transform infrared spectrometry
傅里叶变换光谱仪 Fourier transform spectrometer
傅里叶变换红外光谱学 FT-IR
傅里叶变换近红外分光仪 Fourier transform near infrared spectrometer
傅里叶分析 Fourier analysis
傅里叶分析器 Fourier analyzer
傅里叶级数 Fourier's series
傅里叶镜片 Fourier optics
傅里叶系数 Fourier coefficients
傅里叶展开式 Fourier expansion
覆盖层 cover layer
覆金属箔基材 metal-clad bade material
覆铜箔层压板 copper-clad laminate

G

伽马射线液位测量	gamma ray level measuring device
钆	gadolinium (Gd)
改变	change
改变报警优先级	alarm priority change
改变存储选择	change storage option
改性剂	modifier
钙	calcium (Ca)
盖	cap
盖板	cover
盖板中心部分	cover center section
盖顶	coping
盖顶石	coping stone
盖格尔式测振仪	Geiger type vibrograph
盖螺母	cap nut
盖瓦	cover tile
盖形螺母	acorn nuts, cap nut
概率	probability
概率的	probabilistic
概率分布函数	probability distribution function
概率积分	probabilities integration
概率逻辑	probabilistic logic
概率密度函数	probability density function
概率模式	probabilistic model
概念表达式	conceptual representation
概念设计	concept design
干电池	dry battery, dry cell
干法施釉	dry glazing
干法施釉器	dry glazing dispenser
干粉型灭火器	dry powder type fire extinguisher
干净核弹	clean bomb
干空气	dry air
干路	trunk road
干密度	dry density
干摩擦	dry friction, static friction
干摩擦阻尼	Coulomb damping
干球（温度计）	dry bulb
干球温度计	dry-bulb thermometer
干扰	disturbance, interference
干扰电流	disturbance current, interference current
干扰电平	interference level
干扰理论	perturbation theory
干扰探测	interference detection
干扰误差	interference error
干扰效应	disturbance effect
干扰信号	interfering signal
干扰性的	interfering
干热灭菌器	drying sterilizers
干热试验	dry heat test
干涉点	interference point
干涉滤光片	interference filter
干涉仪	interferometer
干湿球温度计	wet and dry bulb thermometer
干式气体表	dry gas meter
干水收缩	drying shrinkage
干线	trunk route
干燥	dry
干燥计	dry meter
干燥剂	dryer, siccative
干燥剂筒	dehydrating cartridge
干燥器	drier
干燥箱	drying oven
甘汞半电池	calomel half-cell
甘汞电极	calomel electrode
甘特图	Gantt chart
甘油	glycerine
杆螺母	stem nut
杆上变压器	pole-mounted transformer
杆组	Assur group
坩埚	crucible
感测探针	sensing probe
感光性树脂	photosensitive resin
感抗	inductive reactance
感纳	inductive susceptance
感生电流	induced current
感生电流像	induced current image
感温探测器	thermal detector

感性分量　inductive component
感性分量　inductive component
感性负载　inductive load
感烟式探测器　smoke detector
感应　induce, induced, inductance
感应测井仪　induction logger
感应淬火　induction hardening
感应电动机　induction motor
感应电动势　induced EMF, inductive EMF
感应电流　electromagnetic induction
感应电流法　induced current method
感应电路　induced wire
感应电压　induced voltage
感应发电机　induction generator
感应分压器　induction voltage divider
感应光　induction light
感应加热线圈　load coil
感应控制　inductive control
感应脉冲瞬变系统　induced pulse transient system
感应式传感器　inductance transducer, inductosyn
感应式电度表　induction meter
感应式电机　induction machine
感应式电机设计　induction motor design
感应式检出器　inductive pick-off
感应式盐度计　induction salinometer
感应式仪表　induction instrument
感应同步式位移传感器　inductosyn displacement transducer
感应系数　inductance
感应硬化　induction hardening
感知器　perceptron
刚度系数　stiffness coefficient
刚度准则　rigidity criterion
刚体　rigid body
刚体导引机构　body guidance mechanism
刚性冲击　rigid impulse, rigid shock
刚性单面印制板　rigid single-sided printed board
刚性导管　rigid conduits
刚性多层印制板　rigid multilayer printed board, rigid-flex multilayer printed board

刚性构架　rigid frame
刚性航天动力学　rigid spacecraft dynamics
刚性基质电极　rigid matrix electrode
刚性接缝　rigid joint
刚性联轴器　rigid coupling
刚性双面印制板　rigid double-sided printed board, rigid-flex double-sided printed board
刚性印制板　rigid printed board, rigid-flex printed board
刚性轴承　rigid bearing
刚性转子　rigid rotor
缸砖　quarry tile
钢　steel
钢板　steel plate
钢板弹簧　leaf spring
钢材卷料架　coil reel stand
钢锭模　ingot mold
钢管　steel pipe
钢结构　steel structure
钢结构支架配管　steel structure bracket tubing
钢筋　cramp iron, steel reinforcement
钢筋混凝土　reinforced concrete
钢筋混凝土封盖　reinforced concrete cover
钢筋丝　reinforcing wire
钢筋条　reinforcement bar
钢筋网　steel fabric
钢锯　hack saw
钢拉线　steel guy wire
钢片　sheet steel
钢琴丝　piano wire
钢丝　steel wire
钢丝床用钢丝　mattress netting steel wire
钢丝护网　wire guard
钢丝绒　steel wool
钢丝软轴　wire soft shaft
钢丝绳　cable wire, steel wire rope, wire rope
钢丝绳夹　wire rope grips
钢丝绳用普通套环　general purpose thimbles for use with steel wire ropes
钢丝绳用压板　clamping plates for

fixing steel wire ropes
钢丝刷　wire brush
钢丝网　steel wire netting
钢铁工业　steel industry
钢铁冶金　ferrous metallurgy
钢铁制造　steel manufacture
钢芯铝线　steel reinforced aluminium wire, steel-cored aluminium wire
钢芯铜线　steel-cored copper wire
钢制电缆管　rigid steel conduit
钢质焊接气瓶　welded steel gas cylinders
钢质无缝气瓶　seamless steel gas cylinders
钢桩　steel pile
杠杆　lever
杠杆臂　lever arm
杠杆记录式流量计　lever-arm recording flowmeter
杠杆平衡式浮钟压力计　balanced level-type bell gauge
杠杆式测振仪　lever-type vibrograph
杠杆式制动电动机　level-braked motor
杠杆组件　level assembly
杠铃　barbell
高比移值　high Rf value
高冲击聚苯乙烯　high impact polystyrene
高冲击性聚苯乙烯　high impact polystyrene rigidity
高磁测试　electronic magnetism inspect
高次　higher order
高次谐波　higher harmonic, upper harmonic
高次谐波共振　higher harmonic resonance
高低平均算法　high low average algorithm
高低通滤波器　low-and-high-pass filter
高低信号选择器　high-low signal selector
高低选择器　high-low selector
高低作用　high-low action
高碘酸盐　periodate

高电流密度　high current density
高电平过程接口单元　high level process interface unit (HLPIU)
高电平模拟输入　high level analog input (HLAI)
高电热电位差计　high EMF potentiometer
高电压　high-voltage
高电压与绝缘技术　high voltage and insulation technology
高顶值　high limited value
高度计　altimeter, height gauge
高度角　altitude angle
高度冷凝　heavy condensation
高度系列　height series
高度指示器　altitude indicator
高分辨电子显微镜　high resolution electron microscope
高分辨核磁共振波谱仪　high-resolution NMR spectrometer
高分辨率　high resolution visible (HRV)
高分辨率γ探测器　high-resolution gamma detector
高分辨率核磁共振波谱仪　high-resolution NMR spectroscope
高分辨率气相色谱法　high resolution gas chromatography (HRGC)
高分辨率图像仪　high resolution visible image instrument
高分辨衍射附件　high resolution diffraction attachment
高分辨衍射像　high resolution diffraction image
高分辨质谱仪器　high-resolution mass spectroscope
高分散衍射像　high dispersion diffraction image
高硅生铁　glazy pig iron
高合金钢丝　high alloy steel wire
高级的　advanced
高级多功能控制器　advanced multi-function controller (AMC)
高级过程管理综合控制器　advanced process manager (APM)
高级过程控制　advanced process control (APC)
高级控制程序　advanced control pro-

gram (ACP)
高级双重内嵌式内存模块　advanced dual in-line memory modules (AD-IMM)
高级移动电话系统　advanced mobile phone system (AMPS)
高级语言　high-level language
高架电线　overhead wire
高架轨道　elevated trackway
高架结构　elevated structure
高架起重机　overhead crane
高架引道　approach viaduct, elevated approach road
高阶统计表　higher-order statistics
高阶微分方程　higher-order differential equation
高阶系统　higher-order system
高精度　high accuracy, high definition
高精度的　high-precision
高精度脑立体定向仪　high precision stereotaxic
高聚物传声器　high polymer microphone
高抗拉钢　high tensile steel
高抗拉钢缆　high tensile steel tendon
高抗拉螺栓　high tension bolt
高空气象计　aerometeorograph
高灵敏度的　highly sensitive
高炉　blastfurnace
高炉生铁　blast-furnace cast iron
高锰酸钾法　potassium permanganate method
高密度　high density
高密度 I/O 子系统　high density I/O subsystem
高密度聚乙烯　high density polyethylene
高能 X 射线　high energy X-rays
高能电压敏电阻器　high energy varistor
高能电子衍射仪　high electron energy diffractometer (HEED)
高频　high-frequency
高频 X 射线诊断机　high frequency X-rays diagnostic machine
高频变压器　high-frequency transformer
高频测量仪表　high-frequency measuring instrument
高频滴定　high-frequency titration
高频电刀　high frequency electrotome
高频电压敏电阻器　high frequency varistor
高频扼流圈　high-frequency choke coil
高频干扰　high-frequency interference
高频干扰电压　radio influence voltage
高频隔直流电容器　chokon
高频加热　high-frequency heating
高频介电分选仪　high frequency dielectric splitter
高频控制　high-frequency control
高频疲劳试验机　high frequency fatigue testing machine
高频示波器　ondograph
高频调制　high-frequency modulation
高频通信　high-frequency communication
高频性能　high-frequency performance
高频衍射　high-frequency diffraction
高频仪表　high-frequency instrument
高频移动 X 射线机　high frequency mobile X-rays machine
高频噪声　high-frequency noise
高频振荡器　high-frequency oscillator
高强度钢丝　high-tensile steel wire
高强度铸铁　meehanite cast iron
高清晰度电视　high definition television
高山气压表　mountain barometer
高身花槽　raised planter
高水温故障停机　high coolant temperature shutdowns
高斯（磁通密度单位）　gauss
高斯的　Gaussian
高斯定理　Gauss theorem
高斯分布　Gaussian distribution
高斯光学　Gauss optics
高斯过程　Gaussian process
高斯函数　Gaussian function
高斯计　Gauss-meter
高斯-马尔科夫源　Gauss-Markov source
高斯求积法　Gaussian integration method
高斯噪声　Gaussian noise

高速车床 high-speed lathe
高速传递 high-speed transmission
高速打印机 high-speed printer
高速带 high speed belt
高速度工具钢 high speed tool steel
高速缓冲存储器 cache memory
高速绘图仪 high-speed plotter
高速开关 high-speed switch
高速冷冻离心机 high speed refrigerated centrifuge
高速平衡 high speed balancing
高速平衡设备 high speed balancing installation
高速数据通道 data highway (DHW)
高速数据通道接口 highway gateway
高速数据通道接口模块 highway interface module
高速数据通道接口状态显示画面 highway gateway status display
高速数据通道接口字库 highway gateway library
高速数据通道控制状态 highway control state
高速数据通道耦合模块 highway coupler module
高速数据通道通信处理器 highway communication processor
高速数据通道通信指挥器 highway traffic director
高速数据通道状态 highway status
高速数据通道状态显示画面 highway status display
高速数据通道子系统接口 data highway port (DHP)
高速通信指挥器 highway traffic director
高速液相色谱法 high speed liquid chromatography
高速阅读器 high-speed reader
高速钻床 drilling machines high-speed
高通滤波器 high pass filter
高桅（杆）照明 high mast lighting
高温 high temperature device
高温安装线 high-temperature hook-up wire
高温处理电热调节器 high temperature thermistor
高温电阻炉 high temperature electric furnaces
高温钢 high temperature steel
高温高压电站闸阀 high temperature pressure power station gate valves
高温管式电阻炉 high temperature electric resistance tubular furnace
高温计 pyrometer
高温炉加热设备 high temperature furnaces heating apparatus
高温试验 high temperature test
高温试验机 high temperature testing machine
高温试验箱 high temperature test chamber
高温稳定 high-temperature stability
高温应变计 high temperature strain gauge
高限报警 alarm high
高效薄层色谱法 high performance thin layer chromatography
高效二极管 efficiency diode
高效率 high-efficiency
高效毛细管电泳法 high performance capillary electrophoresis (HPCE)
高效液相色谱法 high performance liquid chromatography (HPLC)
高效液相色谱仪 high performance liquid chromatograph
高效液相色谱-质谱联用 high performance liquid chromatography-mass spectrometry (HPLC-MS)
高信号选择器 high-signal selector
高性能的 high-performance
高性能过程管理站 high-performance process manager
高性能计算机与通信规划 high performance computing and communication program
高性能计算与通信 high performance computing and communicating
高性能扫描电子显微镜 high performance scanning electron microscope
高血压 hypertension
高压 high voltage
高压变压器 high-tension transformer

高压并联电抗器 high voltage shunt reactor
高压侧 high side
高压测试器 high voltage tester
高压差式热分析仪 high pressure DTA unit
高压电动机 high-tension motor
高压电缆 high-voltage cable
高压电力 high voltage electricity
高压电桥 high voltage bridge
高压电网 high-voltage fence
高压电线 high tension wire
高压电源 high voltage source
高压电子显微镜 high voltage electron microscope
高压端 high voltage terminal
高压汞灯 high pressure mercury lamp
高压柜 high pressure cabinet
高压回路 high tension loop
高压集成电路 high voltage IC
高压绝缘箱 high voltage insulating tank
高压馈电干线 high-tension supply main
高压缆槽 high voltage cable trough
高压力恢复阀 high-recovery valve
高压炉头 high pressure burner
高压灭菌器 autoclaves sterilizers
高压钠灯 high pressure sodium lamp
高压喷水器 high pressure water jet
高压起动器 high volt starter
高压气动薄膜执行机构 high pressure pneumatic diaphragm
高压清洁器 high pressure cleaner
高压试验 high-voltage test
高压输电电线 high-tension transmission line
高压输入 high voltage injection
高压水推送清洗器 high pressure water propagating cleaner
高压水银灯 high pressure mercury lamp
高压探针 high voltage probe
高压稳定度 high voltage stability
高压线 high-tension line
高压线铁塔 pylon
高压限压器 high pressure limiter
高压液相色谱 high pressure liquid chromatography (HPLC)
高压整流二极管 kenotron
高压直流传输线 HVDC transmission lines
高压直流放大器 high-voltage DC amplifier
高压铸造 squeeze casting
高增益 high-gain
高增益反馈 high-gain feedback
高张力螺帽 heigh strength nuts
高真空 high vacuum
高阻标准电阻器 high value standard resistor
高阻表 earthometer, insulation resistance meter
高阻抗 high impedance
高阻抗差式探头 high-impedance differential probe
割集 cut set
割集矩阵 cut set matrix
格林函数 Green function
格森管 Gershum tube
格式化 format
格线 grid lines device
隔板 isolating plate
隔爆外壳 flame proof enclosure, flameproof enclosure
隔尘网 dust screen
隔断阀 isolating valve
隔离变压器 isolating transformer
隔离放大器 disconnect amplifier, isolated amplifier
隔离管 wet leg
隔离间壁 isolating partition
隔离开关 disconnector, isolating link
隔离模拟输入 isolated analogue input
隔离盘 isolation pad
隔离器 isolator
隔离网络 isolated network
隔离线 shielded wire, shielding wire
隔离液 sealing liquid
隔膜 diaphragm
隔膜阀 diaphragm valve
隔膜浇口 diaphragm gate
隔膜压力表 diaphragm sealed manometer, diaphragm-seal pressure gauge

隔气　trapping
隔气弯管　trap
隔气弯管去水口　trap outlet
隔墙　diaphragm wall
隔热　thermal insulation
隔热屏障　heat shield
隔声板　noise barrier
隔声屏　acoustic screen
隔音盖罩　noise enclosure
隔音屏障　noise barrier
隔直流电容器　blocking condenser
镉　cadmium
镉电池　cadmium cell
镉光电池　cadmium photocell
镉镍蓄电池　cadmium-nickel accumulator
镉铜合金线　cadmium copper wire
个人计算机　personal computer (PC)
个人计算机外设　personal computer enclosure
个人计算机串行接口　personal computer serial interface
个人数字助理　personal digital assistant (PDA)
各向同性　isotropism
各种系统动态特性　dynamic behavior of various system
铬　chromium (Cr)
铬不锈钢　chromium stainless steel
铬钼钒钢　chromium-molybdenum-vanadium steel
铬钼钢　chromium-molybdenum steel
铬镍线　chrome-nickel wire
铬酸盐　chromate
铬铁合金　chromium irons
铬铁矿　chromic iron
铬铜　chrome bronze
铬铜线　chromium-copper wire
铬线　chrome wire
给定的　given
给定与测量机构　set point unit and measurement receive assembly
给定值　set value
给水泵　feed water pump
给水处理池　feed water treatment reservoir
给水用硬聚氯乙烯管件　unplasticized polyvinyl chloride fitting for water supply
给水止回阀　feed water check valve
根　root
根部半径　root radius
根部焊道　root run
根轨迹　root loci, root locus
根轨迹的条数　number of separate loci
根轨迹法　root locus method, root method
根轨迹图　root locus diagram
根灵敏度　root sensitivity
根锥　root cone
根锥顶　root apex
根锥顶至安装基准面距离　root apex to back
根锥顶至相错点的距离　root apex beyond crossing point
跟随控制系统　follow-up control system
跟随模型控制　model-following control
跟随器　follower
跟踪　follow, tracking
跟踪校准　calibration traceability
跟踪器　servomechanism
跟踪特性　tracking characteristic
跟踪误差　tracking error
跟踪系统　tracking system
跟踪线　tracker wire
跟踪应用　tracking application
工厂　plant
工厂底层数据　plant floor data
工厂集成控制　integrated plant control
工厂控制系统　plant control system
工厂通信　factory communication
工厂网络模块　plant network module
工厂信息网络　plant information network
工厂信息协议　factory information protocol
工厂照明　factory illumination
工厂诊断与改善方法　factory diagnosis and improvement method
工厂自动化　factory automation (FA)

工厂总貌显示画面　plant overview display
工程变更次序　engineering change order
工程变更通知　engineering change notice (ECN)
工程标准　engineering standardization
工程单位　engineering unit (EU)
工程仿真器　engineering simulator
工程功能　engineering functions
工程功能的组态　configuration of engineering functions
工程管理综合控制器　process manager (PM)
工程技术人员　engineer
工程技术站　engineerings
工程界限　project limit
工程进度　works programme
工程经纬仪　engineer's transit
工程控制论　engineering cybernetics
工程力学　engineering mechanics
工程模　stage die
工程瓶颈　engineering project difficulty
工程热物理　engineering thermophysics
工程设计自动化　engineering design automation
工程师　engineer
工程师操作站　engineer's operating station
工程师工作站　engineering working station
工程师键盘　engineer keyboard
工程师属性　engineer personality, engineering personality
工程试模材料　material for engineering mold testing
工程塑胶　engineering plastics
工程统计　engineering statistics
工程图　dwg
工程系统仿真　engineering system simulation
工程项目协议　project agreement
工程制图　engineering drawing
工程资料　project data
工件　workpiece
工件夹具　workholding equipment
工件生命质量　quality of work life
工件头座　workhead
工具箱　kit, tool box
工况　application factor
工令　work order
工模焊接　jig welding
工频　power frequency
工频感应电炉　low-frequency induction furnace
工效学　ergonomics
工业　industrial
工业安全　industrial safety
工业标准架构　industry standard architecture (ISA)
工业布线系统　industrial distribution system
工业催化　industrial catalysis
工业辐射高温器　industrial total radiation pyrometer
工业过程　industrial process
工业过程检测控制仪表　industrial process measurement and control instrument
工业环境评估　industrial environment evaluation
工业机器人　industrial robot
工业机箱　industrial enclosure
工业酒精　industrial alcohol
工业控制　industrial control
工业控制系统　industrial control system
工业频率　industrial frequency, power frequency
工业区　industrial area
工业生产系统　industrial production system
工业数据处理　industrial data processing
工业双金属温度计　industrial bimetallic thermometer
工业通信　industrial communication
工业污水　trade effluent
工业用计算机　industrial computer
工业造型设计　industrial moulding design
工业自动化　industrial automation
工业自动化仪表　industrial process measurement and control instrument

工业组织与管理　industrial organizations and management
工艺标准　international workman standard
工艺动态　process dynamics
工艺工程师　process engineer
工艺流程设计　process design
工艺流程图　flowchart
工艺设备　process equipment
工艺设计　technological design
工字钢　H-shaped iron
工字梁　I beam
工字桩　H-pile
工作　work
工作比控制系统　duty factor control system
工作车辆　operational vehicle
工作齿侧　drive side
工作齿高　working depth
工作单　work sheet
工作单画面　worksheet display
工作电压　operating voltage
工作范围　operating range
工作负荷　operating load
工作荷载　working load
工作机构　operation mechanism
工作极限值　limiting value for operation
工作间　work cell
工作进度表　work schedule
工作空间　working space
工作量规　working gauge
工作能力　service ability
工作扭矩　operating torque
工作平台　working platform
工作守则　code of practice
工作台　test table
工作特性　performance characteristic
工作条件影响　operating influence
工作调节器　service regulator
工作通知单　job order
工作温度　working temperature, operating temperature
工作误差极限　limits of operating error
工作循环图　working cycle diagram
工作压力　operating pressure
工作压力角　operating pressure angle
工作允许应力　working stress
工作载荷　external loads
工作站　workstation, work station
工作站处理机　workstation processor
工作周期　duty cycle
工作状态　working order
工作阻力矩　effective resistance moment
公差　tolerance
公称力矩　nominal moment
公称通径　nominal bore
公称压力　nominal pressure
公称应力　nominal stress
公称直径　nominal diameter
公称值　nominal value
公法线　common normal line
公共存储器测试程序　common memory test program
公共的　common
公共端隔离输入　input with isolated common point
公共控制信号　common control signals
公共块　common block
公共区　common area
公共约束　general constraint
公模　male die
公母模　feature die
公顷（面积单位）　hectare (ha)
公升　liter
公用电力系统　municipal power supply system
公用天线电视　communication antenna television
公制齿轮　metric gears
公众电信服务　public telecommunication service
功耗　power dissipation
功角　power-angle
功率　power
功率比　power rate
功率变换器　power transformer
功率表　power meter
功率表校验器　wattmeter calibrator
功率沉积　power deposition
功率沉积特性　power deposition charac-

功率放大　power amplification
功率放大器　power amplifier
功率分布　power distribution
功率分布电路　power distribution circuit
功率分配器　power divider
功率辅助控制　power assisted control
功率管理　power management
功率激励　power drive
功率计　dynamometer
功率角　power angle
功率控制　power control
功率流　power flow
功率密度频谱　power density spectrum
功率平衡　power balance
功率谱　power spectra, power spectrum
功率谱密度　power spectrum density, power spectral density
功率器件　power device
功率绕组　power winding
功率三角形　power triangle
功率输出　power output
功率输入　power input
功率损耗　loss of power, watt loss, power loss
功率调节器　power regulator
功率因数　power factor
功率因数表　factor meter, power-factor meter
功率因数改善　power-factor improvement
功率因数控制　power-factor control
功能　function
功能板测试转接器　card test module
功能的　functional
功能电刺激　functional electrical stimulation (FES)
功能仿真　functional simulation
功能分解　functional decomposition
功能分析　function analysis
功能分析设计　function analyses design
功能规格　functional specification
功能级模块　function class module
功能件定义　function key definition
功能键　function key
功能近似　function approximation
功能绝缘　functional insulation
功能卡　function card
功能块　function block, functional block
功能块的功能列表　list of function blocks
功能块细目定义　function block detail definition
功能链　functional chain
功能模块　function module
功能强大　powerful
功能相似　functional similarity
功能型光纤温度传感器　function type optic-fibre temperature transducer
功能整定　functional adjustment
功能总貌　overview of functions
供单回路调节器用的回路通信卡　loop communication card for single loop controller
供电　electric power supply
供电电缆　feeder cable
供电电路　supply circuit
供电电压　power supply voltage, supply voltage
供电商　electrical supplier
供电设施　electrical service
供电网　supply network
供电系统　power supply system
供电原理图　scheme of power supply
供电子取代基　electron donating group
供料管　feed pipe
供料器　loader
供气出口　air supply outlet
供气阀　air supply valve
供气能力　pneumatic delivery capability
供气压力　supply pressure
供设计优化坐标轴　predominant axis
供水　water supply
供水点　water supply point
供水管　supply pipe
供应中断　interruption of supply
汞　mercury (Hg)
汞滴振幅　mercury drop amplitude
汞浸继电器　mercury-wetted relay
汞齐化法　amalgamation method

拱度	camber
拱门	arch
共磁式振动子	common magnet galvanometer
共轭齿廓	conjugate profile
共轭齿轮	conjugate tooth
共轭的	conjugate
共轭点	conjugate point
共轭复根	complex conjugate root
共轭复数	conjugate complex number
共轭根	conjugate roots
共轭匹配	conjugate match
共轭曲线	conjugate curves
共轭梯度法	conjugate gradient method
共轭凸轮	conjugate cam
共发射极晶体管	grounded-emitter transistor
共基极	common-base
共基极晶体管	grounded-base transistor
共集电极	common-collector
共集电极晶体管	grounded-collector transistor
共价键	covalent bond
共聚合体	copolymer
共模电压	common mode voltage
共模干扰	common mode interference
共模信号	common mode signal
共模抑制	common mode rejection
共模抑制比	common mode rejection ratio (CMRR)
共态变量	costate variable
共同使用	common use
共享	share
共享内存结构	share memory architecture (SMA)
共型表面	conformal surfaces
共用废水管	common waste pipe
共振	resonance
共振频率	resonant frequency
沟道晶体管	channel transistor
沟道作用	channeling
沟渠基底	channel bases
钩头道钉	dog spike
钩头楔键	taper key
钩穴	hook cavity
构架接合	framed joint
构件	structural hardware, structural member
估计	estimation
估计参数	estimation parameter
估计理论	estimation theory
估计量	estimate
估计器	estimator
估计算法	estimation algorithms
孤立系统	isolated system
古典文丘里管	classical Venturi tube
骨架	skeleton
骨料	aggregate
钴	cobalt (Co)
鼓风机	air blower, blower
鼓风炉	blast furnace
鼓风式风扇	forced draught fan
鼓式制动器	drum brake
鼓形齿	crowned teeth
鼓形电枢	drum armature
鼓形记录仪	drum recorder
鼓形控制器	drum controller
固定	fix
固定安装法	fixed setting
固定标度测量仪表	moving index measuring instrument
固定侧模板	fixed bolster plate
固定程序计算机	fixed program computer
固定触点	fixed contact
固定串联电容补偿	fixed series capacitor compensation
固定存储器	fixed storage
固定的	fixed
固定的温彻斯特磁盘驱动器	fixed Winchester disk driver
固定点	fixed point
固定电容	fixed capacity
固定电容器	fixed capacitor
固定电压	fixed voltage
固定电阻	fixed resistance
固定负载	constant load
固定附着物	fixtures
固定格式报告	fixed format log
固定构件	fixed link
固定化相开管柱	immobilized phase open tubular column
固定基础固有频率	fixed-based natu-

ral frequency
固定校正作用　definite corrective action
固定接合　lock seaming
固定连接　permanent connection
固定螺栓　fixing bolt
固定螺丝　set screw
固定命令控制　fixed command control
固定偏压　fixed bias
固定频率　fixed frequency
固定频率振荡器　constant frequency oscillator
固定时间间隔　fixed time lag
固定式测量仪表　fixed measuring instrument
固定式仪表　fixed instrument
固定水泵装置　fixed pump installation
固定水平　fixed level
固定顺序机械手　fixed sequence manipulator
固定误差　fixed error
固定线圈　fixed coil
固定相　stationary phase
固定销　retainer pin
固定液　stationary liquid
固定支座　fixed bearing
固定装置　fixtures
固化　stiffening
固化剂　curing agent
固件　firmware
固结　consolidation
固结系数　consolidation coefficient
固溶退火　solution treatment
固态　solid state
固态电池　solid state cell
固态光电传感器　solid state photosensor
固态激光器　solid state laser
固态逻辑组件　solid state logic card
固态相变　solid state phase changes
固态元件　solid state component
固体　solid
固体光敏元件　solid state photosensor
固体力学　solid mechanics
固体流量计　solid flowmeter
固体润滑剂　solid lubricant
固体微型逻辑元件　solid state micrologic element
固相萃取　solid phase extraction
固有不稳定性　inherent instability
固有常数　fundamental constant
固有的　inherent, intrinsic
固有反馈　inherent feedback
固有非线性　inherent nonlinearity
固有关系　fundamental relation
固有矩阵　fundamental matrix
固有流量特性　inherent flow characteristic
固有滤过当量　inherent filtration
固有模式　intrinsic mode
固有膜片压力范围　inherent diaphragm pressure range
固有黏度　intrinsic viscosity
固有频率　natural frequency
固有弱质失效　inherent weakness failure
固有双稳定性　intrinsic bistability
固有特性　deterministic behaviour, natural characteristic
固有稳定性　inherent stability
固有误差　intrinsic error
固有像散　intrinsic astigmatism
固有振荡　natural oscillations
故障　fail, failure trouble, fault, malfunction, trouble
故障保护　fail safe
故障保险系统　fail-safe system
故障查找　trouble shooting
故障点　fault point, trouble spot
故障定位　fault location
故障分布　fault distribution
故障隔离　failure isolation, fault isolation
故障记录　fault log
故障检测　failure detection, fault detection
故障检测器　failure monitor
故障检修　corrective maintenance
故障开关　breakdown switch
故障切除时间　fault clearing time
故障清除　clearing of fault, fault clearance

故障区间	fault section
故障时间	down time
故障识别	fault identification
故障树分析	fault tree analysis
故障停机时间	downtime
故障维修	breakdown maintenance
故障抑制	fault containment
故障诊断	failure diagnosis, fault diagnosis
故障诊断与分离	fault detection and isolation (FDI)
故障指示灯	fault indication
故障阻抗	fault impedance
刮刀	scraper
刮痕	scratch
刮伤	flaw
拐点	inflection point
拐点电压	knee point voltage
拐点频率	break frequency
关	off
关闭	close, shut down
关闭阀	shut off valve
关键构件	critical element
关键过程控制器	critical process controller
关键路线分析	critical path analysis
关节型操作器	jointed manipulator
关节型机器人	articulated robot, revolute robot
关联画面	associate display
关联矩阵	incidence matrix
关联设备	associate equipment
关联系数	correlation coefficient
关上	shut
关锁	locking device
关系	relationship
关系代数	relational algebra
关系数据库	relational database
关系型数据	relational data
观测器	observer
观测误差	error in observation
观察误差	error of observation
观察仪	visualizer
冠状齿轮	crown gear
管壁效应	wall effect
管道	pipe, pipeline
管道安全阀	piping safety valves
管道泵	piping pumps
管道电缆	conduit cable, duct cable, underground cable
管道敷设过程	pipelining processing
管道工程	pipe works
管道离心泵	piping centrifugal pumps
管道系统	piping
管道预留地	pipeline reserve
管端配件	end fittings
管件	fitting
管接头	pipe joint
管颈	neck
管口盖板	blank flange
管理	management
管理单元	administration unit
管理级	management level
管理级网络	supervisory level network
管理计算机控制	supervisory computer control
管理决策	management decision
管理科学	management science
管理控制	supervisory control
管理控制计算机	supervisory control computer
管理模件测试系统	manager module test system
管理目标	managed object
管理系统	management system
管理信息系统	management information system (MIS)
管理站	manager
管配件	pipe fittings
管钳子	box spanner
管式过滤器	pipe filter
管式罗盘	tubular compass
管筒	pipe barrel
管用螺帽	pipe nuts
管支柱	pipe stanchion
管制标志	regulatory sign
管柱试验	lysimeter experiment
管状暗渠	pipe culvert
管状电极	pipe graphite
管状扶手	tubular handrail
管状栏杆	tubular railing
管状线	hollow wire
管状荧光灯	tubular fluorescent lamp

贯通裂缝　through crack
惯量　inertia factor
惯性　inertia
惯性参考部件　inertial reference unit
惯性测量部件　inertial measurement unit
惯性常数　constant of inertia
惯性传感器　inertial sensor
惯性导航　inertial navigation
惯性的　inertial
惯性负载　inertia load
惯性合成向量　resultant vector of inertia
惯性矩阵　inertial matrix
惯性力　inertia force
惯性力部分平衡　partial balance of shaking force
惯性力矩　moment of inertia
惯性力平衡　balance of shaking force
惯性力完全平衡　full balance of shaking force
惯性轮　inertial wheel
惯性平台　inertial platform
惯性运动　inertial motion
惯性主矩　resultant moment of inertia
惯性姿态敏感器　inertial attitude sensor
惯性坐标系　inertial coordinate system
灌浆　grout
灌浆阀　grout valve
灌浆工程　grouting works
灌浆管　grout pipe, grout tube
灌浆排气管　grout vent pipe
灌浆套筒　grout sleeve
灌注阀　fill valve
光按钮　light button
光暗变压器　dim transformer
光暗控制器　dimmer
光笔　light pen
光标　cursor
光标键　cursor key
光标控制键　cursor control key
光标式仪表　instrument with optical index
光标位置　cursor position
光参量振荡器　optical parametric oscillator
光场　optical field
光冲量　photoimpact
光传感器　optical sensor, optical transducer
光传输　optical transmission
光磁的　photomagnetic
光存储装置　optical storage device
光导发光元件　optron
光导纤维　fiber optic
光点参数放大器　photo parametric amplifier
光点式检流计　galvanometer with optical point
光电　photoelectricity
光电倍增管　photomultiplier, photomultiplier tube
光电倍增晶体管　photomultiplier cell
光电倍增器　photoelectric multiplier, photomultiplier
光电变换器　photoelectric transducer, photoelectric transformer
光电波形发生器　photoformer
光电测距仪　electro-optical distance meter
光电池　optoelectronic cell, photocell
光电传感器　photoelectric sensor
光电磁效应　photo-electromagnetic effect
光电的　photoelectric
光电二极管　photodiode
光电发射　photoemission
光电发射管　photo-emissive cell
光电放大器　optical amplifier
光电管　optoelectronic cell, photocell, phototube, photovalve
光电管放大器　photocell amplifier
光电管控制　phototube control
光电管振荡器　photoformer
光电函数发生器　photoformer
光电继电器　photoelectric relay
光电晶体管　phototransistor
光电开关　photoswitch, photo electric switch
光电耦合器　photocoupler
光电三极管　phototriode
光电设备　optoelectronic device
光电式转速传感器　photoelectric ta-

chometric transducer
光电效应　photoeffect, photoelectric effect
光电信号发生器　photogenerator
光电遥感器　photoelectric sensor
光电元件　photovalve
光电转换器　optical to electrical converter
光电子脉冲放大器　optoelectronic pulse amplifier
光电子学　optoelectronics
光度感应控制　photo electric control
光二极管阵列检测器　photodiode array detector (DAD)
光发射元件　photovalve
光非线性　optical nonlinearity
光隔离器　photocoupler
光焊丝　bare wire
光焊条　bare electrode
光滑剂　slip agent
光化学烟雾　smog
光辉热处理　bright heat treatment
光极化双稳定性　optical polarization bistability
光继电器　light relay
光接收器　optical receiver
光禁带　optical band gap
光开关　optical switch
光控场效应管　photo-FET
光控继电器　photoswitch
光控晶闸管　light triggered thyristor
光控脉冲　photoimpact
光亮钢丝光面线　bright wire
光亮退火钢丝　bright annealed wire
光脉冲　optical pulse
光密度计　densitometer
光面　shiny side
光面钢丝　smoothly surfaced steel wire
光面管状栏杆　plain tubular railing
光敏变阻器　photovaristor
光敏的　photosensitive
光敏电池　light-sensitive cell
光敏电阻　light dependent resistance, photoconductive cell, photoresistance, photosensitive resistance
光敏电阻继电器　photoresistance relay
光敏感器　light sensor
光敏开关　light activated switch
光敏元件　light sensor
光盘　optical disk
光盘只读存储器　CD Rom
光频隔离器　optical isolator
光平均功率计　optical average power meter
光谱　spectra, spectrum
光谱的　spectral
光谱分析　spectral analysis
光谱分析法　spectroscopic analysis
光谱高温计　spectral pyrometer
光谱化学分析仪器　spectro-chemical analysis apparatus
光谱滤器　spectral filter
光谱匹配法　spectral matching
光谱学　spectroscopy
光谱仪　spectrograph
光谱仪器　optical spectrum instrument
光强　intensity of light, light intensity
光强计　lighting-intensity meter
光溶解　optical solution
光散射检测器　light scattering detector
光栅　grating
光栅重合　image registration
光栅单色仪　grating monochromator
光栅摄谱仪　grating spectrograph
光栅式位移传感器　grating displacement transducer
光时分复用　optical time division multiplex (OTDM)
光实现　optical implementation
光束检波计　light-beam galvanometer
光随机控制　optical stochastic control
光调制　light modulation, optical modulation
光调制器　light modulator, optical modulator, photomodulator
光通量　light flux, luminous flux, optical flow
光纤　fiber optic, optical fiber

光纤放大器	fibre amplifier
光纤干涉仪	fibre interferometer
光纤连接器	fibre connector
光纤耦合器	fibre coupler
光纤前置放大器	fibre preamplifier
光纤声传感器	optical fiber acoustic sensor
光纤时钟传送器	fiber optic clock transmitter
光纤时钟接收器	fiber optic clock receiver
光纤式转速表	optic fiber tachometer
光纤通导速度	fibre conduction velocity
光纤通信	fiber communication
光纤图像电缆	fiber optic image cable
光纤网络	fibre network, optical fibre network
光线	ray
光线强度计	actinometer
光线示波器	light beam oscillograph
光响应	optical response
光学薄膜厚度计	optical film thickness meter
光学常数	optical constant
光学的	optical
光学定向耦合器	optical directional coupler
光学反馈	optical feedback
光学高温计	optical pyrometer
光学高温计电测法	electrical measurement method of optical pyrometer
光学工程	optical engineering
光学聚焦转换开关	optical focus switch
光学控制	optical-control
光学流量计	optical flowmeter
光学平晶	optical flat
光学平行	optical parallel
光学摄谱仪	optical spectrograph
光学数据存储	optical data storage
光学双稳定性	optical bistability
光学通信	optical communication
光学显微镜	optical microscope
光学性能测定仪	optical property tester
光学性质	optical property
光学元件	optical component
光学字符辨识	optical character recognition
光源	light source
光泽	brilliance, gloss
光增强屏	intensifying screen
光斩波器	light chopper
光折光	optical birefringence
光制下料加工	finish blanking
光子	photon
光字排	illuminated nameplate
广播	broadcast
广播波段	broadcast band
广播干扰	broadcast interference
广播网	broadcast network
广角	wide angle
广角全辐射高温计	wide angle total radiation pyrometer
广义分析预估器	generalized analytical predictor (GAP)
广义建模	generalized modeling
广义李雅普诺夫函数	generated Lyapunov function
广义连接网络	generalized connection network
广义稳定判据	general stability criterion
广义误差系数	generalized error coefficients
广义线性系统	generalized linear system
广义预测控制	generalized predictive control (GPC)
广义预测控制器	generalized predictive controller
广义状态空间	generalized state space
广义最小二乘估计	generalized least squares estimation
广义坐标	generalized coordinate
归档时间周期	archive time period
归结原理	resolution principle
归纳建模法	inductive modeling method
归一法	normalization method
规范	norms
规范化状态变量	canonical state variable
规则	rule
规则库	rule-base
规则系统	rule-based system

硅	silicon（Si）
硅单结晶体管	silicon unijunction transistor
硅二极管	silicon diode
硅钢	silicon iron
硅钢板	silicon steel sheet
硅光电二极管	silicon photodiode
硅胶	silica gel
硅晶体管	silicon transistor
硅树脂	silicone resin
硅树脂油漆	silicone paint
硅酸钙绝热制品	calcium silicate insulation
硅酸盐复合绝热涂料	silicate compound plaster for thermal insulation
硅酸盐水泥	Portland cement
硅铁	silicon iron
硅酮建筑密封膏	silicone sealant for building
硅土	silica
硅烷	silane
轨道	orbit, track
轨道电路连接线	track circuit wire
轨道电子	orbital electron, planetary electron
轨道和姿态耦合	coupling of orbit and attitude
轨道交会	orbital rendezvous
轨道摄动	orbit perturbation
轨道速限	civil speed limit
轨道陀螺罗盘	orbit gyrocompass
轨迹	locus, trajectory
轨迹对接	joint trajectory
轨迹发生器	path generator
轨迹计划	trajectory planning
轨迹球	trackball
轨迹生成	path generation
辊缝测量仪	roll gap measuring instrument
辊构件枢轴	roller member pivot
辊式电极	roller electrode
辊式送料	roller feed
滚齿机	hobbing
滚刀	hob
滚道	raceway
滚动定心	roll centering
滚动摩擦	rolling friction
滚动速度	rolling velocity
滚动体	rolling element
滚动轴承	rolling bearing
滚动轴承代号	rolling bearing identification code
滚镀	barrel plating
滚光加工	barreling
滚花	knurling
滚花刀	knurling
滚花身	knurled shank
滚筒	roller
滚筒打光	barrel tumbling
滚筒筛滤器	rotary drum screen
滚筒弯曲加工	roll bending
滚型机	roll forming machine
滚修正比	modified roll
滚压	rolling
滚压加工	roll finishing
滚针	needle roller
滚针轴承	needle bearing, needle roller bearing
滚轴	roller
滚珠丝杠	ball screw
滚珠丝杠装配	ball screw assembly
滚珠轴承	ball bearing
滚珠轴承钢丝	ball-bearing wire
滚柱式单向超越离合器	roller clutch
滚柱式输送机	roller conveyor
滚转机	roller
滚子半径	radius of roller
滚子从动件	roller follower
滚子电极	roller electrode
滚子链	roller chain
滚子链条	bead chain
滚子轴承	roller bearing
锅炉	boiler
锅炉房	boiler room
锅炉炉膛安全监控系统	furnace safety supervision system（FSSS）
锅炉水位表	boiler water gauge
国际标准化组织	International Standardization Organization（ISO）
国际标准器	international standard
国际纯粹与应用化学联合会	International Union of Pure and Applied Chemistry（IUPAC）
国际单位制	International System of

Units (SI)

国际电报电话咨询委员会 Consultative Committee International Telegraph and Telephone (CCITT)

国际电工委员会 International Electrotechnical Commission (IEC)

国际电信联盟 International Telecommunications Union (ITU)

国际电信联盟-电信标准部 International Telecommunication Union - Telecommunication Standardization Sector

国际科学组织 International Science Organization (ISO)

国际实用温标 international practical temperature scale (IPTS)

国家标准 national standard

国家电气规范 national electrical code

过程报告 process reports

过程报告菜单画面 process report menu panel

过程报警 process alarm

过程报警窗口 process alarm window

过程报警功能 process alarm function

过程报警信息 process alarm message

过程变量 process variable (PV)

过程变量分类 classification of process variable

过程变量上限 process variable high limit

过程变量下限 process variable low limit

过程变量源选择 PV source selection

过程辨识 process identification

过程参数估计 process parameter estimation

过程操作员 process operator

过程测量 process measurement

过程层析成像 process tomography (PT)

过程窗口 window to the process

过程单元 process unit

过程定时器 process timer

过程断层摄影 process tomography (PT)

过程反应曲线 process reaction curve

过程反应曲线法 process reaction curve method

过程仿真 process simulator

过程负荷 process load

过程工程师 process engineer

过程管理和控制 process management and control

过程管理机 process manager (PM)

过程管理机测试执行 process manager test executive

过程管理机的控制语言 control language for process manager

过程管理机模件 process manager module (PMM)

过程和系统的历史数据 history process and system

过程集成 integrating process

过程计算机 process computer

过程接口 process interface

过程接口单元 process interface unit

过程控制 process control

过程控制工程师控制台 process engineers console

过程控制计算机 process control computer

过程控制系统 process control system

过程控制信息系统 information system for process control

过程控制语言 process-control language

过程模件 process module

过程模件点 process module point

过程模块数据点 process module data point (PMDP)

过程模型 process model

过程模型化和优化 process modeling and optimization

过程区域 process area

过程输入输出 process input output (PIO)

过程数据 process data

过程数据调理 reconciliation of process data

过程顺序控制 process-oriented sequential control

过程网络接口 process network interface

过程网络调制解调器 process network modem

过程温度 process temperature

过程相关 process correlation
过程压力 process pressure
过程仪表 process instrumentation
过程仪表及分析仪器 process instrumentation and analytics
过程优化操作 optimization in process operation
过程优化器 process optimizer
过程增益矩阵 process gain matrix
过程自动化 process automation
过冲 transient overshoot
过电流 excess current, overcurrent
过电流保护 overcurrent protection
过电流继电器 overcurrent relay
过电压 excess voltage, overtension, overvoltage
过电压保护 excess voltage protection
过电压保护装置 overvoltage protective device
过电压熔断器 overvoltage fuse
过度地 overly
过渡电流 transient current
过渡电压 transient voltage
过渡过程 transient process
过渡时间 settling time
过渡转向 oversteer
过多的缺陷 excessive defects
过范围 overrange
过负荷极限 overload limit
过激励 over-excitation
过激励放大器 overdriven amplifier
过老化 overageing
过冷 supercooling
过流 over current

过流阀 restrictor valves
过滤电子透镜 filtered electron lens
过滤电子像 filtered electron image
过滤机型号编制方法 model designation of filters
过滤器 strainer
过滤器滤芯 filter element
过热 overheat, overheating
过热保护继电器 overheat protective relay
过热的 overheated
过热感应器 excessive temperature sensor
过热器 superheater
过烧钢 burnt iron
过失误差 gross error
过压 over voltage
过压外壳 pressurized enclosure
过压型电动仪表 pressurized electrical instrument
过盈配合 interference fit
过载 overloading, over loading
过载保护 overcurrent protection
过载保护装置 overload protection device
过载继电器 overload relay
过载容量 overload capacity
过载试验 overload test
过载特性 overload characteristic
过载系数 overload factor
过早损坏 early failure
过阻尼 overdamping
过阻尼过程 overdamped process
过阻尼响应 overdamped response

H

铪　hafnium（Hf）
哈伯接地电流表　Harber earth current ammeter
哈迪分光光度计　Hardie spectrophotometer
哈考特光度计　Harcourt photometer
哈里多诺夫定理　Kharitonov theorem
哈蒙-佩尔电位差计　Hamon-Pair potentiometer
哈脱莱振荡电路　Hartley circuit
哈脱莱振荡器　Hartley oscillator
海岸带水色扫描仪　coastal zone color scanner (CZCS)
海床　seabed
海堤壕沟　seawall trench
海堤通孔　seawall opening
海底沟管出口处　submarine outfall
海底排污管　submarine outfall
海绵铁　sponge iron
海绵橡胶　sponge rubber
海沙　marine sand
海水　seawater
海水泵　seawater pump
海水冲水系统　salt water flushing system
海水抽水机　salt water pump
海水抽水站　salt water pumping station
海水抽水站入口　seawater pumping station intake
海水隔滤网　sea water screen
海水碱度　alkalinity of seawater
海水溶解氧测定仪　dissolved oxygen analyzer for seawater
海水增压泵　sea water booster pump
海水直接冷却系统　direct seawater cooling system
海水中的溶解氧　dissolved oxygen of seawater
海图基准面　chart datum
海洋采泥区　marine borrow area
海洋沉积土　marine deposit
海洋地震勘探　marine seismic prospecting
海洋调查装备　installation of oceanographic survey
海洋工程　marine works
海洋光泵磁力仪　marine optical pumping magnetometer
海洋数字地震仪　marine digital seismic apparatus
海洋振弦重力仪　marine vibrating-string gravimeter
海洋质子磁力仪　marine proton magnetometer
海洋质子梯度仪　marine proton gradiometer
海洋重力仪　marine gravimeter
亥姆霍兹共振器　Helmholtz resonator
亥姆霍兹线圈　Helmholtz coils
氦　helium（He）
氦气测量　helium survey
氦质谱检漏仪　helium mass spectrometer leak detector
含量均匀度　content uniformity
含偶数个电子　even electron
含硼水贮存箱　borated water storage tank
含奇数个电子　odd electron
含湿量　moisture content
含水率　water content
含油轴承　oil bearing
函数　function
函数的　functional
函数发生器　function generator
函数关系　function relation
函数近似　function approximation
函数卡　signal characterizer card
函数开关　function switch
函数发生　function generation
焓衰减　enthalpy relaxation
寒流捕获器　cooling trap
焊疤　crator

焊粉	welding powder
焊缝	weld bead, welded joint, welding bead, welding line
焊缝量规	weld gauge
焊缝凸起	weld flush
焊根焊道	root run
焊工	welder
焊弧	welding arc
焊剂	welding flux
焊接	weld, welding
焊接变形	welding distortion
焊接电极	welding electrode
焊接方向	welding direction
焊接痕	weld mark
焊接痕迹	welding mark
焊接机	welding machine
焊接接点	welded joint
焊接流痕	weld flow mark
焊接螺帽	weld nuts
焊接面	solder side
焊接区	weld zone
焊接纹	weld line
焊接周期	welding interval
焊料	solder
焊丝	welding wire
焊条	filler rod, welding electrode, welding rod, welding wire
焊条钢丝	electrode wire
焊条极性	electrode polarity
焊条芯	core wire
焊条直径	core diameter
焊透深度	weld penetration
焊锡	tin solder
焊锡面	solder side
焊线	bonding wire
焊渣	welding slag
焊珠	welding bead
行编辑程序	line editor
行扫描	line scanning
行式打印机	line printer
行业技能测验	trade test
航测图	aerial view
航道	navigation channel, navigation waterway
航道净空	navigation clearance
航海系统	marine system
航空磁通门磁力仪	airborne flux-gate magnetometer
航空飞行器	aircraft
航空伽玛辐射仪	airborne gamma radiometer
航空伽玛能谱仪	airborne gamma spectrometer
航空工程	aerospace engineering
航空光泵磁力仪	airborne optical pumping magnetometer
航空轨迹	aerospace trajectory
航空航天空间	aerospace
航空集散站	air terminal
航空计算机控制	aerospace computer control
航空控制	aerospace control
航空气象记录仪	aerograph
航空遥感	aerial remote sensing
航空宇航科学与技术	aeronautical and astronautical science and technology
航空宇航器制造工程	manufacturing engineering of aerospace vehicle
航空宇航推进理论与工程	aerospace propulsion theory and engineering
航空照相机	aerial camera
航空质子磁力仪	airborne proton magnetometer
航空重力测量	aerogravity survey
航空重力仪	air-borne gravimeter
航摄仪	aerial surveying camera
毫（千分之一）	milli (m)
毫安（电流单位）	milliampere
毫安表	milliammeter
毫伏（电压单位）	millivolt
毫伏安培计	millivolt ammeter
毫伏计	millivoltmeter
毫米波磁控管	millimetric wave magnetron
毫升	milliliter (ml)
毫瓦（功率单位）	milliwatt
耗电量	electrical power consumption
耗气量	air consumption
耗散参数	scattering parameter
耗散功率	dissipation power
耗散结构	dissipative structure
耗散问题	scattering problem
耗油量	fuel consumption, oil consumption

耗油量系数　oil consumption factor
合并标准偏差　pooled standard deviation
合并载重　combined load
合成　synthesis
合成半导体　compound semiconductor
合成进给运动　resultant movement of feed
合成切削运动　resultant movement of cutting
合成切削运动方向　direction of resultant movement of cutting
合成双绞线　compound twisted wire
合成弯矩　resultant bending moment
合成物　synthetic
合成误差　composite error
合成纤维　synthetic fibre
合成橡胶　neoprene
合成信号发生器　synthesized signal generator
合成一体的　incorporated
合法证明　formal verification
合格　up to grade
合格标志　mark of conformity
合格率　yield factor
合格人员　qualified person
合格认证　conformity certification
合格证书　certificate of conformity
合计的　total
合金　alloy
合金钢筋条　alloy steel bar
合金钢螺帽　alloy steel nuts
合金工具钢　alloy tool steel
合金结构钢丝　alloy structural steel wire
合金铸铁　alloy cast iron
合力　resultant force
合力矩　resultant moment of force
合模　die assembly, spotting
合模锻造　closed-die forging
合模机　die spotting machine
合模面　land area
合谱　combinations
合适　pertain
合约图纸　contract drawing
合约细则　article of agreement

合闸　switch on
合闸磁铁　closing magnet
合闸动作　feed motion
合闸继电器　closing relay
合闸旋钮　switching knob
合作对策　cooperative game
和子系统通信的可选软件包　option pa-ckages for communicating with subsystem
核　nucleus
核磁共振　nuclear magnetic resonance (NMR)
核磁共振波谱　NMR spectrum
核磁共振波谱法　NMR spectroscopy
核磁共振分光计　nuclear magnetic resonance spectrometer
核磁共振流量计　nuclear magnetic resonance flowmeter
核磁共振装置　NMR equipment
核电　nuclear power
核电站　nuclear power plant, nuclear power station
核对　verification
核对点　check pointing
核对色谱法　iteration chromatography
核反应堆　nuclear reactor
核辐射物位计　nuclear radiation level meter
核工厂　nuclear plant
核技术及应用　nuclear technology and applications
核间双共振法　internuclear double resonance
核科学与技术　nuclear science and technology
核能科学与工程　nuclear energy science and engineering
核燃料循环与材料　nuclear fuel cycle and materials
核实验　nuclear tests
核糖核酸　RNA
核物理　nuclear physics
核准　approved by
核子　nucleon
荷载测试　load testing

荷载强度　loading intensity
荷载状况　loading condition
盒　box
盒式磁带　cartridge tape, cassette
盒式磁盘　cartridge disk
盒式磁盘机　cartridge disk drive
盒式盘或软盘　removable media
赫兹　cycle per second, hertz (Hz)
赫兹公式　hertz equation
赫兹振荡器　hertz oscillator
黑钢皮　black sheet iron
黑光探伤器　black light crack detector
黑色　black
黑色导线　black wire
黑体　blackbody
黑体炉　blackbody furnace
黑体腔　blackbody chamber
黑匣　black box
黑箱　black box
黑箱测试法　black box testing approach
黑箱模型化　black box modeling
黑心可锻铸铁　black-heart malleable iron
黑烟　black smoke
亨利（电感单位）　henry
恒定带宽滤波器　constant bandwidth filter
恒定电压　constant voltage
恒定速率注入法　constant-rate injection method
恒定误差　constant error
恒定压头流量计　constant head flowmeter
恒定状态　constant state
恒荷载　dead load
恒节流孔　fixed restrictor
恒流变压器　constant-current transformer
恒流电源　constant current power supply
恒流可调信号源　constant current adjustable signal source
恒流特性　constant-current characteristic
恒流调节　constant-current regulation
恒流源　constant-current source
恒频振荡器　constant-frequency oscillator
恒容高位发热量　heating quantity of constant capacity on high position
恒湿箱　humidistat
恒湿状态　humidity constant state
恒速操纵器　cruise control
恒温池　constant temperature bath
恒温锻造　isothermal forging
恒温干燥箱　constant temperature drying ovens
恒温恒湿器　constant temperature and humidity chambers
恒温培养器　constant temperature incubators
恒温器　thermostat
恒温式热量计　isothermal point
恒温水槽　constant temperature water baths
恒温箱　temperature chamber
恒温循环泵　constant temperature circulator
恒温装置　constant temperature devices
恒稳直流电　steady direct current
恒向电流　continuous current
恒压电源　constant voltage power supply
恒压恒频　constant voltage constant frequency (CVCF)
恒压器　barostat, manostat
恒压源　constant voltage source
恒液位压头容器　constant-level head tank
恒值控制系统　fixed set-point control system
恒重温度计　weight thermometer
珩磨机　honing machines
桁条　purlin
横隔板　diaphragm plate
横河公司计算机系统　Yokogawa computer
横截面　cross section
横截面积　cross sectional area
横截面内的平均动压　mean dynamic pressure in a cross-section
横梁　cross-beam
横丝　horizontal wire
横纹　cross grain

横向承托工程	lateral support works
横向承托系统	lateral support system
横向尺寸	lateral dimension
横向定位凹槽	lateral positioning groove
横向分解	horizontal decomposition
横向风力	transverse wind
横向干扰	transverse interference
横向滑动闸门	horizontal sliding gate
横向滤波器	transversal filter
横向搜索	breadth-first search
横向稳定性评议	lateral stability assessment
横向限动块	lateral stopper
横向移动	lateral movement
横向约束	horizontal restraint
横向直径	horizontal diameter
横斜度	crossfall
横轴	axis of abscissas
横坐标	abscissa
横坐标轴	abscissa axis
衡准仪	balance level
烘干机	dryer
烘干密度	oven-dried density
红丹	red lead
红外测距仪	infrared distance meter
红外分光光度计	infrared spectrophotometer
红外辐射计	infrared radiometer
红外辐射温度计	infrared radiation thermometer
红外光传感器	infrared light sensor, infrared light transducer
红外光电开关	infrared photoelectric switch
红外光谱法	infrared spectrometry
红外检测器	infrared detector
红外检测仪	infrared radiation detection apparatus
红外湿度表	infrared hygrometer
红外吸收光谱法	infrared spectroscopy
红外吸收光谱湿度计	infrared absorption spectra hygrometer
红外线	infrared ray (IR)
红外线保安系统	infra-red security system
红外线电子测距仪	infra-red electronic distance meter
红外线气体分析器	infrared gas analyzer, infrared type gas analyzer
红外线探伤法	infrared radiometry
红外线照射	infrared radiation
红外遥感	infrared remote sensing
红外照相机	infrared camera
红氧化铁底漆	red oxide primer
红移	red shift
宏	macro
宏汇编语言	macro assembly language
宏指令	macro instruction
虹吸管	siphon
后板	rear plate
后备串级	backup cascade
后备电源	backup power source
后备发电机	standby generator
后备寄存器	background register
后备水泵	standby water pump
后部触点	back contact
后存	post store
后端	rear
后盖	end cap
后固化	post cure
后滚翻	flip-chip
后加拉力	post-tensioning
后角	clearance angle
后模镶针	core pin
后台程序	background program
后台处理	background processing
后台作业	background job
后续的	consequent
后续检定	subsequent verification
后验估计	posteriori estimate
后张钢筋束	post-tensioned tendon
后张混凝土梁	post-tensioned concrete beam
后张预应力	post-tensioned prestressing
后置放大器	post amplifier
后柱反应	post-column reaction
后柱反应器	post-column reactor
厚薄规	feeler gauge, thickness gauge
厚度	thickness
厚度层铁皮	heavy iron
厚度计	thickness meter
厚膜	thick film
厚膜电路	thick film circuit

厚膜放大器　thick film amplifier
厚膜微调电位器　thick film trimming potentiometer
呼叫　calling
呼叫继电器　calling relay
呼叫设备　calling device
呼叫信号　calling signal
呼吸机　respirator
呼吸器　breather
呼吸器具　breathing apparatus
弧齿锥齿轮　spiral bevel gear
弧度　radian (rad)
弧线　circular thickness
弧线厚度　circular thickness
弧形闸门通用技术条件　general technical requirements for radial gates
弧阻　arc resistance
互补测量　complementary measurement
互补单结晶体管　complementary unijunction transistor
互补的　complementary
互补对称电路　complementary symmetry circuit
互补冠状齿轮　complementary crown gears
互补金属氧化物半导体　complementary metal oxide semiconductor
互补晶体管逻辑电路　complementary transistor logic circuit
互补问题　complementarity problem
互导纳　mutual admittance
互电导　mutual conductance
互干扰　mutual interference
互感　mutual-inductor
互感系数　coefficient of mutual inductance
互换性　interchangeability
互换性齿轮　interchangeable gears
互联　interconnection
互联外围设备　peripheral component interconnect (PCI)
互联网络　interconnection network
互联系统　interacted system
互联预估法　interactive prediction approach
互调失真仪　intermodulation distortion meter
互通量　mutual flux
互相参考　crosstalk interference
互相关函数　cross-correlation function
互相关流量计　cross-correlation flowmeter
互易定理　reciprocity theorem
互易原理　principle of reciprocity
互阻抗　mutual impedance
户外　outdoor
户外设备　outdoor equipment
护轨　guard rail
护栏　barrier
护轮轨　check rail
护面罩　face shield
护目镜　goggle, protective goggle, safety goggle, visor
护目面罩　safety visor
护圈　collar
护舷桩　fender pile
护眼屏　eye screen
护眼罩　eye shield
护罩　protective guard
护柱基座　bollard plinth
护桩　fender pile
戽式输送机　scoop conveyor
花键　multiple keys
花线　flexible wire
华氏度　Fahrenheit, degree Fahrenheit
华氏温标　Fahrenheit temperature scale
滑车轮　sheave
滑尺　slide gauge
滑触电阻线　slide wire
滑动　slide, sliding
滑动表面　sliding surface
滑动开关　slide switch
滑动力　sliding force
滑动率　sliding ratio
滑动模板　slip form
滑动摩擦　sliding friction
滑动耦合器　slide coupler
滑动平衡器　slide balancer
滑动曲线　sliding curve
滑动速度　sliding velocity
滑动系数　sliding coefficient
滑动轴承　sliding bearing

滑阀	slide valve
滑轨	slide rail
滑环电动机	slip-ring motor
滑键	feather key
滑接馈线	trolley wire
滑块	slider, sliding block, cam block, die block
滑块固定块	sliding dowel block
滑块联轴器	oldham coupling
滑料架	sliding rack
滑轮	pulley
滑轮组	pulley block
滑模	sliding mode
滑模控制	sliding mode control (SMC)
滑配接头	slip joint
滑石	talc
滑石粉	french chalk
滑台	moving table
滑线	slide wire
滑线变阻器	rheochord
滑线电阻	slide rheostat
滑移扭矩	slip torque
滑枕式滚珠轴承	pillow ball bearing
化工变量控制	chemical variables control
化工过程机械	chemical process equipment
化学测试	chemical test
化学传感器	chemical sensor
化学电池	chemical cell
化学电镀	chemical plating
化学电离	chemical ionization
化学发光检测	chemiluminescence detection
化学发光仪	chemiluminescence apparatus
化学反应器	chemical reactor
化学分析	chemical analysis
化学工程	chemical engineering
化学工程与技术	chemical engineering and technology
化学工业	chemical industry
化学工艺	chemical technology
化学灌浆	chemical grout
化学光度计	actinometer
化学计量点	stoichiometric point
化学计量学	chemometrics
化学剂量	chemical dosing
化学键合相色谱法	bonded phase chromatography
化学键合相填充剂	chemically bonded phase packing
化学垃圾	chemical refuse
化学离子源	chemical ionization source
化学品	chemicals
化学双电层	chemical double layer
化学特性	chemical property
化学推进	chemical propulsion
化学微型传感器	chemical microsensor
化学位移	chemical shift
化学蒸镀	chemical vapor deposition
化学作用	chemical action
化油剂	dispersant
划痕硬度	scratch hardness
划片式流量计	slide vane flowmeter
划线	marking out, scribing
划线刀	marking knife
划线工具	marking tool
划线规	marking gauge
划线轮	stitch marker
划线盘	tosecan
划线器	scriber
画法几何学	descriptive geometry
画面编辑程序	picture editor
画面名称	panel name
画面注释	panel comment
坏料	billet
坏值	bad value
坏值检查器	check for bad
还原剂	reducer
还原性介质	reductant
环称压力计	ring balance manometer
环大小效应	ring size effect
环带状的电热器	band heater
环规	ring gauge
环糊精	cyclodextrin
环糊精动电色谱	cyclodextrins electrokinetic chromatography
环糊精色谱法	cyclodextrin chromatography
环境	environmental
环境保护工程	environmental engineering
环境参数	environmental parameter

环境成分分析仪　CHN Analysis
环境工程　environmental engineering
环境工程学　environmental engineering
环境规范　environmental specification
环境监测　environmental test
环境监测站　environmental monitor station
环境建筑学　environment architectures
环境科学　environmental science
环境科学与工程　environmental science and engineering
环境美化工程　landscaping works
环境气体分析仪　environmental gas analyzer
环境区域　environmental area, environmental location
环境湿度范围　ambient humidity range
环境试验　environmental test
环境条件　environmental condition
环境条件影响　environmental influence
环境调节　environmental regulation
环境温度　ambient temperature
环境温度范围　ambient temperature range
环境稳定　environmental stability
环境污染　environmental pollution
环境误差　environmental error
环境压力　ambient pressure
环境压力误差　ambient pressure error
环境因素　environmental factor
环境应力龟裂试验　environmental stress cracking test
环境影响评估　environmental impact assessment
环境噪声　ambient noise, environmental noise
环境振动　ambient vibration
环流　circular current, loop current, ring current
环路　loop circuit
环路测试　loop test
环路控制　loop control
环面蜗杆　toroid helicoids worm
环耦合型衰减器　loop-couple-mode attenuator
环室　annular chamber

环首螺帽　eye nuts
环首螺栓　eye bolt
环线　loop wire
环形齿轮　ring gear
环形磁路　ring magnetic circuit
环形电流　circulating current
环形电路　ring circuit
环形电线　coil aerial
环形钢筋混凝土电杆　circular reinforced concrete pole
环形间隙　annular space
环形浇口　ring gate
环形螺帽　round nuts
环形盘　annular pad
环形燃烧管灭菌器　loop cinerator
环形弹簧　annular spring
环形线圈　encircling coil
环形线圈间隙　annular coil clearance
环形预应力混凝土电杆　circular prestressed concrete pole
环形展开　circular development
环氧玻璃布基覆铜箔板　epoxide woven glass fabric copper-clad laminates
环氧玻璃布纸复合覆铜箔板　epoxide cellulose paper core glass cloth surfaces copper-clad laminates
环氧玻璃基板　epoxy glass substrate
环氧当量　weight per epoxy equivalent
环氧酚醛　epoxy novolac
环氧合成纤维布覆铜箔板　epoxide synthetic fiber fabric copper-clad laminates
环氧煤焦油　coal-tar epoxy
环氧漆　epoxy paint
环氧树脂　epoxy resin
环氧树脂浆　epoxy resin grout
环氧树脂胶　epoxy glue
环氧树脂密封钢筋　epoxy-coated reinforcement
环氧树脂黏合剂　epoxy adhesive
环氧树脂搪层　epoxy resin coat
环氧值　epoxy value
环氧纸质覆铜箔板　epoxide cellulose paper copper-clad laminates
环状外对流敏感元件　external-convection ring sensitive element
缓冲　shock-absorber

缓冲存储器　buffer storage
缓冲地区　buffer area
缓冲电路　buffer circuit
缓冲放大器　buffer amplifier
缓冲级　buffer stage
缓冲块　bumper block
缓冲路拱　speed hump
缓冲器　bumper, dashpot, impact damper
缓冲区　buffer
缓冲溶液　buffer solution
缓冲弹簧　cushioning spring
缓冲销　cushion pin
缓解阀　graduated release valve
缓凝剂　retarder
缓跑径　jogging track
换流器　inverter
换路定律　switching law
换码误差　commutation error
换模　die change
换能器　energy converter, transducer
换算系数　reduction factor
换算因子　conversion factor
换相器　phase changer
换向磁场　commutating field
换向磁阻电机　switched reluctance motor
换向电流　current of commutation
换向电容滤波器　switched capacitor filter
换向电容器　switched capacitor
换向极　commutating pole
换向开关　reversing switch
换向器　commutator
换向器变频机　commutator frequency changer
换向器电动机　commutator motor
换向器-电刷总线　commutator-brush combination
换向器感应电动机　commutator induction motor
换向绕组　commutating winding
换向损耗　commutation loss
换向线圈　commutating coil
换向状况　commutation condition
黄色闪灯　amber flashing light
黄铜　brass

黄铜闸门阀　brass gate valve
黄炸药　dynamite
灰尘　dust
灰口铁　graphitic pig iron, gray cast iron, gray pig iron
灰口铸铁　gray cast iron, gray pig iron, grey cast iron, grey pig iron
灰泥板　plaster board
灰砂斗　hod
灰体　graybody
灰铸铁　gray cast iron, gray pig iron, grey cast iron
恢复　recovery
恢复供应　reinstatement of supply
恢复归档定义　restore archived definition
恢复归档文件　restore archive file
恢复时间　recovery time
恢复事件描述　restore event descriptor
恢复事件数据　restore event data
恢复已存档案　restore saved archive
挥发法　volatilization method
挥发物　volatile
挥发性　volatility
辉光放电　glow discharge
辉光放电管　glow-tube
辉面电镀　bright electroplating
回比矩阵　return ratio matrix
回波　echo
回波参数放大器　backward-wave parametric amplifier
回差　return difference
回差比　return difference ratio
回差大　exceeded variation
回差矩阵　return difference matrix
回程　return
回程误差　hysteresis error
回风进口　return air inlet
回复时间　reset time
回复原状　reinstatement
回归　regression
回归分析　regression analysis
回归估计　regression estimate
回归关系　regression relationship
回归算法系统　regression algorithm system

回火 tempering	回位弹簧 return spring
回火脆性 temper brittleness	回位销 return pin
回火裂痕 tempering crack	回线数据 retrieve data
回火颜色 temper colour	回响 reverberation
回火制止器 flash back arrester	回旋加速器 cyclotron
回馈制动 regenerative braking	回旋交通 gyratory traffic
回流 contra-flow	回油 return oil
回流层现象 back layering	回油管 oil return tube
回流阀 reflux valve	回原键 reset key
回路 loop	回转泵 rotary pump
回路传递函数 loop transfer function	回转底座 swinging base
回路电流 loop current	回转锻造 rotary forging
回路电流法 loop method	回转接头 swivel joint
回路级模块 loop class module	回转面 surface of revolution
回路连接定义 loop connection definition	回转器 gyrator, gyroscope
回路连接状态显示画面 loop connection status display panel	回转式吊机 slewing crane
	回转式密封装置 rotating seal
回路弱相关 weak interaction of loops	回转体 revolution
回路适配器 loop adapter	回转体平衡 balance of rotors
回路手动 loop manual	回转弯曲机 rotary bender
回路通信单元 loop communication unit	回转压碎机 gyratory crusher
	汇编程序 assembler
回路通信卡 loop communications card	汇编机 assembler
回路显示 loop display	汇编语言 assembler language, assembly language
回路相角 loop phase angle	汇流排 busbar
回路相位特性 loop phase characteristic	汇流条 bus-bar
回路增益 loop gain	会话监督系统 conversational monitoring system
回路增益特性 loop gain characteristic	会刊 transaction
回路整定 loop tune	会议电视系统 video conferphone system
回路转换器 loop transfer	会议记录 meeting minutes
回路状态 loop status	绘图 drawing
回扫脉冲电源 kickback power supply	绘图板 drawing board
回声测深仪 echo sounder	绘图材料 drawing materials
回声测探接收器 echo sounding receiver	绘图格式 drawing format
	绘图机 drafting machine, plotter
回声探测装置 sounding device	绘图技巧 drawing technique
回声验潮仪 acoustic tide gauge	绘图铅笔 drawing pencil
回收箱 receive tank	绘图仪 plotting device
回授 back feed	绘图仪器 drawing instruments
回水箱 recovery tank	绘图纸 drawing paper
回送检验 loopback checking	惠斯通电桥 Wheatstone bridge
回弹性 resilience	惠斯通电桥记录仪 Wheatstone bridge recorder
回填 backfill	
回填材料 backfill material	惠特克·香农采样定理 Whittaker-
回跳硬度 shore hardness	

Shannon sampling theorem
混沌　chaos
混沌控制　chaotic control
混沌理论　chaos theory
混沌时间序列　chaotic time series
混沌特性　chaotic behaviour
混沌系统　chaotic system
混合　commix, mixture
混合 PI 调节器　blending PI controller
混合比　mixing ratio
混合比例控制　blend ration control
混合参数　hybrid parameter
混合长度　mixing length
混合电路　hybrid circuit
混合仿真　hybrid simulation
混合机　kneader
混合计算机　hybrid computer
混合控制　blending control
混合连接　hybrid junction
混合料　compound
混合灵敏度问题　mixed sensitivity problem
混合轮系　compound gear train
混合面板功能块　hybrid faceplate block
混合气体热导率　thermal conductivity of mixture gas
混合驱动汽车　hybrid vehicle
混合绕组电流互感器　compound-wound current transformer
混合润滑　mixed lubrication
混合设计　mix design
混合水泥　blended cement
混合调节器　blending controller
混合温度　mixing temperature
混合物　mixture
混合系统　blending system
混合效应　combined effect
混合信号　composite signal
混合用料比例　mix proportion
混合指示剂　mixed indicator
混合柱　mixed column
混合桩　composite pile
混联　hybrid connection
混流泵　mixed flow pumps
混凝管　concrete pipe
混凝土　concrete
混凝土板　concrete slab
混凝土拌和厂　concrete mixing plant
混凝土保护层　concrete cover
混凝土标号　concrete grade
混凝土剥落　concrete spalling
混凝土测试　concrete test
混凝土掺和料　concrete admixture
混凝土厂砖　ready-mix plant
混凝土衬砌　concrete lining
混凝土导管　tremie
混凝土等级　concrete grade
混凝土工艺　concrete technology
混凝土轨枕　concrete sleeper
混凝土海堤　concrete block seawall
混凝土和钢筋混凝土排水管　concrete and reinforced concrete drainage and sewer pipes
混凝土缓冲　concrete buffer
混凝土混合机　concrete mixer, batching plant
混凝土混合物　concrete mix
混凝土基脚　concrete plinth
混凝土建造物　concrete structure
混凝土浇灌工程　concrete pour works
混凝土结构　concrete structure
混凝土科technology　concrete technology
混凝土立方块　concrete cube
混凝土路面　rigid pavement
混凝土路面振动整实器　concrete vibrator
混凝土耐用程度　concrete durability
混凝土强度　concrete strength
混凝土墙　concrete wall
混凝土外加剂　concrete admixture
混凝土芯　concrete core
混凝土样本　concrete sample
混凝土应力　concrete stress
混凝土再碱性化　concrete re-alkalization
混凝土振动器　concrete vibrator
混凝土砖　concrete block
混凝土桩　concrete pile
混频　mixing
混频器　mixer
混杂模型　hybrid model
混杂系统　hybrid system
混浊度　turbidity
活扳手　adjustable wrench

| 活的　active
| 活动扳手　adjustable spanner
| 活动板　active plate
| 活动臂　lever arm
| 活动标度测量仪表　moving scale measuring instrument
| 活动部分　moving part
| 活动衬套　loose bush
| 活动的　moving
| 活动工作台　moving bolster
| 活动构架　movable frame
| 活动模板　floating platen
| 活动式海底生物呼吸测量器　free vehicle respirometer
| 活动式模具　loose mold
| 活动铁芯式继电器　moving core type relay
| 活度系数　activity coefficient
| 活荷载　live load
| 活化剂　activator
| 活化作用　activation
| 活节联轴器　coupling joint
| 活裂缝　live crack
| 活零件模具　loose detail mold
| 活区　live zone
| 活塞　piston, stopcock
| 活塞阀　piston valve
| 活塞连杆　piston rod
| 活塞式阀　plunger valve
| 活塞式压力计　piston type pressure gauge
| 活塞式压缩机　reciprocating compressor
| 活时间　live time
| 活性腐蚀　active corrosion
| 活性金属指示电极　active metal indicated electrode
| 活性氧化　active oxidation
| 火车控制　train control
| 火花　spark
| 火花点火器　spark lighter
| 火花放电　spark discharge
| 火花间隙　spark gap
| 火警探测　fire detection
| 火炮、自动武器与弹药工程　artillery, automatic gun and ammunition engineering

火漆　sealing wax
火山岩　volcanic rock
火石点火器　flint gun
火线　live conductor
火线带电线　fire wire
火星塞高压线　plug wire
火焰处理　flame treatment
火焰传播　flame propagation
火焰发射光谱法　flame emission spectrometry
火焰光度检测　flame photometric detection
火焰光度检测器　flame photometric detector (FPD)
火焰监测器　flame monitor
火焰警报器　flame failure alarm
火焰离子化检测　flame ionization detection
火焰离子化检测器　flame ionization detector (FID)
火焰切割　flame cutting, torch-flame cut
火焰热离子检测　flame thermionic detection
火焰探测器　flame detector
火焰温度检测器　flame temperature detector
火焰硬化　flame hardening
火灾报警系统　fire alarm system
火灾探测　fire detection
钬　holmium (Ho)
或　OR
或电路　OR circuit
或非　NOR
或非电路　NOR circuit
或非门　NOR gate
或非算子　NOR operation
或非元件　NOR element
或门　OR gate
或运算　OR-operation
货车吊机　lorry crane
货物处理　freight handling
货物装卸区　cargo handling area
货运场　freight yard
货运调车设施　freight marshalling facilities
霍尔　Hall
霍尔常数　Hall constant

霍尔传感器　Hall element
霍尔丹空气分析仪　Haldane's apparatus
霍尔电导检测器　Hall conductivity detector
霍尔电解质电导率检测　Hall electrolytic conductivity detection
霍尔式位移传感器　Hall displacement transducer
霍尔式压力传感器　Hall type pressrue transducer
霍尔探头　Hall probe
霍尔效应　Hall effect
霍尔效应倍增器　Hall effect multiplier
霍尔效应磁力计　Hall effect magnetometer
霍尔效应磁通计　Hall effect fluxmeter
霍尔效应器件　Hall effect device
霍尔效应式转速传感器　Hall effect tachometric transducer
霍尔效应位移传感器　Hall effect displacement transducer
霍尔效应线性检测器　Hall effect linear detector
霍尔元件　Hall element
霍夫曼编码　Huffman code
霍尼韦尔验证测试系统　Honeywell verification test system

J

击穿　disruptive
击穿电压　breakdown voltage, disruptive voltage
击穿二极管　breakdown diode
击穿试验　breaking-down test
机车　locomotive
机车柴油机　locomotive diesel engine
机车动力学　vehicle dynamics
机车仿真器　vehicle simulator
机车管理　energy management
机车空气动力学　vehicle aerodynamics
机车库　locomotive shed
机床　machine tool
机床切削平面　machine plane
机床调整卡　summary of machine settings
机床中心至工件安装基准面　machine center to back
机电的　electromechanical
机电模拟方法　method of electro-mechanical analogy
机电式液位指示器　electromechanical liquid level indicator
机电调节器　electromechanical regulator
机电一体化　mechanical-electrical integration
机电一体化系统设计　mechanical-electrical integration system design
机顶盒　set top box (STB)
机动　engine driven
机动泵　engine driven pump
机动性　manoeuvrability
机房　machine room, plant room
机构分析　analysis of mechanism
机构和零部件　mechanism and component parts
机构平衡　balance of mechanism
机构运动简图　kinematic sketch of mechanism
机构运动设计　kinematic design of mechanism
机构综合　synthesis of mechanism
机构组成　constitution of mechanism
机柜接地　cabinet grounding
机架　frame fixed link
机架安装　rack-mount
机架变换　kinematic inversion
机架检查程序　frame-check sequence
机壳　body
机壳接地　chassis earth, rack earth
机理模型　mechanism model
机器代码　machine code
机器检查指示仪　machine check indicator
机器描述格式　machine description format (MDF)
机器描述格式数据库　MDF database
机器人　robot
机器人编程　robot programming
机器人编程语言　robot programming language
机器人操纵器　robotic manipulator
机器人导航　robot navigation
机器人动力学　robot dynamics
机器人控制　robot control
机器人视觉　robot vision
机器人学　robotics
机器人运动学　robot kinematics
机器识别　machine recognition
机器学习　machine learning
机器语言　machine language
机器智能　machine intelligence
机身构架拉线　fuselage truss wire
机箱　crate
机箱地址　crate address
机箱控制器　crate controller
机械　machine
机械臂　robot arm
机械操纵器　mechanical manipulator
机械测振仪　mechanical vibration meter
机械冲击　mechanical shock

机械创新设计　mechanical creation design
机械的　mechanical
机械电子工程　mechatronic engineering
机械动力分析　dynamic analysis of machinery
机械动力设计　dynamic design of machinery
机械动力学　dynamics of machinery
机械工程　mechanical engineering
机械共振　mechanical resonance
机械加工　machining
机械加工余量　machining allowance
机械加速度计　mechanical accelerometer
机械金属探伤器　mechanical stethoscope
机械锯床　sawing machine
机械利益　mechanical advantage
机械联动装置　mechanical linkage
机械量　mechanical quantity
机械量传感器　mechanical quantity sensor, mechanical quantity transducer
机械零件　mechanical parts
机械零位　mechanical zero
机械零位调节器　mechanical zero adjuster
机械螺丝用六角螺帽　hex machine screw nut
机械能守恒　conservation of mechanical energy
机械平衡　balance of machinery
机械强度　mechanical strength
机械扫查声全息　acoustical holography by mechanical scanning
机械设计　mechanical design
机械设计与理论　mechanical design and theory
机械湿度计　mechanical hygrometer
机械式上皿天平　mechanical top-loading balance
机械式深温计　mechanical bathythermograph（MBT）
机械式试验机　mechanical testing machine
机械式岩矿密度仪　machine type rock ore densimeter
机械式整流器　mechanical rectifier
机械手校准　robot calibration
机械套管　mechanical coupler
机械特性　mechanical behavior
机械调节器　mechanical regulator
机械调速　mechanical speed governors
机械通风　mechanical ventilation
机械脱出　mechanical runout
机械无级变速　mechanical stepless speed changes
机械系统　mechanical system
机械系统设计　mechanical system design
机械效率　mechanical efficiency
机械性能　mechanical capacity, mechanical property
机械性能测定仪　mechanical property tester
机械性能试验　mechanical test
机械压线钳　crimping tools
机械应变　mechanical strain
机械应力　mechanical stress
机械原理　theory of machines and mechanisms
机械振动器　mechanical vibrator
机械制动器　mechanical brake
机械制图　mechanical drawing
机械制造及其自动化　mechanical manufacture and automation
机械中心　machine center
机械装置　machinery
机械阻抗　mechanical impedance
机翼　aerofoil
机油　engine oil
机油平面　oil level
机载红外光谱辐射计　airborne infrared spectroradiometer
机罩　hood
机组运行小时表　genset running hour meter
机组运行正常　normal running
机座　bed
肌电控制　myoelectric control
肌电图　electromyography（EMG）
肌电图传感器　electromyographic sensor, electromyographic transducer
唧嘴口径　sprue diameter

积层磁铁　laminated magnet
积层多层印制板　build-up multilayer printed board
积层印制板　build-up printed board
积分　integral
积分饱和　integral windup
积分表示法　integral representation
积分测量仪表　integrating measuring instrument
积分电路　integrating circuit
积分反馈　integral feedback
积分方程　integral equation
积分方程式　integral equation formulation
积分放大器　integrating amplifier
积分公式　integral formulation
积分检波　quadrature detection
积分检测器　integral detector
积分镜像滤波器　quadrature mirror filter
积分控制　integral control
积分控制器　I controller, integral controller
积分控制作用　integral control action
积分偏差判据　integral error criteria
积分器　integrating instrument, integrator
积分球　integrating sphere
积分时间　integral time
积分时间常数　integral time constant
积分式电动执行机构　integral electric actuator
积分式激发电位仪　integrating induced type polarization potentiometer
积分式记录仪　integrating recording instrument
积分速率　integral action rate
积分调节器　reset controller
积分调节作用　reset control action
积分性能指标　integral performance index
积分性能准则　integral performance criterion
积分转换　integrating conversion
积分作用　integral action
积分作用时间　integral action time
积分作用时间常数　integral action time constant
积分作用系数　integral action coefficient
积分作用限幅器　integral action limiter
积分作用因子　integral action factor
积复励　cumulative compound excitation
积复励电动机　cumulative compound motor
积复励发电机　cumulative compound generator
积复励绕组　cumulative compound winding
积算　integrating
积算法　integration method
积算仪表　integrating meter
积算仪器　integration instrument
基　radical
基本操作站　basic operation station
基本测量单位　base unit of measurement
基本处理单元　basic processing unit
基本处理周期　basic processing cycle
基本点　primary site
基本电路　basic circuit
基本固有振型　fundamental natural mode of vibration
基本过程　fundamental process
基本核磁共振频率　basic NMR frequency
基本控制器　basic controller
基本量　base quantity
基本频率　fundamental frequency
基本输出子系统　basic output subsystem
基本输入子系统　basic input subsystem
基本算法模块　basic arithmetic module
基本物理量测定　basic physics
基本误差　intrinsic error
基本误差限　limit of intrinsic error
基本型端子　basic terminal
基本修正　matrix correction
基本应用软件　base application software (BAS)
基本周期　fundamental period
基波　fundamental wave

基波电流　fundamental current
基波因数　fundamental factor
基波振幅　amplitude of first harmonic
基材　base material
基层岩　bedrock
基础标准　basic standard
基础测量法　fundamental method of measurement
基础单元　base unit
基础灌浆技术　foundation grouting technique
基础机构　fundamental mechanism
基础图则　foundation plan
基带同轴电缆　baseband coaxial cable
基底　substrate
基底结构　floor structure
基地式差压流量变送器　integral flow orifice differential pressure transmitter
基地式调节仪表　local-mounted controller
基尔霍夫电流定律　Kirchhoff's current law
基尔霍夫电压定律　Kirchhoff's voltage law
基尔霍夫定律　Kirchhoff's law
基峰　base peak
基极　base electrode
基极信号发生器　base signal generator
基架　base frame
基架绝缘器　base insulator
基节距　base pitch
基流　background current
基体材料　basis material
基体效应　matrix effect
基线　baseline, base line
基线计划　baseline programme
基线漂移　baseline drift
基线噪声　baseline noise
基因检查仪器　gene pattern analyzer
基于规则控制　rule-based control
基于技巧生产　skill-based production
基于技巧系统　skill-based system
基于模型控制　model-based control (MBC)
基于模型识别　model-based recognition
基于实例设计　case-based design

基于微机控制　microcomputer-based control
基于微机系统　microcomputer-based system
基于知识控制　knowledge-based control
基于知识自适应控制　knowledge-based adaptive control
基圆　base circle
基圆半径　base radius, radius of base circle
基圆齿距　base pitch
基圆螺旋角　base spiral angle
基圆压力角　pressure angle of base circle
基圆直径　base diameter
基圆柱　base cylinder
基圆锥　base cone
基址寄存器　base register
基准　benchmark
基准变量　reference variable
基准尺　reference meter
基准齿　datum tooth
基准齿条　basic rack
基准点　reference point
基准电平　reference level
基准电压　reference voltage, voltage reference
基准电压发生器　pedestal generator
基准记号　datum mark, reference mark
基准结　reference junction
基准结补偿器　reference junction compensator
基准精确度　reference accuracy
基准输入元件　reference input element
基准误差　datum error
基准线　baseline
基准信号　reference signal
基准轴　axis of reference
基座　chassis, plinth
基座坐标系　base coordinate system
畸变功率　distortion power
畸变衰减器　distortion pad
畸峰　distorted peak
激磁电流　exciting current
激磁绕组输入阻抗　input impedance of the magnetizing winding

激发　stimulus
激发光谱　excitation spectrum
激发指数　excitation index
激光　laser
激光测距仪　laser distance meter
激光测云仪　laser ceilometer
激光传感器　laser sensor
激光垂准仪　laser plummet
激光导向仪　laser alignment instrument
激光电离　laser beam ionization
激光二极管　laser diode
激光风速计　laser anemometer
激光干涉测量法　laser interferometry
激光干涉仪　laser interferometer
激光钢板切割机　laser cutting for SMT stensil
激光光谱辐射计　laser spectrum radiator
激光经纬仪　laser theodolite
激光内径测量仪　laser inside diameter measuring instrument
激光喷射打印机　laser jet printer
激光器触发开关　laser triggered switching
激光切割　laser cutting
激光散射计　laser scatterometer
激光式测长仪　laser length measuring instrument
激光束扫描声全息　acoustical holography by laser scanning
激光水准仪　laser level
激光探针质谱计　laser probe mass spectrometer
激光椭圆度测量仪　laser ellipticity measuring instrument
激光外径测量仪　laser outside diameter measuring instrument
激光显微光谱分析仪　laser microspectral analyzer
激光印刷机　laser printer
激光荧光探测仪　laser fluorescence detector
激光诱导荧光检测　laser-induced fluorescence detection
激光原子化器　laser atomizer
激光指向仪　laser orientation instrument

激光重力仪　laser gravimeter
激光转速仪　laser tachometer
激活　activation
激励电压　energizing voltage
激励控制　excitation control
激励频率　energizing frequency
激励绕组　excitation winding
激励线圈　drive wire
激励者　actuator
级　degree
级别　grade
级间负载电阻　interstage load resistance
级联变换器　cascade converter
级联补偿　cascade compensation
级联光电调制器　cascade electrooptic modulator
级联连接　cascade connection
级联式电压互感器　cascade voltage transformer
即热式热水炉　instantaneous water heater
极大规则　maximum rule
极大熵　maximum entropy
极大似然法　maximum likelihood
极大似然估计器　maximum likelihood estimator
极大值原理　maximum principle
极地浮标　arctic buoy
极点　pole
极点定义　poles definition
极点和零点　poles and zeros
极点配置　pole assignment, pole placement
极端温度　extreme temperature
极光　aurora
极化　polarization
极化分析　polarization analysis
极化关系　polarization dependence
极化继电器　electropolarized relay
极化强度　intensity of polarization
极谱法　polarography
极谱仪　polarograph
极数　number of poles
极微时间测定器　chronoscope
极限　limit
极限报警检测器　limit alarm sensor

极限承载故障　ultimate bearing failure
极限承载力　ultimate bearing capacity
极限当量电导　limit equivalent conductance
极限刀顶距　limit point width
极限电流　limited current
极限分辨率　limit resolution
极限工作条件　limiting operating condition
极限可用度　limiting availability
极限控制　limiting control
极限量规　limit gauge
极限频率　limit frequency
极限强度　ultimate strength
极限切除时间　critical clearing time
极限位置　extreme position, limiting position
极限温度　limiting temperature
极限稳定性　marginal stability
极限压力角　limit pressure angle
极限应力　ultimate stress
极限有效波长　limiting effective wavelength
极限载重　ultimate load
极限增益法　ultimate gain method
极限真空度　limiting vacuum
极限值　extreme value
极限值问题　boundary-value problem
极限周期　limit cycle
极限转矩　breakdown torque
极限状态　limit state
极限状态设计　ultimate limit state design
极小极大技术　mini-max technique
极性　polarity
极性变化　change of polarity
极性变换　reversal of pole
极坐标式电位差计　polar coordinate type potentiometer
极坐标图　polar plot
极坐标型机器人　polar robot
急回机构　quick-return mechanism
急回特性　quick-return characteristics
急回运动　quick-return motion
急救人员　first aider
急救箱　first aid box
急弯　sharp bend

急转弯　hairpin turn
棘轮　ratchet
棘轮机构　ratchet mechanism
棘爪　pawl
集成　integrated
集成测试　integration testing
集成传感器　integrated sensor
集成电路　integrated circuit (IC)
集成电路放大器　integrated circuit amplifier
集成电路天线　integrated circuit antenna
集成光学　integrated optics
集成门极换流晶闸管　integrated gate-commutated thyristor
集成汽车高速公路系统　integrated vehicle highway system (IVHS)
集成塑模　family mold
集成陀螺　integrating gyro
集成稳压器　integrated voltage regulator
集成元件　integrating element
集成注射逻辑　integrated circuit injection logic
集电方式　current collecting methods
集电环　collector ring
集电极　collector
集电设备　current collecting device
集肤效应　skin effect
集合管　manifolds
集合式铸模　all core molding
集结矩阵　aggregation matrix
集料　aggregate
集流时间　time of concentration
集气管　air header
集散控制　total distributed control
集散控制系统　distributed control system (DCS)
集束荷载　crowd loading
集水槽　catchwater channel
集水阀　deluge valve
集水沟　gully
集水沟隔气弯管　gully trap
集水坑　sump pit
集体运输路线　mass transit line
集体运输系统　mass transit system
集线器　concentrator, hub
集烟罩　exhaust fume collecting hood

集液器　liquid receiver
集油器　oil interceptor
集中　centralize
集中参数　lumped parameter
集中参数系统　lumped parameter system
集中操作　centralized operation
集中的　collective
集中电阻　constriction resistance
集中负荷　concentrated load, point load
集中供热　district heating
集中管理系统　centralized management system
集中化　centralization
集中控制　centralized control, collective control
集中模型　lumped model
集中器　concentrator
集中式网络　centralized network
集中输入系统　input centralized system
集中数据处理　centralized data processing
集中型过程控制计算机　centralized process control computer
集中性　centrality
集中智能　centralized intelligence
集装箱功能　container function
集装箱起重机　container cranes
集总参数模型　lumped parameter model
集总常数元件　lumped constant element
集总的　lumped
集总电感　lumped inductance
集总电容　lumped capacitance
集总电阻　lumped resistance
几何编码　geometric code
几何的　geometrical
几何分布　geometric distribution
几何理论　geometrical theory
几何特性　geometric property
几何位置　geometrical position
几何相似　geometric similarity
几何像差　geometric aberration
几何学　geometry
几何学的　geometric
几何学方法　geometric approach

挤出成形　extrusion molding
挤出模　extrusion die
挤锻压加工　swaging
挤光模　burnishing die
挤压　extrusion
挤压模　squeezing die
计　meter
计测器框架　instrument rack
计尘器　dust counter
计划　scheme, programme
计划流程　planning process
计量泵　metering pump
计量标准器　measurement standard
计量规　gauge
计量盘　metering panel
计量性能　metrological performance
计量学　metrology
计时　clocking
计时变量　time variable
计时表　time meter
计时电路　timing circuit
计时脉冲　timed pulse
计时员　timer
计数　counting
计数电路　count circuit
计数风速表　counting anemometer
计数鼓　shaft encoder
计数寄存器　counter register
计数开关　counter switch
计数率　counting rate
计数率计　counting rate meter
计数脉冲　count impulse
计数器　counter
计数式超导磁力仪　counter type superconduction magnetometer
计数型浅层地震仪　counting-shallow-layer seismograph
计算　calculate, compute
计算表达式　computational expression
计算操作器　computing station
计算方法　computational method
计算-复位-保持三用开关　computer-reset-hold switch
计算功能　calculation function
计算功能块　calculation block
计算机　computer
计算机编程　computer programming

计算机编码　computer code
计算机程序　computer program
计算机创造　computer recreation
计算机断层摄影　computer tomography
计算机仿真　computer simulation
计算机辅助测试　computer-aided test (CAT)
计算机辅助电路分析　computer-aided circuit analysis
计算机辅助电路设计　computer aided circuit design
计算机辅助反馈控制器设计　computer-aided design of feedback controller (CADFC)
计算机辅助分析　computer aided analysis
计算机辅助工程　computer aided engineering (CAE), computer-aided engineering (CAE)
计算机辅助工作　computer-aided work
计算机辅助故障诊断　computer aided debugging
计算机辅助管理　computer assisted management
计算机辅助规划　computer aided planning
计算机辅助控制工程　computer aided control engineering
计算机辅助控制系统设计　computer-aided control system design (CACSD)
计算机辅助软件工程　computer-aided software engineering
计算机辅助设计　computer aided design (CAD)
计算机辅助设计工作站　work station for computer aided design
计算机辅助生产计划　computer aided production planning
计算机辅助系统设计　computer-aided system design
计算机辅助研究开发　computer aided researching and developing
计算机辅助诊断　computer aided diagnosis
计算机辅助指令　computer aided instruction
计算机辅助制图　computer aided drawing
计算机辅助制造　computer aided manufacturing (CAM)
计算机辅助质量管理　computer aided quality management
计算机辅助质量控制　computer aided quality control
计算机管理说明　computer management instruction
计算机和计算硬件　computer and computing hardware
计算机集成　computer-integrated
计算机集成过程系统　computer integrated process system (CIPS)
计算机集成制造　computer-integrated manufacturing
计算机集成制造系统　computer integrated manufacturing system (CIMS)
计算机监控系统　supervisory computer control system
计算机接口　computer gateway, computer interface
计算机接口操作器　computer interface station
计算机接口单元　computer interface unit
计算机结构　computer architecture
计算机科学与技术　computer science and technology
计算机控制　computer control
计算机控制系统　computer control system, computer controlled system
计算机控制显示　computer controlled display (CCD)
计算机控制应用　control applications of computer
计算机软件　computer software
计算机软件与理论　computer software and theory
计算机设定操作器　computer set station
计算机视觉　computer vision
计算机试验　computer experiment
计算机数控　computerized numerical control (CNC)
计算机通信网络　computer communication network

计算机图形学　computer graphics (CG)
计算机网络　computer network
计算机系统　computer system
计算机系统结构　computer systems organization
计算机应用　computer application
计算机应用技术　computer applied technology
计算机硬件　computer hardware
计算机子程序　computer subroutine
计算机自动备用　computer auto backup
计算机自动测量和控制　computer automated measurement and control (CAMAC)
计算机自动/手动操作器　computer auto-to-manual station
计算机自动-手动操作器　computer automatic-manual station
计算继电器　computing relay
计算结果数据点　calculated results data point (CRDP)
计算力矩　factored moment
计算力矩控制　computed torque control
计算连接　computing linkage
计算模件　computing module (CM)
计算模块状态显示画面　computing module status display
计算器　calculator, numerator
计算器算法　calculator algorithm
计算书　calculation sheet
计算弯矩　calculated bending moment
计算系统　computing system
计算仪表　computing instrument
计算元件　computing element
计算值　calculated value, calculation value
计算装置　computing unit
记录　logging, recording
记录笔　recording pen
记录笔张力　pen tension
记录编码　recording code
记录分析仪　recording analyzer
记录工具　recording medium
记录技术　recording property technology
记录控制台　log console

记录器　logger
记录式报警器　recording alarmer
记录式电流表　recording ammeter
记录式电压表　recording voltmeter
记录头　recording head
记录性能　recording performance
记录仪　recorder
记录仪表　recording instrument
记录纸　chart
记录纸标度尺长度　chart scale length
记录纸分度线　chart lines
记录纸驱动机构　chart driving mechanism
记录纸速度　chart speed
记录装置　marking device, recording device
记账依据　recording media
技术　technology
技术更改指令　engineering change order
技术规格　technical specification
技术过程　technique process
技术经济评价　technical and economic evaluation
技术说明书　engineering description
技术条件　technical conditions, technical specifications
技术系统　technique system
技术要求　technical requirements
技术预报　technological forecasting
技术援助中心　technical assistance center
技术支持网络　technological support network
技术转换　technology transfer
技术资料　technical information
剂量　dose
剂量泵　dosing pump
剂量反应模型　dose-response model
剂量率　dose rate
剂量率计　dose rate meter
继承　inheritance
继电保护　relaying protection
继电保护装置　relay protection
继电控制系统　contactor control systems
继电器　electric relay, relay

继电器-接触器控制	relay-contactor control
继电器控制	relay control
继电器逻辑卡	relay logic card
继电器释放	releasing of relay
继电器输入/输出模块	relay I/O module
继电器特性	relay characteristic
寄存器	register
寄存器配置	register allocation
寄生电容	stray capacitance, stray capacity
寄生反馈	parasitic feedback
寄生干扰	parasitic disturbance
寄生现象	parasitics
寄生元件	parasitic element
寄生振荡	parasitic oscillation, spurious oscillation
加蔽线	drain wire
加标	labelling
加长线圈调谐器	lengthening coil tuner
加成法用层压板	laminate for additive process
加电诊断	power-up diagnostic
加法电路	summing circuit
加法放大器	summing amplifier
加法计算器	totalizer
加法器	adder, summer
加法运算	additive operation
加感电缆	pupin cable
加感线路	loaded line
加工硬化	workhardening
加工余量	finishing allowance
加工制造报文服务	manufacturing message service (MMS)
加工中心	machining center
加固	consolidate, strengthen
加固工程	strengthening works
加固镍	reinforcing wire
加焊硬面法	hard facing
加荷系统	loading system
加厚	thicked
加减器	adder-subtracter
加减时间	add-subtract time
加劲杆	stiffener
加宽	widening
加宽校正	broadening correction
加宽校正因子	broadening correction factor
加料漏斗	charging hopper
加氯器	chlorinator
加气混凝土含水率	moisture of aerated concrete
加气混凝土吸水率	water absorption of aerated concrete
加气剂	air entraining agent
加气水泥	air entrained cement
加强	reinforcement
加强焊接	reinforcement of weld
加权法	weighting method
加权方法	method of weighted
加权函数	weighting function
加权平均	weighted mean
加权移动平均	weighted moving average
加热	heating
加热板	hot plates
加热过程	heat process
加热片	heater band
加热器	calorifier, heater
加热器冷却	heater cooler
加热曲线	heating curves
加热室	heating boxes
加热速率曲线	heating rate curves
加热线	heater wire
加热元件	heating element, heating unit
加热蒸汽管道	steam trace
加润滑油	add lubricating oil
加通风口	additional vent
加湿	humidification
加湿器	humidifier
加速电压	acceleration voltage
加速度	acceleration
加速度传感器	acceleration sensor, acceleration transducer, jerk sensor, jerk transducer
加速度仿真器	acceleration simulator
加速度分析	acceleration analysis
加速度曲线	acceleration diagram
加速度误差系数	acceleration error coefficient
加速管	accelerating tube
加速计	accelerometer

加速键	accelerating key
加速器敏感元件	acceleration sensitive element
加速器踏板	acceleration pedal
加速器响应	acceleration response
加速试验	accelerated test
加速室	accelerating chamber
加压管	pressure pipe
加压密封	pressure seal
加压硬化	press quenching
加油管	oil filler pipe
加有氯气的水	chlorinated water
加运算	add operation
加载装置参数	load device parameter
加装砂芯	embedded core
夹板	plywood, plywood board
夹持进料	gripper feed
夹紧	clamping
夹紧传送	gripper feeder
夹具	clamp, jig
夹具系统	clamping systems
夹口胶	joint sealant
夹模器	die clamper
夹钳	pincers tongs
夹套	jacket
夹头	grip
夹腿机	adductor
夹线板	cord bracket
夹渣生铁	cinder pig iron
家庭自动化	house automation
家用电器	household appliances
镓	gallium (Ga)
甲苯	toluene
甲酚树脂	cresol resin
甲阶树脂	A-stage resin
甲类放大	class A amplification
甲类放大器	class A amplifier
甲类工作状态	class A operation
甲类调制	class A modulation
甲醛	formaldehyde
甲烷	methane
甲烷着火	methane ignition
钾	potassium (K)
驾驶动作分析仪	tell-tale
驾驶控制台	driving console
驾驶盘	steering wheel
驾驶室	cab
驾驶室加温器	cab heater
驾驶室音响指示器	audible cab indicator
架空地线	overhead earth wire, overhead ground wire
架空电话线	overhead telephone line
架空电缆	electric overhead line, hook-up wire
架空电缆系统	cableway system
架空电信线路	overhead telecommunication line
架空管道	overhead line
架空索道	aerial ropeway
架空线	trolley wire
架空线路	overhead line
架空行车路	elevated vehicular link
架模高度	shut height of a die
架装	rack mounting
架装仪表	rack mounted instrument
假的	false
假定	assume
假峰	ghost peak
假设	hypothesis
假天花板	false ceiling
尖底从动件	knife-edge follower
尖脉冲	spike pulse
坚硬不透水物料	hard impervious material
间断分度	intermittent index
间断喷雾	intermittent spray
间隔	interval
间隔臂效应	spacer arm effect
间隔尺寸	granularity
间隔符	blank character
间隔环	spacer ring
间隔块	spacer block
间隔区	spacer
间隔时钟	interval timer
间接被控变量	indirectly controlled variable
间接被控系统	indirectly controlled system
间接测量	indirect measurement
间接测量法	indirect method of measurement
间接传动	indirect drive
间接的	indirect

间接电流变换电路　indirect DC-DC converter
间接电流控制　indirect current control
间接电阻加热　indirect resistance heating
间接控制　indirect control
间接作用记录仪　indirect acting recording instrument
间接作用式测量仪表　indirect acting measuring instrument
间接作用仪表　indirect acting instrument
间距　spacing
间隙　gap
间隙测量　gap measurement
间隙差式气动控制器　differential gap pneumatic controller
间隙动态力矩　gap transient torque
间隙规　gap gauge, clearance gauge
间隙过大　excessive gap
间隙空间　clearance space
间隙特性　backlash characteristics
间隙调节器　gap action controller
间隙元件　gap element
间歇传动装置　intermittent gearing
间歇反应器　batch reactor
间歇工作方式　intermittent duty
间歇过程　batch process
间歇过程控制　batch process control
间歇联用技术　discontinuous simultaneous techniques
间歇模式　batch mode
间歇运动机构　intermittent motion mechanism
肩部螺丝　shoulder bolt
肩峰　shoulder peak
兼容的　compatible
兼容分时系统　compatible time-sharing system
兼容软件　compatible software
兼容性　compatibility
兼容硬件　compatible hardware
监测　monitoring
监测继电器　control relay
监测控制和数据采集　supervisory control and data acquisition (SCADA)
监测站　monitoring station

监督　supervision, supervise
监督程序　monitoring program
监督计算机控制　supervisory computer control
监督继电器　supervisory relay
监督控制　supervisory control
监督人　supervisor
监督学习　supervised training
监工计划书　supervision plan
监控反馈　monitoring feedback
监控环节　monitoring element
监控回路　monitoring loop
监控顺序程序　supervisory sequence program
监控系统　monitored control system
监控信息系统　supervisory information system
监视器　monitor
监视线路　monitoring wire
监视硬件　monitoring hardware
监听器　audio monitor, monitor
监听线　monitoring wire
减低噪声措施　noise abatement measure
减法电路放大器　differential amplifier
减法器　subtracter
减法运算　subtraction
减幅比　decrement ratio
减活化作用　deactivation
减流电阻器　current reducing resistor
减摩性　anti-friction quality
减轻措施　mitigation measure
减色效应　hypochromic effect
减少　reduction, decrease
减少键　decrement key
减速　retardation, deceleration
减速比　reduction ratio
减速车道　deceleration lane
减速齿轮　reduction gear
减速阀门　deceleration valves
减速剂　moderator
减速器　gear reducer, retarder, speed reducer
减速箱　reduction gearbox
减速装置　reduction gear
减损　impair
减缩因数　reduction factor

减压阀 pressure reducing valve, reducing valve
减压拱 relieving arch
减压梁 relieving beam
减压螺钉 bleed screw
减压室 decompression chamber
减压嘴 bleed nipple
减振拉筋 damping wire
减震器 absorber, dashpot, vibration damper
剪边模 shearing die
剪断 shearing
剪角 shear angle
剪力 shearing force
剪力钢筋 shear reinforcement
剪力键 shear key
剪力墙 shear wall
剪切 shear
剪切板 cut to size panel
剪切和粘贴 cut and paste
剪式振动 scissoring vibration
剪线钳 wire cutters
剪应力 shearing stress
剪子 shears
检波 demodulation
检波二极管 detector diode
检波放大器 detect amplifier
检波器 detecting instrument
检测 detection
检测点 measured point
检测技术与自动化装置 detection technology and automatic equipment
检测能力 detectability
检测频率 test frequency
检测器性能 detector performance
检测算法 detection algorithm
检测条件 measurement condition
检测系统 detection system
检测限 limit of detection (LOD)
检测元件 detecting element, detecting instrument
检测者 tester
检查 examine
检查齿轮啮合涂色剂 gear marking compound
检查点 checkpoint
检查接线 connection test
检查井 inspection chamber, inspection manhole, inspection pit
检查箱 hatch box
检查员 inspector
检出器 detecting device
检定炉 furnace for verification use
检管器 pipeline inspection gauge
检流计 galvanometer
检漏 leak detection
检漏器 hydrophone, leak detector
检漏仪 leakage detector
检索方法 index method
检索系统 searching system
检修 overhauling
检修门 access door
检验器 checker
简单网络管理协议 simple network management protocol (SNMP)
简化双重跳线法 easy setting dual jumper (ESDJ)
简谐运动 simple harmonic motion
简易模 plain die
简易水质检查试验纸 water quality tester strips
碱硅反应 alkali-silica reaction
碱化 alkalinization
碱火焰检测 alkali flame detection (AFD)
碱火焰离子化 alkali flame ionization
碱误差 alkaline error
碱性 alkalinity
碱性的 alkali
碱性集料反应 alkaline aggregate reaction
碱性溶剂 basic solvent
碱性铁 basic iron
碱焰离子化检测器 alkali flame ionization detector (AFID)
碱液泵 alkaline pump
碱液缸 alkaline tank
建立 build
建立命令文件 creat command file
建立目录 creat directory
建立时间 build-up time, setting time
建立实体 establish
建模 modeling
建模误差 modeling error

建设阶段　construction stage
建筑技术科学　building technology science
建筑卷扬机　construction winch
建筑物自动化和控制网络　building automation and control net
建筑用硅酮结构密封胶　structural silicone sealants for building
渐近逼近　asymptotic approximation
渐近的　asymptotic
渐近分析　asymptotic analysis
渐近近似　asymptotic approximation
渐近稳定性　asymptotic stability
渐近线　asymptote
渐近性　asymptotic property
渐近中心　asymptote centroid
渐开螺旋面　involute helicoid
渐开线　involute
渐开线齿　involute teeth
渐开线齿廓　involute profile
渐开线齿轮　involute gear
渐开线发生线　generating line of involute
渐开线方程　involute equation
渐开线干涉点　involute interference point
渐开线函数　involute function
渐开线花键　involute spline
渐开线螺旋角　involute spiral angle
渐开线蜗杆　involute worm
渐开线压力角　pressure angle of involute
溅射电离　ionization by sputtering
鉴别　discriminate
鉴别力　discrimination
鉴别灵敏度　discrimination threshold
鉴别器　descriminator
鉴别试验　identification test
鉴别阈　discrimination threshold
鉴频　frequency discrimination
鉴相器　phase discriminator
键槽　keyway
键槽离合器　key clutch
键合相柱　bonding phase column
键控穿孔　keypunch
键控放大器　keyed amplifier
键盘　key panel, keyboard
键盘处理器　keyboard processor
键盘发送　keyboard send
键盘接收　keyboard receive
键盘控制键　keyboard control key
键盘控制器　keyboard controller (KBC)
键盘输入　key in
键锁　keylock
键锁级别　key level
键形开关　key switch
键组态　button configuration
浆液阀　parallel slide valves
降低灵敏度　desensitization
降低泄漏电流　leakage current reduction
降阶观测器　reduced order observer
降阶模型　reduced-order model
降落张线　landing wires
降水量　amount of precipitation
降压变换器　buck converter
降压变压器　step down transformer, step-down transformer
降压室　decompression chamber
降压斩波电路　buck chopper
降雨强度　intensity of rainfall
交变磁场　alternating magnetic field
交变负荷　alternating load
交变应力　repeated stress
交变载荷　repeated fluctuating load, repeated load
交叉　interleaved, intersect
交叉编译　cross compiling
交叉存储器　interleaved memory
交叉带传动　cross-belt drive
交叉点　cross point, intersection
交叉干扰　crosstalk interference
交叉汇编　cross assembling
交叉矩阵　interactor matrix
交叉设计　interdisciplinary design
交叉相位解调　cross-phase modulation
交磁放大机　amplidyne
交错点　crossing point
交错线　stagger wire
交错轴斜齿轮　crossed helical gears
交叠分解　overlapping decomposition
交叠滤波器　cross over filter
交互服务站　interacting service station

交互式多媒体协议　interactive multimedia association
交互式制图设计　interactive drawing design
交互式终端　interactive terminal
交互主磁通　mutual flux
交换记忆　memory swap
交换律　commutative law
交换容量　exchange capacity
交汇处　interchange
交联度　degree of cross linking
交联柱　crosslinking column
交流变换器　AC converter machine, AC-frequency converter
交流测速发电机　AC tachometer generator
交流差动变压器式位移传感器　AC differential transformer displacement transducer
交流触发器　AC flip-flop
交流电　alternating current
交流电导率　AC conductivity
交流电动机　AC motor, alternating current motor
交流电机　AC machine
交流电力电子开关　AC power electronic switch
交流电力网　alternating current mains
交流电路　AC circuit, alternating current circuit
交流电路稳压器　AC circuit constant voltage regulator
交流电桥　alternating current bridge
交流电压　alternating voltage
交流电源的谐波含量　harmonic content of AC power supply
交流电源控制器　AC power controller
交流电阻　alternating current resistance
交流发电机　alternating current generator
交流功率控制器　AC power controller
交流换向器电动机　alternating-current commutator motor
交流换向器电机　AC commutator machine

交流继电器　AC relay, alternating-current relay
交流检流计　AC galvanometer
交流声　hum
交流输电　alternating current transmission
交流数字电压表　AC digital voltmeter
交流数字伏特计　AC digital voltmeter
交流-数字转换器　AC to digital converter
交流调功电路　AC power controller
交流调压电路　AC voltage controller
交流稳压电源　AC regulated power source, AC regulated power supply, AC voltage-stabilized source
交流稳压器　AC voltage stabilizer
交流整流子电机　AC commutator machine
交替脉冲发生器　alternate pulse generator
交调失真　crosstalk
交通标志　traffic sign
交通灯　traffic light
交通灯号　traffic light signal
交通灯控制行人过路处　signal-controlled pedestrian crossing
交通管制及监视系统　traffic control and surveillance system
交通交汇处　transport interchange
交通控制　traffic control
交通控制中心　traffic control centre
交通容量　traffic volume
交通信号（灯）　traffic signal
交通信息工程及控制　traffic information engineering & control
交通运输工程　communication and transportation engineering
交通运输规划与管理　transportation planning and management
交通噪声　traffic noise
交通走廊　traffic corridor
交越频率　crossover frequency
交直流差动电压表　AC/DC differential voltmeter
交直流差动式伏特计　AC/DC differential voltmeter
交直流发电机　double current generator

交直流两用电动机 universal motor
交直流网络分析仪 AC and DC network analyzer
浇封 encapsulation
浇注 bottom pouring
浇注包 casting ladle
浇注底板 bottom board
浇注法 pouring process
浇注锚固 cast-in anchorage
胶合板 plywood, plywood board
胶合指数 scoring index
胶筐 plastic basket
胶膜 film adhesive
胶泥 puddle
胶泥封合剂 mastic sealant
胶黏剂 adhesive
胶黏剂面 adhesive face
胶黏水泥 mastic
胶束动电毛细管色谱法 micellar electrokinetic capillary chromatography (MECC)
胶束动电色谱 micellar electrokinetic chromatography (MEKC)
胶束色谱法 micellar chromatography (MC)
焦点 focus
焦点尺寸 focus size
焦耳（能量、热量、功的单位）Joule (J)
焦耳定律 Joule's law
焦耳-楞次定律 Joule-Lenz's law
焦痕 gas mark
焦距 focal distance, focus-to-film distance
焦平面 focal plane
焦深 depth of focus
焦炭生铁 coke pig iron
角编码器 angular encoder
角侧隙 angular backlash
角动量 angular momentum
角度 angle, angular
角度传感器 angle sensor
角度针 angle pin
角阀 angle valve
角分辨电子谱法 angle resolved electron spectroscopy (ARES)
角钢 angle iron, edge iron
角钢梁测试 angle beam testing
角焊 fillet welding
角焊接 fillet weld
角加速度 angular acceleration
角加速度传感器 angular acceleration sensor, angular acceleration transducer
角接触推力轴承 angular contact thrust bearing
角接触向心轴承 angular contact radial bearing
角接触轴承 angular contact bearing
角节距 angular pitch
角灵敏度 angular sensitivity
角偏移 angular deviation
角频率 angular frequency, circular frequency
角式节流阀 angle throttle valves
角式截止阀 angle type globe valves
角速比 angular velocity ratio
角速度 angular velocity
角速度传感器 angular velocity sensor, angular velocity transducer
角铁 angle iron, edge iron
角铁切割机 angle cutter
角铁支架 angle iron bracket
角位 corner
角位移 angular displacement
角位移传感器 angular displacement transducer
角位移光栅 angular displacement grating
角位置 angular position
角铣刀 angle cutter
角销 angular pin
角行程阀 rotary motion valve
角形断流阀 angle stop valves
角形控制极 corner gate
角形外螺纹切断阀 angle external threaded block valve
角型 angle form
角应变 angle strain
绞车 winch
绞合电缆 stranded cable
绞合天线 radio wire
绞合线 strand wire, twisted wire
绞孔 fraising

矫顽力　coercive force
矫顽力计　coercivity meter
矫直退火　straightening annealing
脚本　scripting
脚手架　falsework, scaffolding
脚踏泵　foot pump
脚踏冲床　foot press
脚线　leg wire
铰刀　reamer
铰孔机　broaching machine
铰孔加工　reaming
铰链　hinge
铰链杆　hinge bar
铰链销　hinge pin
搅拌棒　stirring rod
搅拌鼓　mixing drum
搅拌滚筒　mixing drum
搅拌器　agitator, mixer, stirrer
较低的　lower
校对规　check gauge
校对机　interpolator
校平机　leveller
校验报告　calibration report
校验和　checksum
校验寄存器　check register
校验检查值　check value
校验器　tester
校验仪　check meter
校正　calibrate, correcting, regulating
校正变量　correcting variable
校正单元　correcting unit
校正动作　corrective action
校正动作报告　corrective action report
校正范围　correcting range
校正方法　method of correction
校正率　correction rate
校正平面　correcting plane
校正气　calibrating gas
校正时间　correction time, settling time
校正数据　correction data
校正衰减器　calibrated attenuator
校正条件　correcting condition
校正系数　correction coefficient
校正因子　correction factor
校正用仪表　reference instrument
校正元件　correcting element
校正针　adjusting pin

校正值　correction value
校准　calibration
校准层次　calibration hierarchy
校准点　calibration point
校准电压　calibrating voltage
校准工作台　calibration table
校准公式　calibration equation
校准函数或校准曲线　universal calibration function or curve
校准混合气　calibration gas mixture
校准记录　calibration record
校准量　calibration quantity
校准器　regulator
校准曲线　calibration curve
校准特性　calibration characteristics
校准信号发生器　calibrated signal generator
校准液　calibration solution
校准用测量仪表　calibration instrument
校准用电池　calibration battery
校准用调节器　calibration regulator
校准振荡器　alignment oscillator
校准周期　calibrating period
校准组分　calibration component
阶段放气阀　graduated release valve
阶梯光栅　echelon grating
阶梯刨刀　corrugated tool
阶梯信号发生器　step generator
阶跃　step
阶跃法　step change method
阶跃函数　step function
阶跃函数输入信号　step-function input signal
阶跃输入　step input
阶跃误差常数　step error constant
阶跃响应　step function response, step response
阶跃响应模型　step response model
阶跃响应时间　step response time
阶跃作用　step action
接触　tangency
接触比　contact ratio
接触不良　loss contact
接触导线　trolley wire
接触点　contact points, point of contact

接 jie

接触点电压降　contact drop
接触电动势　contact electromotive force
接触电压　contact voltage
接触电阻　contact resistance, constriction resistance
接触垫　contact pads
接触法　contact process
接触放电加工　electro-arc contact machining
接触感测模块　contact sense module
接触过程　contact process
接触迹　path of contact
接触角　angle of contact
接触开关　contact switch
接触面　mating face
接触疲劳　contact fatigue
接触器控制　contactor control
接触式电子调温计　electronic temperature contact controller
接触式风速表　contact anemometer
接触式密封　contact seal
接触图形　contact pattern
接触线　line of contact
接触液体的部件　wetted parts
接触应力　contact stress
接地　earthing, ground connection, grounding
接地板　earth plate
接地棒　earth rod, ground rod
接地保护　ground protection
接地保护继电器　ground relay
接地保护装置　earth-fault protection
接地测量仪　earthometer
接地层　ground plane
接地带　earth tape
接地导板　earth plate
接地导线　earthed conductor
接地的　grounded
接地点　earthed point, ground point
接地电极　earth electrode, grounding electrode
接地电缆　earthing cable
接地电路　earthed circuit, grounded circuit
接地电容　capacity ground
接地电阻　earth resistance
接地电阻表　earth resistance meter, earth-resistance meter
接地端　grounded junction
接地故障　earth fault
接地绞线　earthing twisted pair
接地开关　ground switch
接地控制　ground control
接地跨接线　earthing jumper
接地漏电断路器　earth leakage circuit breaker (ELCB)
接地漏电检示器　earth leakage detector
接地母线　earth bus, earthing bus, ground strap
接地片　earth tag
接地事故　ground fault
接地输出　earthed output, grounded output
接地输入　earthed input, grounded input
接地探针　grounded probe
接地体　earth electrode
接地铜织带　earth copper strap
接地系统　earthed system, earthing system, ground system
接地线　bonding wire, earth wire
接地型电压互感器　earthed voltage transformer
接地噪声　grounded noise
接地中点　grounded neutral
接地中端　earthed neutral conductor
接地终端　earthing terminal
接地转换开关　earthing switch
接地自耦变压器　earthing autotransformer
接地总线入口　ground bus inlet
接地阻抗测试　earth loop impedance test
接点　connection point
接点输出　contact output
接点输入　contact input
接管式防水密封接头　water-proof packing gland for connecting pipe
接合　seaming
接合部件　mating part
接合剂　jointing compound
接户线　drop wire

接近距离	approach distance	节点对数	node pair number
接近锁定	approach lock	节点接口单元	node interface unit
接近效应	proximity effect	节点数量	node number
接口处理器	gateway processor	节点总线	node bus
接口单元	interface unit	节点总线上的个人工作站	personal workstation on node bus
接口垫片	gasket joint	节距	pitch
接口管	mouthpiece	节距圆	pitch circle
接口通信处理机	interface message processor	节流阀	throttle, throttle valve
接口状态	interface state	节流式仪表截止阀	throttling type instrument block valve
接口状态发生	interface state generation	节面	pitch plane
接零	connect to neutral	节面母线	pitch element
接入继电器	cut into operation	节曲面	pitch surfaces
接上	connect up	节热器	economizer
接收	receive	节线	pitch line
接收波段	receiving wave range	节线跳动	pitch-line runout
接收池	reception tank	节圆半径	pitch radius
接收的	receiving	节圆齿厚	thickness on pitch circle
接收电路	acceptor circuit, receiving circuit	节圆夹具	pitch-line chuck
接收机	receiver	节圆直径	pitch diameter
接收机单元	receiver element	节圆锥角	pitch cone angle
接收器	acceptor	节锥	pitch cone
接收器狭缝	collector slit	节锥顶	pitch apex
接收通道	receiving channel	节锥顶至轮冠	pitch apex to crown
接通	switch on	节锥角	pitch angle
接通的	connective	洁净	cleansing
接通开关	engaged switch	洁净恒温器	clean ovens
接头电阻测试仪	bond tester	结电容器	junction capacitor
接线	wiring	结构	configuration
接线板	connection box	结构参数	structural parameter
接线电容	lead capacitance	结构的	structural, structured
接线端子	terminal with switch	结构分析	structural analysis, structured analysis
接线端子板	terminal block	结构改善工程	structural improvement works
接线盒	patera, terminal box	结构钢	structural iron, structural steel
接线检测仪	wire-space detector	结构钢构件	structural steel member
接线盘	junction panel	结构工程	structural engineering, structural works
接线图	hookup	结构构架	structural frame
接线箱	joint box	结构构件	structural element
接线柱	post	结构骨架	structural skeleton
接中线	connect to neutral	结构化程序设计	structured programming
街道照明	street lighting	结构检定	structural appraisal
街道照明设施	public lighting		
节点	node, pitch point		
节点安排	node assignment		
节点对	node pair		

结构勘测　structural survey
结构勘查　structural investigation
结构奇异值　structured singular value
结构强度　structural strength
结构设计　physical design, structural design
结构试验　structure test
结构松弛　structural relaxation
结构图　dwg
结构完整程度　structural integrity
结构稳定性　configuration stability, structural stability
结构稳定性分析　structural stability analysis
结构系统　structure system
结构形式　form of structure
结构型传感器　mechanical structure type sensor, mechanical structure type transducer
结构性能　structural behaviour, structural property
结构硬件　constructional hardware
结构用途　structural use
结构优化　structural optimization
结构约束　structural constraint
结构状况勘测　structural condition survey
结构自重　self-weight of structure
结果　result
结果传达　read out
结合　bond
结合板　joint plate
结合垫片　joint sheet
结合剂　bonding agent
结合能　binding energy
结晶粒粗大化　coarsening
结晶粒度　grain size
结晶水　water of crystallization
结晶紫　crystal violet
结论　conclusion
结面　nodal plane
结束字节　end byte
结霜点　hoarfrost point
结型场效应晶体管　junction field effect transistor
结型晶体管积分器　junction transistor integrator
结型压敏电阻器　junction type varistor
结型氧化锌电压敏电阻器　junction type zinc oxide varistor
截断　chopping
截断放大器　chopper amplifier
截流阀　shut off valve
截流沟　intercepting drain
截流件　closure member
截流污水管　intercepting sewer
截面　section
截面分析　section analysis
截面积分　integral cross section
截面面积　sectional area
截片法　stripping method
截水沟　intercepting channel, intercepting ditch
截尾文丘里管　truncated Venturi tube
截止　cut off
截止电压　cut-off voltage
截止阀　shutoff valve, stop valves
截止频率　cut-off frequency
截止频率增益　gain cut-off frequency
截止速率　cut-off rate
截止状态　cut-off state
截锥铆钉　pan head rivet
解除　relief
解冻　defrost
解裂　disaggregation
解码器　decoder
解耦参数　decoupling parameter
解耦控制　decoupling control
解耦问题　decoupling problem
解释程序器　interpreter
解释语言　interpretive language
解锁　unlock
解调器　demodulator, detuner, modulation eliminator
解调制　demodulation
解吸化学电离　desorption chemical ionization (DCI)
解析测图仪　analytical plotter
解析近似　analytic approximation
解析设计　analytical design
解相关　decorrelation
介电常数　permittivity, specific inductive capacity
介电强度　dielectric strength

介电原片　dielectric sheets
介质　medium
介质常数　dielectric constant
介质存取控制　media access control
介质电导　dielectric conductance
介质放大器　dielectric amplifier
介质损耗　dielectric loss
介质损耗角　dielectric loss angle
介子　meson
界面　interface
界面反射波　interface echo
界面润滑　boundary lubrication
金　gold (Au)
金刚砂　carborundum, emery
金刚石切割器　diamond cutters
金刚石镶嵌瓣　diamond pad
金刚烷　adamantane
金工锯　metal saw
金连接线　gold bonding wire
金融系统　financial system
金相显微镜　metallurgical microscopy
金属　metal
金属板　metal plate
金属编织电缆　metal braid
金属表面斜层初始疲劳破裂　subsurface initiated fatigue breakage
金属材料试验机　metallic material testing machine
金属电弧焊　metal arc welding
金属电极电位　electrode potential
金属垫片　metallic gasket
金属钉　metal spike
金属粉末　metal dust
金属封盖　metal cover
金属工艺学　technology of metals
金属护皮电线　sheathed wire
金属基覆铜层压板　metal base copper-clad laminate
金属基印制板　metal base printed board
金属基指示电极　metal base indicated electrode
金属加工　metalwork
金属夹边　metal gripping rim
金属壳　metal clad, metal casing
金属拉网　metal lath
金属卤化物灯　metal halide lamp, metallic halide lamp

金属裸露　bare metal, exposed metal
金属密封蝶阀　hard seal butterfly valves
金属膜电阻　metalster
金属膜片　metallic diaphragm
金属-难溶盐指示电极　metal-insoluble salt indicated electrode
金属喷镀法　metallikon
金属屏蔽线　metal-shielded wire
金属嵌钉　metal stud
金属切削　metal cutting
金属熔液　molten metal
金属丝网　gauze, wire mesh
金属碎片　metal chip
金属陶瓷 X 射线管　metal-ceramic X-ray tube
金属陶瓷电位计　cermet potentiometer
金属套管　metal casing
金属弹簧重力仪　metal-spring gravimeter
金属网线　mesh wire
金属线　wire
金属屑　metal filing
金属芯覆铜箔层压板　metal core copper-clad laminate
金属芯印制板　metal core printed board
金属氧化物半导体　metal-oxide semiconductor (MOS)
金属氧化物半导体场效应晶体管　MOSFET
金属氧化物气体传感器　metal-oxide gas sensor, metal-oxide gas transducer
金属氧化物湿度传感器　metal-oxide humidity sensor, metal-oxide humidity transducer
金属预置扭矩式螺帽　all-metal prevailing torque type nuts
金属元素分析仪　metal elemental analysis
金属圆片　metal disc
紧边　tight-side
紧固板　securing plate
紧固定位销　securing dowel
紧固钩　securing hook
紧固环　securing ring
紧固连杆　securing rod

紧固螺钉	securing screw
紧固螺帽	securing nut
紧固索带	securing strap
紧合封盖	close fitting cover
紧合配件	close fittings
紧急按钮	emergency push button
紧急报警	emergency alarm
紧急出口	emergency exit
紧急出口门	emergency exit door
紧急栏障	panic barrier
紧急频道	emergency channel
紧急切断阀	emergency cut-off valves
紧急情况	emerg
紧急人手绞动器	emergency hand-winding equipment
紧急事故控制中心	emergency control centre
紧急事故手册	emergency manual
紧急事件	emergency
紧急跳闸系统	emergency trip system
紧急停车	emergency shutdown
紧急停车条件处理程序	emergency condition handler
紧急停止保护装置	emergency stop protection
紧急通道	emergency access
紧急照明设备	emergency lighting
紧急止动按钮	emergency stop button
紧急制动器	emergency brake
紧接的	back-to-back
紧邻	immediate vicinity
紧密的	tight
紧密耦合	close coupling
近代极谱法	modern polarography
近红外光谱	near infra-red spectrum (NIRS)
近距离效应	proximity effect
近似等效电路	approximate equivalent circuit
近似分析	approximate analysis
近似绝对温标	approximate absolute temperature scale
近似推理	approximate reasoning
近似值	approximate value
进场速度	approach speed
进程模型	process model
进	ingress and egress
进出路径	access lane
进出途径	means of access
进刀	set-in
进度表	progress chart
进风槽	air intake duct
进风口	air inlet, air intake
进港航道	approach channel
进给齿轮	feed gears
进给阀	inlet valve
进给方向	direction of feed
进给凸轮	feed cam
进给运动	feed motion, feed movement
进化系统	evolutionary system
进料品质管制人员	incoming quality control (IQC)
进气百叶窗	air inlet louver
进气阀	inlet valve, intake valve
进气风扇	intake fan
进气管道隔滤器	air line strainer
进气过滤器	air intake filter
进气孔	air intake
进气口	air inlet port
进气歧管	intake manifold
进气扇	supply fan
进入	access
进入同步	coming into step, falling into step
进入位	gate location
进水管	pressure pipe
进位	carry
进位存储	carry storage
进位寄存器	carry storage register
进相器	phase advancer
进行	execution
进样系统	inlet system
浸胶基应变	impregnated base strain gauge
浸蜡铜线	paraffin copper wire
浸蜡线	paraffin wire
浸没式冷却器	immersion cooler
浸入深度	immersion depth
浸入式反射计	immersion reflectometer
浸入误差	immersion error
浸润剂含量	size content
浸碳钢	blister steel

浸透压测定表 osmotic pressure meters
浸渍薄层色谱法 impregnated thin layer chromatography
浸渍绝缘纸 impregnating insulation paper
浸渍试验 immersion testing
禁区 closed area
禁区许可证 prohibited zone permit
禁止 disable
禁止潮湿 guard against damp
禁止令 injunction
禁止线 inhibit wire
经典控制理论 classical control theory
经典信息模式 classical information pattern
经济控制 economic control
经济控制理论 economic control theory
经济控制论 economic cybernetics
经济模型 economic model
经济数据 economic data
经济系统模型 economic system model
经济性设计 economic design
经年变形 secular distortion
经碾压碎石底层 rolled hardcore
经批准的图则 approved plan
经批准的物料 approved material
经纱 warp yarn
经纬测角仪 pantometer
经纬仪 theodolite
经验分布 empirical distribution
经验模型 empirical model
经验数据 empirical evidence
经验温标 experimental temperature scale
晶片检测计 wafer prober
晶体 crystal
晶体变换器 crystal converter
晶体传声器 crystal microphone
晶体的 crystalline
晶体电极 crystalline electrode
晶体二极管 crystal diode
晶体放大器 crystal amplifier
晶体管 transistor
晶体管参数测试仪 transistor parameter tester
晶体管电路 transistor circuit
晶体管多级放大器 transistor multistage amplifier
晶体管放大器 transistor amplifier
晶体管毫伏表 transistorized millivoltmeter
晶体管化表面膜厚计 transistorized surface film thickness meter
晶体管-晶体管逻辑 transistor-transistor logic
晶体管稳压电源 transistorized voltage stabilizer
晶体光电池 crystal cell
晶体光电元件 crystal photoelement
晶体光栅 crystal grating
晶体混频器 crystal mixer
晶体频率 crystal frequency
晶体切片 crystal cut
晶体压力传感器 crystal pressure transducer
晶体扬声器 crystal speaker
晶体振荡器 crystal oscillator, piezoelectric oscillator
晶体整流器 crystal rectifier
晶体钟 crystal clock
精度标称值 accuracy rating
精度测量仪表 accurate measuring instrument
精加工 finish machining
精加工机床 finishing machines
精炼铁 metallic iron
精馏控制 distillation control
精馏柱 distillation column
精密测量 precision measurement
精密冲裁 fine blanking
精密低温恒温水槽 precision low constant temperature water baths
精密度 precision
精密锻造 precision forging
精密恒温器 precision constant temperature ovens
精密计时表 chronometer
精密衰减器 precision attenuator
精密下料冲床 fine blanking press
精密下料加工 fine blanking
精密小孔测定器 bore check
精密压铸 accurate die casting
精密仪表 precision instrument
精密仪器及机械 precision instrument

and machinery
精确 accurate
精确度 accuracy
精确度等级 accuracy class
精筛 fine screening
精实生产 lean manufacturing
精研机 lapping machines
精整加工 finishing
精整加工线 finishing machines
鲸醇 kitol
井 well
井底水窝水泵 sump pump
井盖 manhole cover
井水泵 well water pump
井温仪 borehole thermometer
井下γ能谱仪 gamma spectrometer in borehole
井中γ辐射仪 gamma radiometer in borehole
井中五分量磁测井仪 five-component borehole magnetometer
井中重力仪 borehole gravimeter
颈缩加工 necking
景深 depth of focus
警报 alarm
警报电路 warning circuit
警报蜂鸣器 alarm buzzer
警报器 alarm buzzer
警报信号 warning signal
警告标志 caution sign, warning sign
警告灯 warning lantern
警钟 alarm bell
净高 clear height
净开口 clear opening
净空高度 headroom
净空间 clear space
净跨距 clear span
净水厂 water treatment plant
净水器 water purifier
净有效长度 clear effective length
净重 net weight
净重测试仪 dead weight tester
径 path
径节 diametral pitch
径向 radial direction
径向泵 radial pump
径向当量动载荷 dynamic equivalent radial load
径向当量静载荷 static equivalent radial load
径向的 radial
径向定位表面 radial locating surface
径向基本额定静载荷 basic static radial load rating
径向基函数 radial basis function (RBF)
径向基函数网络 radial base function network
径向畸变系数 coefficient of radial distortion
径向接触轴承 radial contact bearing
径向平面 radial plane
径向平面绘图仪 radial planimetric plotter
径向前角 radial rake angle
径向跳动 runout
径向跳动公差 runout tolerance
径向游隙 radial internal clearance
径向载荷 radial load
径向载荷因素 radial load factor
径向展开 radial development
竞争离子 competing ions
静电八极透镜 electrostatic octupole lens
静电表 electrometer
静电场干扰 electrostatic field interference
静电除尘器 static precipitator
静电的 electrostatic
静电电子显微镜 electrostatic electron microscope
静电发射 field emission
静电分析器 electrostatic analyzer
静电感应 electrostatic induction, static induction
静电功率表 electrostatic wattmeter
静电过滤器 electrostatic filter
静电荷 electrostatic charge
静电荷测量仪 static charge measuring instrument
静电激发器 electrostatic actuator
静电计式射气仪 electrostatic emanometer
静电力 electrostatic force
静电耦合 electrostatic coupling

静电排放　electro-static discharge (ESD)
静电偏转　electrostatic deflection
静电屏蔽　electrostatic screen
静电屏蔽笼　Faraday cage
静电屏蔽仪表　instrument with electrostatic screening
静电四极透镜　electrostatic quadrupole lens
静电透镜　electrostatic lens
静电吸引定律　law of electrostatic attraction
静电系仪表　electrostatic instrument
静电显示记录仪　electrostatic display recorder
静电印刷机　electrostatic printer
静合接点　back contact
静寂地带　silent zone
静力　static force
静密封　static seal
静摩擦　static friction
静平衡　static balance
静态　quiescent state
静态测量　static measurement
静态充电　static electrification
静态存储器　static storage
静态感应半导体　static induction transistor
静态工作点　quiescent point
静态解耦　static decoupling
静态精度　static accuracy
静态控制器　static controller
静态灵敏度　static sensitivity
静态模型　static model
静态偏差　steady state deviation
静态随机存取存储器　static RAM
静态特性曲线　static characteristics curve
静态调节　static regulation
静态稳定性　static stability
静态误差　static error
静态增益　static gain
静压　static pressure
静压头　hydrostatic head
静压影响　influence of static pressure
静载重　static load
镜面光泽度试验方法　test method for specular gloss
镜铁　mirror iron, spiegel iron

镜像　mirroring
镜像法　image method
镜像开关　mirror image switch
纠错剖析　error-correction parsing
久期方程　secular equation
酒精温度表　alcohol thermometer
救生掣　exit switch
就地安装　local mounting
就地控制单元　local control unit
就地控制柜　local control panel
就地批量操作站　local batch operator station
压密试验　consolidation test
居里　Curie
居里温度　Curie temperature
锔　curium (Cm)
局部磁化　local magnetization
局部淬火　selective hardening
局部的　partial
局部反馈　local feedback
局部封闭　partial closure
局部故障　partial failure
局部回路　local loop
局部渐近稳定性　local asymptotic stability
局部结构　local structure
局部抗磁屏蔽　local diamagnetic shielding
局部可控性　local controllability
局部控制　local control
局部控制网络　local control network (LCN)
局部控制网络段　local control network segment
局部控制网络扩展器　local control network extender (LCNE)
局部控制网络练习程序　local control network exerciser
局部控制者　local controller
局部渗漏　localized seepage
局部手动　local manual
局部数据库　local data base
局部坍塌　partial collapse
局部稳定性　local stability
局部主参考地　local master reference ground
局部自动化　local automation

局部自由度	passive degree of freedom

局部自由度　passive degree of freedom
局部最优　local optimum
局地闪电计数器　local lightning counter
局间的　interoffice
局限性投标　restricted tender
局限应力　confining stress
局域　local
局域操作网络　local operating networks (LON)
局域计算机系统　local computer system
局域网　local area network (LAN)
菊花链　daisy chain
菊花链总线　daisy chain bus
举模器　die lifter
矩形　rectangle
矩形波　rectangular wave
矩形波导　rectangular waveguide
矩形波发生器　square wave-form oscillator
矩形波信号发生器　square wave signal generator
矩形波整形电路　squaring circuit
矩形的　rectangular
矩形扶手　rectangular handrail
矩形钢丝　oblong steel wire
矩形铝线　rectangular aluminum wire
矩形螺纹　square threaded form
矩形填充　rectangle filling
矩形牙嵌式离合器　square-jaw positive-contact clutch
矩阵　matrice, matrix
矩阵代数　matrix algebra
矩阵多项式方程　matrix polynomial equation
矩阵方程　matrix equation
矩阵方法　matrix method
矩阵公式　matrix formulation
矩阵黎卡提方程　matrix Riccati equation
矩阵求逆　matrix inversion
矩阵式打印机　matrix printer
矩阵式放大器　matrix amplifier
矩阵式加法器　matrix adder
矩阵元素　matrix element
巨砾　boulder
巨系统　huge system
巨型计算机　supercomputer

拒绝　rejection
具体结构　concrete structure
具有附加金属膜盒密封的气动调节阀　pneumatic control valve with additional metal bellows seal
距离　distance
距离变换　distance transformation
距离常数　distance constant
距离传感器　range sensor
距离刻度　distance marker
距离扫描发生器　range-sweep generator
距离速度滞后　distance velocity lag
距离振幅补偿　distance amplitude compensation (DAC)
距离振幅曲线　distance amplitude curve
锯　saw
锯齿波　saw-tooth wave
锯齿波发生器　saw-tooth wave generator
锯齿波截止二极管　saw turn-off diode
锯齿波振荡器　saw-tooth oscillator
锯齿形螺纹　buttress thread form
锯片　saw blade, blade saw
锯削　sawing
锯屑　saw dust
聚氨酯　polyurethane (PU)
聚苯醚　polyphenylene oxide
聚苯乙烯　polystyrene (PS)
聚丙烯　polypropylene (PP)
聚丙烯薄板　acrylic sheet
聚丙烯酸　polyacrylic acid
聚丙烯酰胺凝胶电泳　polyacrylamide gel electrophoresis
聚尘器　precipitator
聚醋酸乙烯　polyvinyl acetate
聚芳酰胺纤维纸　aromatic polyamide paper
聚氯乙烯　polyvinyl fluoride
聚光灯　spotlight
聚光镜　condenser lens
聚光器　condenser
聚光透镜　condenser lens
聚硅酮橡胶　silicon rubber
聚合树脂　polymerization resin

聚合物　polymer
聚合作用　polymerization
聚集日照计　Campbell-stokes sunshine recorder
聚甲基丙烯酸甲酯　polymethyl methacrylate（PMMA）
聚焦　focusing
聚焦探头　focusing type probe
聚焦调整器　focusing regulator
聚焦透镜　condenser lens
聚类分析　cluster analysis
聚氯乙烯　polyvinyl chloride（PVC）
聚氯乙烯防水卷材　polyvinyl chloride plastic sheets for waterproofing
聚氯乙烯绝缘屏蔽软线　PVC insulated shielded soft wire
聚氯乙烯绝缘线　thermoplastic-covered wire
聚氯乙烯用户引入线　polyvinyl chloride drop wire
聚偏二氯乙烯　polyvinylidene chloride
聚全氟乙烯丙烯薄膜　fluorinated ethylene-propylene copolymer film
聚水器　water trap
聚四氟乙烯　polytetrafluoroethylene（PTFE）
聚四氟乙烯绝缘电线　teflon insulated wire
聚碳酸酯　polycarbonate（PC）
聚酰胺　polyamide（PA）
聚酰胺环氧漆　polyamide epoxy paint
聚酰胺纤维　polyamide fiber
聚酰亚胺玻璃布覆铜箔板　polyimide woven glass fabric copper-clad laminates
聚酰亚胺薄膜　polyimide film, PI film
聚酰亚胺树脂　polyimide resin
聚亚胺酯　polyurethane（PU）
聚氧化乙烯　polyoxyethylene
聚乙二醇　polyethyleneglycol
聚乙烯　polyethylene（PE），polythene
聚乙烯醇　polyvinyl alcohol（PVAC）
聚乙烯醇缩丁醛　polyvinyl butyral
聚乙烯绝缘电缆　polyethylene insulated cable
聚乙烯绝缘线　formal insulated wire, polyethylene insulated wire

聚乙烯内搪层　polyethylene lining
聚乙烯铜线　formale copper wire
聚酯　polyester
聚酯玻璃布覆铜箔板　polyester woven glass fabric copper-clad laminates
聚酯树脂　polyester resin
涓流充电　trickle charge
卷边工具　crimping tools
卷边加工　hemming
卷边弯曲加工　curl bending
卷材进料　coil cradle
卷簧　coil spring
卷积积分　convolution integral
卷积模型　convolution model
卷缆柱　cable column
卷料　coil stock, roll material
卷曲加工　curling
卷刃　wired edge
卷弯成形机　rotary bender
卷圆压平　reel stretch
卷圆压平冲子　reel-stretch punch
卷闸　roller shutter
卷纸　roll chart
卷纸机构　roll chart drive unit
决策　decision making
决策表　decision table
决策程序　decision program
决策电路　decision circuit
决策反馈　decision feedback
决策分析　decision analysis
决策合成　decision fusion
决策空间　decision space
决策理论　decision theory
决策模型　decision model
决策树　decision tree
决策支持系统　decision support system
诀窍数据　recipe data
绝对测量　absolute measurement
绝对尺寸　absolute dimension, absolute size
绝对尺寸系数　absolute dimensional factor
绝对磁导率　absolute permeability
绝对磁道地址　absolute track address
绝对单元地址　absolute cell address
绝对　absolute

绝对地址　absolute address
绝对分辨率　absolute resolution
绝对静电计　absolute electrometer
绝对零点　absolute zero
绝对偏差　absolute deviation
绝对偏差积分　absolute deviation integral, integral of the absolute error (IAE)
绝对频率计　absolute frequency meter
绝对切向速度　absolute tangential velocity
绝对湿度　absolute humidity
绝对速度　absolute velocity
绝对位址　absolute address
绝对温标　absolute scale, absolute temperature scale
绝对温度　absolute temperature
绝对稳定性　absolute stability
绝对误差　absolute error
绝对误差积分准则　integral of absolute value of error criterion
绝对误差判据　absolute error criterion
绝对压力变送器　absolute pressure transmitter
绝对压力表　absolute pressure indicator
绝对压力传感器　absolute pressure sensor, absolute pressure transducer
绝对压强　absolute pressure
绝对盐度　absolute salinity
绝对仪器　absolute apparatus (AA)
绝对运动　absolute motion
绝对噪声计　objective noise-meter
绝对值　absolute value
绝对值报警　absolute alarm
绝对值偏差乘时间积分　integral of the time and absolute error (ITAE)
绝对重力仪　absolute gravimeter
绝热扫描热量计　adiabatic scanning calorimeter
绝缘　isolation
绝缘棒　insulating rod
绝缘被覆线　covered wire
绝缘材料　insulating material, insulation material
绝缘测试　megger test
绝缘测试器　megger tester
绝缘层　insulating layer, insulation lagging
绝缘层信移连接件　insulation displacement connection
绝缘等级　class of insulation, insulation class
绝缘电缆　insulated cable
绝缘电线　covered wire
绝缘电阻　dielectric resistance, insulation resistance
绝缘电阻测试器　insulating-resistance tester
绝缘垫片　insulation spacer
绝缘端　isolated junction
绝缘隔板　isolating partition
绝缘护片　insulating shield
绝缘护套　insulation sleeve
绝缘击穿电压　insulation voltage breakdown
绝缘浇道方式　insulated runner
绝缘跨接线　bridle wire
绝缘铝线　insulated aluminum wire
绝缘破坏　electrical breakdown
绝缘强度　dielectric strength, insulating strength
绝缘强度测试　breakdown voltage testing
绝缘栅双极型晶体管　insulated gate bipolar transistor
绝缘试验　insulation test
绝缘试验电压　insulating test voltage
绝缘损坏检示仪表　insulation fault detecting instrument
绝缘套圈　insulated ferrule
绝缘体　dielectric, insulator
绝缘型铠装热电偶　isolated junction type sheathed thermocouple
绝缘障　insulating barrier
掘进机　heading machine
掘井　sink well
军事化学与烟火技术　military chemistry and pyrotechnics
军用标准　military-standard (MIL-STD)
均布荷载　uniformly distributed load
均方根　root mean square
均方根值　root mean square value
均方谱密度　mean squared spectral density

均方误差 mean-square error
均方误差准则 mean-square error criterion
均衡 equalization, equilibrium
均衡槽 balance tank
均衡负载的 equally loaded
均衡器 equalizer
均衡增长 equilibrium growth
均化剂 leveling agent
均相膜电极 homogeneous membrane electrode
均压器 voltage balancer
均一化调整 adaptive equalization
均匀磁场 uniform magnetic field
均匀地 evenly, homogeneous
均匀电路 uniform electric field
均匀反应堆 homogeneous reactor
均匀冷却 even cooling
均匀性突变脉冲 homogeneity spoiling pulse
均值分析 mean value analysis
竣工图纸 as-built drawing

K

卡槽　card slot
卡车起重机　truck crane
卡尔曼-布西滤波器　Kalman-Bucy filter
卡尔曼滤波器　Kalman filter
卡尔逊应变计　Carlson type strain gauge
卡箍　hoop
卡规　caliper gauge, calipers, snap gauge
卡件安装程序　card installation procedure
卡件插座　card connector
卡具　fixtures
卡口式盖环　cam bezel ring
卡（热量单位）　calorie (cal)
卡曼漩涡流量计　Karman swirlmeter
卡诺图　Karnaugh map
卡盘　chuck
卡片测试延伸板　card test extender
卡片穿孔机　card punch
卡片穿孔装置　card punch unit
卡片复制机　card reproducer
卡片阅读机　card reader
卡片阅读装置　card read unit
卡式磁带　cartridge tape, cassette
卡套式　ferrule-type
卡套式球阀　globe valve with sealing bushing
开［尔文］（温度单位）　Kelvin
开尔文电桥　Kelvin bridge
开尔文电桥式欧姆表　Kelvin bridge ohmmeter
开尔文绝对静电计　Kelvin absolute electrometer
开尔文双臂电桥　Kelvin double bridge
开发者　developer
开方器　square root extractor
开放程序　unpacking procedure
开放界面　open interface
开放系统　operating system
开放系统互联　open system interaction (OSI)
开关　switching, contactor, on-off, switch
开关闭合信号　switch closure
开关变量　switching variable
开关表面　switching surface
开关代数　switching algebra
开关单元　switch unit
开关点　switching point
开关定律　switching law
开关函数　switching function
开关控制　bang-bang control, car-switch control, on-off control
开关控制器　on-off controller
开关理论　switching theory
开关脉冲发生器　switching pulse generator
开关时间　switching time
开关时间比例控制器　on-off time proportional controller
开关伺服机构　on-off servo mechanism
开关算法　switch algorithm, switching algorithm
开关随动机构　on-off servo mechanism
开关台　switch-desk
开关特性　switching characteristic
开关网络　switching network
开关位置标志　switch position label
开关位置选择器　switch position selector
开关误操作　faulty switching
开关箱　switchbox
开关仪表　switch instrument
开关仪表功能块　switch instrument block
开关元件　on-off element
开关站　switch station
开关整流器　switching rectifier
开关值　switching value
开管柱　open tubular column (OTC)

开环　open loop, open-loop
开环传递函数　open loop transfer function
开环极点　open loop pole
开环结构矩阵　open structure matrix
开环控制　open loop control
开环控制系统　open-loop control system
开环频率响应　open loop frequency response
开环增益　open loop gain
开孔　tapping
开口皮带传动　open-belt drive
开口销　retainer pin
开链机构　open chain mechanism
开路　open circuit
开路参数　open-circuit parameter
开路插座　open-circuit jack
开路电磁流量计　open loop electromagnetic flowmeter
开路电压　open-circuit voltage
开路试验　open-circuit test
开路状态　open-circuit position
开启式感应电动机　open type induction motor
开始操作　commencement of operation
开氏温标　Kelvin temperature scale, absolute scale, absolute temperature scale
开式链　open kinematic chain
开位置　on position
开支模式　expenditure pattern
锎　californium (Cf)
凯尔文电桥　double bridge
凯撒效应　Kaiser effect
铠装电缆钢丝　armoured cable wire
铠装（电缆用）钢丝　armouring wire
铠装热电偶　armored thermocouple, sheathed thermocouple
铠装线　sheathed wire
勘探　exploration
勘探工程　exploratory works
坎贝尔电桥　Campbell bridge
坎贝尔-斯托克斯日照计　Campbell-Stokes sunshine recorder
看门狗　watchdog
看门狗定时器　watchdog timer
康铜　constantan, konstantan
抗差估计　robust estimation
抗冲击性能　properties of impact resistance
抗臭氧试验　ozone resistance test
抗磁材料　diamagnetic material
抗磁的　diamagnetic
抗磁效应　diamagnetic effect
抗磁性　diamagnetism
抗断强度　breaking strength
抗风化　weather resistance
抗干扰　anti-interference, disturbance rejection
抗干扰的　jamproof
抗干扰滤波器　anti-interference filter
抗干扰能力　antijamming capability
抗干扰设备　anti-interference equipment
抗剪钢筋　shear reinforcement
抗剪键　shear key
抗剪试验　shearing test
抗静电措施　countermeasures against static electricity
抗静电链　anti-static chain
抗静电轮胎　anti-static tyre
抗拉钢筋　tensile reinforcement
抗拉强度　tensile strength
抗拉性能　properties of direct tension
抗流电路　choke circuit
抗硫酸盐硅酸盐水泥　sulfate resistance Portland cement
抗磨铸铁　antifriction cast iron
抗耦双频激电仪　anti-coupling bi-frequency induced polarization instrument
抗蠕变强度　creep strength
抗蚀油脂　corrosion inhibiting grease
抗碎强度　crushing strength
抗弯性能　flexural properties
抗咸水物料　salt water resistant material
抗压测试　compression test
抗压强度　compression strength, the compressive strength
抗压强度试验　test of compressive strength

抗压应力　compressive stress
抗氧化漆　anti-oxidizing paint
抗噪声对策　countermeasures against noise
抗折强度试验　test of bending strength
抗震墙　shear wall
钪　scandium (Sc)
烤炉　grill
烤漆不到位　lack of painting
烤漆厂　painting factory
烤箱　oven
靠模　former
靠模板　master plate
苛性钾　caustic potash
苛性钠　caustic soda
科恩-库恩控制器整定　Cohen and Coon controller setting
科伐丝　kovar wire
颗粒　granule, particle
颗粒材　granular material
颗粒尺寸测量　particle size measurement
壳　shell
壳模铸造　shell casting
壳体试验压力　shell test pressure
可编程的　programmable
可编程的逻辑块　programmable logic block
可编程的只读存储器　programmable read-only memory
可编程功能键分配　programmable function key assignment
可编程控制器　programmable controller
可编程逻辑控制器　Programmable Logic Controller (PLC)
可编程逻辑控制器接口　programmable logic controller gateway
可编程脉冲宽度输出指示调节器　programmable indicating controller with pulse width output
可编程序控制器　programmable logic controller (PLC)
可编程运算器　programmable computing
可编程增益放大器　programmable gain amplifier

可编程只读存储器　programmable read only memory (PROM)
可编程指示调节器　programmable indicating controller
可变长度编码　variable-length code
可变电感　variable inductance
可变电容　variable capacity
可变电容器　adjustable capacitor, adjustable condenser, variable capacitor, variable condenser
可变电阻　adjustable resistance, variable resistance
可变电阻器　adjustable resistor
可变阀门时间控制　variable valve timing control
可变结构　variable structure
可变结构控制　variable-structure control
可变结构系统　variable-structure system
可变孔径流量计　variable orifice flow indicator
可变口径式流量计　variable aperture type flowmeter
可变面积流量计　variable area flowmeter
可变全息干涉仪　variable holographic interferometer
可变性　variability
可变压差流量计　variable head flowmeter
可变延迟多谐振荡器　variable delay multivibrator
可变增益　variable gain
可变增益放大器　variable gain amplifier
可变阻抗功率表　variable impedance power meter
可辨识性　identifiability
可擦可编程的只读存储器　erasable programmable read-only memory
可测试性　testability
可拆链环　connecting link
可沉浸的污水泵　submersible sewage pump
可程式低温培养器　low temperature program type incubators

可导址远程传感器数据公路　highway addressable remote transducer (HART)
可定位寻址　addressable location addressing
可定误差　determinate error
可动侧模板　moving bolster plate
可动法兰普通式铂热电阻温度计　platinum resistance thermometer with movable flange
可动空间　motion space
可断拼板　break-away panel
可锻化处理　malleablizing
可锻铸铁　malleable iron
可分解搜索　decomposable searching
可分解搜索问题　decomposable searching problem
可分配空间　configuration space enable
可分性　divisibility
可关断晶闸管　gate turn-off thyristor (GTO)
可观测规范型　observable canonical form
可观测指数　observability index
可观的　observable, observability
可互操作系统协议　interoperable system protocol (ISP)
可互换附件　interchangeable accessory
可互换终端　interchangeable terminal
可计算一般均衡模型　computable general equilibrium model
可见光检测　visible detection
可接受的质量标准　acceptable quality level
可开闭的吊桥　drawbridge
可靠的　reliable
可靠度　degree of reliability
可靠度工程　reliability engineering
可靠性　certainty, dependability, reliability
可靠性测试系统　reliability test system
可靠性分析　reliability analysis
可靠性理论　reliability theory
可靠性评估　reliability evaluation
可靠性设计　reliability design

可控规范型　controllable canonical form
可控性　controllability
可控雪崩整流器　controlled avalanche rectifier
可控指数　controllability index
可扩充性　augment ability
可扩展系统配置数据　extended system configuration data (ESCD)
可逆的　reversible
可逆电动机　reversible motor, reversing motor
可逆电气传动　reversible electric drive
可逆控制器　reversing controller
可逆启动器　reversing starter
可逆式水表　reversible water meter
可逆系统　reversible system
可逆性　reversibility
可凝性气体　condensable gas
可配置的　configurable
可膨胀性　expansibility factor
可切削铸铁　machine cast pig iron
可燃成分　combustible component
可燃气　combustible gas
可燃气体的爆炸限　explosive limit of flammable gas
可燃烧物料　combustible material
可燃物品　combustible goods
可燃性　combustibility
可设计的　programmable
可实现性　realizability
可视对讲门铃　video interphone
可塑剂　plasticizer
可调变压器　transtat
可调测量范围上限　adjustable higher measuring range limit
可调测量范围下限　adjustable lower measuring range limit
可调电阻接线端子　terminal with adjustable resistance
可调角度试验机　angular testing machine
可调脉冲计数器　adjustable impulse counter
可调启动变阻器　adjustable starting rheostat
可调前置放大器　tunable preamplifier

可调式长度量规　adjustable length gauge
可调手钳　adjustable pliers
可调系数　adjustability coefficients
可调谐滤波器　tunable filter
可调型楔块　adjusting wedge
可调性　adjustability
可调增益　adjustable gain
可调整的　adjustable
可弯曲导管　pliable conduit
可吸入的悬浮颗粒　respirable suspended particulate
可卸式平衡混频仪　removable balanced mixer
可行协调　feasible coordination
可行性　feasibility
可行性研究　feasibility study
可行域　feasible region
可修改性　modifiability
可选的　optional
可寻址的便携测试仪表　addressable portable test instrument
可寻址火灾监视系统　addressable fire monitor system
可延伸护栏　extendible barrier
可移动介质软盘　removable media
可以忽略的　negligible
可以立即启用的电脑系统　turnkey system
可用材料　available material
可用时间　available time
可用性　availability
可用状况　serviceable condition
可再生的　reproducible
可再生能源系统　renewable energy system
可折叠悬臂平台　collapsible cantilever platform
可制造性设计　design-for-manufacturability
可转换元素　fertile element
可追踪的审计数据　traceability and audit records
可组态的控制功能　configurable control function
克拉天平　carat balance
克努森移液管　Knudsen pipette
克原子　gram atom
刻槽机　grooving machine
刻度　graduation of scale, scale division
刻度范围　graduated range, scale range
刻度盘　graduated dial
刻度盘进给装置　dial feed
刻度盘式指示表　dial indicator
刻度吸管　graduated pipettes
刻痕　notch
刻印　stamp mark
氪　krypton (Kr)
空洞率　void content
空腹桁架　Vierendeel truss
空盒气压计　aneroidograph
空间　space
空间波形　spatial waveform
空间传动机构　spatial mechanism
空间电荷　space charge
空间飞船自主性　spacecraft autonomy
空间构架　space frame
空间结构相互作用　space structure interaction
空间连杆机构　spatial linkage
空间排阻色谱法　steric exclusion chromatography
空间凸轮机构　spatial cam
空间运动副　spatial kinematic pair
空间运动链　spatial kinematic chain
空间运载工具　space vehicle
空气　air
空气安全阀　air safety valve
空气出口栅格　air outlet grille
空气储存器　air receiver
空气磁导率　air permeability
空气磁芯　air core
空气电容器　air capacitor
空气动力平衡　aerodynamic balance
空气断路器　air break switch, air circuit breaker
空气阀　air valve
空气阀门　air valves
空气分配器　air distributor
空气干燥器　air dryer
空气格栅　air grill
空气隔断旋塞　air isolating cock
空气过滤器　air cleaner, air filter

空气过滤室　air filter chamber
空气过滤调压器　air pressure regulator-filter
空气换向阀　air shuttle valve
空气回路　air circuit
空气活度监测器　air activity monitor
空气活塞　air piston
空气加压系统　air pressurization system
空气节流器　air restrictor
空气开关　air switch
空气冷却器　air cooler
空气冷却系统　air cooling system
空气粒子　air particle
空气联结阀　air coupling valve
空气流量计　air flowmeter
空气幕送风定温恒温器　forced convection constant temperature ovens with air curtain
空气枪　air gun
空气韧化　air patenting
空气湿度　humidity of the air
空气湿度指示器　air humidity indicator
空气弹簧　air spring
空气调节　air conditioning
空气通风器　air ventilator
空气污染　air pollution
空气箱　air tank
空气压力试验　air pressure test
空气压力天平　air pressure balance
空气压缩机　air compressor
空气压缩机调压器　air compressor governor
空气中凝固　air set
空气阻尼　air damping
空腔稳频振荡器　cavity-stabilized oscillator
空速控制　idle speed control
空调场所　air conditioned location
空调区　air condition area
空调设备　air conditioner
空投式极地浮标　air-deployable buoy
空投自动气象站　air-drop automatic station
空心　air core
空心板　voided slab

空心钢栏　hollow section steel fencing
空心桥板　voided slab
空心铜线　hollow copper wire
空心线圈　air-core coil
空心阴极灯　hollow-cathode lamp
空心阴极原子化器　hollow-cathode atomizer
空心柱　open tubular column (OTC)
空穴　positive carrier, positive hole
空穴电流　hole current
空穴迁移率　hole mobility
空穴载流子　hole carrier
空穴噪声　cavitation noise
空穴作用　cavitation
空压机　air compressor
空载　no load
空载电流　no-load current
空载电压　no-load voltage
空载和短路法　no-load and shot-circuit method
空载启动　no-load starting
空载曲线　no-load curve
空载试验　no-load test
空载损耗　no-load loss
空载特性　no-load characteristic
空载运行　no-load operation
空载转速　no-load speed
空载状态　no-load state
空站　idle stage
空指令　skip
空中交通控制　air traffic control
空转　idle running
空转轮　idler
孔板　orifice plate
孔板流量计　hole-plate flowmeter, orifice flowmeter, orifice plate flowmeter
孔环　annular ring
孔加工　spot facing machining
孔径规　calliper gauge
孔密度　hole density
孔-塞式可变面积流量计　orifice and plug-type variable area flowmeter
孔图　hole pattern
孔位　hole location
孔隙　void
孔隙量　void content

空白　blank
空白半模拟仪表盘　blank semi-graphic instrument panel-board
空白格式　blank form
空隙度　void content
空隙时间　aperture time
控制　controlling
控制板　control board
控制变量　control variable, controlled variable
控制变压器　control transformer
控制标志　control mark
控制参数　control parameter
控制策略　control strategy
控制处理机　control processor
控制触点　control contact
控制单元的启动　control unit start
控制单元的停止　control unit stop
控制单元定义　control unit definition
控制单元状态显示画面　control unit status display panel
控制等级　control hierarchy
控制点　control point, control station
控制电动机　control motor
控制电极　control electrode
控制电缆　control cable
控制电流　control current
控制电路　control circuit, pilot channel
控制电器　control apparatus
控制电压　control voltage
控制阀　control valve, regulating valve
控制范围　control range, regulating range
控制方案　control scheme
控制方程　control equation
控制方式　control mode
控制工程　control engineering
控制功能　control function
控制规则　control law
控制柜　control cabinet
控制函数　control function
控制回路　control loop
控制回路导向图　control loop guideline
控制回路故障排除　control loop troubleshooting
控制回路交互作用　control loop interaction
控制机构　control gear
控制机械　controlling machine
控制级别　levels of control
控制计数器　control counter
控制计算机　process computer
控制技术　control technology
控制继电器　control relay, pilot relay
控制监视器　control monitor
控制键　control key
控制接触点　control contactor
控制结构　control structure
控制精度　control accuracy, control precision
控制井口数量　controlled wellhead number
控制开关　control switch
控制科学与工程　control science and engineering
控制理论与控制工程　control theory and control engineering
控制力矩陀螺　control moment gyro
控制论　cybernetics
控制论系统　cybernetic system
控制命令字　control command word
控制模件冗余　control module redundancy
控制目标　control object
控制盘　control panel
控制盘开关　panel switching
控制屏　control board
控制器　control device, control unit, controller
控制器饱和　controller saturation, saturation of controller
控制器的比例带　proportion band of a controller
控制器设计　controller design
控制器输出　controller output
控制器优化整定　optimization in controller tuning
控制器增益　controller gain
控制器整定　controller setting
控制器子系统　controller subsystem
控制区　controlled area
控制绕组　control winding
控制上限　upper control limit

控制设备　control appliance, control equipment
控制时间范围　control time horizon
控制式自整角机　control synchro
控制室　control room
控制室管理系统　control room management system
控制室环境　control room environment
控制室区　control room area
控制室设计　control room design
控制手柄　control handle
控制输出　control output
控制输出指定站　control output destination
控制输出组件　control output module
控制输入　control input
控制输入源　control input source
控制算法　control algorithm
控制算法初始化　initialization of control algorithm
控制台　console, control console, control desk
控制台打印机　console printer
控制台监视器　console monitor
控制台显示器　console display, console scope
控制台信号处理机　console message processor
控制台用备用仪表板　backup panel for console
控制特性　control characteristic
控制条件　controlled condition
控制图　control drawing
控制图窗口　control drawing window
控制图定义　control drawing definition
控制图举例　control drawing example
控制图之间的自由信号流　free signal flow between control drawing
控制误差　control error
控制系统　control system, controlled system
控制系统的控制特性　control characteristic of a control system
控制系统分析　control system analysis
控制系统计算机辅助设计　computer aided design of control system
控制系统设计　control system design
控制系统仪表化　control system instrumentation
控制系统整定　tuning control systems
控制系统综合　control system synthesis
控制下限　lower control limit (LCL)
控制线　control wire
控制线圈　control winding
控制相关因数　control interaction factor
控制箱存储器卡　nest memory card
控制箱公用卡　nest common card
控制箱体　control console body
控制箱微处理器卡　nest processor card
控制信号　control signal, pilot signal
控制信号显示　control message display
控制信息协议　internet control message protocol (ICMP)
控制要求　control requirement
控制仪表　control instrument, controlling instrument
控制因子　controlling element
控制应用　control application
控制硬件　control hardware
控制用计算机　control computer
控制语言　control language (CL)
控制运算　control computation
控制站　control kiosk
控制证书　certificate of control
控制职能　control function
控制中心　control center
控制装置　controlling device
控制状态显示功能　control status display function
控制组合　control combination
控制组显示块　control group display block
控制作用　control action
口径　caliber
口径测量器　calibrator
口令　password
扣件　fastener
扣紧　tighten
寇乌气压表　Kew pattern barometer
库秤　hopper scale
库存管理系统　inventory management

system
库存控制　inventory control
库仑（电量电位）　Coulomb
库仑表　Coulometer
库仑滴定法　coulometric titration
库仑定律　Coulomb's law, law of electrostatic attraction
库仑摩擦　Coulomb friction
库仑阻尼　Coulomb damping
库仑作用　Coulomb interaction
跨导　transconductance
跨度　span
跨交　crossover
跨式铣刀　straddle cutter
跨线电阻　link resistance
跨装　straddle mounting
块　block
块传递　block transfer
块对角化　block diagonalization
块规　block gauge
块检验　block check
块检验字符　block check character
快变模态　fast mode
快的　fast
快干油漆　quick-drying paint
快恢复二极管　fast recovery diode
快恢复外延二极管　fast recovery epitaxial diodes
快开阀　quick-opening valve
快开式　quick-open type
快凝固环氧胶黏剂　fast-setting epoxy adhesive
快凝水泥　quick-setting cement
快速编程　rapid programming
快速并行算法　fast parallel algorithms
快速处理器　fast processor
快速定时法　fast timing method
快速仿真　high speed simulation
快速傅里叶变换　fast Fourier transforms
快速换模系统　quick die change system
快速继电器　fast relay
快速晶闸管　fast switching thyristor
快速卡尔曼算法　fast Kalman algorithms
快速开关　quick switch
快速排污阀　quick draining valves

快速熔断器　fast acting fuse
快速扫描单色仪　rapid-scan monochromator
快速原子轰击　fast atom bombardment (FAB)
快速运输系统　rapid transit system
快行车道　fast lane
快硬　rapid hardening
快硬硅酸盐水泥　rapid harding Portland cement
快硬水泥　rapid hardening cement
宽标尺自动平衡记录仪　wide scale self-balancing recorder
宽波段　broad band spectrum
宽波段运行　broad-band operation
宽带差动直流放大器　wide band differential DC amplifier
宽带放大器　wide band amplifier
宽带光电调制器　broad band electro-optic modulator
宽带毫伏计　broad band millivoltmeter
宽带随机振动　broad-band random vibration
宽带网　wide area network
宽带运行　broad-band operation
宽度　width
宽度系列　width series
宽度优先搜索　breadth-first search
宽范围温度控制器　wide range temperature controller
宽量程　wide range
宽量程交流毫伏表　wide range AC millivoltmeter
宽量程仪表　wide meter, wide range meter
宽频带　broad band
宽限期　grace period
宽压插孔板流量计　wide range orifice meter
矿产普查与勘探　mineral resource prospecting and exploration
矿场　mine
矿井　pit
矿山罗盘仪　mining compass
矿物　mineral
矿物加工工程　mineral processing en-

gineering
矿物油　mineral oil
矿样　core sample
矿业工程　mineral engineering
框架　framework
框架结构　frame construction
框架劲度　frame stiffness
框架式仪表盘　frame type instrument panel
亏损　loss
馈电　feeding
馈电盘　feeder panel
馈送　feed
馈线　feeder line
馈线吊线　feeder messenger wire
馈线自动化　feeder automation
捆扎用带材　tie band
捆扎用丝（锁口丝）　bag tie wire
扩充的工业标准结构　extended industry standard architecture
扩径模　expander die
扩孔　counterbore
扩孔钻　reamer
扩散　diffusion
扩散泵　diffusion pump
扩散电流　diffusion current
扩散硅式测力计　diffused silicon semiconductor force meter
扩散硅式力传感器　diffused silicon semiconductor force transducer
扩散面结型晶体管　diffused junction transistor
扩散器　diffuser
扩散声场　diffuse field
扩散退火　diffusion annealing
扩散系数　diffusion coefficient
扩散型半导体应变计　semiconductor strain gauge
扩音器　amplifier
扩音系统　public-address system
扩展 V 网　extending the V net
扩展标度　expanded scale
扩展标度尺仪表　expanded scale instrument
扩展槽　accessory slot
扩展存储器　expanded memory
扩展的额定电流　extended rating current
扩展的额定型电流互感器　extended rating type current transformer
扩展卡尔曼滤波器　extended Kalman filter
扩展控制器　extended controller
扩展网络　extended network
扩展总线　expansion bus（EB）

L

垃圾袋	garbage bag
垃圾焚化炉	incinerator
拉拔	draw out
拉拔模入口	die approach
拉出器	puller
拉床	broaching machine
拉钉枪	rivet gun
拉杆	draw bar
拉格朗日对偶性	Lagrange duality
拉格朗日法	Lagrangian method
拉孔	broaching, draw hole
拉力	drawing force, tensile force
拉力的	tensile
拉力试验机	tensile testing machine
拉曼分光光度计	Raman spectrophotometer
拉曼散射光	Raman scattering light
拉普拉斯变换	Laplace transform
拉普拉斯反变换	inverse Laplace transform
拉普拉斯方程式	Laplace's equation
拉普拉斯-高斯分布	Laplace-Gauss distribution
拉伸	pulling, stretching
拉伸成形模	stretch form die
拉伸冲击试验	tensile impact test
拉伸矫直机	stretcher leveler
拉氏变换	Laplace transform
拉丝模板	wortle plate
拉索	guy
拉条组件	tension assembly
拉铁丝	bracing wire
拉系索	lashing wire
拉线	tightening wire
拉线井	draw pit
拉线开关	ceiling switch, cord switch
拉线箱	draw box
拉应力	tensile stress
拉制钢丝	drawn wire
喇叭线	horn wire
来自窗口启动菜单	from windows start menu
来自系统生成菜单的帮助	from help menu of builder
铼	rhenium (Re)
拦截围墙	catch fence
栏板锁杆	locking lever
栏栅	barrier
蓝宝石硅片	silicon on sapphire
蓝色	blue
蓝烟	blue smoke
蓝移	blue shift
镧	lanthanum (La)
揽货	cargo collection
缆绳	rope
缆索紧固锚	rope anchorage
缆索卷筒槽	rope drum groover
缆索起重机	cable cranes
浪涌电压	surge voltage
劳斯表	Routh tabulation
劳斯-赫尔维茨判据	Routh-Hurwitz criterion
劳斯近似判据	Routh approximation method
劳斯稳定判据	Routh stability criterion
铹	lawrencium (Lr)
老虎钳	pliers, vice, vise
老化	decrepitation
老化处理	ageing
老化性能测定仪	aging property tester
铑	rhodium (Rh)
酪蛋白	casein
勒克斯（光照度单位）	lux (lx)
雷达	radar
雷达控制	radar control
雷达网	radar network
雷达显示器	radar display
雷达制导系统	radar guidance system
雷电计	ceraunograph
雷诺方程	Reynolds's equation
镭	radium (Ra)
泪滴盘	teardrop pad
泪孔	weep hole
类模器	analog-mode device

累积时差　accumulated time difference
累积误差　accumulated error
累积雨量器　accumulative raingauge
累计光能计　integrating actinometer
累计误差　cumulative error
累计仪表　totalizing measuring instrument
累加　accumulate
累加器　accumulator
累进误差　progressive error
楞次定律　Lenz's law
冷　cold
冷藏库　cold storage
冷藏瓶　vacuum flask
冷冲螺母　cold punched nut
冷处理　cold treatment
冷冻机　chiller
冷冻溶剂焊接　cold solvent welding
冷冻水泵　chilled water pump
冷冻温度　cryogenic temperature
冷端　cold junction
冷端参考点　cold junction reference
冷锻　cold forging
冷锻模用钢　cold work die steel
冷镦钢丝　cold-heading wire
冷风风扇　chilled air fan
冷风生铁　cold blast pig iron
冷挤压制模　cold hobbing
冷加工　cold machining
冷拉铜　hard drawn copper
冷拉线　hard drawn wire
冷料渣　cold slag
冷流道　cold runner
冷凝　condensation
冷凝器　condenser
冷凝容器　condensate pot
冷凝装置　refrigeration plant
冷却　cooling
冷却板　chill plate
冷却除湿盘管　cooling and dehumidifying coil
冷却风扇　cooling fan
冷却管　cooling pipe
冷却罐　cooling tank
冷却剂　coolant, cooling agent
冷却流体　cooling fluid
冷却螺管　cooling spiral

冷却盘管　cooling coil
冷却器　cooler
冷却水泵　cooling water pump
冷却水温度表　coolant temperature gauge
冷却水循环器　cooling water circulators
冷却塔　cooling tower
冷却系统　cooling system
冷却液　cooling fluid
冷却液体循环器　cooling liquid circulators
冷式压铸　cold chamber die casting
冷态电阻　cold resistance
冷阴极离子源　cold-cathode source
冷硬用铸模　chill mold
冷硬铸铁　chilled cast iron
冷油　tack coat
冷轧钢丝　cold reduced steel wire, cold-rolled steel wire
厘　centi
厘米波校准仪　centimeter-wave calibrating instrument
离岸　off shore
离地净高　ground clearance
离合器　clutch
离合器覆盖　clutch lining
离合器轮壳　clutch boss
离合器制动器　clutch brake
离解　dissociation
离散　discrete
离散测量　discontinuous measurement
离散的　discontinuous
离散傅里叶变换　discontinuous Fourier transformer
离散化　discretization
离散积分仪　discrete integrator
离散控制系统　discrete control system
离散类型　discrete type
离散量输出　discrete output
离散量输入　discrete input
离散时间　discontinuous time
离散时间检测　discontinuous time detection
离散时间模型　discrete-time model
离散时间模型时域响应　time-domain

responses for discrete time model
离散时间系统　discontinuous time system, discrete-time system
离散时间响应　discrete-time response
离散时间信号　discrete-time signal
离散事件动态系统　discontinuous-event dynamic system, discrete event dynamic system
离散事件系统　discontinuous-event system
离散数字动态控制　discontinuous digital dynamic control
离散系统　discontinuous system, discrete system
离散系统仿真　discrete system simulation
离散系统仿真语言　discrete system simulation language
离散系统模型　discrete system model
离散信号　discrete signal
离散余弦变换　discrete cosine transformer
离散状态　discrete state
离线　off line, off process
离线编程　off-line programming
离线测量　off-line measurement
离线测试　off-process test (OPT)
离线环境　off-process environment
离线控制　off-line control
离线诊断　off-line diagnostic
离线装载数据　down line loading (DLL)
离心泵　centrifugal pump
离心泵技术条件　technical specifications for centrifugal pumps
离心的　centrifugal
离心风机　centrifugal fan
离心过滤器　centrifugal filter
离心荷载　centrifugal load
离心机　centrifuge
离心机型号编制方法　model designation of centrifuges
离心开关　centrifugal switch
离心力　centrifugal force
离心力式平衡机　centrifugal balancing machine
离心密封　centrifugal seal

离心式积分器　centrifugal integrator
离心式离合器　centrifugal clutch
离心式启动开关　centrifugal starting switch
离心式转速表　centrifugal tachometer
离心调速器　centrifugal governor
离心型冻结干燥器　centrifugal freeze dryers
离心应力　centrifugal stress
离子　ion
离子泵　ion pump
离子传感器　ion sensor, ion transducer
离子传输效率　ion transmission efficiency
离子氮化　plasma nitriding
离子的　ionic
离子缔合物　ion pair, ion-pair
离子电镀　ion plating
离子动能谱　ion kinetic energy spectra
离子对　ion-pair, ion pair
离子对色谱法　ion pair chromatography, paired ion chromatography
离子对试剂　ion pair reagent
离子对提取法　ion pair extraction method
离子丰度　ion abundance
离子光学　ion optics
离子轰击二次电子像　ion bombardment secondary electron image
离子化损失谱法　ionization lose spectroscopy (ILS)
离子计　ion-activity meter
离子计数器　ion counter
离子加速电压　ion accelerating voltage
离子减薄机　ion beam thinner
离子检测器　ion detector
离子交换薄层色谱法　ion exchange thin layer chromatography
离子交换低压色谱法　ion-exchange low pressure chromatography
离子交换动电色谱　ion-exchange electrokinetic chromatography
离子交换剂　ion exchanger
离子交换容量　ion exchange capacity

离子交换色谱法　ion exchange chromatography (IEC)
离子截面积检测器　cross-section ionization detector
离子流风速表　ion flow anemometer
离子排斥极　ion repeller
离子排斥色谱法　ion-exchange chromatography
离子强度　ion strength
离子散射谱　ion-scattering spectrum
离子散射谱法　ion-scattering spectroscopy (ISS)
离子散射谱仪　ion scattering spectrometer
离子色谱法　ion chromatography
离子渗碳处理　ion carburizing
离子渗碳氮化　ion carbonitriding
离子淌度　ion mobility
离子淌度光谱仪　ion mobility spectrometer
离子透镜　ion lens
离子选择测量系统　ion-selective measuring system
离子选择场效应管　ISFET
离子选择电极　ion selective electrode, ion-selective electrode
离子选择电极分析　ion-selective electrode analysis
离子选择电极式气体传感器　ion-selective electrode gas sensor, ion-selective electrode gas transducer
离子液相色谱法　ion liquid chromatography
离子抑制色谱法　ion suppression chromatography
离子源　ion source
离子中和　ion neutralization
离子中和谱法　ion neutralizing spectroscopy (INS)
离子中和谱仪　ion neutralization spectrometer
黎卡提　Riccati
黎卡提微分方程　Riccati equation
李雅普诺夫　Lyapunov
李雅普诺夫方程　Lyapunov equation
李雅普诺夫方法　Lyapunov method
李雅普诺夫函数　Lyapunov function
李雅普诺夫渐近稳定性定理　Lyapunov theorem of asymptotic stability
李雅普诺夫稳定性　Lyapunov stability
里程表　odometer
理论　theory
理论的　theoretical
理论模型　theoretical models
理论啮合线　theoretical line of action
理论斜率因数　theoretical slope factor
理想变压器　ideal transformer
理想导线　ideal wire
理想的　ideal
理想低通滤波器　ideal low-pass filter
理想电流源　ideal current source
理想电压放大器　ideal voltage amplifier
理想电压源　ideal voltage source
理想电源　ideal source
理想发动机　hysteresis motor
理想回转器　ideal gyrator
理想控制器　ideal controller
理想气体温标　ideal gas temperature scale
理想条件　ideal condition
理想系统　idealized system
理想元件　ideal element
理想值　ideal value
理想终值　ideal final value
锂　lithium (Li)
力　force
力标准机　force standard machine
力测量仪表　force measuring instrument
力传感器　force sensor, force transducer
力多边形　force polygon
力封闭型凸轮机构　force-closed cam mechanism
力矩　torque
力矩电动机　torque motor
力矩法　moment method
力矩放大器　torque amplifier
力矩控制　torque control
力矩同步　torque synchro
力控制　force control

力量　strength
力量训练　strength training
力敏元件　mechanical sensor
力偶　couple
力偶不平衡　couple unbalance
力偶矩　moment of couple
力平衡　force balance
力平衡变送器　force-balance transmitter
力平衡式比例调节器　force balance proportional controller
力平衡式加速度传感器　force-balance acceleration transducer
力平衡式加速度计　force-balance accelerometer
力平衡原理　force-balance principle
力学　mechanics
力学的　mechanical
力学性能　mechanical property
力学性能测定仪　mechanical property tester
历史　history
历史处理器　history processor
历史模块　history module
历史模块处理单元　history module processing unit
历史模块状态显示画面　history module status display
历史趋势　historical trend
历史趋势画面　historical trend panel
历史事件　event history
历史信息报告　historical message reports
立方米（体积单位）　cubic metre
立管式压力计　vertical-tube manometer
立剖面图　sectional elevation
立式车床　vertical lathe
立式加工制造中心　vertical machine center
立式加工中心　vertical machining centers
立式健身车　upright cycle
立式铣床　vertical milling machines
立式止回阀　vertical lift check valves
立式钻床　drilling machines vertical
立体剖面图　sectional axonometric drawing
立体视角　stere vision

利萨如图形　Lissajou's figures
利益函数　profit function
励磁变阻器　field rheostat
励磁电流　exciting current, field current
励磁电路　exciting circuit
励磁电压　excitation voltage, exciting voltage
励磁放电　field discharge
励磁机　exciter
励磁绕组　exciting winding
励磁系统　excitation system
励磁线圈　field coil
沥滤液　leachate
沥青　asphalt, bitumen
沥青衬里　bitumen lining
沥青的防水度　water impermeability of asphalt
沥青的拉力　tensile strength of asphalt
沥青的耐热度　heat resistance of asphalt
沥青的吸水度　water absorption of asphalt
沥青防水膜　bituminous waterproof membrane
沥青混凝土　asphaltic concrete, bituminous concrete
沥青喷洒机　asphalt distributor
沥青摊铺机　asphalt paver
沥青涂层　asphaltic coating
沥青外衬　bitumen coating
沥铸成形法　slush molding
例外制作　exception build (EB)
例行的　routine
隶属度　membership
隶属函数　membership function (MF)
粒度分析仪　particle size analyzer
粒间体积　interstitial volume
粒径分析　particle size analysis
粒料　pellet
粒子探测器　particle detector
连分式　continued fraction
连分数　continued fraction
连杆　connecting rod
连杆曲线　coupler-curve
连杆调节螺钉　connection screw

连接 conjunction, connecting
连接到子系统 connection to subsystem
连接端 end connection
连接杆 connecting link
连接管线 connecting pipeline
连接环节 connecting link
连接机制 connectionism
连接技术 interconnection technology
连接矩阵 interconnection matrix
连接链节 connecting link
连接片 connecting link
连接器 connection box, connector
连接器插头 connector plug
连接数据点 bound data point
连接匣 connection box
连接线误差 connecting wire error
连接形式 connecting format
连锁编辑 linkage editor
连通性 connectivity
连续 sequential
连续变换 continuous transformation
连续变量 continuous variable
连续操作 continuous operation
连续测试结果 consecutive test result
连续成形加工 progressive forming
连续冲击台 bump testing machine
连续带搅拌反应器 continuous stirred tank reactor (CSTR)
连续到离散时间变换 continuous to discrete-time conversion
连续的 continuous
连续分度 continuous index
连续负荷 continuous duty
连续工作 continuous duty
连续轨道控制 continuous path control
连续过程 continuous process
连续荷载 continuous load
连续护栏 continuous barrier
连续检查 consistency check
连续搅拌釜［槽］式反应器 continuous stirred tank reactor (CSTR)
连续开关 sequential switching
连续控制 continuous control
连续控制功能 continuous control function
连续控制功能块 continuous control block
连续控制算法 sequential control algorithm
连续控制系统 continuous control system
连续离散事件混合系统仿真 continuous discrete event hybrid system simulation
连续梁 continuous beam
连续溶解保温炉 aluminum continuous melting and holding furnaces
连续生产系统 continuous system
连续时间滤波器 continuous time filter
连续时间系统 continuous time system
连续时间信号 continuous time signal
连续使用 continuous duty
连续数据点 continuous data point
连续送料 progressive
连续弯曲加工 progressive bending
连续系统 continuous system
连续下料加工 progressive blanking
连续相位调制 continuous phase modulation
连续信号 continuous signal
连续信号重组 reconstruction of continuous signal
连续性 continuity
连续性试验 continuity test
连续循环方法 continuous cycling method
连续引伸加工 progressive drawing
连续语音识别 continuous speech recognition
连续运算 continuous operation
连续运行 continuous duty
连续展开 continuous development
连续自动送料 transfer feed
连续作用 continuous action
连续作用控制器 continuous controller
联苯 biphenyl
联动机构 linkage
联动开关 gang switch, linked switch
联动扫描 linked-scanning
联二苯 biphenyl

联合基脚　combined footing
联机的　on line
联机设备　online equipment
联机诊断　online diagnosis
联结概率　joint probability
联结翼板　coupling flange
联络单　liaison
联络点　contact points
联锁　interlock, interlocking
联锁按钮　block push-button
联锁触点　interlocking contact
联锁继电器　interlocking relay
联锁顺序试验　sequence test
联锁装置　interlocking device
联想记忆模型　associative memory model
联轴器连接　coupling joint
炼钢生铁　conversion iron
炼后熔渣　finishing slag
炼油厂　petroleum refinery
链　chain
链测长度　chainage
链传动　chain drive
链接编辑器　linkage editor
链轮　chain wheel, sprocket gear, sprocket-wheel
链码校准　captive chains calibration
链式表　chained list
链式打印机　chain printer
链式反应　chain reaction
链式集结　chained aggregation
链式码　chain code
良品　accepted goods, accepted parts, good parts, good product, qualified products
粮食、油脂及植物蛋白工程　cereals, oils and vegetable protein engineering
两倍的　double
两极式　two plate
两维凸轮　two-dimensional cam
两位式调节器　differential gap controller
两位作用　two-step action
两相的　two phase
两相电动机　two-phase motor
两相电路　two-phase circuit
两相感应电动机　two-phase induction motor
两相伺服电动机　two-phase servomotor
两相瓦特计　two-phase wattmeter
两性溶剂　amphoteric solvent
两用表　dual meter
两爪钉　cramp iron
两爪铁扣　dog iron
亮度　brightness, intensity, luminance
亮度传感器　luminance sensor, luminance transducer
亮度计　brightness meter
亮度调节器　brightness regulator
量程　measuring span
量程迁移　span shift
量程上限　upper range-limit
量程上限值　upper range-value
量程调整器　span adjustment
量程误差　span error
量程下限　lower range limit
量程允差　span tolerance
量纲系统　dimensional system
量纲转换函数　dimensional transfer function
量规定位板　gauge plate
量规因数　gauge factor
量化　quantization
量化编码器　quantizing encoder
量化的　quantized
量化频率　sampling frequency
量化器　quantizer
量化器设计　quantizer design
量化误差　quantization error
量化信号　quantization signal, quantized signal
量化噪声　quantization noise, quantized noise
量化状态　quantized state
量块　gauge block
量气计　aerometer
量筒　graduated cylinder
量筒量杯　graduated flask
量值　value of quantity
量制　system of quantities
量子数　quantum number
钌　ruthenium (Ru)

料斗　hopper
料斗秤　hopper weigher
料斗送料　hopper feed
料片厚度　material thickness
料套式模具　loading shoe mold
列　column
列车自动保护速限　automatic train protection speed limit
列车自动控制装置　automatic train control（ATC）
列出归档目录　list archive catalog
列出命令文件清单　list command file
列出目录清单　list directory
列举　enumeration
列奇偶校验字段　column-parity field
烈性炸药　trinitrotoluene
裂边　edge crack
裂变　fission
裂变产物　fission products
裂变物质　fissionable material
裂缝　crack
裂缝分布图　cracking pattern
裂纹　fissure
裂纹扩展应变计　crack propagation strain gauge
裂纹图形　cracking pattern
邻接　abutment
邻接端　abutting end
邻近　neighbourhood
邻近效应　proximity effect
邻频功率比　adjacent-channel power ratio（ACPR）
林产化学加工工程　chemical processing engineering of forest products
林业工程　forestry engineering
临床控制系统　clinical control system
临床药物浓度仪　analyzer for clinic medicine concentration
临界的　critical，marginal
临界点　critical point
临界点干燥器　critical evaporator
临界电流密度　critical current density
临界电压　critical voltage
临界电阻　critical resistance
临界构件　critical element
临界荷载　critical load
临界激发　critical stimulus

临界胶束浓度　critical micelle concentration（CMC）
临界流　critical flow
临界流量测量　critical flow measurement
临界流喷嘴　critical flow nozzle
临界区域　critical area
临界缺陷　critical defect
临界衰减　critical damped
临界速度　critical speed
临界条件　critical condition
临界通道分析　critical path analysis
临界温度热敏电阻器　critical temperature thermistor
临界温度系数热敏电阻器　critical temperature coefficient thermistor
临界稳定性　critical stability，marginal stability
临界稳定状态　critical stable state
临界物质　critical mass
临界压差　critical pressure differential
临界压差比　critical pressure differential ratio
临界压力比　critical pressure ratio
临界黏性阻尼　critical viscous damping
临界值　threshold
临界转速　critical speed
临界状态模式　critical state model
临界阻尼　critical damping
临时措施　interim measure
临时改善工程　interim improvement works
临时基底应变计　temporary base strain gauge
临时结构　temporary construction
临时支架　falsework
磷　phosphorus（P）
磷化处理　bonderizing
磷硫火焰离子化检测　phosphorus sulfur flame ionization detection
磷青铜　phosphorous bronze
磷青铜线　phosphor-bronze wire
磷酸盐　phosphate
磷酸盐皮膜处理　phosphating
灵敏　sensitive
灵敏度　degree of sensitivity，limita-

tion of sensibility, sensitivity
灵敏度分析　sensitivity analysis
灵敏度函数　sensitivity function
灵敏度控制　sensitivity control
灵敏度控制器　sensitivity controller
灵敏度时间控制　sensitivity-time control
灵巧控制　agile control
铃　bell
菱形盘　diamond pad
菱形阵　diamond array
零　zero
零标度标记　zero scale mark
零标度线　zero scale mark
零地址　zero-address
零点检定器　calibrator for ice-point
零点校正　zero correction
零点漂移　zero shift
零点迁移　zero shift
零点提升　zero elevation
零点提升范围　elevated-zero range
零点允差　zero tolerance
零电阻电流表　zero-resistance ammeter
零基线性度　zero-based linearity
零基一致性　zero-based conformity
零极点配置　pole zero assignment
零极点相消　pole-zero cancellation
零件　part
零件表　parts list
零件分散图　exploded drawing
零件名称　name of parts
零件图　parts drawing
零件详图　details drawing
零交叉电平检测器　zero-cross-level detector
零阶保持器　zero-order holder
零解耦　decoupling zero
零静差　absence of offset, zero offset
零距触发脉冲发生器　zero-range trigger generator
零漂　zero drift
零频率　zero frequency
零频增益　zero frequency gain
零上检定器　calibrator above ice-point
零时间调节　zero time reference
零输入响应　zero input response

零位　zero bit
零位调节　zero control
零位调整　zero adjustment
零位调整器　zero adjuster
零位误差　zero error
零位指示器　zero indicator
零下检定器　calibrator below ice-point
零线　naught line
零相电流继电器　zero phase current relay
零相序继电器　zero phase-sequence relay
零序　zero sequence
零序对称分量　zero-sequence symmetrical component
零序分量　zero-sequence component
零序系统　zero-sequence system
零序阻抗　zero sequence impedance
零值　zero value
零状态响应　zero state response
领示控制器　pilot controller
领域　domain
领域知识　domain knowledge
令牌传递方式　token passing
令牌传送　token pass
令牌规程　token-ring protocol
令牌环　token ring
溜槽　chute
流变学　rheology
流程图　flow diagram, flow chart, flow sheet
流程图窗口　graphic window
流程图定义　graphic definition
流程图对象　graphic object
流出式黏度计　efflux viscometer
流出液　eluate
流道　runner, shoot
流道顶出器　runner ejector set
流道拉销　runner lock pin
流道模板　runner plat
流道平衡　runner balance
流道脱料板　runner stripper plate
流道系统　runner system
流动　flow
流动剖面　flow profile
流动气象站　mobile weather station
流动调整器　flow conditioner
流动相　mobile phase

流动相流速　flow rate of mobile phase
流动相平均线速　mean linear velocity of mobile phase
流动相前沿　mobile phase front
流动载体电极　electrode with a mobile carrier
流动整直器　flow straightener
流动注射分析　flow injection analysis
流关　flow to close
流痕　flow mark
流经管道横截面的流体流量　flow rate of a fluid through a cross section of a conduit
流开　flow to open
流量　flow, flow rate, flux
流量比　flow ratio
流量变送器　flow transmitter
流量补偿算法　flow compensation algorithm
流量测量　flow measurement
流量测量校准　flow measurement calibration
流量测量仪器　flow measuring device
流量传感器　flow sensor, flow transducer
流量范围　flow-rate range
流量积算仪　flow integrator
流量计　flowmeter, flow meter, flow instrument
流量计二次仪表　flowmeter secondary device
流量计一次仪表　flowmeter primary device
流量监测　flow monitoring
流量开关　flow switch
流量控制　flow control
流量控制阀　flow control valve
流量喷嘴　flow nozzle
流量特性　flow characteristic
流量特性曲线　flow characteristic curve
流量调节阀　flow control valve
流量弯管　flow elbow
流量稳定器　flow stabilizer
流量系数　discharge coefficient, flow coefficient
流量信号　flow signal
流量修正器　flow corrector

流量指示控制器　flow indicator controller
流量指示仪　flow rate indicator
流明（光通量单位）　lumen
流式细胞仪　flow cytometer
流水板　flow board
流速　current velocity
流速计　current meter
流速型流量计　velocity flowmeter
流体　fluid
流体单元动压　dynamic pressure of fluid element
流体动力学　fluid dynamics
流体机械及工程　fluid machinery and engineering
流体绝对静压　absolute static pressure of the fluid
流体力学　fluid mechanics
流体力学体积　hydrodynamic volume
流体摩擦　fluid friction
流线　current line
流线型张线　streamline wire
留隙角　clearance angle
硫　sulphur (S)
硫化氢　hydrogen sulfide
硫化橡胶　vulcanized rubber
硫磺　sulphur (S)
硫酸　sulphuric acid
硫酸灰分　sulphated ash
硫酸铈法　cerium sulphate method
硫酸盐　sulphate
硫酸酯　sulphate
馏分收集器　fraction collector
六分仪　sextant
六杆机构　six-bar linkage
六角车床　turret lathe
六角盖头螺帽　hex cap nuts
六角钢丝　hexagonal steel wire
六角割沟螺帽　hex slotted nut
六角锯齿螺帽　hex serrated nuts
六角立式铣床　turret vertical milling machines
六角轮缘螺帽　hex flange nuts
六角螺帽　hexagon nut
六角头铰制孔用螺栓　hexagon fit bolts
六角头螺杆带孔铰制孔用螺栓　hexagon fit bolts with split pin hole on

shank
六角头螺杆带孔螺栓　hexagon bolts with split pin hole on shank
六角头螺栓　hexagon headed bolt
六角头头部带槽螺栓　hexagon bolts with slot on head
六角重型螺帽　heavy hex nuts
六面锻造　six sides forging
龙格-库塔积分方法　Runge-Kutta integration method
龙门刨床　double housing planer
龙门起重机　gantry crane
龙头　stop cock
笼式感应电动机　squirrel cage induction motor
笼形转子　cage rotor
楼面荷载　floor load
漏磁场　leakage magnetic field, stray field
漏磁探伤法　leakage magnetic field inspection
漏磁探伤仪　leakage magnetic flow detector
漏磁通　leakage flux
漏电　short circuit
漏电保护断路器　earth leakage circuit breaker (ELCB)
漏电感　leakage inductance
漏电继电器　leakage relay
漏电抗　leakage reactance
漏电流　drain current
漏电指示器　leakage detector
漏电阻　leakage resistance
漏斗　hopper
漏斗秤　hopper scale
漏斗式泥浆黏度计　funnel-shaped mud viscometer
漏件　missing part
漏件式落料模　blank through dies
漏水报警器　water leakage alarm meter
露点　dew point, dew-point
露点传感器　dew point sensor, dew point transducer
露点计　dew-point hygrograph
露点湿度计　dew point hygrometer
露点温度　dew point temperature

露点仪　dew-point meter
露端型铠装热电偶　exposed junction type sheathed thermocouple
露光计　actinometer
露量计　drosometer
露天场地　open area
炉底灰　furnace bottom ash
炉灶面　cooktop
炉渣水泥　blast-furnace slag cement
颅内压传感器　intracranial pressure sensor, intracranial pressure transducer
卤素　halogen
卤素灯　halogen light
卤素计数器　halogen counter
卤素检漏仪　halogen leak detector
鲁棒　robust
鲁棒变换　robust transmission
鲁棒估计　robust estimation
鲁棒估计器　robust estimator
鲁棒控制　robust control
鲁棒稳定　robust stability
鲁棒稳定性　stability robustness
鲁棒系统　robust system
鲁棒性　robustness
鲁棒性能　robust performance
录像机　videocorder
录音磁头　recording head
录音钢丝　magnetic wire
录音机　recorder
录音介质噪声　recording media noise
录音通道　recording channel
路灯电箱　road lighting pillar box
路灯具　road lighting lantern
路径冲突　conflicting routes
路径规划　path planning
路径可重复性　path repeatability
路径问题　routing problem
路口容量　junction capacity
路面排水　surface drainage
路由选择　routing
路闸　barrier gate
路障　barrier block
铝　aluminium (Al)
铝带　aluminium tape
铝合金　aluminium alloy
铝合金线　aluminium alloy wire

铝螺帽 aluminum nuts
铝塑复合板 aluminium-plastic composite panel
铝线 aluminium wire
履历卡 career card
绿色设计 design for environment, green design
氯 chlorine (Cl)
氯仿 chloroform
氯化聚乙烯防水卷材 chlorinated polyethylene plastic sheets for waterproofing
氯化锂湿敏电阻器 lithium chloride humidity-dependent resistor
氯化氢 hydrogen chloride
氯化物 chloride
氯化物含量 chloride content
氯化物扩散 chloride diffusion
氯离子 chloride ion
氯离子含量 chloride ion content
滤波 filtering, smoothing
滤波电路 smoothing circuit
滤波电容器 filter capacitor, smoothing capacitor
滤波放大器 filter amplifier
滤波技术 filtering technique
滤波理论 filtering theory
滤波器 wave filter, smoother
滤波器电路 filter circuit
滤波器设计 filter design
滤波器特性 filter characteristic
滤波器通带 filter transmission band
滤波器稳定性 filter stability
滤波器元件 filter cell
滤波器阻带 filter stop band
滤波式气体分析仪 optical filter gas analyzer
滤波问题 filtering problem
滤波装置 filter device
滤罐型呼吸器 cartridge type respirator
滤光镜 glass filter
滤光片 optical mask, spectral filter
滤网压差 filter differential pressure
滤油罐 oil filter canister
滤油器 oil filter
滤油器滤芯 fuel filter element

滤质器的接收容限 acceptance of the mass filter
卵形（蒸馏）瓶 matrass
伦琴 roentgen
轮齿表面 tooth surface
轮齿接触斑点 tooth contact pattern
轮齿接触分析 tooth contact analysis
轮齿接触面 tooth bearing
轮齿剖面图 tooth layout
轮齿收缩 tooth taper
轮机工程 marine engine engineering
轮壳螺帽 wheel nuts
轮廓 contour, profile
轮廓光学投影仪 profile projector
轮廓锯床 contouring machine
轮廓模 profile die
轮系 gear train
轮轴 wheel axle
罗盘 compass
罗盘经纬仪 compass theodolite
罗西-皮克斯流动试验 rossi-peakes flow test
逻辑 logic
逻辑部件 logical unit
逻辑操作 logical operation
逻辑操作功能块 logical operation block
逻辑槽数量 number or logic slot
逻辑乘 logical product
逻辑代数 logic algebra
逻辑单元 logic unit
逻辑的 logical
逻辑地址 logical address
逻辑点 logic point
逻辑电路 logic circuit, logical circuit
逻辑分析器 logic analyser
逻辑分析仪 logic state analyzer
逻辑管理综合控制器 logic manager (LM)
逻辑和 logical sum
逻辑环 logical ring
逻辑混合 logic mix
逻辑节点 logical node
逻辑控制 logic control, logical control
逻辑控制器 logic controller
逻辑控制压力 logic control pressure
逻辑块 logic block (LB)

逻辑门　logic gate
逻辑门电路　logic gate circuit
逻辑模拟　logic simulation
逻辑模型　logical model
逻辑设备标识符　logical device identifier
逻辑设计　logic design
逻辑设计自动化　logic design automation
逻辑数据点　logic data point
逻辑算法　local algorithm
逻辑梯形图　ladder logic diagram
逻辑图　logic diagram
逻辑应用　logic application
逻辑运算　logic operation, logical operation
逻辑阵列　logic array
逻辑最小化　logic minimization
螺钉　screw
螺杆泵　screw pumps
螺距　pitch of thread, thread pitch
螺口插座　screw socket
螺母　nut, nutsert, screw nut
螺栓　bolt
螺丝打头机　heading machine
螺丝刀　screw driver
螺丝滑头　slipped screwhead, slippery screw head
螺纹　thread
螺纹加工　thread processing
螺纹磨床　grinders thread
螺纹切削　thread cutting
螺纹升角　lead angle
螺纹栓　threaded bolt
螺纹套管接头　screw joint
螺纹效率　screw efficiency
螺线管　solenoid
螺线管执行机构　solenoid actuator
螺旋　helix, screw
螺旋波纹管　helical bellows
螺旋传动　power screw
螺旋方向　hand of spiral
螺旋钢筋　spiral reinforcement
螺旋规　helicograph
螺旋机构　screw mechanism
螺旋桨式流量计　propeller flowmeter

螺旋角　helix angle, spiral angle
螺旋接合　screw joint
螺旋冷却栓　cooling spiral
螺旋流动试验　spiral flow test
螺旋密封　spiral seal
螺旋塞　screw plug
螺旋伞齿轮　spiral bevel gear
螺旋式回旋通道　access spiral loop
螺旋式熔断器　screw base fuse
螺旋式水龙头　screw tap
螺旋弹簧　coil spring, helical spring, spiral spring
螺旋线　helical line
螺旋压缩弹簧　helical compression spring
螺旋运动　helical motion
螺旋桩　screw pile
螺旋锥齿轮　helical bevel gear
螺旋钻孔桩　bored pile
裸板　bare board
裸露电极　bare electrode
裸露式热电偶温度传感器　bare thermocouple temperature transducer
裸铜线　bare copper wire
裸线　bare wire, open wire, skinned wire
洛仑兹电子　Lorentz electron
洛氏表面硬度计　superficial Rockwell hardness tester
洛氏硬度　Rockwell hardness
洛氏硬度计　Rockwell apparatus
洛氏硬度试验　Rockwell hardness test
络合滴定　compleximetry
络合色谱法　complexation chromatography
落锤　blocking hammer
落锤式冲击试验机　landing impact testing machine
落地式　floor model
落地式数控铣床　CNC floor-type milling and boring machine
落点　drop point
落料模　punching die
落球冲击试验　falling ball impact test
落球黏度计　falling sphere viscometer

M

麻点　pock
麻花钻　twist drill
麻黄碱　ephedrine
麻醉机　anesthetic equipment
马达　motor
马达启动器　motor starter
马顿斯耐热试验　Martens heat distortion temperature test
马尔特机构间歇传动轮　geneva wheel
马弗炉　Muffle Furnaces
马赫数　Mach number
马可夫参数　Markov parameter
马可夫决策过程　Markov decision process
马可夫决策问题　Markov decision problem
马可夫模型　Markov model
马口铁　tin, tinplate
马力　horsepower
马氏体　martensite
马蹄形磁铁　horseshoe magnet
马蹄形电磁铁　horse-shoe electromagnet
码变换　code conversion
码分多址　code-division multiple access (CDMA)
码序发生器　code sequence generator
埋藏的　buried
埋地电线　buried wire
埋电阻板　buried resistance board
埋孔　buried via hole
埋入地下的混凝土　buried concrete
埋入砂芯　filling core
埋入式应变仪　embedded strain gauge
埋头孔　counterbore, countersink
埋头螺丝　countersunk socket set screw
麦克风　microphone
麦克斯韦电桥　Maxwell bridge
麦克斯韦方程　Maxwell equation
麦氏真空计　McLeod vacuum gauge
脉冲　impulse, pulse
脉冲（导压）管线　impulse pipe
脉冲比率　ratio rate

脉冲编码器　pulse encoder
脉冲变压器　pulse transformer
脉冲波　impulse wave
脉冲波形　pulse form, pulse shape
脉冲持续时间　pulse duration
脉冲传递函数　pulse transfer function
脉冲电动机　pulse motor
脉冲电流发生器　impulse current generator
脉冲电流检测　pulsed-amperometric detection
脉冲电路　impulse circuit, pulse circuit
脉冲多路输入卡　pulse multiple input card
脉冲发射　pulse radiation
脉冲发生器　impulse generator
脉冲放大器　pulse amplifier
脉冲分配器　pulse distributor
脉冲分频器　countdown frequency divider
脉冲分析仪　impulse analyzer
脉冲幅度　impulse amplitude
脉冲函数　delta function, impulse function
脉冲计数器　impulse counter
脉冲继电器　impulse relay
脉冲尖峰　pulse spike
脉冲键控器　pulse keyer
脉冲精密声级计　impulse precision sound level meter
脉冲控制　impulse control
脉冲控制的　pulse controlled
脉冲控制器　pulse manipulator
脉冲宽度　pulse width
脉冲宽度控制　pulse duration control
脉冲宽度输出附件端子　additional terminals for pulse width output
脉冲宽度调制　pulse-width modulation
脉冲累加值　pulse accumulation value
脉冲频率控制　pulse frequency con-

trol
脉冲群 pulse train
脉冲群发生器 group pulse generator
脉冲声 impulsive sound
脉冲式测距仪 impulse distance meter
脉冲输入 impulse input
脉冲输入模件 pulse input module
脉冲输入模拟量输出卡 pulse input analog output card
脉冲速率指示仪 pulse rate indicator
脉冲条件 impulse condition
脉冲调节器 impulse regulator
脉冲调宽控制系统 pulse width modulation control system
脉冲调频控制系统 pulse frequency modulation control system
脉冲调制 pulse modulation
脉冲调制的 pulse modulated
脉冲调制法 pulse modulation method
脉冲调制器 pulse modulator
脉冲调制系统 variable duration impulse system
脉冲同步示波器 pulse synchroscope
脉冲位置调制 pulse position modulation
脉冲响应 impulse response, impulse-forced response, pulse response
脉冲响应模型 impulse response model
脉冲削波电路 peaker
脉冲信号 impulse signal, pulse signal
脉冲形式 pulse shape
脉冲形状合成 pulse shape synthesis
脉冲序列 impulse sequence, pulse train
脉冲周期 clock cycle
脉动 pulsation
脉动的 pulsating
脉动电流 pulsating current
脉动电压 pulsating voltage
脉动无级变速 pulsating stepless speed changes
脉动信号 intermittent signal
脉动循环应力 fluctuating circulating stress
脉动载荷 fluctuating load
脉宽键控器 pulse-width keyer

脉宽控制器 pulse duration controller
脉宽调制 pulse duration modulation (PDM), pulse-width modulation (PWM)
脉宽调制逆变器 PWM inverter
满标度流量 full scale flow rate
满量程 full scale
满载 full load
满载测试 full load test
满载电流 full-load current
满载试验 full-load test
满载转矩 full-load torque
曼彻斯特编码 Manchester encoding
曼哈顿距离 Manhattan distance
慢变子系统 slow subsystem
慢燃线 slow-burning wire
慢速处理器 slow processor
漫射 diffusion
忙 busy
忙碌状态 busy state
忙闲度 duty cycle
盲板 blinding plate
盲段 dead section
盲孔 blind hole, blind via
盲目搜索 blind search
盲区 dead zone
毛发湿度表 hair hygrograph, hair hygrometer
毛发湿度记录仪 hair hygrograph
毛圈长 feather length
毛细管超临界流体色谱法 capillary supercritical-fluid chromatography
毛细管等电聚焦 capillary isoelectric focusing
毛细管电色谱法 capillary electrochromatography
毛细管电泳 capillary electrophoresis
毛细管离子分析 capillary ion analysis
毛细管黏度计 capillary viscometer
毛细管凝胶电泳 capillary gel electrophoresis
毛细管气相色谱法 capillary gas chromatography
毛细管气相色谱仪 capillary gas chromatograph
毛细管区带电泳法 capillary zone e-

lectrophoresis (CZE)
毛细管柱　capillary column
毛细管柱气相色谱法　capillary column gas chromatography
毛细裂缝　hair crack
毛细提升高度　height of capillary rise
毛细现象　capillary phenomenon
锚　anchor
锚定板　anchor plate, plate anchor
锚固长度　anchorage length
锚固强度　anchoring strength
锚链垫板　flash plate
锚栓　anchor bolt
锚索　anchoring wire
锚销　anchor pin
锚座　anchor bearing
铆钉　rivet
铆钉机　riveting machine
铆合不良　defective to staking, poor staking
铆合模　riveting die
冒烟　smoke
冒烟测试　smoke test
梅森增益公式　Mason's gain formula
煤气表　gas meter
煤气耗用量　gas consumption
煤气热水炉　gas geyser
煤气用具　gas appliance
煤气总管　gas main
煤油　kerosene
酶式葡萄糖传感器　glucose enzyme sensor, glucose enzyme transducer
酶电极　enzyme electrodes
酶多联免疫测定技术　enzyme-multiplied immunoassay technique
酶联免疫吸附测定　enzyme-linked immunosorbent assay (ELISA)
酶免疫测定　enzyme immunoassay
酶敏电极　enzyme substrate electrode
镅　americium (Am)
每分钟转数　revolution per minute
每相电流　current by phase
美国大众运输协会　American Public Transit Association
美国国家标准学会　American National Standards Institute (ANSI)
美国信息交换标准码　American Standard Code for Information Interchange (ASCII)
镁　magnesium (Mg)
镁铝合金　magnalium
门电路　gate circuit
门槛　door sill
门捷列夫称量法　Mendeleev weighing
门开关　door switch
门框　door frame
门联锁触点　door contact
门联锁装置　door interlock
门限电流　threshold current
门限电压　threshold voltage
门限消失　threshold decomposition
门限元件　threshold element
门阵列　gate array
钔　mendelevium (Md)
蒙乃尔铜-镍合金　monel
蒙特卡罗　Monte Carlo
蒙特卡罗法　Monte Carlo method
蒙特卡罗仿真　Monte Carlo simulation
蒙特卡罗计算　Monte Carlo calculation
锰　manganese (Mn)
锰钢　manganese steel
锰铁　manganese iron
锰铜线　manganin wire
弥散　dispersion
迷宫式密封　labyrinth seal
米（长度单位）　metre (m)
米汉纳铁　meehanite metal
米汉纳铸钢　meehanite cast iron
泌尿学　urology
秘诀　know how
密闭空间　confined space
密闭性检查　leak detection
密度　density
密度测量　density measurement
密度传感器　density sensor, density transducer
密度和含水率　density and water content
密度计　density meter
密度修正　density correction
密耳（千分之一英寸）　mil
密封　seal
密封棒　sealing bar
密封带　seal belt, sealing tape

密封的	hermitic
密封垫圈	seal washer, sealing washer
密封罐	sealing pot
密封环	sealing ring
密封剂	sealant, jointing compound
密封件	sealing
密封胶	seal gum
密封胶条	sealing rubber strip
密封接头	sealing fitting
密封膜片	diaphragm seal
密封配件	sealing fitting
密封圈	ball seat
密封式电动机	hermitic motor
密封式晶体管	packaged transistor
密封室	seal chamber
密封型热敏电阻器	enveloped thermistor
密封元件	potted component
密封装置	sealing arrangement
密缝錾	calking tool
密合封板平台	close-boarded platform
密集安装式输入/输出模件	densely-installed I/O module
免疫血型传感器	blood-group immune transducer
面板	bezel panel
面板插座	panel connector
面板功能块	faceplate blocks
面板功能块功能	faceplate block function
面板开关	panel switching
面板设定	panel set
面板式仪表	panel meter
面层装饰	facing decorations
面对面安装	face-to-face arrangement
面内弯曲振动	in-plane bending vibration
面漆	finish paint, finishing coat
面扫描	line-by-line scanning
面外弯曲振动	out-of-plane bending vibration
面外摇摆振动	wagging vibration
面向产品生命周期设计	design for product's life cycle
面向对象的程序设计	object-oriented programming
面向过程	procedure-oriented
面向过程的仿真	process-oriented simulation
面向机器语言	machine-oriented language
面向商业的通用语言	common business-oriented language
面向问题语言	problem-oriented language
面锥角	face angle
面锥母线	face cone element
描述	description
描述符	descriptor
描述函数	describing function
描述顺序的方法	methods of representing sequences
描述系统	descriptor system
秒表	stop watch
灭弧	extinction of arc
灭火器	extinguisher, fire extinguisher, fire-extinguisher
灭声器	muffler, silencer
灭音套	noise suppressing case
民用煤气	town gas
民用燃气	town gas
敏感性	susceptibility
敏感元件	sensing element, sensing probe
敏捷	agile
敏捷制造	agile manufacturing
名义应力	nominal stress
明场电子像	bright field electron image
明杆	rising stem
明沟	ditch
明火	naked flame
明火直热式汽化器	direct-fired vaporizer
明挖回填法	cut and cover method
明线	open wire, surface wiring
明装	unfold install
铭牌	nameplate
命令	command
命令报文	command message
命令不符	command disagree
命令操作	command operation
命令接受	command accepted
命令输出	command output

命令消息　command message
命令字典　command data dictionary
命中　hit
命中质量系数　hit quality index (HQI)
模板　die plate, shuttering
模板库　template base
模板印痕　plate mark
模衬　die insert
模唇　die lip
模锻法　drop forging
模工　die worker
模糊　fuzziness
模糊辨识器　fuzzy identifier
模糊传感器　fuzzy sensor
模糊的　fuzzy
模糊对策　fuzzy game
模糊关系　fuzzy relation
模糊规则　fuzzy rule
模糊规则库　fuzzy rule base
模糊化　fuzzification
模糊混杂系统　fuzzy hybrid system
模糊基函数　fuzzy basis functions (FBF)
模糊集　fuzzy set
模糊集理论　fuzzy-set theory
模糊监控　fuzzy supervision
模糊决策　fuzzy decision
模糊控制　fuzzy control
模糊控制器　fuzzy controller
模糊理论与应用　fuzzy theory and application
模糊逻辑　fuzzy logic
模糊逻辑控制器　fuzzy logic controller
模糊逻辑系统　fuzzy logic system
模糊模型　fuzzy model
模糊模型化　fuzzy modelling
模糊评价　fuzzy evaluation
模糊神经网络　fuzzy-neural network, fuzzy neural network (FNN)
模糊输出　fuzzy output
模糊输入　fuzzy input
模糊数据　fuzzy data
模糊推理　fuzzy inference, fuzzy reasoning
模糊推理矩阵　fuzzy reasoning matrix
模糊误差　ambiguity error
模糊系统　fuzzy system

模糊信息　fuzzy information
模糊专家系统　fuzzy expert system
模糊子集　fuzzy subset
模糊自适应滤波　fuzzy adaptive filter
模架　formwork
模具　die, mold
模具备品　spare dies
模具备品仓　spare molds location
模具工程　die engineering
模具固定用零件　die fastener
模具缓冲垫　die cushion
模具结构图　die structure dwg
模具孔　die opening
模具零件　mold components
模具寿命　die life
模具叶状模槽　die button
模口挤痕　shock line
模口角度　die approach
模块　die block, module
模块化　modularization
模块化程序设计　modular programming
模块化设计　modular design
模块控制　modular control
模块式传动系统　modular system
模块总貌画面　module summary display
模拟　analog, analogy, simulation
模拟比较计算仪表　analogue comparing calculating hardware
模拟比较控制仪表　analogue comparing control hardware
模拟操作功能块　analog operation blocks
模拟测量仪表　analogue measuring instrument
模拟测试　dummy test
模拟乘法器　analog multiplier
模拟除法器　analog divider
模拟传感器　analog transducer
模拟磁带地震记录仪　analog seismograph tape recorder
模拟磁带记录强震仪　analog magnetic tape record type strong-motion instrument
模拟单元　analog unit (AU)
模拟点　analog point
模拟定位机　simulated positioner

模拟多路转换器　analog multiplexed
模拟发生器　simulative generator
模拟仿真　analog simulation
模拟积分器　analog integrator
模拟计算单元　analogue computing unit
模拟计算机　analog computer, analogue computer
模拟计算机控制　analog computer control
模拟计算元件　analog computing element
模拟开关　analog switch
模拟刻度盘　analog dial
模拟控制　analog control
模拟控制板　mimic control panel
模拟量控制器　analog controller
模拟量控制系统　modulating control system
模拟量输入数据点　analog input data point
模拟-脉冲变换器　analogue-to-pulse converter
模拟面板功能块　analog faceplate block
模拟模型　analog model
模拟器　imitator, simulator
模拟曲线描绘器　analog curve plotter
模拟深层地震仪　analog deep-level seismograph
模拟实验　simulation experiment
模拟示值　analogue indication, analogue read-out
模拟式超导磁力仪　analog superconduction magnetometer
模拟输出　analog output (AO)
模拟输出数据点　analog output data point
模拟输入　analog input (AI)
模拟输入/输出　analog input/output
模拟数据　analog data, analogue data, simulation data
模拟数字适配器　analog digital adapter
模拟-数字瞬态记录仪　analog to digital transient recorder
模拟/数字译码器　A/D encoder
模拟/数字转换　A/D conversion
模拟/数字转换器　A/D converter
模拟数字转换器　analog digital converter
模拟-数字转换器　analog-to-digital converter (ADC)
模拟调节仪表　analog regulator
模拟通道　analog channel
模拟系统　analog system
模拟信号　analog signal, analogue signal
模拟遥测系统　analog telemetering system
模拟中断　simulated interrupt
模胚　mold base
模式　mode
模式报警　pattern alarm
模式辨识　pattern identification
模式的　modal
模式分析　mode analysis
模式基元　pattern primitive
模式结构　mode structure
模式理论　mode theory
模式识别　pattern recognition (PR)
模式识别与智能系统　pattern recognition and intelligent systems
模数　modulus
模/数与数/模转换　A/D and D/A conversion
模-数转换　analog-digital conversion, analogue-to-digital conversion
模-数转换分辨率　resolution in analog-to-digital
模-数转换精确度　analog-to-digital conversion accuracy
模-数转换速度　analog-to-digital conversion rate
模塑电路板　molded circuit board
模态变换　modal transformation
模态集结　modal aggregation
模态矩阵　modal matrix
模态控制　modal control
模态耦合器　modal coupler
模头料道　die approach
模网　lay wire
模型　model
模型逼真度　model fidelity
模型变换　model transformation
模型变量　model variable
模型参考　model reference

模型参考控制　model reference control
模型参考控制系统　model reference control system
模型参考自适应　model reference adaptive（MRA）
模型参考自适应控制　model reference adaptive control
模型参考自适应控制系统　model reference adaptive control system
模型参数拟合　fitting model parameter
模型测试　model test
模型产生　pattern generation
模型分解　model decomposition
模型分析　model analysis
模型跟踪控制器　model following controller
模型跟踪控制系统　model following control system
模型管理　model management
模型画面　templet display
模型夹头　die holder
模型简化　model simplification
模型降价　model reduction
模型降价法　model reduction method
模型近似　model approximation
模型精确度　model accuracy
模型库　model base（MB）
模型库管理系统　model base management system（MBMS）
模型目录　templet directory
模型评价　model evaluation
模型确认　model validation
模型设计　model design
模型试验　mock up test, model experiment
模型算法控制　model algorithm control（MAC）
模型协调法　model coordination method
模型修改　model modification
模型验证　model verification
模型预测控制　model-predictive control（MPC）
模型预测启发控制　model predictive heuristic control（MPHC）
模型置信度　model confidence
模型装载　model loading
模型状态反馈　model state feedback

（MSF）
模修　die repair
模穴托板　cavity retainer plate
模压机　stamping press
模压印制线路板　stamped printed wiring board
模用板　mold platen
模组　die set
膜盒　diaphragm capsule
膜盒气压表　aneroid barograph
膜盒气压计　capsule aneroid
膜盒式差式流量计　bellows differential flowmeter
膜盒式绝对压力表　bellows absolute-pressure gauge
膜盒式孔板流量计　bellows-type orifice meter
膜盒式煤气表　diaphragm gas meter
膜盒式微压计　capsule-type micromanometer
膜盒式压力计　capsule type manometer
膜盒式制动器　diaphragm actuator
膜盒压力表　capsule pressure gauge
膜盒元件　diaphragm-box element
膜片　membrane
膜片式差压计　membrane differential pressure gauge
膜片式电阻温度计　wafer-type-resistance thermometer
膜片式压力表　diaphragm gauge
膜片式压力传感器　diaphragm type pressure sensor
膜片式应变计　diaphragm strain gauge
膜片压力表　diaphragm pressure gauge
膜片压力量程　diaphragm pressure span
膜片有效面积　effective diaphragm area
膜片真空表　membrane vacuum-gauge
膜式压敏电阻器　film varistor
摩擦　friction
摩擦负荷　friction load
摩擦角　friction angle
摩擦力　friction force, frictional force
摩擦力矩　friction moment
摩擦刹车　friction brake
摩擦速度　friction velocity
摩擦误差　friction error, frictional error

摩擦系数	coefficient of friction, friction coefficient
摩擦学设计	tribology design
摩擦圆	friction circle
摩擦圆碟	friction disc
摩擦轧光	friction glazing
摩擦制动器	friction brake
摩擦桩	friction pile
摩擦阻力	frictional resistance
摩托车	motor cycle
磨	grind
磨床	grinding machine
磨床工作台	grinder bench
磨刀石	hone
磨光	polish
磨光钢丝	ground steel wire
磨光机	flint glazing machine
磨光器	polisher
磨耗层	wearing course
磨耗环	wear ring
磨耗试验	rattler test
磨痕	grinding defect
磨机	mill
磨具	abrasive tool
磨粒	abrasive particle
磨轮	grinding wheels
磨盘	abrasive disk
磨石	grinding stone
磨损	abrasion, wear
磨损硬度	abrasion hardness
磨削工具	grinding tools
磨削裂纹	grinding cracks
蘑菇云	mushroom cloud
末端	end
末端吸收	end absorption
末端执行器	end-effector
末级放大器	final amplifier, last amplifier
莫氏锥度量规	Morse taper gauge
默认区域	default area
母板	mother board
母离子	parent ion
母模	female die
母式	matrice
母栓	female plug
母线	busbar
木槌	mallet
木料堆置场	timber yard
木料间隔	timber partition
木桩	timber pile
目标	goal, object, target
目标辨认	object recognition
目标程序计算机	target computer
目标方针	quality policy
目标跟踪	target tracking
目标跟踪滤波器	target tracking filter
目标函数	objective function
目标控制	target control
目标模型技术	object modelling technique
目标位置指示器	target position indicator
目标协调法	goal coordination method
目的论	teleology
目的码	object code
目的系统	teleological system
目的站	destination
目录	catalog, catalogue, directory, list
目录编辑表	edit table of contents
目录和索引的应用表	using table of contents and index
目录文件	directory file
目视的	visual
目视模式识别	visual pattern recognition
目视移动	visual motion
钼	molybdenum (Mo)
钼钢	molybdenum steel
钼丝	molybdenum wire
钼系高速钢	molybdenum high speed steel
幕墙	curtain wall
幕墙承托物	curtain wall supports

N

镎　neptunium（Np）
耐酸滤水器　acid-resistant water purifier
内板　inner plate
内半径　internal radius
内包装　inner package, interior package
内标法　internal standard method
内标式玻璃温度计　enclosed-scale liquid-in-glass thermometer
内标式温度计　enclosed-scale thermometer
内标线　internal standard line
内标准　internal standard
内表面　internal surface
内不等率　inherent regulation
内的　inner, internal
内部电话　house telephone
内部钢丝　inner wire
内部检查　inner parts inspect
内部校验　built-in check
内部校准器　internal calibrator
内部矩阵　inner matrix
内部面积　internal area
内部气孔　internal porosity
内部设定　internal setpoint
内部通信系统　intercommunication system
内部状态　internal dynamics
内部状态开关　internal status switch
内部自行分包制　domestic sub-contracting system
内部组件　internal component
内参比电极　internal reference electrode
内参比试样　internal reference sample
内插近似　interpolation approximation
内插算法　interpolation algorithm
内插误差　interpolation error
内插线圈　inside coil
内插振荡器　interpolating oscillator
内衬　lining

内齿轮　internal gear
内冲头　inner punch
内存储器　built-in storage
内存唤醒　suspend to ram
内存转换中心　memory transfer hub
内挡块　internal stop
内导柱　inner guiding post
内地段　inland lot
内对角接触　bias in
内对流敏感元件　internal-convection sensitive element
内反射光谱法　internal reflection spectrometry
内反射元件　internal reflection element
内分泌控制　hormonal control
内环　inner loop
内接圆面　inscribed circle
内径　clear width
内径尺　inside calliper
内径千分尺　dial bore gauge, inside micrometer
内卡钳　inside calipers
内力　internal force
内六角螺钉　inner hexagon screw
内螺模　male die
内螺纹　inside thread
内螺纹截止阀　internal threaded block valve
内模控制　internal model control（IMC）
内模预测控制　internal model predictive control（IMPC）
内模原理　internal model principle
内摩擦　internal friction
内切刀齿　inside blade
内切刀尖直径　inside point diameter
内圈　inner ring
内燃机　internal combustion engine
内扰　internal disturbance
内容定址存储器　associate storage
内生变量　endogenous variable
内锁信号　internal lock signal
内索　inner wire

内脱料板　inner stripper
内拓扑学　internal topology
内外径比　diameter ratio
内稳态　homeostasis
内镶玻璃　inside glazing
内销包装　domestic package
内斜齿轮　internal bevel gear
内引线　internal lead
内圆磨床　internal cylindrical machine
内圆磨削　internal grinding
内在自励磁电机　machine with inherent self-excitation
内直径　internal diameter
内置金属　embedded metal
内置驱动器　native driver
内置泄压阀　internal relief valve
内重心转子　inboard rotor
内柱塞　inner plunger
内转换　internal conversion
内装式检流计　built-in galvanometer
内锥齿轮　internal bevel gear
内阻　internal resistance
纳什最优性　Nash optimality
钠　sodium (Na)
钠线　sodium wire
氖　neon (Ne)
氖管振荡器　neon oscillator
奈奎特　Nyquist
奈奎斯特滤波器　Nyquist filter
奈奎斯特判据　Nyquist's criterion
奈奎斯特图　Nyquist diagram, Nyquist plot
奈奎斯特稳定判据　Nyquist stability criterion
耐高温的　fire resistant
耐火玻璃　fire-retarding glazing
耐火材料　fire resistant material
耐火导线　fire-resistant wire
耐火混凝土　fire proof concrete
耐火结构　fire resisting construction
耐火内衬　refractory lining
耐火性　fire resistance
耐久极限　endurance limit
耐久寿命　endurance life
耐久物料　durable material
耐久性　durability
耐久性试验　endurance test
耐久性系数　durability factor
耐力训练　endurance training
耐磨板　wedge wear plate
耐磨度　wear resistance
耐磨性　wear resistant
耐热试验　heat test
耐蚀性　corrosion resistance
耐酸铸铁　acid proof cast iron
耐特压润滑剂　extreme pressure lubricant
耐压　proof pressure
耐震壁　shear wall
挠曲临界转速　flexural critical speed
挠曲强度　flexural strength
挠曲试验　deflection test
挠曲试验机　deflection testing machine
挠曲应力　flexural stress
挠曲主振型　flexural principl mode
挠性扁平电缆　flexible flat cable (FFC)
挠性单面印制板　flexible single-sided printed board
挠性多层印制板　flexible multilayer printed board
挠性覆铜箔绝缘薄膜　flexible copper-clad dielectric film
挠性机构　mechanism with flexible elements
挠性双面印制板　flexible double-sided printed board
挠性印制板　flexible printed board
挠性印制电路　flexible printed circuit (FPC)
挠性印制线路　flexible printed wiring
挠性转子　flexible rotor
脑电图　electroencephalogram (EEG)
脑电图传感器　electroence-phalographic sensor, electroence-phalographic transducer
脑力负荷　mental workload
脑模型　brain model
能达性　reach ability
能当量　energy equivalent
能耗制动　dynamic braking
能级　energy level
能级传送器　level transmitter

能级图　energy level diagram
能见度　visibility
能量　energy
能量储存　energy storage
能量处理组件　energy processor module
能量分布　energy distribution
能量分散　energy spread
能量过滤器　energy filter
能量加权搜索　energy weighted acquisition
能量控制　energy control
能量平衡　energy balance
能量色散　energy dispersion
能量守恒　conservation of energy, energy conservation
能量守恒定律　law of conservation of energy
能量守恒原理　principle of conservation of energy
能量输送　power transfer
能量损耗　energy dissipation, energy loss
能量损失谱仪　energy loss spectrometer
能量相关　energy dependence
能量转换　energy transformation
能谱　energy spectra
能谱仪　energy disperse spectroscopy
能源　source of power
能源管理系统　energy management system
能源危机　energy crisis
能源消耗　energy expenditure
能源效率　energy efficiency
尼柯尔斯　Nichols
尼柯尔斯图　Nichols diagram
尼柯尔斯图表　Nichols chart
尼龙　polyamide（PA）
尼龙过滤器　nylon filter
尼龙绝缘块　nylon insulator
尼龙嵌入防松螺帽　nylon insert lock nuts
尼龙塑料　nylon
尼龙套圈　nylon ferrule
泥饼黏滞性测定仪　cake adhesive retention meter

泥刀　trowel
泥钉　soil nail
泥浆　mud
泥浆波　mud wave
泥浆比重计　mud hydrometer
泥浆电阻仪　mud logger, mud resistance meter
泥浆含砂量测定仪　mud sand content meter
泥浆流量计　mud flow meter
泥浆润滑性测定仪　mud lubrification meter
泥浆失水量测定仪　mud water loss meter
铌　niobium（Nb）
霓虹灯　neon light
逆变　inversion
逆变换　inverse transfer
逆变换轨迹　inverse transfer locus
逆变器　invertor
逆电流器　spacer
逆动力学问题　inverse kinematic problem
逆动态控制　inverse dynamics control
逆动态问题　inverse dynamic problem
逆火　back-fire
逆加热速率曲线　inverse heating rate curves
逆矩阵　inverse matrix
逆蒙特卡罗　inverse Monte Carlo
逆奈奎斯特阵列　inverse Nyquist array
逆奈奎斯特图　inverse Nyquist diagram
逆散射　inverse scattering
逆散射问题　inverse scattering problem
逆时针方向的　anticlockwise, counterclockwise counter-clockwise
逆向偏压　reversed bias
逆序　inverted sequence
逆转换函数　inverse transfer function
腻子　sealing compound
年负荷变化　annual load variation
年负荷曲线　annual load curve, yearly load curve
年负荷系数　annual load factor

黏度　viscosity
黏度测定槽　kinematic viscosity baths
黏度计　viscometer, viscous meter, viscosimeter
黏合层　bond coat
黏合剂　emulsifier, binder
黏合漆包线　bonded wire
黏合强度　bonding strength
黏合应力　bond stress
黏结层　bonding layer
黏结力试验　bond test
黏结膜　film adhesive
黏结片　bonding sheet
黏土　clay
黏性　stickiness
黏性的　viscous
黏性力　viscous force
黏性摩擦　viscous friction
黏性阻尼系数　viscous damping coefficient
黏滞摩擦　viscous friction
黏滞阻尼　viscous damping
黏着磨损　adhesive wear
尿沉渣分析器　urine sediments analyzer
尿液分析仪　medical equipments urine analyzer, urine analyzer
啮出　engaging-out, recess action
啮出弧　arc of recess
啮合　engagement, meshing
啮合长度　length of action
啮合点　contact points, mesh point
啮合点轨迹　path of action
啮合角　working pressure angle
啮合面　surface of action
啮合平面　plane of action
啮合曲线　pitch curve
啮合系数　contact ratio
啮合线　line of action
啮合线长度　length of line of action
啮接　bridle joint
啮入　approach action, engaging-in
啮入弧　arc of approach
镊子　nipper, tweezers
镍　nickel (Ni)
镍白口铁　nickel white iron
镍钢丝　nickel steel wire
镍铬电阻线　nichrome resistance wire
镍铬钢　Atbas metal, nickel chromium steel
镍铬合金　chromel alloy
镍铬合金线　nichrome wire
镍铬-康铜热电偶　chromel-constantan thermocouple
镍铬-考铜补偿导线　chromel-copel extension wire
镍铬-镍硅热电偶　chromel-nisiloy thermocouple
镍铬-镍铝热电偶　chromel-alumel thermocouple
镍铬丝　nickel chrome wire
镍-钼热电偶　nickel-molybdenum thermocouple
镍银　nickel silver
拧紧型真空快卸法兰　vacuum technology screwed type quick release flange
凝固　curdle
凝固点　freezing point, solidifying point
凝固热　freezing heat
凝固值　set value
凝胶过滤色谱法　gel filtration chromatography (GFC)
凝胶色谱法　gel chromatography
凝胶渗透色谱法　gel permeation chromatography (GPC)
凝胶渗透色谱仪　gel permeation chromatograph
凝胶体　gel
凝结　condensation
牛顿（力量单位）　newton (N)
牛头刨床　shaping machine
扭簧　helical torsion spring
扭矩　moment of torque
扭矩常数　torque constant
扭力　torsion, torsional force, twisting force
扭力扳手　torque wrench
扭力丝　torsion wire
扭曲　twist
扭曲试验　torsion test
扭振频率　torsional vibration frequency
扭振阻尼器　torsional vibration damper
扭转强度　torsional strength
扭转应力　torsion stress

| 扭转载荷 torsion load
| 钮子开关 toggle switch
| 农村电话线 rural distribution wire
| 农业 agriculture
| 农业电气化与自动化 agricultural electrification and automation
| 农业工程 agricultural engineering
| 农业机械化工程 agricultural mechanization engineering
| 农业生物环境与能源工程 agricultural biological environmental and energy engineering
| 农业水土工程 agricultural water-soil engineering
| 农业用汽车 motor vehicle for agricultural use
| 农用分析仪 agricultural analyzer
| 浓差电池 concentration cell
| 浓度 concentration
| 浓度饱和 bottoming
| 浓度变送器 consistency transmitter
| 浓度极限 limit of concentration
| 浓度计 lysimeter
| 浓度敏感型检测器 concentration sensitive detector
| 浓缩 enrichment
| 浓缩薄层板 concentrating thin layer plate
| 浓缩导磁板 concentrating flux plate
| 浓缩导磁套 concentrating flux sleeve
| 浓缩器 evaporators
| 浓缩铀 enriched uranium
| 浓缩柱 concentrating column
| 努普硬度压头 Knoop hardness penetrator
| 努普硬度值 Knoop hardness number
| 钕 neodymium（Nd）
| 暖风机 fan heater
| 暖气片 heating unit
| 暖气系统 heating system
| 诺顿定理 Norton's theorem
| 诺模图 nomogram
| 诺司卡品 noscapine
| 锘 nobelium（No）

O

欧几里得距离　Euclidean distance
欧姆（电阻单位）　ohm
欧姆的　ohmic
欧姆定律　Ohm's law
欧姆计　ohmer
欧姆接触　ohmic contact
欧氏联轴节　Oldham coupling
欧文电桥　Owen bridge
偶发故障　chance failure
偶极子　dipole
偶然负荷　accidental loading
偶然荷载　accidental loading
偶然误差　accidental error, random error
耦合　coupling, linking
耦合变压器　coupling transformer
耦合的　coupled
耦合电感　coupling inductance
耦合电抗　coupling reactance
耦合电路　coupled circuit
耦合电容　coupling capacity
耦合电容器　blocking condenser, coupling capacitor
耦合方式　coupled modes
耦合函数　coupling function
耦合剂　couplant
耦合控制回路　interacting control loop
耦合滤波器　coupling filter
耦合模型　coupling model
耦合模型分析　coupled mode analysis
耦合模型理论　coupled mode theory
耦合器　coupler
耦合腔法互易校准　coupled chamber method reciprocity calibration
耦合损失　coupling loss
耦合系数　coefficient of coupling, coupling coefficient, coupling factor
耦合系统　coupled system, interacting system
耦合谐振器　coupled resonator
耦合元件　coupling element
耦合振荡电路　coupled oscillatory circuit
耦合振动　coupled vibration
耦合装置　coupled device
耦合阻抗　coupling impedance

P

爬坡速率　creeping speed
帕累托最优　Pareto optimality
帕［斯卡］（压力单位）　pascal（Pa）
拍扣式电缆线夹　claplock cable clamp
拍频　beat frequency
拍频振荡器　beat frequency oscillator
拍子　beat
排出阀　outlet valve
排除故障　fault clearing
排除极限　exclusion limit
排代泵　positive displacement pump
排队　queue, queuing
排队论　queueing theory
排队网络模型　queuing network model
排废管汇　exhaust manifold
排风机　exhaust blower
排空　empty
排料逃孔　opening
排列　rank
排列图　pareto diagram
排列组合　permutation and combination
排气　breathing
排气阀　drain cock, exhaust valve
排气隔板　exhaust baffle
排气管　exhaust stack
排气管道　exhaust duct
排气净化　exhaust purification
排气口　air exhaust
排气螺钉　vent screw
排气排放物控制系统　exhaust emission control system
排气扇　exhaust fan
排气通风机　suction fan
排气系统　exhaust system
排气消声器　exhaust muffler, exhaust silencer
排气循环　exhaust gas recirculation
排气烟度　exhaust smoke
排气有害成分　poisonous exhaust composition
排气肘管　exhaust elbow
排水泵　drainage pump
排水层　drainage layer
排水措施　drainage measure
排水阀　blow down valve, drainage valves
排水干管　main drain
排水工程　drainage works
排水管　drain tube
排水井　catchpit
排水孔　drain hole
排水入口　drainage inlet
排水物料　drainage material
排水系统　drainage system
排污阀　waste valves
排线　flat cable
排泄管　drain pipe
排泄孔　drain holes
排屑输送机　chip conveyor
排序与排程　sequencing and scheduling
排液槽　sump pit
排液旋塞　drain cock
排油软管　oil drain hose
排障器　cowcatcher
盘点数量　physical inventory
盘簧　coil spring
盘面安装　surface mounting
盘塞式流量计　float flowmeter
盘式离合器　plate clutch
盘铣刀　disc-mill cuter
盘形凸轮　disk cam
盘形芯　disc plug
盘形转子　disk-like rotor
盘址　anchoring spur
盘装仪表　shelf mounted instrument
判别函数　discriminant function
判定表　decision table
庞德里亚金极大值原理　Pontryagin's maximum principle
旁管　side pipe
旁路　bypass, by-pass

旁路电容	feed through capacitor	配电器	distributer, distributor
旁路电容器	shunt capacitor	配电网	distribution network
旁路集管	bypass manifold	配电系统	distribution system
旁路接头	by-pass joint	配电线	distribution line
旁路连接	shunt connection	配电线路	distribution feeder
旁路热处理	bypass heat treatment	配电箱	distribution box
旁路位移电流	shunt displacement current	配电站	power sub-station, transformer room
旁频	side frequency	配电装置	distribution equipment
旁热式热敏感电阻器	indirectly heated type thermistor	配方细目显示画面	recipe detail display
		配合公差	fit tolerance
旁通阀	bypass valve, by-pass valve	配基交换剂	ligand exchanger
旁通管式流量计	bypass type flowmeter	配件	accessory
旁通进样器	by-pass injector	配接等化	adaptive equalization
抛光	buffing, burnishing, polishing	配料	ingredient
抛光粉	abrasive grain	配料工厂	batching plant
抛光轮	buffing wheel	配流轴	pintle valve
抛射体	projectile	配气系统	air distribution system
抛物面天线	parabolic antenna	配水池	header tank
抛物线的	parabolic	配位滴定法	complexmetry
抛物线输入	parabolic input	配位体色谱法	ligand chromatography
抛物线运动	parabolic motion	配置	deploy
刨	plane	配置问题	allocation problem, assignment problem
刨齿	gear shaping		
刨齿机	gear shaper	喷出	ejection
刨床	planer, planing machine	喷淋器	drencher
刨刀	plane iron	喷淋头	drencher head
刨尖头	spear head	喷墨式彩色绘图机	spray plotter
炮铜	gun metal	喷墨式记录仪	jet recorder
跑表	timer	喷墨印刷机	ink jet printer
泡浸	pickling	喷漆	spraying painting
泡沫混凝土	foam concrete	喷气推进	jet propulsion
培德近似	Pade approximation	喷枪	spray gun
培养箱	incubator	喷洒器	sprinkler
锫	berkelium (Bk)	喷砂	sand blasting
配电	electric power distribution, power distribution	喷砂处理	sand blast
		喷射	ejected
配电板	electrical panel	喷射泵	ejector pump
配电板式仪表	panel meter	喷射除草剂	weed killer spraying
配电变压器	distribution transformer, substation transformer	喷射管	injection pipe
		喷射混凝土	shotcrete, sprayed concrete
配电馈线	distribution feeder		
配电盘	distribution panel, power distribution panel, switchboard	喷射器	ejector
		喷射润滑	jet lubrication
配电盘式仪表	panel-type instrument	喷射造型法	injection molding
配电盘装配	panel mounting	喷水泵	water jet pump
配电屏	distribution board	喷水孔	spout

喷水器　water jet
喷水系统　water spray system
喷涂器　air sprayer
喷丸处理　shot blast
喷雾干燥器　spray dryer
喷雾控制器　airgun controller
喷雾器　atomizer
喷釉　spraying glazing
喷釉法　spray glazing
喷釉枪　glazing spray gun
喷嘴　nozzle, spray nozzle, spraying nozzle
喷嘴分离器　jet orifice separator
硼　boron
硼酸　boric acid
膨润土　bentonite
膨胀　expansion, inflation
膨胀法　dilatometry
膨胀剂　expanding agent
膨胀率　expansion factor
膨胀石墨　expanded graphite
膨胀系数　coefficient of expansion, expansion coefficient, expansion factor
膨胀珍珠岩绝热制品　expanded perlite insulation
碰　bump
碰撞电离　collision ionization
碰撞激活　collisional activation
碰撞激活质谱计　collisional activation mass spectrometer
批处理　batch processing
批处理仿真　batch processing simulation
批量控制　batch control
批量控制器　batch controller
批量控制站　batch control station
批量历史模型数据点　batch history prototype data point
批量历史数据点　batch history data point
批量趋势　batch trend
批量设定　batch set
批量设定单元　batch set unit
批量设定器　batch set station
批量数据采集器　batch data acquisition unit
批量数据设定单元型端子　batch data set unit terminal
批量状态指示器　batch status indicator
批内精密度　within-run precision
批准　approval
铍　beryllium（Be）
劈裂抗拉强度试验　test of tensile splitting strength
皮带　belt, strap
皮带秤　belt weigher
皮带传动　belt drive
皮带惰轮锯齿板　toothed belt idler plate
皮带护罩　belt guard
皮带拉力　belt tension
皮革化学与工程　leather chemistry and engineering
皮拉尼真空计　Pirani vacuum gauge
皮托管　Pitot tube
皮托静压管　Pitot static tube
疲劳　fatigue
疲劳测试　fatique test
疲劳极限　fatigue limit
疲劳破裂　fatigue breakage
疲劳强度　fatigue strength
疲劳失效　fatigue failure
疲劳试验机　fatigue testing machine
疲劳寿命　fatigue life
疲劳特性　fatigue characteristic
疲劳应变计　fatigue strain gauge
匹配　matching
匹配变压器　matching transformer
匹配测量仪　adapt meter
匹配电路　matching circuit
匹配滤波器　matched filter
匹配指数　match index
匹配准则　matching criterion
片料　sheet stock
片选　chip enable
片状阀　flapper valve
偏差　deviation
偏差报警　deviation alarm
偏差报警检测器　deviation alarm sensor
偏差变量　deviation variable
偏差低限报警点　deviation low alarm trip point

偏差高限报警点　deviation high alarm trip point
偏差继电器　biased relay
偏差检查　deviation check
偏差平方积分　integral of the squared error (ISE)
偏差限值　deviation limit
偏距　offset distance
偏距圆　offset circle
偏离　departure
偏离线性度　deviation from linearity
偏流源　current bias source
偏微分　partial differential
偏微分方程　partial differential equation
偏微分方程离散化　discretization of partial differential equation
偏向　deflection
偏心　off centre
偏心角　eccentric angle
偏心率　eccentricity ratio
偏心旋转阀　rotary eccentric plug valve
偏心载荷　eccentric load
偏心质量　eccentric mass
偏心轴　eccentric shaft
偏压电路　biased circuit
偏压调整电位计　bias control potentiometer
偏移电压　offset voltage
偏置电压　bias voltage
偏置滚子从动件　offset roller follower
偏置尖底从动件　offset knife-edge follower
偏置连接盘　offset land
偏置误差　bias error
偏转　deflecting
偏转电压　deflecting voltage
偏转公差　runout tolerance
偏转角　deflecting angle
偏转力矩　deflecting torque
偏转系数　deflection coefficient
偏转线圈　yoke
偏最小二乘法　partial least squares method
漂流浮标　drifting buoy
漂流卡片　drift card
漂流瓶　current bottle, drift bottle
漂移　drift
漂移电流　drift current
漂移校正放大器　drift corrected amplifier
漂移率　drift rate
漂移迁移率　drift mobility
漂移速度　drift velocity
漂移载流子　drift carrier
拼合锻模　split forging die
拼合接头　splice joint
拼合模　sectional die
拼合铸模　split mold
拼接　splice
拼接板　splice bar
频差　frequency difference
频带　frequency band
频带宽度　band width
频道　frequency channel
频道选择器　channel selector
频分多路传输　frequency division multiplexing
频分多址　frequency division multiple access
频分遥测系统　telemetering system of frequency division type
频率　frequency
频率变换　frequency conversion, frequency transformation
频率标准　frequency standard
频率补偿器　frequency compensator
频率测量　frequency measurement
频率测深仪　frequency sounding instrument
频率成分　frequency component
频率范围　frequency range
频率分布　frequency distrbution
频率分析　frequency analysis
频率分析仪　frequency analyzer
频率跟踪　frequency tracking
频率估计　frequency estimation
频率轨迹　frequency locus
频率计　cymometer, frequency meter
频率校正　frequency correction
频率校准　frequency calibration
频率控制　frequency control

频率扫描　frequency sweeping
频率失真　frequency distortion
频率输出　frequency output
频率特性　frequency characteristic
频率特性函数　transfer function in the frequency domain
频率调整　frequency regulation
频率调制　frequency modulation (FM)
频率-温度系数　frequency-temperature coefficient
频率稳定性　frequency stability
频率相关　frequency-dependent
频率相关特性　frequency-dependent characteristic
频率响应　frequency response
频率响应的最大值　maximum value of the frequency response
频率响应法　frequency response method
频率响应范围　frequency response range
频率响应方法　frequency response method, frequency-response method
频率响应轨迹图　frequency response locus
频率响应曲线　frequency response curve
频率响应特性　frequency response characteristic
频率响应显示仪　frequency response tracer
频率信号分析　frequency signal analysis
频率选择　frequency selection
频率指数　frequency index
频率转换积算器　frequency transducer with integrator
频率转换开关　frequency change-over switch
频谱　frequency spectrum
频谱发生器　spectrum generator
频谱分析　spectral analysis
频谱分析仪　spectrum analyzer
频谱估计　spectral estimation
频谱激电仪　frequency spectra induced polarization instrument
频谱密度　spectral density
频谱密度函数　spectral density function
频谱特性　spectral characteristic
频谱相关　spectral correlation
频谱仪　frequency spectrograph
频移磁强计　frequency shift magnetometer
频移键控　frequency shift keying (FSK)
频域　frequency domain
频域法　frequency domain method
频域分析　frequency domain analysis, frequency-domain analysis
频域模型降阶法　frequency domain model reduction method
频域设计　frequency-domain design
频域稳定　frequency stabilization
品牌　brand
品质　quality
品质保证　quality assurance (QA)
品质保证计划　quality assurance scheme
品质改善　quality improvement
品质改善活动　quality ameliorate notice
品质改善团队　quality improvement team (QIT)
品质工程　quality engineering
品质管理技术和实践　quality management techniques and practice
品质目标　quality target
品质圈　quality control circle (QCC)
品质因数　quality factor
平板　slab
平板色谱法　plane chromatography
平板压光　plate glazing
平板闸阀　flat gate valves
平带　flat belt
平带传动　flat belt driving
平底板　base slab
平底从动件　flat-face follower
平方误差积分准则　integral of squared error criterion
平方滞后　quadratic lag
平分线　bisector
平箍钢　flat hoop iron
平刮铸模　sweeping mold
平衡　balancing
平衡变换器　balance converter
平衡锤　balance weight
平衡的　balanced
平衡点　equilibrium point

平 ping

平衡电动机　balancing motor
平衡电桥　balanced bridge
平衡电源　balanced supply
平衡阀　balance valves
平衡放大器　balanced amplifier
平衡机　balancing machine
平衡机灵敏度　balancing machine sensitivity
平衡继电器　balancing relay
平衡检波器　balanced detector
平衡解调器　balanced demodulator
平衡孔　equalizing orifice
平衡品质　balancing quality
平衡气　balance gas
平衡器　balancer
平衡桥式取样器　balanced bridge sampler
平衡设备　balancing equipment
平衡式差压记录仪　balance-type differential pressure recorder
平衡式限噪器　balanced noise limiter
平衡水管　balancing pipeline
平衡调制器　balanced modulator
平衡指示器　balance indicator
平衡质量　balancing mass
平衡转速　balancing speed
平衡状态　balance condition, balance state, equilibrium state
平滑滤波器　smoothing filter
平滑判据　smoothness criterion
平均场强　average field intensity
平均出厂品质　average output quality (AOQ)
平均出厂品质水平　average output quality level
平均刀尖直径　average cutter diameter
平均电流　average current
平均电压　average voltage
平均风速　average wind speed
平均负荷　mean load
平均高度　mean height
平均功率　average power
平均故障间隔时间　mean time between failures (MTBF)
平均故障时间　mean time to failure (MTTF)
平均绝对值　average absolute value
平均可用度　average availability
平均流量　mean flow-rate
平均强度　average strength, mean strength
平均声级　average sound level
平均时间　mean time
平均寿命　mean life
平均水平　mean level
平均速度　average velocity
平均无故障时间　mean time between failure (MTBF)
平均误差　average error, mean error
平均修理时间　mean time to repair (MTTR)
平均压应力　average compressive stress
平均应变　mean strain
平均应力　average stress, mean stress
平均有效波长　mean effective wavelength
平均值　average value, mean, mean value
平均值的实验标准偏差　experimental standard deviation of the mean
平均值检波器　average value detector
平均值控制　averaging control
平均中径　mean screw diameter
平流扼流圈　smoothing choke
平炉生铁　open-hearth iron
平面波　plane wave
平面电路　plane circuit
平面规　surface gauge
平面机构　planar mechanism
平面节流阀　flat throttling valve
平面连杆机构　planar linkage
平面磨床　surface grinder
平面磨削　plane grinding
平面曲线　horizontal curve
平面色谱法　planar chromatography
平面凸轮　planar cam
平面凸轮机构　planar cam mechanism
平面图则　horizontal plan drawing
平面位置显示器　plane position indicator
平面运动副　planar kinematic pair
平面轴承　plain bearing
平面轴斜齿轮　parallel helical gears

平纹组织　plain structure
平稳随机过程　stationary random process
平行尺　drafting machine
平行垫块　parallel block
平行度　parallelism
平行计算　parallel computation
平压式印刷机　stamping press
评定标准　evaluation standard
评估　assessment
评价　evaluation
评价技术　evaluation technique
评价与决策　evaluation and decision
屏蔽　shielding
屏蔽常数　shielding constant
屏蔽电缆　braided cable, screened cable, shielded cable
屏蔽线　screening wire
屏蔽效应　screening effect
屏蔽装置　shield assembly
屏蔽作用　shielding action
屏幕　screen
屏障　barrier
钋　polonium (Po)
坡度　inclination
坡口焊　groove welding
坡莫合金　permalloy
坡莫合金膜　permalloy film
泼釉　glazing by splashing
钷　promethium (Pm)
迫切危险　imminent danger
破坏性测试　destructive test
破坏性点蚀　destructive pitting
破坏性磨损　destructive wear
破坏压力　rupture pressure

破裂　breakage
破裂压力　burst pressure
破碎强度　crushing strength
剖面显示功能　profiler display function
剖面线　hatching
剖视图　profile chart
铺缆井　cable draw pit
铺设　lay
铺设煤气总管工程　gas main laying works
葡聚糖　sephadex
葡糖苷酸　glucuronides
镤　protactinium (Pa)
普朗克常数　Plank constant
普通操作输入　general purpose inputs
普通齿轮系　gear train with fixed axes, ordinary gear train
普通二极管　general purpose diode
普通硅酸盐水泥　ordinary Portland cement
普通混凝土小型空心砌块　normal concrete small hollow block
普通六角螺帽　hex nuts
普通模式噪声　common-mode noise
普通平键　parallel key
普通强度钢丝　common-strength steel wire
普通水泥　Portland cement
普通卧式压力蒸汽灭菌器　horizontal pressurized steam sterilizer
普通型接线端子　ordinary terminal
谱带扩展　band broadening
谱库检索　library searching
镨　praseodymium (Pr)

Q

期望值　desired value
漆包绝缘线　enamel insulated wire
漆包线　enamel covered wire, enamelled wire, lacquered wire
齐格勒-尼科尔斯方法　Ziegler-Nichols method
齐格勒-尼科尔斯整定　Ziegler-Nichols setting
齐纳二极管　Zener diode
齐纳二极管校准器　Zener diode regulator
齐平导线　flush conductor
齐平印制板　flush printed board
其他故障显示及输入　other common fault alarm display and input
其他特殊螺帽　special nuts
奇点　singularity
奇异　singular
奇异点　singular point
奇异分解　singular value decomposition
奇异控制　singular control
奇异扰动方法　singular perturbation method
奇异位置　singular position
奇异吸引子　singular attractor
奇异系统　singular system
奇异值　singular value
歧管　manifold
歧管挡块　manifold block
歧管仪表　manifold gauge
旗标点　flag point
旗号员　flagman
企业　enterprise
企业标准　company standard
企业管理　enterprise management
企业集成　enterprise integration
企业技术解决　enterprise technology solutions
企业模型　enterprise modelling
企业资源规划　enterprise resource planning (ERP)
启动　start up, start-up
启动按钮　start button
启动变阻器　starting rheostat
启动补偿器　starting compensator
启动程序　start-up sequence
启动次数　number of starts
启动电流　starting current
启动电阻　starting resistance
启动方式　startup mode
启动接触器　starting contactor
启动绕组　starting winding
启动条件　starting conditions
启动元件　trigger element
启动周期　starting period
启动转矩　torque at starting
启动装置　starting equipment
启发式搜索　heuristic search
启发式推理　heuristic inference
起步阻力　breakaway force
起吊试验　hoisting test
起吊止挡　hoisting stopper
起动扭矩　starting torque
起动失败停车　fail to start shutdowns
起落架控制　chassis control
起模板　draw plate
起模长针　draw spike
起模杆　ejector pin, rapping rod
起模装置　die lifter
起始啮合点　beginning of contact, initial contact
起始温度　initial temperature
起重臂　jib
起重电磁铁　lifting electromagnet
起重电动机　crane motor
起重吊钩　lifting hooks
起重杆　lifting rod
起重钢索　hoisting wire
起重滑车　lifting tackle
起重机　chain block, crane, hoist
起重机接触导线　crane trolley wire
起重机控制器　crane controller
起重机设计规范　design rules for cranes
起重机械　lifting appliance

起重机械安全规程　safety rules for lifting appliances
起重机支架　sheer leg
起重力矩限制器　load moment limiters
起重链　hoisting chain, lifting chain
起重马达　hoisting motor
起重绳　hoisting rope
起重轴承　hoisting bearing
起重装置　lifting gear
起子插座　screwdriver holder
起子头　head of screwdriver
气窗　transom window
气电插座　air electric connecting plug
气电转换器　pneumatic to current converter
气-电转换器　pneumatic-electrical converter
气垫　air spring, gas cushion
气垫板　air cushion plate
气动泵　pneumatic pump
气动变送器　pneumatic transmitter
气动薄膜控制阀　diaphragm control valve
气动薄膜调节阀　pneumatic diaphragm control valve
气动薄膜执行机构　pneumatic loading diaphragm actuator
气动差压变送器　pneumatic differential pressure transmitter
气动长行程执行机构　pneumatic long-stroke actuating mechanism
气动串级指示记录调节器　pneumatic cascade indicating and recording
气动单元组合仪表　instruments of the pneumatic aggregate
气动的　pneumatic
气动电开关　air operated electrical switch
气动蝶阀　pneumatic butterfly valve
气动二针记录仪　pneumatic double-pen recorder
气动阀门定位器　pneumatic positioner, pneumatic valve positioner
气动法兰式差压变送器　pneumatic flanged differential pressure transmitter
气动放大器　pneumatic amplifier

气动风闸　air-operated damper
气动副线板　pneumatic by-pass station
气动管　pneumatic tube
气动管线外部连接系统图　scheme of external pneumatic piping connections
气动活塞切断阀　pneumatic piston cut-off valve
气动机构　pneumatic mechanism
气动积分器　pneumatic integrating counter, pneumatic integrator
气动积算器　pneumatic integrating counter, pneumatic integrator
气动继电器　pneumatic relay
气动加法器　pneumatic summing unit
气动加减器　pneumatic adder-subtractor
气动夹紧　pneumatic lock
气动绝对压力变送器　pneumatic absolute pressure transmitter
气动开关　air pressure switch
气动控制　pneumatic control
气动控制阀　pneumatic control valve
气动控制液压泵　pneumatic control hydraulic pump
气动力平衡式传感器　pneumatic force-balance transducer
气动模拟操作器　pneumatic analog station
气动排气能力　pneumatic exhaust capability
气动旁路遥控板　pneumatic by-pass remote control station
气动起子　pneumatic screw driver
气动色带指示仪　pneumatic colour-strip indicator
气动条形指示仪　pneumatic strip-type indicator
气动调节及安全停止阀　pneumatic control and safety shutoff valve
气动温度变送器　pneumatic temperature transmitter
气动系统　pneumatic system
气动限幅器　pneumatic limit operator
气动限位开关　pneumatic limit switch
气动信号　pneumatic signal

气动旋转执行器　pneumatic rotary actuator
气动压力变送器　pneumatic pressure transmitter
气动压力表　pneumatic pressure gauge
气动遥控板　pneumatic remote control station
气动液位调节变送器　pneumatic level controller transmitter
气动一针记录仪　pneumatic single-pen recorder
气动仪表校验台　testing stand for pneumatic instruments
气动执行机构　piston actuator, pneumatic actuator
气动指示记录调节器　pneumatic indicating and recorder controller
气动指示调节器　pneumatic indicating controller
气动制动系统　pneumatic brake system
气动自锁阀　pneumatic self-locking valve
气缸　air cylinder
气缸效率　cylinder efficiency
气固色谱法　gas-solid chromatography
气关　air to close
气管　air pipe
气焊　gas welding
气化　gasification
气开　air to open
气孔　gas vent
气控　air-control
气口　gas port
气流表　airmeter
气流换向器　flow switch
气密的　gas-tight
气密盖　airtight cover
气密式压力计　gas-enclosed pressure gauge
气密式仪表　air tight instrument
气密式仪器仪表　air-tight instrument
气密外壳　hermetically sealed enclosure
气敏电极　gas sensing electrode
气敏元件　gas-sensitive element
气囊筒　bellow pot
气泡　bubble

气泡室　bubble chamber
气泡贮存器　bubble accumulator
气瓶减压器　gas cylinder regulator
气瓶颜色标记　coloured mark for gas cylinders
气瓶用易熔合金塞　fusible plug for gas cylinders
气蚀　cavitation corrosion
气锁　air lock
气胎压路机　pneumatic tyred roller
气态　gaseous form
气体　gas
气体测微计　air micrometer
气体传感器　gas sensor, gas transducer
气体等离子灰化机　gas plasma asher
气体等离子清洗机　gas plasma dry cleaner
气体等离子蚀刻机　gas plasma etcher
气体动力噪声　aerodynamic noise
气体发生器　gas generator
气体放电　gas discharge
气体放电源　gas-discharge source
气体分配站　gas insulated substations
气体分析　gas analysis
气体分析器　gas analyzer
气体隔离　gas insulated
气体激光器　gas laser
气体继电器　Buchholz relay, gas relay
气体检测继电器　gas detector relay
气体流量计算器　gas flow computer
气体密度计　gas densitometer, gas density gauge
气体温度计　gas thermometer
气体氧化法　gaseous cyaniding
气体硬化　air hardening
气体遮蔽　gas shield
气体正比检测器　gas proportional detector
气体轴承　gas bearing
气体状态方程　equation of state of gas
气味　odor
气温　air temperature
气隙　air gap
气隙磁场　air-gap field

气隙磁化线　air-gap line
气隙磁通　air-gap flux, mutual flux
气隙磁通分布　air-gap flux distribution
气隙磁阻　gap reluctance
气相分离计　vapour fractometer
气相色谱法　gas chromatography
气相色谱仪　gas chromatograph
气相色谱-质谱法　gas chromatorgaphy-mass
气相色谱-质谱联用仪　gas chromatograph-mass spectrometer (GC-MS)
气象观测　meteorological observation
气象火箭　meteorological rocket
气象计　meteorograph
气象雷达　meteorological radar
气象数据　meteorological data
气象塔　meteorological tower
气象卫星　meteorological satellite
气象仪器　meteorological instrument
气象指示器　weather indicator
气穴现象　cavitation
气压　pneumatic pressure, air pressure
气压表　air pressure gauge
气压表零点高度　elevation of zero point of barometer
气压冲击体内碎石机　ballistic intracorporeal lithotrite
气压高度计　atmospheric pressure altimeter
气压计　barograph, barometer
气压继电器　barometer relay
气压开关　air pressure switch
气压温度计　barothermograph
气压修正　barometric correction
气压制动器　air brake
气液分离器　gas-liquid separator
气液色谱法　gas-liquid chromatography
气源球阀　ball valve for air supply
气远传转子流量计　pneumatic remote transmitting rotameter
汽车　automobile, motor vehicle
汽车车身　automotive body
汽车的　automotive
汽车的转向轮安装角测定仪　aligner

汽车底盘　automotive chassis
汽车电气设备　automotive electrical equipment
汽车发动机　automotive engine
汽车竞赛　auto-car racing
汽车力学　mechanics of motor vehicle
汽车排气污染　exhaust pollution of motor vehicle
汽车试验　testing of motor vehicle
汽车噪声　noise of motor vehicle
汽车制造工业　automobile industry
汽缸　cylinder
汽缸效率　cylinder efficiency
汽化表　atmometer
汽化器　carburettor
汽轮发电机组　turbo-generator set
汽轮机　steam turbine
汽轮机监测仪表　turbine supervisory instrumentation
汽油　petrol
汽油表　gasoline gauge
汽油发动机　petrol engine
砌箱造模法　brick molding
器件换流　device commutation
器具　appliance
千　kilo (K)
千吨　kiloton
千分表　dial indicator
千分表检查仪　micrometer checker
千分尺　micrometer
千分卡尺　micrometer calipers
千伏安　kilovolt-ampere (KVA)
千伏表　kilovoltmeter
千斤顶　lifting cylinder, lifting jack, jack
千兆 (十亿)　giga (G)
千字节 (1024字节)　kilobyte
迁移　migration
迁移率　mobility
迁移水管　diversion of water main, diversion of water pipe
牵力　towing force
牵入同步　lock in synchronism
牵索　towing cable
牵引的　tractive
牵引机　tractor
牵引力　traction

牵引力扭矩　tractive effort torque
牵引线　fishing wire
铅　lead (Pb)
铅包线　lead-covered wire, lead-sheathed wire
铅垂线　plumb line
铅锤　plumb bob
铅封丝　seal wire, sealing wire
铅管　lead pipe
前板　front plate
前刀面　rake face
前端　front end
前端处理机　front end processor
前后两车时间间隔　headway
前级放大器　prime amplifier
前角　front angle, rake angle
前馈　feedforward
前馈补偿　feedforward compensation
前馈-反馈控制　feedforward-feedback control (FFC)
前馈校正　correcting feedforward
前馈控制　feedforward control
前馈通路　feedforward path
前馈信号　feedforward signal
前伸峰　leading peak
前式加法器　look-ahead adder
前台　foreground
前台程序　foreground program
前台处理　foreground processing
前向差分　forward difference
前向控制　forward control
前向信号　forward signal
前沿　leading edge
前置放大器　head amplifier, preamplifier, prime amplifier
前置警告标志　advance warning sign
前置滤波器　prefilter
前轴定位器　aligner
前锥　front cone
前锥齿冠　front crown
前锥齿冠至交错点　front crown to crossing point
钱币计数及袋装设备　cash counting and bagging equipment
钳工　locksmith
钳口　jaw
钳式电流表　tong-test ammeter

钳位电路　clamping circuit
钳位二极管　clamping diode, catching diode
钳位交流毫安表　clamp-on alternating-current milliammeter
钳位脉冲发生器　clamp pulse generator
钳位器　clamper
钳形流量计　clamp-on flowmeter
钳子　pinchers, pliers
潜热　latent heat
潜入式浇口　submarine gate
潜水电泵　submerged motor pumps
潜水钟　diving bell
潜在危险安装　potentially hazardous installation
欠电压保护装置　under-voltage protection
欠实时仿真　slower-than-real-time simulation
欠压　undervoltage
欠载继电器　underload relay
欠阻尼　underdamped, underdamping
欠阻尼系统　underdamped system
欠阻尼响应　underdamped response
嵌件销　insert pin, retainer pin
嵌入开关　flush switch
嵌入式温度计　embedded thermometer
嵌入式仪表　flush-type instrument
嵌入系统　embedded system
嵌线卷边机　wiring press
嵌装表　flush mounted gauge
嵌装压力表　flush mounted pressure gauge
强大的操作与管理功能　powerful operation and monitoring function
强大的控制与通信功能　powerful control and communication function
强带　strong band
强电工程　heavy current engineering
强度　intensity
强度不够　insufficient rigidity
强度测试　strength test
强度改变　intensity change
强度控制　intensity control
强度试验压力　strength testing pressure

强度调制	intensity modulation
强度调制方法	intensity modulation method
强度系数	strength factor
强跟踪滤波	strong tracking filter (STF)
强健控制	robust control
强烈噪声	intensity noise
强耦合系统	strongly coupled system
强迫换流	forced commutation
强迫振荡	force oscillation
强迫振动	forced vibration
强行励磁	reinforced excitation
强阳离子交换	strong cation exchange
强阴离子交换	strong anion exchange
强制	constrain
强制对流	force convection
强制函数	forcing function
强制励磁	constrained current operation
强制性标准	mandatory standard
强制振荡	forced oscillation
羟化物	hydroxide
跷板开关	rocker switch
敲打成形	beatening
敲击法	hammering method
乔丹日照计	Jordan sunshine recorder
桥秤	weighbridge
桥大梁	bridge girder
桥墩	bridge pier
桥接	bridge connection
桥接电路	bridge circuit
桥接器	bridge
桥接阻抗	bridged impedance
桥梁工程	bridgeworks
桥梁与隧道工程	bridge and tunnel engineering
桥路电阻	bridge resistance
桥门构架	portal frame
桥面板	bridge deck
桥式吊车	bridge crane, traversing crane
桥式吊机	bridge crane
桥式断点	bridge-contact
桥式高阻表	bridge megger
桥式可逆斩波电路	bridge reversible chopper
桥式起重机	bridge crane, overhead crane
桥式整流器	bridge rectifier
桥台	bridge abutment
桥形滤波器	lattice type filter
桥型接触斑点	bridged contact pattern
翘曲	warpage
翘曲试验	warpage test
切边	side cut
切边冲子	trimming punch
切边模	trimming die
切边碎片	edge
切变强度	shear strength
切槽刀	slotting tool
切齿安装距	cutting distance
切齿深度	depth of cut
切断	cutting out
切断机	cutting-off machines
切断开关	cut-out switch
切断器	disconnector
切割	cut
切割机	dicing saw
切割锯	dicing saw
切割器	cutter
切换	cut-over, switched
切换差	differential gap
切换阀	switch-over valve
切换机构	switching mechanism
切换开关型端子	switch terminal
切换值	switching value
切换中值	mean switching point
切机	generator tripping
切孔	cutting opening
切口冲模	blanking die
切口模	lancing die
切口效应	notch effect
切平面	tangent plane
切线	tangent
切向点	tangent point
切向负荷	tangential load
切向力	tangential force
切削深度	cutting depth
亲和色谱法	affinity chromatography
侵蚀防治	erosion control
青熟脆性	blue shortness
青铜	bronze

氢　hydrogen（H）
氢弹　hydrogen bomb
氢电极　hydrogen electrode
氢离子（浓度的）负指数指示剂　pH indicator
氢硫酸　hydrosulphuric acid
氢气减压器　hydrogen regulator
氢酸　hydracid
氢压力表　hydrogen pressure gauge
氢焰离子化检测器　hydrogen flame ionization detector
氢氧化钙　hydrated lime
氢氧化物　hydroxide
轻便式配电屏　portable distribution board
轻便数字测井仪　light digital logging instrument
轻便铁路系统　light rail system
轻工技术与工程　light industry technology and engineering
轻填料　lightweight filler
轻型上盖　lightweight cover
轻油　light oil
倾覆　overturning
倾覆力矩　overturning moment
倾角　angle of inclination, dip
倾角计　inclinometer
倾角罗盘　inclinometer
倾斜　declination
倾斜计　inclinometer
倾斜面　inclined plane
倾斜式压力计　tilting manometer
倾斜误差　inclination error
倾泻阀　dump valve
倾卸斗车　tipper
清仓品　clearance goods
清除　clear, purging
清除目录　purge directory
清除氧化皮　descale
清洁的　clean
清洁剂　detergent
清理棒　cleaning rod
清理孔　cleaning eye
清零　null
清扫　cleanness
清扫器　sweeper
清晰度　clearness

清晰显示　clear display
擎住效应　latching effect
请求　demand, request
请求函　request letter
请求信息结构　solicited message structure
请求注意　look-at-me (LAM)
请求注意信号　look-at-me signal
请勿用钩　no hooks
求导　derivation
求和放大器　summing amplifier
求和器　summator
求积　quadrature
求积仪　planimeter
求解矩阵　resolvent matrix
球　ball
球差系数　coefficient of spherical aberration
球阀　ball valve, globe valve
球化处理　spheroidizing
球面副　spheric pair
球面滚子　spherical roller
球面渐开线　spherical involute
球面渐开线齿　spherical involute teeth
球面运动　spherical motion
球墨铸铁　ductile iron, ductile cast iron, nodular cast iron, spheroidal graphite cast iron
球墨铸铁管　ductile iron pipe
球塞滑块　ball slider
球窝接头　ball-and-socket joint
球销副　sphere-pin pair
球形支承　spherical bearing
球轴承　ball bearing
球状碳化铁　globular cementite
球坐标操作器　polar coordinate manipulator
区带　zone
区带脱尾　zone tailing
区分效应　differentiating effect
区间分化　range splitting
区间平均运算器　cumulative average unit
区位指示器　zone-position indicator
区域　area
区域报警总貌画面　area alarm sum-

mary display
区域定位　area location
区域发电厂　regional power station
区域分析　domain analysis
区域改变　area change
区域供暖　district heating
区域规划模型　regional planning model
区域交通控制系统　area traffic control system
区域名　area name
区域名称组态　area name configuration
区域数据库　area data base
曲板　crank plate
曲柄　crank
曲柄滑块机构　crank-slider mechanism, slider-crank mechanism
曲柄销轴承　crank pin bearing
曲柄摇杆机构　crank-rocker mechanism
曲柄轴　crank shaft
曲管地温表　bent stem earth thermometer
曲径密封垫片　labyrinth seal
曲率　curvature
曲率半径　radius of curvature
曲面从动件　curved-shoe follower
曲线　curve
曲线的　curvilinear
曲线发生器　curve generator
曲线绘制仪　plotting device
曲线拼接　curve matching
曲线运动　curvilinear motion
曲线族　family of curves
曲线坐标记录仪　curvilinear ordinates recording instrument
曲轴平衡机　crank balancing machine
驱动　driving
驱动齿轮箱　driving gearbox
驱动磁铁　drive magnet
驱动电动机　driving motor
驱动电路　driving circuit
驱动电压　driving voltage
驱动放大器　drive amplifier
驱动绞车　driving winch
驱动力　driving force
驱动力矩　driving moment

驱动器　driver
驱动器模式　driver model
驱动器特性　driver behavior
驱动文件　drive file
驱动信号　actuating signal
驱动转矩　driving torque
屈光度　diopter
趋势　trend
趋势窗口　trend window
趋势点窗口　trend point window
趋势分析　trend analysis
趋势功能　trend function
趋势画面　trend panel
趋势记录定义　trend record definition
趋势记录仪　trend recorder
趋势仪表　trend instrument
取料机　taker
取料装置　take out device
取消　cancel
取压点　pressure tapping point
取压分接管　pressure tap
取样比率　rate of sampling
取样触发器　sample flip flop
取样及保持放大器　sample-and-hold amplifier
取样空间　sample space
取样频率　sampling frequency
取样扰动　sampling disturbance
取样显示器　sampling scope
取样用涡轮流量计　sampling type turbine flowmeter
去磁　demagnetization
去磁器　degausser, demagnetizer
去加重电路　deemphasis circuit
去角刀具　chamfering tool
去角斜切　chamfering
去卷积　deconvolution
去毛刺　debur
去耦　decouple
去耦的　decoupling
去耦电路　decoupling circuit
去耦合子系统　decoupled subsystem
去耦滤波器　decoupling filter
去水管　outlet pipe
去铜箔面　foil removal surface
去相关　decorrelation

去脂溶剂　degrease solvent
全包合约　turn-key contract
全波　full wave
全波的　all wave
全波段　all band, all-wave band
全波分析　full wave analysis
全波滤波器　all-wave filter
全波整流　full-wave rectification
全波整流器　full wave rectifier
全部钻孔　all drilled hole
全尺寸测量　absolute dimension measurement
全齿高　whole depth
全齿高齿　full-depth teeth
全磁通　fluxoid
全点　full point form
全镀锌钢丝　fully-galvanized wire
全封闭式电动机　fully enclosed motor
全工序循环　completing cycle
全集成能源管理　totally integrated power
全减器　full-subtracter
全阶观测器　full order observer
全解　complete solution
全景放大器　panoramic amplifier
全局的　global
全局定位系统　global positioning system
全局观测　international survey
全局渐进稳定性　global asymptotic stability
全局说明文件　global description file (GDF)
全局稳定　global stability
全局优化　global optimization
全局最优　global optimum
全绝缘电流互感器　fully insulated current transformer
全跨度　full span
全面大修　complete overhaul
全面地　all sidedly
全面定向道路交汇处　full directional interchange
全面功能试验　complete function test
全面品质管理　total quality management (TQM)
全面生产维护　total productive maintenance (TPM)
全面质量管理　total quality control (TQC)
全屏模式　full-screen mode
全屏幕编辑　full-screen editing
全屏幕处理　full-screen processing (FSP)
全桥测量　full bridge measurement
全桥电路　full bridge converter
全桥整流电路　full bridge rectifier
全球定位系统　global position system (GPS)
全球概览　global overview
全球数据库　global data base
全球移动通信系统　global system for mobile communications
全容量内件　full capacity trim
全世界的　international
全数字分析器　all digital analyzer
全套程序污水处理厂　full sewage treatment plant
全特色批量控制软件包　full-featured batch control package
全天候风向风速计　all-weather wind vane and anemometer
全天空照相机　all-sky camera
全通　all pass
全通滤波器　all pass filter
全通元件　all pass element
全图形显示器　full graphic display
全息地震仪　holographic SDRS
全息干涉仪　holographic interferometer
全息光栅　holographic grating
全息照相　holography
全响应　complete response
全向方位变换器　omni-bearing convertor
全血　whole blood
全压式模具　positive mold
全质量双聚焦　double focusing at all masses
全自动操作　fully automatic operation
全自动的　full automatic
全自动血细胞分析仪　automatic blood cell analyzer
权衡分析　trade-off analysis

权限　authority
权因子　weighting factor
权重集　weight sets
醛　aldehyde
缺点　defect
缺口　nick
缺口修整加工　shaving
缺少的　absent
缺省选择　default option
缺省值　default value
缺陷的回波信号　flaw echo
缺陷分辨力　flaw resolution
缺陷回波　flaw echo
缺陷回声　flaw echo
缺陷灵敏度　flaw sensitivity
缺相　phase not together

缺氧安全装置　oxygen-deficiency safety device
缺油的　oilless
确定　make sure
确定的　definite, deterministic
确定矩阵　matrix determinant
确定性系统　deterministic system
确定性自动机　deterministic automaton
确认　validation
确认操作　acknowledge operation
确认中断　acknowledge interrupt
群　bunch
群延时均衡器　group delay equalizer
群验潮杆　multiple tide staff

R

燃-空比控制	fuel-air ratio control

燃料 fuel
燃料安全系统 fuel safety system
燃料电池 fuel cell
燃料费用 cost of fuel
燃料控制 fuel control
燃料喷射控制 fuel control injection
燃料消耗量 fuel consumption
燃气轮机 gas turbines
燃气涡轮 gas turbines
燃烧 combustion
燃烧法 burning method
燃烧极限 limit of inflammability
燃烧器 burner
燃烧器管理系统 burner management system
燃烧器控制系统 burner control system
燃烧热 heat of combustion
燃烧室 combustion chamber
燃烧性 combustibility
燃烧性能测定仪 combustion property tester
燃油泵 fuel pump
燃油表 fuel gauge
燃油阀 fuel valve
燃油分配泵 fuel dispensing pump
燃油供应 fuel supply tap
燃油供应系统 fuel supply system
燃油管 fuel pipe
燃油管线 fuel line
燃油过滤器 fuel filter
燃油紧急截断杆 fuel emergency cut-off lever
燃油喷嘴 oil nozzle
燃油喷射 fuel injection
燃油喷射泵 fuel injection pump
燃油喷射器 fuel injector
燃油输送泵 fuel transfer pump
燃油吸入装置 fuel intake device
燃油液位控制开关 fuel level control switch
燃油溢出 oil spillage
燃油运送控制系统 fuel transfer control system
染料渗透剂 dye penetrant
染料渗透检查 dye penetrant inspection
染料渗透试验 dye penetrant test
扰动 perturbation
扰动补偿 disturbance compensation
扰动参数 disturbance parameter
扰动分析 perturbation analysis
扰动幅度 range of disturbance
扰动量 disturbance variable
扰动系数 perturbed coefficient
扰动信号 disturbance signal
绕绳滑轮 rope round pulley
绕线管 bobbin
绕行的 circuitous
绕组 winding
绕组铜损耗 winding loss
绕组温度指示器 winding temperature indicator
绕组线 winding wire
绕组展开图 developed winding diagram
热斑 hot mark
热波电子显微镜 thermal wave electron microscope
热波电子像 thermal wave electron image
热冲击 thermal shock
热冲击试验 thermal shock test
热冲击试验箱 thermal shock test chamber
热处理 heat treatment
热处理玻璃观察窗 heat treated glass window
热传导 conduction of heat, heat conduction
热传导检测器 thermal conductivity detector (TCD)
热传导式流量计 heat transfer flowmeter

热磁式氢气分析器　thermal-conductivity hydrogen analyzer
热磁式氧分析器　thermal magnetic oxygen analyzer
热单位　heat unit（HU）
热导池　thermal conductivity cell
热导计　conductometer
热导率　heat conductivity, thermal conductivity
热导率计　thermal conductivity meter
热导式气体传感器　thermal conductivity gas sensor, thermal conductivity gas transducer
热导式气体分析器　thermal conductivity gas analyzer
热导式湿度传感器　thermal conductivity humidity sensor, thermal conductivity humidity transducer
热导体　heat conductor
热的　hot
热滴定　thermal titration
热电　thermal electricity
热电变换器　thermal convertor
热电导线　thermoelectric wire
热电动势　thermal electromotive force
热电高温计　thermoelectric pyrometer
热电离质谱计　thermal ionization mass spectrometer
热电偶　thermal couple, thermocouple, thermoelement
热电偶比较检定法　comparison method of calibrating thermocouple
热电偶电路　thermocouple circuit
热电偶继电器　thermo-relay
热电偶烧断　thermocouple burnout
热电偶输入模块　thermocouple input module
热电偶双极标定法　double-polarity method for calibrating thermocouple
热电偶套管　thermowell
热电偶微差检定法　differential method of calibrating thermocouple
热电偶组件　thermocouple assembly
热电式温度检测器　pyroelectric thermodetector
热电效应　thermoelectric effect
热电站系统　hydrothermal power system
热电子的　thermionic
热电阻　resistance thermometer, resistance thermometer sensor
热电阻传感器　resistance thermometer sensor
热电阻输入信号　RTD input signals
热电阻组件　resistance thermometer assembly
热动式疏水器　thermo-operated steam trap
热镀锌钢丝　hot-dip galvanized steel wire
热端　hot junction
热锻　hot forging
热锻模用钢　hot work die steel
热对流　heat convection
热反应堆　thermal reactor
热分析　thermal analysis
热分析范围　thermal analysis range
热分析曲线　thermal analysis curve
热分析系统　thermo system
热分析仪　thermal analyzer
热分析仪器　thermal analysis instrument
热辐射　heat radiation, thermal radiation
热辐射体　thermal radiator
热固性　thermosetting
热固性树脂　thermosetting resin
热核的　thermonuclear
热红外遥感范围　thermal infrared range remote sensing
热化学式气体分析器　thermal chemical gas analyzer
热继电器　thermal relay
热加工　hotwork
热降解　thermal degradation
热交换器　heat exchanger
热浇道　hot runner
热浸镀　hot dipping
热浸镀锌　hot-dip galvanizing
热精炼　thermal refining
热空气消毒箱　hot air sterilizer
热扩散率　thermal diffusivity
热扩散系数　thermal diffusivity
热拉钢丝　hot drawing wire

热离子检测器　thermal ion detector
热力学温标　thermodynamic scale (of temperature)
热量　heat
热量的　thermal
热量集成　heat integration
热量计　calorimeter
热量计式测试仪表　calorimeter instrument
热量流　heat flow
热灵敏度漂移　thermal sensitivity drift
热零点漂移　thermal zero drift
热流传感器　heat flux sensor, heat flux transducer
热流道模具　hot-runner mold
热流计　heat flow meter, heat-flow meter
热流密度　density of heat flow
热流型差示扫描量热仪　heat-flux differential scanning calorimeter
热敏电缆　heat sensing cable
热敏电阻　critesister, thermistor
热敏电阻测温链　thermistor chain
热敏电阻辐射测量仪　thermistor bolometer
热敏电阻流量计　thermistor flowmeter
热敏印刷机　thermal printer
热敏元件　thermo-sensitive element
热能分析仪　thermal energy analyzer (TEA)
热能工程　thermal power engineering
热能勘测任务　heat-capacity mapping mission (HCMM)
热偶式仪表　thermocouple instrument
热喷雾　thermo spray
热膨胀　thermal expansion, thero expansion
热膨胀系数　thermal expansion coefficient
热膨胀仪　thermal dilatometer
热疲劳试验机　thermal fatigue testing machine
热平衡　heat balance, thermal equilibrium
热容量　calorific capacity, thermal capacity
热容量制图卫星　heat-capacity mapping satellite
热容式热量计　thermal capacitance calorimeter
热熔机　fuse machine
热渗氮　thermal nitridation
热式记录仪　thermal recorder
热式流量计　thermal flowmeter
热式质量流量计　thermal mass flowmeter
热室压铸　hot chamber die casting
热输出　thermal output
热输出系数　thermal output coefficient
热水锅炉　hot water boiler
热水炉　heater
热水循环泵　hot water circulation pump
热丝流量测量仪表　hot wire flow measuring device
热丝式呼吸流量传感器　hot-wire respiratory flow sensor, hot-wire respiratory flow transducer
热塑材料　thermoplastic material
热塑路标漆　thermoplastic road marking paint
热塑性树脂　thermoplastic resin
热塑性塑料　thermoplastic
热损耗　heat loss
热缩式电力电缆终端头　pyrocondensation power cable terminal
热态启动　hot start
热探测器　heat detector
热特性　thermal property
热天平　thermobalance
热通量计　heat flux meter
热同步　hot sync
热弯试验　hot bend test
热稳定性　thermal stability
热物理性能测定仪　thermal physical property tester
热系仪表　thermal instrument
热线风速表　hot wire anemometer
热线流量传感器　hot wire flow sensor, hot wire flow transducer
热线式仪表　hot-wire instrument
热线湍流计　hot-wire turbulence me-

ter
热响应时间 thermal response time
热效率 heat efficiency
热效式延时继电器 thermal time-delay relay
热学分析 thermal analysis
热循环 thermal cycle
热循环试验 heat cycle test
热压配合 shrink fit
热应变 thermal strain
热应力 thermal stress
热浴淬火 hot bath quenching
热噪声 thermal noise
热轧钢筋条 hot rolled steel bar
热值 calorific value, heat value
热滞现象 thermal hysteresis
热中子 thermal neutron
热重分析法 thermogravimetric analysis
热嘴 hot sprue
人工操作 manual operation
人工的 artificial
人工费 cost of labor
人工环境 artificial environment
人工监督控制 human supervisory control
人工可靠性 human reliability
人工控制 manual control
人工气候室 artificial atmospheric phenomena simulator
人工缺陷 artificial defect
人工神经网络 artificial neural network (ANN)
人工调节 manual regulation
人工系统交互 human system interaction
人工系统接口 human system interface
人工照明 artificial lighting
人工智能 artificial intelligence (AI)
人工智能工作站 artificial intelligent work station
人机 man-machine
人机对话 man-machine interaction
人-机对话方法 interactive approach
人-机机车动力学 interactive vehicle dynamics
人-机机车控制 interactive vehicle control
人机交互 human-computer interaction (HCI)
人机交互程序 interactive program
人机接口 man-machine interface (MMI)
人机接口处理器 man-machine interface processor
人机界面站 human interface station (HIS)
人机界面站设定窗口 HIS setting window
人机通信 man-machine communication
人机系统 man-machine system
人机系统设计 human-machine system design
人机协调 man-machine coordination
人机与环境工程 man-machine and environmental engineering
人类信息处理 human information processing
人类知觉 human perception
人力资源管理 human resource management
人脑 human brain
人体系统诊断与改善 human system diagnosis and improvement
人为误差 human error
人为因素 human factor
人行道指示灯 Belisha beacon
人因工程学 human factors engineering
人造卫星 satellite artificial
人字齿轮 herringbone gear
刃口冲裁模 finish blanking
刃口余隙角 cutting edge clearance
刃用锉刀 edge file
认可标准 recognized standard
认可规范 certification specifications
认证 certification
认证体系 certification system
认知 cognitive
认知机 cognitron
认知科学 cognitive science
认知系统 cognitive system
任务程序 tasking program
任意波形发生器 arbitrary waveform

generator
任意层内部导通孔　any layer inner via hole（ALIVH）
韧青铜　tough bronze
韧铜　tough pitch copper
韧性　toughness
韧致辐射　bremsstrahlung
日报　daily report
日本工业标准　Japanese Industrial Standard（JIS）
日变化　diurnal variation
日常的　routine
日常检查　routine inspection
日常维护　routine maintenance
日负荷　daily load
日负荷曲线　daily load curve
日记　journal
日内精密度　within-day precision
日射　insolation
日照计　heliograph
容错　fault tolerance
容错软件　fault tolerant software
容错设计　fault tolerant design
容错系统　fault tolerant system
容积　volume
容积泵　positive displacement pump
容积流量计　volume flowmeter
容积式流量测量仪表　positive displacement flow measuring device
容积式水表　water displacement meter
容积式液体流量计　liquid positive displacement flowmeter
容抗　capacitive reactance, captance
容量　capacity, containment
容量滴定法　volumetric precipitation method
容量分析法　volumetric analysis
容量控制阀　capacity control valve
容量修正　capacity correction
容量因子　capacity factor
容纳　capacitive susceptance
容器　container, receptacle, vessel
容许电流　allowable current
容许负载　allowable load
容许极限　permissible limit
容许误差　admissible error, allowable error
容许值　allowable value
溶度计　lysimeter
溶剂　solvent
溶剂萃取法　solvent extraction
溶解　dissolution, solution
溶解氧分析器　dissolved oxygen analyzer
溶剂回收单元　solvent recovery unit
溶体处理　solution treatment
溶胀因子　expansion factor
溶质性能检测器　solute property detector
熔点　melting point
熔点测定仪　melting point measuring instrument
熔点型消耗式温度计　melting point type disposable fever thermometer
熔断　fusing
熔断电流　fusing current
熔断片　fuse link
熔断器　fuse
熔断器接线端子　terminal with fuse
熔断时间　fusing time
熔敷焊道　welding bead
熔焊　fusion welding
熔合　fuse together
熔化　fusion
熔接气炬　welding torch
熔接应力　welding stress
熔接硬面法　hard facing
熔解　fusion melting
熔解热　melting heat
熔炉　furnace
熔模铸造　investment casting
熔热处理炉　heating treatment furnaces
熔融金属　molten metal
熔融石英电容器　melted quartz capacitor
熔融石英开管柱　fused silica open tubular column
熔丝　fuse wire, safety wire
熔丝开关　fuse-switch
熔态金属　molten metal
熔渣　molten slag, slag
冗余　redundancy

冗余操纵器	redundant manipulator
冗余的	tolerant
冗余化控制模块	redundancy control module (RCM)
冗余简约	redundancy reduction
冗余控制	redundancy control
冗余控制模件	redundant control module
冗余信息	redundant information
冗余自由度	redundant degree of freedom
柔轮	flexible gear
柔性	flexible
柔性臂	flexible arm
柔性波导管	flexible waveguide
柔性冲击	flexible impulse, soft shock
柔性路面	flexible pavement
柔性行车道	flexible carriageway
柔性制造系统	flexible manufacturing system (FMS)
柔性转子	flexible rotor
柔性自动化	flexible automation
铷	rubidium (Rb)
蠕变	creep
蠕变断裂强度	creep rupture strength
蠕变试验机	creep testing machine
蠕变速度	creeping speed
蠕动	creeping motion
蠕动泵	peristaltic pump
蠕缓放电	creeping discharge
乳化剂	emulsifier
乳剂	emulsion
乳胶漆	emulsion paint
乳胶液	latex solution
入轨姿势	injection attitude
入口	adit, entrance, ingress point, inlet
入口管	inlet pipe
入库	be put in storage
入块	insert
入射	incidence
入射波	incident wave
入射的	incident
入射角	angle of incidence
软板材及自由发泡板机组	soft and free expansion sheet making plant
软测量	soft measurement
软传感器	soft sensor
软磁材料	magnetically soft material
软磁盘	diskette, flexible disk
软磁盘机	floppy disk drive (FDD)
软磁性材料	non-retentive material
软氮化	nitrocarburizing
软电缆	flexible cable
软垫	cushion
软钢管	mild steel pipe
软钢围栏	mild steel barrier
软管	flexible conduit, flexible hose, flexible pipe, hose
软管接头	hose coupler
软管卷盘	hose reel
软件	software
软件安全性	software safety
软件产率	software productivity
软件度量	software metrics
软件工程	software engineering
软件工具	software tool
软件规格	software specification
软件环境	software environment
软件可靠性	software reliability
软件体系结构	software architecture
软件项目管理	software project management
软件心理学	software psychology
软件性能	software performance
软键	soft key
软键盘	soft key
软金属线	soft-annealed wire
软拉钢丝	mild drawn wire
软磨石	flex-hone
软盘	floppy disk
软盘驱动器	flexible disk driver
软铜线	annealed copper wire
软线	flexible cord
软性绞合线	flexible stranded wire
软轴	flexible shaft
软轴用钢丝	flexible shaft wire
锐边	sharp edge
锐角效应	corner effect
瑞利散射光	Reyleith scattering light
润滑	lubrication
润滑机油	lubricating oil
润滑剂	lubricant
润滑膜	lubricant film

润滑系统 lubrication systems	润滑装置 lubrication device
润滑性 lubricity	弱带 weak band
润滑油 lub oil	弱耦合器 loss coupler
润滑油泵 lub oil pump	弱阳离子交换 weak cation exchange
润滑与磨损学 tribology	弱阴离子交换 weak anion exchange

S

洒水系统　water sprinkler system
塞缝片　spline
塞焊　plug welding
塞环线　ring wire
塞孔熔接　plug welding
塞曼效应激光仪　Zeeman laser
塞套引线　sleeve wire
塞子　plug
三倍倍压器　voltage tripler
三倍器　tripler
三次谐波　third harmonic
三重逻辑　ternary logic
三重调制遥测系统　triple modulation telemetering system
三等标准测力计　grade Ⅲ standard dynamometer
三点插头　three-point plug
三点手性识别模式　three-point chiral recognition model
三阀组　three valve bypass manifold
三极管　triode
三角变换　trigonometric transformations
三角波谐振器　triangular resonator
三角锉　triangular file
三角符号　triangular symbol
三角矩阵　matrix triangularization
三角皮带　V belt，V-belt
三角形花键　serration spline
三角形接法　delta connection
三角形隶属度函数　triangle-shape grade of membership function
三角形螺纹　V thread screw
三角椅　utility bench
三脚架　tripod
三聚氰胺甲醛树脂　melamine formaldehyde resin
三片式模具　three plates mold
三绕组变压器　three-column transformer
三条螺纹　three start screw
三通　T-junction
三通电磁阀　three-way solenoid valve
三通阀　three-way valve，triple valve
三通分流阀　diverging three way valve
三维凸轮　three-dimensional cam
三位控制　three-step control
三位控制器　three state controller
三位式开关调节器　three-position on/off controller
三位式调节器　three-position controller
三位数字显示直流电压表　three-digit DC digital voltmeter
三线电源电缆　three-conductor power cable
三线制　three-wire system
三相　three-phase，triphase
三相半波整流电路　three-phase half wave rectifier circuit
三相变压器　three-phase transformer
三相电动机　three-phase motor
三相电力传输　three-phase power transmission
三相电力网　three-phase power network
三相电路　three-phase circuit
三相电压　voltage by three phase
三相发电机　three-phase generator
三相负载　three-phase load
三相故障　three phase fault
三相交流电　three-phase AC
三相笼式电动机　three-phase squirrel cage motor
三相四线电度表　three-phase four-wire watt-hour meter
三相四线制　three-phase four-wire system
三相同步发电机　three-phase synchronous generator
三相异步电动机　three-phase asynchronous motor
三相制　three-phases system
三硝基甲苯　trinitrotoluene

三要素法	method of three key factors
三叶浮标	cloverleaf buoy
三轴姿态稳定	three-axis attitude stabilization
三作用	three-term action
三作用控制	three-term control
三作用控制器	three-mode controller, three-term controller
伞齿轮底角	root angle
伞形齿轮	bevel gear
散布图	scatter diagram
散粒噪声	signal shot noise
散列法	hashing
散绕	random-wound
散热	heat dissipation
散热片	cooling fin, radiator fin
散热器	radiator
散热扇片	cooling fan blade
散热设备	heat sink device
散射光	scattering light
散射剂量	dispersion dose
散射离子的实验强度	experimental intensity of scattered ion
散射器	scatter
散射全息干涉仪	scattering holographic interferometer
散线印制板	discrete wiring board
散装货	bulk cargo
扫描	scan
扫描电路	scanning circuit
扫描监测器	scanning monitor
扫描频率	sweep frequency
扫描器	scanner
扫描探针电子显微镜	scanning probe microscope
扫描探针显微镜	scanning probe microscopy
扫描填充	scan filling
扫描装置	scanister
扫频标志发生器	marker sweep generator
色斑	color mottle
色饱和度	saturation
色标管	chromatron
色层分离谱	chromatogram
色差	aberration
色差系数	coefficient of chromatic aberration
色带	colour bar
色度计	colorimeter
色码	colour code
色母料	color masterbatch
色谱法	chromatography
色谱分析	chromatographic analysis
色谱峰	chromatographic peak
色谱联用技术	hyphenated techniques chromatography
色谱图	chromatogram
色谱响应函数	chromatographic response function
色谱优化函数	chromatographic optimization function
色谱柱	chromatographic column
色散本领	dispersive power
色散红外线气体分析器	dispersive infra-red gas analyzer
色散计	dispersion meter
铯	caesium (Cs), cesium (Cs)
森林工程	forest engineering
杀虫液体	insecticidal fluid
杀菌器	sterilizer
沙	sand
沙尔皮冲击试验	Charpy impact test
沙浆底层	screeding
沙漏	hourglass, hour-glass
纱包漆包线	cotton covered enamel wire
纱包线	cotton-covered wire
纱线	yarn
刹车距离	braking distance
刹车踏脚板	brake pedal
砂布	emery cloth
砂砾	grit
砂轮	grinding wheel
砂轮机	grinding machine
砂轮切断机	abrasive cut-off machine
砂轮越程槽	grinding wheel groove
砂轮整修机	dresser
砂磨工具	abrasive tool
砂芯模板	core template
砂芯排气孔	core vent
砂纸	glass-paper
筛子	sifter
山泥倾泻补救工程	landslip remedial

works
删除 delete
删除功能 delete function
删除控制点 control point deletion
删除命令文件 delete command file
删除事件 event deletion
删除一个文件 delete a file
钐 samarium (Sm)
栅电压 gate voltage
栅格 grille
栅极电容 grid capacitance
栅极电阻 grid resistance
栅极控制 grid control
栅极引线 grid lead wire
栅控 X 射线管 grid-controlled X-ray tube
栅偏压 grid bias
栅偏压电源 grid bias supply
栅状测云器 grid nephoscope
闪灯 flashing light
闪灯继电器 flasher relay
闪电 lightning
闪电电流磁检示器 magnetic detector for lightning currents
闪动标灯 flashing beacon
闪动灯具 flashing lantern
闪光 flash
闪光测频 frequency measurement by stroboscope
闪光灯 flash light
闪光电焊机 flash butt welding machine
闪光对接焊机 flash butt welding machine
闪光焊 flash butt weld
闪光继电器 flasher relay, flicker relay
闪光器 lamp flasher
闪光信号报警器 flashing annunciator
闪光信号灯 blinker lamp
闪光信号装置 blinker signal equipment
闪弧继电器 flasher relay
闪络 flashover
闪频观测仪 stroboscope
闪熔镀层 flash plate
闪烁 flicker
闪烁灯光 blinker light
闪耀光栅 blazed grating
扇尾形模具 fantail die
扇形磁场质谱计 sector magnetic field mass spectrometer
扇形防护网架 catch fan
扇形浇口 fan gate
扇叶 fan blade
商用硬件与软件 commercial hardware and software
熵 entropy
上部 upper
上层 upper storey
上层大气 upper atmosphere
上层构造 superstructure
上层平台 upper platform
上层网 upper wire
上承板 upper supporting blank
上齿面 addendum surface
上电 power-up
上垫板 punch pad
上垫脚 top block
上段集水区 upper catchment area
上盖 roof cover
上盖面积 site coverage
上光机 glazing machine
上光剂 glazing agent
上光涂料 glazing paint
上级问题 upper level problem
上卷 scroll up
上毛毯 glazing felt
上模板 upper plate
上模座 upper die base
上内模 cavity insert
上切换值 upper switching value
上升 rise
上升力 uplift force
上升时间 rise time
上网 top wire
上下文无关文法 context-free grammar
上下文状态记录 context state record
上限 ceiling, upper limit
上限调整 high-limit adjustment
上限越界 off-normal upper
上限值范围 rangeability
上限值控制 high limiting control

上行电动扶梯　up-escalator
上行电缆　rising main
上行展开法　ascending development
上压块　upper holder block
上游　upstream
上釉　glazing
上釉陶器　glazing pottery
上釉窑　glazing kiln
上中间板　upper mid plate
烧杯　beaker
烧结板　sintered plate
烧结锻造　sinter forging
烧磨砂　chamotte sand
烧瓶　flask
勺皿　casserole
少灰混凝土　lean mix concrete
少数载流子　minority carrier
舍弃式刀头　disposable toolholder bits
舍入误差　round-off errors
舍入噪声　round-off noise
设备　fittings, equipment
设备控制　plant control
设备控制点　device control point
设备露点温度　apparatus dew point temperature
设备驱动程序　device driver
设备型号规格　instrument type and specification
设定　setting
设定重定操作　set-reset operation
设定点控制　set point control
设定轨迹　set point trajectory
设定盲插头　blind set plug
设定值　set point
设定值变化　set point change
设定值范围　range of set value
设计　design
设计变化　design modification
设计变量　design variable
设计常数　design constant
设计方法学　design methodology
设计工程　project engineering
设计规则检查　design rule checking
设计荷载　design load
设计后处理　post design processing
设计计算资料　design calculation
设计假定　design assumption
设计浸入深度　designed immersion depth
设计距离　design distance
设计流量　design flow
设计品质保证　design quality assurance
设计使用年限　design life
设计数据表　design data sheet
设计数据库　design database
设计速度　design speed
设计算法　device degradation
设计土压力　design earth pressure
设计系统　design system
设计压力　design pressure
设计应力　design stress
设计原点　design origin
设计约束　design constraints
设计状态　design condition
设计准则　design code
设计自动化　design automation
设施　facilities
设施规划　facilities planning
设置归档指针　set archive pointers
设置缺省目录　set default directory
射出循环　shot cycle
射极电压　emitter voltage
射极电阻　emitter resistance
射极输出器　emitter follower
射料浇口　sprue gate
射流流量计　fluidic flowmeter
射频　radio-wave frequency
射频变压器　radio frequency transformer
射频电流计　radio frequency ammeter
射频发生器　RF generator
射频放大　radio frequency amplification
射频放大器　radio frequency amplifier
射频干扰　radio frequency interference
射频毫伏计　RF millivoltmeter
射频脉冲发生器　radio frequency pulse generator
射频敏感器　radio frequency sensor
射频微电位器　radio frequency micro-potentiometer
射气测量　emanation survey
射气仪　emanometer

射线　ray
摄动理论　perturbation theory
摄谱仪　spectrograph
摄氏　Celsius
摄氏度（温度单位）　degree Celsius
摄氏温标　Celsius temperature scale
摄氏温度　Celsius temperature
摄氏温度计　Celsius
摄像管　image pickup tube
摄影测量与遥感　photogrammetry and remote sensing
伸臂　extension jib
伸长率　elongation
伸拉角　stretch corner
伸拉中心　stretch zone
伸缩接头　expansion joint
伸缩栓　expansion bolt
伸缩套筒　extension sleeve
伸缩性量规　telescopic gauge
伸缩振动　stretching vibration
砷　arsenic (As)
砷化铁　arsenic iron
深波纹管　depth bellows
深槽　deep groove
深槽滚珠轴承　deep groove ball bearing
深槽式笼形感应电动机　deep-slot squirrel-cage induction motor
深层地基　deep foundation
深层压实　deep compaction
深层振荡式压实　deep vibration compaction
深的　deep
深度计　depth gauge
深度控制器　depth controller
深度游标卡尺　vernier depth gauge
深沟挖掘　deep trench excavation
深海仪器舱　deep sea instrument capsule
深井水泵　deep well water pump
深冷处理　subzero
深陷式　submarine
深钻井泵　deep bore well pump
神经的　neural
神经动力学　neural dynamics
神经激活　neural activity
神经集合　neural assembly
神经控制　neural control
神经网　neural net
神经网络　neural network
神经网络计算机　neural network computer
神经网络模型　neural network model
神经元　neuron
审核　approval examine and verify
甚高频调制信号发生器　VHF modulation signal generator
渗氮法　nitriding
渗流力　seepage force
渗漏　leakage
渗漏率　leak rate
渗滤系数　permeability coefficient
渗入　penetration
渗蚀层　eluvial
渗水坑　soakaway
渗碳　carburization, carburizing, cementation
渗碳层深度　carburized case depth
渗碳体　cementite
渗透　percolating
渗透极限　permeability limit
渗透试验　penetration test
渗透系数　permeability coefficient
升（容积单位）　litre (L)
升高　raise
升级　upgrade
升降机　lift
升降机顶升销　lift pin
升降机工程　lift works
升降机井　lift shaft
升降机坑　lift pit
升降机厢　lift car
升降平台　lifting platform
升降器　lifter
升降式止回阀　lift check valves
升温和降温封闭场所　heated and cooled enclosed location
升温或降温封闭场所　heated or cooled enclosed location
升温曲线测定　heating curve determination
升温曲线测定仪　heating curve determination apparatus
升温速率　heating rate

升压　increase of voltage
升压泵　boost pump
升压变换器　boost converter
升压变压器　booster transformer, step-up transformer
升压二极管　booster diode
升压放大器　booster amplifier
升压继电器　booster relay
升压水泵　water booster pump
升压斩波电路　boost chopper
升油泵　fuel lift pump
生产工艺　processing technique
生产和物料控制　product material control（PMC）
生产能力　throughput
生产线喷雾干燥器　spray dryer for product line
生产线主管　line supervisor
生产阻力　productive resistance
生成　builder
生成函数　generation function
生化分析仪　biochemical analyzer
生化量传感器　biochemical quantity sensor, biochemical quantity transducer
生冷处理　subzero treatment
生理模型　physiological model
生理学　physiology
生命系统　living system
生命周期　life cycle
生色团　chromophore
生态学　ecology
生物安全柜　biohazard safety equipment, bio-safety cabinet
生物传感器　biosensor
生物反馈系统　biological feedback system
生物反应器　bio-reactor
生物分子间相互作用分析系统　biomolecular interaction analysis system
生物工艺学　biotechnology
生物化工　biochemical engineering
生物技术分析　biochemical analysis
生物技术关连仪器　bio-technology related instruments
生物碱　alkaloid
生物控制论　biocybernetics
生物量传感器　biological quantity sensor, biological quantity transducer
生物医学　biomedical
生物医学分析仪　biomedical analyzer
生物医学工程　biomedical engineering
生物医学控制　biomedical control
生物医学系统　biomedical system
生物自显影法　bioautography
生锈　rust
声波测井仪　acoustic logging instrument
声波幅度记录器　acoustic amplitude logger
声波全波列测井仪　full-wave logger
声波示波器　acoustic oscillograph
声程　beam path distance
声传输线　acoustic line
声发射　acoustic emission
声发射分析系统　acoustic emission analysis system
声发射光谱　acoustic emission spectrum
声发射换能器　acoustic emission transducer
声发射技术　acoustic emission technique
声发射检测系统　acoustic emission detection system
声发射检测仪　acoustic emission detector
声发射率　acoustic emission rate
声发射脉冲发生器　acoustic emission pulser
声发射脉冲分析仪　AE pulse analyzer
声发射能量　acoustic emission energy
声发射前置放大器　acoustic emission preamplifier
声发射事件　acoustic emission event
声发射事件计数　count of acoustic emission event
声发射信号处理器　acoustic emission signal processor
声发射源　acoustic emission source
声发射源定位及分析系统　acoustic emission source location and analysis system

声发射源定位系统 acoustic emission source location system
声发射振幅 acoustic emission amplitude
声放大器 acoustic amplifier
声辐射计 acoustic radiometer
声干涉仪 acoustic interferometer, acoustical interferometer
声光调制器 acoustical light modulator
声光信号 acoustic and light signals
声换能器 acoustic transducer
声级计 sound level meter, sound-level meter
声宽网络 bandwidth voice network
声敏控制装置 voice-activated control unit
声明 declaration
声呐压力计 sonar manometer
声耦合剂 acoustic couplant
声耦合器 acoustic coupler
声疲劳 acoustic fatigue
声匹配层 acoustic matching layer
声频 audible frequency, audio-frequency
声频变压器 audioformer
声频测量仪表 AF measuring instrument
声频带通放大器 audio band amplifier
声频发生器 acoustic frequency generator
声频频谱仪 audio-frequency spectrograph
声谱仪 sound spectrograph
声强度 sound intensity
声全息术 acoustical holography
声全息图 acoustical hologram
声扫描器 acoustic scanner
声失效 acoustic malfunction
声释放器 acoustic releaser
声束比 beam ratio
声速偏转式超声流量计 beam-deflection ultrasonic flowmeter
声透镜 acoustical lens
声学比 acoustic ratio
声学多普勒定位系统 acoustic Doppler system

声学分析仪 acoustic analyzer
声学海流计 acoustic current meter
声学流量计 sonic flowmeter
声学温度计 acoustic thermometer
声学元件 acoustic element
声音 voice
声应答器 acoustic transponder
声阻 acoustic resistance
声阻法 acoustical impedance method
声阻抗 acoustic impedance
绳端套环 rope thimble
绳夹 rope clamp
绳索包皮线 snake wire
剩磁 remanence
剩余强度 residual strength
剩余熔渣 remaining slag
失步 desynchronizing
失步保护 out-of-step protection
失配 mismatch
失去同步 loss of synchronization
失调 off-tune
失效 invalidation
失效安全 fail safe
失效机理 failure mechanism
失效率 failure mode
失效模式分析 failure model effectiveness analysis (FMEA)
失效识别 failure recognition
失效树分析 fail tree analysis
失效诊断 failure diagnosis
失协调 discoordination
失修 out of repair
失压保护 no-voltage protection
失真 distort, distortion
失真波 distorted wave
失真测试仪 distortion tester
失真度测量仪 distortion factor meter
失真交变电流 distorted alternating current
失真系数 distortion factor
施工缝 construction joint
施工用升降机 builder's lift
施控系统 controlling system
施塔克尔贝格决策理论 Stackelberg decision theory
施压阀 pressure valve
施主 donor

施主杂质　donor impurity
湿度　humidity
湿度表检定箱　hygrometer calibration chamber
湿度波动度　humidity fluctuation
湿度测定点　measuring point for the humidity
湿度传感器　humidity sensor, humidity transducer
湿度范围　humidity range
湿度分析器　moisture analyzer
湿度计　hygrograph, hygrometer
湿度均匀性　humidity uniformity
湿度控制器　humidity controller
湿空气　humid air
湿敏电容器　humicap
湿敏电阻器　humidity-dependent resistor
湿敏开关　humidity sensitive switch
湿敏元件　dew cell, moisture sensor
湿强度保留率　wet strength retention
湿热试验　humid heat test
湿式气体流量计　wet type gas flowmeter
十二烷基磺酸钠　sodium dodecyl sulfate (SDS)
十分之一　deci
十进电阻箱　decade resistance box
十进位电阻箱　decad type resistance box
十进制编码　coded-decimal notation
十进制的　decimal
十进制计数管　decatron
十进制数　decimal number
十字滑块联轴器　oldham coupling
十字接头　cross joint
十字路口　cross road
十字丝　cross wire
十字头联轴节　Oldham coupling
石膏灰泥　gypsum plaster
石膏模　plaster mold
石膏铸模　gypsum mold
石化工业　petroleum industry
石灰　lime
石蜡　paraffin
石锚　rock anchor
石棉拆除工程　asbestos abatement works

石棉垫片　asbestos gasket
石棉绝缘线　asbestos covered wire
石棉水泥　asbestos cement
石棉水泥井管　asbestos-cement pipe for well casing
石棉水泥输煤气管　asbestos-cement pressure pipe for gas transmission
石墨　graphite
石墨加工机　graphite machine
石墨模子　graphite mould
石墨润滑脂　graphite grease
石蕊　litmus
石蕊试纸　litmus paper
石英　quartz
石英补偿元件　quartz compensating element
石英晶体　quartz crystal
石英晶体滤波器　quartz crystal filter
石英晶体压电式加速度计　quartz piezoelectric accelerometer
石英晶体振荡器　quartz crystal oscillator
石英砂　quartz sand
石英弹性元件　quartz elastic element
石油产品　petroleum products
石油化工装置　petrochemical plant
石油截流隔　petrol intercepting trap
石油气　liquefied petroleum gas (LPG)
石油与天然气工程　oil and natural gas engineering
时报　hourly report
时变参数　time-varying parameter
时变对象　time-varying plant
时变系统　time-varying system (TVS)
时标比较法测频　frequency measurement by comparison with time scale
时不变的　time invariant
时不变系统　time invariant system
时差计时单元　delta-T timing unit
时基扫描多谐振荡器　time base sweep multivibrator
时基线性　linearity of time base
时间　time
时间比例开关调节器　time proportioning on/off controller
时间比例控制　time proportioning con-

trol
时间标签分辨率 time stamp resolution
时间常数 time constant
时间常数实验测定 experimental determination of time constant
时间程序 time programme
时间串联分析 time-series analysis
时间淬火 time quenching
时间积分判据 time-integral criteria
时间继电器 time relay
时间控制 time control
时间设定对话 time setting dialog
时间顺序控制 time-oriented sequential control
时间相位 time-phase
时间响应 time response
时间延迟 time delay
时间延迟模型 time delay model
时间优化控制 time-optimal control
时间整定 time setting
时间整定范围 time setting range
时间滞后 time lag
时间滞后补偿 time delay compensation
时间周期控制器 time cycle controller
时速限制 speed limit
时限 time limit
时限保护 time-limit protection
时限继电器 timing relay
时效硬化 age hardening
时序 sequence in time
时序机 sequential machine
时序控制 sequential control
时序控制器 time schedule controller
时序模拟 timing simulation
时延估计 time delay estimation
时延扩张 time delay spread
时域 time-domain
时域 Petri 网 time Petri-net
时域不变对象 time-invariant plant
时域不变式 time-invariant
时域反射仪 time-domain reflectometer
时域方法 time-domain method
时域分光光度计 time-domain spectroscopy

时域分析 time domain analysis, time-domain analysis
时域共享控制 time shared control
时域共用程序 time-sharing program
时域共用系统 time-sharing system
时域规划控制 time schedule control
时域计算 time-domain calculation
时域逆变器 time reversal
时域频率 time-frequency
时域频率表示 time-frequency representation
时域频率定位 time-frequency localization
时域通道分配 time-slot assignment
时域同步 time synchronization
时域系统 time system
时域相关 time-domain correlation
时域响应 time-domain response
时域信号 time signal
时域综合 time-domain synthesis
时滞 delay
时滞补偿 delay compensation
时滞分析 delay analysis
时滞估计 delay estimation
时滞解调 delay demodulation
时滞延伸 delay spread
时滞元件 delay element
时钟 clock
时钟点 timers point
时钟开关 clock switch
时钟控制调节器 clock-controlled governor
时钟练习程序 clock exerciser
时钟脉冲 clock pulse
时钟脉冲频率 clock frequency
时钟同步 clock synchronization
时钟系统重复器 clock system repeater
识别 recognition
识别标志 identification mark
识别分析 discriminant analysis
识别器 recognizer
实部 real part
实地测量 field surveying
实地测试 field testing, on-site test
实地勘测 field investigation
实际测量范围上限值 actual higher

measuring range value
实际测量范围下限值　actual lower measuring range value
实际布置图　physical layout
实际尺寸　physical dimension
实际电流源　practical current source
实际电压比　actual voltage ratio
实际电压源　practical voltage source
实际观测时间　actual time of observation
实际啮合线　actual line of action
实际位移　positive displacement
实际位姿漂移　attained pose drift
实际限制　physical constraint
实际值　actual value
实践论　realization theory
实时报警　real alarm
实时部分　real component
实时操作系统　real-time operating system
实时的　real-time
实时计算机　real-time computer
实时计算机系统　real-time computer system
实时趋势　real time trend
实时人工智能　real-time AI
实时任务　real-time task
实时日志　real time journal
实时时钟　real-time clock
实时数字相关器　real-time digital correlator
实时通信　real-time communication
实时网络操作系统　real-time network operating system
实时系统　real-time system
实时谐波分析仪　real-time harmonic analyzer
实时遥测　real time telemetry
实时优化程序　real-time optimizer
实时语言　real-time language
实时专家系统　real-time expert system
实体　entity
实体筏基　solid raft
实体量器　material measure
实体模型　solid model
实体设计　physical design

实物校准　actual material calibration
实现　realization
实线　soild wire
实心导线　solid conductor
实验标准偏差　experimental standard deviation
实验测试　experimental testing
实验方差　experimental variance
实验技术　laboratory technique
实验教学　laboratory education
实验模型　experimental modeling
实验室　laboratory
实验室玻璃器皿清洗机　laboratory glassware washers
实验室测试　laboratory testing
实验室可靠性试验　laboratory reliability test
实验室盐度计　laboratory salinometer
实验室应用资料　laboratory application data
实验台　laboratory furniture
实验台用附属器具　carts and laboratory table attachments
实轴　real axis
实轴上的根轨迹段　root locus segments on the real axis
蚀刻机　etching machines
蚀刻术　etching
食品分析仪　food analyzer
食品分析仪器　food analysis instruments
食品工业　food processing
食品科学　food science
食品科学与工程　food science and engineering
食水泵　potable water pump
食水抽水站　fresh water pumping station
食指　forefinger
史密斯方法　Smith's method
史密斯预估器技术　Smith predictor technique
矢量　vector
矢量法　vector method
矢量方向　direction of vector
矢量和　vector sum
矢量积　cross product

矢量量化　vector quantization
矢量量化器　vector quantizer
矢量图　vectogram, vector diagram
使能够　enable
使弯曲　buckle
使无效　deactivate
使用环境　suitable surrounding
使用率　utilization rate
使用期限　life span
使用人　user
使用寿命　service life, working life
使用说明书　instruction manual
使自动　automate
市场与行销　market and marketing
市政工程　municipal engineering
示波管　oscilloscope tube
示波计　oscillometer
示波器　oscillograph, oscilloscope
示差热曲线　differential thermal analysis curve
示差扫描量热仪　differential scanning calorimeter
示差折光检测器　refractive index detector
示教编程　teaching programming
示教再现式机器人　playback robot
示速器　speed indicator
示意图　schematic
示值　indication
示值范围　indication range
示踪物浓度　concentration of tracer
示踪原子　tracer element
事故　accident
事故备用按钮　emergency push-button switch
事故备用电源　emergency power supply
事故记录仪　event recorder
事故排放阀　dump valve
事故切断　emergency switching-off
事故手动装置　emergency hand-drive
事故停机　disaster shutdown
事故照明器　emergency light
事件　event
事件不良　poor incoming part
事件初始点　event initiated point
事件触发处理　event initiated processing (EIP)
事件记录程序包　event logger package
事件链　event chain
事件脉冲　event pulse
事件序列开关量输入　digital input-sequence of event
事件选项　event option
事实　fact base
势能　potential energy
视差　parallax error
视场光阑　field stop
视场角　angle of field of view
视角　view angle
视觉与色彩　vision and colors
视力　vision
视频　image frequency
视频第二检波器　video second detector
视频点播　video-on-demand (VOD)
视频电压放大器　video voltage amplifier
视频放大器　video amplifier
视频放大器频宽　band width of video amplifier
视频监视器　video display monitor
视频双芯电缆　video pair cable
视频信号　video signal
视频组件　video module
视图　view
视网膜电图传感器　electroretinographic sensor, electroretinographic transducer
视线　sight line
视野　field of view
视野深度　depth of field
视在的　apparent
视在功率　apparent power
视在功率消耗　apparent power consumption
视在阻抗　apparent impedance
视轴　collimation axis
视准线　collimation line
试差整定　trial and error tuning
试车　commissioning, trial run
试电表　neon tester
试管　test tube
试剂　reagent

试剂瓶　reagent bottles
试件　test piece
试孔　trial hole
试料混合器　blender
试切前调整　undeveloped settings
试切调整　developed setting
试算表　spreadsheet
试台　testing bench
试探法　method of trial and error
试验　test
试验泵　test pump
试验长度　terminal length
试验程序标准　test procedure standard
试验电压　testing voltage
试验发生　terminal generation
试验方法　test methods
试验管加热板　heating blocks
试验机　testing machine
试验机附件　accessories of testing machine
试验机最小负荷　minimum load of the testing machine
试验计划　pilot project, pilot scheme
试验结果　test result
试验空间　test space
试验溶液　test solution
试验设备　test equipment, testing facility
试验数据　test data
试验数据充分度　test data adequacy
试验顺序　testorder
试验系统的柔度　testing system flexibility
试验线圈　test coil
试验质量　test mass
试验装置　testing device
试样空间　sample space
试样探针　sample probe
饰　decorative finish
室内土工试验　consolidation test
室温　ambient temperature
适配器　adapter, adaptor
适配器检验　adapter check
适配器控制块　adapter control block
适应　adaptation
适应层　adaptation layer

适用介质　applicable medium
适用温度　applicable temperature
铈　cerium (Ce)
释放　release
释放线　release wire
释放延迟　hand-over
释药系统　drug delivery system (DDS)
收发器　transceiver
收集　collect
收集时间　acquisition time
收敛　convergence
收敛的　convergent
收敛分析　convergence analysis
收敛级数　convergent series
收敛数值方法　convergence of numerical method
收敛因子　convergence factor
收敛原理　conservation principle
收敛证明　convergence proof
收缩　shrinkage
收缩接缝　contraction joint
收缩配合　shrinkage fit
收缩系数　contraction coefficient, shrinkage coefficient
收益　profitability
手　hand
手、足放射性污染监测仪　hand-and-feet contamination monitor
手柄　handle
手持测角器　hand goniometer
手持风速仪　hand anemometer
手持罗盘　hand compass
手持式风钻　jack-hammer
手持式模具　handle mold
手持式数字转速表　handy digital tachometer
手持水准仪　hand pattern level
手持通话器　handset
手持研磨机　hand-held grinder
手电钻　hand electric drill
手动　handling
手动操作　hand operation
手动操作阀　hand-operated valve
手动操作器　manual control unit, manual station
手动齿轮齿条式冲床　hand rack pinion press

手动冲床　hand press
手动阀　hand operated valve
手动方式回路状态转换键　manual mode loop status switching key
手动复位　hand reset
手动滚动试验机　hand-rolling tester
手动开关　manual switch
手动控制　manual control
手动控制盘　manual control panel
手动螺旋式冲床　hand screw press
手动模式　manual mode
手动扫查　manual scanning
手动设定　manual set
手动输出操作站　manual output station
手动输出卡　manual output card
手动输入穿孔机　hand-feed punch
手动数据进入组件　manual data entry module
手动数据输入编程　manual data input programming
手动调节器　hand-operated regulator
手动吸气式湿度计　hand-aspirated psychrometer
手动信号　hand signal
手动运转方式　manual operating mode
手动/自动　manual/automatic
手动-自动的切换　manual to automatic transfer
手动/自动切换站　manual/automatic station
手工的　manual
手工具　hand tool
手工修润　hand finishing
手控操作　manual mode operation
手轮　hand wheel, handwheel
手套　gloves
手提灯　hand lamp
手提灭火筒　portable fire extinguisher
手提式　portable
手提式比重计　hand hydrometer
手提式分光光度计　hand spectrophotometer
手提式逻辑电路故障探测器　hand-held logic circuit probe
手提式模具　portable mold
手提式数字温度计　hand-held digital thermometer
手提式温度计　handy type thermometer
手提式小型万用表-示波器两用仪　hand-held multimeter-oscilloscope
手提钻孔机　portable driller
手推车　trolley
手握面罩　hand face shield
手写字符　hand printed character
手性分离色谱　chiral separation chromatography
手性固定相　chiral stationary phase
手性离子对络合剂　chiralion pair complex
手性色谱法　chiral chromatography
手性衍生试剂　chiral derivatization reagent
手摇泵　hand pump
手摇试压泵　wobble pump for pressure testing
手摇油泵阀　manual oil pumps valves
手摇钻　hand brace
手制动器　hand brake
首标字节　header byte
首次检定　initial verification
首件确认　first article assurance
寿命　lifetime
寿命系数　life factor
受端　receive-side
受控变量　controlled variable
受控大气　controlled atmosphere
受控源　dependent source
受役系统　slaved system
售后服务　after service
授权代码　authoritarian code
授权检查　authoritarian checking
授权书　power of attorney
授权信息　authoritarian message
枢轴　pivot
梳板　comb plate
梳齿型滤波器　comb filter
梳形刨刀　tooth plane iron
疏忽　negligence
疏忽误差　spurious error
疏水　drain
疏水层　drainage blanket
疏水器　steam trap

疏水作用色谱法　hydrophobic interaction chromatography
输出　output
输出变化率检查　output velocity check
输出变压器　output transformer
输出补偿　output compensation
输出操作站　output station
输出传递函数　output transfer function
输出打印机　output printer
输出电路　output circuit
输出电刷　output brush
输出电压　output voltage
输出电阻　output resistance
输出反方向作用　output reverse action
输出反馈　output feedback
输出方程　output equations
输出放大器　output amplifier
输出负载电阻　output load resistance
输出跟踪　output tracking
输出功　output work
输出功率　output power
输出构件　output link
输出机构　output mechanism
输出级　output stage
输出矩阵　output matrix
输出开关　out switch
输出开路报警检查　output open alarm check
输出力矩　output torque
输出量　output variable
输出滤波器　output filter
输出绕组　output winding
输出容量选择　output capacity selection
输出数据　data-out
输出调节器　output regulator
输出调整　output regulation
输出网络　output network
输出误差辨识　output error identification
输出限　output limit
输出向量　output vector
输出信号　output signal
输出信号处理　output signal processing
输出信号类型　output signal type
输出预估法　output prediction method
输出振荡　vibrating of output
输出值　output value
输出轴　output axis, output shaft
输出轴最大转数　maximum revolutions of output shaft
输出注射　output injection
输出装置　output device
输出阻抗　output impedance
输电　electricity transmission
输电网　transmission network
输电系统　power transmission system
输电线　power line
输电线路故障　transmission line malfunction
输入　entry, input
输入变量　input variable
输入补偿　input compensation
输入参数　input parameter
输入导纳　input admittance
输入的　incoming
输入电流　incoming current
输入电路　incoming circuit, input circuit
输入电阻　input resistance
输入动态　input dynamics
输入端　input terminal
输入构件　input link
输入估计　input estimation
输入环节　input element
输入矩阵　input matrix
输入开路检查　input open check
输入量-频率转换型　input quantity to frequency conversion type
输入脉冲　input impulse
输入设备　input device
输入矢量　input vector
输入/输出　input/output (I/O)
输入/输出报告　I/O report
输入/输出操作　input/output operation
输入-输出处理器　input-output processor
输入/输出单元　I/O unit

输入/输出单元的装置 mounting of I/O units
输入/输出点 I/O point
输入/输出端口 input/output port
输入-输出接口 input-output interface
输入/输出模件箱 I/O module nest
输入/输出模块 I/O module (IOM)
输入-输出设备 input-output device
输入-输出通道 input-output channel
输入-输出线性化 input-output linearization
输入-输出装置 input-output equipment
输入数据 data-in
输入向量 input vector
输入信号 input signal
输入预估法 input prediction method
输入指示器 input indicator
输入阻抗 input impedance
输水管改移工程 water main diversion works
输水设施 water carrying services
输送带 conveyer belt
输送电力 transmission of electricity
输送机 conveyer
输送架 transmission rack
输送链 conveying chains
输送延迟 transport delay
输血器 blood transfusion set
输油管 pipeline
熟石灰 hydrated lime
熟铁 dug iron, wrought iron
熟铁块 bloomery iron
鼠标 mouse
鼠笼 squirrel cage
术语标准 terminology standard
束帆索 gasket
束缚电荷 bound charge
束缚电子 bound electron
树 tree
树形检索 tree search
树形结构 tree structure
树脂 resin
树脂胶合 resinoid bond
树脂胶浆 resin mortar
树脂绝缘器 resin insulator
树脂流纹 resin streak
树脂漆 resin emulsion paint
树脂射出法 resin injection
树脂脱落 resin wear
竖板 riser
竖井通道 access shaft
竖向净空 vertical clearance
竖向曲线 vertical curve
竖直度盘指标补偿器 automatic vertical index
竖直度盘指标补偿误差 compensating error of automatic vertical index
竖直面板 vertical panel
数传机 data set
数据 data
数据保持 data hold
数据保密 data privacy
数据采集 data acquisition
数据采集设备 data acquisition equipment
数据采集算法 data acquisition algorithm
数据采集与处理系统 data acquisition system
数据采集站 data acquisition station
数据采样开关 data sampling switch
数据采样系统 sampled-data system
数据采样系统的稳定性 stability of a sampled-data system
数据储存 data storage
数据处理 data handling, data processing
数据处理机子系统 data processor subsystem
数据处理器 data processor
数据处理系统 data handling system, data processing system
数据传递 data transfer
数据传输 data transmission
数据传输接口 data transmission interface
数据传输系统 data communication system
数据传输指令指示器 data link command indicator
数据传送率 data transfer rate
数据传信率 data signalling rate
数据存取 data access

数 shu

数据点　data point (DP)
数据点方式　data point mode
数据点方式属性　data point mode attribute
数据电路　data circuit
数据电路终端设备　data circuit-terminating equipment (DCE)
数据定义表　data definition table
数据读出设备　data-taking equipment
数据段说明　data segment description
数据发生器　data generator
数据发送器　data source
数据范围　range data
数据分配器　data distributor
数据浮标系统　data buoy system
数据符　data symbol
数据复制　data replication
数据公路　data highway
数据公路单元　highway unit
数据公路协议　highway protocol
数据公路帧　highway frame
数据和时间设定画面　data and time setting panel
数据汇集器　data concentrator
数据集　data set
数据集中　data concentration
数据集中分配器　data concentrator
数据记录　data logging
数据记录装置　data logger
数据加密　data encryption
数据简化　data reduction
数据简化系统　data reduction system
数据接收器　data sink
数据结构　data structure
数据库　data base (DB), data pool
数据库管理系统　data base management system
数据库结构　database structure
数据库系统　database system
数据捆绑功能　data binding functions
数据类型　data type
数据连接　data connection
数据链路　data link
数据链路层　data link layer (DLL)
数据链路服务定义　data link service definition
数据链路协议规范　data link protocol specification
数据流　data flow, data stream
数据流程图　data flow diagram (DFD)
数据流分析　data flow analysis
数据流式磁带驱动器　streaming tape drive
数据拟合　data fitting
数据驱动　data driven
数据设定开关　data set switch
数据设定器　data set unit
数据升级　data upgrading
数据实体编制程序　data entity builder (DEB)
数据适配器　data adapter unit
数据输出电位计　data potentiometer
数据输入面板　data entry panel
数据所有者　data owner
数据通道　data channel
数据通信　data communication
数据完整性　data integrity
数据网络　data network
数据显示模块　data display module
数据线路　data circuit
数据协调　data reconciliation
数据压缩　data compression
数据压缩算法　data compression algorithm
数据译码系统　data encoding system
数据预处理　data preprocessing
数据源　data source
数据站　data station
数据指示器　data pointer
数据终端设备　data terminal equipment (DTE)
数据转移　data symbol transmission
数据字　data word
数据字回波　data word echo
数据字回信　data word echo
数据总线　data bus, data highway (DHW)
数据总线接口　data highway interface
数据总线子系统接口　data highway port (DHP)
数控　computerized numerical control (CNC)
数控车床　CNC lathe, CNC turning machine

数控车刀　CNC turning tool
数控成形外圆磨床　CNC formed cylindrical grinder
数控齿轮精加工机床　CNC gear hard finishing machine
数控冲模回转头压力机　CNC turret punch press
数控单柱立式车床　CNC single column vertical turning and boring mill
数控单柱移动式立式车床　CNC movable single column vertical turning and boring mill
数控单柱坐标镗床　CNC single column jig boring machine
数控电火花加工机床　CNC electrical discharge machine
数控电火花线切割机床　CNC wire-cut electric discharge machine
数控电极磨床　CNC electrode grinding machine
数控电解工具磨床　CNC electrolytic tool and cutter grinder
数控端面外圆磨床　CNC angular approach cylindrical grinding machine
数控仿形铣床　CNC copy milling machine
数控杆料进料匣　CNC bar loading magazine
数控高速双砂轮端面外圆磨床　CNC high speed double-wheel angular approach cylindrical grinding machine
数控高速转塔铣床　CNC high speed turret milling machine
数控高效深孔镗床　CNC deep hole boring machine
数控工具曲线磨床　CNC tool profile grinding machine
数控工作台不升降立式铣床　CNC bed-type vertical milling machine
数控工作台不升降卧式铣床　CNC bed-type horizontal milling machine
数控滑枕式铣床　CNC ram-type milling machine
数控激光加工机　CNC laser processing machine
数控精密导轨磨床　CNC precision guideway grinder

数控卡盘车床　CNC chucking lathe
数控控制器　CNC controller
数控立式车床　CNC vertical turning machine
数控立式升降台铣床　CNC knee type vertical milling machine
数控立式外拉床　CNC vertical external broaching machine
数控立柱移动工作台不升降铣床　CNC bed-type milling machine with travelling column
数控龙门铣床　CNC planer type milling machine
数控磨床　CNC grinding machine
数控内螺纹磨床　CNC grinder for internal thread
数控内圆端面磨床　CNC internal thread grinding machine
数控刨台卧式铣镗床　CNC planer horizontal milling and boring machine
数控前载机车床　CNC front loaded machine
数控三维测量机　CNC three dimensional measuring machine
数控砂带平面磨床　CNC abrasive belt surface-grinding machine
数控砂线切割机床　CNC abrasive wire sawing machine
数控深孔钻镗床　CNC deep-hole drilling and boring machine
数控双轴卡盘式车床　CNC twin-spindle chucker-type turning machine
数控双柱立式车床　CNC double-column vertical turning & boring mill
数控双柱坐标镗床　CNC double column jig boring machine
数控水力切割机　CNC water cutting machine
数控丝锥磨床　CNC tap grinding machine
数控凸轮轴铣床　CNC camshaft milling machine
数控外螺纹磨床　CNC grinder for external thread
数控弯管机　CNC tube bending machine

数控万能工具铣床　CNC universal tool milling machine
数控万能铣床　CNC universal milling machine
数控卧式矩台平面磨床　CNC horizontal surface grinding machine with rectangular table
数控卧式拉床　CNC horizontal broaching machine
数控卧式升降台铣床　CNC knee type horizontal milling machine
数控卧式铣镗床　CNC horizontal milling and boring machine
数控铣刀　CNC milling cutter
数控系统　numerical control system
数控线切割机床　CNC wire-cut machine
数控斜角全面进磨式外圆磨床　CNC external cylindrical angular plunge-cut grinder
数控斜切外圆专用磨床　CNC oblique cutting cylindrical special grinding machine
数控斜式车床　CNC slant-bed turning machine
数控液压折弯机　CNC hydraulic press brake
数控轧辊磨床　CNC roll grinder
数控重型车床　CNC heavy turning machine
数控重型卧式车床　CNC heavy-duty horizontal lathe
数控轴承内圈沟磨床　CNC race grinding machine for ball bearing inner ring
数控轴承套圈内圆磨床　CNC internal grinding machine for bearing ring
数控轴承外圈沟磨床　CNC race grinding machine for ball bearing outer ring
数控专用铣床　CNC special purpose milling machine
数控自动车床　CNC automatic turning machine
数控自动多轨车床　CNC automatic multi-slide lathe
数控自动铆接机　CNC automatic riveting machine
数控纵切车床　CNC sliding headstock auto-lathe
数控坐标测量机　CNC coordinate measuring machine
数控坐标磨床　CNC jig grinding machine
数量　quantity
数模　digital analogy
数/模转换　digital-to-analog (D/A)
数/模转换器　D/A converter
数位式测微仪　digital micrometer
数学的　mathematical
数学规划　mathematical programming
数学模型　mathematic model
数学系统理论　mathematical systems theory
数学相似　mathematical similarity
数值　numerical value
数值变量　numeric variable
数值点　numeric point
数值方法　numerical method
数值方法稳定性　stability of numerical methods
数值仿真　numerical simulation
数值分析　numerical analysis
数值解　numerical solution
数值算法　numerical algorithm
数字　digit
数字 I/O 接口　digital I/O interface
数字补偿　digital compensation
数字补偿器　digital computer compensator
数字测井仪　digital logging instrument
数字测量仪表　digital measuring instrument
数字车用无线电　digital mobile radio
数字传感器　digital sensor, digital transducer
数字磁带记录强震仪　digital magnetic tape record type strong-motion instrument
数字磁通表　digital fluxmeter
数字的　digital, numerical
数字地震仪　digital seismic recording system
数字电流表　digital ammeter

数字电路 digital circuit
数字电压表 digital voltmeter
数字电液控制系统 digital electro-hydraulic (DEH) control system
数字电阻表 digital ohmmeter
数字读出 digital readout
数字阀门 digital valve
数字反馈控制回路 digital feedback control loop
数字反馈系统 digital feedback system
数字仿真 digital simulation
数字仿真计算机 digital simulation computer
数字仿真器 digital simulator
数字复合点 digital composite point
数字功率表 digital power driver
数字化 digitization, digitizing
数字化误差 digitalization error, digitization error
数字混合数据点 digital composite data point
数字积分电路 digital integrating circuit
数字积分器 digital integrator
数字积分式磁通表 digital integrating fluxmeter
数字集成电路测试仪 digital integrated circuit tester
数字计数器 digital counter
数字计算机 digital computer
数字计算机控制 digital computer control
数字计算机应用 digital computer application
数字记录仪表 digital recording instrument
数字胶片室 digital film room
数字警报器 data alarm
数字控制 digital control, numerical control
数字控制器 digital controller
数字控制算法 digital control algorithm
数字控制系统 digital control system
数字量输入/输出模块 digital I/O module

数字滤波处理器 digital filter processor
数字滤波器 digital filter
数字滤波器结构 digital filter structure
数字脉宽调制 digital pulse duration modulation
数字模拟仿真器 digital-analog simulator
数字模拟混合计算机 digital-analog hybrid computer
数字-模拟译码器 digital-analog decoder
数字-模拟转换 digital-analog conversion
数字-模拟转换器 digital-analog converter
数字模式 digital pattern
数字黏度计 digital viscometer
数字频率表 digital frequency meter
数字设备公司 digital equipment corporation
数字深层地震仪 digital deep-level seismograph
数字示值 digital indication
数字式滴定管 digital burette
数字式电动执行机构 digital electric actuator
数字式定位器 digital positioner
数字式高斯计 digital gaussmeter
数字式光度计 digital photometer
数字式热红外扫描仪 digital thermal infrared scanner
数字式微分积分器 digital differential integrator
数字式位移测量仪 digital displacement measuring instrument
数字式位移传感器 digital displacement transducer
数字式位置发送器 digital position transmitter
数字式无线电 digital radio
数字式显示仪表 digital display instrument
数字式综合记录仪 combination digital logger
数字输出 digital output (DO)

数字输出数据点 digital output data point
数字输入 digital input (DI)
数字输入数据点 digital input data point
数字数据 digital data
数字数据采集系统 digital data acquisition system
数字数据处理系统 digital data processing system
数字伺服机构 digital servomechanism
数字通信 digital communication
数字图像 digital image
数字万用表 digital multimeter
数字网络结构 digital network architecture
数字微分电路 digital differential circuit
数字微分分析器 digital differential analyser
数字温度计 digital thermometer
数字系统 digital system, number system
数字显示模拟测量仪表 analogue measuring instrument with digital presentation
数字相位表 digital phase meter
数字信号 digital signal
数字信号处理 digital signal processing
数字信号处理器 digital signal processor
数字信号分析仪 digital signal analyzer
数字信息处理系统 digital information processing system
数字压力表 digital pressure gauge
数字遥测系统 digital telemetering system
数字应变传感器 digital tension sensor
数字应变仪 digital strain indicator
数字震动计 digital vibration meter
数字转换技术 digital conversion technique
数字转换器 digital converter, digitiser, digitizer
数组 array
数组处理机 array processor
数组点 array point
衰减 attenuation, decay
衰减比 attenuation ratio, decay ratio
衰减波 damped wave
衰减长度 attenuation length
衰减程度 degree of attenuation
衰减的 damped
衰减观察器 attenuation observation
衰减交流 damped alternating current
衰减校正 attenuation correction
衰减控制 convergent control
衰减器 attenuator, dimmer
衰减系数 attenuation coefficient, coefficient of attenuation, damping coefficient
衰减振荡频率 relaxation oscillation frequency
双倍密度格式 double-density format
双倍字长 double-length
双臂电桥 double bridge
双波纹管差压计 differential pressure gauge with double bellows
双波纹管差压元件 double-bellows differential element
双层玻璃 double-glazing
双层玻璃空隙 interspace of double glazing
双层的 double layer
双层模具 double stack mold
双差值报警卡 duplex difference alarm card
双程分隔车道 dual carriageway
双重的 double
双重方式控制系统 dual-mode control system
双重绝缘 double insulation
双重控制 duplex control
双重螺旋法 duplex helical
双重模式控制 dual-mode control
双重双面刀齿 duplex spread blade
双重双面法 duplex
双重调制遥测系统 dual modulation telemetering system

双重用途电压互感器　dual purpose voltage transformer
双刀双掷开关　double-pole double-throw switch
双等离子体离子源　duoplasmatron ion source
双电流滴定法　double amperometric titration
双端电路　double end converter
双端馈电　dual-feed
双放大倍率成像法　double-magnification imaging
双分度　double index
双峰　doublet
双干式热量计　double-dry calorimeter
双杆扭秤　double-beam torsion balance
双工传输　duplex transmission
双工压力表　duplex pressure gauge
双股线　twin wire
双管式管道　twin pipe
双管水银压力表　double tube mercury manometer
双光束光谱辐射计　double beam spectrum radiator
双轨　double track
双轨铁路　twin track railway
双滚子链联轴器　double roller chain coupling
双滑块机构　double-slider mechanism
双换向器式电动机　double commutator motor
双火焰离子化检测器　dual-flame ionization detector
双基型二极管　double-base diode
双极电动机　bipolar motor
双极结型晶体管　bipolar junction transistor（BJT）
双极型晶体管　bipolar transistor
双计气压表　duplex air gauge
双计算机系统　duplexed computer system
双焦点X射线管　double-focus X-ray tube
双绞线　twisted wire
双接触导线　double-contact wires
双接收器　double collectors

双金属　bimetal
双金属的　bimetallic
双金属继电器　bimetal relay
双金属式仪表　bimetallic instrument
双金属丝　composite wire
双金属温度传感器　bimetallic temperature transducer
双金属温度计　bimetallic thermometer
双金属线　bimetallic wire
双金属元件　bimetallic element
双金属转子　bimetallic rotor
双进样系统　double inlet system
双晶体衍射计　double-crystal diffractometer
双晶现象　dimorphism
双聚焦　double focusing
双聚焦分析器　double focusing analyzer
双聚焦质谱仪器　double-focusing mass spectroscope
双口排气阀　double opening exhaust valves
双馈　double-fed
双馈推斥电动机　double-fed repulsion motor
双量程电压表　double range voltmeter
双列轴承　double-row bearing
双笼式电动机　double cage motor, double squirrel cage motor
双逻辑系统　binary logic system
双马达的　bimotored
双马来酰亚胺三嗪树脂　bismaleimide-triazine resin
双面处理铜箔　double treated foil
双面刀盘　alternate blade cutter
双面电路板　double sided board
双面覆铜箔层压板　double-sided copper-clad laminate
双面印制板　double-sided printed board
双盘天平　double pan balance
双片边刨刨刀　dual edge trimming iron
双偏差报警卡　duplex deviation alarm card
双频谱　bispectrum

双频谱估计　bispectrum estimation
双氰胺　dicyandiamide
双曲柄机构　double crank mechanism
双曲柄轴冲床　double crank press
双曲面齿轮　hyperboloid gear
双曲轴压力机　double crank press
双绕组变压器　double-column transformer, two-winding transformer
双三极管　dual triode
双三角接法　double-delta connection
双时标系统　two-time scale system
双束质谱计　double-beam mass spectrometer
双丝包线　double-silk covered wire
双搜索树　binary search tree
双速控制器　two-speed controller
双速离合器　two speed clutch
双探头法　double probe technique
双弹性散射峰　binary elastic scattering peak
双弹性散射过程　binary elastic scattering event
双通道比色温度计　double-path ratio thermometer
双通道标度变换卡　dual sealer card
双通道高灵敏度电压表　dual-channel high-sensitive voltmeter
双通道绝对值报警卡　dual absolute alarm card
双通道载波放大器　two-channel carrier amplifier
双通内反射元件　double-pass internal reflection element
双头扳子　double-ended spanner
双头螺柱　studs
双头砂轮机　double-ended grinder
双万向联轴节　double universal joint
双位继电器　two-position relay
双位控制　on-off control
双温恒温器　dual temperature thermostat
双稳定装置　bistability device
双稳态触发器　flip and flop generator
双稳态触发元件　bistable trigger element
双稳态电路　bistable circuit
双稳态多谐振荡电路　flip-flop circuit

双稳态多谐振荡器　bistable multivibrator
双稳态逻辑多谐振荡器　bistable logic multivibrator
双稳态元件　bistable element
双稳态装置　two-state device
双稳性　bistability
双涡轮质量流量计　twin-turbine mass flowmeter
双线的　bifilar
双线电缆　paired cable
双线规　mortise gauge
双线同步检定器　dual-beam synchroscope
双线无感绕组　bifilar winding
双线无感线圈　bifilar coil
双线性　bilinear
双线性变换　bilinear transformation
双线性控制　bilinear control
双线性系统　bilinear system
双线压力传感器　two-wire pressure transmitter
双相的　biphase
双相平衡电动机　two-phase balancing motor
双向传动　two-way drive
双向的　two-way
双向电动机　reversing motor
双向电路　bilateral circuit
双向二极管　bilateral diode
双向风向标　bivane
双向滚动　double roll
双向计数器　back-forward counter
双向晶闸管　bi-directional triode thyristor
双向开关　bilateral switch
双向控制器　reversible controller
双向脉冲　doublet impulse
双向脉冲列　bidirectional pulse train
双向伺服机构　bilateral servomechanism
双向天线　bilateral antenna
双向推力轴承　double-direction thrust bearing
双向稳流器　bilateral current stabilizer
双向旋转电动机　reversing motor

双像速测仪　double-image tacheometer
双芯导线　duplex wire
双芯电缆　duplex cable
双信号选择器　dual signal selector
双星形接法　double star connection
双压力湿度计　two-pressure humidity apparatus
双摇杆机构　double rocker mechanism
双液柱平衡压力计　double liquid balance manometer
双映像　binary image
双元阵　binary array
双闸板　double disc
双折射偏光器　birefringent polarizer
双针指示器　dual indicator
双振幅　double vibration amplitude
双振子探头　double crystal probe
双支热电偶　twin thermocouple
双指示剂滴定法　double indicator titration
双指数滤波器　double-exponential filter
双轴数控车床　CNC double spindle lathe
双转子式流量计　duplex rotor type flowmeter
双转子式容积式流量计　displacement flowmeter of duplex rotor pattern
双锥黏度计　double-cone viscometer
双自旋稳定　dual spin stabilization
双组分控制　dual composition control
双作用定位器　double acting positioner
双座阀　double-ported globe valve
霜点湿度计　frost point hygrometer
水　water
水泵　water pump
水标尺　water gauge
水表　water meter
水车　water bowser
水锤　water hammer
水淬火　water quenching
水的三相点　triple point of water
水底照明　underwater lighting
水电的　hydroelectric
水电站　hydropower station
水分　moisture
水分渗入　penetration of moisture
水封　water seal
水封式旋转气体流量计　water-sealed rotary gas meter
水封闸阀　water seal gate valves
水负载功率计　water load power meter
水工结构工程　hydraulic structure engineering
水管　water pipe
水合物　hydrate
水灰比　water cement ratio
水解　hydrolysis
水浸　flooding
水浸探头　immersion type probe
水口大小　gate size
水口形式　gate type
水库　reservoir
水力发电系统　hydroelectric system
水力蓄力器　hydraulic accumulator
水力学及河流动力学　hydraulics and river dynamics
水力直径　hydraulic diameter
水利工程　hydraulic engineering
水利水电工程　hydraulic and hydropower engineering
水帘　drencher curtain
水流量计　water flowmeter
水流速计　water current meter
水龙头　stopcock
水轮泵名词术语及定义　terms and definitions of water-turbine pumps
水轮泵试验方法　testing methods of water-turbine pump
水轮发电机　hydraulic generator
水轮机　hydraulic turbine
水落管　down pipe
水密　watertight
水泥　cement
水泥管　concrete pipe
水泥含量　cement content
水泥灰泥　cement plaster
水泥浆　laitance
水泥强度试验用标准砂　standard sand for cement strength test

水泥砂浆　cement mortar, cement sand mix
水泥质成分　cementitious content
水泥砖　concrete block
水泡　blister
水平表面　horizontal surface
水平玻璃横格条　horizontal glazing bar
水平测长仪　horizontal length measuring machine
水平尺寸　horizontal dimension
水平导面　horizontal guide face
水平分量　horizontal component
水平横楣梁　lintel beam
水平极限　horizontal limit
水平减震器　horizontal damper
水平净空　horizontal clearance
水平面积　horizontal area
水平能见度　horizontal visibility
水平偏置　horizontal offset
水平式热天平　beam-loading thermobalance
水平位移　horizontal displacement
水平线向　horizontal alignment
水平信号放大器　X axis amplifier
水平翼　horizontal tail
水平铸造　casting on flat
水汽　aqueous vapour
水汽含量　moisture content
水色计　colour meter
水声工程　underwater acoustics engineering
水声信标　acoustic beacon
水塔　water tower
水听器校准器　hydrophone calibrator
水头压力　head pressure
水位　water level
水位计　water level meter
水温表　water-temperature gauge
水文测量　hydrographic survey
水文绞车　hydrographic winch
水文学及水资源　hydrology and water resources
水污染　water pollution
水洗　rinse
水下试验　underwater test
水箱　water tank
水性漆　acrylic paint

水锈　incrustation
水压力　water pressure
水压气动系统　hydro-pneumatic system
水压试验　hydrostatic test
水翼升力　lift of hydrofoil
水翼型拖曳体　hydrofoil shape vehicle
水银灯具　mercury lamp
水银电机式仪表　mercury motor meter
水银电极　mercury pool electrode
水银继电器　mercury-wetted relay
水银开关　mercury switch
水银气压表　mercury barometer
水银温度表　mercury thermometer
水银压力计　mercury manometer
水藻　alga
水阀　sluice valve
水闸门　sluice gate
水诊器　hydrophone
水蒸气　water vapour
水质分析仪　water test kits
水中听音器　hydrophone
水柱压力计　water manometer
水准仪　level gauge, spirit level
水渍　water spots
顺磁体　paramagnetic body
顺磁物质　paramagnetic substance
顺磁性　paramagnetism
顺磁性材料　paramagnetic material
顺时针　clockwise
顺序　sequence
顺序（程序）控制系统　sequence control system
顺序表　sequence table
顺序表参考　sequence table reference
顺序表窗口　sequence table window
顺序表功能块　sequence table block
顺序表状态　sequence table status
顺序表状态报告　sequence table status report
顺序槽路　sequence slot
顺序层压多层印制板　sequentially-laminated multilayer
顺序程序　sequence program
顺序传感器　sequential transducer
顺序定向的过程语言　sequence orien-

ted procedural language（SOPL）
顺序分解　sequential decomposition
顺序辅助功能块　sequence auxiliary block
顺序计算机　sequential computer
顺序控制　sequential control
顺序控制的类型　types of sequence control
顺序控制功能　sequence control function
顺序控制功能块　sequence control function block
顺序控制器　sequence controller
顺序面板功能块　sequence faceplate block
顺序模　progressive die
顺序调试画面　sequence debug display
顺序相位控制　sequential phase control
顺序信息输出　sequence message output
顺序元素　sequence element
顺序元素参考画面　sequence element reference panel
顺序元素搜索画面　sequence element search panel
顺序执行方式　sequence execution mode
顺序执行状态　sequence execution state
顺序注射分析法　sequential injection analysis
瞬间的　instantaneous
瞬时电功率　instantaneous electric power
瞬时荷载　transient load
瞬时机械功率　instantaneous mechanical power
瞬时接触斑点　instantaneous contact pattern
瞬时可用度　instantaneous availability
瞬时力矩　transient torque
瞬时流量计　instantaneous flowmeter
瞬时声压　instantaneous sound pressure
瞬时系统误差　transient system deviation
瞬时振动　transient vibration
瞬时值　instantaneous value

瞬时状态　transient state
瞬态　transient
瞬态超调　transient overshoot
瞬态电源扰动　transient power disturbance
瞬态辐射　transient radiation
瞬态光栅　transient grating
瞬态过电压　transient overvoltage
瞬态控制　control instant
瞬态能量传递　transient energy transfer
瞬态偏差　transient deviation
瞬态散射　transient scattering
瞬态响应　transient response
瞬态振荡　transient oscillation
瞬心　instantaneous center
说明　notes
说明程序　daemon declaration program
说明符　declarator
说明文件　daemon declaration file
丝包漆包线　enamel slik-covered wire
丝杠　screw rod
丝棉包线　silk-and-cotton-covered insulated wire
司机座椅　operater seat
私人专用电报线　private wire
斯托克斯位移　Stokes shift
锶　strontium（Sr）
死层　dead layer
死点　dead point
死区　dead band
死区误差　dead band error, dead zone error
死锁　dead lock
死体积　dead volume
四倍器　quadrupler
四层　four-layer
四重峰　quartet
四端标准电阻器　four-terminal standard resistor
四端口网络　four terminal network
四端器件　four-terminal device
四端网络　electrical quadripole
四环素类　tetracyclines
四极质量分析仪　quadrupole type mass analyzer
四极质谱仪　quadrupole mass spec-

trometer
四连杆机构 four-bar linkage
四轮 four-wheel
四轮操纵 four-wheel steering
四轮驱动 four-wheel drive
四芯电缆 four core cable
四芯线 four-core wire
四叶式交汇处 cloverleaf interchange
四元反馈 quaternion feedback
伺服 servo
伺服传动 servo drive
伺服电动机 pilot motor, servomotor
伺服电机 actuating motor
伺服电机执行机构 servomotor actuator
伺服放大器 servoamplifier
伺服机构 servomechanism
伺服机械控制 servomechanism control
伺服控制 servo control
伺服平衡式液位计 servo-balance type tank gauge
伺服驱动电位差计 servo-operated potentiometer
伺服水力学 servo hydraulics
伺服问题 servo problem
伺服系统 servo, servo system
伺服自整角机系统 servo-selsyn system
似然 likelihood
似然函数 likelihood function
松弛 relaxation
松弛分析 relaxation analysis
松垂 sagging
松的 loose
松动的 untight
松件 loose piece
松开 loosen
送风定温干燥器 forced convection constant temperature drying ovens
送风定温恒温器 forced convection constant temperature ovens
送风机 forced draught fan
送料不到位 feeding is not in place
送料长度 feed length
送料高度 feed level
送料机 feeder

搜索 search
搜索方法 search method
搜索引擎 search engine
苏打 soda
速闭阀 quick-closing valve
速测水准仪 quick-setting level
速动 quick-action
速动继电器 quick-acting relay
速动开关 quick-action switch
速动熔断器 quick-action fuse
速动调节器 quick-acting regulator
速 rate, speed, velocity
速度饱和 vehicle saturation
速度比 speed ratio
速度变送器 speed transmitter
速度表 speedometer
速度波动 speed fluctuation
速度不均匀系数 coefficient of speed fluctuation
速度测量 speed measurement, vehicle measurement
速度超调 vehicle overshoot
速度传感器 velocity transducer
速度反馈 vehicle feedback
速度范围 speed range
速度积分器 velocity integrator
速度计 speed indicator, velocimeter
速度检查 velocity check
速度控制 speed control
速度控制系统 speed control system
速度曲线 velocity diagram
速度瞬心 instantaneous center of velocity
速度伺服电动机 speed control servomotor
速度探测器 speed probe
速度调节 speed regulation, speed governing
速度调整器 speed regulator
速度误差 vehicle error
速度误差常数 velocity error constant
速度限值控制 velocity limiting control
速度限制器 velocity limiter
速度转矩特性 speed-torque characteristic
速断开关 quick-break switch

| 速率反馈 rate feedback
| 速率积分陀螺 rate integrating gyro
| 速率控制 rate control
| 速脱联轴节 quick-release coupling
| 塑胶 thermosetting plastic
| 塑胶覆面钢板 vinyl tapped steel sheet
| 塑胶管 plastic tube
| 塑胶片 sheet
| 塑孔栓 core pin
| 塑料 plastics
| 塑料变形测试 plastic yield test
| 塑料的 plastic
| 塑料工业 plastics industry
| 塑料护套 PVC-sheathed
| 塑料绝缘电线 thermoplastic-covered wire
| 塑料绝缘控制电缆 PVC-insulated control cable
| 塑料绝缘线 plastic-insulated wire
| 塑料零件 plastic parts
| 塑限 plastic limit
| 塑性变形 plastic deformation, plastic distortion
| 塑性沉降 plastic settlement
| 塑性流动 plastic flow
| 塑性系数 plasticity coefficient
| 酸 acid
| 酸酐 anhydride
| 酸碱滴定法 acid-base titrations
| 酸碱度值 pH value
| 酸碱控制器 pH controller
| 酸碱探测器 pH detector
| 酸碱指示剂 acid-base indicator
| 酸洗钢丝 pickled steel wire
| 酸洗设备 acid plant
| 酸性平炉 acid open-hearth furnace
| 酸性溶剂 acid solvent
| 酸性熔铁炉 acid lining cupola
| 酸性转炉 acid converter
| 酸液泵 acid pump
| 酸液缸 acid tank
| 算法 algorithm, arithmetic algorithm
| 算法的 algorithmic
| 算法语言 algorithmic language
| 算术 arithmetic
| 算术加权平均值 arithmetic weighted mean

算术逻辑运算单元 arithmetic logic unit(ALU)
算术平均值 arithmetic mean
算术运算器 calculation unit
算子演算 operational calculus
随动控制 servo control
随动系统 servo system
随机 random
随机编程 stochastic programming
随机变量 random variable, stochastic variable
随机不确定度 random uncertainty
随机参数 stochastic parameter
随机存取 random access
随机存取存储器 random access memory(RAM)
随机存取装置 random access device
随机的 stochastic
随机电报噪声 random telegraph noise
随机仿真 probabilistic simulation
随机负载流 probabilistic load flow
随机复杂性 stochastic complexity
随机过程 random process, stochastic process
随机函数 random function
随机函数发生器 randomizer
随机检查 random inspection
随机介质 random media
随机近似 stochastic approximation
随机控制 stochastic control
随机控制系统 stochastic control system
随机理论 stochastic theory
随机领域 random field
随机模型 stochastic modelling
随机漂移 random drift
随机扰动 random disturbance, random perturbation
随机实现 stochastic realization
随机输入 random input, stochastic input
随机数 random number
随机数生成程序 random number generator
随机数字发生器 random number generator
随机搜寻 random search

随机特性	stochastic property
随机跳变过程	stochastic jump process
随机误差	accidental error, random error
随机系统	stochastic system
随机信号	random signal
随机有限自动机	stochastic finite automaton
随机噪声	random noise
随机自动化	stochastic automation
随时可燃烧	readily combustible
碎裂过程	fragmentation
碎裂图形	fragmentation pattern
碎片	chippings
碎片离子	fragment ion
碎屑	debris
隧道	tunnel
隧道衬砌	tunnel lining
隧道导洞	pilot tunnel
隧道电缆	tunnel cable
隧道二极管	esaki diode
隧道工程	tunnelling works
隧道口	tunnel portal
隧道连接	tunnel junction
隧道连接器	tunnel junction receiver
隧道通风风扇	tunnel ventilation fan
损害	damage
损耗角	loss angle
损耗角正切	loss tangent
损失极小化	loss minimization
榫钉接缝	dowel joint
榫规	mortise gauge
榫锯	tenon saw
榫眼	mortise
榫凿	mortise chisel
缩回	retract
缩颈砂芯	break-off core
缩孔	shrinkage hole
缩裂	contraction crack
索赔	claim
索引控制	traction control
索引偏航控制	traction yaw control
索引平面	key plan
索引图	key diagram
锁闭齿轮	locking gear
锁定	locking
锁定电路	latch circuit
锁定块	clamping block
锁定转子	locked-rotor
锁定转子转矩	locked-rotor torque
锁固垫圈	lock washer
锁紧螺母	lock nut
锁模块	lock plate
锁模力	mould clamping force
锁模器	die locker
锁气装置	air-lock device
锁线	locking wire
锁相	phase locking
锁相回路	phase-locked loop
锁相阵列	phase-locked array
锁信号	lock signal

T

他激式发电机　separately excited generator
他励的　separately excited
他励电动机　separately excited motor
他励多谐振荡器　separately excited multivibrator
他励发电机　separately excited generator
铊　thallium(Tl)
塌陷　collapse
塔　tower
塔轮　step pulley
塔式起重机　tower crane
踏脚板　foot rest
踏脚开关　foot switch
台秤　platform balance
台阶器　stepper
台面式晶体管　mesa transistor
台式电子检流计　electronic galvanometer table model
台式计算机　desk top
台式记录器　desk-top type recorder
台式轮廓投影仪　table profile projector
台式砂轮机　desk grinding wheel machine
台式设备　desktop equipment
台式手绕线机　bench type hand winding machine
太空机器人　space robot
太阳　solar
太阳电池板　solar array
太阳电池组件　solar module
太阳轮　sun gear
太阳能　solar energy
太阳能板指向控制　solar array pointing control
太阳能电池　solar battery, solar cell
太阳能电池阵　solar array
太阳能发电厂　solar power plant
钛　titanium(Ti)
钛线　titanium wire

泰勒级数近似　Taylor series approximation
泰勒控制语言　Taylor control language
泰勒梯形逻辑　Taylor ladder logic
泰勒远程 I/O　Taylor remote I/O
弹出压力　stripping pressure
弹簧　spring
弹簧常数　spring constant
弹簧钢　spring steel
弹簧管　Bourdon tube
弹簧管式压力计　spring-tube manometer
弹簧管压力表　Bourdon tube pressure gauge, Bourdon-tube manometer
弹簧管压力检测元件　Bourdon pressure sensor
弹簧离合器　spring clutch
弹簧螺帽　spring nuts
弹簧片　spring test
弹簧式电流表　spring ammeter
弹簧箱　spring box
弹力　elastic force
弹性　elasticity
弹性变形　elastic deformation
弹性地基　elastic foundation
弹性垫片　resilient pad
弹性分析　elastic analysis
弹性构架　flexible frame
弹性后效　elastic after-effect
弹性滑动　elasticity sliding motion
弹性机垫　resilient mounting
弹性极限　elastic limit
弹性接缝　flexible joint
弹性联轴器　elastic coupling, flexible coupling
弹性模量　modulus of elasticity
弹性散射　elastic scatter
弹性体　elastomer
弹性系数　elastic coefficient, elastic modulus
弹性系统　elastic system
弹性元件压力计　elastic pressure gauge

弹性闸板　flexible disc
弹性支座　elastic support
钽　tantalum(Ta)
钽丝　tantalum wire
探测　probing
探测　detect
探测器　detector
探测器饱和　detector saturation
探测器性能　detector performance
探测算法　detection algorithm
探测系统　detection system
探测元件　detecting element
探伤　fault detection
探伤面　test surface
探伤仪　defectoscope
探头　probe
探头传感器　probe sensor
探头放大器　probe amplifier
探头式磁流量计　probe type magnetic flowmeter
探头引线　probe lead
探向　direction finding
探照灯　cloud searchlight
探针臂　feeler arm
碳　carbon(C)
碳电阻器　carbon resistor
碳钢　carbon steel
碳化　carbonization
碳化硅　carborundum
碳化过程　carbonation process
碳化深度　carbonation depth
碳化铁　cementite
碳化物　carbide
碳结构钢丝　carbon structural wire
碳氢化合物　hydrocarbon
碳氢元素分析仪　carbon and hydrogen analysis meter
碳湿敏电阻器　carbon humidity-dependent resistor
碳刷　carbon brush
碳丝灯泡　carbon-filament lamp
碳素电极　carbon pole
碳素钢铸件　steel casting iron
碳素工具钢　carbon tool steel
碳酸钾　potassium carbonate
碳酸钠　sodium carbonate
碳烟　soot

炭环　carbon ring
炭条　carbon strip
汤浸釉　glazing by dipping
汤釉　glazing by rinsing
汤蘸釉　glazing by immersion
羰基铁　carbonyl iron
镗床　boring machine
镗孔头　boring heads
镗瓷铁　enamelled iron
烫印　hot stamping
逃料孔　slug hole
陶瓷　ceramics
陶瓷传声器　ceramic microphone
陶瓷的　vitrified
陶瓷基覆铜箔板　ceramics base copper-clad laminates
陶瓷金属　ceramic metal
陶瓷滤波器　ceramic filter
陶瓷式振动位移计　ceramic type vibration displacement meter
陶瓷印制板　ceramic substrate printed board
陶管　earthenware pipe
陶土管　vitrified-clay pipe
套管　sleeve, thimble
套管阀　cage valve
套管法　pipe-in-pipe method
套管式铂电阻温度计　capsule platinum resistance thermometer
套管式电流互感器　bushing type current transformer
套管温度表　double tube thermometer
套接　muff coupling
套接管　socketed pipe
套圈　ferrule
套筒扳手　socket wrench
套筒导向　cage guiding
套线　sleeve wire
特别规格　particular specification
特点　features
特殊处理　process special
特殊的　special
特殊断面钢丝　special cross section steel wire
特殊工具钢　low alloy tool steel
特殊工作需求　special work request

特殊命令信号　special semaphore command
特殊事件日记管理　journal manager special event
特殊用途计算机　special purpose computer
特殊运动链　special kinematic chain
特殊指令　special instruction
特［斯拉］（磁通量密度单位）　tesla
特效晶体管结构　effect transistor structure
特效装置　effect devices
特效装置电源　effect device power
特性　characteristics
特性变更　feature change
特性导纳　characteristic admittance
特性方程式　characteristic equation
特性曲线　characteristic curve
特性曲线图示仪　characteristic curve tracer
特性阻抗　characteristic impedance
特征X射线　characteristic X-ray
特征抽取　feature extraction
特征多项式　characteristic polynomial
特征方程　secular equation
特征分析　signature analysis
特征根　characteristic root
特征轨迹　characteristic locus
特征函数　eigenfunction
特征化运算器　characterizer
特征检测　feature detection
特征频率　characteristic frequency, eigen frequency
特征强度　characteristic strength
特征时间　characteristic time
特征矢量　characteristic vector
特征数据点　flag data point
特征向量　eigenvector
特征值　eigenvalue
特征值低限　eigenvalue lower bound
特征值赋值　eigenvalue assignment
特征值替换　eigenvalue replacement
特征值问题　eigenvalue problem
特制弯管　purpose-made bend
铽　terbium(Tb)
梯度　gradients
梯度薄层板　gradient thin layer plate
梯度磁强计　gradient magnetometer
梯度法　gradient method
梯度计　gradiometer
梯度洗脱　gradient elution
梯度展开　gradient development
梯形　ladder
梯形波　trapezoidal wave
梯形钢丝　trapezoidal steel wire
梯形畸变　keystone distortion
梯形滤波器　ladder filter
梯形逻辑　ladder logic
梯形螺纹　acme thread form
梯形渠　trapezoidal channel
梯形衰减器　ladder attenuator
梯形算法　ladder algorithm
梯形图　ladder diagram
梯形网络　ladder-type network
梯状平台　terraced platform
梯状斜坡　terraced slope
锑　antimony(Sb)
提案改善　proposal improvement
提出问题　posed problem
提杆开关　rocker switch
提高功率因数用电容　power-factor capacitor
提交任务　submit job
提前点火控制　spark advance control
提取器　extractor
提升钩　hoisting hook
提升模板技术　climb form technique
体积　physical dimension
体积标记器　volume marker
体外冲击碎石机　extracorporeal shock wave lithotrite
体温传感器　body temperature transducer
体系间跨越　inter-system crossing
体型半导体应变计　bulk type semiconductor strain gauge
体型氧化锌电压敏电阻器　bulk zinc oxide varistor
替代方法　method of substitution
替代机构　equivalent mechanism
替代路线　alternative route
替代设计　alternative design
替换　replace
天电　atmospherics

天花板　ceiling slab
天花吊钩　ceiling suspension hook
天平　balances
天平动阻尼　libration damping
天气　weather
天然气　natural gas
天然石英　natural quartz
天然水压供水　gravity water supply
天然通风　natural ventilation
天然照明　natural lighting
天台贮水箱　roof tank
天线　aerial line, aerial wire, antenna
天线架空线　air wire
天线馈线　aerial feeder
天线匹配　antenna matching
天线系统　antenna system
天线线圈　aerial coil
天线引下线　antenna down-lead
天线指向控制　antenna pointing control
天线转换开关　aerial change-over switch
添加剂　additive
填表式　fill in the form
填充床反应器　packed bed reactor
填充毛细管柱　packed capillary column
填充物　padding
填充因数　fill factor
填充域　region filling
填充柱　packed column
填料　filler, infill
填砂　filling in
填土材料　fill material
填土斜坡　fill slope
填隙合金　caulking metal
填隙料　caulking compound, caulking material
条　bar
条件　condition
条件不稳定性　conditional instability
条件的　conditional
条件电位　conditional potential
条件概率　conditional probability
条件码　condition code
条件数　condition number
条件稳定　conditional stability
条件稳定系统　conditional stable system
条件信号　condition signal
条件属性　conditional behavior

条码技术　barcode
条码扫描器　barcode scanner
条纹状组织　banded structure
条形磁铁　bar magnet
条形码　barcode
条形自动平衡显示仪　self-balancing strip indicator
条状信号发生器　bar generator
调幅　amplitude modulation（AM）
调幅波　modulating wave
调幅的　amplitude-modulated
调幅检波器　amplitude modulation detector
调幅器　amplitude modulator, modulator
调幅器带通滤波器　modulator band filter
调幅/调频函数发生器　AM/FM function generation
调高旋钮　height adjustment knob
调和函数　harmonic function
调校　adjustment
调节　conditioning
调节变压器　regulating transformer
调节波纹管　adjustment bellows
调节池　balance tank
调节单元　controlling unit, regulating unit
调节PV点　regulation PV, regulation PV point
调节电动机　regulating motor
调节电路　regulating circuit
调节电阻　regulating resistance
调节阀　modulating valve, regulating valve
调节范围　range of regulation, regulating range
调节方式　control mode, type of control
调节放大器　conditioning amplifier, regulating amplifier
调节回路　regulating loop
调节机构　regulating mechanism
调节继电器　regulating relay
调节精度　control accuracy, control precision
调节控制点　regulatory control point
调节控制功能块　regulatory control function block
调节控制数据点　regulatory control data point
调节能量　regulating energy
调节器　adjuster, governor, regula-

ting apparatus
调节器的比例带　proportion band of a controller
调节器控制　regulator control
调节器理论　regulator theory
调节器装置　adjustment device
调节设备　regulating device
调节时间　settling time
调节弹簧　regulating spring
调节特性　regulating characteristic
调节系统　control system, controlled system, regulating system
调节元件　regulating element
调节值　regulated value
调节装置　controlling device
调节作用　control action
调零　zero set
调零移相器　zero-setting phase shifter
调频　frequency modulation(FM)
调频波　FM wave
调频测试　frequency modulation inspect
调频广播　frequency-modulation broadcasting
调频激光器　chirping laser
调频激励机　FM exciter
调频接收机　FM receiver
调频器　frequency modulator
调频-调幅倍增器　FM-AM multiplier
调色　color matching
调试　debug
调试程序　debugger, debugging routine
调试程序包　debugging package
调试器　debugger
调试实用程序　debugging utility
调速　speed governing
调速电动机　adjustable speed motors
调速发电机　velodyne
调速器　regulator governor, speed governor
调速器电动机　governor motor
调速器开关　governor switch
调速系统　speed control system
调温器　thermoregulator
调相　phase modulation
调相机　phase regulator
调相开关　phasing switch
调相器　phase modifier
调谐　tune

调谐放大器　resonance amplifier
调谐器　tuner
调心球轴承　self-aligning ball bearing
调心轴承　self-aligning bearing
调压变压器　voltage regulating transformer
调压螺钉　pressure adjusting screw
调压器　pressure governor, pressure regulator, voltage regulator
调整保留时间　adjusted retention time
调整保留体积　adjusted retention volume
调整变量　regulated variable
调整参数存储功能　tuning parameter save function
调整参数打印输出　tuning parameter printout
调整电压　regulating voltage
调整电源放大器　tuned power amplifier
调整画面　tuning panel
调整机构　adjuster
调整螺钉　adjusting screw
调整器　aligner, speed regulator
调整趋势　tuning trend
调制　modulation
调制边带　modulation sideband
调制波　modulating wave
调制程度　degree of modulation
调制的　modulating
调制的脉冲宽度　pulse width modulated
调制度　modulation degree
调制放大器　modulation amplifier
调制分析　modulation analysis
调制分子束质谱计　molecular beam mass spectrometer
调制解调器　modem
调制控制　modulation control
调制控制器　controller modulator
调制器　modulator, keyer
调制深度　modulation depth
调制失真　modulation distortion
调制作用　modulating action
调质处理　hardening and tempering, thermal refining
跳变过程　jump process
跳齿分度　skip index
跳刀　jump
跳焊法　skip welding process
跳火　flashover

跳线　bridle wire, jumper wire
跳闸线路　breaker coil
铁　iron
铁板　iron plate
铁棒　pontil
铁笔　stencil pen, crowbar
铁饼　discus
铁磁的　ferromagnetic
铁磁流体密封　ferrofluid seal
铁磁物质　ferromagnetic material
铁磁谐振　ferro-resonance
铁磁性　ferromagnetism
铁搭　cramp iron
铁-酚试剂　iron-phenol reagent
铁钩　dog iron
铁合金　alloy iron
铁剪　snips
铁-柯柏试剂　iron-Kober reagent
铁矿石　iron ore
铁片　sheet iron
铁钳　pincers
铁撬棍　grab iron, crowbar
铁丝网　wire mesh, woven wire
铁素体　ferrite
铁素体的　ferritic
铁损　iron loss, iron-loss
铁索　iron chain
铁-碳合金相图　iron-carbon equilibrium diagram
铁-铜镍热电偶　iron/copper-nickel thermocouple
铁桶　metal pail
铁屑　scrap iron, filings
铁芯　iron core
铁芯的　iron-cored
铁芯面积　core area
铁芯线圈　iron-core coil
铁氧磁体　magnetized ferrite
铁氧体磁芯　ferrite core
铁氧体磁芯存储器　ferrite-core memory
铁砧　anvil
听诊器　stethoscope
烃化　alkylation
停车阶段　stopping phase
停电　power failure
停电管理系统　outage management system(OMS)
停工时间　down time
停机坪　apron
停机时间　down time
停机指令　halt instruction
停留时间　dwell time
停用文件　dead file
停止　stop
停止按钮　stop button
停止点开关　breakpoint switch
停止供电　cut-off of supply
停止位置　closed position
停止旋塞　stop cock
停止运行　decommissioning
停滞时间　dead time
停转转差率　breakdown slip
通带　pass-band
通带宽带限制开关　pass-band limiting switch
通带宽度　pass-band width
通道　access road, channel, gangway
通道带宽　channel bandwidth
通道地址字　channel address word
通道多路转换器　channel multiplexer
通道范围　channel range
通道解调器　channel demodulator
通道量程　channel span
通道门　access door
通道适配器　channel adapter
通道总线控制器　channel bus controller
通地泄漏电路断电器　earth leakage circuit breaker(ELCB)
通电　energization
通电线路　hot circuit
通断开关　on-and-off switch
通断控制　on-off control
通断作用　on-off action
通风　ventilation vent
通风管　air ventilator, ventilating pipe
通风管道　air duct
通风管旋塞　air vent cock
通风柜　fume hoods
通风机房　fan room
通风机空气动力性能试验方法　test methods of aerodynamic performance for fans
通风孔　air ventilator, vent
通风口　ventilation opening

通风螺塞	vent plug
通风扇	ventilation fan
通风设备	ventilator
通风竖井	airshaft
通风塔	ventilation shaft
通风系统	ventilating system
通风装置	aerator
通管孔	cleaning eye, rodding eye
通航宽度	navigation span
通话按钮	talk-listen button
通孔形式	through-hole form
通量空间矢量	flux space vector
通量密度	flux density
通路	access road, thoroughfare
通路测试	continuity test
通路耦合器	access coupler
通气阀	air vent valve, breather valve
通气管	vent pipe
通气孔	air vent
通气性模具	porous mold
通信	communication
通信处理机	communication processor(CP)
通信电缆	cable for communication
通信多路转接器	communication multiplexer
通信方式控制	communication mode control
通信功能	communication function
通信环境	communication environment
通信缓冲器	communication buffer
通信激光器	communication laser
通信接口适配器	communication interface adapter
通信卡	communication card
通信控制器	communication control unit
通信控制协议	communication control protocol
通信控制应用程序	communication control application program
通信门单元	communication gateway unit
通信模件	communication module
通信区域	communication zone
通信网	communication net
通信网络	communication network
通信卫星	communication satellite
通信系统	communication system
通信线接口	communication line interface
通信线路	communication line
通信协议	communication protocol
通信与信息系统	communication and information systems
通信终端	communications terminal
通信子网	communication subnet
通行权	wayleave
通用扳手	allen wrench
通用比较仪	universal comparator
通用操作和监视功能	common operation and monitoring function
通用操作站	universal station
通用测量仪表	universal measuring instrument
通用测试电流表	universal test ammeter
通用穿线盒	universal fitting
通用串行总线	universal serial bus(USB)
通用的	general, universal
通用仿真器	general simulator
通用非周期波	general nonperiodic wave
通用钢梁	universal beam
通用工作站	universal work station
通用广义双线性变换	general bilinear transformation
通用化	generalization
通用计算机	general purpose computer
通用计算机接口	general purpose computer interface
通用技术条件	general technical requirements
通用加工中心	general machining centers
通用接口总线	general purpose interface bus
通用卡存放器组件	common card file assembly
通用控制	universal control
通用控制网接口	universal control network interface
通用控制网络	universal control network
通用控制网调制解调器	universal control network modem
通用模具	universal mold
通用模型	allround die holder
通用模座	all round die holder
通用软件	common software
通用数据压缩	universal data compression

通用数字转换器　generalized quantizer
通用同步示波器　universal synchroscope
通用线性化算法　general linearization algorithm
通用斜坡发生器　universal ramp generator
通用型片状模塑料　sheet moulding compounds for general purposes
通用异步发送器　universal asynchronous transmitter
同步　synchronization
同步保护　synchronous security
同步辨识　synchronous identification
同步补偿机　synchronous compensator
同步操作计算机　simultaneous computer
同步传递函数　synchronous transfer function
同步传动　synchro drive
同步带　synchronous belt
同步带传动　synchronous belt drive
同步的　synchronous
同步电动机　synchronous motor
同步电机系统　synchronous motors system
同步电抗　synchronous reactance
同步发电机　synchronous generator
同步分解器　synchro resolver
同步分析　synchronous analysis
同步工程　concurrent engineering
同步故障　synchronous failure
同步故障与修复　synchronous failure and recovery
同步回旋加速器　synchrocyclotron
同步机　synchro
同步机器　synchronous machine
同步积分　synchronous integration
同步加速器　synchrotron
同步建筑　synchronous architecture
同步降阶　synchronous order reduction
同步校准器　synchronous regulator
同步阶次　synchronous order
同步结合　synchronous synthesis
同步矩阵　synchronous matrix
同步开关　synchro switch
同步可靠性　synchronous reliability
同步控制　synchronization control
同步控制变压器　synchro control transformer
同步控制发射机　synchro control transmitter
同步控制接收器　synchro control receiver
同步理论　synchronous theory
同步力矩变送器　synchro torque transmitter
同步力矩接收器　synchro torque receiver
同步灵敏度　synchronous sensitivity
同步模型　synchronous model
同步起动电路　trigger circuit
同步驱动皮带　synchronous belt
同步示波器　oscilosynchroscope
同步数据流　synchronous data flow
同步调相机　synchronous condenser
同步文件　synchronous documentation
同步误差　synchro error
同步运行　synchronous operation
同步噪声　synchronous noise
同步诊断　synchronous diagnosis
同步指示灯　lamp synchroscope
同步指示器　synchro indicator
同步转速　synchronous speed
同步装置　synchronizer
同步状态估计　synchronous state estimation
同步自耦变压器　synchronous compensator
同分异构物　isomer
同核锁信号　homonuclear lock signal
同类量　homogeneous quantities
同频抑制比　co-channel rejection ratio
同时操作　concurrent operation
同时式自动光谱仪　meantime autospectrometer
同时稳定　simultaneous stabilization
同态系统　homomorphic model
同位素　isotope
同位素 X 荧光光谱仪　isotope X-ray fluorescence spectrometer
同位素丰度测定　isotope ratio measurement
同位素峰　isotope peak
同位素离子　isotopic ion
同位素稀释质谱法　isotope dilution mass spectrometry
同位素质谱计　isotope mass spectrometer
同相的　inphase, in-phase
同相电压　in-phase voltage

同相分量	inphase component
同心度	concentricity
同心度检查仪	concentricity tester
同心孔板流量计	concentric orifice plate flowmeter
同心绕组	concentric winding
同心锐孔隔板	concentric orifice plate
同心线	concentric line
同心电缆	concentric cable
同轴-波导变换器	coaxial-to-waveguide transformer
同轴测辐射热计	coaxial bolometer
同轴的	coaxial
同轴电缆	coaxial cable, concentric line, concentric cable
同轴度	coaxiality
同轴继电器	coaxial relay
同轴开关	gang switch
同轴控制	gang control
同轴匹配渐变器	coaxial matched taper
同轴调谐器	coaxial tuner
同轴线变换器	coaxial line transformer
同轴圆筒下落黏度计	falling coaxial cylinder viscometer
铜	copper(Cu)
铜包钢丝	weld wire
铜包线	copper binding wire
铜箔	copper foil
铜箔面	copper-clad surface
铜触点	copper contact
铜电极	copper electrode
铜管	copper tube
铜焊钢丝	copper weld steel wire
铜焊线	brazing wire
铜合金	cuprum alloy
铜螺帽	brass nuts
铜皮	copper sheet
铜片	copper sheet
铜丝刷	brass wire brush
铜损	copper loss
铜条嵌镶玻璃	copper glazing
铜芯电力电缆	copper core
铜芯聚氯乙烯绝缘软线	PVC insulated copper soft wire
铜芯聚氯乙烯绝缘线	PVC insulated copper wire
铜芯线	copper core
铜阻尼器	copper damper
统计	statistics
统计的	statistical
统计方法	statistical method
统计分析	statistical analysis
统计过程控制	statistical process control(SPC)
统计决策理论	statistical detection theory
统计模式识别	statistic pattern recognition
统计设计	statistical design
统计推理	statistical inference
统计质量控制	statistical quality control(SQC)
统一数据处理	integrated data processing
统一系统监测管理器	unified system diagnostic manager(USDM)
桶形畸变	barrel distortion
筒夹	collet
筒式柴油打桩锤	diesel pile hammer, tubular diesel pile hammer
筒型防毒面具	cartridge type respirator
头塞	screw plug
投弃式探海温度测量器	expendable bathythermograph
投入产出表	input-output table
投入产出分析	input-output analysis
投入产出模型	input-output model
投影磨床	profile grinding machine
投资决策	investment decision
透度计	penetrameter
透光率	light transmittance
透过紫外线的滤光片	black light filter
透镜测微器	lens micrometer
透明漆	transparent lacquer
透明塑胶	perspex
透明性	transparency
透气系数	permeability coefficient
透气性试验	permeability test
透射式电子显微镜	transmission electron microscope
透声压力容器	acoustically transparent pressure vessel

透视图　perspective drawing
凸规划　convex optimization, convex programming
凸极　salient poles
凸计划　convex projection
凸轮　cam
凸轮倒置机构　inverse cam mechanism
凸轮机构　cam mechanism
凸轮开关　cam operated switch
凸轮廓线　cam profile
凸轮廓线绘制　layout of cam profile
凸轮设定控制器　cam set controller
凸轮弯曲加工　cam die bending
凸锣刀片　convex cutter
凸面　convex side
凸面体　convex
凸模　male die
凸榫　tenon
凸形铣刀　convex cutter
凸缘　flange
凸缘管　flanged pipe
凸缘加工　flanging
凸缘联轴器　flange coupling
凸缘尼龙盖帽　flange nylon insert lock nuts
凸缘起皱　flange wrinkle
凸缘套筒　flange bush
突变论　catastrophe theory
突出　protrude
突触可塑性　synaptic plasticity
突跃　hop
图案　pattern
图表　diagram, graph
图表记录器　chart recorder
图表式指示器　card type indicator
图画纸　cartridge paper
图解　diagram, graphic
图解打印　graphic printer
图解的　graphic
图解法　diagram method, graphic method
图解计算法　graphic calculation
图解平面　diagrammatic plan
图框　drawing frame
图例　legend
图灵机　Turing machine
图论　graph theory
图示板　graphic panel
图搜索　graph search
图像　image, picture
图像编码　image coding
图像插入　image interpolation
图像重构　image reconstruction
图像处理　image processing, picture processing
图像处理器　image processor
图像传感器　image sensor, image transducer
图像传意　graphical communication
图像电磁位移　image electro-magnetic shift
图像放大　image amplification
图像分割　image segmentation
图像分析　image analysis
图像模型化　image modelling
图像配准　image registration
图像匹配　image matching
图像平滑　image smoothing
图像失真　image distortion
图像识别　image recognition
图像修复　image restoration
图像压缩　image compression
图像移动补偿　image motion compensation
图像增强　image enhancement
图像增强器　image intensifier
图像转换器　image converter
图形　graphs
图形表达技巧　graphic presentation technique
图形符号　graphical symbols
图形核心系统　graphical kernel system
图形画面　schematic display
图形加速接口　accelerated graphics port(AGP)
图形库　graphic library
图形显示　graphic display, graphical display
图形显示板　graphic display panel
图形显示文件　graphic display file
徒手切割　free hand cutting
涂布器　spreader

涂层	coating
涂层材料	coating material
涂胶催化层压板	adhesive-coated catalyzed laminate
涂胶黏剂绝缘薄膜	adhesive coated dielectric film
涂胶铜箔	adhesive coated foil
涂胶无催层压板	adhesive-coated uncatalyzed laminate
涂胶脂铜箔	resin coated copper foil
涂料疤	blacking scab
涂料孔（铸疵）	blacking hole
涂漆钢	painted steel
涂色剂	marking compound
涂树脂铜箔	resin coated copper foil
涂油脂	greasing
土地沉降	ground settlement
土地勘测	ground investigation
土方工程	earthworks
土方机械	earth moving machinery
土力测量	geotechnical survey
土木工程	civil engineering
土壤稳定工程	soil stabilization works
土芯	concrete core
土芯样本	core sample
土芯样本资料库	core data bank
土压力	earth pressure
钍	thorium(Th)
湍流	turbulence
湍流对流	turbulent convection
推	push
推斥电动机	repulsion motor
推斥式仪表	repulsion type meter
推导式流量计	inferential flowmeter
推断控制	inferential control
推断问题	prediction problem
推杆	push bar
推杆式送料	pusher feed
推荐的备件清单	recommended spare parts list
推进控制	propulsion control
推进式套筒接头	push-fit spigot and socket joint
推拉开关	push and pull switch
推理	inference
推理策略	inference strategy
推理过程	inference process
推理机	inference engine
推理模型	inference model
推力器	thruster
推力球轴承	thrust ball bearing
推力矢量控制系统	thrust vector control system
推力销	thrust pin
推力轴承	thrust bearing
推土机	bulldozer, loading shovel
推挽的	push-pull
推挽电路	push-pull circuit
推挽式放大器	differential amplifier, push-pull amplifier
推挽输出放大器	push-pull output amplifier
推移图	transition diagram
退磁	degaussing, demagnetize
退刀槽	tool withdrawal groove
退格	backspace
退格符	backspace character
退化	degeneration, degradation, deterioration
退火	anneal
退火铝线	annealed aluminum wire
退火铜箔	annealed copper foil
退货单	material reject bill
退扣式模具	unscrewing mold
退入界	setback line
退色	fading
托板	support plate
托杯形电机	drag cup motor
托架	mounting bracket, shelf
托料板	retainer plate
托盘叉式起重车	pallet fork lift truck
拖板	extension socket
拖板车	long vehicle
拖车	trailer
拖尾峰	tailing area
拖尾因子	tailing factor
拖曳体升阻比	lift-drag of towed vehicle
脱轨器	derailer
脱辊	roll release
脱机设备	off-line equipment
脱机诊断	off-line diagnostic
脱蜡铸造	lost wax casting
脱离	deviate
脱料背板	stripper pad

脱料螺栓　stripper bolt
脱磷生铁　dephosphorized pig iron
脱模衬套　stripper bushing
脱模杵　knockout bar
脱模剂　release agent
脱色剂　discoloring agent
脱水　dehydration
脱水器　dehydrator
脱水物　anhydride
脱水作用　dewatering
脱碳　decarbonizing, decarburization, malleablizing
脱碳退火　decarburizing
脱脂　degrease
陀螺测斜仪　gyro balancing machine
陀螺经纬仪　gyro-theodolite
陀螺漂移率　gyro drift rate
陀螺水平仪　gyro-level
陀螺体　gyrostat
陀螺型质量流量计　gyroscopic mass flowmeter
陀螺仪　gyroscope
椭圆齿轮流量测量仪表　oval wheel flow measuring device
椭圆齿轮流量计　oval gear flowmeter
椭圆度　ovality
椭圆钢丝　oval wire
椭圆极化仪　elliptical polarization instrument
椭圆振动　elliptical vibration
拓扑结构　topological structure
拓扑图　topology graph

W

挖沟机　ditching machine
挖掘工程　excavation works
挖掘许可证　excavation permit
挖坑机　trench excavator
挖泥船　dredge
挖泥工程　dredging works
挖填设计　cut and fill design
挖土机　excavator
瓦管　clay field pipe
瓦楞玻璃屋面　corrugated roof glazing
瓦斯　methane gas
瓦斯继电器　gas relay
瓦特（功率单位）　watt(W)
瓦特计　power meter, wattmeter
瓦特数　wattage
歪轮　skew bevel gears
外包装　exterior package
外保护层　external protection
外标法　external standard method
外标式玻璃温度计　external-scale liquid-in-glass thermometer
外标式温度计　external-scale thermometer
外标准化　external standardization
外表裂缝　external crack
外表面　external surface, outside surface
外部的　external, outer
外部方式键允许状态　external mode switching option state
外部干扰　outside interference
外部工程　external works
外部巩固工程　external strengthening works
外部加法　external summing
外部接缝　external joint
外部界面　external interface
外部临界阻尼电阻　external critical damping resistance
外部支撑　external bracing
外部支撑结构架　external structural bracing frame
外部总线　front side bus(FSB)

外参比试样　external reference sample
外侧的　outboard
外差变频器　heterodyne converter
外差法　heterodyne
外差检波器　heterodyne detector
外差接收机　heterodyne receiver
外差频率　heterodyne frequency
外差式分析仪　heterodyne analyzer
外差式收音机　beat receiver
外差振荡器　heterodyne oscillator
外齿轮　external gear
外存储器　external store
外导柱　outer guiding post
外电路　external circuit
外端齿顶圆　crown circle
外对角接触　bias out
外反馈　external feedback
外干路　external distributor road
外观不良　cosmetic defect
外观检查　cosmetic inspect, visual inspection
外护罩　outer shield
外环　outer loop
外汇　foreign exchange
外激式电动机　separately excited motor
外加电压　applied voltage
外加负载　applied load
外加覆盖物　applied covering
外加荷载　imposed load
外加剂　admixture
外径　outer diameter, outside diameter
外径规　snap gauge
外径卡规　caliber gauge
外径千分尺　outside micrometer
外壳　external shell
外壳极限温度　housing limit temperature
外壳接地　earthing of casing
外力　external force
外临界电阻　external critical resistance
外露端　exposed junction

外露面　exposed face
外螺纹　male thread, outside thread
外墙　external wall
外切刀尖直径　outside point diameter
外切刃点　outside blade
外圈　outer ring
外扰　external disturbance
外生变量　exogenous variable
外锁信号　external lock signal
外特性　external characteristic
外脱料板　outer stripper
外围的　peripheral
外围设备　peripheral equipment
外形　profile
外形尺寸　boundary dimension, physical dimension
外圆车削　cylindrical lathe cutting
外圆磨床　cylindrical grinding machine, grinding machines cylindrical
外圆磨削　external grinding
外缘　outer edge
外直径　external diameter
外指示剂　external indicator, outside indicator
外转换　external conversion
外装玻璃法　outside glazing
外锥距　outer cone distance
弯管机　pipe bender, pipe bending machine
弯矩　bending moment
弯曲　bend
弯曲度　bending
弯曲刚性　flexible rigidity
弯曲机　bending machines
弯曲力　bending force
弯曲疲劳　bending fatigue
弯曲强度　bending strength
弯曲应力　beading stress
弯曲振动　bending vibration
弯铁剪　curved snips
弯头　elbow
弯头接合　toggle mechanism
完美控制　perfect control
完全　full
完全的　complete
完全焊透　full penetration, full welding
完全解耦　complete decoupling

完全可观测性　complete observability
完全可控性　complete controllability
完全能观测性　complete observability
完全能控性　complete controllability
完全熔接　complete fusion
完全失效　complete failure
完全停机　dead halt
完全退火　full annealing
完整　integrity
顽磁性铁磁材料　retentive ferromagnetic material
烷基化　alkylation
万测仪　pantometer
万能工具磨床　universal tool grinding machine
万能控制网络　universal control networks(UCN)
万能磨床　grinding machines universal
万能手钳　multipurpose pliers, universal pliers
万能铣床　universal milling machines
万能属性　universal personality
万向接头　universal joint
万向联结轴　universal coupling shaft
万向球节　universal ball joint
万向轴　cardan shaft
万向轴节　universal coupling
万用表　multimeter
万用槽钢　versatile U-steel
万用电桥　universal bridge
万有引力　gravitational attraction
网钢丝　netting wire
网格　grid
网格纸　grid paper
网关　gateway
网孔　mesh
网孔电流　mesh current
网络　network
网络版　network version
网络表　net list
网络布置　network layout
网络层　network layer
网络方程　equation of network
网络分析　network analysis
网络分析仪　network analyser
网络故障　network fault
网络管理系统　network management

system(NMS)
网络函数 network function
网络缓冲器 network buffer
网络监控系统 net control system
网络接口模块 network interface module(NIM)
网络结构 lattice
网络可观察能力 network observability
网络可实现性 network reliability
网络控制 network control
网络滤波器 lattice filter
网络数量 network number
网络损耗 network loss
网络拓扑 network topology
网络组态文件 network configuration file(NCF)
网筛 mesh sieve
网上设计 on-net design
网上协作工程 concurrent engineering on network
网纹板 chequered plate
网纹钢 checkered iron
网线 screen wire
网状滤器 gauze strainer
网状污水渠系统 sewerage reticulation system
往复泵 reciprocating pump
往复活塞空气压缩机 reciprocating piston air compressors
往复活塞流量计 reciprocating piston flowmeter
往复螺杆 reciprocating screw
往复式密封 reciprocating seal
往复式压缩机 reciprocating compressor
往复移动 reciprocating motion
往复运动唇形密封圈 reciprocating seal
望远镜 telescope
望远镜视场角 angle of view of telescope
望远镜探测器 telescope detector
危险 danger, risk
危险标志 danger signal
危险程度评估研究 hazard assessment study
危险的 dangerous
危险概率评估 probabilistic risk assessment

危险建筑物 dangerous building
危险警告灯 hazard warning lantern
危险品 dangerous goods
危险品包装 dangerous articles package
危险品包装标志 hazardous substances mark
危险品仓库 dangerous goods store
危险区 danger zone
危险信号 danger signal
危险信号灯 danger light
威森伯格效应 Weissenberg effect
微气压计 microbarograph
微安计 microammeter
微波 microwave
微波测距仪 microwave distance meter
微波传输带 microstrip
微波磁盘操作系统 microsoft-disk operating system
微波等离子体光谱仪 microwave inductive plasma emission spectrometer
微波等离子体检测器 microwave plasma detector
微波段 microwave band
微波多普勒流量计 microwave Doppler meter
微波辐射计 microwave radiometer
微波管 microwave tube
微波厚度计 microwave thickness meter
微波检测仪 microwave detection apparatus
微波雷达 microwave radar
微波滤波器 microwave filter
微波谱 microwave spectrum
微波散射仪 micro-wave scatterometer
微波收发两用机 microwave transmitter-receiver
微波探伤法 microwave distance method
微波网络分析仪 microwave network analyzer
微波遥感 microwave remote sensing
微差测量 differential measurement
微差测量法 differential method of measurement

微场扩流发电机　metadyne
微处理器　microprocessor
微处理器控制　microprocessor control
微滴　droplet
微滴乳状液动电色谱法　microemulsion electrokinetic chromatography
微电脑野外检测系统　micro-computer field measuring system
微电子器件　microelectronic device
微电子学　microelectronics
微电子学与固体电子学　microelectronics and solid state electronics
微动开关　microswitch
微动开关式变送器　microswitch transmitter
微动螺旋机构　differential screw mechanism
微动装置　microinching equipment
微法［拉］（电容单位）　microfarad
微分的　differential
微分电路　differentiator
微分动力学系统　differential dynamical system
微分对　differential pair
微分对策　differential game
微分反馈　derivative feedback
微分方程　differential equations
微分方程式　differential equation
微分分析器　differential analyzer
微分环节　differentiation element
微分几何　differential geometry
微分控制　derivative control
微分控制器　D controller
微分器　differentiator
微分时间　rate time
微分时间常数　derivative time constant
微分调节作用　derivative control action
微分型检测器　differential detector
微分增益　differential gain
微分作用　derivative action
微分作用时间　derivative action time
微分作用时间常数　derivative action time constant
微分作用系数　derivative action coefficient

微分作用增益　derivative action gain
微观结构　micro-structure
微观经济模型　micro-economic model
微观经济系统　micro-economic system
微机化交流电阻率仪器　microcomputer alternating current resistivity instrument
微机激电仪　microcomputer induced polarization instrument
微机系统　microcomputer system
微积分学　calculus
微径柱　microbore column
微库仑检测器　micro coulometric detector
微粒柱　microparticle column
微量分析　micro analysis
微量高速离心机　high speed microcentrifuges
微量天平　micro balance
微量吸附热检测器　micro-heat of adsorption detector
微米　micron
微密度计　micro-densitometer
微生物谷氨酸传感器　glutamate microbial transducer, glutamic acid microbial sensor
微生物简单测试仪　simple germ test
微生物培养箱　microbiological incubator
微生物自动分析系统　automatic analyzer for microbes
微缩影片　microfilm
微填充柱　microbore packed column, micro-packed column
微调尺寸　inching
微调电容器　alignment capacitor
微调电阻箱　fine turning resistance box
微通道结构　microchannel architecture(MCA)
微系统　microsystem
微线印制板　micro wire board
微型分光光度计　microplate spectrophotometer
微型计算机　microcomputer
微音器　capacitor microphone
微阻缓闭止回阀　tiny drag slow shut

check valves
韦伯（磁通量单位） weber
韦布尔分布　Weibull distribution
围板　hoarding
围压应力　confining stress
围堰坝　cofferdam
违反　contravention
桅杆起重机　derrick crane
唯一性　uniqueness
唯一作用　unique action
维恩电桥　Wien bridge
维护　maintenance
维护出口　outlets for maintenance
维护建议信息　maintenance recommendation message
维护试验　maintenance test
维护支持软件　maintenance support software
维纳滤波　Wiener filtering
维纳滤波器　Wiener filter
维生素　vitamin
维氏硬度试验　Vickers hardness test
维修保养　corrective maintenance
维修车辆　recovery vehicle
维修工程　maintenance works
维修工程师　maintenance engineer
维修轨道　maintenance track
维修建议　maintenance recommendation
维修坑　maintenance pit
维修系统　maintenance system
维修仪器　maintenance device
伪速率增量控制　pseudo-rate-increment control
伪随机序列　pseudo random sequence
纬纱　weft yarn
纬向　weft-wise
卫星　satellite
卫星控制　satellite control
卫星控制应用　satellite control application
卫星联络线　satellite order wire
卫星通信　satellite communication
未保护的　unprotected
未定义单元　off unit
未加工铸件　unworked casting
未接地寻线器　absence-of-ground search selector

未经处理的污水　raw sewage
未经净化水　raw water
未修正结果　uncorrected result
位　bit
位并行　bit parallel
位串行　bit serial
位串行信息公路　bit-serial highway
位面计　level instrument
位/秒　bit per second(bps)
位向量　bit vector
位移测量仪表　displacement measuring instrument
位移传感器　displacement pickup, displacement sensor, displacement transducer
位移串级　displacement cascade
位移电流　displacement current
位移定律　displacement law
位移曲线　displacement diagram
位移通量　displacement flux
位移误差　displacement error
位移相　displaced phase
位移因数　displacement factor
位移振幅传感器　displacement vibration amplitude transducer
位/英寸　bit per inch
位置　location, position
位置变送器　position transmitter
位置测量仪　position measuring instrument
位置度　true position
位置反馈　position feedback
位置估计　position estimation
位置继电器　positioning relay
位置精度　position accuracy
位置刻度　position scale
位置控制　position control
位置控制器　positioner
位置灵敏度　position-sensitive
位置速度　position velocity
位置图　location plan
位置误差　position error
位姿　position and orientation
位姿过调量　pose overshoot
胃肠内压传感器　gastrointestinal inner pressure sensor, gastrointestinal inner pressure transducer

谓词逻辑　predicate logic
温标　temperature scale
温差电动势　thermal electromotive force
温差电效应　thermoelectric effect
温差电元件　thermoelement
温差式比重计　thermo hydrometer
温度　temperature
温度变化　temperature variation
温度变化试验　change of temperature test
温度变换器　temperature converter
温度变送器　temperature transmitter
温度补偿　temperature compensation
温度补偿静电计　temperature compensated electrometer
温度补偿装置　temperature compensating device
温度测定点　measuring point for the temperature
温度测量　temperature measurement
温度测量仪表　temperature measuring instrument
温度传感器　temperature sensor, temperature transducer
温度分布　temperature distributions
温度轨线　temperature profile
温度计　temperature meter, thermometer
温度计算　temperature calculation
温度计套管　thermometer well
温度记录仪　temperature recorder
温度交变试验箱　high-low temperature chamber
温度均匀性　temperature uniformity
温度可调范围　adjustable temperature range
温度控制　temperature control
温度控制回路　temperature control loop
温度控制继电器　temperature control relay
温度调节器　thermoregulator
温度调节仪表　temperature controller
温度稳定度　temperature stability
温度误差　temperature error, thermal error
温度系数　temperature coefficient, temperature factor
温度指示　temperature indicators
温度锥　pyrometric cone, thermal cone
温度自动记录器　automatic temperature recorder
温锻　warm forging
温升　temperature rising
温湿度红外辐射计　temperature humidity infrared radiometer
温湿计　hygrothermograph
温室　conservatory
文本编辑程序　text editor
文本输入口　text input port
文法推断　grammatical inference
文件　document, file
文件编码程序　file builder
文件分类机　document sorter
文件夹　document folder
文件结构　document structure
文件结束　end of file(EOF)
文件结束指示器　end of file indicator
文件证据　documentary evidence
文卷维护　file maintenance
文丘里管　venturi, Venturi tube, Venturi air gauge
文丘里管流量计　Venturi flowmeter
文丘里空气压力计　Venturi air gauge
文丘里喷嘴　Venturi nozzle
纹理　texture
稳定变压器　stabilizing transformer
稳定的　stable, steady
稳定电源　stabilized power supply
稳定度　degree of stability
稳定度判据　criterion for stability
稳定反馈　stabilizing feedback
稳定范围　stability range
稳定化　stabilization
稳定化控制器　stabilizing controller
稳定极限　stability limit
稳定剂　stabilizer, stabilizing agent
稳定开关　stable switch
稳定力矩　stabilizing moment
稳定连接　connective stability
稳定平衡　stable equilibrium
稳定前馈　stabilizing feedforward
稳定条件　stability condition
稳定网络　stabilization network

稳定系数　coefficient of stability, coefficient of stabilization
稳定系统　stable system
稳定性　stability, steadiness
稳定性测试　stability test
稳定性的确定　determination of stability
稳定性分析　stability analyses
稳定性判据　stability criterion
稳定性性质　stability property
稳定因数　stability factor
稳定域　stability domain
稳定运转阶段　steady motion period
稳定状态　stable state
稳幅器　amplitude stabilizer
稳固平台　solid platform
稳健设计　robust design
稳流电源　constant-current power supply
稳流器　constant-current stabilizer
稳频电源　constant frequency power supply
稳频器　frequency stabilizer
稳态　steady state
稳态电源条件　steady-state power condition
稳态偏差　steady-state deviation
稳态热阻的测定　determination of steady-state thermal resistance
稳态条件　steady-state condition
稳态稳定性　steady-state stability
稳态误差　steady-state error
稳态误差系数　steady state error coefficient
稳态响应　steady-state response
稳态有效性　steady-state availability
稳态值　steady-state value
稳压电源　regulated power supply
稳压二极管　Zener diode
稳压二极管校准器　Zener diode regulator
稳压管　voltage-regulator tube
稳压集成电路　integrated regulator
稳压器　manostat, voltage regulator, voltage stabilizer
问题求解器　problem solver
涡动速度　eddy velocity
涡流　eddy, eddy current, eddy-current
涡流分析　eddy current analysis
涡流技术　eddy current technique
涡流扩散　eddy diffusion
涡流流量计　eddy-current flowmeter, vortex flowmeter
涡流模拟速度传感器　vortex analog speed sensor
涡流式传感器　eddy-current type transducer
涡流式流量计　swirlmeter
涡流式转速表　eddy-current revolution counter
涡流损耗　eddy current loss
涡流探伤法　eddy current testing method
涡流探伤仪　eddy current inspection instrument
涡流问题　eddy current problem
涡轮发电机　turbine generator, turbo-generator
涡轮机　turbine
涡轮流量测量仪表　turbine flow measuring device
涡轮流量计　turbine flowmeter, turbine meter
涡轮式泵　turbine pump
涡轮式间接流量计　turbine type inferential flowmeter
涡轮式流量传感器　turbine type flow transducer
蜗杆　worm
蜗杆传动机构　worm gearing
蜗杆驱动　worm drive
蜗杆头数　number of threads
蜗杆蜗轮机构　worm and worm gear
蜗杆形凸轮步进机构　worm cam interval mechanism
蜗杆旋向　hands of worm
蜗杆直径系数　diametral quotient
蜗轮　worm gear
蜗轮传动蝶阀　butterfly valves with gear actuator
蜗轮流量变送器　turbine flow transmitter
沃斯回火法　austempering
卧式电机　horizontal machine
卧式锻造机　impacter

卧式及立式加工中心　horizontal and vertical machining centers
卧式加工制造中心　horizontal machine center
卧式加工中心　horizontal machining centers
卧式平衡机　horizontal balancing machine
卧式镗床　horizontal boring machine
卧式镗孔机　horizontal boring machine
卧式铣床　horizontal milling machines
握固长度　grip length
握手　handshake
污染　contaminate, pollution
污染监测系统　contamination monitoring system
污染性物质　polluting substance
污水　effluent, foul water
污水泵　foul water pump, sewage pump
污水泵房　sewage pump house
污水池　cesspool
污水处理　sewage disposal, sewage treatment
污水处理厂　sewage treatment plant
污水干管　trunk sewer
污水管　sewage pipe
污水坑　sewage sump
污水排放管　effluent outfall
污损　deface
屋顶荷载　roof load
乌氏黏度计　Ubbelohde viscometer
钨　tungsten(W)
钨-铼热电偶　wolfram-rhenium thermocouple
钨丝　tungsten wire
钨丝灯　tungsten filament lamp
无差控制　deadbeat control
无差调节　floating control
无承梁转向架　bolsterless bogie
无触点传感器　contactless pickup, non-contacting pickup
无触点的　contactless
无触点继电器　no touch relay
无触点开关　non-contacting switch, contactless switch
无触点运动　brushless
无窗框安装法　glazing without frame
无磁性的　nonmagnetic
无电流的　currentless
无定位控制器　floating controller
无定位速度　floating speed
无定位作用　astatic action, floating action
无端铜网　endless wire
无法兰阀　flangeless valve
无法兰连接端　flangeless ends
无缝　seamless, smls
无缝锻造　seamless forging
无干扰供电　clean supply
无感的　non-inductive
无感电路　non-inductive circuit
无感电容器　non-inductive capacitor
无感电阻　non-inductive resistance
无功补偿　reactive power compensation
无功部分　reactive component
无功电度表　reactive kilovolt-ampere-hour meter, wattless component watt-hour meter
无功电流　idle current, reactive current
无功电流发生器　reactive current generator
无功电流分量　reactive current component
无功电压分量　reactive voltage component
无功分量　idle component
无功伏安　var
无功负载　reactive load
无功功率　reactive power
无功功率表　wattless power meter
无功功率因数　reactive factor, reactive power factor
无功损耗　reactive loss
无轨电车　trolley bus
无滚动粗切　no-roll roughing
无互作用控制系统　non-interacting control system
无换向器电机　commutatorless machine
无火花冲击式电铃　non-spark im-

无	pulse type alarm bell	无线电测距	radio distance
无机化学	inorganic chemistry	无线电测量仪器	radio instrument
无机泥土	non-organic soil	无线电测向	radio direction finding
无级变速	stepless speed changes devices	无线电传真	radio facsimile
无级变速传动	variable-speed drive	无线电干扰	radio interference
无极作用	stepless action	无线电工程	radio engineering
无极放电灯	electrodeless-discharge lamp	无线电广播	radiocast
无记忆	memoryless	无线电话	radiophone, radiotelephony
无记忆信源	memoryless source	无线电控制	radio control
无甲烷碳氢化合物	non-methane hydrocarbons	无线电频率干扰	radio frequency interference
无接地搜索选择器	absence-of-ground search selector	无线电通信	telecommunication
无静差控制	astatic control	无线电线路	wireless link
无孔	imperforate	无线电遥控	wireless remote control
无连接盘导通孔	landless via hole	无线电自动风向风速仪	automatic radio wind wane and anemometer
无连接盘孔	landless hole	无相互干扰	non-interference
无量纲量	dimensionless quantity	无效事件	invalid event
无流道冷料模具	runnerless mould	无效自由度	passive degree of freedom
无墨水式记录仪	inkless recorder	无心精研机	lapping machines centerless
无内搪层镀锌铁管	unlined galvanized iron pipe	无心磨床	grinding machines centerless
无捻粗纱	rovings	无压继电器	no-voltage relay
无偏估计	unbiased estimate	无液气压计	aneroid barometer
无偏置测量	freedom from bias of measurement	无液气压记录器	aneroidograph
无铅汽油	unleaded petrol	无液晴雨表	aneroid barometer
无穷大	infinite	无液自动气压计	aneroidograph
无穷大电压增益	infinite voltage gain	无引线元件	leadless component
无穷控制	infinity control	无影灯	shadowless lamp
无穷维系统	infinite dimensional system	无用数据	garbage
无曲柄式	crankless	无油灰镶玻璃法	puttyless glazing
无扰动启动	bumpless start	无源补偿	passive compensation
无熔丝断路器	free fuse breaker	无源测距	passive ranging
无声链	silent chain	无源的	passive
无失真	distortionless	无源电路元件	passive circuit elements
无损耗	lossless		
无损耗网络	ideal network	无源滤波器	passive filter
无损检测仪	instrument for nondestructive testing	无源四端网络	passive four-terminal network
无损线路	lossless line	无源网络	passive network
无铁芯的	ironless	无源线性二端口网络	passive linear two-port network
无误差	error-free		
无线的	wireless	无源悬挂	passive suspension
无线电	radio	无源元件	parasitic element, passive

component, passive element
无源组件　passive block
无噪声的　noise-free
无噪声电路　quiet circuit
无遮蔽露台　open balcony
无支撑胶黏剂膜　unsupported adhesive film
无质子溶剂　aprotic solvent
无中断自动控制　uninterrupted automatic control(UAC)
无中心的　centreless
无阻塞式数字流量计　obstructionless digital flowmeter
无阻行车速度　free flow operating speed
无阻直通路线　free flow through route
五重峰　quintet
五次谐波　quintuple harmonic
五分仪　quintant
五金器件　hardware
五孔气插座　five-hole pneumatic socket
五坐标数控铣镗床　CNC 5-coordinate milling-boring machine
武器系统与运用工程　weapon systems and utilization engineering
物理测量　physical measurement
物理层　physical layer
物理程序库　physical properties library
物理电子学　physical electronics
物理分析　physical analysis
物理符号系统　physical symbol system
物理化学分析　physicochemical analysis
物理节点　physical node
物理可行性　physical realizability
物理量的模拟表示　analogue representation of a physical quantity
物理量的数字表示　digital representation of a physical quantity
物理模型　physical model
物理气相沉积　physical vapor deposition
物理设计　physical design
物理特性　physical property
物理维度　physical dimension
物理学　physics
物料　materials
物料检查表　material check list
物料控制　material control
物料平衡　material balance
物料平衡控制　material balance control
物料清单　bill of material
物料统计明细表　material statistics sheet
物料系统　material system
物料需求计划　material requirements planning(MRP)
物料质量　quality of material
物流自动化　material flows automation
物位传感器　level sensor, level transducer
物形反光镜　shaped reflector
物性分析　physical property analysis
物质守恒　conservation of matter
误操纵　mishandle
误操作　faulty operation
误差　error
误差补偿　error compensation
误差传递　propagation of error
误差传递函数　error transfer function
误差带宽　error bandwidth
误差分析　error analysis
误差估计　error estimation
误差极限　limits of error
误差检测　error detection
误差检测码　error-detecting code
误差检测器　error detector
误差校正　error correction
误差校正码　error-correcting code
误差控制　error control
误差率　error probability
误差敏感元件　error-sensing element
误差判据　error criteria
误差系数　error coefficient
误差信号　error signal
误读　misreading
误码率　bit error rate(BER)
雾度　haze
雾化　atomizing
雾化空气风扇　atomization air fan
雾量器　fog-gauge
雾浊　blushing

X

西格示温熔锥　Seger cone
西林电桥　Schering bridge
西门子（电导单位）　siemens
吸持电流　holding current
吸持线圈　locking coil
吸电子取代基　electron-withdrawing group
吸附　adsorption
吸附剂　adsorbent
吸附色谱法　adsorption chromatography
吸力　suction
吸起电流　operating current
吸起线　pick-up wire
吸气阀　aspirating valves
吸气器　aspirator
吸热反应　endothermic reaction
吸热峰　endothermic peak
吸入管过滤器　suction strainer
吸入水　water of imbibition
吸收　absorption
吸收X射线度谱术　absorption X-ray spectrometry
吸收比　absorptance
吸收比色计　absorptiometer
吸收边沿　absorption edge
吸收池　absorption cell
吸收带　absorption band
吸收带的强度　intensity of absorption band
吸收电路　absorber circuit
吸收光谱　absorption spectrum
吸收率　absorptance, absorption factor, absorptivity
吸收能力　absorptivity
吸收频谱　absorption spectrum
吸收湿度表　absorption hygrometer
吸收式红外光谱仪　absorption infrared spectrometer
吸收式监测器　absorption type monitor
吸收式探伤仪　absorption flaw detector
吸收式探头分析仪　absorption probe analyzer
吸收体　absorber
吸收系数　absorption coefficient
吸收限　absorption edge
吸收性　absorptivity
吸收修正　absorption correction
吸水测试　water absorption test
吸水阀　suction valve
吸水干管　suction main
吸水性　water absorptivity
吸液管　pipette
吸引材料的吸收率　absorptivity of an absorbing
吸引子　attractor
希尔波特　Hilbert
希尔波特变换器　Hilbert transformer
希尔波特空间　Hilbert space
希望特性　expected characteristics
希望值　desired value
析出硬化　precipitation hardening
硒　selenium(Se)
稀释比　dilution ratio
稀释比例　dilution rate
稀释法　dilution methods
稀释剂　thinner
稀释空气取样探头　sample probe for dilution air
稀释气　diluent gas
稀释速率　dilution rate
稀释因数　dilution factor
锡　tin
锡包线　solder-covered wire
熄火保险　flame failure protection
席式基脚　mat footing
洗　wash
洗净器　washers
洗瓶　plastic wash bottle
洗脱（淋洗）　elution
洗脱色谱法　elution chromatography
铣床　milling machine

铣刀　milling cutter
铣头　milling heads
系船柱　bollard
系紧螺母　anchor nuts
系紧线　stay wire
系列滤波器　filter bank
系留螺帽　anchor nuts
系数　coefficient
系数矩阵　coefficient matrix
系统　system
系统安装图号　system mounting drawing No.
系统报警　system alarm
系统报警窗口　system alarm window
系统报警通知　system alarm notification
系统报警信息画面　system alarm message panel
系统不确定度　systematic uncertainty
系统菜单画面　system menu display
系统传递函数　system transfer function
系统存储　system save
系统等级　system scale
系统电压　system voltage
系统方法论　system methodology
系统仿真　systems simulation
系统概念　system concept
系统工程　system engineering
系统功能键名　system function key name
系统构成　system configuration
系统固有特性　inherent characteristic of a system
系统观察　system view
系统规格说明　system specification
系统环境　system environment(SE)
系统控制　system control
系统偏差　systematic deviation
系统品质保证工程　system quality assurance engineering
系统评价　system assessment
系统设备　system equipment
系统设计　system design
系统生成菜单　system builder menu
系统数据库　system database
系统同构　system isomorphism
系统同态　system homomorphism
系统完整性组件　system integrity module
系统维护　system maintenance
系统维护画面　system maintenance panel
系统维护控制中心　system maintenance control center
系统稳定性　system stability
系统误差　systematic error
系统误差处理　system error handler
系统信息窗口　system message window
系统性能评估　system performance evaluation
系统学　systematology
系统应用　system utility
系统增益　system gain
系统状态画面　system status display
系统状态执行程序　system state executive
系统状态总貌窗口　system status overview window
系统总貌　system overview
细胞的　cellular
细胞电位传感器　cell potential sensor, cell potential transducer
细胞计数分析仪　cell scalar analyzer
细胞生死判别系统　cell vital analyzer
细分　subdivide
细粉尘　fine dust
细钢丝　small-gauge wire
细骨料　fine aggregate
细木工作　joinery
细目丝网　gauze wire
细目文件　detail file
细目显示　detail display
细目显示调出键　detail key
细水口　pin-point gate
细弹簧　hairspring
细牙螺纹　fine threads
隙缝　aperture
下半　low-half
下边带　lower side-band
下齿面　dedendum surface
下齿面加工　flanking
下传动式压力机　bottom slide press

下垂线　faller wire
下垫板　die pad
下垫脚　bottom block
下盖　bottom flange
下滑块板　lower sliding plate
下降率　droop rate
下降时间　fall time
下降特性　drooping characteristic
下井仪器　downhole instrument
下卷　scroll down
下料模　blanking die
下模　lower die
下模板　lower plate
下模座　die holder
下批批量设定值　next batch set value
下切换值　lower switching value
下投式探空仪　dropsonde
下推自动机　pushdown automaton
下托板　bottom plate
下脱料板　lower stripper
下网　lower wire
下限　lower limit
下限截止频率　lower-cut-off frequency
下限控制　low limiting control
下限调整　low-limit adjustment
下限越界　off-normal lower
下限值控制　low-limiting control
下向力　downward force
下行展开　descending development
下压模　dip mold
下游　down stream
下游阀　downstream valve
下游区　catchment area
下载　download
夏时制　daylight savings time
先进操作　advanced operation
先进磁盘操作系统　advanced diskette operating system
先进的控制接口数据点　advanced control interface data point(ACIDP)
先进控制　advanced control
先行缓冲器　look-ahead buffer
先行控制点标志　advanced control point ID
先行控制功能　advanced control function

先行控制技术　advanced control technique
先行控制接口数据点　advanced control interface data point(ACIDP)
先行控制系统　advanced control system
先验估计　priori estimate
纤度计　deniermeter
纤维断口铁　fibrous iron
纤维光学　fibre optics
纤维灰泥　fibrous plaster
纤维绝缘线　fibre-insulated wire
纤维强化热固性　fiber reinforcement
氙　xenon(Xe)
酰基化　acylation
鲜风供应风扇　fresh air supply fan
弦齿高　chordal addendum
弦齿厚　chordal thickness
弦丝电位计　string potentiometer
咸水　salt water
衔铁线圈　armature
衔铁行程　armature stroke
显谱　visualization
显色剂　color agent
显示　display
显示板　display board
显示打印任务单　show print queue
显示单元　display unit
显示管　display tube
显示过程本身的参数　show own process parameter
显示画面　display schematic
显示画面请求　display request
显示记录　display log
显示控制台　display console
显示逻辑参数　show logical parameter
显示批任务单　show batch queue
显示器　visual display unit(VDU)
显示器下盖　display lower
显示器支撑杆　display stem
显示缺省目录　display default directory
显示设备　display device
显示生成程序　display generator
显示数据库　display data base
显示系统清单　show system

显示一个文件　list a file
显示仪表　display instrument
显示元件　display element
显示属性　display attribute
显微光度计　microphotometer
显微镜　microscope
显微硬度值　microhardness number
显像管　charactron
显眼地方　conspicuous place
显眼位置　conspicuous position
显影罐　developing tank
显影剂　developer
显影溶剂　developing solvent
显著性水平　level of significance
现场安装　field mounting
现场萃取采水器　in-situ extraction sampler
现场电缆区域　field cable area
现场端子组件　field termination assemblies
现场含水量　in-situ moisture content
现场混凝土强度　in-situ concrete strength
现场检测　witnessed inspections
现场检修　local repair
现场浇铸　cast-in-place
现场可靠性试验　field reliability test
现场控制单元　field control unit
现场控制的晶闸管　field controlled thyristor
现场控制系统　field control system
现场控制站　field control station
现场配电盘　local panel
现场平衡　field balancing
现场平衡设备　field balancing equipment
现场渗透度　in-situ permeability
现场试验　field test
现场数据　field data
现场水色计　in-situ colour meter
现场物料测试　in-situ material testing
现场信号　field signal
现场整定　field tuning
现场值班员　local attendant
现场总线　field bus(FB)
现场总线控制系统　field bus control system(FCS)
现场总线模件　field bus module
现代控制理论　modern control theory
现有库存　on-hand inventory
线　line
线电流　line current
线电压　line voltage
线电压调整率　regulation of line voltage
线定位　linear location
线对　line pair
线分辨力　line to line resolution
线割　wire EDM
线加速度　linear acceleration
线加速度传感器　linear acceleration sensor, linear acceleration transducer
线夹型探头　clip-type probe
线间的　line-to-line
线间电压　voltage between lines
线间距离　distance between conductors
线焦点　line focus
线坑　trench
线拉伸　line pulling
线灵敏度　line sensitivity
线路补偿器　line drop compensation
线路故障报警器　circuit alarmer
线路换向　line turnaround
线路继电器　line relay
线路监听多次存取检测　carrier sense multiple access detect
线路监听多次碰撞检测　carrier sense multiple collision detect
线路均衡器　line equalizer
线路开关　line switch
线路耦合器　line coupler
线路配置图　wiring layout
线路试验器　line tester
线路图　wiring diagram
线路陷波器　line trap
线路协议　line protocol
线路帧　line frame
线路阻抗　line impedance
线切割　linear cutting, wire-cutting
线驱动器　line driver
线圈　coil
线圈布置　coil arrangement
线圈法　coil method
线圈绝缘测试器　coil insulator tester

线圈绕组　coil winding
线圈式振动子　coil galvanometer
线绕式转子　wound rotor
线扫描曲线　linescan
线色散　linear dispersion
线速度　linear velocity
线速度传感器　linear velocity sensor, linear velocity transducer
线损　line loss
线条图　Gantt chart
线网　net
线位移　linear displacement
线性编码　linear code
线性标度　linear scale
线性差分变换　linear differential transformer
线性超前　linear lead
线性代数　linear algebra
线性的　linear
线性电机　linear motor
线性电位计　linear potentiometer
线性电阻　linear resistance
线性迭代　linear regression
线性定常控制系统　linear time-invariant control system
线性独立　linear independence
线性度　linearity
线性度误差　linearity error
线性多变量系统　linear multivariable system
线性二次高斯控制　LQG control
线性二次高斯控制方法　LQG control method
线性二次调节器　linear quadratic regulator(LQR)
线性二次调节器控制方法　LQR control method
线性二次调节器问题　linear quadratic regulator problem
线性范围　linear range
线性方程　linear equation
线性方程组　linear equations
线性放大　linear amplification
线性放大器　linear amplifier
线性分析　linear analysis
线性分组码　linear block code
线性功率放大器　linear power amplifier
线性估计　linear estimation
线性关系　linear relation
线性规划　linear programming
线性化　linearization
线性化方法　linearization technique
线性化模型　linearized model
线性化系统　linearizable system
线性集成光学　linear integrated optics
线性检波器　linear detector
线性可变磁阻传感器　linear variable reluctance transducer
线性控制　linear control, linearity control
线性控制系统　linear control system
线性控制系统的绝对稳定性　absolute stability of a linear control system
线性控制系统的条件稳定性　conditional stability of a linear control system
线性控制系统的稳定性　stability of a linear control system
线性控制系统理论　linear control system theory
线性理论　linear theory
线性滤波　linear filter
线性模型　linear model
线性黏性阻尼　linear viscous damping
线性黏性阻尼系数　linear viscous damping coefficient
线性排列　linear array
线性膨胀系数　coefficient of linear expansion
线性平衡变换器　line balance converter
线性区　linear zone
线性热敏电阻器　linear thermistor
线性扫描振荡器　linear sweep generator
线性失真　linear distortion
线性适应元　adaline
线性输出反馈　linear output feedback
线性双斜型　linear dual slope type
线性调节器　line conditioner, linear regulator
线性调制　linear modulation
线性网络　linear network

线性位移传感器　linear displacement transducer
线性吸收系数　linear absorption coefficient
线性系统　linear system
线性系统的条件稳定性　conditional stability of a linear system
线性系统仿真　linear system simulation
线性相关　linear dependence
线性相角　linear phase
线性斜坡型　linear ramp type
线性优化　linear optimal
线性与范围　linearity and range
线性预报　linear prediction
线性元件　linear element
线性滞后　linear lag
线性转换　linear conversion
线性转换器　linear transducer
线性最优控制系统　linear optimal control system
线应变　line strain
线与中性点间的　line-to-neutral
限动环　stop collar
限峰器　peak limiter
限幅二极管　limiter diode
限幅放大器　limiting amplifier
限幅器　amplitude limiter, amplitude lopper, limiter
限流电抗器　current limiting reactor
限流电阻　limiting resister
限流电阻器　current limiting resistor
限流阀　drain valve
限流继电器　current-limit relay
限流晶体管　current limiting transistor
限流器　current limiter
限流阈值　current limiting threshold
限期　deadline
限速继电器　speed relay
限速器　speed governor
限位阀　limit valves
限位开关　limit switch
限压器　voltage limiter
限制　limiting
限制手动存取　restricted manual access
限制通气外壳　limited breathing enclosure
陷波电路　trap circuit
陷波滤波器　notch filter
陷波器　trapper
相对保留值　relative retention value
相对比移值　relative Rf value
相对标准差　relative standard deviation (RSD)
相对呈感性　reactive in respect to
相对呈阳性　active in respect to
相对丰度　relative abundance
相对间隙　relative gap
相对论　relativistic
相对密度　relative density
相对平均偏差　relative average deviation
相对曲率半径　relative radius of curvature
相对湿度　relative humidity
相对湿度传感器　relative humidity sensor
相对湿度可调范围　adjustable relative humidity range
相对速度　relative velocity
相对位移　relative displacement
相对稳定性　relative stability
相对误差　relative error
相对运动　relative motion, relative movement
相对增益　relative gain
相对增益阵　relative gain array (RGA)
相对值　relative value
相对阻尼　relative damping
相干检测　coherent detector
相关　membership
相关表达式　relational expression
相关的　coherent, relative
相关度　membership degree
相关对比法感应冲瞬变系统　correlative method input system
相关干涉仪　correlation interferometer
相关工程　associated works
相关函数　membership function (MF)
相关联控制回路　interacting control loop

相关频率响应综合　interactive frequency response synthesis
相关器　correlator
相关式测长仪　correlation length measuring instrument
相关吸收带　correlation absorption band
相关系数　correlation coefficient
相关显示画面　related display panel
相关因子　correlator
相关指数　interaction index
相互的　mutual
相互作用　interaction
相互作用的　interactive
相互作用机理　interaction mechanism
相加　summing
相加单元　summing unit
相加点　summing point
相加环节　summing element
相邻水平　adjacent level
相邻信道　adjacent channel
相邻信道干扰　adjacent-channel interference
相容性　consistency
相似变换　similarity transformation
相似定理　correspondence theorem
相似性　similarity
相同的　interphase
香农采样定理　Shannon's sampling theorem
箱　case
箱控制状态　box control state
箱式浮子液位计　cage-type float-operated gauge
箱形大梁　box girder
箱型退火　box annealing
箱状态显示画面　box status display
镶玻璃条　glazing bead
镶块　embedded lump, panel board
镶片刀盘　inserted blade cutter
镶嵌　inlay
镶嵌玻璃槽口　rebate of glazing
镶入式圆形凹模　button die
镶有玻璃　glazed
镶铸法　inlay casting
详图　detailed plan
详细描述　detailed description

详细设计区　layout area
响应　response
响应测量　response measurement
响应函数　response function
响应曲线　response curve
响应时间　response time
相　phase
相变　phase transition
相补角　phase margin
相电流　phase current
相电压　phase voltage
相负载　phase load
相轨迹　phase locus, phase trajectory
相机长度　camera length
相间电压　voltage between phases
相角　phase angle
相角差　phase-angle difference
相界电位　phase boundary potential
相空间　phase space
相控阵　phased array
相量　phasor
相量和　phasor sum
相量图　phasor diagram
相敏整流器　phase sensitive rectifier
相频特性　frequency-phase characteristic, phase-frequency characteristic
相平衡　phase balance
相平面　phase plane
相数　number of phases
相图　phase diagram
相位比较器　phase comparator
相位变化　phase change
相位表　phase meter
相位补偿　phase compensation
相位差　phase difference
相位差仪　phase difference instrument
相位差指示器　phase difference indicator
相位超前　phase lead
相位超前控制器　phase-lead controller
相位干扰技术　phase perturbation technique
相位计　phasometer
相位交越频率　phase crossover frequency
相位校正　phase correction

相位校正器　phase corrector
相位解调　phase demodulation
相位介电感应测井仪　dielectric phase induction logger
相位控制　phase control
相位轮廓线　phase contour
相位逆变器　phase inverter
相位排列　phased array
相位容限　phase margin
相位失真　phase distortion
相位特性　phase characteristic
相位调制　phase modulation
相位外延　phase epitaxy
相位稳定度　phase stability
相位稳定性　phase stability
相位系统　phase system
相位系统辨识　phase system identification
相位系统分析　phase frame analysis
相位响应　phase response
相位旋转　phase rotation
相位移　phase displacement
相位移前器　phase advancer
相位滞后　phase lag
相位滞后控制　phase-lag control
相位滞后控制器　phase-lag controller
相位中心　phase centre
相线　phase line
相序　phase sequence
相序继电器　phase sequence relay, phase-rotation relay
相序控制　control of phase-sequence
相移　phase shift
相噪声　phase noise
向导　guide
向后传播的波　backward-travelling waves
向后退入　setback
向量方程　vector equation
向量李雅普诺夫函数　vector Lyapunov function
向前传播的波　forward-travelling waves
向前　forward
向上　up
向上加载　up line loading(ULL)
向下　down, downward
向下辐射　downward radiation
向下滚动　down roll

向下全辐射　downward total radiation
向心力　centripetal force
向心展开　centripetal development
向心轴承　radial bearing
象限　quadrant
像场弯曲　curvature of field
像点　image point
像频干扰　image frequency interference
像平面　image plane
像素　picture element, pixel
像旋转　image rotation
橡胶　rubber
橡胶避震垫　rubber anti-vibration mounting
橡胶成形　rubber molding
橡胶垫圈　rubber gasket
橡胶管套　rubber sleeve
橡胶喉管　rubber tube
橡胶缓冲器　rubber bump stop
橡胶绝缘线　India-rubber wire
橡胶孔环　rubber grommet
橡胶密封垫圈　rubber sealing washer
橡胶密封环　rubber seal ring
橡胶软管　rubber hose
橡胶石墨板　rubber graphite board
橡胶弹簧　rubber spring
橡胶条　rubber strip
橡胶外包线　rubber sheathed wire
橡胶支座　rubber boot
橡皮绝缘低压腊克线　rubber insulated low voltage lacquer wire
橡皮绝缘聚氯乙烯护套控制电缆　PVC sheathed rubber insulated control cable
橡皮泥　plasticine
削波　clipping
削波电路　clipping circuit
削土斜坡　cut slope
消磁电压敏电阻器　anti-magnetized varistor
消磁器　degausser
消磁线圈　demagnetizing coil
消电离　deion
消毒灭菌设备　sterilization and disinfection equipment
消毒箱　disinfectant tank
消防泵　fire pump

消防泵房　fire pump room
消防车　fire fighting engine
消防街井　fire hydrant
消防局　fire station
消防卷闸　fire roller shutter
消防控制台　fire console
消防龙头　hydrant
消防龙头出口　hydrant outlet
消防气阀　fire safety gas valve
消防设备　fire fighting equipment, fire services equipment
消防设备检查　fire services inspection
消防设施　fire public device
消防水管　fire main
消防系统　fire fighting system
消防员紧急开关　fireman's emergency switch
消防员升降机　fireman's lift
消防装置　fire services installation
消费函数　consumption function
消费性电子产品　consumer electronics
消耗　consumption
消弧电压敏电阻器　arc suppressing resistor
消弧线圈　extinction coil
消极约束　passive constraint
消流充电　trickle charge
消散　scattering
消声　silence
消声器　noise remover, silencer
消声室　anechoic chamber
消声水池　anechoic tank
消逝时间调节器　elapsed time controller
消旋体　despinner
消音　silence
消音器　noise remover
消隐画面　erase panel
消震器　shock eliminator
硝酸　aqua fortis, nitric acid
硝酸纤维素　nitrocellulose
销尖　pin nose
销售包装　consumer package
销售及服务品质保证　sales and service quality assurance
销套　pin bush

销形阀　pintle valve
销子　dowel pin
小包　packet
小波变换　wavelet transformation
小册子　pamphlet
小齿轮　pinion
小齿轮线坯　pinion wire
小端槽宽　inner slot width
小端齿顶高　inner addendum
小端齿根高　inner dedendum
小端螺旋角　inner spiral angle
小端窄大端宽接触　narrow-tow-wide-heel
小端锥距　inner cone distance
小功率放大器　miniwatt amplifier
小夹　clip
小检修　line check
小块修补　patch repair
小六角头导颈螺栓　bolts small hexagon head with fit neck
小六角头螺杆带孔导颈螺栓　bolts small hexagon head with fit neck and hole through the shank
小六角头螺杆带孔螺栓　bolts small hexagon head with hole through the shank
小六角头头部带孔螺栓　bolts small hexagon head with holes through the head
小轮粗切机　pinion rougher
小轮后端后轴承　pinion rear bearing
小轮偏置距　pinion offset
小轮前端轴承　pinion front bearing
小轮止端前轴承　pinion head bearing
小轮轴向位移　pinion axial displacement
小轮锥距　pinion cone
小脑模型连接计算机　cerebellar model articulation computer (CMAC)
小钳子　tweezers
小时平均画面　hourly average display
小塔　turret
小条型指示仪　small strip-type indicator
小条型自动平衡指示仪　small strip-type self-balancing indicator
小心搬运　handle with care

小信号模式　small signal mode
小型的　compact
小型化　miniaturization
小型计算机　minicomputer
小型计算机系统接口　small computer system interface
小型结构工程　minor structural works
小型示波器　minioscilloscope
小型通用继电器　small ordinary relay
小型现场控制站　compact FCS
小修　minor overhaul
小直径　minor diameter
肖氏硬度　shore hardness
肖式硬度计　shore hardness tester
肖特基势垒二极管　Schottky barrier diode
肖特基效应　Schottky effect
效果　effect
效率　efficiency
效率曲线　efficiency curve
效率提高　effective enhancement
效能驱动　performance drive
效益理论　effectiveness theory
效用函数　utility function
楔式　wedge
楔式闸阀　wedge gate valves
楔形光谱仪　wedge spectrometer
协处理器　coprocessor
协处理器子板　coprocessor daughter board
协方差　covariance
协方差矩阵　covariance matrix
协调　co-ordination
协调策略　coordination strategy
协调控制　co-operative control
协调控制系统　coordinated control system
协调器　coordinator
协同学　synergetics
协议　protocol
协议工程　protocol engineering
协作　co-operation
斜撑　inclined strut
斜齿轮的当量直齿轮　equivalent spur gear of the helical gear
斜齿圆柱齿轮　helical-spur gear

斜导边　angle pin
斜顶杆　angle ejector rod
斜度　gradient
斜键　taper key
斜交锥齿轮　angular bevel gears
斜角　bevel
斜角法　angle beam technique
斜率　slope
斜率的微分误差　differential error of the slope
斜面　chamfer
斜面冲击试验　incline impact test
斜坡　ramp
斜坡函数　ramp function
斜坡函数响应　ramp function response
斜坡输入　ramp input
斜坡误差常数　ramp error constant
斜坡响应　ramp response
斜坡响应时间　ramp response time
斜坡延时电路　ramp delay circuit
斜伞齿轮　skew bevel gears
斜视规　bezel
斜束　diagonal beam
斜探头　angle probe
斜网　inclined wire
斜针　angle pin
斜桩　batter pile
谐波　harmonic, harmonic wave
谐波齿轮　harmonic gear
谐波传动　harmonic driving
谐波电流　harmonic current
谐波发生　harmonic generation
谐波发生器　harmonic generator
谐波放大器　harmonic amplifier
谐波分量　harmonic component
谐波分析　Fourier analysis, harmonic analysis
谐波分析器　Fourier analyzer, harmonic analyzer
谐波功率　harmonic power
谐波含量　harmonic content
谐波滤波器　harmonic filter
谐波平衡分析　harmonic balance analysis
谐波平衡技术　harmonic balance technique

谐波平衡器　harmonic balancer
谐波驱动　harmonic drive
谐波失真　harmonic distortion
谐波响应　harmonic response
谐波响应特性　harmonic response characteristic
谐振波长计　resonance frequency wavemeter
谐振的　resonant, resonanting
谐振电路　resonance circuit
谐振频率　resonance frequency, resonant frequency
谐振腔频率计　cavity-resonator frequency meter
谐振曲线　resonance curve
谐振扫描　harmonic sweep
谐振式仪表　resonant type instrument
谐振状态　resonance state
谐振追踪　harmonic search
携带式光谱仪　lineman's spectrometer
携带式直流单臂电桥　portable single-arm DC bridge
携带式直流电位差计　portable DC potentiometer
写　write
写入电路　write circuit
泄放电路　bleeder circuit
泄放喷嘴　release nozzle
泄漏　leakage, leak
泄漏电流　leakage current
泄漏电流屏蔽　leakage current screen
泄漏特性　leakage property
泄漏压力　leak pressure
泄气阀　release valve
卸料阀　discharge valve
卸料机　unloader
卸泥区　dumping area
卸去尾部检查　unload audit trail
卸下档案　unload archive
卸压阀　pressure relief valve
卸载　unload
心电图(仪)　electrocardiograph(ECG)
心电图传感器　electro-cardiography sensor, electro-cardiography transducer
心率计　heart-rate meter

心线　core wire
心形销　core pin
心音传感器　heart sound sensor, heart sound transducer
心轴距　arbor distance
心轴组件　arbor assembly
芯片　chip
芯片安装面积　die pad
芯片测试　chip testing
芯铁　core iron
锌铁　galvanized iron
新版　new version
新浇混凝土　fresh concrete
新品首件检查　first article inspection
信道　communication channel
信道移频器　channel shifter
信号　signal
信号报警　alarming
信号变换　conversion of signal
信号表征器　signal characterizer
信号重构　signal reconstruction
信号处理　signal processing
信号处理器　signal processor
信号处理算法　signal processing algorithm
信号传输　signal transmission
信号灯　signal lamp, signal light
信号电缆端头　signal cable termination
信号电路　signal circuit
信号电平　signal level
信号电压　signal voltage
信号发生器　signal generator
信号发送器　signal transmitter
信号放大器　signal amplifier
信号分配组件　signal distribution component
信号分析　signal analysis
信号幅度排序　signal amplitude sequencing
信号隔离　signal isolation
信号-工位号显示块　single-tag display block
信号公共端　signal common(SC)
信号合成　signal synthesis
信号畸变　signal distortion
信号级　signal level

信号继电器　signal relay
信号间隔　signal duration
信号检测　signal detection
信号检测和估计　signal detection and estimation
信号交换　handshaking
信号流图　flow chart, flow diagram, signal flow diagram, signal flow graphs
信号脉冲　signal impulse
信号屏　signal screen
信号器　annunciator
信号失真　signal distortion
信号式样　signal aspect
信号调节　signal conditioning
信号调节放大器　signal conditioning amplifier
信号线　signal line
信号线路　signal circuit
信号相关　signal correlation
信号形态　signal aspect
信号选择单元型端子　single selector terminal
信号选择器　signal selector
信号压力　signal pressure
信号延迟　signal delay
信号与信息处理　signal and information processing
信号源　signal source
信号噪声比　signal to noise ratio (SNR)
信号指示器　signal indicator
信号转换器　signal converter
信号状态编码　signal state code
信息　information
信息采集　information acquisition
信息参数　information parameter
信息处理　information processing
信息传递线　information-carrying wire
信息分析　information analysis
信息公路驱动器　highway driver
信息集成　information integration
信息技术　information technology
信息检索　information retrieval
信息结构　information structure
信息流　information flow

信息流图　information flow diagram
信息论　information theory
信息模式　information pattern
信息屏幕　message screen
信息容量　information capacity, message capacity
信息深度　information depth
信息网　information network
信息网络接口　information network interface
信息系统　information system
信息线　information wire
信息与通信工程　information and communication engineering
信息总貌画面　message summary display
信噪比　signal noise ratio (SNR), signal to noise ratio (SNR)
星形连接　star connection
星形轮　chain gearing
星形网络　star network
行波　travelling wave
行波保护　traveling wave protection
行波放大器　travelling wave amplifier
行波继电器　travelling wave relay
行波调制器　travelling wave modulator
行波信号　travelling wave signal
行波应答器　travelling-wave transponder
行车　travelling crane
行程速度变化系数　coefficient of travel speed variation
行程止销　stroke end block
行为科学　behavioural science
行星齿轮变速箱　planetary transmission
行星轮　planet gear
行星轮变速器　planetary speed changing devices
行星轮系　planetary gear train
形成生命时间　generation lifetime
形式　form
形式神经元　formal neuron
形式语言理论　formal language theory
形态　type
形态观察分析系统　mapping analyzer

形状	shape

形状 shape
形状描述 shape description
形状区别 shape discrimination
形状因数 form factor
型底 die bed
型钢 figured iron
型号及规格 type and specification
型数 type number
型铁 figured iron
型芯材料 core material
性能 performance
性能度量 performance measure
性能分析 performance analysis
性能函数 performance function
性能极限 performance limit
性能价格比 cost performance
性能监视 performance monitoring
性能扭矩 performance torque
性能判据 performance criteria
性能评估 performance evaluation
性能指标 performance index
修补 patching
修订 amendment
修复 recondition, restore
修复工程 reinstatement works, restoration works
修改 modify
修改的 modified
修剪 trim
修理 repair
修理工 repairer
修整表面缺陷 chipping
修整机 finisher
修正 correction
修正结果 corrected result
修正曲线 correction curve
修正梯形加速度运动规律 modified trapezoidal acceleration motion
修正系数 coefficient of correction
修正正弦加速度运动规律 modified sine acceleration motion
修正总重合度 modified contact ratio
袖珍仪表 packet instrument
锈斑 rust stain
溴 bromine(Br)
溴酚蓝 bromophenol blue
溴化环氧树脂 brominated epoxy resin
溴量法 bromine method
溴酸钾法 potassium bromate method
虚部 imaginary part
虚根 imaginary root
虚假应变 false strain
虚拟测试功能 virtual test function
虚拟现实 virtual reality
虚拟现实技术 virtual reality technology
虚拟现实设计 virtual reality design
虚拟仪器 virtual instrument
虚数 imaginary number
虚数部分 reactive component
虚数的 imaginary
虚线 dotted line
虚约束 passive constraint
虚轴 imaginary axis
需求 demand and supply
需求弹性 elasticity of demand
需要归档 demand archive
许用不平衡量 allowable amount of unbalance
许用压力角 allowable pressure angle
许用应力 permissible stress
序贯最小二乘估计 sequential least squares estimation
序列 sequence
序列估计 sequence estimation
蓄电池 accumulator, accumulator cell, battery, storage battery
蓄电池电压 battery cell volt
蓄料井 well type
悬臂 cantilever
悬臂地基 cantilever foundation
悬臂吊机 cantilever crane
悬臂基脚 cantilever footing
悬臂结构 cantilever structure
悬臂梁 cantilever beam, outrigger
悬臂桥 cantilever bridge
悬臂支架 cantilever support
悬浮机车 vehicle suspension
悬汞电极 hanging mercury electrode
悬挂式桁架 suspended truss
悬挂式湿度计 sling hygrometer
悬架 suspension
悬缆线 messenger wire

悬式经纬仪　hanging theodolite
悬式倾斜计　hanging clinometer
悬式水准仪　hanging level
悬丝式检流计　filar suspended galvanometer
悬索　span wire
悬索高架桥　cable supported viaduct
悬索桥　suspension bridge
悬索式流速计　rope suspended current meter
旋臂吊机　jib crane
旋臂起重机　slewing crane
旋磁比　gyromagnetic ratio
旋桨　impeller
旋进流量计　vortex precession flowmeter, swirlmeter
旋路　loop road
旋片式真空泵　rotary vane type vacuum pump
旋启式止回阀　swing check valves
旋塞　cock
旋塞阀　plug valve
旋涡　eddy
旋压成形机　spin forming machine
旋翼式冷水水表　rotating vane type cold water-meter
旋转　rotation
旋转泵　rotary pump
旋转变压器　rotating transformer
旋转变阻器　revolving rheostat
旋转成型　rotational molding
旋转磁场　revolving magnetic field, rotating magnetic field
旋转磁盘　rotating disk
旋转磁强计　spinner magnetometer
旋转的　revolving, rotating
旋转电机　rotating machine
旋转换向器　rotating commutator
旋转畸变系数　coefficient of rotational distortion
旋转开关　revolution switch, rotary switch
旋转刻度自动平衡指示仪　self-balancing indicator with rotating scale
旋转力矩　running torque
旋转门　revolving door
旋转面　rotary surface

旋转模塑　rotational molding
旋转平面　plane of rotation
旋转矢量　rotating vector
旋转式泵　rotary pump
旋转式变相机　rotary phase converter
旋转式磁场放大机　metadyne
旋转式磁力仪　spinner magnetometer
旋转式密封　rotating seal
旋转式汽化器　rotary evaporators
旋转式容积流量计　rotary displacement type flowmeter
旋转式水表　rotary water meter
旋转式直流电阻箱　rotary DC resistance box
旋转相量　rotating phasor
旋转叶片式流量计　rotary vane flowmeter
旋转运动　rotary motion
旋转整流子　rotating commutator
漩涡　vortex
漩涡流量计　swirl-meter
选标准则　tender selection criteria
选送放大器　take-off amplifier
选型　type selection
选用设备　optional device
选择　select, option
选择的　selective
选择电路　selective circuit
选择开关　option switch, select switch
选择开关接线端子　select switch terminal
选择器　selector
选择曲线　selectivity curve
选择线　selection wire
选择项　pick
选择性　selectivity characteristic
选择性电离真空计　selective ionization gauge
选择性离子监测　selected ion monitoring, selective ion monitoring
选择性网络　selective network
选择硬化　selective hardening
眩光　glare
穴型导板　pass guide
学习　learning
学习控制　learning control

学习算法　learning algorithm
学习系统　learning system
雪崩　avalanche
雪崩二极管　avalanche diode
雪崩三极管　avalanche transistor
雪崩型光电二极管　avalanche-type photodiode
雪深　depth of snow
血钙传感器　blood calcium ion transducer
血钾传感器　blood potassium ion transducer
血浆　plasma
血流传感器　blood flow transducer
血流速度计　rheometer
血氯传感器　blood chlorine ion transducer
血钠传感器　blood sodium ion transducer
血气传感器　blood gas transducer
血气分析仪　blood-gas analyzer
血球计数器　hematocyte counter
血容量传感器　blood-volume transducer
血糖分析仪　blood sugar analyzer
血细胞计数器　hematocyte counter
血压传感器　blood-pressure transducer
血氧传感器　blood oxygen transducer
血液 pH 传感器　blood pH transducer
血液电解质传感器　blood electrolyte transducer
血液二氧化碳传感器　blood carbon dioxide transducer
寻呼机　pager
寻优技术　optimal search technique
巡边器　edge finder
驯服的装配机器人臂　selective compliance assembly robot arm (SCARA)
循环　circulating, cycle
循环泵　recirculating pump
循环存储器　cyclic memory
循环的　cyclic
循环伏安法　cyclic voltammetry
循环负荷　cyclic load
循环功率流　circulating power load
循环功能　circular function
循环荷载　cyclic loading
循环缓冲　circular buffering
循环计数　cycle count
循环寄存器　circulating register
循环模式　circulation mode
循环器　circulator
循环冗余校验　cyclic redundancy check (CRC)
循环湿热试验　cyclic damp heat test
循环时间　cycle length, cycle time
循环式研磨带　endless grinding belt
循环寿命　cycling life
循环水泵　circulating water pump
循环文件　circular file
循环遥控　cyclic remote control
循环应变　cyclic strain
循环应力　cyclic stress
训练　training

Y

压　press
压板　binder plate, pressure plate
压边浇口　corner gate
压扁丝　flattened wire
压差变送器　pressure difference transmitter
压差光圈　aperture of pressure difference
压差式流量计　pressure differential flowmeter
压出粒涂层法　extruded bead sealing
压磁效应　magnetoelastic effect
压电电阻加速度计　piezoresistance accelerometer
压电晶体　piezocrystal, piezoelectric crystal
压电晶体稳频器　piezoelectric stabilizer
压电控制　piezo-electric control
压电石英　piezoelectric quartz
压电式测力传感器　piezoelectric force transducer
压电式振动仪　piezoelectric type vibration gauge
压电谐振器　piezo-resonator
压电压力敏感元件　piezoelectric pressure sensor
压电振荡器　piezoelectric oscillator
压电振动片　piezoelectric vibrator
压风机　forced draught fan
压感膜片　pressure-sensitive diaphragm
压焊　pressure welding
压痕　dents, indentation
压痕加工　indenting
压花刀　knurling
压花身　knurled shank
压花铁　embossing iron
压花纹　knurling
压挤工具　compressing tool
压筋冲子　ribbon punch
压紧多股绞合线　compact-stranded wire
压紧螺栓　holding down bolt
压紧模　pressure die
压块　lock block
压力　pressure
压力泵　pressure pump
压力变送器　pressure transmitter
压力表接头　gauge connector
压力补偿器　pressure compensator
压力补偿式流量计　pressure-compensated flowmeter
压力补偿式转子流量计　balanced-pressure rotameter
压力测量　pressure measurement
压力测量仪表　pressure measuring instrument
压力抽风机　forced draught fan
压力传感器　pressure transducer
压力传送器　pressure transmitter
压力端口　pressure port
压力范围　pressure range
压力灌浆　pressure grouting
压力计　manometer, pressure gauge
压力记录仪　pressure recorder
压力角　pressure angle
压力开关　pressure switch
压力控制　pressure control
压力控制阀　pressure valve
压力控制回路　pressure control loop
压力流量控制器　pressure and flow controller
压力敏感电极　pressure-sensitive probe
压力泡　pressure bulb
压力气流　forced draft
压力容器　pressure vessel
压力润滑　pressure lubrication
压力式温度计　filled system thermometer
压力试验机　compression testing machine
压力水冷反应堆　pressurized water reactor
压力损失　pressure loss

压力调节器　pressure regulator
压力温度控制器　pressure and temperature controller
压力仪表　pressure instrument
压力元件　pressure element
压力增压阀　pressure valve
压力真空表　combined pressure and vacuum gauge
压力真空关系　pressure volume relationship
压力中心　center of pressure
压毛边冲子　deburring punch
压敏电阻　varistor, voltage dependent resistor
压模嵌入件　die insert
压强计　piezometer
压入　pressfit
压入桩　jacking pile
压实　compaction
压实测试　compaction test
压水堆　pressurized water reactor
压碎强度　crushing strength
压缩　compression, condensation, hitting
压缩成型　compresion molding
压缩光谱　compact spectra
压缩荷载　compression load
压缩机　compressor
压缩机分类　compressors classification
压缩空气　compressed air, pressurized air
压缩空气开挖隧道法　compressed air tunnelling method
压缩力　compression force
压缩零位仪表　instrument with suppressed zero
压缩气体　compressed gas
压缩强度　compressive strength
压缩强度试验方法　test methods for compressive strength
压缩塌毁　compressive failure
压缩因子　compressibility factor
压套式仪表截止阀　instrument block valve with pressed coupling
压条装配玻璃法　bead glazing
压凸　belling

压弯曲加工　compression bending
压纹校平　waffle die flattening
压纹装置　marking device
压线　groove
压线冲子　groove punch
压延成形　calendaring molding
压延铜箔　rolled copper foil
压延退火铜箔　rolled annealed copper foil
压印加工　coining
压应力　compressive stress
压铸冲模　die casting dies
压铸机　die casting machines
压阻效应　piezoresistive effect
雅可比　Jacobian
雅可比矩阵　Jacobian matrix
亚稳分解　metastable decomposition
亚稳峰　metastable peak
亚稳离子　metastable ion
亚稳扫描　metastable scanning
亚硝基化滴定法　nitrosation titration
亚硝基化反应　nitrosation reaction
亚硝酸钠法　sodium nitrite method
氩　argon(Ar)
氩电离检测器　argon ionization detector
氩焊　argon welding
氩弧焊　argon arc welding
氩离子枪　argon-ion gun
烟囱盖顶　chimney coping
烟囱效应　chimney effect
烟道　flue
烟道排气管　flue pipe
烟度传感器　smoke density sensor
烟度计　opacimeter
烟浓度　level of smoke
烟气　flue gas
烟雾报警器　smoke detector
烟雾吸收器　smog absorber
淹没式泄水阀　submerged discharge valve
延长导线　extension lead
延长导线法　extension lead method
延迟喷嘴方式　long nozzle
延迟变换　delayed transformation
延迟触发器　delay flip-flop
延迟电路　delay circuit

延迟动作　retarding action
延迟器　delayer
延迟时间　delay time
延迟线　delay line
延迟自动增益控制　delay automatic gain control
延期　extension of time
延伸　extend
延伸率　elongation rate
延伸翼　extension wing
延时电缆　delay cable
延时动作　delay action, delay-action
延时断路器　delay-action circuit-breaker
延时放大器　delay amplifier
延时继电器　time delay relay, time lag relay, time-lag relay
延时开关　delay switch
延时调节器　timer
延时调制　delay spread modulation
延时遥测　delayed telemetry
延时装置　time delay device
延性　ductility
延展部分　extension
延展时限　extension of time limit
严重破坏　havoc
严重事故　major accident
岩石　rock
岩土工程　geotechnical engineering
岩土纪录　geotechnical record
岩土评估　geotechnical assessment
岩土数据　geotechnical data
岩心测试　core testing
岩心试验　core testing
岩心样本　core sample
沿面放电　creeping discharge
研究方向　research direction
研磨　lapping
研磨膏　grinding paste
研磨工具　abrasive tool
研磨机　finishing machine, grinding machine
研磨料　abrasive
研磨轮　glazing wheel
研磨盘　grinding disc
盐　salt
盐水　saline water
盐水溶液自动补给器　automatic feeder for brine
盐酸　hydrochloric acid
盐釉　salt glazing
盐浴淬火　salt bath quenching
颜色　colour
颜色笔　colour pencil
颜色灯号　colour light signal
颜色滤光片　colour filter
颜色水泥　coloured cement
檐沟　eaves gutter
衍射分辨力　diffraction resolution
衍射光栅　diffraction grating
衍射透镜　diffraction lens
衍生物　derivative
掩蔽场所　sheltered location
掩蔽区　sheltered area
眼电图　electro-oculogram(EOG)
演绎技术　rendering technique
演绎与归纳混合建模法　deductive-inductive hybrid modeling method
验电器　electroscope
验收测试　acceptance testing
验收检验　acceptance inspection
验收准则　acceptance criteria
堰　weir
堰板　weir plate
堰式流量测量仪表　weir type flow measuring device
燕尾锯　dovetail saw
扬声器　loudspeaker, speaker
阳电子　positron
阳极　anode, positive pole
阳极电镀　anodize
阳极电压　anode voltage
阳极电源　B-power
阳极区　positive column
阳极效率　plate efficiency
阳极效应　anode effect
阳极氧化处理　anodizing
阳离子　cation, positive ion
阳模　positive mold
洋铁　tinplate
养护　curing
养护剂　curing compound
氧　oxygen(O)
氧化　oxidation, oxidization
氧化氮　nitrogen oxide

氧化还原滴定法　oxidation-reduction titration
氧化-还原电位变送器　oxidation-reduction potential transmitter
氧化-还原电位电极装置　oxidation-reduction potential electrode assembly
氧化膜电容器　oxide-film capacitor
氧化铅漆　lead oxide paint
氧化物　oxide
氧化性介质　oxidant
氧瓶燃烧法　oxygen flask combustion method
氧气压力表　oxygen pressure gauge
氧气压力表校验仪　calibrator for oxygen pressure gauge
氧气乙炔焰焊接　oxy-acetylene welding
样板　gauge board
样板　master plate
样本集合　sample set
样本空间　sample space
样品　specimen
样品处理　sample handling
样气处理系统　sample handling system
样条　splines
摇摆　swing
摇摆式防滑试验仪　pendulum type skid resistance tester
摇臂　rocker arm
摇臂开关　rocker switch
摇臂钻床　drilling machines radial
摇表　megger
摇动　shake
摇动锻造　rocking die forging
摇动加工　joggling
摇杆　rocker
摇杆式开关　rocker switch
遥测　telemetering
遥测技术　telemetry
遥测水位计　long distance water level recorder
遥测涡流探伤仪　telemetering eddy current detector
遥测系统　telemetering system
遥测指示仪表　remote-indicating instrument
遥测装置　telemetering equipment
遥感仪表　distant-action instrument
遥控　distant control, remote controlling, telecontrol
遥控变量　remote variable
遥控操纵　telemanipulation
遥控操作器　remote manipulator
遥控的　contactless, remote-controlled
遥控机构　telemechanism
遥控机器人　telerobotics
遥控开关　teleswitch
遥控器　remote-controller
遥控设备　remote-control apparatus, remote-control equipment
遥控调节器　distant regular
遥控系统　remote-control system
遥示转子流量计　remote-indicating rotameter
遥调　remote regulating
咬入　bite
药物的鉴别试验　identification test
药物动力学数据　pharmacokinetic data
要点　dominant point
冶金工程　metallurgical engineering
冶金物理化学　physical chemistry of metallurgy
冶金学　metallurgy
冶金自动化　metallurgical automation
野外作业仪器　field instrument
叶轮泵　vane pump
叶轮流速计　propeller-type current meter
叶轮式风速表　fanning mill anemometer
叶轮式流量计　bladewheel type flowmeter
叶轮-涡轮式质量流量计　impeller-turbine mass flowmeter
叶片　blade, vane
叶片式流量计　vane flowmeter
曳进导梁　launching nose
曳进吊梁机　launching girder
曳进架设法　launching erection
曳引系数　drag coefficient

页式打印机　page printer
液滴闪烁计数　liquid scintillation counting
液动节流控制压力　control pressure of hydraulic choke
液动执行器　hydraulic actuator
液封　liquid seal
液封转筒式气体流量计　liquid sealed drum gas flowmeter
液固色谱法　liquid solid chromatography(LSC)
液-固提取法　liquid-solid extraction
液固吸附色谱法　liquid-solid adsorption chromatography
液固柱萃取　column liquid-solid extraction
液化过程　liquefaction
液化石油气钢瓶　liquefied petroleum gas cylinders
液接电位　liquid junction potential
液接界面　liquid junction boundary
液晶温度计　liquid crystal thermometer
液晶显示屏　liquid crystal display (LCD)
液晶显示数字温度计　LCD thermometer
液力传动　hydrodynamic drive
液力耦合器　hydraulic couplers
液面控制　level control
液面控制器　level controller
液面声全息　liquid surface acoustical holography
液面探测管　dip tube
液面系统模型　liquid level system model
液膜强度　film strength
液体比重计　specific gravity hydrometer
液体动力噪声　hydrodynamic noise
液体恒温槽　liquid thermostatic bath
液体静力水准仪　hydrostatic level
液体静水压试验　hydrostatic pressure test
液体控制阀　liquid control valve
液体冷冻机　liquid chiller
液体密度传感器　liquid density sensor, liquid density transducer
液体密度计　liquid densitometer
液体喷砂法　liquid honing
液体渗透探伤　liquid penetrant examination
液体视膨胀　liquid visual expansion coefficient
液体弹簧　liquid spring
液体温度表　liquid thermometer
液体压力表　liquid manometer
液体压力恢复系数　liquid pressure recovery factor
液体置换法　liquid displacement technique
液体置换系统　liquid displacement system
液体阻尼振动子　fluid damping galvanometer
液位　liquid level
液位报警器　liquid level alarm
液位玻璃管　gage glass
液位测量仪表　fluid level measuring instruments
液位计　liquid indicator, liquid level meter, tank gauge
液位控制器　fluid level controller
液位调节器　level regulator
液位指示器　level indicator
液下泵　under water pumps
液相　liquid phase
液相色谱法　liquid chromatography
液相色谱仪　liquid chromatograph
液相色谱-质谱法　liquid chromatography-mass spectrometry(LC-MS)
液相色谱-质谱联用仪　liquid chromatograph-mass spectrometer (LC-MS)
液相载荷量　liquid phase loading
液压　hydraulic pressure
液压泵　hydraulic pump
液压波纹管　hydraulic-formed bellows
液压步进马达　hydraulic step motor
液压传动柴油机车　diesel hydraulic locomotive
液压传动机构　fluid drive mechanism
液压传动控制系统　hydraulic transmission control system

液压的　hydraulic
液压动力工具　hydraulic power tools
液压动力元件　hydraulic power units
液压动力源　hydraulic power supply
液压放大器　hydraulic amplifier
液压缸　hydraulic cylinder
液压固有频率　frequency of the natural hydraulic mode, hydraulic natural frequency
液压回转缸　hydraulic rotary cylinders
液压机构　hydraulic mechanism
液压继电器　hydraulic relay
液压继动阀　hydraulic relay valves
液压减压阀　hydraulic relief valve
液压减震器　hydraulic shock absorber
液压控制　hydraulic control
液压离合器　fluid clutch
液压联轴节　fluid clutch
液压马达　hydraulic motor
液压千斤顶　hydraulic jack
液压驱动泵　fluid clutch
液压润滑　hydraumatic lubrication
液压伸缩筒　hydraulic ram cylinder
液压升降工作台　hydraulic platform
液压式试验机　hydraulic testing machine
液压式挖土机　hydraulic excavator
液压伺服电动机　hydraulic servo-motor
液压无级变速　hydraulic stepless speed changes
液压油　hydraulic oil
液压元件　hydraulic components
液压振动器　hydraulic vibrator
液压转向系统　hydraulic steering system
液液色谱法　liquid-liquid chromatography(LLC)
液-液提取法　liquid-liquid extraction
液柱压力计　liquid column manometer
一般工位号　general tag
一般故障错误　general failure error
一般规格　general specification
一般化运动链　generalized kinematic chain
一般鉴别试验　general identification test
一般均衡理论　general equilibrium theory
一般力学与力学基础　general and fundamental mechanics
一般描述　general description
一般模型控制　generic model control (GMC)
一般输出　general output
一般输入　general input
一般系统理论　general system theory
一般压力表　general-purpose pressure gauge
一半　half
一齿度量中心距变量　tooth-to-tooth composite variation
一次测量元件　primary measuring element
一次电池　one-shot battery
一次绕组　primary winding
一次性静脉输液针　disposable venous infusion needle
一次性使用输液器　disposable infusion set
一次性无菌注射针　disposable sterile injector
一次仪表　primary instrument
一贯单位　coherent of unit
一贯单位制　coherent system of unit
一级参比电极　primary reference electrode
一级光谱　first order spectrum
一阶保持元件　first-order hold element
一阶超前　first-order lead
一阶电路　first-order circuit
一阶加纯滞后　first-order plus time delay
一阶微分器　first order lead unit
一阶谓词逻辑　first order predicate logic
一阶系统　first-order system
一阶滞后　first-order lag
一阶滞后器　first order lag unit
一阶滞后系统　first-order lag system
一维搜索　one-dimensional search
一行程　one stroke

一氧化氮　nitric oxide
一氧化碳　carbon monoxide
一致　coincidence
一致的　uniform
一致渐近稳定性　uniformly asymptotic stability
一致性　conformity
一致性测试　conformance test
一致性检验　consistency check
一致性误差　conformity error
医学应用　medical application
医用导管　medical catheter
医用电子直线加速器　medical electronic linear accelerator
医用灌注泵　medical injection pump
医院信息系统　hospital information system(HIS)
铱　iridium(Ir)
仪表保护箱　instrument protecting box
仪表常数　instrument constant
仪表导线　instrument lead
仪表的安全因数　instrument security factor
仪表电阻　meter resistance
仪表度盘　meter dial
仪表端子盒　meter terminal box
仪表阀　gauge valves
仪表法兰　instrument flange
仪表化　instrumentation
仪表记录笔　instrument pen
仪表检验设备　calibrating devices for instruments
仪表截止阀　instrument block valve
仪表精度　instrumentation accuracy
仪表灵敏度　meter sensitivity
仪表流量　meter flow-rate
仪表面板　instrument faceplate
仪表面板数据输入　instrument faceplate data entry
仪表盘开孔　panel-board cut-out
仪表盘盘后布置图　arrangement on the back of instrument panel-board
仪表盘盘后接管图　piping diagram on the back of instrument panel-board
仪表盘盘后接线图　wiring diagram on the back of instrument panel-board
仪表盘盘内接线图　instrument panel-board inside wiring diagram
仪表盘正面布置图　instrument panel-board layout
仪表屏　instrument panel
仪表台　instrument desk
仪表通信网络　instrument communication network
仪表外壳　meter case
仪表位号　instrument tag No.
仪表误差　instrument error
仪表箱　instrument box, instrument case
仪表针形截止阀　meter needle type globe valves
仪表正面　instrument front
仪表指针　gauge pointer
仪器　apparatus, instrument
仪器本底　instrumental background
仪器分析　instrumental analysis
仪器附件　instrument accessory
仪器科学与技术　instrument science and technology
仪器屏幕　instrument screen
仪器误差　instrumental error
仪用互感器　instrument transformer
仪用自耦互感器　instrument auto transformer
胰岛素敏感　insulin sensitivity
移点器　moving point device
移动标准器　travelling standard
移动参数　motion parameter
移动磁场　moving field
移动从动件　reciprocating follower
移动估计　motion estimation
移动横梁　movable cross-beam
移动机器人　mobile robot
移动接缝　movement joint
移动目标　moving object
移动平均滤波器　moving-average filter
移动平均模型　moving average model
移动平均运算器　moving average unit
移动式X射线探伤机　mobile X-ray detection apparatus
移动式高架起重机　overhead travel-

ling crane
移动式工作台　moving bolster
移动式起重机　mobile crane，travelling crane
移动式转臂起重机　travelling derrick crane
移动式子卡　mobile daughter card
移动凸轮　wedge cam
移位　displacement
移位操作　shifting function
移位寄存器　shift register
移位脉冲　shift pulse
移位脉冲驱动器　shift pulse driver
移相器　phase shifter
移液管　elongated glass bulb
遗传　inheritance
遗传算法　genetic algorithm(GA)
乙电源　B-power
乙二胺四乙酸　ethylene diamine tetraacetic acid(EDTA)
乙基纤维素　ethyl cellulose
乙类放大　class B amplification
乙类放大器　class B amplifier
乙类调制　class B modulation
乙醛　aldehyde
乙炔　acetylene
乙炔减压器　acetylene regulator
乙炔汽缸　acetylene cylinder
乙炔压力表　acetylene pressure gauge
已充电的　charged
已分配的逻辑存储器　allocated logical storage
已分配的物理存储器　allocated physical storage
已调波　modulated wave
已调节的　regulated
已整流的　rectified
已转运　transferred
以百分比表示的 PV 参数　PV in percent
以百分比表示的输出低限　output low limit in percent
以百分比表示的输出高限　output high limit in percent
以棒条通渠　rodding
以人为中心设计　human-centered design
以太网　Ethernet
以太网接口　Ethernet interface
以微处理器为基础的中央处理单元　microprocessor-based central processor unit
异步变频器　asynchronous frequency changer
异步传输　asynchronous transmission
异步传输模式　asynchronous transfer mode (ATM)
异步的　asynchronous
异步电动机　asynchronous motor
异步电机　asynchronous machine
异步电路　asynchronous circuit
异步多路转换器　asynchronous multiplexed
异步发送　asynchronous send
异步计算机　asynchronous computer
异步接收　asynchronous receive
异步启动　asynchronous starting
异步输入　asynchronous input
异步数据传递　asynchronous data transfer
异步数据流　asynchronous flow
异步调制　asynchronous modulation
异步通信接口适配器　asynchronous communication interface adapter
异步通信　asynchronous communication
异步序列　asynchronous sequential
异步序列逻辑　asynchronous sequential logic
异步运行　asynchronous operation
异常处理　abnormal handling
异常情况　abnormal condition
异常释放条件　abnormal release condition
异常条件处理程序　abnormal condition handler
异常停止　emergency stop
异常运转工况　abnormal operating condition
异常终止　abnormal termination
异常终止程序　abort routine
异核锁信号　heteronuclear lock signal

异或	xor	溢流阀	relief valve
异或门	xor gate	溢流管	overflow pipe
异径管	reducer	溢流容量	overflow capacity
异物	foreign matter	溢流式模具	flash mold
异形冲子	special shape punch	溢流水位	flood level

异或　xor
异或门　xor gate
异径管　reducer
异物　foreign matter
异形冲子　special shape punch
抑流电阻器　current suppressing resistor
抑止　choke
抑制　suppression
抑制放大器　rejective amplifier
抑制柱　suppression column
译码　decipher, decode
译码继电器　decoding relay
译码器　code converter, code translator, deciphering machine, transcoder
易腐物品　perishable goods
易接近性　accessibility
易起燃性　ease of ignition
易切削钢丝　free cutting steel wire
易燃　inflammable
易燃产品　flammable products
易燃的　combustible
易燃气体　inflammable gas
易燃下限　lower flammable limit
易燃性极限　limit of flammability
易燃性下限　lower limit of flammability
易燃液体　flammable liquid
易燃蒸气　flammable vapour
易受影响的　susceptible
易碎　fragile
易损件　easily damaged parts
易于移走的封盖　readily removable cover
逸出气分析　evolved gas analysis (EGA)
逸出气分析仪　evolved gas analysis apparatus
逸出气检测　evolved gas detection (EGD)
逸出气检测器　evolved gas detection apparatus
意外坍塌　accidental collapse
溢出　overflow
溢出物　spillage
溢流保护装置　excess flow protection device
溢流阀　relief valve
溢流管　overflow pipe
溢流容量　overflow capacity
溢流式模具　flash mold
溢流水位　flood level
溢流水位计　weir meter
溢流污水渠　relief sewer
溢流装置　excess flow device
溢水道　spillway
翼轮式水表　rotary vane water meter
翼式流量计　airfoil flow meter
因变量　dependent variable
因式分解　factorization
因式分解方法　factorization method
因数　factor
阴极　cathode
阴极保护　cathodic protection
阴极耦合　cathode coupling
阴极耦合器　cathode follower
阴极射线管　cathode ray tube (CRT), Braun-tube
阴极射线管显示器　cathode ray tube display
阴极射线示波器　cathode-ray oscillograph
阴极射线指零仪　cathode ray null indicator
阴极输出放大器　cathode follower amplifier
阴极引线　cathode lead wire
阴离子　anion
音叉式液位控制开关　tuning-fork level switch
音量控制　volume control
音频　audible frequency, audio-frequency
音频变压器　audio-frequency transformer
音频带　voice-frequency band
音频带宽　audio bandwidth (AB)
音频发生器　audio-frequency generator
音频放大器　audio amplifier, audio-frequency amplifier, speech amplifier, voice amplifier
音频轨道电路　audio frequency track circuit

音频检波器　aural detector
音频接收　audio-frequency reception
音频设备　audio equipment
音频调制　voice modulation
音频信号　audio signal
音速喷嘴　sonic nozzle
音速文丘里喷嘴　sonic Venturi-nozzle
音响信号　audible signal
铟　indium (In)
银　silver (Ag)
银-氯化银电极　silver silver-chloride electrode
引出透镜　extraction lens
引出线　leading-out wire
引导衬套　guide bushing
引导柱　guide post
引导装入　bootload
引道　approach road
引道坡　approach ramp
引力　attraction
引擎　engine
引擎排气系统　engine exhaust system
引入电缆　leading-in cable
引入线　lead-in wire, drop wire
引伸计　extensometer
引示线　pilot wire
引水渠　approach channel
引缩加工　ironing
引用误差　fiducial error, error expressed as a percentage of the fiducial value
引用值　fiducial value
隐蔽喉管　concealed piping
隐蔽振荡　hidden oscillation
隐藏式安装　concealed installation
隐含反馈回路　hidden feedback loop
隐含算子　implication operator
隐极　non-salient pole
隐极同步发电机　non-salient pole synchronous generator
隐式系统　implicit system
隐丝式光学高温计　disappearing-filament optical pyrometer
印记冲头　marking iron
印刷　printed
印刷电路　printed circuit
印刷电路板　printed circuit board (PCB)
印刷电路板装配　printed circuit board assembly (PCBA)
印刷电路天线　printed circuit antenna
印刷工业　printing industry
印刷偶极　printed dipole
印刷天线　printed antenna
印制　printing
印制板　printed board
印制板装配　printed board assembly
印制板组装图　printed board assembly drawing
印制接点　printed contact
印制线路　printed wiring
印制线路板　printed wiring board (PWB)
印制线路布设　printed wire layout
印制线路装配　printed wired assembly
印制元件　printed component
应收票据　bills receivable
应用　application
应用编码程序和执行　application builder and executive
应用层　application layer (AL)
应用层协议规范　application layer protocol specification
应用程序　application program
应用处理器　application processor
应用函数　utility function
应用化学　applied chemistry
应用控制语言　application control language
应用模块冗余测试　application module redundancy test
应用模块使用的控制语言　control language for application module
应用模块状态显示画面　application module status display
应用软件　application software
应用神经控制　applied neural control
应用生物力学　applied biomechanics
应用室服务定义　application layer service definition
应用数据点　special data point
应用组件　application module
英寸　inch
英特尔　Intel
迎角传感器　angle-attack sensor, an-

gle-attack transducer
迎头色谱法　frontal chromatography
盈亏功　increment or decrement work
荧光　fluorescence
荧光薄层板　fluorescence thin layer plate
荧光磁粉　flurescent magnetic particle
荧光磁粉探伤机　fluorescent magnetic particle inspection machine
荧光灯管　fluorescent lamp tube；fluorescent light tube
荧光分析法　fluorometry
荧光光谱　fluorescence spectrum
荧光计　fluorometer
荧光检测　fluorescence detection
荧光检测器　fluorescence detector
荧光量子产率　fluorescence quantum yield
荧光面漆　fluorescent finish paint
荧光偏振免疫测定法　fluorescence polarization immunoassay
荧光屏　fluorescent screen
荧光屏前面防护玻璃　cover plate
荧光渗透探伤法　fluorescent penetrant testing method
荧光寿命　fluorescence life time
荧光熄灭法　fluorescence quenching method
荧光像　fluorescent image
荧光效率　fluorescence efficiency
荧光效应　fluorescence effect
营运条件　serviceable condition
影线　hatch
影响量　influence quantity
影响特性　influence characteristic
影响误差　influence error
影响系数法　influence coefficient method
影像信号　camera signal
应变　strain
应变计电测技术　electric measurement technique of strain gauge
应变计灵敏系数　gauge factor
应变计形式　model of strain gauge
应变量　strain capacity
应变时效　strain ageing
应变式称重传感器　strain gauge load cell
应变系数　gauge factor
应变仪　strain gauge, strain measuring instrument
应答　answering
应答信号　answerback
应急备用电源　emergency supply source
应急待命　emergency standby
应急断电　emergency-off
应急发电机　emergency generator
应急计划　emergency plan
应急开关　emergency switch
应急排放阀　dump valve
应急停机开关　emergency stop switch
应急演习　emergency exercise
应急准备　emergency preparedness
应力　stress
应力龟裂　stress crack
应力分析　stress analysis
应力幅　stress amplitude
应力集中　factor of stress concentration, stress concentration
应力集中系数　stress concentration factor
应力图　stress diagram
应力消除退火　stress relieving annealing
应力-应变图　stress-strain diagram
映像分析　image analysis
硬齿面精加工　hard finishing
硬磁材料　retentive material
硬底层　hardcore
硬度　hardness, rigidity
硬度比　hardness ratio factor
硬度传感器　hardness sensor, hardness transducer
硬度计　hardness tester, sclerometer
硬化　hardening
硬化混凝土　hardened concrete
硬化剂　hardener, hardening agent
硬化强度　hardening strength
硬化深层　depth of hardening
硬化深度　depth of hardening
硬化时间　curing time
硬化性　hardenability
硬化性曲线　hardenability curve
硬件的配置　configuration of hardware

硬件检验测试系统　hardware verification test system
硬胶　polystyrene（PS）
硬金属制的喉管　rigid metallic pipework
硬拷贝　hard copy
硬铝线　hard-drawn aluminium wire
硬膜　hard film
硬盘驱动器　hard disk drive（HDD）
硬铜绞线　hard-drawn copper strand wire
硬性塑胶　rigid plastic
硬约束　hard constraint
硬支承平衡机　hard bearing balancing machine
硬质聚氯乙烯管　rigid polyvinyl chloride pipe
硬转换　hard switching
永磁　permanent magnetism
永磁波动器　permanent magnet undulator
永磁材料　magnetically hard material
永磁电动机　permanent magnet motor
永磁动圈式电流表　permanent-magnet moving-coil ammeter
永磁发电机　dynamo magneto
永磁开放式磁共振系统　permanent magnet open magnetic resonance system
永磁式低速同步电动机　low-speed magneto synchronous motor
永动机　hysteresis motor
永久变形　permanent deformation
永久磁铁　permanent magnet
永久的　permanent
永久间隔　permanent partition
永停滴定法　dead-stop titration
涌流　inrush current
涌流阀　deluge valve
用尺寸标明的位置　dimensioned location
用电装置　power device
用户　consumer
用户计算功能级模块　user calculation function class module
用户技术建议　engineering custom engineering proposal
用户接口　user interface
用户卷　user volume
用户可定义的键　user assignable key
用户名称　user name
用户平均值　user average
用户屏幕　user screen
用户数据点　custom data point
用户数据段　custom data segment
用户调整　user adjustment
用户图形显示　custom graphic display
用户许可　user permission
用户组　user group
用利萨如图形测频　frequency measurement by Lissajou's figure
用前检查　pre-operational check
用前馈的 PID 算法　PID with feedforward algorithm
用数字频率计测频　frequency measurement by digital meter
用途　usage
用外部重置反馈的 PID 算法　PID with external reset feedback algorithm
用于多功能控制器的控制语言　control language for the multifunction controller
用于连续控制点的控制语言　control language for continuous data points
优化操作点　optimum operating point
优化的　optimal
优化负载流　optimal load flow
优化功率流　optimal power flow
优化估计　optimal estimation
优化轨迹　optimal trajectory
优化控制　optimum control
优化滤波　optimal filtering
优化设计　optimal design, optimization design
优化试验设计　optimal experiment design
优化调节器　optimal regulator
优化稳态控制　optimum steady state control
优化系统　optimal system
优化抑制　optimal rejection
优先存取设备　preferred access device
优先的　preferential
优先级　priority

优先设备	preferred device
优质碳素钢铸件	fine steel casting iron
油杯	oil bottle
油泵	oil pump
油表	fuel meter
油槽	oil baths
油槽润滑	sump lubrication
油池泵	sump pump
油淬化	oil quenching
油断路器	oil breaker
油封	oil seal
油沟密封	oily ditch seal
油管压力表	tubing gauge
油壶	oil can
油灰	putty
油灰抹子	jointing compound
油灰镶玻璃法	glazing with putty, putty glazing
油回火钢丝	oil temper wire
油浸变压器	oil-immersed transformer
油浸电力变压器	oil-immersed power transformer
油浸启动器	oil-immersed starter
油浸琴键	oil-immersed key switch
油浸式多点切换	oil-immersed multi-point switch
油浸调整变压器	oil-immersed regulating transformer
油井水泥	oil well cement
油开关	oil switch
油库	oil depot
油冷却系统	oil cooling system
油门	accelerator
油漆	paint
油漆层	paint coating
油气储运工程	oil-gas storage and transportation engineering
油气井工程	oil-gas well engineering
油气田开发工程	oil-gas field development engineering
油枪	oil gun
油熔断器	oil-break fuse
油位	fuel level
油污	grease stains, oil stains
油污截流井	oil interceptor
油箱	fuel tank
油箱容积	oil tank volume
油箱液位计	tank level meter
油压板车	hydraulic handjack
油压表	oil pressure gauge
油压断路器	oil circuit breaker
油压机	hydraulic machine
油脂	grease
油脂分离器	grease trap
油脂枪	grease gun
油脂室	grease chamber
油阻尼器	oil damper
柚木	teak
铀	uranium (U)
游标	cursor, vernier
游标尺	vernier scale
游标卡尺	nonius, slide caliper, slide gauge, vernier caliper
游标千分尺	vernier micrometer
游离	free
游离铁	free iron
游丝	hairspring
有补偿的感应电动机	compensated induction motor
有承梁转向架	bolster bogie
有触点电动执行机构	contact controlled electric actuator
有抵抗力的	resistive
有毒物质	poisonous substance
有功电度表	active energy meter
有功分量	active component
有功负载	active load
有功功率	active power
有关经验	relevant experience
有关特性	related properties
有轨电车	electric tramway
有害气体	noxious gas
有害杂质	deleterious substance
有害阻力	detrimental resistance
有机玻璃	perspex
有机定量分析仪	organic quantitative analyzer
有机化学	organic chemistry
有机溶剂	organic solvent
有机酸	organic acid
有界的	bounded
有界扰动	bounded disturbance

有界输入有界输出 bounded input and bounded output (BIBO)
有界噪声 bounded noise
有理矩阵 rational matrix
有色金属 nonferrous metal
有色金属冶金 non-ferrous metallurgy
有色噪声 coloured noise
有损耗同轴电缆 lossy coaxial cable
有线高频通信 high-frequency wire communication
有线载波通信 line radio
有限编码 limited code
有限差分 finite difference
有限场 finite field
有限的 finite, limited
有限分布 limiting distribution
有限互换附件 accessory of limited interchangeability
有限积分器 limited integrator
有限阶跃响应 finite step response (FSR)
有限空间 confined space
有限控制作用 limiting control action
有限脉冲响应 finite impulse response (FIR)
有限数据 limited data
有限线段 finite arc segment
有限元 finite element
有限元场仿真 finite field simulation
有限元分析 finite analysis
有限元计算 finite computation
有限增益放大器 limited gain amplifier
有限状态机 finite state machine
有限自动化 finite automata
有限自动机 finite automaton
有箱造模法 flask molding
有效波道长度 effective channel length
有效补偿 active compensation
有效出射度 effective radiation exitance
有效纯滞后 effective deadtime
有效磁场 effective magnetic field
有效带宽 effective bandwidth
有效的 effective, efficient, valid
有效电感 effective inductance

有效电抗 effective reactance
有效电流 effective current
有效电阻 effective resistance
有效发射率 effective emissivity
有效范围 effective range
有效工程 efficient engineering
有效功率 real power
有效功率消耗 true power consumption
有效光程长度 effective path length
有效基长 active gauge length
有效基宽 active gauge width
有效激振力 effective excitation force
有效加权 weighting efficient
有效截止波长 effective cut-off wavelength
有效孔径 effective aperture
有效控制 active control
有效拉力 effective tension
有效面积 effective area
有效评估 efficient evaluation
有效声压 effective sound pressure
有效时间 available time
有效时间常数 effective time constant
有效数据传送率 effective data transfer rate
有效数字 significant figure
有效衰减系数 effective attenuation factor (EAF)
有效算法 efficient algorithm
有效性 effectiveness, validity
有效元件 active component
有效圆周力 effective circle force
有效噪声控制 active noise control
有效窄带悬挂 active narrow band suspension
有效值 effective values
有效质量 effective mass
有效阻抗 effective impedance
有序参数 order parameter
有益阻力 useful resistance
有用 utility
有用的 useful
有源传感器 active transducer
有源电路 active electric circuit
有源电路元件 active circuit elements
有源滤波器 active filter

有源器件　active device
有源四端网络　active four-terminal network
有源线性二端网络　active linear two-port network
有源组件　active block，active element
有噪信道　noisy channel
有噪映像　noisy image
有噪语言　noisy speech
有证标准物质　certified standard material
铕　europium（Eu）
右　right
右手定则　right-hand rule
右手螺旋定则　right-hand screw rule
釉瓷耐火黏土　enamelled fire clay
釉瓷瓦　vitreous tile
釉面陶土　glazed earthenware
釉面物料　glazed ware
釉面砖　glazed brick
釉色　glazing color
淤积物　deposit
淤泥收集器　silt trap
淤泥水　sullage water
余函数　complementary function
余数系统　residue number system
余隙孔　access hole
余下工程　remaining works
余弦　cosine（cos）
余弦变换　cosine transform
余弦定理　cosine law
余弦辐射体　cosine radiator
余弦加速度运动　cosine acceleration motion
余弦收集器　cosine collector
鱼雷形分流板　torpedo spreader
鱼探仪　fish finder
鱼尾形　fishtail
鱼尾形模具　fishtail die
鱼眼　fish eye
与　AND
与……同相　coincide in phase with
与操作运算　AND operation
与电路　AND circuit
与非　NAND
与非门　dispersion gate，AND NOT gate
与非算子　NAND operation
与非元件　NAND element
与或非门　AND OR NOT gate
与门　AND gate
与元件　AND element
宇宙　universe
宇宙飞船　interplanetary spacecraft
宇宙射线　cosmic rays
宇宙线　cosmic rays
雨滴静电干扰　precipitation statics
雨淋阀　deluge valve
雨水槽　gutter
雨水排放主渠　stormwater main drain
雨水渠　stormwater drain
雨水渠排水口　stormwater outfall
雨水疏导系统　stormwater drainage system
雨水溢流室　stormwater overflow chamber
语言变量　linguistic variable
语言的　linguistic
语言信息　linguistic information
语言支持　linguistic support
语言综合　linguistic synthesis
语义网络　semantic network
语音　speech
语音分析　speech analysis
语音干扰分析系统　voice interference analysis system
语音控制　speech control
预备退火　pre-annealing
预测　prediction
预测报告　forecast
预测方法　prediction method
预测函数控制　predictive functional control（PFC）
预测控制　predictive control
预测区间　prediction interval
预测误差方法　prediction error method
预处理　preprocessing
预处理器　preprocessor
预定完成日　pre-fixed finishing date
预锻　dummying
预防　prevention
预防措施　precautionary measure

预防性维修	preventive maintenance
预激励	pre-excitation
预加荷载	preloading
预紧力	preload
预浸材料	prepreg
预警系统	early warning system
预聚物	prepolymer
预滤器	prefilter
预期的长期稳定性	expected long term stability
预期值	desired value
预取	prefetch
预燃室	precombustion chamber
预热	preheating
预热器	preheater
预热时间	warm-up period
预调时间	rate time
预应力	prestressing
预应力钢丝	reinforcement wire
预应力构件	prestressed element
预应力混凝土梁	prestressed concrete beam
预制	precast
预制混凝土梁	precast concrete beam
预制混凝土砌块	precast concrete segment
预制件工场	casting basin
预制式水箱	prefabricated water tank
预置脉冲计数器	predetermining impulse counter
预组构件	prefabricated parts
阈电平	threshold level
阈值	threshold value
阈值分辨率	threshold of resolution
阈值函数	threshold function
阈值逻辑	threshold logic
阈值选择	threshold selection
裕度	margin
遇热收缩软管	heat shrinkable tubing
元件	component, parts
元件安置	component positioning
元件孔	component hole
元件密度	component density
元件面	component side
元素	element
元素分析	element analysis
元知识	meta-knowledge
原点	origin
原电池	galvanic cell, primary cell
原动机	prime motor
原动机扭矩	prime mover torque
原动件	driving link
原动力	prime mover
原来图则	original plan
原理	principle, theorem
原理图	schematic diagram
原料	feed stock, raw materials
原模型控制	control oriented model
原绕组	primary winding
原设备制造	original equipment manufacture
原始机构	original mechanism
原始设计	original design
原始数据	raw data
原位定量	in situ quantitation
原因分析	cause analysis
原因说明	cause description
原油	crude petroleum
原子	atom
原子弹	atom bomb
原子动力反应堆	power reactor
原子发射分光光度法	atomic emission spectrophotometry
原子发射光谱术	atomic emission spectrometry (AES)
原子发射光谱学	atomic emission spectroscopy
原子发射检测	atomic emission detection (AED)
原子共振磁强计	atomic resonance magnetometer
原子光谱	atomic spectrum
原子光谱法	atomic spectroscopy
原子锅炉	atomic boiler
原子化	atomization
原子力显微镜	atom force microscope
原子量	atomic weight
原子能	atomic power
原子能的	nuclear
原子束磁共振仪	atomic beam magnetic resonance apparatus
原子吸收分光光度法	atomic absorption spectrophotometry (AAS)
原子吸收分光光度计	atomic-absorp-

tion spectrophotometer
原子吸收光谱学　atomic absorption spectroscopy（AAS）
原子序数　atomic number
原子序数修正　atomic number correction
原子荧光测定术　atomic fluorometry
原子荧光分光光度法　atomic fluorescence spectrophotometry
原子荧光光度计　atomic fluorescence spectrophotometer
原子荧光光谱法　atomic fluorescence spectrometry
原子荧光光谱学　atomic fluorescence spectroscopy（AFS）
原子质量　atomic mass
原子质量单位　atomic mass unit（AMU）
圆冲子　round punch
圆带　round belt
圆带传动　round belt drive
圆顶　dome
圆钢　round steel
圆弧齿厚　circular thickness
圆弧圆柱蜗杆　hollow flank worm
圆基脚　circular footing
圆角半径　fillet radius
圆螺母　round nuts
圆盘形记录仪　disc recorder
圆盘端面铣刀　circular face-mill
圆盘摩擦离合器　disc friction clutch
圆盘式容积流量计　disc type positive displacement flowmeter
圆盘铣刀　circular peripheral-mill
圆盘制动器　disc brake
圆缺孔板　segmental orifice plate
圆筒　cylinder
圆筒形喉部文丘里喷嘴　cylindrical throat Venturi nozzle
圆筒直尺　cylinder square
圆线弹簧　wire spring
圆形测量刻度盘　circular measuring dial
圆形齿轮　circular gear
圆形加工　rounding
圆形均压环取压的孔板　orifice plate with circular pressure-equalizing
圆形盘　round pad

圆形心线　round core wire
圆柱齿轮　cylindrical gear
圆柱滚子　cylindrical roller
圆柱滚子轴承　cylindrical roller bearing
圆柱螺旋拉伸弹簧　cylindroid helical-coil extension spring
圆柱螺旋扭转弹簧　cylindroid helical-coil torsion spring
圆柱螺旋压缩弹簧　cylindroid helical-coil compression spring
圆柱体抗压强度　cylinder crushing strength
圆柱凸轮　cylindrical cam
圆柱蜗杆　cylindrical worm
圆柱销　straight pin
圆柱坐标操作器　cylindrical coordinate manipulator
圆柱坐标型机器人　cylindrical robot
圆锥齿轮机构　bevel gears
圆锥滚子　tapered roller
圆锥滚子轴承　tapered roller bearing
圆锥角　cone angle
圆锥螺旋扭转弹簧　conoid helical-coil compression spring
圆锥入口孔板　conical entrance orifice plate
圆锥轴承　conical bearing
源　source
源程序　source program
源极输出电路　source follower circuit
源码　source code
源语言　source language
源装置参数　source device parameter
远场　far field
远程端子板　remote terminal panel
远程控制器　remote manipulator
远程耦合　long range coupling
远程屏蔽效应　long range shielding effect
远程设定点调整器　remote set point adjuster
远程通告　remote notify
远程通信　telecommunication
远程显示协议　remote display protocol
远方设定　remote set

远方设定控制器　remote set controller
远方手动操作器　remote manual loader
远红外分光光度计　far-infrared sprectrophotometer
远红外辐射　far infrared radiation
远红外辐射加热　heating by far infrared radiation
远红外辐射元件　far infra-red radiant element
远红外干涉仪　far-infrared interferometer
远距离的　remote
远距离调节　remote regulating
远距离无线电通信　distant radio communication
远离锅炉　keep away from boiler
约当标准型　Jordan canonical form
约当范式　Jordan canonical form
约当块　Jordan block
约定真值　conventional true value
约束　constraint
约束补偿　constraint satisfaction
约束补偿问题　constraint satisfaction problem
约束参数　constrained parameter
约束的　constrained
约束方法　bounding method
约束极点　constrained pole
约束控制　constraints on control
约束力　constraining force
约束条件　constraint condition
约束优化　constrained optimization
约束指令集　restricted instruction set
月　month
月报　monthly report
月负荷率　monthly load factor
月负荷曲线　monthly load curve
钥匙开关　key switch
跃度　jerk
跃度曲线　jerk diagram
越野汽车　cross-country vehicle
云滴取样器　cloud-drop sampler
云底　cloud base
云底记录仪　cloud-base recorder
云顶　cloud top

云辐射仪　cloudiness radiometer
云高计　ceilometer
云高仪　cloud height meter
云高指示器　cloud height indicator
云量　cloud amount
云母　talc
云母电容器　mica capacitor, mica condenser
云母膜片　mica diaphragm
云石　marble
云室　cloud chamber
云速　cloud speed
云向　cloud direction
匀染剂　leveling agent
匀速滚动　uniform roll
匀速试验机　uniform velocity tester
允许　permit
允许的API中断状态　permit API interrupt
允许的负载阻抗　allowable load impedance
允许相位失真　phase margin
允许压差　allowable pressure differential
允许应力　allowable stress
陨铁　corrugated iron, undulated sheet iron
运筹学　operations research
运筹学模型　operational research model
运动部件　moving element
运动部件的漂移运动　erratic motion of the moving element
运动部件电谐振频率　electrical resonance frequency of the moving element
运动部件机械共振频率　mechanical resonance frequency of the moving element
运动部件悬挂动刚度　dynamic stiffness of the moving element suspension
运动部件有效质量　effective mass of the moving element
运动倒置　kinematic inversion
运动方案设计　kinematic precept design
运动分析　kinematic analysis

运动副　kinematic pair
运动构件　moving link
运动简图　kinematic sketch
运动控制　motion control
运动链　kinematic chain
运动密封　kinematical seal
运动黏度　kinematic viscosity
运动曲线　motion curves
运动设计　kinematic design
运动学　kinematics
运动周期　cycle of motion
运动装置　telemechanical apparatus
运动综合　kinematic synthesis
运输　transportation
运输和储存条件　transportation and storage condition
运输极限值　limiting value for transport
运输控制　transportation control
运输模式　mode of transport
运输能力　traffic capacity
运输器　conveyer
运输设施　transportation facilities
运输系统　traffic system
运输需求　transport demand
运数计　operameter
运算电路　operation circuit
运算放大器　operational amplifier
运算微积分　operational calculus
运算阻抗　operational impedance
运行　run
运行程序　run program
运行的　operating
运行方式　method of operation
运行记录　operational log
运行时编译执行的技术　just-in-time (JIT)
运行时间　running time
运行试验　commissioning test, operation test
运行速度　running speed
运行-停机-遥控启动选择键　local run-stop-remote starting selector switch
运行在线手动观察窗　running online manual viewer
运行噪声　running noise
运行中断　outage
运行转速　running rpm
运行状态　operating condition
运转不均匀系数　coefficient of velocity fluctuation
韵律　rhythm

Z

匝数比　turns ratio
杂化影响　hybridization affect
杂散　stray
杂散电磁场　stray magnetic field
杂散电流　stray current
杂散电容　stray capacitance, stray capacity
杂散电压　stray voltage
杂散光　stray light
杂散损耗　stray loss
杂色　mixed color
杂讯容限　noise margins
杂质　impurity
载波　carrier wave
载波磁放大器　carrier magnetic amplifier
载波带局域网络　carrier band LAN
载波电缆　carrier wire
载波电平　carrier level
载波发生器　carrier generator
载波和边带　carrier and sideband
载波检波器　carrier detector
载波频率　carrier frequency
载波频率放大器　carrier frequency amplifier
载波设备　carrier equipment
载波通信　carrier-current communication
载波同步　carrier sync
载波侦听　carrier sense
载荷-变形曲线　load-deformation curve
载荷-变形图　load-deformation diagram
载荷能力　loading capacity
载荷预测　load forecasting
载货升降机　goods lift
载客量　passenger capacity
载客升降机　passenger hoist
载流导体　current-carrying conductor
载流量　current-carrying capacity
载流容量　current carrying capacity
载流子　carrier, current carrier
载气　carrier gas
载人潜水器　manned submersible
载体箔　carrier foil
载芯片板　chip on board
载运工具运用工程　vehicle operation engineering
载重勘测　loading survey
载重量　load carrying capacity
再充电　recharge
再次研磨　regrinding
再结晶　recrystallization
再入控制　reentry control
再生　regeneration
再生的　retroactive
再生反馈　regenerative feedback
再生放大　retroactive amplification
再生式放大器　regenerative amplifier
再生式收音机　regenerative receiver
再生制动　regenerative braking
再填　refilling
再调质　reconditioning
再现性　reproducibility
再循环　recirculation
再循环阻尼器　recirculation damper
在规定特性曲线　specified characteristic curve
在后台编译和连接　compile and link in background
在连接盘中导通孔　via-in-pad
在线安全分析　online security analysis
在线帮助　on-line assistance
在线闭环回路　online closed loop
在线操作　online operation
在线测试　on-process tests
在线重组态　online reconfiguration
在线的　on line, online
在线分析　on-process analysis (OPA)
在线估计　online estimation
在线控制　online control
在线手动　online manual
在线维护功能　online maintenance function

在线文件	online documentation
在线优化	online optimization
在线诊断	online diagnosis
在线整定	online tuning
在一般个人计算机上运行	run on generic PC
在制品	work in progress product
暂时的	temporal
暂态电流	transient current
暂态电压	transient voltage
暂态放电	transient electrical discharges
暂态分量	transient state component
暂态分析	transient analysis
暂态解	transient solution
暂态逻辑	temporal logic
暂态推理	temporal reasoning
暂态稳定	transient stability
暂态稳定性分析	transient stability analysis
暂态稳定性评估	transient stability assessment
暂态误差	transient error
暂态信号	transient signal
暂态行波	transient state travelling wave
凿	chisel
凿子	broach
早期故障	early failure
早期失效	early failure
造价	construction cost
造链机	chain making tools
造模刮板	sweep template
造线机	cable making tools
造纸工业	paper industry
噪声	noise
噪声安全系数	noise margins
噪声电流	noise current
噪声电压	noise voltage
噪声分析	noise analysis
噪声干扰	noise interference
噪声功率谱	noise power spectrum
噪声计	audio noise meter, noise meter
噪声控制	noise control
噪声滤波器	noise filter
噪声容限	noise margins
噪声声级	noise level
噪声试验	sound test
噪声特性	noise characteristic
噪声特性化	noise characterization
噪声污染	noise pollution
噪声吸收系数	noise absorption factor
噪声系数	noise factor
噪声信号比	jam-to-signal
噪声抑制	noise suppression
噪声抑制器	noise suppressor
增安型	increased safety
增安型电动仪表	increased safety electrical instrument
增安型电气设备	increased safety electrical apparatus
增补计划	additional plan
增广系统	augmented system
增加	increase
增加工程能力的成分和模件	increasing engineering productivity component and module
增加键	incremental key
增加物	accretion, addition
增力丝	reinforcing wire
增量	increment
增量测试	incremental testing
增量积分器	incremental integrator
增量加法器算法	incremental summer algorithm
增量控制	incremental control
增量运动控制	incremental motion control
增量运动控制系统	incremental motion control system
增强	enhancement
增强板材	stiffener material
增强材料	reinforcing material
增强剂	reinforcing agent
增强器	booster
增强效应	enhancement effect
增强型操作台	excellent operator console
增强型操作站	enhanced operator station
增强型工业标准架构	enhanced industry standard architecture (EISA)
增强型计算机	excellent computer

增强型接口单元　excellent gateway unit
增强型现场监视站　excellent field monitoring station
增强型现场接口　excellent field gateway
增强型现场控制站　excellent field control station
增色效应　hyperchromic effect
增塑剂　plasticizers
增压泵　booster pump
增压变压器　boost transformer, booster transformer
增压抽水机　booster water pump
增压抽水站　booster pumping station
增益　gain
增益饱和　gain saturation
增益动力学　gain dynamics
增益规划　gain scheduling
增益交越频率　gain crossover frequency, gain-crossover frequency
增益控制　gain control
增益控制器　gain controller
增益时间控制　gain time control
增益特性　gain characteristic
增益调节分压器　gain setting divider
增益抑制　gain suppression
增益裕度　gain margin
增益增强方法　gain enhancement method
增益转换放大器　gain-switching amplifier
增殖反应堆　breeder reactor
轧锻　roll forging
轧钢　rolled steel
轧制表面　rolled surface
轧制钢槽　rolled steel channel
闸　brake
闸板　wedge disc
闸板式流量控制开关　paddle type flow switch
闸刀　knife-switch
闸刀开关　knife switch
闸流管　thyristor
闸门阀　gate valve
闸门式流量计　gate-type flowmeter
闸式阀　gate valve, sluice valve

炸药　explosive
炸药贮藏室　explosive store
摘取　pick off
摘要　abstract
窄V带　narrow V belt
窄带传输　narrow-band transmission
窄带晶体滤波器　narrow-band crystal filter
窄带滤波器　narrow-band filter
窄带调制　narrow-band frequency modulation
窄带制　narrow-band system
窄频带　narrow band
窄频道放大器　narrow band amplifier
沾染　contamination
沾湿台　wet station
毡圈　felt ring
毡圈密封　felt ring seal
毡条　felt strip
粘贴式应变计　bonded strain gauge
斩波电路　chopper circuit
斩波放大器　chopper amplifier
斩波器　chopper
斩波稳压器　chopper voltage stabilizer
斩波信号　chopping signal
展成齿轮　generating gear
展成传动键　generating train
展成法齿轮　generated gear
展成法刨齿机　planing generator
展成凸轮　generating cam
展开　development
展开槽　developing tank
展开剂　developing solvent
展开室　development chamber
展开图　extension dwg
占空度　duty cycle
占空因数　duty factor, fill factor
占线信号　busy-back signal
站　station
站控制箱　station control nest
站默认存取级别　station default access level
张弛电路　relaxation circuit
张紧轮　tension pulley
张力　tension, tension force
张力的　tensile

张力计　tensile gauge, tensiometer
张力试验　tension test
张力调整杆　draw bead
张索　guy wires
章动敏感器　nutation sensor
长霉试验　mould growth test
胀缩心轴　expanding arbor
障碍　obstacle
障碍探测　obstacle detection
障碍探测器　obstacle detector
障碍探测系统　obstruction detection system
障碍物　barrier
招牌　signboard
爪牙夹头　dog chuck
沼气含量　methane content
沼气检查　methane detection
沼气-空气混合物　methane-air mixture
沼气泄出　methane emission
兆欧表　insulation resistance meter, megger, earthometer
兆欧计　earthometer
兆字节　megabyte (MB)
照度　illuminance
照度传感器　illuminance sensor, illuminance transducer
照度计　illumination meter, illuminometer, lumeter
照明　illuminate, lighting
照明标志　illuminated sign
照明的　illuminating
照明电缆　lighting cable
照明度　illuminance level, illumination level
照明开关　light switch
照明孔径角　illuminating aperture angle
照明亮度　illuminating brightness
照明器　illuminator
照明设备　luminaire
照明调节器　light-regulator
照明系统　illuminating system
照相机　camera
照相温度表　film recording thermograph
照准丝　sighting wire

照准仪　alidade
罩极式电动机　shaded-pole motor
遮挡板　baffle plate
遮挡墙　baffle wall
遮盖力计　cryptometer
遮光剂　opaquer
遮光效果　shading effect
遮光眼罩　eyeshade
折尺　folding rule
折刀　bending block
折叠式记录纸　folding paper
折反射望远镜　catadioptric telescope
折光率　refractive index detection
折光率检测器　refractive index detector
折扣　discount
折射　refraction
折射的　refracted
折射角　angle of refraction
折铁器　folding bar
折铁砧　hatchet stake
折弯块　folded block
折线函数发生器　function fitter
折线函数运算器　line-segment function unit
折纸机构　scan-fold chart
折纸机构驱动装置　scan-fold chart drive unit
锗　germanium (Ge)
锗二极管　germanium diode
锗三极管　germanium triode
锗温度计　germanium thermometer
褶合光谱法　convolution spectrometry
褶合积分　convolution integral
针尖浇口　pin gate
针角焊接　stitch welding
针孔摄影法　aperture photographic method
针孔试验机　pinhole test
针头销毁器　needle destroyer
针形阀　pintle valve
真空　vacuum
真空泵　vacuum air pump, vacuum pump
真空表校验表　calibrator for vacuum gauges
真空淬火　vacuum hardening

真空氮化　vacuum nitriding
真空低温保持器　vacuum cryostat
真空法兰　vacuum flanges
真空干燥箱　vacuum drying ovens
真空管　vacuum tube
真空光电二极管　vacuum photodiode
真空计　vacuometer, vacuum gauge
真空技术　vacuum technology
真空接触式膨胀计　vacuum contacting dilatometer
真空解除阀　vacuum relief valve
真空气体置换炉　gas replacement vacuum furnaces
真空热处理　vacuum heat treatment
真空热离子检测计　vacuum thermionic detector
真空渗碳处理　vacuum carburizing
真空式气动测微计　vacuum type pneumatic micrometer
真空涂膜　metallizing
真空自动控制电源　automatic control source of vacuum
真实的　true
真圆度　out of roughness
真值　true value
真值表　matrice, truth table
诊断　diagnosis
诊断测试　diagnostic test
诊断程序　diagnostic program, diagnostor
诊断错误的处理程序　diagnostic error handler
诊断的　diagnostic
诊断功能　diagnostic function
诊断功能测试程序　diagnostic function test
诊断模型　diagnostic model
诊断推理　diagnostic inference
枕木　cross tie
阵风效应　gust effect of wind
阵列排列　array configuration
振荡　oscillation
振荡倍频器　oscillator doubler
振荡的　oscillating, oscillatory
振荡电路　oscillating circuit
振荡放电　oscillatory discharge
振荡晶体　oscillation crystal

振荡培养器　shaking incubators
振荡频率　oscillation frequency
振荡器　oscillator
振荡器石英　oscillator quartz
振荡式低温水槽　low constant temperature shaking baths
振荡线圈　oscillating coil
振荡周期　oscillating period, period of oscillation
振动　vibration
振动测量　vibration measurement
振动测量计　vibration meter
振动荷载　vibrating load
振动计　vibrometer
振动力矩　shaking couple
振动量的测量　measurement of vibration quantity
振动烈度　vibrational severity
振动模具　swing die
振动模态　mode of vibration
振动频率　frequency of vibration
振动式检流计　vibration galvanometer
振动式黏度计　vibro viscometer
振动试验系统　vibration test systems
振动试验方法和限值　vibration testing method and limiting levels
振动送料机　vibration feeder
振动影响　influence of vibration
振动桩锤　vibratory pile hammer
振幅　amplitude of vibration
振幅检测组件　amplitude detector module
振幅鉴别器　amplitude discriminator
振幅滤波器　amplitude filter
振幅取样器　amplitude sampler
振幅误差　amplitude error
振幅因数　crest factor
振簧式微静电计　vibrating-reed micro quantity electrometer
振纹　rippling
振弦式测力传感器　vibrating wire force transducer
振形　mode shape, mode of vibration
镇定方法　stabilization method
镇流电阻　ballast resistance
镇流电阻器　barretter
镇流器　ballast, current stabilizer

镇流阻抗　ballast impedance
蒸发　vaporization
蒸发成像仪　evaporograph
蒸发度　evaporativity
蒸发光散射检测器　evaporative light scattering detector
蒸发冷却系统　evaporative cooling system
蒸发皿　evaporating dish
蒸发器　evaporator
蒸发器蛇形管　evaporator coil
蒸发器旋管　evaporator coil
蒸发仪　atmidometer, evaporimeter
蒸馏　distillation
蒸馏水　distilled water
蒸汽　steam
蒸汽发生器　steam generator
蒸汽工厂　steam plant
蒸汽锅炉　steam generator, steam boiler
蒸汽流量计　steam flow meter
蒸汽凝结水　steam condensate
蒸汽热交换器　steam-heated exchanger
蒸汽压力式感温系统　vapor pressure thermal system
蒸汽养护混凝土　steam-cured concrete
蒸汽转换阀　steam converting valve
蒸散表　evapotranspirometer
蒸压灰砂砖　autoclaved lime-sand brick
蒸压加气混凝土板　autoclaved aerated concrete slabs
蒸压加气混凝土砌块　autoclaved aerated concrete blocks
拯救设备　rescue equipment
整定特性　tuning characteristic
整个深度　entire depth
整流　commutate, rectifying
整流变压器　rectiformer
整流电路　rectification circuit
整流光电管　rectifier photocell
整流级　rectifier stage
整流极　commutating pole
整流滤波器　rectifier filter
整流器　commutator, rectifier
整流器用变压器　rectifier transformer

整流器元件　rectifier element
整流式仪表　rectifier instrument
整流子频率变换器　commutator frequency changer
整流作用　rectifying action
整面机　finishing machines
整平层　levelling course
整平机　uncoiler and straightener
整数　integer
整数规划　integer programming
整体尺寸　overall dimensions
整体刀盘　solid cutter
整体锻模　solid forging die
整体分析　global analysis
整体工厂控制　control of total plant
整体化　totalization
整体进度　overall progress
整体坍塌　total collapse
整体稳定性　overall stability
整体运输研究　comprehensive transport study
整体铸件　one piece casting
整体铸造　inblock cast
整形　reshaping
整形电路　shaping circuit
整形滤波器　shaping filter
整形器　shaper
整形网络　shaping network
整修　renovate
整直器　straightener
正、余弦电位计　sine-cosine potentiometer
正常　normal
正常齿厚收缩　normal thickness taper
正常电流　normal current
正常方式　normal mode
正常负载　normal load
正常工作条件　normal operating condition
正常计算流量系数　calculated normal flow coefficient
正常磨损　normal wear
正常收缩　standard taper
正常输出　normal output
正常运行　normal operation
正常照明　normal illumination
正常值　normal value

正常值状态　normal value status
正常状态特性　normal mode attribute
正齿轮　spur gear
正的　positive
正电荷载流子　positive carrier, positive hole
正电性的　electropositive
正电柱　positive column
正电子　positron
正反馈　positive feedback, regenerative feedback
正反馈控制　positive feedback control
正负三位作用　positive-negative three step action
正负作用　positive-negative action
正规方法　formal method
正规语言　formal language
正回授　regenerative feedback
正火　normalizing
正极　anode, positive pole
正极导线　positive wire
正交电位差计　quadratic potentiometer
正离子　cation, positive ion
正面图　front view, elevation
正面压印　tick-mark nearside
正排量泵　positive displacement pump
正迁移　positive transfer
正迁移比　suppression ratio
正切　tangent
正切电流计　tangent galvanometer
正切机构　tangent mechanism
正式的　formal
正式模图设计　final mold design
正式说明规格　formal specification
正弦　sine
正弦波　sine-wave
正弦波发生器　sine-wave generator
正弦波输出　sine wave output
正弦波数字信号发生器　digital sine wave generator
正弦磁密度　sinusoidal density wave
正弦电流　sinusoidal current
正弦电压　sine voltage
正弦发生器　sine generator
正弦量规　sine bar
正弦时间函数　sinusoidal time function
正弦态最大激振力　maximum rated force under sinusoidal conditions
正弦稳态　sinusoidal steady state
正弦响应　sine-forced response, sinusoidal response
正弦信号　sinusoidal signal
正弦振荡器　sinusoidal oscillator
正相　normal phase
正相序　positive phase-sequence
正相液相色谱法　normal phase liquid chromatography
正向变流器　forward converter
正向传递函数　forward transfer function
正向电流　forward current
正向电压　forward voltage
正向电阻　forward resistance
正向环节　forward element
正向设置　forward construction
正向通道　forward channel
正向通路　forward path
正向推理　forward reasoning
正向运动学　direct kinematics, forward kinematics
正向主控元件　forward controlling element
正像望远镜　erecting telescope
正序阻抗　positive sequence impedance
正压室　high pressure chamber
正样　engineering prototype
正应力　normal stress
正则化　regularization
正作用　direct action
正作用调节控制器　direct acting actuator controller
正作用执行机构　direct acting actuator
证明　demonstrate
证明书　certificate
证实　authentication
证实试验　confirmatory test
帧　frame
帧缓冲缓存器　frame buffer, frame buffer cache (FBC)
帧同步　frame synchronization

支撑　bracing
支撑结构　bracing structure
支撑孔　supported hole
支撑销　support pin
支撑支柱　support pillar
支撑轴　back shaft
支承垫板　bearing plate
支承垫片　bearing pad
支承构件　supporting member
支承面　bearing surface
支承栓钉　bearing pin
支承应力　bearing stress
支承桩　bearing pile
支持　support
支持系统　holding systems
支点　fulcrum
支管　branch pipe
支架　bracket
支路　branch, branching, circuit branch
支路导纳矩阵　branch admittance matrix
支路电流　branch current
支路电流法　branch current method
支路阻抗矩阵　branch impedance matrix
支线　branch line
支线电缆　branch cable
支轴销　fulcrum pin
支柱　stanchion, strut
支座垫板　holder base plate
知识　knowledge
知识表达　knowledge representation
知识产权法　intellectual property laws
知识辅助协议自动化　knowledge aided protocol automation
知识工程　knowledge engineering
知识工具　knowledge tool
知识获取　knowledge acquisition
知识库　knowledge base
知识库管理系统　knowledge base management system (KBMS)
知识库系统　knowledge based system (KBS)
知识模型　knowledge model
知识同化　knowledge assimilation
知识推理　knowledge inference
知识转换器　knowledge transfer

织布机　loom
织物经纬密度　thread count
织物组织　weave structure
脂肪酸　fatty acid
执行　execute
执行步骤　step execution
执行程序　executive routine
执行单元　execution unit
执行构件　executive link
执行机构　actuator
执行机构波纹管　actuator bellows
执行机构动力部件　actuator power unit
执行机构负载　actuator load
执行机构刚度　actuator stem force
执行机构输出杆　actuator stem
执行机构输出轴　actuator shaft
执行机构弹簧　actuator spring
执行机构行程特性　actuator travel characteristic
执行器传感器接口　actuator sensor interface (ASI)
执行时间　execution time
执行元件　executive component, correcting element
直槽刨刀　straight grooving iron
直尺　straight edge
直齿圆柱齿轮　straight toothed spur gear
直齿锥齿轮　straight bevel gear
直齿锥齿轮粗拉法　revex
直动式电磁阀　direct action solenoid valve
直读式测流计　direct reading current meter
直读式频率计　magmeter
直读式声频计　direct reading audio-frequency meter
直读式陶瓷电容器测试仪　direct reading ceramic condenser tester
直读式仪表　direct-reading instrument
直读式仪器　direct reading instrument
直度　straightness
直交直电路　DC-AC-DC converter
直角边薄孔板　square edged thin orifice plate
直角尺　square

直角三角形　right triangle
直角压陷　corner shear drop
直角坐标操作器　Cartesian coordinate manipulator
直角坐标型机器人　Cartesian robot
直角坐标轴　quadrature axis
直接　direct
直接安装压力表　direct mounting gauge
直接被控变量　directly controlled variable
直接被控系统　directly controlled system
直接比较测量法　direct-comparison method of measurement
直接测量法　direct method of measurement
直接成像质量分析仪　direct-imaging mass analyser
直接重写　direct overwrite
直接存储器内存存取　direct memory access
直接存储器内存访问　direct memory access
直接电流控制　direct current control
直接电位测定法　direct potentiometry
直接电阻加热　direct resistance heating
直接动力学问题　direct dynamic problem, direct kinematic problem
直接放大　straight amplification
直接傅里叶重构　direct Fourier reconstruction
直接记录式强震仪　direct record strong-motion instrument
直接记录式仪表　direct-recording
直接浇口　direct gate
直接控制　direct control
直接控制层　direct control layer
直接连接　direct connection
直接耦合的　direct coupled
直接耦合放大器　direct coupled amplifier
直接耦合晶体管电路　directly-coupled transistor circuit
直接喷射　direct injection
直接频率调制　direct frequency modulation

直接启动　direct-on starting
直接驱动机器人　direct-drive robot
直接驱气　direct purging
直接射出法　direct injection
直接示值　direct indication
直接数字控制　direct digital control (DDC)
直接数字控制站　direct digital control station
直接探头进样　direct probe inlet
直接图形　directed graph
直接协调　direct coordination
直接制动阀　direct brake valve
直接制动器　direct brake
直接注入　direct injection
直接注入燃烧器　direct injection burner
直接综合法　direct synthesis method
直接作用测量仪表　direct acting actuator measuring instrument
直接作用式调节　direct-operated regulator
直接作用仪表　direct acting instrument
直径　diameter
直径比　diameter ratio
直径的　diametral
直径螺距比　diameter ratio
直径系列　diameter series
直径系数　diametral quotient
直立墙　upright wall
直立式检流计　vertical galvanometer
直立式落水管　vertical down pipe
直流　DC，direct current
直流比较仪式电桥　DC comparator type bridge
直流比较仪式电位差计　DC comparator potentiometer
直流标准电压发生器　DC standard voltage generator
直流标准电阻　DC standard resistance
直流部分　direct component
直流传动　DC drive
直流传输　direct-current transmission
直流串励电动机　DC series motor
直流等离子体发射光谱仪　direct current plasma emission spectrometer
直流电　direct current（DC）
直流电的　galvanic

直流电动机　DC motor
直流电焊发电机　DC welding generator
直流电机　DC machine
直流电流表　DC ammeter
直流电流互感器　DC current transformer
直流电路　DC circuit, direct current circuit
直流电位差计　DC potentiometer
直流电压校准器　DC voltage calibrator
直流电源　DC electrical source, direct current power supply
直流电源电压纹波　DC power voltage ripple
直流电阻分压箱　DC resistor volt ratio box
直流电阻箱　DC resistance box
直流动圈式仪表　direct-current moving coil meter
直流发电机　DC generator
直流发电机-电动机组传动　DC generator-motor set drive
直流分量　DC component
直流复励电动机　DC compound generator
直流复射式检流计　DC reflecting galvanometer
直流高压　high direct voltage
直流高压输电　DC high voltage transmission
直流高阻电桥　DC bridge for measuring high resistance
直流供电　DC supply
直流毫安表　DC milliammeter
直流积分表　DC integrating meter
直流-交流变换器　DC-AC converter
直流进样器　direct injector
直流控制回路　DC control circuit
直流控制系统　DC control system
直流馈电屏　direct current feed control panel
直流励磁绕组　DC exciting-winding
直流配电装置　DC distributing equipment
直流无触点电位计　direct-current contactless potentiometer
直流信号　DC signal
直流斩波　DC chopping
直流斩波电路　DC chopping circuit
直流-直流变换器　DC-DC converter
直流作用记录仪　direct acting recording instrument
直热式热敏电阻器　direct heated type thermistor
直升机　helicopter
直升机动力学　helicopter dynamics
直升机控制　helicopter control
直通交通　through traffic
直线天线阵　linear array
直线位移光栅　linear displacement grating
直线运动　linear motion
直线运动电气传动　linear motion electric drive
直行程阀　linear motion valve
直形橱式操纵台　console with wardrobe
直形外螺纹截止阀　straight external threaded block valve
直轴　direct axis, straight shaft
直轴瞬变时间常数　direct axis transient time constant
值　value
值状态　value status
止动板　check plate
止动环　stop ring
止动螺钉　check screw
止动位置　rest position
止动销　stop pin
止回阀　check valve
止口镶嵌方式　counter lock
止逆阀　non-return valve
只读存储器　read only memory (ROM)
只读光碟　compact disk read-only memory
纸板箱　carton
纸带　paper tape
纸带穿孔机　paper tape punch
纸带机　paper tape unit
纸带阅读机　paper tape reader
纸浆工业　pulp industry
纸色谱分析法　paper chromatography

纸箱　carton box
纸张厚度计　caliper profiler
指挥站　director
指令　instruction
指令和控制系统　command and control system
指令级语言　instruction level language
指令集体系结构　instruction set architecture (ISA)
指令计数器　instruction counter, location counter
指令寄存器　instruction register
指令监视译码器　command monitoring decoder
指令控制　command control
指令码　command code
指令信号　command variable
指示　indicate
指示标志　indicative mark
指示表支撑座　indicator anchorage
指示灯　display lamp, indicating lamp, indicator lamp, indicator light
指示点　indication point
指示电极　indicator electrode
指示符　designator
指示剂常数　indicator constant
指示角度　indicated angle
指示空速　equivalent air speed (EAS)
指示控制器　indicating controller
指示器　indicator
指示曲线　index profile
指示误差　indicating error, indication error
指示压力调节器　indicating pressure controller
指示仪表　indicating instrument
指示仪表行程　indicator travel
指示应变　indicated strain
指示值　indicated value
指示装置　indicating device
指数　index
指数函数　exponential function
指数滤波器　exponential filter
指数滞后　exponential lag
指向角　angle of spread
指向性　directivity
指向性响应图案测量　measurement of directional response pattern
指形护罩　finger guard
指形销　finger pin
指样订货　sample order
指针　cursor, needle, pointer
指针式流量计　dial flowmeter
指针式热电偶　needle thermocouple
指针式仪表　pointer instrument
指针式指示器　needle indicator
制备薄层色谱法　preparative thin layer chromatography
制备液相色谱法　preparative liquid chromatography
制备液相色谱仪　preparative liquid chromatograph
制表　prepared by
制程　manufacture procedure
制程不良　deficient manufacturing procedure
制导系统　guidance system
制钉铁　nail iron
制动　braking
制动保护继电器　brake protecting relay
制动底板　backing plate
制动电磁铁　brake magnet, braking magnet
制动电动机　brake motor
制动阀　brake valves
制动功率　brake power
制动继电器　braking relay
制动距离　braking distance
制动开关　brake switch
制动力矩　rake moment
制动马达　actuating motor
制动马力　brake horse power
制动器　arrester brake
制动器摩擦衬片　brake lining
制动器试验　brake test
制动时间　braking time
制动系统　brake system
制动系统测试器　brake tester
制动元件　braking element
制动转矩　retarding torque
制浆造纸工程　pulp and paper engineering
制冷及低温工程　refrigeration and cry-

ogenic engineering
制冷剂　refrigerant
制冷剂回路　refrigerant circuit
制冷设备　chiller plant
制冷压缩机　refrigeration compressor
制冷装置试验　testing of refrigerating systems
制糖工程　sugar engineering
制图机　drafting machine
制药设备　pharmaceutical equipments
制造　fabrication, manufacturing
制造工程　manufacturing engineering
制造工业　products industry
制造工艺　fabrication process, processing technique
制造管理　manufacturing management
制造规范　manufacturing specifications
制造过程　manufacturing process
制造品质保证　manufacture quality assurance
制造图纸　fabrication drawing
制造文件　manufacturing documentation
制造系统　manufacturing system, production system
制造系统与管理　manufacturing systems and management
制造执行系统　manufacturing execution systems
制造自动化　manufacturing automation
制造自动化协议　manufacturing automation protocol (MAP)
制造作业规范　standard operation procedure
制作图　fabrication drawing
治疗模型　therapy model
治疗药物浓度监测　therapeutic drug monitoring
质荷比　mass-to-charge ratio
质径积　mass-radius product
质量　mass
质量保证　quality assurance (QA)
质量定心　mass centering
质量定心机　mass centering machine
质量范围　mass range

质量分析　quality analysis
质量分析离子动能谱仪　mass analyzed ion kinetic energy spectrometer (MIKES)
质量分析器　mass analyzer
质量峰　mass peak
质量跟踪　quality tracking
质量估计　quality estimation
质量管理　quality control (QC)
质量检测　mass detection
质量检测系统　quality measurement system
质量流量　mass flow-rate
质量流量变送器　mass flow transmitter
质量流量计　mass flowmeter
质量流量计算机　mass flow computer
质量流量敏感型检测器　mass flow rate sensitive detector
质量逻辑测试　quality logic test
质量平衡　mass balance
质量平衡式　mass balance equation
质量歧视效应　mass discrimination effect
质量扫描　mass scanning
质量色谱法　mass chromatography (MC)
质量色散　mass dispersion
质量数　mass number
质量碎片谱法　mass fragmentography (MF)
质量弹簧系统　mass-spring system
质量体系　quality system
质量稳定性　mass stability
质量吸收系数　mass absorption coefficient
质量选择检测　mass selective detection
质量指示器　mass indicator
质量中心　centre of mass
质谱　mass spectrum
质谱的可迭加性　additivity of mass spectra
质谱法　mass-spectrometric method
质谱分析法　mass spectrometry (MS)
质谱计　mass spectrometer
质谱学　mass spectroscopy

质谱仪　mass spectrograph, mass spectroscope
质谱仪器　mass spectroscope
质谱-质谱法　mass spectrometry-mass spectrometry (MS-MS)
质谱-质谱法扫描　MS-MS scanning
质心　center of mass
质子　proton
质子核磁共振　proton magnetic resonance (PMR)
质子核磁共振谱　proton magnetic resonance spectrum
质子加速器　bevatron
质子平衡式　proton balance equation
质子溶剂　protonic solvent
质子自递常数　autoprotolysis constant
质子自递反应　autoprotolysis reaction
致冷剂　refrigerant
致冷剂回路　refrigerant circuit
致冷设备　chiller plant
致冷压缩器　refrigeration compressor
致死的　lethal
致死性　lethality
智力　intelligence
智能变送器　intelligent transmitter
智能变送器接口　smart transmitter interface
智能变送器接口模件　smart transmitter interface module
智能传感器　intelligent sensor
智能的　intelligent
智能电源应用　smart power application
智能阀　smart valve
智能仿真　intelligent simulation
智能功率模块　intelligent power module
智能管理　intelligent management
智能管理系统　intelligent management system
智能化仿真软件　intellectualized simulation software
智能化设计　intelligent design
智能化现场仪表　smart field instrumentation
智能机器　intelligent machine
智能机器人　intelligent robot
智能计算机　intelligent computer
智能建筑管理系统　building management system
智能键盘系统　intelligent keyboard system
智能决策支持系统　intelligent decision support system (IDSS)
智能控制　intelligent control
智能控制器　intelligent controller
智能控制系统　intelligent control system
智能温度变送器　intelligent temperature transmitter
智能系统　intelligent system
智能系统模型　intelligent system model
智能现场通信器　smart field communicator
智能巡航控制　intelligent cruise control
智能仪表　intelligent instrument, smart instrument
智能站　intelligent station
智能知识库　intelligent knowledge-base
智能制造系统　intelligent manufacturing system
智能终端　intelligent terminal
智能自动化系列　intelligent automation series
滞后　lag
滞后补偿　lag compensation
滞后补偿器　lag compensator
滞后-超前补偿　lag-lead compensation
滞后-超前控制器　lag-lead controller
滞后-超前网络　lag-lead network
滞后电流　lagging current
滞后功率因数　lagging power-factor
滞后环节　lag element
滞后距离　delay distance
滞后时间补偿器　dead time compensation unit
滞后时间器　dead time unit
滞后网络　lag network
滞后误差　hysteresis error
滞后相位　lagging phase
滞后组件　lag module

滞环　B-H loop, hysteresis curve, hysteresis loop
滞环比较方式　hysteresis comparison
滞环误差　hysteresis error
置换　substitution
置位时间　setting time
置信度　confidence
置信级　confidence level
置信区间　confidence interval, confidence limits
置于甲板　on deck
中波　medium wave
中波段　medium-wave band
中点　mid-point
中点半径　mean radius
中点测量齿顶高　mean measuring addendum
中点测量齿高　mean measuring depth
中点测量厚度　mean measuring thickness
中点齿槽宽　mean slot width
中点齿顶高　mean addendum
中点齿根高　mean dedendum
中点法向基节　mean normal base pitch
中点法向径节　mean normal diameter pitch
中点法向模数　mean normal module
中点截面　mean section
中点径节　mean diametral pitch
中点螺旋角　mean spiral angle
中点锥距　mean cone distance
中断　interrupt
中断点功能　breakpoint function
中断屏蔽　interrupt mask
中断器　interrupter
中断优先权　interrupt priority
中断优先权系统　interrupt priority system
中断源　interrupt source
中高频　medium-high frequency
中国电力　China Power
中和池　neutralizing tank
中和电阻　neutralizing resistance
中和区控制　neutral zone control
中和线　neutralizing wire
中红外　mid infrared range (MIR)

中红外光谱　mid infra-red spectrum (MIRS)
中红外吸收光谱　mid-infrared absorption spectrum
中红外遥感　mid infrared range remote sensing
中继器　repeater
中继线　relay line
中继自整角机　relay selsyn motor
中间带　neutral zone
中间的　medial, mid, middle
中间分配线架　intermediate distribution frame
中间继电器　intermediate relay
中间镜　intermediate lens
中间孔　interstitial hole
中间平面　mid-plane
中间数据文件　intermediate data file (IDF)
中间退火　process annealing
中间文件　intermediate file
中径　mean diameter
中空部分　cavity
中空吹出成形　hollow blow molding
中控室　central control room
中跨　mid-span
中跨接头　mid-span joint
中跨距　central span
中频　intermediate frequence, median frequency
中频变压器　intermediate frequence transformer
中频带　mid-frequency band
中频放大器　IF amplifier, intermediate frequence amplifier
中频接收机　intermediate frequence receiver
中碳钢丝　bullet wire
中位知识　metal-level knowledge
中温应变计　medium temperature strain gauge
中线　neutral wire
中线电流　current in middle wire, neutral current
中心　center
中心齿轮　sun gear
中心抽样检验　core testing

中心导体法　central conductor method
中心浇口式模具　center-gated mold
中心距　center distance
中心距变动　center distance change
中心零位式仪表　zero-center instrument
中心轮　central gear
中心区　intermediate zone
中心调整电位器　centering potentiometer
中心线　centre line
中心销　center pin, core pin, king pin
中心指示器　direction indicator
中心轴　center pin
中心主惯性轴　central principal inertia axis
中性操作　neutral steer
中性的　neutral
中性点　neutral point
中性接地　neutral earthing, neutral grounding
中性线　neutral conductor, neutral line
中性轴　neutral axis
中压　middle voltage
中压配电　medium-power distribution
中央处理单元　central processing unit (CPU)
中央处理器　central processor
中央动力机械　central power-driven machine
中央分隔带　central dividing strip
中央分隔栏　central divider
中央计算机　central computer
中央控制室　central control room
中央路栏　median barrier
中央预留带　central reserve
中值　median
中值滤波器　median filter
中止　discontinue
中转仓库　transshipment depot
中子　neutron
中子束瞄准仪　neutron beam collimator
中子通量　neutron flux

终点电压　terminal voltage
终点可实现性　terminal reliability
终点控制　end point control
终点品质管制人员　final quality control (FQC)
终端电缆　terminal cable
终端机　terminal
终端接头　end connector
终端控制　terminal control
终端控制元件　final controlling element
终端连接　terminal connection
终端面板　terminal panel (TP)
终端线　tag wire
终端指示器　terminal indicator
终极定理　limit theorem
终检　final inspection
终结器　terminator
终了温度　final temperature
终饰工程　finishing works
终态　final state
终值　final value
终值原理　final value theorem
终止啮合点　final contact
钟铜　bell metal
钟罩校准器　bell prover
钟罩式压力计　inverted bell manometer
钟罩压力计　bell manometer
重大紧急事件　major emergency
重铬酸钾法　potassium dichromate method
重金属排液处理　heavy metal eliminator
重力　gravity
重力测量　gravity survey
重力垂向梯度测量　gravity vertical gradient survey
重力荷载　gravity load
重力加速度　acceleration of gravity
重力平台　gravity platform
重力剖面　gravity profile
重力式测波仪　floating accelerometer
重力式海堤　gravity type seawall
重力式平衡机　gravitational balancing machine
重力式取样管　gravity corer

重力水平梯度测量　gravity horizontal gradient survey
重力梯度测量　gravity gradient survey
重力梯度力矩　gravity gradient torque
重力梯度仪　gravity gradiometer
重力修正　gravity correction
重力仪　gravimeter
重力异常　gravity anomaly
重力铸造　gravity casting
重力铸造机　gravity casting machines
重量　weight
重量分析法　gravimetric analysis
重量流量计　weight-flow meter
重燃料油　heavy fuel oil
重水　heavy water
重心距　distance of centre of gravity
重心铊　gravity nut
重型　heavy duty
重型车床　heavy-duty lathe
重型固土机　heavy compaction plant
重要工位号　important tag
重油引擎　heavy oil engine
重载　heavy load
重症监护仪　ICU monitor
周波变流器　cycle convertor
周波表　cycle counter
周出货需求　weekly delivery requirement
周节　circular pitch
周界　perimeter
周期　period
周期波　periodic wave
周期的　periodic
周期电磁阀　cycling solenoid valve
周期电流　periodic current
周期工作制　periodic duty
周期函数　periodic function
周期结构　periodic structure
周期脉冲列　periodic pulse train
周期时间　duration of cycle
周期数　cycling life
周期替换　periodic replacement
周期性　periodicity
周期性负载　periodic load
周期性速度波动　periodic speed fluctuation

周期运动　periodic motion
周期阻尼　periodic damping
周围的　ambient
周围压力　ambient pressure
周转轮系　epicyclic gear train
轴　axis
轴承　bearing, axle bearing
轴承高度　bearing height
轴承合金　bearing alloy
轴承加工机　bearing processing equipment
轴承宽度　bearing width
轴承内径　bearing bore diameter
轴承配件　bearing fittings
轴承寿命　bearing life
轴承套圈　bearing ring
轴承外径　bearing outside diameter
轴承销　pin bearing
轴承有效间距　effective bearing spacing
轴承预负荷　bearing preload
轴承侦测器　bearing detector
轴承中心线　bearing axis
轴承座　bearing block
轴端挡圈　shaft end ring
轴封　shaft seal
轴环　shaft collar
轴肩　shaft shoulder
轴角　shaft angle
轴颈中心　journal centre
轴颈中心线　journal axis
轴流泵　axial pump, axial-flow pump
轴流式风扇　axial fan
轴套　axis guide
轴瓦　bearing bush
轴线　axes
轴向　axial direction
轴向齿距　axial pitch
轴向齿廓　axial tooth profile
轴向重合度　face contact ratio
轴向当量动载荷　dynamic equivalent axial load
轴向当量静载荷　static equivalent axial load
轴向定位面　axial locating surface
轴向分力　axial thrust load
轴向基本额定动载荷　basic dynamic

axial load rating
轴向基本额定静载荷　basic static axial load rating
轴向间隙　axial clearance
轴向接触轴承　axial contact bearing
轴向力　axial force
轴向灵敏度　axial sensitivity
轴向平面　axial plane
轴向倾角　axial rake angle
轴向通电法　axial current flow method
轴向推力　axial thrust
轴向位移　axial displacement
轴向移动　end movement
轴向应力　axial stress
轴向游隙　axial internal clearance
轴向载荷　axial load
轴向载荷系数　axial load factor
轴向振动　axial vibration
轴心抗压强度试验　test of axial compressive strength
轴针式喷油器　pintle valve
肘杆式锁模装置　toggle type mould clamping system
肘杆式压力机　toggle press
肘形机构　toggle mechanism
皱纹　wrinkle
珠光体铸铁　pearlitic iron
珠击法　shot peening
珠粒喷击清理　shot blasting
蛛网模型　cobweb model
竹节钢　knotted bar iron
烛光（发光强度单位）　candela（cd）
逐步精化　stepwise refinement
逐次比较法　method of successive comparison
主变电站　bulk supply substation
主变压器　main transformer
主参考地　master reference ground
主成分分析　principal component analysis（PCA）
主成分回归法　principal component regression（PCR）
主程序　master program
主处理机　host processor
主触点　main contact
主窗口　main window

主磁通　main flux
主从触发器　master-slave flip-flop
主从鉴别　master/slave discrimination
主从系统　master-slave system
主存储器　main memory, main storage
主存储器数据库系统　main memory database system（MMDBS）
主导极点　dominant pole
主导轴　predominant axis
主电路　main circuit
主动齿轮　driving gear
主动带轮　driving pulley
主动土压力　active earth pressure
主动寻址　active addressing（AA）
主动遥感　active remote sensing
主动姿态稳定　active attitude stabilization
主阀　main valve
主反馈通路　main feedback path
主钢筋　main reinforcement
主杠杆　force-bar
主根　dominant route
主观频率　subjective probability
主航线　trunk route
主回路　major loop
主机板　mainboard
主机架　main frame
主级门路　primary gate
主集流脉　main manifold
主计算机　host computer
主剪力线　main shear wire
主接触器　main contactor
主控　master control
主控板　master control board
主控接线机　master switch
主控站　master station
主控制回路　primary loop
主控制模件　primary control module
主控制屏　master control panel
主控制器　primary controller
主控制室　main control room, master control room
主跨　main span
主楼层　main storey
主楼梯　main staircase
主配线架　main distribution frame

（MDF）
主频率　dominant frequency
主频区　primary frequency zone
主时间常数　dominant time constant
主栓　king bolt
主水平基准　principal datum（PD）
主送风方向横截面积　area of cross section of the main air flow
主体　main block
主体合约　main contract
主调节　primary regulation
主文卷　master file
主销衬套　king pin bush
主信号　master signal
主要的　main，major，primary
主要废水管　main waste pipe
主要缺陷　major defect
主运动　main movement
主运动方向　direction of main movement
主轴　spindle
主轴箱　headstock
主轴旋转角　spindle rotation angle
属性　attribute
属性标号　attribute label
属性字节　attribute byte
助动型重力仪　auxiliary type gravimeter
助熔剂　complete fusion
助色团（用以使染料固着在织物上）　auxochrome
助听器　audiphone，hearing aid
注道导套　sprue bushing guide
注道定位衬套　sprue lock bushing
注解列　comment column
注口衬套　sprue bushing
注入成形　encapsulation molding
注入横截面　injection cross-section
注入口　sprue
注射泵　syringe pump
注射管嘴　injection nozzle
注射活塞　injection plunger，injection ram
注射模　injection moulding
注射器　injector
注射柱塞　injection plunger
注视　watch

注释　comment
注释行　comment line
注塑冷料　cold slug
注意　attention，caution
注油机　lubricators
注油器　oiler
贮存极限值　limiting value for storage
贮存量　storage capacity
贮存器　container
贮料区　stockpiling area
贮水箱　water storage tank
贮油装置　oil storage installation
驻波　standing wave
驻极体传声器　electret microphone
柱基脚　column footing
柱架　column frame
柱老化　aging of column
柱流失　column bleeding
柱帽　column cap
柱切换　column switching
柱塞阀　plunger valve
柱塞截止阀　plunger globe valves
柱塞流模型　plug flow model
柱塞式阀　plunger valve
柱色谱法　column chromatography
柱上安装　pole mounting
柱寿命　column life
柱填充剂　column packing
柱头　column head
柱外效应　extra-column effect
柱效能　column capacity
柱形蜗杆　cylindrical worm
柱液相色谱法　liquid column chromatography
柱状图　histogram
柱状外对流敏感元件　external-convection column sensitive element
铸锭　ingot
铸钢　cast steel
铸灰口铁　cast gray pig iron
铸件凹痕　clamp-off
铸件气孔　blow hole
铸件清理　cleaning of casting
铸铝　cast aluminium
铸铝黄铜　casting aluminium brass
铸铝青铜　casting aluminium bronze
铸铝转子　cast-aluminum rotor

铸模 mould
铸模成形 moulding
铸模紧固夹 mold clamp
铸模座 die body
铸坯 ingot blank
铸砂烧贴 sintering of sand
铸铁 cast iron
铸铁导管 cast iron conductor
铸铁管 cast iron pipe
铸铜 cast copper
铸造 casting
铸造合模记号 assembly mark
铸造设备 foundry equipment
铸造生铁 foundry iron
铸造凸缘 casting flange
铸造叶片 cast blade
专家控制系统 expert control system (ECS)
专家系统 expert system
专家系统的壳 shell of expert system
专家自适应控制器整定 expert adaptive controller tuning
专利 patent
专题制图仪 thematic mapper
专线 individual line
专用 dedication
专用标准 specialized standard
专用测试仪表 special test unit
专用尺寸 special size
专用存储区 dedicated storage
专用的 dedicated
专用分组交换机 private branch exchange (PBX)
专用附件 non-interchangeable accessory
专用工具 special tool
专用集成电路 application-specific integrated circuit (ASIC)
专用节点 owning node
专用实体 reserved entity
专用通道 dedicated channel
专用通信接口 private communication interface
专用线 dedicated line
专用箱 dedicated cabinet
专用仪表 dedicated instrument
转变 transition
转变点 transition point

转发器 transponder
转化器 convertor
转换 conversion, transform, transformation
转换触点 change-over contact
转换阀 change-over valve
转换矩阵 transformation matrix
转换开关 diverter switch, transfer switch
转换开关盒 switchbox
转换设备 conversion equipment
转换室 switch room
转换文法 transformation grammar
转换装置 transfer device
转接点 switching point
转接控制 car-switch control
转录器 transcriber
转向架 bogie
转向架清洗设备 bogie washer
转向器 diverter
转移函数 transfer function
转移和连接 branch and link
转移矩阵 transition matrix
转移模式 transition mode
转移时间 transition time
转移系统 transition system
转移噪声 transition noise
转运站 transfer station
转杯风速表 cup anemometer
转差率 slip ratio, slip speed
转动关节 revolute joint
转动惯量 moment of inertia
转动面温度元件 roll surface temperature element
转动时间系统 run-time system
转动铁芯 moving iron
转动组件 runner assembly
转/分 revolutions per minute
转角频率 corner frequency
转矩变换器 torque variator
转矩计 torquemeter
转炉 convertor
转/秒 revolutions per second
转盘 dial
转式测速仪 rotameter
转数 number of revolutions
转数指示器 revolution indicator

转速　rotating speed revolution
转速表式流量计　tachometer type flowmeter
转速波动　fluctuation of speed
转速传感器　revolution speed transducer
转速计　revolution indicator, tachometer
转速力矩特性曲线　speed-torque curve
转塔车床　turret lathe
转塔冲床　turret punch press
转塔式数控立式铣床　CNC turret vertical miller
转桶式水表　drum water meter
转轴　revolving shaft
转子　rotor
转子电路　rotor circuit
转子电阻　rotor resistance
转子发动机　rotor generator
转子感应电动机　rotor induction motor
转子流量计　rotameter
转子漏磁电抗　rotor leakage reactance
转子偏心距　eccentricity of rotor
转子频率　rotor frequency
转子平衡　balance of rotor
转子平衡精度　balance quality of rotor
转子启动器　rotor starter
转子绕组　rotor winding
转子损耗　rotor loss
转子铁芯　rotor core
转子铜损　copper loss of rotor
转子线圈　rotor coil
装配　assemblage
装配工　assembler
装配机器人　assembly robot
装配条件　assembly condition
装配图　assembly drawing
装配线　assemble line
装饰照明　decorative lighting
装箱单　packing list
装修工程　fitting out works
装有隔气弯管的集水沟　trapped gully
装载归档文件　load archive file

装置　device
装置仿真　device simulation
装置仿真器　device simulator
装置式自动开关　cluster automatic switch
状态　state
状态变化检测器　state scintillation detector
状态变量　state variable
状态变量法　state variable method
状态触点输出连接　status contact output connection
状态反馈　state feedback
状态方程　state equations
状态方程模型　state equation model
状态赋值　state assignment
状态估计　state estimation
状态观测器　state observer
状态监控　state monitoring
状态空间　state space
状态空间方法　state-space method
状态空间公式　state-space formula
状态空间轨迹　state trajectory
状态空间描述　state space description
状态空间模型实现　state-space model realization
状态空间实现　state-space realization
状态输出　status output
状态输入-输出卡　status input-output card
状态图　state diagram
状态显示　status display
状态向量　state vector
状态序列估计　state sequence estimation
状态转移方程　state transition equation
状态转移矩阵　state transition matrix
撞击　bump
锥-板黏度计　cone-plate viscometer
锥齿轮　bevel gear
锥齿轮的当量直齿轮　equivalent spur gear of the bevel gear
锥齿轮冠圆　crown circle
锥顶　common apex of cone
锥度　taper
锥度车削　taper turning

锥距　cone distance
锥面包络圆柱蜗杆　milled helicoids worm
锥形变压器　tapered transformer
锥形螺母　tapered nut
锥形筒　cone
锥形销　taper pin
锥形轴承　conical bearing
准备工作　preliminary
准光学元件　quasi-optical component
准距线视距丝　stadia wire
准确度　degree of accuracy
准确度等级　class of accuracy
准确度极限　accuracy limit
准双曲面齿轮　hypoid gear
准双曲面齿轮偏置距　hypoid offset
准线　alignment wire
准线性特性　quasilinear characteristics
准线仪　alignment instrument
准许地积比率　permitted plot ratio
准许上盖面积　permitted site coverage
准则　criterion
桌面管理接口　desktop management interface (DMI)
浊度计　turbidity meter
浊气　foul air
着火性能　fire behaviour
着色剂　colorant
着色渗透探伤法　dye-penetrant testing method
咨询　consultation
姿态　attitude
姿态捕获　attitude acquisition
姿态传感器　attitude sensor, attitude transducer
姿态轨道控制系统　attitude and orbit control system (AOCS)
姿态机动　attitude maneuver
姿态角速度　attitude angular velocity
姿态控制　attitude control
姿态扰动　attitude disturbance
姿态算法　attitude algorithm
资料夹　file folder
资料库　database
资料预处理　data preprocessing

资源分配　resource allocation
资源计划　resource plan
资源利用状态显示　resource usage status display
子程序　subprogram
子回路　subsidiary loop
子机构　sub-mechanism
子空间　subspace
子空间方法　subspace method
子口刨　rebate plane
子系统　subsystem
子系统通信功能　subsystem communication function
子系统综合功能　subsystem integration function
紫铜　tough pitch copper
紫外光电倍增管　ultraviolet photomultiplier
紫外光检测　ultraviolet detection
字符　character
字符边界　character boundary
字符串　alphabetic string
字符串点　string point
字符打印机　character printer
字符集　character set
字符码　character code
字符识别　character recognition
字节　byte
字节串行　byte serial
字块长度　block length
字块结束　end off block
字模　character die
字模冲子　stamped punch
字母　alpha
字母代码　alphabetic code
字母的　alphabetic
字母换挡　alphabetic shift
字母数字的　alphanumeric
字母数字键盘　alphameric keyboard, alphanumeric keyboard
字母数字字段　alphameric field
字母字符　alphabetic character
字母字符集　alphabetic character set
字数据块　block word data
字数组　alpha-numeric
自保持　self-hold
自保持继电器　lock-up relay

自 zi

自备电厂　self-supply power plant
自备电源　auxiliary power supply
自补偿　self-compensation
自操作控制　self-operated control
自定义功能键　soft key
自动　automatic
自动安平水准仪　compensator level
自动包装机　automatic packer
自动曝光装置　automatic exposure device
自动编码　automatic coding
自动辨识　automata recognition
自动操作　automata operation，automatic operation
自动操作站　automatic station
自动测量　automatic measurement
自动测试　automata testing
自动测试设备　auto test equipment
自动抄表　automatic message recording
自动车床　automatic lathe
自动程序设计　automatic programming
自动程序中断　automatic program interrapt
自动重复启动　automata restart
自动重合闸　auto reclose，automatic reclosing
自动导向机车　automated guided vehicle
自动的　automotive
自动滴定仪　automatic titrator
自动点火装置　automatic ignition device
自动电平控制器　automatic level controller
自动电压调整器　automatic voltage regulator（AVR）
自动断路器　automatic circuit breaker
自动发电控制　automatic generation control
自动发射　automotive emission
自动反馈放大器　self-feedback amplifier
自动方式回路状态转换键　auto mode loop status switching key
自动防松螺帽　self-locking nuts

自动复位继电器　automatic reset relay
自动跟踪　automatic following
自动跟踪仪　automatic tracking
自动功率控制　automatic power control
自动过程控制　automata process control
自动戽斗定量秤　hopper scale
自动化　automation
自动化的　automatic
自动化工程　automatic engineering
自动化检验　automated inspection
自动化交互系统　automation system interaction
自动化设备　automation equipment
自动化系统　automatic system
自动化系统接口　automation system interface
自动化与驱动　automation and drives
自动化装配系统　automatic assembly system
自动绘图仪　autographic apparatus
自动火花点火装置　automatic spark ignition device
自动机　automaton
自动机理论　automata theory
自动极化补偿器　auto-polarization compensator
自动记录的　self-recording
自动记录器　self-recorder
自动加热　self-heating
自动驾驶装置　servomechanism
自动减压阀　automatic pressure reducing valve
自动检测　self-verifying
自动交换区　dial exchange
自动校验　automatic check
自动校正　autocorrection
自动开关　automatic switch
自动空气断路器　automatic air circuit breaker
自动控制　autocontrol，automatic control
自动控制开关　automatic control switch
自动控制器　automata controller

679

自动控制系统　regulating system, automatic control system
自动控制装置　servomechanism
自动联锁装置　automatic interlocking device
自动零平衡电位计　automatic null-balancing potentiometer
自动灭火材料　self-extinguishing material
自动模式　automatic mole
自动喷水灭火系统　sprinkler system
自动偏压　automatic bias
自动频道开关　automatic channel switch
自动频率控制放大器　AFC amplifier
自动频域控制　automata frequency control
自动平衡电路　automatic balancing circuit
自动平衡机　automatic balancing machine
自动启动　automatic start
自动启动器　automatic starter
自动气动断路器　autopneumatic circuit breaker
自动气象站　automatic weather station
自动切断　automatic break
自动倾卸车　dumper
自动洒水系统　automatic sprinkle system
自动洒水装置　sprinkler system
自动扫查　automatic scanning
自动试验机　automatic testing machine
自动收发机　automatic send-receive
自动收费系统　automatic fare collection system
自动-手动操作器　automatic manual station
自动手动控制站算法　auto manual station algorithm
自动手动转换开关　automatic and hand operated change-over switch
自动售票机　automatic ticket issuing machine（ATIM）
自动输出操作站　automation output station

自动数据处理　automatic data processing
自动顺序控制　automata sequence control
自动顺序控制仪表　instrument for automatic sequence control
自动弹簧式吸液管　automatic spring pipette
自动调焦和消像散　automatic focus and stigmator
自动调节　self-adjustment, automatic regulation
自动调节的　self-regulating
自动调节控制　automatic modulation control
自动调节器　automata regulator, automatic regulator
自动调温器　thermostat
自动调整仪　control gear
自动同步电动机　auto synchronous motor
自动同步机　autosyn
自动同步接收机　receiving selsyn
自动脱扣　automatic release
自动稳压器　constac
自动相位同步　automatic phase synchronization
自动选择器　auto selector
自动选择系统　auto selector system
自动研光　self-glazing
自动遥控开关　automatic teleswitch
自动移相器　automatic phase shifter
自动引导装入程序　automation bootstrap loader
自动增益控制　automata gain control, automatic gain control（AGC）
自动轧管机　plug mill
自动闸门　automatic gate
自动站　auto station
自动诊断程序　automatic diagnostic program
自动装置　automata
自动准直机　autocollimator
自返式取样器　boomerang grab
自感　self-inductance
自感电动势　self-induced EMF
自感体　self-inductor

自 zi

自感系数　coefficient of self-inductance
自感应线圈　self-inductor
自给偏压　auto-bias
自攻螺丝　self tapping screw
自回归模型　autoregressive model
自回归移动平均　autoregressive moving-average
自回归移动平均参数估计　ARMA parameter estimation
自回归移动平均模型　ARMA model
自激励绕组　self-excitation winding
自激励振荡　self-excited oscillation
自激振荡　auto-excitation
自激振荡器　self-excited oscillator
自校开关　self-check key
自校正　self-correction
自校正调节器　self-tuning regulator
自举电路　bootstrap circuit
自均衡结构　self-aligning structure
自控设备表　list of equipment for process control
自力式控制器　self-operated controller
自励　self excited, self-excitation
自励的　self-excited, self-exciting
自励电机　self-excited machine
自励多谐振荡器　static multivibrator
自励过程　process of self-excitation, self-excitation process
自铆机　tox machine
自耦变压器　autoformer
自耦变压器启动器　compensator starter
自耦磁放大器　autotransductor
自耦式混合绕组电流互感器　auto-compound current transformer
自偏差放大器　self biased amplifier
自偏置电阻　self-bias resistor
自平衡　inherent regulation
自平衡滑线电位差计　self-balancing slide wire potentiometer
自平衡静电计　self-balancing electrometer
自平衡仪表　self-balancing instrument
自启动　self-starting
自然　natural
自然函数发生器　natural function generator
自然冷却　natural cooling
自然冷却式电机　machine with nature cooling
自然频率　natural frequency
自然特性　natural characteristic
自然无阻尼频率　natural undamped frequency
自然行宽度　natural line width
自然语言　natural language
自然语言生成　natural language generation
自然振荡　natural oscillations
自燃　spontaneous combustion, spontaneous ignition
自上而下测试　top-down testing
自上向下方法　top-down method
自适应补偿器　adaptive equalizer
自适应的　adaptive
自适应规则　adaptive law
自适应矩阵　adaptive array
自适应均衡　adaptive equalization
自适应均衡器　automatic adaptive equalizer
自适应控制　adaptive control, self-adaptive control
自适应控制器　adaptive controller
自适应控制系统　adaptive control system
自适应滤波器　adaptive filter
自适应模糊控制　adaptive fuzzy control
自适应模糊系统　adaptive fuzzy system
自适应数字滤波器　adaptive digital filter
自适应算法　adaptive algorithm, self-adapting algorithm
自适应系统　adaptive system
自适应遥测系统　adaptive telemeter system
自适应一般模型控制　adaptive generic model control（AGMC）
自适应预报　adaptive prediction
自适应整定　adaptive tuning
自锁　self-hold
自锁螺母　locknut

自锁条件　condition of self-locking
自调节　self-regulation
自调整系统　self-adjusting system
自调制过程　self-regulating process
自同步系统　automatic synchronized system
自同构　automorphism
自位滚柱轴承　self-aligning roller bearing
自下而上开发　bottom-up development
自相关　adaptive correlation
自相关函数　auto correlation function, autocorrelation function
自相位　self phase
自相位调制　self-phase modulation
自行调节　inherent regulation
自旋体　spinner
自旋轴　spin axis
自学习　self-learning
自优化控制　self-optimizing control
自优化系统　self-optimizing system
自由场修正曲线　free field correction curves
自由电荷　free charge
自由电子　free electron
自由度　degree of freedom
自由感应衰减信号　free induction decay signal
自由格式报表　free format log (FFL)
自由声场　free field
自由声场互易校准　free field reciprocity calibration
自由运动　free motion
自由振荡　free oscillation
自由振动　free vibration
自由振动周期　free oscillating period
自运行　self-running
自诊断　self diagnostic
自振荡　self oscillation
自整定控制　self-tuning control
自整定自适应控制器　self-adaptive controller
自整角机　selsyn
自整角机系统　selsyn system
自整角接收机　receiving selsyn, selsyn motor
自整流　self-rectifying

自制动　self braking
自治的　autonomous
自治控制　autonomous control
自治系统　autonomous system
自主机车　autonomous vehicle
自主移动机器人　autonomous mobile robot
自阻尼　self-damping
自组织策略　self-organizing strategy
自组织系统　self-organizing system
综合布线　generic cabling
综合测试　integration testing
综合重建区　comprehensive redevelopment area
综合发展区　comprehensive development area
综合服务数字网　integrated services digital network (ISDN)
综合交通交汇设施　comprehensive transport interchange facilities
综合试验　combined test
综合试验箱　combined test cabinet
综合预负荷　combined preload
综合载荷试验机　combined load testing machine
综合照明　combined lighting
棕色　brown
总沉降　total settlement
总重合度　total contact ratio
总电流　total current
总电压　total voltage
总电阻　total resistance
总度量中心距变动量　total composite variation
总反虹吸管　main anti-syphonage pipe
总分度变动量　total index variation
总分度公差　total index tolerance
总辐射　global radiation
总功率　total power
总管　main pipe
总和　summation
总荷载　total load
总开关　main switch
总控制屏　general control panel
总离子强度调节缓冲剂　adjust buffer total ion strength
总貌　overview

总貌画面　overview display
总貌画面定义　overview panel definition
总面积　aggregate area
总排水管　main drain
总配电板　main switchboard
总平面图　general layout plan
总热传送值　overall thermal transfer value
总实用楼面空间　aggregate usable floor space
总水管　water main
总速度　sum velocity
总损耗　total losses
总碳分析　total carbon analysis
总体布置　general arrangement
总体设计　overall design
总误差　overall error, total error
总线　bus line
总线从设备　bus slave
总线多处理过程　bus multiprocess
总线母板　bus mother board
总线拓扑　bus topology
总线网　bus network
总线主设备　bus master
总线转发器　bus repeater
总线转换器　bus converter
总压毕托管　total pressure Pitot tube
总匝数　total number of turns
总质量　gross mass
总重量　gross weight, total wt.
总综合公差　total composite tolerance
纵波　longitudinal wave
纵波法　longitudinal wave technique
纵波探头　longitudinal wave probe
纵截面　longitudinal section
纵向重合度　overlap contact ratio
纵向磁动势　longitudinal magnetomotive
纵向磁化　longitudinal magnetization
纵向磁通　longitudinal magnetic flux
纵向的　longitudinal
纵向分解　vertical decomposition
纵向风力分析　longitudinal wind analysis
纵向干扰　longitudinal interference
纵向滑动速度　lengthwise sliding velocity
纵向坡度　longitudinal gradient
纵向失配　lengthwise mismatch
纵轴　axis of ordinates, longitudinal axis
纵坐标　ordinate
走火通道　fire escape
走纸偏差　driving chart paper, driving chart swaying
阻带滤波器　stop band filter
阻挡板　barrier plate
阻挡层　barrier layer, blocking layer
阻挡器　stopper
阻抗　impedance
阻抗变换器　impedance transformer
阻抗分析　impedance analysis
阻抗矩阵　impedance matrix
阻抗控制　impedance control
阻抗匹配　impedance matching
阻抗平面图　impedance plane diagram
阻抗三角形　impedance triangle
阻抗式呼吸频率传感器　impedance respiratory frequency sensor, impedance respiratory frequency transducer
阻抗式血容量传感器　impedance blood volume sensor, impedance blood volume transducer
阻抗头　impedance head
阻抗系数　resistance coefficient
阻抗线圈　impedance coil
阻抗值　impedance value
阻力系数　drag coefficient
阻力张线　drag wire
阻流体　bluff body
阻尼　damping
阻尼比　damping ratio
阻尼波　damped wave
阻尼常数　damping constant
阻尼的　damped
阻尼电路　buffer circuit
阻尼固有频率　damped natural frequency
阻尼拉筋　damping wire
阻尼力矩　damping torque
阻尼力矩系数　damping torque coefficient

阻尼频率	damped frequency
阻尼器	amortisseur, damper
阻尼曲线	damping curve
阻尼绕组	damping winding
阻尼时间	damping time
阻尼特性	damping characteristic
阻尼系数	damping coefficient
阻尼线圈	damping coil
阻尼因数	damping factor
阻尼振荡	damped oscillation, damping oscillation
阻尼振荡器	damped oscillator
阻尼作用	damping action
阻燃的	fireproof
阻容耦合	capacitance-resistance coupling
阻容振荡器	RC oscillator
阻抑剂	inhibitor
阻滞剂	retarder
组成控制	composition control
组成原理	theory of constitution
组分平衡	component balance
组合	combination
组合安装	stack mounting
组合爆炸	combinatorial explosion
组合测井仪	combination logging instrument
组合齿形	composite tooth form
组合的	combinational
组合电路	combinational circuit
组合阀	combination valves
组合放大器	group amplifier
组合恒载	combined dead load
组合机构	combined mechanism
组合决策	combination decision
组合开关	combinational switching
组合空间	configuration space
组合控制	composite control
组合滤波器	array filter
组合逻辑	combinatory logic
组合逻辑元件	combinational logic element
组合模	segment mold
组合式放大器	composition amplifier
组合式键盘	modular keyboard
组合式模具	modular mold
组合试验	composite test
组合数学	combinational mathematic
组合网络	combinational network
组合温度补偿应变计	composite-temperature-compensation strain gauge
组合性	modularity
组合组	packed group
组合作用	compound action
组件	module
组件式信号器	modular type annunciator
组件选择	packaging option
组趋势画面	group trend display
组态标识表	configuration identification table
组态表	configuration list
组态管理	configuration management
组态控制	configuration control
组态说明书	configuration specification
组织与管理	organization and management
组装	assembly
组装线	assembly line
组装在一起	fit together
钻床	drilling machines
钻床工作台	drilling machines bench
钻孔机	drilling machine
钻孔记录	borehole log
钻孔图	drill drawing
钻取土芯	core drilling
钻石	diamond
钻石刀具	diamond cutters
钻石轮划片机	dicing saw
钻台	drill stand
钻头	drill bit, drills
钻挖的隧道	bored tunnel
最大标度	maximum scale value
最大超调量	maximum overshoot
最大称量	maximum capacity
最大穿透力	maximum penetration power
最大风速	maximum wind speed
最大浮置电压	maximum floating voltage
最大幅值	maximum amplitude
最大负荷试验机	maximum load of the testing machine

最 zui

最大负载　peak load
最大工作压差　maximum operating pressure differential
最大工作压力　maximum working pressure
最大功耗　maximum power consumption
最大荷载　maximum load
最大横向负荷　maximum transverse load
最大回路电阻　max loop resistance
最大激励　maximum excitation
最大计算流量　calculated maximum flow coefficient
最大加速度　maximum acceleration
最大控制器增益　maximum controller gain
最大利润规划　maximum profit programming
最大流量　maximum flow-rate
最大排水量　max discharging capacity
最大平坦响应滤波器　Butterworth filter
最大容许负载量　maximum permitted load
最大施工负载量　maximum safe working load
最大试验负荷　maximum load of the test
最大输出　maximum output
最大输出电感　maximum output inductance
最大输出电阻　maximum output resistance
最大似然估计　maximum likelihood estimation
最大速度　maximum velocity
最大特征值　largest singular value
最大位移　maximum displacement
最大线性偏转　max deflection of linearity
最大需量电度表　meter with maximum demand indicator
最大循环负荷　maximum cyclic load
最大循环应力　maximum cyclic stress
最大应变　maximum strain
最大应力　maximum stress
最大盈亏功　maximum difference work between plus and minus work
最大允许连续工作电流　max allowable continuous working current
最大允许偏差　maximum allowed deviation
最大值　maximum
最大周向磁化电流　maximum rated circumferential magnetizing current
最低储备　minimum reserve
最低电源电压　minimum power supply voltage
最低航速　creeping speed
最低收益率　minimum rate of benefit
最低温度　minimum temperature
最低温度计　minimum thermometer
最短距离　minimum distance
最高表面温度　highest face temperature
最高/低电流及电压　max/min current and voltage
最高电源电压　maximum power supply voltage
最高工作温度　max operating temperature
最高交通流量　peak traffic flow
最高静水压力　maximum static pressure
最高楼层　uppermost storey
最高速度　maximum safe speed
最高温度　maximum temperature
最高温度计　maximum thermometer
最高温升　maximum temperature-rise
最高允许温度　max allowable temperature
最高允许压力　max allowable pressure
最高允许噪声声级　maximum allowable noise level
最佳的　optimized
最佳化控制　optimizing control
最佳可换单元　optimum replacement unit
最佳滤波器　optimum filter
最佳条件　optimum
最近邻　nearest-neighbor
最经济观测　most economic observing (MEO)
最经济控制　most economic control (MEC)

685

最速下降 steepest descent
最小标度 minimum scale value
最小电源电压 minimum power voltage
最小二乘辨识 least-squares identification
最小二乘法 least square (LS), method of least square
最小二乘方法 least-squares method
最小二乘估计 least-squares estimation (LSE)
最小二乘近似 least-squares approximation
最小二乘拟合 least-squares fitting (LSF)
最小二乘算法 least-squares algorithm
最小二乘问题 least-squares problem
最小负载电阻 min load resistance
最小化 minimization
最小目标函数 minimum of objective function
最小能量问题 minimum-energy problem
最小偏差控制 minimum variance control

最小燃料问题 minimum-fuel problem
最小冗余 minimum redundancy
最小时间控制 minimum-time control
最小时间问题 minimum-time problem
最小输出 minimum output
最小相位系统 minimum phase system
最小值 minimum, minimum value
最小值原理 minimum principle
最优更换单元 optimal replacement unit
最优化 optimization
最优化问题 optimal problem
最优控制 optimal control
最优性 optimality
最优装置 optimal device
最终被控变量 ultimately controlled variable
最终控制元件 final control element
最终用户 end user
左右偏位指示器 right-left bearing indicator
作废字符 cancel character
作业程序 sequence of operation
作用选择 activity selection
坐标系 co-ordinate
坐标原点 origin of coordinate

其他

1/4 圆孔板　quarter circle orifice plate
Ⅱ型单回路控制器　single strategy controllerⅡ（SSCⅡ）
Ⅲ型高性能操作站　enhanced operator stationⅢ（EOSⅢ）
A 级绝缘　class A insulation
A 形骨架　A-frame
B-H 曲线　B-H curve
CRT 显示器　CRT display device
CRT 显示终端　CRT display terminal
CRT 终端　CRT terminal
DNA 测序仪　DNA sequencers
DNA 及蛋白质的测序和合成仪　sequencers and synthesizers for DNA and protein
DTA 范围　DTA range
D 型端口　D-port
EC 链　EC-link
FCS 控制功能的组态　configuration of FCS control functions
FCS 状态显示窗口　FCS status display window
FF 协议智能温度变送器　FF protocol intelligent temperature transmitter transmitter
FF 总线　foundation fieldbus (FF)
H∞　H-infinity
H∞ 控制　H-infinity control
H∞ 优化　H-infinity optimization
HART 协议直行程电动执行器　HART protocol intelligent linear electric actuator
HART 协议智能压力变送器　HART protocol intelligent pressure transmitter
HF 总线　HF-bus
H 无穷　H-infinity
I/O 连接扩展器　I/O link extender
IOP 点的类型　point's IOP type
ISA1932 喷嘴　ISA 1932 nozzle
I 控制器　integral controller
LCN 子系统测试程序　LCN subsystem test program
NCF 工作文件　NCF working file
NPN 晶体管　N-P-N transistor
N 型半导体　N-type semiconductor
PD 作用　proportional plus derivative action
Petri 网　Petri-net
pH 变送器　pH transmitter
pH 测量系统　pH measurement system
pH 电极装置　pH electrode assembly
pH 计　pH meter
pH 控制　pH control
pH 控制系统　pH control system
pH 值指示剂　pH indicator
PID 控制器　PID controller, proportional-integral-derivative controller
PID 作用　proportional plus integral plus derivative action
PI 控制器　proportional-integral (PI) controller, proportional plus integral controller, PI cotroller
PI 作用　proportional plus integral action
PLC 过程控制　PLC process control
PM 箱数据点　PM box data point
PV 标识参数　PV flag parameter
PV 标志　PV flag
PV 低量程　PV low range in engineering units
PV 高量程　PV high range in engineering units
PV 来源　PV source
PV 行程的负向变化率　PV negative rate of change trip point
PV 行程的正向变化率　PV positive rate of change trip point
PV 原始值　PV raw value
PV 正常状态　PV normal state
PV 值低报警点　PV low alarm trip point
PV 值低低报警点　PV low low alarm

trip point	
PV 值高报警点	PV high alarm trip point
PV 值高高报警点	PV high high alarm trip point
PV 自动值	PV auto value
RC 振荡器	RC oscillator
RIO 总线	RIO bus
RIO 总线装配	RIO bus arrangement
RL 总线	RL bus
SFC 窗口	SFC window
SFC 块功能	SFC block function
s 域	s-domain
T 形连接	T connection
U 形钉	staple U
U 形管密度传感器	U-tube density transmitter
U 形管压力计	U-tube manometer
U 形夹	clevis
U 形螺帽	U nuts
U 形螺栓	U bolt, U-bolt
U 形金属管压力计	metal U-tube manometer
V 带	V belt, V-belt
V 网	V net
V 网电缆	V net cable
V 网耦合器安装	V net coupler installation
V 形盘	V-shaped pad
V 形刀具	V-tool
V 形皮带	V belt
W 平面	W-plane
X-Y 绘图仪	X-Y plotter
X-Y 记录器	X-Y recorder
X-Y 示波器	X-Y oscilloscope
X 波段微波干涉仪	X-band microwave interferometer
X 频带微自旋波谱仪	X-band microspin spectrometer
X 射线	X-ray
X 射线报警器	X-ray warning device
X 射线测角仪	X-ray goniometer
X 射线度谱术	X-ray spectrometry
X 射线发射光谱仪	X-ray emission spectrometer
X 射线发生器	X-ray generator
X 射线辐射圆锥角	angle of X-ray projection
X 射线干扰	X-ray interference
X 射线管灯丝特性	heater characteristic of X-ray tube
X 射线管寿命试验	life of X-ray tube
X 射线检查	X-ray examination
X 射线气体密度计	X-ray gas density gauge
X 射线探伤仪	X-ray detection apparatus
X 轴信号放大器	X axis amplifier
Y-△启动器	Y-△ starter, star-delta starter
Y 形连接	Y-connection
Y 轴信号放大器	Y axis amplifier
Z 变换	Z- transform
Z 传递函数	Z transfer function
Z 轴信号放大器	Z-axis amplifier
△接法	delta connection
γ 测井仪	gamma ray logger
γ 处理像	gamma processing image
γ 定向辐射仪	gamma directioned radiometer
γ 光谱测量	gamma-spectrometry
γ 能谱仪	gamma ray spectrometer
γ 闪烁辐射仪	gamma scintillator radiometer
γ 射线	gamma rays, gamma-rays
γ 射线探伤机	gamma-ray detection apparatus
γ 射线物位测量仪表	gamma ray level measuring device

欢迎订阅自控类图书

书　名	定价/元	书号
自控软件		
MATLAB实用教程——控制系统仿真与应用	48	9787122049889
MATLAB语言与控制系统仿真实训教程（附光盘）	38	9787122060617
LabVIEW虚拟仪器程序设计与应用	48	9787122103321
组态软件应用指南——组态王Kingview和西门子WinCC	68	9787122104748
仪器仪表		
自动抄表系统原理与应用	29	978712212683
LabWindows/CVI虚拟仪器测试技术及工程应用（附光盘）	85	9787122113702
热工仪表与自动控制技术问答	35	9787122052735
有毒有害气体检测仪器原理和应用	20	9787122038463
污水处理在线监测仪器原理与应用	28	9787122029218
仪表工程施工手册	98	9787502567415
传感器手册	75	9787122014702
在线分析仪表维修工必读	55	9787122009319
在线分析仪器手册	148	9787122024398
过程控制技术及施工		
从新手到高手——过程控制系统基础与实践	48	9787122117342
从新手到高手——自动调节系统解析与PID整定	48	9787122138200
集散控制系统应用技术	48	9787122110343
现场总线系统监控与组态软件	28	9787122030030
油库电气控制技术读本	28	9787122027122
过程控制系统及工程（二版）	30	9787502538514
过程自动检测与控制技术	20	9787502593322
工业过程控制技术——应用篇	65	7502577955
石油化工自动控制设计手册（三版）	138	9787502526962
自动化及仪表技术基础	28	9787122026316
电视监控系统及其应用	36	9787122012579
人机界面设计与应用	36	9787122014016

续表

书名	定价/元	书号
PLC 技术		
PLC 现场工程师工作指南	59	9787122121868
西门子 S7-300/400PLC 快速入门手册	58	9787122138545
PLC 编程和故障排除	30	9787122037947
西门子 S7 系列 PLC 电气控制精解	46	9787122083708
西门子 S7-200 系列 PLC 应用实例详解	36	9787122072313
可编程控制器使用指南	38	9787122036407
可编程序控制器原理及应用技巧(二版)	30	7502544062
PLC 技术及应用	18	9787122012203
西门子 PLC 工业通信网络应用案例精讲(附光盘)	48	9787122099655
变频器技术丛书		
变频器实用手册	68	9787122103338
变频器故障排除 DIY	38	9787122032973
变频器使用指南	40	9787122036520
变频器应用问答	22	9787122038890
变频器应用技术丛书——变频器的使用与维护	29	9787122047649
变频器应用技术丛书——变频器应用实践	29	9787122047625
变频器应用技术丛书——电气传动与变频技术	30	9787122093271
变频器的使用与节能改造	28	9787122102515
变频器应用技术及实例解析	23	9787122023469
单片机系列		
单片机应用入门——AT89S51 和 AVR	33	9787122029515
单片机系统设计与调试(吉红)	29	9787122087010
单片机 C51 完全学习手册(附光盘)	68	9787122035820

以上图书由化学工业出版社电气分社出版。如需以上图书的内容简介、详细目录以及更多的科技图书信息，请登录 www.cip.com.cn。

邮购地址：(100011) 北京市东城区青年湖南街 13 号化学工业出版社

服务电话：010-64519685，64519683（销售中心）

如要出版新著，请与编辑联系。联系方法：010-64519262，sh_cip_2004@163.com